Lecture Notes in Networks and Systems

Volume 243

The series "Lecture Notes in Networks and Systems" publishes the latest developments in Networks and Systems—quickly, informally and with high quality. Original research reported in proceedings and post-proceedings represents the core of LNNS.

Volumes published in LNNS embrace all aspects and subfields of, as well as new challenges in, Networks and Systems.

The series contains proceedings and edited volumes in systems and networks, spanning the areas of Cyber-Physical Systems, Autonomous Systems, Sensor Networks, Control Systems, Energy Systems, Automotive Systems, Biological Systems, Vehicular Networking and Connected Vehicles, Aerospace Systems, Automation, Manufacturing, Smart Grids, Nonlinear Systems, Power Systems, Robotics, Social Systems, Economic Systems and other. Of particular value to both the contributors and the readership are the short publication timeframe and the world-wide distribution and exposure which enable both a wide and rapid dissemination of research output.

The series covers the theory, applications, and perspectives on the state of the art and future developments relevant to systems and networks, decision making, control, complex processes and related areas, as embedded in the fields of interdisciplinary and applied sciences, engineering, computer science, physics, economics, social, and life sciences, as well as the paradigms and methodologies behind them.

Indexed by SCOPUS, INSPEC, WTI Frankfurt eG, zbMATH, SCImago.

All books published in the series are submitted for consideration in Web of Science.

More information about this series at http://www.springer.com/series/15179

Duc-Tan Tran · Gwanggil Jeon ·
Thi Dieu Linh Nguyen ·
Joan Lu · Thu-Do Xuan
Editors

Intelligent Systems and Networks

Selected Articles from ICISN 2021, Vietnam

 Springer

Editors
Duc-Tan Tran
Phenikaa University
Hanoi, Vietnam

Thi Dieu Linh Nguyen
Hanoi University of Industry
Hanoi, Vietnam

Thu-Do Xuan
University of Transport Technology
Hanoi, Vietnam

Gwanggil Jeon
Department of Embedded Systems
Engineering
Incheon National University
Incheon, Korea (Republic of)

Joan Lu
University of Huddersfield
Huddersfield, UK

ISSN 2367-3370   ISSN 2367-3389   (electronic)
Lecture Notes in Networks and Systems
ISBN 978-981-16-2096-6   ISBN 978-981-16-2094-2   (eBook)
https://doi.org/10.1007/978-981-16-2094-2

This Springer imprint is published by the registered company Springer Nature Singapore Pte Ltd.
The registered company address is: 152 Beach Road, #21-01/04 Gateway East, Singapore 189721,
Singapore

Preface

The First International Conference on Intelligent Systems & Network (ICISN 2021) was held on March 19–20, 2021, at the University of Transport Technology (UTT), Hanoi, Vietnam. The first edition of ICISN 2021, at UTT given an international forum, where academician, researchers as well as industry practitioners who were actively involved in the research fields of mobile and wireless networks, software engineering, computational intelligence, signal processing, IoT applications, or any other emerging trends related to the theme covered by the conference.

This conference was in hybrid nature, was offline and online technical paper sessions, invited talks, and panels organized around the relevant theme. ICISN 2021 received a massive response in terms of the submission of papers across the globe. ICISN 2021 received 275 reports from various countries outside Vietnam, such as the USA, India, China, Russia, Australia, Japan, Philippines, Bangladesh, Malaysia, Romania, and Iraq.

The organizing committee of ICISN 2021 constituted a robust international program committee for reviewing papers. The manuscript submission was managed through EasyChair, and among all, only 78 papers have been selected after a thorough review process. The conference proceedings will be published in lecture note in network systems (ISSN: 2367-3370), Springer, currently indexed in SCOPUS, Google Scholar, and Springerlink. We convey our sincere gratitude to the authority of Springer for providing the opportunity to publish the proceedings of ICISN 2021.

To realize this conference in 2021, we appreciate the University of Transport Technology (UTT), Hanoi, Vietnam, to agree to host the conference and continuously support the organization team during the preparation of two days of the forum. Without their support, this conference would not have been successful as the first time being held in Vietnam.

Our sincere gratitude to all keynote address presenters, invited speakers, session chairs, and high officials from Vietnam for their gracious presence on the campus on this August occasion. We want to thank the keynote speakers as Prof. Zhongyu Lu, UK; Prof. Marco Anisett, Italy; Prof. Gwanggil Jeon, Korea, for giving their excellent knowledge conference. At the same time, our sincere thanks to the reviewers for completing a reviewing task timely.

The Team ICISN conveys sincere thanks to all the program committee members and honored dignitaries, among few, are Dr. Vu Ngoc Khiem, Chairman of UTT's Council; Dr. Tran Duc Lai, President of the Radio and Electronics Association of Vietnam; Dr. Nguyen Hoang Long, Rector of UTT; Dr. Nguyen Manh Hung, Vice-rector of UTT; Dr. Doan Quang Hoan, Vice President of the Radio and Electronics Association of Vietnam; Dr. Ngo Quoc Trinh, Head of Science & Technology, UTT; Dr. Tran Ha Thanh, Dean of IT Department, UTT; and Team ICISN for their efforts to make the congress a huge success.

Last but not least, we convey our sincere thanks to all the authors who submitted papers to ICISN 2021 and made a high-quality technical program possible. Finally, we acknowledge the support received from the faculty members, Science and Technology Department of the UTT, officers, staff, and University of Transport Technology Authority. We hope that the article published through ICISN21 will be helpful for the researchers pursuing research in computer science, information technology, and related areas. Practicing technologists would also find this volume to be a good source of reference.

Tran Duc Tan
Jeon Gwanggill
Nguyen Thi Dieu Linh
Joan Lu
Do Xuan Thu

Contents

About the Editor

Nguyen Thi Dieu Linh, Ph.D. is Dy. Head of Science and Technology Department, Hanoi University of Industry, Vietnam (HaUI). She has more than 20 years of academic experience in electronics, IoT, smart garden, and telecommunication.

She has authored or co-authored many research articles that are published in journals, books, and conference proceedings. She teaches graduate and postgraduate-level courses at HaUI, Vietnam.

She received Ph.D. in Information and Communication Engineering from Harbin Institute of Technology, Harbin, China, in 2013; Master of Electronic Engineering from Le Quy Don Technical University, Hanoi, Vietnam, in 2006; and Bachelor of Electronic Engineering from HCMC University of Technology and Education, Vietnam, in 2004.

She completed two edited books, "Artificial Intelligence Trends for Data Analytics Using Machine Learning and Deep Learning Approaches" and "Distributed Artificial Intelligence: A Modern Approach," publisher by Taylor & Francis Group, LLC, USA. She is also Board Member of the International Journal of Hyperconnectivity and the Internet of Things (IJHIoT) IGI-Global, USA, Information Technology Journal, Mobile Networks and Application Journal, and some other reputed journals and international conferences.

She has attended two conferences in India as Keynote Speaker in the year 2019 and January 2020. In December 2020, she participated at the first International Conference on Research in Management and Technovation (ICRMAT - 2020) as Guest Honor. Apart from that, she has chaired many technical events in different universities in Vietnam.

She was Organizing Chair for the 4th International Conference on Research in Intelligent and Computing in Engineering. Hanoi University of Industry, Vietnam (RICE-2019), and working actively as Chair in the 5th International Conference on

Research in Intelligent and Computing in Engineering, Thu DAU Mot University, Vietnam (RICE-2020). She is Core Member of the International Conference on Intelligent Systems & Networks (ICISN 2021).

Proposal System for Monitoring Parameters in an Industrial Incubator Incorporating SVM Model to Detect Diseased Chickens

Phat Nguyen Huu[(⊠)]

School of Electronics and Telecommunications,
Hanoi University of Science and Technology, Hanoi, Vietnam
phat.nguyenhuu@hust.edu.vn

Abstract. The development of artificial poultry eggs has facilitated farming that allows it to become not only higher productive industry but also lower production costs in agriculture. In recent years, a lot of advanced technology has been applied in high-tech agriculture. The type of automatic incubator can adjust the temperature and humidity factors to achieve high efficiency. In the paper, we present details design of the parameter monitoring system in an industrial incubator with support vector machine (SVM) model in detecting sick chickens based on voice processing. These parameters sending to the central server for monitoring and training models to classify sounds will be used to predict healthy. Experimental results show that the proposed model meets real-time requirements with accuracy over 90% and processing time of less than one minute.

Keywords: Support vector machine · Acorn RISC machine · Voice processing · IoT · Monitoring system

1 Introduction

In order to have good hatching results, people often choose female poultry capable of hatching eggs that have high efficiency in small businesses. In manual incubation methods, one can use hot paddy or rice husks as ingredients. In this process, people combine to expose eggs and then incubate them with rice husks or hot paddy and then use incubated eggs with new eggs. Besides, people still use hot water to incubate geese and ducks.

Currently, the incubators used in large-scale farms are all high-cost automatic incubators. Besides, one can use a semi-automatic furnace. However, incubation capacity is not high due to cumbersome in the use stage. This results in a high incidence of hatching eggs causing economic losses. This is also the disadvantage that households are facing.

Currently, there are a number of incubators available on the market [1–4]. In [2], the authors propose the smart egg incubator system integrating IoT

© The Author(s), under exclusive license to Springer Nature Singapore Pte Ltd. 2021
D.-T. Tran et al. (Eds.): ICISN 2021, LNNS 243, pp. 1–10, 2021.
https://doi.org/10.1007/978-981-16-2094-2_1

technology. The system includes two types of sensors, namely temperature and humidity sensors. In [3], the authors propose a set of procedures for detecting the development of chicken eggs using Self-Organizing Mapping (SOM) and K-means clustering algorithm. The results show that proposal method is more efficient than traditional method with the learning rate from 0.1 to 0.5. The authors [1] described the eggs smart incubator that is able to control the temperature, humidity, and reversal automatically. The results show that the system has been successfully hatch normally with 87.55%, 0.41% hatch and 1.84% dead hatch. In [4], the authors performed to evaluate the effect of a 12 h light and dark of green light during 1st day to 18th day of incubation time. The results show that green light effects to embryo development and earlier hatching.

The disadvantage of these machines is the high cost and not fully suitable with conditions in Vietnam. Therefore, the objective of the paper is to understand the poultry incubation system, to improve understanding and provide more material for the livestock industry as well as related industries. The study aims to apply it appropriately in Vietnam's cultural, economic, political and social conditions.

The main contribution of the paper is the design of the system to automatically adjust the temperature for the incubator and website to monitor the temperature and humidity. In addition, the system also integrates machine learning (ML) to monitor the sound of chickens and to give an alert if the chicken sounds abnormal.

The rest of the paper include four parts and organize as follows. Section 1 will discuss the related work. In Sect. 2, we present the proposal system. Section 3 will evaluate the proposal system and analyze the results. In final section, we give conclusions and future research directions.

2 Proposal System

2.1 Effective Poultry Incubation Conditions

Poultry is the common name for the two-legged species belonging to the group of winged animals that are kept and propagated by humans for the purpose of producing eggs, meat or feathers. Typical poultry species include chickens, ducks, swans and geese.

Parameters that affect the incubation process include temperature, incubation time, humidity, and air quality. Incubation temperature is important for embryo formation and development. It affects seed quality. If we adjust the temperature wrongly, it can lead to the chickens being shelled, not hatching, or chicks having their legs crossed. Different types of eggs, poultry, birds, and waterfowl have different incubation temperatures.

As well as incubation temperatures, different types of poultry eggs have different times. Incubation period is 28 days duck, goose and goose is 30 days, chicken eggs are from 20 to 21 days and quail eggs are from 16 to 17 days. The time for eggs to hatch can range from 5 to 10 h since larger eggs usually hatch later and smaller eggs usually hatch earlier.

Humidity greatly affects the metabolism and development of egg embryos. Moisture affects the evaporation of the eggs since it affects their metabolism. In addition, the moisture also helps lower the temperature of the egg to hatch easily. During incubation, air exchange is required to provide oxygen for good embryo development. Therefore, the incubator must have ventilation holes and a volume suitable for the number of eggs. For large machines, there should be a cooling fan to exchange air in the machine and the outside environment to ensure sufficient fresh air [1–4].

As analyzing above, the best incubation temperature for chickens is from 37.4 to 38.4°, the ideal number of incubation days is 21 days, and the humidity is from 55% to 65% [1–4].

2.2 Sound and Related Diseases

If poultry is raised, it will lead to sick chickens causing great economic damage. Several diseases can be easily detected based on their sound as in Table 1 [5–9].

Table 1. Respiratory diseases of chickens [5].

Disease name	Symptoms of disease
ORT	Breathing neck up to yawn, gasping for air
CRD	Shortness of breath, whistling sound when extending to breathe
IC	Breathing breath sounds
Avian influenza	Difficulty breathing and open their mouths to breathe
ILT	The chicken shook its head to sneeze, breathless, yawn, raise its neck to breathe
Fungal lung	Shortness of breath, whistling sound when extending to breathe
CORYZA	Runny nose, trouble breathing

Above is a table of some respiratory diseases of chickens. Symptoms are noticeable through the sound of a pet.

2.3 Proposal System

Hardware Design. The design system includes the following functions: measuring and adjusting parameters of incubator (temperature and humidity); displaying results on LCD; sending parameters to web-server; monitoring and displaying the parameters on web-server; and training model to determine animal health. An overview of the proposed system is shown in Fig. 1.

The system consists of 7 main blocks:

1. **Power block:** maintains the operation of the remaining 3 blocks
2. **Processor block:** This is the central processing unit of the system using STM32F10C8T6

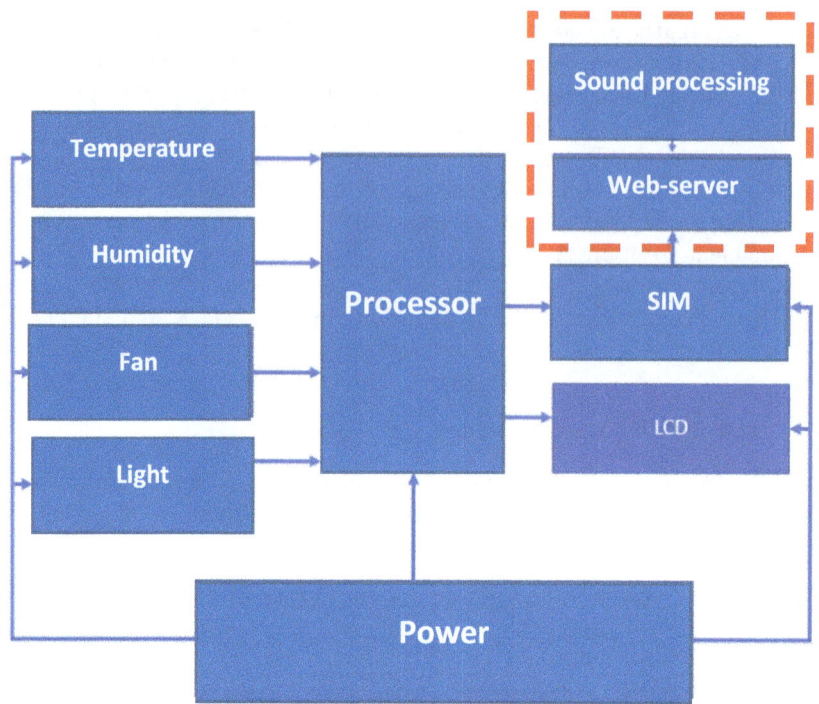

Fig. 1. Diagram of the proposed system.

3. **Equipment block:** This block has the function of measuring the parameters of the environment. It consists of four devices: temperature, humidity, fan, and light bulb to adjust the parameters of the incubator.
4. **Display block:** This block includes a 20 × 4 LCD that displays parameters of temperature, humidity, and user interaction.
5. **Communication block:** use SIM module to transfer data to web-server.
6. **Web-server block:** monitors the temperature and humidity values that are sent from restoring communication.
7. **Sound processing unit:** captures the sound of chirping chickens and puts into the trained model for predicting pet healthy.

Figure 2 is the principle diagram designed on Altium. It includes power, humidity temperature sensor, light and fan equipment, data communication (module SIM 900a), LCD connecting to I2C, and microprocessor blocks.

In the power block, we use LM7805 IC to make voltage from 5 V to 12 V for micro-controller and LCD. Capacitor C_3 (1000 uF/16 V) performs to filter the input source. Capacitors C_2, C_4, C_5, C_6 filter noise of out input and output. LED_2 and LED_4 are mounted in series with resistors R_3 and R_6 that have a notification when the circuit is turned on.

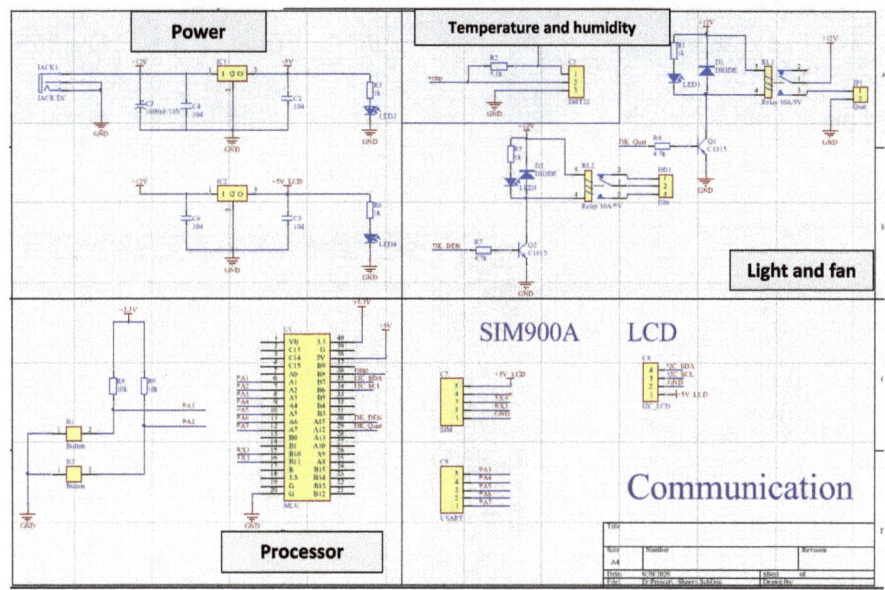

Fig. 2. Principle circuit diagram of the proposed system.

In communication block, SIM module 900A communicates with the micro-controller via USART3 standard of STM32F103C8T6. TX3 pin of the processor is connected to RX of SIM module. RX3 pin of the micro-controller is connected to TX of module SIM. The bit rate is 9600 bits/s.

In display unit, it shows value via a 20 × 04 LCD connected to the micro-controller through the I2C communication standard. The microprocessor acts as master and LCD acts as slave. Data is transmitted via I2C_SDA and synchronized with the clock signal from the I2C_SCL.

Block of equipment includes temperature sensor, humidity DHT22, fan, and lamp where:

DHT22 sensor relays with micro-controller through 1 wire communication standard and 3.3 V power. The data pin is connected to a resistor (5.1 kΩ) and is pulled onto the source.

Light and fan are controlled by micro-controller and two transistors (Q_1, Q_2) act as on/off switch. Diodes D_1 and D_2 have the role of short-circuit inductance when the relay is closed for protecting the transistors. Led_1 and Led_3 in series with resistors (R_1 and R_5) light up when the switch is turned on.

Microprocessor block includes STM32F103C8T6; IC AMS1117 to make from 3.3 V to 5 V; external clock; reset button; header to connect STLINK to load and debug programs; and main flash; system memory; and embedded SRAM.

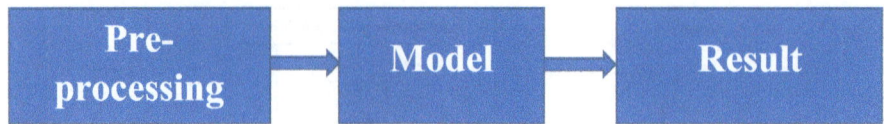

Fig. 3. Sound classification model.

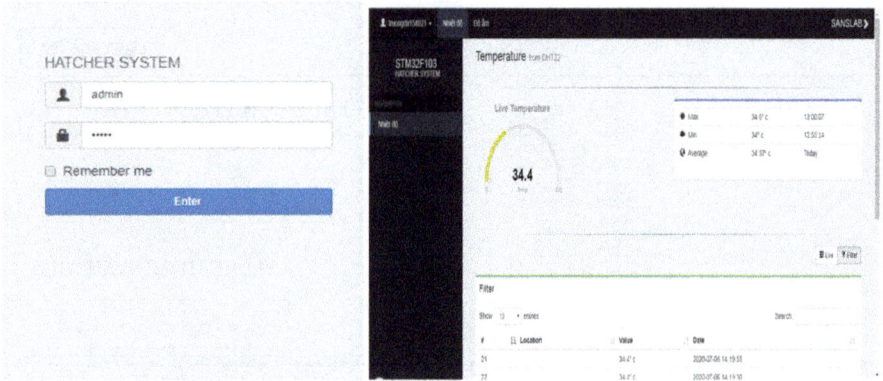

Fig. 4. Result of real-time data.

Sound Processing Design. To handle the sound, we propose to use the support vector machine (SVM) model with two steps, create data for the model and training the model. Details of SVM model can be found in [10,11].

In step 1, we select the sounds of strong and sick chickens from [7,12–14,18] to create data and put into training model.

In step 2, we will choose the training model. The language training model implementing for the newly created data set is SVM. We train with two classes where the strong chicken class is class 0 and the ill chicken class is class 1 [17,22–24].

Through the pre-processing step, the input data that are collected from [19, 20] will be extracted for their characteristics and then passed to the training model (SVM). The predicting sound model is shown in Fig. 3.

3   Result

The system includes incubator, web-server, and pet health prediction. The results are shown in Fig. 4. In Fig. 4, there are two types of account, namely user and admin accounts. The admin account is used to manage a database including users, devices, and other parameters.

In Fig. 4, we also see the instantaneous, maximum, minimum, the average of day, and the 10 closest values. Besides, the filter contains all of the values that have been submitted.

3.1   Result of Temperature and Humidity

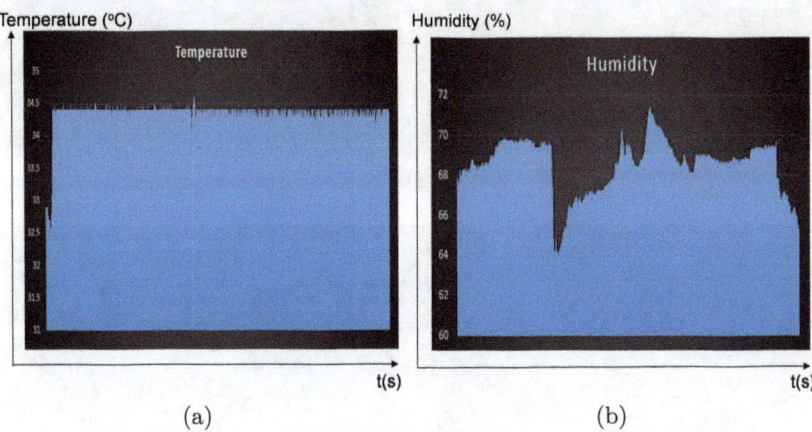

Fig. 5. Evaluating of the stability of measurement results with a) temperature and b) humidity.

We perform system measurements with temperature and humidity in real conditions. The results are shown as in Table 2 and Fig. 5. In Table 2, the average error of temperature is 0.6% while humidity is stable. This results show that the system is suitable for real environment.

Table 2. Measurement results with temperature and humidity.

Number	Temperature (°C)	Humidity %	DS18B20
1	34.4	66.0	34.363
2	34.3	66.1	34.388
3	34.4	66.2	34.375
4	34.5	66.1	34.463
5	34.4	66.1	34.425
6	34.4	66.2	34.438
7	34.3	66.1	34.413
8	34.3	66.2	34.375
9	34.4	66.2	34.513
10	34.5	66.1	34.463
Average value	34.39	60.12	34.42
Error	0.6%		

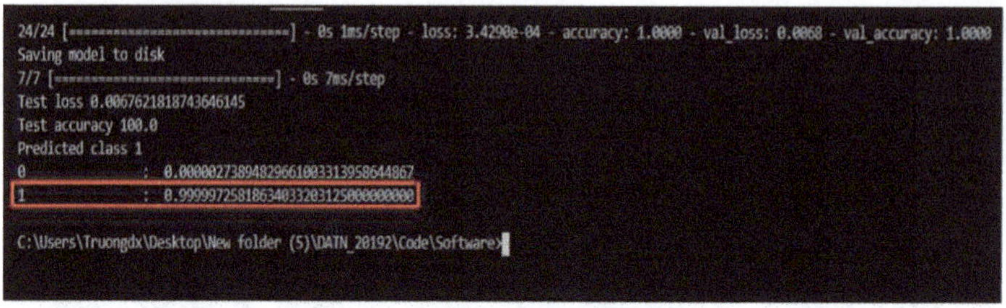

(a)

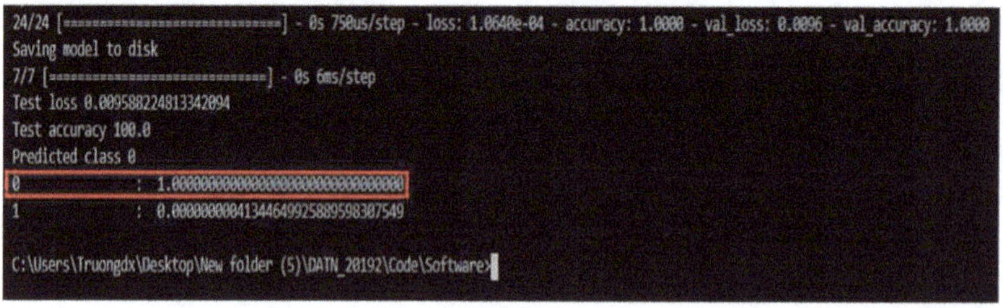

(b)

Fig. 6. The predicting results for two cases: a) sick and b) healthy chickens.

3.2   Result of Sound Processing Model

We perform for a real system in two cases of sick and healthy chickens. Figure 6(a) and (b) shows the predicted results for two sick and healthy chickens, respectively. Both cases have an accurate result of nearly 99% of which class 0 corresponds to the case of healthy chickens and class 1 to the case of sick chickens.

4   Conclusion

In the paper, we have successfully implemented the design of the incubator that includes a server to monitor and classify sounds to make pet health predictions. The system has achieved as high precision (over 90%). However, it still has limitations, for example the cooling system for the system still has defects and it has not integrated audio processing model into the web-server.

Therefore, the future development direction is to create data management on mobile applications and integrate audio processing model into web-server for monitoring of our IoT systems that is [15,16,21].

Acknowledgment. This research is carried out in the framework of the project funded by the Ministry of Education and Training (MOET), Vietnam under the grant B2020-BKA-06. The authors would like to thank the MOET for their financial support.

References

1. Sanjaya, W.S.M., Maryanti, S., Wardoyo, C., Anggraeni, D., Aziz, M.A., Marlina, L., Roziqin, A., Kusumorini, A.: The development of quail eggs smart incubator for hatching system based on microcontroller and internet of things (IoT). In: 2018 International Conference on Information and Communications Technology (ICOIACT), pp. 407–411 (2018)
2. Aldair, A.A., Rashid, A.T., Mokayef, M.: Design and implementation of intelligent control system for egg incubator based on IoT technology. In: 2018 4th International Conference on Electrical, Electronics and System Engineering (ICEESE), pp. 49–54 (2018)
3. Lumchanow, W., Udomsiri, S.: Chicken embryo development detection using self-organizing maps and k-mean clustering. In: 2017 International Electrical Engineering Congress (iEECON), pp. 1–4 (2017)
4. Wang, X.C., Li, B.M., Tong, Q.: Manipulation of green led in chicken egg incubation. In: 2017 14th China International Forum on Solid State Lighting: International Forum on Wide Bandgap Semiconductors China (SSLChina: IFWS), pp. 84–87 (2017)
5. Ali, M.Z., Park, J.E., Shin, H.J.: Serological survey of avian metapneumovirus infection in chickens in Bangladesh. J. Appl. Poult. Res. **28**(4), 1330–1334 (2019)
6. Wigle, W.: Respiratory diseases of gallinaceous birds. The veterinary clinics of North America. Exotic Animal Pract. **3**(2), 403–421 (2000)
7. Rizwan, M., Carroll, B., Anderson, D., Daley, W., Harbert, S., Britton, D., Jackwood, M.: Identifying rale sounds in chickens using audio signals for early disease detection in poultry, pp. 55–59, December 2016
8. Rohollahzadeh, H., Nili, H., Asasi, K., Mokhayeri, S., Najjari, A.H.A.: Respiratory and git tract immune responses of broiler chickens following experimental infection with Newcastle disease's virus. Comp. Clin. Pathol. **3**(27), 1241–1255 (2018)
9. Barbosa, E., Cardoso, C., Silva, R., Cerqueira, A., Liberal, M., Castro, H.: Ornithobacterium rhinotracheale: an update review about an emerging poultry pathogen. Vet. Sci. **3**(7), 1–13 (2020)
10. Bishop, C.M.: Pattern Recognition and Machine Learning (Information Science and Statistics). Springer, Heidelberg (2006)
11. Duda, R., Hart, P., Stork, D.G.: Pattern Classification, vol. xx, p. 688. Wiley Interscience, November 2000
12. Carpentier, L., Vranken, E., Berckmans, D., Paeshuyse, J., Norton, T.: Development of sound-based poultry health monitoring tool for automated sneeze detection. Comput. Electron. Agric. **162**, 573–581 (2019)
13. Fontana, I., Tullo, E., Carpentier, L., Berckmans, D., Butterworth, A., Vranken, E., Norton, T., Berckmans, D., Guarino, M.: Sound analysis to model weight of broiler chickens. Poul. Sci. **96**(11), 3938–3943 (2017)
14. Fontana, I., Tullo, E., Carpentier, L., Berckmans, D., Butterworth, A., Vranken, E., Norton, T., Berckmans, D., Guarino, M.: Sound analysis to model weight of broiler chickens. Poul. Sci. **96** (2017)
15. Huu, P.N., The, H.L.: Proposing recognition algorithms for hand gestures based on machine learning model. In: 2019 19th International Symposium on Communications and Information Technologies (ISCIT), pp. 496–501, September 2019
16. Huu, P.N., Thu, H.N.T.: Proposal gesture recognition algorithm combining CNN for health monitoring. In: Proceedings of 7th NAFOSTED Conference on Information and Computer Science (NICS), pp. 209–213. Vietnam, December 2019

17. Wang, J.-C., Wang, J.-F., He, K.W., Hsu, C.-S.: Environmental sound classification using hybrid SVM/KNN classifier and MPEG-7 audio low-level descriptor. In: The 2006 IEEE International Joint Conference on Neural Network Proceedings, pp. 1731–1735 (2006)
18. Liu, L., Li, B., Zhao, R., Yao, W., Shen, M., Yang, J.: A novel method for broiler abnormal sound detection using WMFCC and HMM. J. Sens. **2020**, 7 (2020)
19. Sadeghi, M., Banakar, A., Khazaee, M., Soleimani, M.: An intelligent procedure for the detection and classification of chickens infected by clostridium perfringens based on their vocalization. Revista Brasileira de Ciência Avícola **17**, 537–544 (2015)
20. McEnnis, D., McKay, C., Fujinaga, I., Depalle, P.: jAudio: an feature extraction library, pp. 600–603, January 2005
21. Nguyen Huu, P., Tran-Quang, V., Miyoshi, T.: Video compression schemes using edge feature on wireless video sensor networks. J. Electr. Comput. Eng. **2012**, 1–20 (2012). https://doi.org/10.1155/2012/421307
22. Piczak, K.J.: Environmental sound classification with convolutional neural networks. In: 2015 IEEE 25th International Workshop on Machine Learning for Signal Processing (MLSP), pp. 1–6 (2015)
23. Sameh, S., Lachiri, Z.: Multiclass support vector machines for environmental sounds classification in visual domain based on log-gabor filters. Int. J. Speech Technol. **16**, 203–213 (2013)
24. Uzkent, B., Barkana, B., Cevikalp, H.: Non-speech environmental sound classification using SVMs with a new set of features. Int. J. Innov. Comput. Inf. Control **8** (2012)

Designing a Remote Monitoring System for Lakes Based on Internet of Things

Quang-Trung Hoang[1], Nguyen Canh Minh[2], and Duc-Tan Tran[3(✉)]

[1] Faculty of Electrical and Electronic Engineering, Thuyloi University,
175 Tay Son, Dong Da, Hanoi, Vietnam
trunghq@tlu.edu.vn
[2] University of Transport and Communications, Hanoi, Vietnam
[3] Faculty of Electrical and Electronic Engineering, Phenikaa University,
Hanoi 12116, Vietnam
tan.tranduc@phenikaa-uni.edu.vn

Abstract. The problem of the environment has been a topic considered by many researchers in recent years. The Internet of Things (IoT) is an effective solution to collect relevant data from the environment. This paper proposes a system that allows remote data collection to support environmental parameters monitoring and provide warnings about water levels for lakes based on the IoT platform. The research results show that our proposed system works well with remote information connections, creating the foundation for the analysis and processing the lake environment data.

Keywords: Network architecture · Wireless sensor network · IoT

1 Introduction

With the development of the industrial revolution 4.0, almost all fields of production, operation, and management systems are rapidly shifting to a new form of organization based on the internet of things [1]. The basis for creating the internet of things platform is wireless sensor networks (for collecting data from surrounding environments), communication connection systems, and data management systems on the internet. Many solutions apply IoT in different fields of management and production organization. In practical system design, each selected IoT model will have different effects and specific optimization parameters. For example, for the same problem related to data collection or environmental parameters, the designs in [2, 3] are based on the Zigbee communication standard, while the system proposed in [4] uses the LoRa communication standard for the sensor devices.

Water resource management and water quality monitoring can now be moved entirely to a new way, based on IoT [5]. This solution effectively reduces the cost of human effort, and improves accuracy, ensures timeliness in the data processing. However, as analyzed above, each design system may already exist in different types of models, and therefore the effect is not the same. Therefore, a flexible system model is required. This paper proposes a practical system model that supports remote monitoring

D.-T. Tran et al. (Eds.): ICISN 2021, LNNS 243, pp. 11–17, 2021.
https://doi.org/10.1007/978-981-16-2094-2_2

of environmental parameters for lakes on the IoT platform. The system allows flexible operation according to the actual conditions of information connection and transmission distance.

2 Proposal for Remote Monitoring System for Lakes

An IoT-based monitoring system of environmental parameters for lakes is shown in Fig. 1. In which, the system includes three main subsystems: (1) Wireless sensor network subsystem to collect data, (2) Internet connection subsystem, (3) Data management subsystem (data center). The system allows the monitoring of multiple lakes.

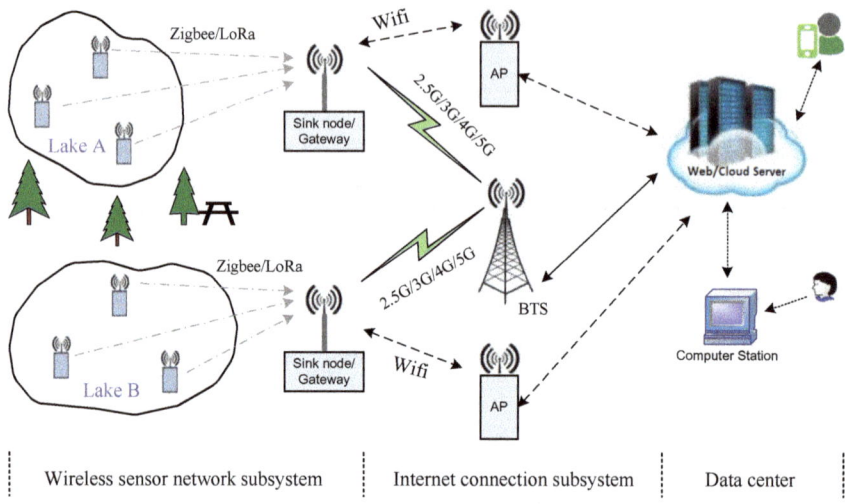

Fig. 1. Diagram of the proposed system.

2.1 Wireless Sensor Network Subsystem

The wireless sensor network elements here include the sensor nodes and sink/gateway nodes linked together in a star topology or tree topology [7–16]. Each sensor node collects and packages data of environment parameters monitored (for example, pH concentration, salinity, dissolved oxygen in water, temperature, water level, etc.). The data packet is transmitted to the sink/gateway node through the radio link. Each sensor node can be designed to collect many lakes surrounding environment parameters by integrating with multiple sensor probes. In this case, we call the multi-parameter sensor node. Sink/Gateway nodes aggregate data and send control commands to the sensor nodes (if any). The data is first preprocessed at the sink/gateway node and then forwarded to the remote data center via the internet connection. Figure 2 presents the flowchart of these nodes.

2.2   Internet Connecting Subsystem

There are two modes of internet connection: (1) Using the mobile network via the base station (BTS), (2) Using the WiFi network with the radio access point (AP). Depending on the existing communication networks in the lake area to be monitored, we can choose one of these two modes of internet connection. It is noted that using BTS will create longer distance connections than AP.

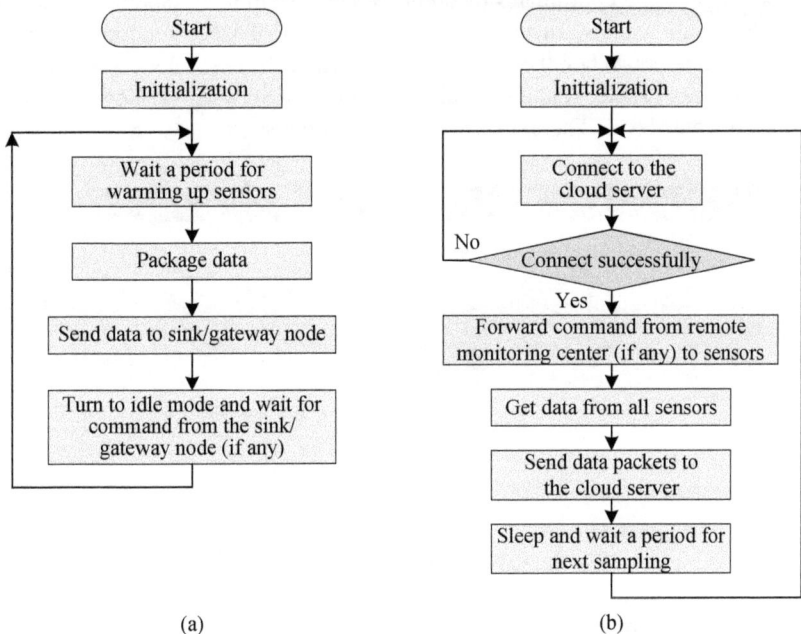

(a)   (b)

Fig. 2. Flowchart of the sensor node (a) and the sink/gateway node (b).

2.3   Data Management Subsystem

This subsystem is also called the data center. The data center components are web servers or cloud servers integrated with the IoT platform, capable of storing data sent from sensor networks. Cloud servers integrate an IoT-enabled platform (such as ThingsBoard [6]), which allows implementation algorithms that analyze, compute and form data, and then send the results to a web interface for the user to observe lake on a smart mobile phone or a computer.

3   Experiment on the Proposal System

To run the test with the proposed system, we give the following scenario:

- Cloud server: ThingsBoard
- Number of lakes to be monitored: 02 (setup on a cloud server)

- Number of sensor nodes per lake: 02 (setup on a cloud server)
- Communication protocol for sensor networks: LoRa (on testbed device)
- Internet connection method: GPRS/3G mobile network (on testbed device)
- Monitoring parameters: water level, pH concentration, surrounding temperature (embedded programming to create data packet on testbed device).

The interface of the proposed system is shown in Fig. 3. The small window in the upper left (District map) shows the measuring areas corresponding to the two lakes. It is possible to get these coordinates by adding a GPS positioning module on the sensor device. The small window in the upper right is a list of lakes for monitoring environmental parameters. When the user wants to know each sensor node's information in any lake, he clicks to select the lake's label. The third small window shows the warning status of the water level. The expert sets the alarm threshold.

Fig. 3. Lake monitoring system interface.

3.1   Hardware Devices in Sensor Networks

We created two hardware devices using components: Nano Arduino Kit; Transceiver module 433 MHz E32-TTL-100 LoRa; sensing elements: water level, pH concentration, and temperature; and SIM808 R14.18 module. The block diagram of the sensor node is shown in Fig. 4. We do embedded programming in Arduino language with AT instruction set and SIM module library for the hardware devices to work.

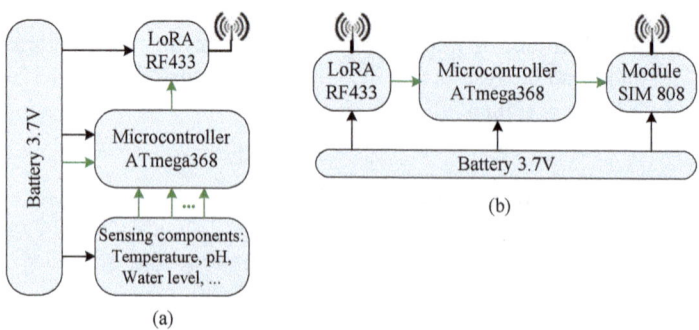

Fig. 4. (a) Diagram of the sensor node; (b) Diagram of the sink/gateway node.

Figure 5 shows an experimental design system for sending data from the sensor node to the ThingsBoard cloud server.

Fig. 5. The photo of our experimental system.

3.2   Reliability of Transmission

To evaluate the transmission reliability of the LoRa protocol used for wireless sensor networks, we measure the packet-transmission success rate (or packet distribution rate: PDR) of each sensor node at various distances. The main parameters are used as follows:

- Number of sent packets: 360
- Period for sending data packet: 5 s
- Payload size: 42–46 bytes
- Baudrate: 9600
- Radio power: 100 mW
- SMA antenna: RF433, 3 dBi, TX433-JK-20
- Environment: LoS or NLoS.

We use the radio path model for testing PDR (see Fig. 6). In the NLoS path, each wall is 40 cm thick and higher than the transceiver antennas.

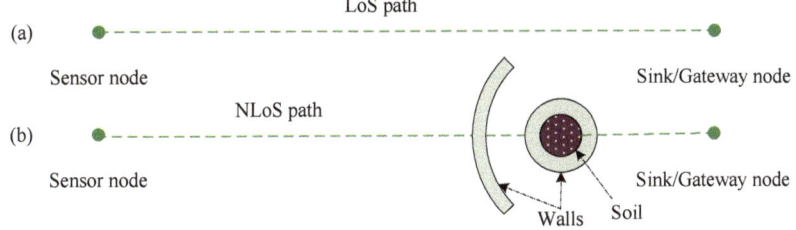

Fig. 6. The radio paths: (a) Line-of-Sight, (b) Non-Line-of-Sight.

The measuring results are shown in Table 1. We find that the transmission reliability is quite high in the transmission line without obstruction (LoS path). Especially at a distance of 400 m, reliability reaches 100%. At long distances, more than 400 m, the reliability begins to decrease. However, in the range of 600 m–1000 m, the reliability shown in Table 1 is still quite good. In the case of NLoS, we perform a transmission reliability measurement with obstruction of walls and soil block (as shown in Fig. 6-b). The measurement results are shown in Table 1. Observing Table 1, we see that the transmission reliability of the sensor network is greatly affected by obstacles. However, with a distance of below 400 m, the reliability is acceptable for the sensor network. At a distance below 300 m, the reliability is so good. From the actual measurement results on the link quality of WSN, we can calculate the Number of packets required to send how the sink/gateway can receive environment data well. If PDR is p then the Number of the packet needs to send by a sensor node during a receiving period of the sink/gateway node is $N = 1/p$. For example: if $p = 65.3\%$ at a distance 400 m (NLoS), then N is proximately 1.54. So during a receiving period of the sink/gateway node, if a sensor node repeatedly sends a data packet, twice the sink/gateway can receive data well from the sensor node. So, the above experimental results have necessary implications in making recommendations for selecting the lake monitoring system's working parameters.

Table 1. Results on Line of Sight (LoS).

	Line of Sight (LoS)			Non-Line of Sight (NLoS)		
Distance (m)	400	600	1000	200	300	400
Number of received packets	360	355	326	340	334	235
PDR (%)	100	98.6	90.1	94.4	92.8	65.3

4 Conclusions

In this paper, we focus on designing a holistic system that approaches the internet of things. The system monitors general environmental parameters and collects environmental data around the lake area through a wireless sensor network with an internet connection to the remote data center. Experimental results show that the system works well, can be deployed in practice, and developed on a larger scale. In particular, the hardware system used in the design is popular in the market and has a low cost, meeting small, medium, and large scale system development requirements.

Acknowledgments. The project is partly funded by Thuyloi University, 175 Tay Son, Dong Da, Hanoi, Vietnam.

References

1. Wortmann, F., Kristina, F.: Internet of things. Bus. Inform. Syst. Eng. **57**(3), 221–224 (2015)
2. Aarti, R.J., Karishma, K., Pankaja, P., Kiran, M., Gauri, S.: Environmental monitoring using wireless sensor networks (WSN) based on IOT. IRJET **4**(1), 1371–1378 (2017)
3. Xiao, J., Li, T.J.: Design and Implementation of Intelligent Temperature and Humidity Monitoring System Based on ZigBee and WiFi. In: 3rd International Conference on Mechatronics and Intelligent Robotics, Published by Elsevier (2019)
4. Botero-valencia, J., Castano-Londono, L., Marquez-Viloria, D., Rico-Garcia, M.: Data reduction in a low-cost environmental monitoring system based on LoRa for WSN. IEEE Internet Things J. **6**(2), 3024–3030 (2019)
5. Deshmukh, S., Barapatre, P.: Internet of things based system for water resource engineering. In: International Conference on Emanations in Modern Technology and Engineering, ISSN 2321-8169 (2017)
6. De Paolis, L.T., De Luca, V., Paiano, R.: Sensor data collection and analytics with thingsboard and spark streaming. In: IEEE Workshop on Environmental, Energy, and Structural Monitoring Systems, pp. 1–6 (2018)
7. Ta, T.D., Tran, T.D., Do, D.D., Nguyen, H.V., Vu, Y.V., Tran, N.X.: GPS-based wireless ad hoc network for marine monitoring, search and rescue (MSNR). In: Second International Conference on Intelligent Systems, Modelling and Simulation, pp. 350–354, IEEE (2011)
8. Do, D.D., Nguyen, H.V., Tran, N.X., Ta, T.D., Tran, T.D., Vu, Y.V.: Wireless ad hoc network based on global positioning system for marine monitoring, searching and rescuing (MSnR), In: Asia-Pacific Microwave Conference, pp. 1510–1513, IEEE (2011)
9. Duc-Tuyen, T., Duc-Tan, T.: Efficient and reliable GPS-based wireless ad hoc for marine search rescue system. In: Multimedia and Ubiquitous Engineering, pp. 911–918. Springer, Dordrecht (2013)
10. Nguyen, C.D., Tran, T.D., Tran, N.D., Huynh, T.H., Nguyen, D.T.: Flexible and efficient wireless sensor networks for detecting rainfall-induced landslides. Int. J. Distrib. Sensor Netw. **11**(11), (2015)
11. Tran, D.T., Nguyen, D.C., Tran, D.N., Ta, D.T.: Development of a rainfall-triggered landslide system using wireless accelerometer network. Int. J. Adv. Comput. Technol. **7**(5), 14 (2015)
12. Chinh, N.D., Tan, T.D.: Development and implementation of a wireless sensor system for landslide monitoring application in Vietnam. Int. J. Inform. Commun. Technol. **13**(2), 227–242 (2018)
13. Nguyen, D.C., Tran, D.T., Tran, D.N.: Application of compressed sensing in effective power consumption of WSN for landslide scenario, In: Asia Pacific Conference on Multimedia and Broadcasting, pp. 1–5, IEEE (2015)
14. Tan, T.D., Anh, N.T., Anh, G.Q.: Low-cost Structural Health Monitoring Scheme Using MEMS-based Accelerometers. In: Second International Conference on Intelligent Systems, Modelling and Simulation, pp. 217–220, IEEE (2011)
15. Duc, T.T., Duc, T.T., Do, D.D., Nguyen, H.V.: Innovative WiMAX broadband internet access for rural areas of Vietnam using TV broadcasting ultra-high frequency (UHF) bands. In: TENCON IEEE Region 10 Conference, pp. 526–529, IEEE (2011)
16. Gian Quoc, A., Nguyen Dinh, C., Tran Duc, N., Tran Duc, T., Kumbesan, S.: Wireless technology for monitoring site-specific landslide in Vietnam. Int. J. Electr. Comput. Eng. **8**(6), 4448–4455 (2018)

The Role of Blockchain Security in the Internet of Things

Omar Michael$^{(\boxtimes)}$, Doan Trung Tung, and Nguyen Canh Khoa

FPT University, Hanoi, Vietnam
{omar2, tungdt27, khoanc4}@fe.edu.vn

Abstract. Blockchain may be a new technology within the Internet of Things (IoT). It functions with the public, distributed, decentralized, and instantaneous ledger to stockpile various transactions within the nodes of IoT. A Blockchain comprises of sequences of blocks with each component linked to the previous block sequences. Individual blocks contain previous block hash, cryptography hash code, and stored sets of knowledge. The transactions, which happen within the Blockchain are the elemental elements, which transfer data from one IoT node to a different one. The nodes vary in physical composition, though all of them have actuators, sensors, and programs; thus, they will communicate with one another. The function of Blockchain in IoT is to ensure a secure process of transferring data through the nodes. The technology is efficient and usable publicly without worrying about knowledge pilferage. The blockchain is important for IoT because it enables the secure transfer of data between its nodes in an environment, which is heterogeneous. The communications in blockchain are traceable to a person who has the permission to exchange information within the IoT. The system may help in IoT to reinforce the safety of communication data. This paper, reconnoitered the weakness of IoT security, the importance of Blockchain technology in IoT, the integration of blockchain in IoT, and therefore the design architecture of the integrating of IoT and Blockchain.

Keywords: Authentication · Decentralization · Blockchain · Internet of things IoT · Hash · Node · Cloud

1 Introduction

The growth of the Internet of Things is rapid, and each and every year, the technology aims to embrace 5G technologies for various functions, including e-health, smart cities, smart homes, and distributed intelligence, among others [1]. However, IoT has security challenges, and it is linked in a decentralized manner, thus difficult to use the existing typical systems while communicating between IoT nodes. Blockchain technology enhances security in transactions, which occur in IoT devices. The technology avails a decentralized, distributed, and sharable ledger that members of the public can access to keep the blocks' data, which are processed then get verified in a network of IoT. The information, which is held in the ledger, is managed in an automated manner using the peer-to-peer-based topology. Blockchain technology entails processes in which transactions are sent in the form of a block, which is connected with every device and each other. The IoT and Blockchain together function in the cloud integration and IoT frameworks. In a decentralized market, technology enables investors to deal directly with each other instead of operating from within a centralized exchange. The block body depicts all the records of transactions, which occur on the Blockchain. The maximum amounts of transactions rely upon the size of the blocks [2]. In this regard, the paper discusses security issues and solutions regarding IoT and Blockchain.

1.1 Paper Outline

The manuscript is sectioned as follows: The first section represents the introduction, while the literature review is depicted in the second section. The third part covers weaknesses in IoT security as the fourth part talks about Blockchain as a solution to IoT Security problem. The fifth section focuses on the architecture of the IoT and Blockchain integration. The sixth section explains the code design and implementation along with the execution procedure. The seventh section is all about the implementation processes and the last section is the summary of the whole paper.

2 Literature Review

Since 2018, communication experts are paying much attention to the security of IoT devices used in communication. Many researchers have done their publications regarding the topic. Stuart Harber is among the first individuals to publish a literal work on the transfer of information among parties in a secure manner without keeping it in a time stamping service the concept of Blockchain originated from the duo. However, Satoshi Nakamoto was the first person to present Blockchain in 2008 [3, 20]. The scientist demonstrated the technology by showing a paper containing an arrangement of blocks in chains that would be used to authenticate the exchange of data between the IoT network's nodes with a proposed algorithm for information exchange in blockchains and IoT.

Though as connectivity to IoT increases, the architecture of computing devices also gets complex. The nature of the communication devices makes them more vulnerable to cyber-attacks. IoT physical appliances are often kept in unsecured entities that make them prone to hacking; hence the hackers can manipulate information to suit their interest as they pass through the network. Consequently, the information route and device authorization is a critical issue in IoT, therefore, Blockchain becomes an important aspect of security. The IoT is an expansive organization of things entrenched with computer programs, sensors, and comparative technologies with the intention of exchanging data and information with other frameworks and gadgets over the web [4]. The IoT also entails a set of methods given by particular identifiers that empower the movement of data and information over enormous systems without the help of people. The term IoT includes an assortment of gadgets and systems that incorporate Wi-Fi, cellular, Bluetooth gadgets, health wearable, LPWANs, and ZigBee [5]. The IoT is prone to vulnerability, the strategy and the intent to hack diverse IoT frameworks may not solely target the consumer information, but to sabotage the data owners or users in terms of financial incapacitating and putting their lives at risk by virtue the victims use the gadgets every day to accomplish fundamental tasks. Also, the devices with the IoT framework are hacked and abused to assault the web foundation of the provider companies [6].

The common assault or the risks of most IoT frameworks is the denial-of-service attack, which is abridged as DDOS (Distributed Denial of Service). In this assault, the culprit disrupts temporarily or completely to terminate the services the users offer to their clients, and it is common among competing business entities [7]. DDOS attacks happen when the cyber attacker sends a superfluous number of requests to the targeted machine or internet resource with the intention to obstruct or halt the delivery of services to consumers by overloading the system [8]. Consequently, IoT framework becomes prone to an array of assorted security shortcomings and challenges that make the framework unsustainable in the conveyance of critical services.

The process Blockchain execute its functions entails various steps. First, nodes and blockchain network communicate through public and private keys. The nodes utilize their private keys to sign their respective transactions digitally then get the network access through the public key. The transactions, which have been signed by the nodes are broadcasted by the same nodes then undergo verification by all the notes, which are within reach of the blockchain network with the exception of the node, which does the transaction [9]. At that stage, any transaction, which is not valid, is discarded, and it is referred to as verification. The activity that takes place at the stage in each and every node is choosing a block, which has been initially publicized, and authenticating it if it holds transactions, which are legal. Also, it mentions the precise block present on blockchains via the hash. After the verification, the nodes add the blocks to respective blockchains then apply the activities they hold to update the blockchains. In the event the transactions are not authentic, the block projected is discarded, thus terminating the mining round which exists [10].

Notably, the security weakness of the IoT arises when a device such as a surveillance camera, smart car, or intelligent lamp, which is connected to the internet, has a system and does a specific task [11]. The IoT is prone to hacking due to its weak system and easy to infiltrate-infrastructure. The system's weaknesses entail privacy problems, which may risk the information of the users such as a violation of the location data use [12]. Vulnerabilities exist in the interfaces of communication between integrated devices, which depicts a serious challenge to user and the IoT. The inability to identify authorized users results in an infiltration, which may pose a serious threat to data validity.

In the IoT settings, it is not easy to know the source of data generated from a device, which is kept in the whole network and prone to alteration by attackers.

Regarding scalability, the IoT links several devices and sensors so that information can be shared among the parties of interest. The process entails many types of applications. Therefore, meeting scalability is difficult due to the complex nature and rapid growth of the system. The IoT also represents a cluster of techniques, in which specific identifiers enable large data and information to be conveyed over wide networks in the absence of human beings to provide assistance, thus facilitating cyber-attacks [13].

The security and protection in the correspondence among IoT gadgets gave an excess consideration in 2017 and 2018. Haber and Stornet, proposed an article on trading with security without putting away any data on the time-stepping administration.

The possibility of blockchains comes from yet the first blockchains introduced by Satoshi Nakamoto in 2008. The author's introduced the "IoTChain" for verification of data traded between two hubs in an IoT network with the introduction of a calculation to trade the data in IoT and blockchains. The article was centered around the approval part of the security in the IoTChain structure that investigates the cloud system to associate the shrewd gadgets in middleware IoT gadgets which give a smart thought to give secure correspondence to the pleasant structure called web cloud system security [14].

3 Characteristics of Blockchain in Relations to IoT Security

Weakness in IoT Security. The biggest weakness of any IoT device is security. IoT devices can be compromised by attackers. The following bullets below explain the components of weakness found in IoT.

Token security
Client privacy
Denial-of -service
Bootstrapping Key Server
Secure Communications

Decentralization. Each transaction should be validated in a processing setting, which is centralized. The process occurs by means of the centralized party that is trusted like the banking system and results in the performance and cost decrees at the central location. Regarding the model of centralized IoT, the third party does not need Blockchain. The data consistency and integrity are maintained in the Blockchain [15]. Other characteristics such as the listed bullets below are said to have enhanced the blockchain technology security system and is efficient usage.

Persistency
Anonymity
Resilient backend
High efficiency
Transparency

Scalability. Blockchain has an address space of 60-bit. The IPv6 (Internet Protocol version 6) address has an address space of 128-bit, and the general public key has a 160-bit hash. The spacing is produced by the Elliptic Curve Digital Signature Algorithm (ECDSA). Finally, Blockchain has 4.3 billion address compared to IPv6 [16].

Smart Contract. This is a component of Ethereum, Nick Szabo introduced the concept 1994. Several languages of programming are given support by Ethereum, like Solidity, the most utilized compiler and language in programming the smart contract with enabled various policies and access rights to be defined in blockchain IoT [17].

4   Proposed Design and Architecture of IoT and Blockchain Integration

This project focuses on direct communication and the IoT devices classified as actuators and sensors. Sensors collect data like humidity level while actuators do specific tasks like turning on the light depending on the commands they get from the users. Figure 1 shows a conceptual scenario of an IoT Blockchain platform. Figure 2 depicts the IoT Blockchain (Layer-Based). The architecture is modular as every layer is not coupled with other layers to enable us to delete or add novice modules without tampering with the entire system. Figure 3 diagram gives a description of the workflow of the system and provides an in-depth understanding of each unit of the IoT Blockchain platform proposed. Finally, Fig. 4 gives an illustration of the in-depth process of executing transactions in the Blockchain network.

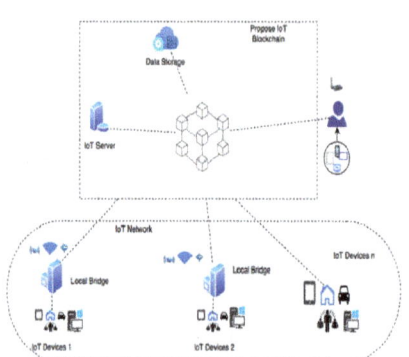

Fig. 1. IoT Blockchain platform conceptual design

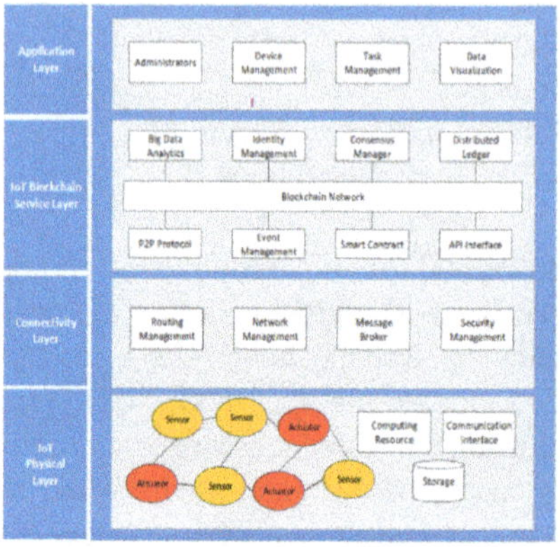

Fig. 2. Layer-based IoT Blockchain platform

4.1   Merits

The merits will lessen the expense since it conveys straightforwardly without the outsider. It takes out all the outsider hubs between the sender and the beneficiary. It gives direct correspondence.

Diminish the Cost,
Reduce Time,
Security & privacy,
Social Services,

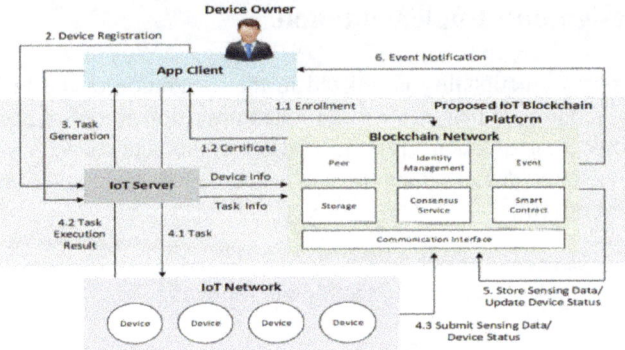

Fig. 3. System work flow of the proposed IoT blockchain platform

4.2   Demerits

The blockchain can become hang due to its hefty heap of the exchange. The Bitcoin stockpiling is getting in excess of 197 GB stockpiling in envision in the event that IoT incorporates with blockchain, at that point the heap will be heavier than the current circumstance (Fig. 5). Depicts the heterogenous blockchain network interaction with smart contract.

Adaptability.
Capacity.
Absence of Skills.
Disclosure and Integration.
Protection.
Interoperability.

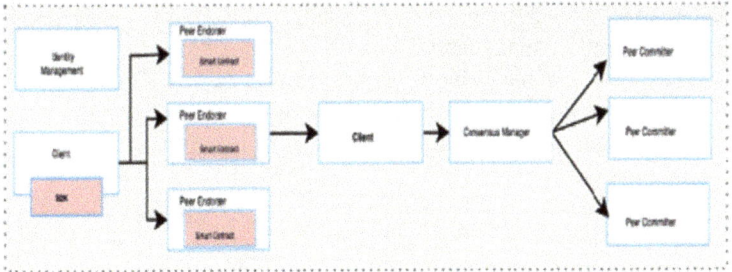

Fig. 4. Detail execution flow

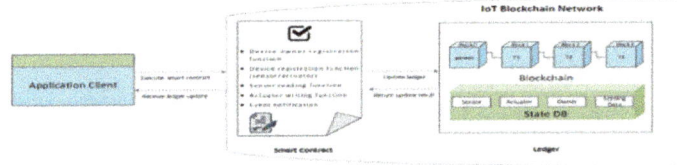

Fig. 5. Smart contract for IoT blockchain network interaction

5   Code Design and Implementation

Transaction process functionality is utilized in the integration of the IoT and Blockchain. Depicted in the code below, the task of the transaction processor has a JavaScript function; the processor function gives an update on the actuator value based on the readings from the input devices, then the new value gets updated in the registry before departing the event [18].

```
BEGIN
define function ActuatorWriting that do the following
get assetRegistry
then
Get actuator by its ID
Then
        If tx's states is available
                Set actuator's state as Tx's state
        If tx's enable is valid
                Set he global t_actuator as actuator
        Update the actuator using assetRegistry
then
        create new actuator event from the factory
        set the event's id as device id
        set the event's state as the actuator's state
        set the event's ownerId as the actuator's ownerId
        set message for the actuator event
END
```
Code 1: Transaction Processor Function for Actuator Writing Transactions
```
Define object variable query
BEGIN
    declare object key description with value
    declare object key statement with value
        If key statement has value
                Select field from table [user define]
        If passed query param match a return set from table
        Return value
else
    return no value
END query
```
Code 2. Query definition in smart contract

6 Implementation

The hosting of the IoT device server uses the Raspberry Pi that is integrated with different actuators and sensors. Then hyper ledger fabric framework is used to develop the blockchain network in which an ordered node and four peers run images in the ducker container. All the peers have data storage and smart contract to write on trans- actions on a ledger using blocks. The couch DB (database) is utilized to act as the state database, which gives an in-depth query report and gives support to smart contract data that is shown as JavaScript Object Notation. It gives the support a variety of query types like put, get, and delete together with a state key that allows the application to invoke a smart contract to have access through APIs (Application Programming Interface).

The couch Database (CDB) gives support to a modest state value with a composite state value, single key-value pair, and several key-value pairs. Contrary to a state database, the blockchain is incorporated physically in a given file, and the structure of the blockchain data utilized to keep records of small data set of operations that are not complex. The REST server gives different restful APIs, which disclose the functions the blockchain network defines. A physical gadget or a web client can invoke the entire service directly. Further, it gives a host to Fabric client that uses system calls of Google remote procedure to engage Hyperledger Fabric network in communication. The blockchain plays the role of a transaction log, which keeps records of all the state changes, the transactions are further gathered into blocks, which are joined together cryptographically to generate a chain sequence in which all the ledger transactions are classified in order of time. Therefore, the user can tell the history of the transactions. Finally, ordered node I and PBFT (Practical Byzantine Fault Tolerance) algorithm is incorporated to make sure there is consistency in all the copies of the parent ledger.

7 Performance/Conclusion

Blockchain provides secure data storage solutions to individuals and organizations. Further, it also avails a chance for the complex IoT systems to function optimally in a secure setting [2]. The technology enriches the network and System of IoT by availing the service that enables the sharing of the network by a third party. In the System, information is easily traceable and reliable. The IoT can immensely benefit from the communication functionality blockchain provides [19]. Further, the integration of the Blockchain and IoT has several benefits, including scalability and Decentralization, which means the move to peer service from a centralized architecture will deter data insecurity . Scalability and Decentralization will also happen by reducing the scenario where a few powerful enterprises control a huge amount of data of individual consumers.

References

1. Atzori, L., Iera, A., Morabito, G.: The Internet of Things: a survey. Comput. Netw. **54**(15), 2787–2805 (2010). https://doi.org/10.1016/j.comnet.2010.05.010
2. Panarello, A., Tapas, N., Merlino, G., Longo, F., Puliafito, A.: Blockchain and IoT integration: a systematic survey. Sensors **18**(8), 2575 (2018)

3. Giusto, L., Iera, D., Morabito, A., Atzori, G.: Focus on middleware and networking issues RFID and sensor networks technology covered as well as network security and privacy issues, vol. 49, no. D, p. 4419 (2010)
4. Hang, L., Ullah, I., Kim, D.H.: A secure fish farm platform based on blockchain for agriculture data integrity. Comput. Electron. Agric. **170**, 105251 (2020). https://doi.org/10.1016/j.compag.2020.105251
5. Ahmad, S., Hang, L., Kim, D.: Design and implementation of cloud-centric configuration repository for DIY IoT applications. Sensors **18**(2), 474 (2018). https://doi.org/10.3390/s18020474
6. Bhattacharjee, S., Salimitari, M., Chatterjee, M., Kwiat, K., Kamhoua, C.: Preserving data integrity in IoT networks under opportunistic data manipulation. In: Proceedings of the 2017 IEEE 15th International Conference on Dependable, Autonomic and Secure Computing. 2017 IEEE 15th International Conference Pervasive Intelligence and Computing 2017 IEEE 3rd International Conference on Big Data Intelligence and Computing, vol. 2018-Janua, pp. 446–453 2018. https://doi.org/10.1109/DASC-PICom-DataCom-CyberSciTec.2017.87
7. Sicari, S., Rizzardi, A., Grieco, L.A., Coen-Porisini, A.: Security, privacy and trust in Internet of Things: the road ahead. Comput. Netw. **76**, 146–164 (2015). https://doi.org/10.1016/j.comnet.2014.11.008
8. Khan, M.A., Salah, K.: IoT security: review, blockchain solutions, and open challenges. Fut. Gener. Comput. Syst. **82**, 395–411 (2018). https://doi.org/10.1016/j.future.2017.11.022
9. Mbarek, B., Jabeur, N., Pitner, T., Yasar, A.U.H.: MBS: multilevel blockchain system for IoT. Pers. Ubiquit. Comput. (2019). https://doi.org/10.1007/s00779-019-01339-5
10. Wei, J.: Association for Information Systems AIS Electronic Library (AISeL) The Adoption of Blockchain Technologies in Data Sharing : a State of the Art Survey (2019)
11. Fernández-Caramés, T.M., Fraga-Lamas, P.: A review on the use of blockchain for the Internet of Things. IEEE Access **6**, 32979–33001 (2018). https://doi.org/10.1109/ACCESS.2018.2842685
12 Zheng, Z., Xie, S., Dai, H., Chen, X., Wang, H.: Blockchain challenges and opportunities: a survey. Int. J. Web Grid Serv. **14**(4), 352–375 (2018). https://doi.org/10.1504/IJWGS.2018.095647
13. Tawalbeh, L., Muheidat, F., Tawalbeh, M., Quwaider, M.: IoT privacy and security: challenges and solutions. Appl. Sci. **10**(12), 1–17 (2020). https://doi.org/10.3390/APP10124102
14 Stuart Haber, W., Stornetta, S.: How to time-stamp a digital document. In: JM, Alfred, AV, Scott (eds.) Advances in Cryptology-CRYPT0' 90, pp. 437–455. Springer Berlin Heidelberg, Berlin, Heidelberg (1991). https://doi.org/10.1007/3-540-38424-3_32
15. De Filippi, P.: The interplay between decentralization and privacy: the case of blockchain technologies, J. Peer Prod. **7**, 18 (2016)
16. Christidis, K., Devetsikiotis, M.: Blockchains and smart contracts for the Internet of Things. IEEE Access **4**, 2292–2303 (2016)
17. Wang, S., Ouyang, L., Yuan, Y., Ni, X., Han, X., Wang, F.: Blockchain-enabled smart contracts: architecture, applications, and future trends. IEEE Trans. Syst. Man Cybern. Syst. **49**(11), 2266–2277 (2019). https://doi.org/10.1109/TSMC.2019.2895123
18. Sompolinsky, Y., Zohar, A.: Secure high-rate transaction processing in Bitcoin. Financial Cryptography (2015)
19. Bruneo, D., et al.: An IoT service ecosystem for smart cities: the #SmartME project. Internet Things **5**, 12–33 (2019). https://doi.org/10.1016/j.iot.2018.11.004
20. Nakamoto, S.: Bitcoin: a peer-to-peer electronic cash system (2008)

Challenges and Merits of Wavelength-Agile in TWDM Passive Optical Access Network

Nguyen-Cac Tran[1], Ngoc-Dung Bui[2(✉)], and Tien-Thanh Bui[3]

[1] Genexis B.V., Eindhoven, The Netherlands
k.tran@genexis.eu
[2] Faculty of Information Technology, University of Transport
and Communications, Hanoi, Vietnam
dnbui@utc.edu.vn
[3] Faculty of Civil Engineering, University of Transport
and Communications, Hanoi, Vietnam
btthanh@utc.edu.vn

Abstract. Passive optical networks are the main technology behind the surge of Gigabit broadband services. As the bandwidth demand continues to increase, the wavelength dimension is added on top of the time dimension. Multiple wavelengths pose a challenge but also an opportunity to further optimize the network operation because the dynamic traffic pattern in which there is significant difference between peak and off-peak hours. Wavelength-agility enables dynamic sharing of both wavelengths and timeslots but requires tunable transceivers in ONUs with tuning times below 150 ms. It increases traffic handling capacity by 36.9% and saves 33% power as compared to wavelength-fixed because the traffic load can spread among wavelengths in peak time while some wavelengths can be shut down in off-peak time.

Keywords: Optical communications · Access networks · Wavelength-agility

1 Introduction

The Full Service Access Network consortium (FSAN) has made a major decision for next-generation passive optical access networks stage 2 (NG-PON2). Time and wavelength division multiplexing (TWDM) was selected after narrowing down from a wide range of technologies [1]. The wavelength dimension is added to further scale up the capacity of a single feeder fiber since increasing the line rate is no longer cost-effective beyond 10 Gbps. TWDM-based NG-PON2 is expected to stack four XG-PON-like TDM PONs in a wavelength plan that allows coexistence with GPON and XG-PON. The optical distribution network (ODN) is untouched as only power splitters are employed. Therefore, the ONU in NG-PON2 has to select the downstream and upstream wavelength among four available wavelength channels.

There are many options for selecting the working channel. Colored optical network units (ONU) can be used with a minimal inventory cost by using colored fiber cords or colored plugs based on Fiber Brag Grating (FBG). The ONU box including the transceiver is common for any working channel and the wavelength filter will be

D.-T. Tran et al. (Eds.): ICISN 2021, LNNS 243, pp. 27–32, 2021.
https://doi.org/10.1007/978-981-16-2094-2_4

installed on site. Using set-and-forget tunable transceivers is another cost-effective option in which the wavelength is selected by mechanically locked tuning. Once tuned, no electrical current is needed to hold the wavelength [2]. The working channel in these options is expected to be unchanged for very long time because the long wavelength switching time, either by filter swap or mechanical tuning, requires a temporary interruption of services and a manual wavelength calibration.

Besides the wavelength preset options, the working channel may be dynamically changed during the course of operation without interruption of services. Networks with such capability are referred to as wavelength-reconfigurable or wavelength-agile networks [3]. Relatively fast tunable transceivers are required to avoid interruption of ongoing services when the wavelength channel handover is performed while data are temporarily buffered during wavelength switch-over. Therefore, it is expected to be more expensive than other options of wavelength selection. However, wavelength-agility enables advanced functions such as load balancing in busy hours and power saving in non-busy hours. In this paper, we will discuss and quantify requirements and merits of wavelength-agile TWDM-based NG-PON2.

2   Challenges of Wavelength-Agility

The schematic representation of wavelength-agile NG-PON2 is shown in Fig. 1. a) in which a tunable filter and tunable laser are used for downstream and upstream wavelength selection, respectively. Wavelength-agility requires the wavelength handover latency to be low enough to avoid any perceptible interruption of on-going services, especially ones associated with two-way speech such as telephony or telepresence. The tolerable latency bound for such real-time services derived from the well-known E-model was specified to be 150 ms in ITU-T G.114 recommendation [4]. Users experience essentially transparent interactivity when the latency is kept within this bound. In fact, while handover is relatively new in optical networks, it is a popular procedure in cellular mobile networks. In mobile networks, cell handover latency needs to be compliant with G.114, i.e., 150 ms to guarantee a seamless switching from one cell to another cell [5]. Therefore, the wavelength tuning time, as the largest component of wavelength handover latency, should be lower than 150 ms. It is even desirable to have a lower tuning time to avoid large buffer sizes given the high bit-rate in access networks. To switch an ONU with 1 Gbps on-going connections, 18.75 MB of data needs to be buffered for one direction during 150 ms. Nevertheless, without the buffer size constraint, the tuning time of 150 ms is a relatively relaxed requirement in comparison to the corresponding requirement in core networks. The maximum tuning time of an integratable tunable laser assembly (ITLA) specified for protection purposes in SONET/SDH networks is 10 ms. However, the relaxed requirement should not necessarily lead to cheaper devices since the cost of laser itself is a minor contributor to the target module cost [6]. In fact, one of main cost drivers is the wavelength control that includes wavelength stability control and wavelength tuning control. Wavelength stability requires temperature control of the laser chip and wavelength tuning needs to be addressed not within the ONU but in the network-wide. The OLT has to assist the ONU to fine-tune the wavelength in order to maximize the received power and to avoid interference [7]. Furthermore, bias currents to the laser electrodes (>10 electrodes) typically need to be set by a microcontroller that also significantly contributes to the cost. The micro controller needs to map input parameters, e.g., output power and wavelength index, to the actual laser current values which vary from chip to chip.

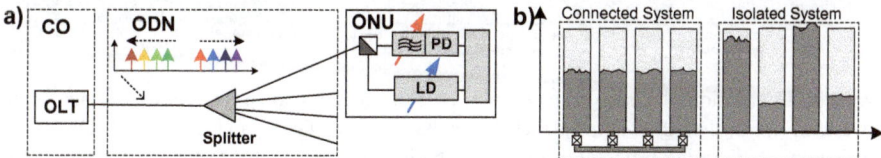

Fig. 1. a) Schematic representation of wavelength-agile TWDM NG-PON2, b) Wavelength-agile vs. wavelength-fixed as system of connected water buckets vs. system of isolated water buckets

The history of fiber to the home (FTTH) has proved that the main requirement for any access technology is low cost, especially in the ONU. The user equipment is very cost-sensitive since it makes up the largest part of access network equipment cost. Therefore, the ultimate challenge for wavelength-agility is to bring the cost of tunable TRx down to the cost point of an access solution. Much progress for both low-cost tunable lasers and tunable filters has been made recently. Uncooled tuning methods were studied to remove expensive cooling but still maintain wavelength stability [7, 8]. External cavity tuning is another promising method to have a low-cost tunable laser [9]. Technologies for the tunable filter, which were developed for core networks, were redesigned for the access application as the tuning range is the main cost driver of filters [10].

3   Merits of Wavelength-Agility

Beyond the cost challenge, wavelength-agility has as a clear principal advantage of the "soft" wavelength sharing factor among the ONUs. The sharing factor can be adjusted on the fly to follow instantaneous user demands [3]. One or more ONUs can be moved out of a busy wavelength and allocated to a lightly-loaded one in order to reduce the sharing factor of the busy wavelength. Thus, the sustainable bandwidth per ONU increases in the busy wavelength. A high-level comparison between the wavelength-agile network and the wavelength-fixed network is shown in Fig. 1. b) as the comparison between a system of connected buckets and a system of unconnected buckets. Although they have the same total network load, moving the load among wavelengths allows load balancing to avoid any local overload.

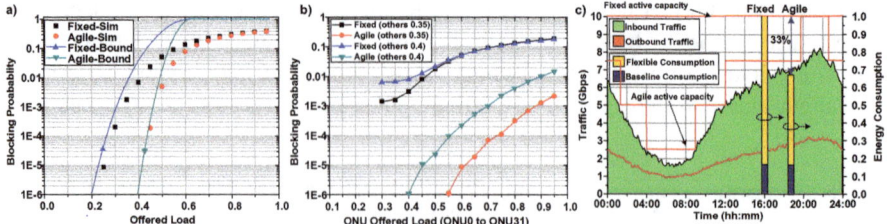

Fig. 2. a) Connection blocking probability in case of offered load speads out evenly across the network of 4 wavelength channels, each has capacity of 10 Gbps, and initially 32 ONUs per wavelength, b) Blocking probability in case of high load from ONU0 to ONU31, other ONUs offered load is maintained at 0.35 or 0.4, c) Traffic trace from a high-take-rate KPN access interface recorded in 2012 in Amsterdam and corresponding OLT energy consumption in cases of the traffic was carried by wavelength-fixed or wavelength-agile NG-PON2 networks.

Nevertheless, wavelength-agility can still exhibit a better statistically multiplexing gain than the wavelength-fixed counterpart in case of the load being distributed evenly over the network. The connection blocking probability for wavelength-fixed and –agile networks is shown in Fig. 2. a). Each network contains four wavelength channels with a capacity of 10 Gbps per channel, 128 ONUs, and an ONU requests a 500 Mbps connection when active. An OPNET simulation is used to yield the blocking probability when requests come in a Poisson arrival process. The upper bound for the blocking probability, when no assumption of request arrival process is applied, is computed based on Chernoff's bound [11]. The results show that wavelength-agility yields a better traffic handling capacity, expressed as the maximum allowable load for certain level of blocking. Wavelength-agility improves 36.9% traffic capacity at 1% blocking probability in comparison to wavelength-fixed. In the case of a more stringent threshold for blocking probability, e.g., 1‰, wavelength-agility can provide a better improvement.

The popularity of video services and mobile devices creates higher temporal and spatial variations in network traffic patterns [12]. Traffic hotspots occur more frequently with higher load magnitudes leading to the local congestion in a wavelength. A representative scenario was simulated in which the load by ONU0 and ONU31 was varied while others (ONU32 to ONU127) maintain the same level of load. Since ONU0 and ONU31 are initially assigned to $\lambda 0$, a local congestion occurs when these ONUs raise their demands. Two different baseline load levels for ONU32 to ONU127 were investigated. The offered load levels of 0.35 and 0.4 are implying that $\lambda 1$, $\lambda 2$, and $\lambda 3$ operate at 56% and 64% of channel capacity, respectively. The results, shown in Fig. 2. b), reveal that wavelength-fixed cases cross 1% blocking threshold at an offered load of 0.45 while wavelength-agile cases can accommodate at least an offered load of 0.9 from the group of ONU0 to ONU31. The high blocking probability in wavelength-fixed cases results from the congestion in $\lambda 0$ because demanding ONUs cannot be reallocated to other λs which still have free bandwidth available.

During non-busy hours when the network load goes down, wavelength-agility allows to save power consumption by powering down unused capacity. ONUs are concentrated to one (or more) wavelength to permit other wavelengths going to standby mode. To have a concrete figure, we compute the OLT daily energy consumption based on a traffic profile recorded from a high-take-rate access interface of the Dutch royal KPN telecom network operator in Amsterdam. This aggregate traffic, shown in Fig. 2. c), is generated by both business and consumer users, which is clearly pronounced in the pattern as 9 am marks the start of the traffic increase and the peak period is in the evening. The wavelength-agile network saves 33% of OLT energy in comparison to the wavelength-fixed network since wavelength-agility can adjust the active capacity by stepwise (de-) activating one or more wavelengths (25% change of aggregate capacity per wavelength). In fact, this traffic pattern provides a conservative saving figure because a typical take rate of FTTH deployments is lower than 50% leading to a lower load level and the mix of user types also makes busy hours longer. For a fair comparison, we assume 20% of baseline consumption that is constant regardless of the number of active channels; the detailed analytical model can be found in [13].

Wavelength-agility can decouple the service provisioning and the physical connectivity provisioning. Services can be added, removed, or modified without touching the physical connectivity. The service provider can offer services in a timely manner, for example, a full wavelength capacity can be provisioned for a single ONU without blocking traffic from other ONUs [3]. Therefore, wavelength-agility allows more flexible service/bandwidth provisioning which is in-line with the anytime and any-where concept. Wavelength agility also can bring network virtualization to a higher level. In the fixed scheme, the ONU is restricted within a wavelength channel with only dynamic sharing of timeslots while the wavelength boundary is removed by wavelength-agility.

4   Conclusion

We have discussed requirements, challenges, and benefits of wavelength-agility in NGPON2. To have a seamless wavelength handover, the laser and filter tuning time should be lower than 150 ms, which is not a demanding technical requirement for such optical devices. Therefore, the main challenge is to bring down the cost of tunable lasers and tunable filters. The low-cost for tunable lasers is more challenging since the cost is not only associated with the laser chip but also the wavelength stability and tuning control.

Concrete figures demonstrate that wavelength-agility yields better performance in terms of blocking probability and power saving in comparison to wavelength-fixed. The local congestion in a wavelength is completely resolved when free bandwidth is available in other wavelengths. Wavelength-agile NGPON2 indicates a saving of 33% energy in a conservative evaluation using a captured access network traffic profile. The flexible service provisioning and network virtualization with the support of wavelength-agility was also discussed. Therefore, we believe that wavelength-agility is a must-have feature for NGPON2 deployments.

Acknowledgement. This work was supported by the Newton Fund Institutional Links through the U.K. Department of Business, Energy, and Industrial Strategy and managed by the British Council under Grant 429715093.

References

1. Peter, V.: Tutorial next generation optical access technologies. In: ECOC, p. Tu.3.G.1 (2012)
2. Ben, K.: DWDM-PON solutions this year. In: OFC/NFOEC 2012, NGPON2 Workshop (2012)
3. Ton, K.: Fiber to the home/fiber to the premises: what, where, and when? Proc. IEEE **94**, 911–934 (2006)
4. ITU-T: G.114 - One-way transmission time. ITU-T Recommendation, May 2003
5. TR 22.976 - Study on Release 2000 services and capabilities. Third Generation Partnership Project (3GPP) (2000)
6. Wale, M.J.: Options and trends for PON tunable optical transceivers. In: ECOC 2011, p. Mo.2.C.1 (2011)
7. Roppelt, M., et al.: Tuning methods for uncooled low-cost tunable lasers in WDM-PON. In: OFC/NFOEC, p. NTuB1 (2011)
8. Pöhlmann, W., et al.: Performance of wavelength-set division multiplexing PON upstream in the O-band with optical preamplification. In: ECOC, p. Tu.4.B.2 (2012)
9. Oh, S.H., et al.: Tunable external cavity laser by hybrid integration of a superluminescent diode and a polymer Bragg reflector. J. Sel. Top. Quantum Electron. **17**(6), 1534–1541 (2011)
10. Murano, R., Cahill, M.J.L.: Low cost tunable receivers for wavelength agile PONs. In: ECOC, p. We.2.B.3 (2012)
11. Koonen, A.M.J., et al.: The merits of reconfigurability in WDM-TDM optical in-building networks. In: OFC/NFOEC, p. JWA63 (2011)
12. Cisco Visual Networking Index: Usage Study, Cisco white paper (2010). https://cisco.com
13. Tran, N.C., et al.: Flexibility level adjustment in reconfigurable WDM-TDM optical access networks. J. Lightwave Technol. **30**(15), 2542–2550 (2012)

Enabling Technologies and Services for 6G Networks

Mohamed El Tarhuni$^{(\boxtimes)}$ and Ahmed I. Salameh

American University of Sharjah, Sharjah, UAE
mtarhuni@aus.edu, asalameh@sharjah.ac.ae

Abstract. We are witnessing a tangible movement into 5G deployment across the globe with approved standards and services that have been developed throughout the last decade. However, there are some concerns that 5G specifications and capacity are insufficient to accommodate the exponential growth in the number of network nodes and to meet the rigorous operation requirements for new applications. Subsequently, researchers are currently interested in depicting beyond 5G network, i.e., 6G. In this paper we aim to describe the main technologies, services, operation scenarios, requirements, and challenges that are expected to be developed.

Keywords: 6G · Artificial intelligence · Edge computing · Mixed reality · Terahertz

1 Introduction

As we step ahead into the 5G era in wireless communications, rising attention into networks beyond 5G is being recorded worldwide recently. The term 6G has been coined for networks beyond 5G, and its expected to become a reality by the end of this decade. Multiple reasons are driving a number of growing research activities into this direction; most importantly are the perceived shortcomings of the envisioned 5G networks. We can note a few, such as 5G's expected down-link data rate ceil of only 20 Gbps [1], which is not enough for next-gen applications and services as will be argued in the rest of the paper. Moreover, 5G's floor of latency is estimated to be 1 ms at best [2], however in many applications, such as robotic arms communication in an industrial facility, the intended latency is sub-ms (0.1–1 ms) [3]. Additionally, in the upcoming few years, it is projected that hundreds of billions of devices will make up the global internet [4], while 5G is characterized to support only a few billion devices [5]. All indicators are pointing at igniting serious research efforts towards facilitating 6G networks over a 10-year plan as 5G is expected to reach its limits by 2030 [6].

It is envisioned that 6G will be able to push way beyond 5G capacity by offering data rates in the order of Terabits per seconds and utilize the frequency spectrum up to few TeraHertz. It is worth to note that 5G uses millimeter waves

D.-T. Tran et al. (Eds.): ICISN 2021, LNNS 243, pp. 33–42, 2021.
https://doi.org/10.1007/978-981-16-2094-2_5

in the range of 20–100 GHz to offer high speeds, however there are hardware limitations still to be overcome, such as poor analong-to-digital conversion (ADC) resolution and increased phase noise [7]. 6G is being studied to offer higher communication speeds than 5G by jumping to the higher frequencies range (Tera-Hertz), where the spectrum is more abundant compared to millimeter waves [8]. Higher speeds are coupled with developing hardware capable of working efficiently at very high frequencies. One of the aspirations of 6G is the focus on efficient green networking by enhancing wireless energy harvesting to an extent of the creation of battery-free devices [1].

It is, therefore, important to explore some of the possible 6G network concepts including technologies, services, applications, deployment scenarios, and challenges.

The rest of this paper is organized as follows: Sect. 2 discusses possible 6G technologies. Section 3 examines 6G services and some deployment scenarios, and the main challenges to be considered. Lastly, in Sect. 4, we share our conclusions.

2   6G Technologies

In this section, we discuss what is expected from 6G networks in terms of technological improvements. We highlight that a higher tier of data rates, lower latencies, and more reliability metrics are anticipated to unlock a new class of services and applications for next-gen networks. Some of the estimates are in the order of Terabytes per second data transfer rates [9], (< 1) ms latencies [3], and 10^{-9} reliability figures [10]. In the course of securing robust communication services, it is essential to realize hardware infrastructure with diversified capabilities. One of the promising approaches to achieve this is via utilizing heavy cell deployment (cell densification) [11,12]. It is worth noting that the co-existence of the different network cell tiers is mandatory for efficient frequency spectrum utilization and the realization of heterogeneous networks, which will be an integral part of 6G. However, this will pose its own set of challenges such as interference management, base stations power management, and optimal cell layout. Another key hardware technique that can be coupled with cell densification is ultra-massive multi-input-multi-output (MIMO) base stations [13,14]. Increasing the number of antennas (up to hundreds) attached to the communicating nodes draws its importance from multiple angles. From one side, the large set of antenna elements enhances the diversity gain in dense urban areas cluttered with many obstacles by receiving the same signal over multiple paths allowing for more reliable signal transmission. Data-hungry users can receive multiple streams over multiple antennas to enhance the throughput. On another side, as the number of devices connected to the network increases, a given base station will be able to serve more nodes in a small area, hence, offering improved spectral efficiency and coverage [15]. However, this further adds to the complexity of the network in the sense of maintaining low cost and energy consumption per antenna and implementing powerful channel estimation and beamforming techniques [16].

2.1 *Intelligent Reflective Surfaces* (IRS)

Many recent works [17–19] suggest the incorporation of another pivotal hardware element into 6G, widely known as *Intelligent reflective surfaces* (IRS). It is proposed to use IRS to control the channel for signal propagation, which is the least controllable part in a given network. This novel approach adjusts the channel between communicating nodes by artificially producing multiple paths via IRS and subsequently enhancing the overall performance of wireless systems. IRS implementation is important for 6G networks more than for any older generation of networks, such as 5G, to cover dead-spots arising from very short-length waveforms [19]. IRS can be looked at as signal-scattering elements that are implanted between the communicating nodes in a network. IRS can be classified into two categories, passive and active. The passive type is mainly concerned with the texture and placement of the IRS to reflect rays that will constructively improve the signal-to-noise ratio (SNR) at the receiver. The active IRS class consumes power and is adjustable based on the deployment scenario to maintain optimal operational conditions. Furthermore, IRS can be conjoined with artificial intelligence (AI) for better adaptation based on the network requirements. Not far from ultra-massive MIMO, IRS can compliment its operation by procuring control over the signal phase shifts at the receiver's antennas. More uses for IRS can be found elsewhere, such as integrating it into walls to create smart indoor radio propagation environments by carrying out frequency-selective signal energy penetration insulation [20], or reinforce wireless power transfer [21] efficiency via executing passive beamforming.

2.2 *Optical Wireless Communications* (OWC)

Optical wireless communications (OWC) is considered as a 6G enabler by many recent works [10,22,23]. OWC is best utilized in indoor communications for short-range high-speed communications, this is possible by targeting unlicensed spectrum resources (> 0.3 THz) that are available in abundance and allowing for extremely higher data rates. OWC is particularly appealing over regular radio frequency (RF) communications for multiple reasons, one is that it does not suffer from electromagnetic interference that is usually faced in RF. Another reason is that the short range of OWC (due to light confinement) can be considered as a merit by allowing high reuse factor. There exists multiple practical scenarios for OWC, such as the tight integration of OWC into vehicle-to-vehicle (V2V) and vehicle-to-infrastructure (V2I) communications on roads [24]. This is made possible by installing communication units in light poles along the roads. Furthermore, some works such as [25], have studied the different wavelengths, sources, and modulation schemes to essentially create different OWC transmission schemes.

2.3 *Distributed Network Computations* (DNC)

The continuous cost-cutting of computation components allows for manufacturing powerful processing units with low power consumption. This has positive implications on 6G by shifting towards *Distributed Network Computations* (DNC), examined in [26]. DNC is foreseen to strengthen the resilience of a network by adapting to the different end devices requirements and applications. For example, critical processes for delay-sensitive applications can be allocated to the geographically closest network component, such as a router. This offers a clear advantage over conventional execution at the network core, by reducing undesired time delays. DNC also gains significant importance in the emerging Internet-of-Everything (IoE) networks, where machine-to-machine (M2M) communications dominates, especially in manufacturing plants and smart homes. DNC at its core is considered to be safer than conventional central computations, as data is being spread over multiple network nodes, thus, contributing to pioneering personalized network security. Additionally, introducing a multi-level network computations scheme allows for offloading non-urgent data from network core to other network components, known as *edge computing* [27,28]. Offloading data can also take place in cloud-based networks, where fog computing is considered and regarded as a modern form of *edge computing*. Fog computing takes place between the end users and the cloud, for reduced latency and increased bandwidth [29]. *Edge computing* allows for manufacturing cheap brainless end devices that offloads all the processing required to the fog while maintaining high battery life. Furthermore, some of the next-gen services demand high processing and battery life, which limits the end user capabilities without *edge computing*.

2.4   *Blockchain and Distributed Ledger*

Not far from DNC, *blockchain and distributed ledger* are essential components for functioning many of the distributed network services as well near-uncompromisable security [30] because they are inherently based on decentralization and sharing data between network nodes. Network security in 6G is projected to be even more salient due to the massive amounts of data about individuals that is being exchanged with the network, including private sensory data. The centralized cloud and distributed *blockchain* concepts can function interchangeably to add versatility to the network, by creating multi-layer network model to serve the different use scenarios, as suggested by [31]. For instance, if a block chain was destroyed at a level of networking for whatever reason, then other layers can still provide the blockchain records. Moreover, creating multiple smaller blockchains allows for achieving savings in time and space that are used for keeping transaction records.

2.5   *Quantum Computing and Communication*

An increasing global interest in *Quantum Computing and Communication* has been demonstrated recently. Quantum mechanics origins date back to the 30's of the previous century [32], recently, several decades after its inception, the Q-bit concept has emerged. In essence, Q-bit is based on the electron state in wires to decide the encoded data by the transmitter [33]. The nature of performing computations in quantum manner is substantially different, due to the uncertainty and entanglement concepts. Theoretically, quantum computations are exponentially faster than conventional processors as a result of being able to perform computations in very high Hilbert space dimensions [34]. The inherent difficulty in using Q-bits for computers is simply to be able to control them, as quantum mechanical waves can be easily disrupted and cause failures. Beside that, more challenges are concealed in how to deal with environmental noises, which can affect the Q-bit state and the photons interactions in Q-bit gates [35]. This sort of unprecedented performance metrics, expected from quantum computers, are projected to enable a multitude of new services in 6G, such as holographic communications, advanced security measures, and decentralized banking. All of these services are hard to be implemented safely and efficiently with the current form of 5G networks and its performance metrics.

2.6 *Molecular Communication*

On the rise is another grade of communications, i.e., *Molecular Communication*, or internet of bio-nano-things [36]. Its principle of operation is based on natural biological processes and implanted sensors, in humans or livestock. *Molecular Communication* is envisioned to revolutionize the concept of communications, to be between body parts (locally) and with the outside world. This newly-formed type of communications is expected to have many applications, most prominent is remote health monitoring and surgeries, which is expected to play a significant role in 6G era. *Molecular Communication* requires to have a very low connection latency between devices facilitating this type of communication, this is not possible with current 5G latency standards. *Molecular Communication* is largely polarizing and attention-grabber for researchers, as its surrounded by challenges and hurdles from multiple angles. From one side, this field is considered multidisciplinary that requires the close coalition of engineers and biologist to produce safe, feasible, and efficient communication units. On a different front, the data transmission mechanisms are fundamentally different from regular RF signals, owing to the dependency on body cells and chemical molecules as the communication medium [37]. In particular, one of the main obstacles is achieving reliable transmission with minimalist error rates, as health-related information necessitates high accuracy in decoding, this requires robust error correction code schemes. Furthermore, studying the channel capacity of *Molecular Communication* is of a great interest to understand the theoretical information transfer rate limit [38], this includes a close inspection of coding schemes, diffusion channels, and noise factors.

3   6G Services

It is worthwhile to explore some of the main 6G services that will play a great part of the future day-to-day communication. We present a few of them in Fig. 1, from (A) to (F). In Fig. 1A, V2V and V2I communication links are established under a smart city setup. It is easy to see that in cases, such as medical emergencies, where ambulances need priority over other vehicles on the road, the signaling between the ambulance, other vehicles, and nearby infrastructure, can bring forward the ambulance arrival time. By looking at Fig. 1B, we can see a deployment scenario for OWC in a small room, where two users are using portable computers to access the network. Switching attention to Fig. 1C, we can observe the capabilities of enabling *Holographic Communication* in a network. This unprecedented type of communication is set to bolster the human-network integration by permitting 3D virtual presence of other humans. This as well can be expanded for other applications, such as demonstrating concepts in 3D, for educational purposes or projects discussion. *Holographic Communication* is largely anticipated to be integrated with another service, i.e., *Mixed Reality* (XR), with both its forms, augmented reality (AR) and virtual reality (VR). XR promises many new features, such as while driving, road signs and features can be filtered into textual warning signs and 3D visual instructions. Moreover, virtual educational institutions that are wholly based on live remote teaching are expected to benefit from such technology, and allowing for a wider access to education at a lower cost.

Fig. 1. 6G services deployment scenarios. (Some icons are taken from dribbble.com/dreamstory, vecteezy.com, and all-free-download.com.)

We can notice in Fig. 1D the integration of AI into the cloud network. As discussed earlier, AI will facilitate many network functionalities and will work on handling the extensive network complexities. For example, base stations can report back to the cloud the information related to end users locations at different times, to increase network acquaintance about users' speed and movement trajectories throughout the day. This can help in cell offloading and resources management for efficient network deployment. In Fig. 1E, we can remark the introduction of Internet of Things (IoT) and IoE into households for various smart home applications. For instance, in appliances coordination system, day energy savings can be achieved, and in activating emergency reporting system during residents absence, higher precaution levels can be reached. Lastly, in Fig. 1F, we can see a vertical stack of different congregated connections that are forming different clusters of sub-networks. Each one of the groups is predicted to be employed at a certain level (as illustrated), to enable the future of non-terrestrial communications integration. We can see that submarines will have their own sub-network that can be based on acoustics or very low frequencies (for water penetration and propagation), while airborne objects, from satellites to drones, can similarly have their own network based on laser communication. The different tiers of the network can communicate with each other via special links. Clustering connections into groups does also scale up horizontally, to accommodate massive multi-tier heterogeneous networks. In that manner, we can have small networks of trains on railways, vehicles on the road, and autonomous robots and sensors in factories, to name a few, in the proximity of each other cooperating without interference.

4   6G Challenges

The movement forward in establishing new network services and applications does however come at a price, i.e., mainly, exacerbated network complexity. In order to ease this predicament and its significance, a motion towards smart networks is taking place. For instance, in smart factories [39], a part of the network is dedicated to support self-organization, coordination, and optimization of IoT devices communication, for optimal operational conditions, such as on conveyor systems. This is where AI is expected to take a major role in IoT and IoE, to fullfill self-sustainability and management goals. AI will not be limited to sensors and will also be integrated into the network core, to actuate many of the 6G services and applications, and for the overall network optimization and stability. On top of that, the network inclusion of advanced AI will set the path to detail human sentiments, for a diverse set of purposes. For example, facial feedback can act as a mechanism for better user-targeted advertisements and online content delivery, on smart phones [40]. Furthermore, AI can aid in human-bot chat scenarios which are becoming increasingly common.

Moreover, most of the mentioned 6G technologies and services are in need for high-precision communication that will unlock seamless and integrated connectivity. This creates a new set of challenges that demand significant research and development efforts. For instance, the shifting towards higher frequencies for achieving unprecedented data rates is faced with atmospheric absorption due to the small wavelength of the transmitted signals [41]. This destabilizes the connectivity between devices at very high frequencies, especially under MIMO formation, by hardening the tasks related to beam management. Another challenging aspect of the move to 6G is the serious predicaments faced by hardware manufacturers of network components, such as designing transceivers capable of operating at $(>THz)$ for lightning data transfer rates, and hardware compatibility between communicating devices in IoT/IoE setups. Moreover, increased network complexity is not easily dealt with, as controlling network resources and reducing interference, becomes demanding with multi-tier heterogeneous networks. This is especially true when working on maintaining stringent metrics for some 6G applications, such as XR and remote surgeries. Much of the underlined technologies are envisaged to become a reality by the year 2030 [10], leaving enough time to create efficient proactive network administration systems.

5   Conclusions

The move to 6G has been started although 5G deployments have just started worldwide. The key performance metrics for the currently-portrayed 5G networks are already looking outdated for the future of networks 10 years away from now, this ultimately energized the move to 6G. In this paper, we discussed in detail this move along with an overview of enabling technologies and services expected to accelerate the move to 6G. Furthermore, we looked at potential challenges and solutions for 6G deployment.

References

1. David, K., Berndt, H.: 6G vision and requirements: is there any need for beyond 5G? IEEE Veh. Technol. Mag. **13**(3), 72–80 (2018)
2. Parvez, I., Rahmati, A., Guvenc, I., Sarwat, A.I., Dai, H.: A survey on low latency towards 5G: RAN, core network and caching solutions. IEEE Commun. Surv. Tutor. **20**(4), 3098–3130 (2018)
3. Berardinelli, G., Mahmood, N.H., Rodriguez, I., Mogensen, P.: Beyond 5G wireless IRT for Industry 4.0: design principles and spectrum aspects. In: 2018 IEEE Globecom Workshops (GC Wkshps), pp. 1–6. IEEE (2018)
4. Andrews, J.G., Buzzi, S., Choi, W., Hanly, S.V., Lozano, A., Soong, A.C., Zhang, J.C.: What will 5G be? IEEE J. Sel. Areas Commun. **32**(6), 1065–1082 (2014)
5. Yang, P., Xiao, Y., Xiao, M., Li, S.: 6G wireless communications: vision and potential techniques. IEEE Netw. **33**(4), 70–75 (2019)
6. Tariq, F., Khandaker, M., Wong, K.K., Imran, M., Bennis, M., Debbah, M.: A speculative study on 6G. arXiv preprint arXiv:1902.06700 (2019)
7. Xiao, M., Mumtaz, S., Huang, Y., Dai, L., Li, Y., Matthaiou, M., Karagiannidis, G.K., Björnson, E., Yang, K., Chih-Lin, I., et al.: Millimeter wave communications for future mobile networks. IEEE J. Sel. Areas Commun. **35**(9), 1909–1935 (2017)
8. Kürner, T., Priebe, S.: Towards THz communications-status in research, standardization and regulation. J. Infrared Millimeter Terahertz Waves **35**(1), 53–62 (2014)

9. Zhang, Z., Xiao, Y., Ma, Z., Xiao, M., Ding, Z., Lei, X., Karagiannidis, G.K., Fan, P.: 6G wireless networks: vision, requirements, architecture, and key technologies. IEEE Veh. Technol. Mag. **14**(3), 28–41 (2019)

10. Strinati, E.C., Barbarossa, S., Gonzalez-Jimenez, J.L., Ktenas, D., Cassiau, N., Maret, L., Dehos, C.: 6G: the next frontier: from holographic messaging to artificial intelligence using subterahertz and visible light communication. IEEE Veh. Technol. Mag. **14**(3), 42–50 (2019)

11. Khaled, A.M., Hassaneen, S., Elagooz, S., Soliman, H.Y.: Small cell network densification using a new space-time block code (STBC). Port-Said Eng. Res. J. **24**(1), 102–107 (2020)

12. Bahlke, F., Ramos-Cantor, O.D., Henneberger, S., Pesavento, M.: Optimized cell planning for network slicing in heterogeneous wireless communication networks. IEEE Commun. Lett. **22**(8), 1676–1679 (2018)

13. Saad, W., Bennis, M., Chen, M.: A vision of 6G wireless systems: applications, trends, technologies, and open research problems. IEEE Netw., 1–9 (2019, in press)

14. Rodrigues, V.C., Amiri, A., Abrao, T., de Carvalho, E., Popovski, P.: Low-complexity distributed XL-MIMO for multiuser detection. arXiv preprint arXiv:2001.11879 (2020)

15. Björnson, E., Hoydis, J., Sanguinetti, L.: Massive MIMO networks: spectral, energy, and hardware efficiency. Found. Trends Signal Process. **11**(3–4), 154–655 (2017)

16. Molisch, A.F., Ratnam, V.V., Han, S., Li, Z., Nguyen, S.L.H., Li, L., Haneda, K.: Hybrid beamforming for massive MIMO: a survey. IEEE Commun. Mag. **55**(9), 134–141 (2017)

17. Gong, S., Lu, X., Hoang, D.T., Niyato, D., Shu, L., Kim, D.I., Liang, Y.C.: Towards smart wireless communications via intelligent reflecting surfaces: a contemporary survey. arXiv preprint arXiv:1912.07794 (2019)

18. Zhao, J., Liu, Y.: A survey of intelligent reflecting surfaces (IRSs): towards 6G wireless communication networks. arXiv preprint arXiv:1907.04789 (2019)

19. Gong, S., Lu, X., Hoang, D.T., Niyato, D., Shu, L., Kim, D.I., Liang, Y.C.: Toward smart wireless communications via intelligent reflecting surfaces: a contemporary survey. IEEE Commun. Surv. Tutor. **22**(4), 2283–2314 (2020)

20. Subrt, L., Pechac, P.: Controlling propagation environments using intelligent walls. In: 2012 6th European Conference on Antennas and Propagation (EUCAP), pp. 1–5. IEEE (2012)

21. Pan, C., Ren, H., Wang, K., Elkashlan, M., Nallanathan, A., Wang, J., Hanzo, L.: Intelligent reflecting surface enhanced MIMO broadcasting for simultaneous wireless information and power transfer. Accepted in IEEE J. Sel. Areas Commun. arXiv preprint arXiv:1908.04863 (2019)

22. Dat, P.T., Kanno, A., Inagaki, K., Umezawa, T., Yamamoto, N., Kawanishi, T.: Hybrid optical wireless-mmWave: ultra high-speed indoor communications for beyond 5G. In: IEEE INFOCOM 2019-IEEE Conference on Computer Communications Workshops (INFOCOM WKSHPS), pp. 1003–1004. IEEE (2019)

23. Ghassemlooy, Z., Zvanovec, S., Khalighi, M.A., Popoola, W.O., Perez, J.: Optical wireless communication systems (2017)

24. Wang, Z., Chen, J.: Networked multiple-input-multiple-output for optical wireless communication systems. Philos. Trans. R. Soc. A **378**(2169), 1–17 (2020)

25. Koonen, T.: Indoor optical wireless systems: technology, trends, and applications. J. Lightwave Technol. **36**(8), 1459–1467 (2017)

26. Destounis, A., Paschos, G.S., Koutsopoulos, I.: Streaming big data meets back-pressure in distributed network computation. In: IEEE INFOCOM 2016-The 35th Annual IEEE International Conference on Computer Communications, pp. 1–9. IEEE (2016)
27. Mahmood, N.H., Alves, H., López, O.A., Shehab, M., Osorio, D.P.M., Latva-aho, M.: Six key enablers for machine type communication in 6G. arXiv preprint arXiv:1903.05406 (2019)
28. Singh, S.: Optimize cloud computations using edge computing. In: 2017 International Conference on Big Data, IoT and Data Science (BID), pp. 49–53. IEEE (2017)
29. Mohan, N., Kangasharju, J.: Edge-fog cloud: a distributed cloud for internet of things computations. In: 2016 Cloudification of the Internet of Things (CIoT), pp. 1–6. IEEE (2016)
30. Jiang, T., Fang, H., Wang, H.: Blockchain-based internet of vehicles: distributed network architecture and performance analysis. IEEE Internet Things J. **6**(3), 4640–4649 (2018)
31. Li, C., Zhang, L.J.: A blockchain based new secure multi-layer network model for internet of things. In: 2017 IEEE International Congress on Internet of Things (ICIOT), pp. 33–41. IEEE (2017)
32. Jaeger, G.: Entanglement, Information, and the Interpretation of Quantum Mechanics. Springer, Heidelberg (2009)
33. Bertoni, A., Bordone, P., Brunetti, R., Jacoboni, C., Reggiani, S.: Quantum logic gates based on coherent electron transport in quantum wires. Phys. Rev. Lett. **84**(25), 5912–5915 (2000)
34. Arute, F., Arya, K., Babbush, R., Bacon, D., Bardin, J.C., Barends, R., Biswas, R., Boixo, S., Brandao, F.G., Buell, D.A., et al.: Quantum supremacy using a programmable superconducting processor. Nature **574**(7779), 505–510 (2019)
35. Ladd, T.D., Jelezko, F., Laflamme, R., Nakamura, Y., Monroe, C., O'Brien, J.L.: Quantum computers. Nature **464**(7285), 45–53 (2010)
36. Akyildiz, I.F., Pierobon, M., Balasubramaniam, S., Koucheryavy, Y.: The internet of bio-nano things. IEEE Commun. Mag. **53**(3), 32–40 (2015)
37. Farsad, N., Yilmaz, H.B., Eckford, A., Chae, C.B., Guo, W.: A comprehensive survey of recent advancements in molecular communication. IEEE Commun. Surv. Tutor. **18**(3), 1887–1919 (2016)
38. Pierobon, M., Akyildiz, I.F.: Capacity of a diffusion-based molecular communication system with channel memory and molecular noise. IEEE Trans. Inf. Theory **59**(2), 942–954 (2012)
39. Wang, S., Wan, J., Li, D., Zhang, C.: Implementing smart factory of industrie 4.0: an outlook. Int. J. Distrib. Sens. Netw. **12**(1), 1–10 (2016). Article ID: 3159805
40. Cambria, E.: Affective computing and sentiment analysis. IEEE Intell. Syst. **31**(2), 102–107 (2016)
41. Rappaport, T.S., Xing, Y., Kanhere, O., Ju, S., Madanayake, A., Mandal, S., Alkhateeb, A., Trichopoulos, G.C.: Wireless communications and applications above 100 GHz: opportunities and challenges for 6G and beyond. IEEE Access **7**, 78729–78757 (2019)

An Enhanced Model for DGA Botnet Detection Using Supervised Machine Learning

Xuan Dau Hoang[1(✉)] and Xuan Hanh Vu[2]

[1] Posts and Telecommunications Institute of Technology, Hanoi, Vietnam
dauhx@ptit.edu.vn
[2] Hanoi Open University, Hanoi, Vietnam
hanhvx@hou.edu.vn

Abstract. Recently, DGA botnet detection has been the research interest of many researchers all over the world because of their fast widespread and high sophistication. A number of approaches using statistics and machine learning to detect DGA botnets based on classifying botnet and legitimate domain-names have been proposed. This paper extends the machine learning-based detection model proposed by [7] by adding new classification features in order to improve the detection accuracy as well as to minimize the rates of false alarms. Extensive experiments confirm that our enhanced detection model outperforms the original model [7] and some other previous models. The proposed model's overall detection accuracy and the F1-score are both at 97.03%.

Keywords: DGA botnet detection · Fast-flux botnet detection · DGA botnet detection model · Machine learning-based botnet detection

1 Introduction

Over last decade, botnets have been seen one of serious threats to the security of Internet-based information systems, individual connected devices and Internet users [1–3]. This is because botnets have been associated with many types of Internet attacks and misuses, such as large-scale DDoS attacks, email spamming, malware transmitting, virtual click generating and sensitive information stealing. For example, The Telegram newspaper has suffered heavy losses in 2019 due to a large-scale DDoS attack believed to have originated in China and involved in the protests in Hong Kong [3]. According to Symantec, botnets produced about 95% of spam emails in the Internet in 2010 [4]. In addition, other botnet-based attacks include web-page injection, URI spoofing, DNS spoofing and collection of sensitive data. The main targets of botnet- based attacks usually are financial and governmental organizations.

During the development history, botnets have continuously been developing on the Internet in terms of the size and complication of controlling techniques [5, 6]. Generally, *a botnet* is a network of Internet connected devices that have been infected with a special type of malware, called *bot* [5, 6]. Bots are usually created by hacking groups, called *botmasters*. A bot running on an Internet connected device allows a botmaster to control the device remotely. The bot-infected device can be a computer, a smartphone, or an IoT device. Bots are highly autonomous and capable of communicating with their

D.-T. Tran et al. (Eds.): ICISN 2021, LNNS 243, pp. 43–51, 2021.
https://doi.org/10.1007/978-981-16-2094-2_6

controller to get updates and commands. In addition, bots can periodically send their operating status to their controller. On the other side, the botmaster sends updates and commands to bots through the controller of botnet, or the C&C (Command & Control) server [5, 6].

In order to connect to the C&C server to download updates and commands, bots send DNS queries containing C&C server's full-qualified names to local DNS services to get the IP of the server. The bots' DNS queries are similar to queries sent by any legitimate applications. However, to avoid the C&C server from being scanned, detected and blocked if using a static name and IP, botmaster constantly changes and updates its name and IP to the nominated DNS server using special methods, such as Fast Flux (FF), or Domain Generation Algorithms (DGA). Bots are also programmed to have the capacity to generate names for the C&C server automatically using the same methods. Therefore, they can still find the C&C server's IP address by generating the server's names, putting them into queries and then sending to DNS service. Since server's names generated by bots are usually different from legitimate domain names, we can monitor, capture DNS queries and then extract domain names in DNS queries for analysis to find signatures of bots and botnet activities [5, 6].

This paper enhances the DGA botnet detection model proposed in [7] by adding new domain classification features in order to increase the detection accuracy and lower the rates of false alarms. The remaining of this paper is structured as the following: Sect. 2 discusses closely related works. Section 3 describes our enhanced model for DGA botnet detection. Section 4 explains our experimental dataset, scenarios and results. Finally, the paper's conclusion is in Sect. 5.

2   Related Works

2.1   Introduction to DGA Botnets

As mentioned in Sect. 1, many botnets use DGA techniques to automatically generate and register different domain names for their C&C servers in order to prevent these servers from being controlled and blacklisted [5–7]. The main reason of using DGA techniques is to increase the complexity of the control and revocation of registered domain names. These botnets are called *DGA-based botnets* or just *DGA botnets*. DGA techniques use operators on variables that have constantly changing values, such as year, month and day to generate random domain names. For example, a form of DGA techniques is implemented by a function that consists of 16 rounds. Each round randomly generates a character as shown in Fig. 1 [5, 6].

year = ((year ^ 8 * year) >> 11) ^ ((year & 0xFFFFFFF0) << 17)
month = ((month ^ 4 * month) >> 25) ^ 16 * (month & 0xFFFFFFF8)
day = ((day ^ (day << 13)) >> 19) ^ ((day & 0xFFFFFFFE) << 12)
domain += chr(((year ^ month ^ day) % 25) + 97)

Fig. 1. Example of a DGA technique to generate domain names

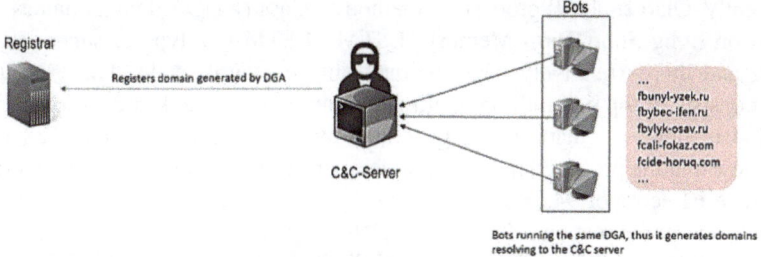

Fig. 2. Example of a botnet using DGA to generate, register and query C&C server names

Figure 2 describes a DGA technique in the botnet operation, in which on one side the botmaster randomly generates domain names using an algorithm, registers these domain names to the DNS system and then assigns them to the botnet's C&C server. On the other side, a bot of the botnet also generates a domain name for the C&C server using the same method and then send a query to the local DNS service to look for the IP address corresponding to the generated domain name. If the IP address of the server is found, the bot creates a connection to the server to get updates and commands. If the DNS query fails, or no valid IP address is found, the bot generates another domain name and repeats the procedure to search for the server's IP address.

2.2   Review of Proposals for DGA Botnet Detection

In this subsection, we discuss some closely related proposals for the detection of DGA botnets, which consist of Truong et al. [8], Hoang et al. [7], Qiao et al. [9] and Zhao et al. [10]. Truong et al. [8] proposes a method to detect domain-flux botnets using DNS traffic features. They use DNS domain features, including the domain length and expected value to distinguish between legitimate and pseudo-random domain names (PDN) generated by several botnets. The experimental dataset consists of top 100,000 legitimate domain names ranked by Alexa [11] and about 20,000 PDN domain names generated by Conficker botnet and Zeus botnet. Several machine learning algorithms, such as naive bayes, kNN, SVN, decision tree and random forest have been used to construct and validate the proposed detection model. Experimental results show that decision tree is the algorithm that gives the highest overall detection accuracy rate and the false positive rate of 92.3% and 4.8%, respectively.

Similarly, Hoang et al. [7] proposes a method to detect botnets based on the classification of legitimate and botnet generated domain names using supervised machine learning techniques. They propose to use 18 classification features, including 16 n-gram features and 2 vowel distribution features to construct and validate the proposed botnet detection model. The experimental dataset consists of 30,000 top legitimate domain names listed by Alexa [11] and 30,000 DGA botnet domain names [12]. Common supervised learning methods, such as kNN, naive bayes, decision tree and random forest have been used to build and validate the proposed model. Various experiments have been conducted using different scenarios and the results confirm that supervised learning methods can be used effectively to detect botnets with the overall detection rate of over 90% using random forest.

Recently, Qiao et al. [9] proposes a method to classify DGA domain names, which is based on Long Short-Term Memory (LSTM). LSTM is a type of supervised deep learning and this is relatively new approach in the security field. The experimental dataset consists of top one million legitimate domain names ranked by Alexa [11] and 1,675,404 malicious domain names generated by various DGA botnets [12]. Experiments show that the proposed model has better performance than existing methods with the average F1-score of 94.58%.

Using another direction, Zhao et al. [10] proposes a method based on n-gram statistics to detect malicious domain names. Each domain name in the training set of legitimate domains is first divided into sequences of substrings using 3, 4, 5, 6 and 7-g method. Then, the statistics and weight values of substrings of all training domains are calculated to form the 'profile'. To determine if an input domain name is legitimate or malicious, it is also first divided into sequences of substrings using 3, 4, 5, 6 and 7-g technique. Then, the statistics of domain name substrings are calculated and then it is used to calculate the 'reputation value' of the domain name based on the 'profile'. A domain name is legitimate if its reputation value is greater than a preset threshold. Otherwise, it is malicious. It is reported that the proposed method produces the high detection accuracy of 94.04%.

In our paper, we extend the model for botnet detection proposed in Hoang et al. [7], by adding new classification features in order to improve the detection accuracy as well as minimize the rates of false alarms. The random forest supervised learning method is used to construct the new model because the algorithm is relatively fast and it gives the best classification performance over other methods, such as kNN, naive bayes, decision tree and SVM [5–7].

3   The Enhanced Model for DGA Botnet Detection

3.1   The Model Description

Figure 3 presents the enhanced model for DGA botnet detection. The proposed model consists of two phases, including the (a) *training phase* to build the detection model from the training data and (b) the *detection phase* to classify a test domain name if it is a legitimate or botnet domain name. The training phase as shown in Fig. 3 (a) is carried out using the following steps:

- Feature extraction: The training set of legitimate and DGA botnet domain names is put into the feature extraction process, in which 24 classification features of each domain name are extracted. Each domain name is transferred to a vector of 24 features and a class label. The final result of the feature extraction step is a 2-dimensional training data matrix of M domains and N features;
- Training: The training data matrix is used to build the 'Classifier', or the detection model using the random forest algorithm. In addition, the constructed model is also validated using the 10-fold cross-validation, in which the training subset and the testing subset account for 80% and 20% of the full data set, respectively.

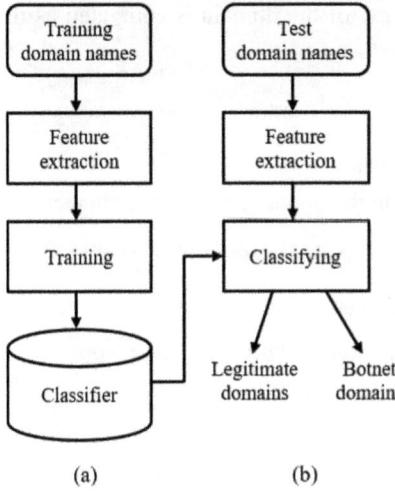

Fig. 3. The proposed detection model: (a) training phase and (b) detection phase

The detection phase as described in Fig. 3 (b) is done using the following steps:

- Feature extraction: The test domain names are put into the feature extraction process using the same procedure as done in the training stage. Each test domain name is converted to a vector of 24 features;
- Classifying: In this step, the vector of each test domain is classified using the 'Classifier' constructed in the training phase. This step's result is the predicted label of the test domain name of either legitimate or botnet.

3.2 Classification Feature Extraction

As mentioned in Sect. 3.1, 24 classification features are extracted for each domain name (we called it as 'domain' for short) as follows:

- 17 features proposed in Hoang et al. [7] include 8 2-g statistical features, 8 3-g statistical features and 1 vowel distribution feature;
- 1 feature proposed in Truong et al. [8] is expected value of a domain;
- 6 new features are as follows:
  - The digit distribution of the domain is computed as the number of digits (0 to 9) in the domain divides by the domain length;
  - The special character distribution of the domain is computed as the number of special characters ('.' and '-') in the domain divides by the domain length;
  - The first character of the domain is a digit or not. If it is a digit the feature's value is 1, otherwise feature's value is 0;
  - The consonant distribution of the domain is computed as the number of consonants in the domain divides by the domain length;
  - The hexadecimal character distribution of the domain is computed as the number of hexadecimal characters in the domain divides by the domain length;

 – The character entropy of the domain is computed as follows:

$$ent = \frac{-\sum\limits_{x} D(x)\log D(x)}{\log len} \tag{1}$$

where x is a character in the domain, $D(x)$ is the character distribution and len is the length of the domain.

3.3   Classification Measures

We use 6 measurements, including TPR, FPR, FNR, PPV, F1 and ACC to measure the enhanced model's performance as follows:

$$PPV (Positive\ Predictive\ Value)\ or\ Precision = \frac{TP}{TP + FP} \tag{2}$$

$$TPR(True\ Positive\ Rate)\ or\ Recall = \frac{TP}{TP + FN} \tag{3}$$

$$FPR(False\ Positive\ Rate) = \frac{FP}{FP + TN} \tag{4}$$

$$FNR(False\ Negative\ Rate) = \frac{FN}{FN + TP} \tag{5}$$

$$F1(F1 - score) = \frac{2TP}{2TP + FP + FN} \tag{6}$$

$$ACC(Overall\ Accuracy) = \frac{TP + TN}{TP + FP + FN + TN} \tag{7}$$

where TP, TN, FP and FN are True Positives, True Negatives, False Positives, False Negatives, respectively [7].

4   Experiments and Results

4.1   Experimental Dataset

Our experimental dataset consists a subset of legitimate domain names and another subset of DGA botnet domain names as follows:

- Top 100,000 legitimate domain names listed by Alexa [11]. We manually download and validate to remove duplicated domain names;
- 153,200 C&C server domain names of various DGA botnets listed in Netlab 360 [12]. These domain names were generated and used by common DGA botnets, such

as *banjori, emotet, gameover* and *murofet*. In this subset, we use 100,000 domain names for training and 53,200 domain names for testing.

The subset of 100,000 legitimate domain names and 100,000 botnet domain names is used for training to construct and validate the 'Classifier'. Another subset of 53,200 botnet domain names is only used for model testing, which is not in training subset.

4.2   Experimental Results

As mentioned in Sect. 4.1, the training set of 200,000 domain names is used for building and validating the detection model using the random forest algorithm. The ratio of data used for training and testing is 8:2 and we use the 10-fold cross-validation to compute the average result, as given in Table 1. Table 2 shows the model's detection accuracy (ACC) on 15 DGA botnets that generate and use large number of malicious domain names.

Table 1. The proposed model's detection performance versus other proposals

Approaches	PPV	TPR	FPR	FNR	ACC	F1
Truong et al. [8]-J48	94.70		4.80		92.30	
Hoang et al. [7]-RF	90.70	91.00	9.30		90.90	90.90
Qiao et al. [9]	95.05	95.14				94.58
Zhao et al. [10]			6.14	7.42	94.04	
Our model-RF 37 trees	97.08	96.98	2.92	3.02	97.03	97.03

Table 2. The proposed model's detection accuracy on various DGA botnets

No.	Botnet names	Total domains	Correctly detected	ACC
1	*banjori*	*4000*	*0*	*0*
2	emotet	4000	3994	99.85
3	gameover	4000	4000	100
4	murofet	4000	3994	99.85
5	necurs	4000	3947	98.67
6	ramnit	4000	3888	97.20
7	ranbyus	4000	3993	99.82
8	rovnix	4000	4000	100
9	shiotob	4000	3892	99.55
10	tinba	4000	3951	98.77
11	qadars	2000	1970	98.50
12	locky	1158	1098	94.81
13	cryptolocker	1000	990	99.00
14	chinad	1000	1000	100
15	dyre	1000	980	98.00
	Overall			**90.33**

4.3 Discussion

From the experimental results given in Table 1 and Table 2, we can draw the following comments:

- The proposed detection model outperforms previous proposals in all measures, in which our model produces much higher overall accuracy and F1-score than previous models and much lower FPR and FNR. For example, the F1-scores of Hoang et al. [7], Qiao et al. [9] and our model are 90.90%, 94.58% and 97.03%, respectively. The FPR and FNR of Zhao et al. [10] and our model are 6.14%, 6.42% and 2.92%, 3.02%, respectively.
- Our model is capable of effectively detecting most DGA botnets, as shown in Table 2. Out of 15 DGA botnets, domain names used by 14 botnets are detected with the accuracy of over 94%. However, it fails to detect '*banjori*' botnet because this botnet's domain names are very similar to legitimate domain names.

5   Conclusion

This paper proposes an enhanced model for detecting DGA botnets based on the random forest algorithm. The proposed model improves the model proposed by Hoang et al. [7] by adding 6 new classification features in order to improve the detection accuracy as well as lower the rates of false alarms. Experimental results confirm that our model produces a much better performance than that of the original model and other proposals based on supervised learning [7, 8], deep learning [9] and statistical method [10]. Moreover, our model can effectively detect most DGA botnets in the test set.

For future work, we will continue to improve our model so that it is able to detect DGA botnets that generate domain names similar to legitimate domain names.

References

1. Spamhaus Botnet Threat Report 2019. https://www.spamhaus.org/news/ article/793/spamh aus-botnet-threat-report-2019. Accessed 19 Aug 2020
2. Kaspersky Lab - Bots and Botnets in 2018. https://securelist.com/bots-and-botnets-in-2018/ 90091/. Accessed 19 Aug 2020
3. Radware Blog - More Destructive Botnets and Attack Vectors Are on Their Way. https:// blog.radware.com/security/botnets/2019/10/scan-exploit-control/. Accessed 19 Aug 2020
4. The Business Journal. https://www.bizjournals.com/sanjose/stories/2010/08/23/daily29.html. Accessed 19 Aug 2020
5. Alieyan, K., Almomani, A., Manasrah, A., Kadhum, M.M.: A survey of botnet detection based on DNS. Nat. Comput. Appl. Forum **28**, 1541–1558 (2017)
6. Li, X., Wang, J., Zhang, X.: Botnet detection technology based on DNS. J. Future Internet **9**, 55 (2017)
7. Hoang, X.D., Nguyen, Q.C.: Botnet detection based on machine learning techniques using DNS query data. J. Future Internet **10**, 43 (2018). https://doi.org/10.3390/fi10050043

8. Truong, D.T., Cheng, G.: Detecting domain-flux botnet based on DNS traffic features in managed network. Secur. Commun. Netw. **9**, 2338–2347 (2016)
9. Qiao, Y., Zhang, B., Zhang, W., Sangaiah, A.K., Wu, H.: DGA domain name classification method based on long short-term memory with attention mechanism. Appl. Sci. **9**, 4205 (2019). https://doi.org/10.3390/app9204205
10. Zhao, H., Chang, Z., Bao, G., Zeng, X.: Malicious domain names detection algorithm based on N-gram. J. Comput. Netw. Commun. **2019** (2019). https://doi.org/10.1155/2019/4612474
11. DN Pedia – Top Alexa one million domains. https://dnpedia.com/tlds/topm.php. Accessed 03 Aug 2020
12. Netlab 360 – DGA Families. https://data.netlab.360.com/dga/. Accessed 10 Aug 2020

Developing Vietnamese Sentiment Lexicon from Social Reviews Corpus Based on Support Measurement

Ha Nguyen Thi Thu[1]([✉]), Vinh Ho Ngoc[2], An Nguyen Nhat[3],
and Hiep Xuan Huynh[4]

[1] Information Technology Faculty, Electric Power University, Hanoi, Vietnam
hantt@epu.edu.vn
[2] Information Technology Faculty, Vinh University of Technology Education,
Vinh, Vietnam
[3] Information Technology Faculty, Army Academy of Science Technology,
Ho Chi Minh City, Vietnam
[4] Information Technology Faculty, Can Tho University, Can Tho, Vietnam
Hxhiep@ctu.edu.vn

Abstract. The sentiment dictionary plays an important role in analyzing or identifying opinion of users. A Sentiment dictionary is widely applicable to many different domains. Therefore, many researchers are interested in and building sentiment dictionaries. However, most of these dictionaries were built based on Vietnamese lexicon, when applied to social reviews often have low accuracy, because of the way social media is used different from Vietnamese lexicon. In this paper, we present a methodology of constructing Vietnamese sentiment dictionary with scoring for analyzing opinion of social reviews. We used training set with 5,200 labeled sentences that were collected from customer's reviews about electronic product domain on electronic commerce websites. After that, we extracted nouns, adverbs and adjectives and then applied support measurement to calculate weight of them. The experimental results with 249 sentences have an accuracy of approximately 92% compared to 87% of dictionary that is developed based on Vietnamese lexicon, showing that our dictionary has a higher accuracy when applied to social sentiment analysis.

Keywords: Vietnamese sentiment lexicon · Social network mining · Social reviews · Association rule

1 Introduction

Opinion mining and sentiment analysis is gaining a lot of attention today because of its application in various fields such as e-commerce, referendum on new policies, banking, finance, entertainment, etc. [6, 12]. The information society is more developed, people like to talk more on social networks. Producers want to mine customers' opinions about their product by collecting their reviews on social media [11, 12]. And then, analyzing customer's sentiment to know how satisfaction about the products and services. In global business, products are distributed everywhere on the world, a hundred of e-

© The Author(s), under exclusive license to Springer Nature Singapore Pte Ltd. 2021
D.-T. Tran et al. (Eds.): ICISN 2021, LNNS 243, pp. 52–58, 2021.
https://doi.org/10.1007/978-981-16-2094-2_7

commerce website offer them, so that the collecting of reviews becomes more difficult because of the large number of customers and the number of reviews [16]. Collecting data has been difficult, analyzing data to understand customer's emotion is even more difficult. When data is becoming huge day by day, we cannot use rudimentary tools or manual analysis to accomplish this task [16, 22].

Sentiment classification is a subfield in the natural language processing field [5, 8]. It performs analyzing text to understand the writer's feelings are negative or positive, which can help e-commerce companies understand the feelings of customers through reviews that customers write on their shopping sites [11, 12]. There are four main approaches as: machine learning, lexicon-based, satistical and rule -based. In this, the lexicon-based and machine learning approaches are the most effective methods. Many dictionary-based methods are developed from a known dictionary such as WordNet, Word2Vec [2, 4, 13, 19] and others [10, 18].

For English, WordNet, SentiWordNet has been built. Other languages have been proposed method or developed sentiwordnet [11, 13, 15]. For the current Vietnamese language, SentiWordNet (VSWN) is also considered as a lexical resource to support perspective analysis in Vietnamese opinion mining applications. A SentiWordNet is usually created from WordNet or Vietnamese dictionary [9, 19], in which each word is also built with positive and negative scores to indicate the polarity of comments [14].

The Vietnamese sentiment lexicon currently has built from dictionaries or WordNet [19]. However, social users often use their own social language (called free word) that are not in dictionary. Therefore, exploiting opinions from social networks through reviews of customer is becoming difficult because the built dictionary does not cover slang words on social networks [16], In addition, when analyzing, it can avoid semantic ambiguity, so it is necessary to build a domain lexicon [1, 3, 17, 21]. In this paper, we present a method to build a Sentiment lexicon by use social reviews of user for training…. The data set is assigned manually with positive and negative label, then extract the nouns, adjectives and adverbs. These words are calculated score by support measurement. With this method, we can build an automated dictionary for any domain.

The rest of the paper is structured as follows: Sect. 2 discusses related work, Sect. 3 describes the process of building a Sentiment lexicon with scoring. Section 4 presents the results of experiment, and finally is concludes.

2   Related Work

To build SentiWordNet, there are two main approaches: dictionary-based and learning-based [2, 13, 19]. Additional, hybrid method that combinations based on machine learning [7, 20, 23]. Rao et al. [16] proposed an algorithm to construct an emotion lexicon based on word-level from dictionary. Their approach was based on Emotion Latent Dirichlet Allocation (LDA). Dictionary based on word levels obtained by maximum likelihood estimation. The data set that used for building this lexicon are topical words that are collected in the real world. Hong Nam Nguyen et al. [9] have proposed a method to build SentiWordNet, to develop for the less popular language is Vietnamese, and also calculate score of all words with TF-IDF in their collected data set. They also adress that a term in Vietnamese can be composed of a single word or a compound word. In their research, firstly, they use a list of all English adjectives, adverbs and verbs extracted from SentiWordnet 3.0 to build. And the next step, they use the collected data set to enriching the Vietnamese sentiment dictionary. In the approach of Xuan-Son Vu [19] used Vietnamese dictionary to build VSWN with 39,561 synsets. Their experiment results compared with English SentiWordNet that is 0.066 and 0.052 differences for positivity and negativity sets respectively.

There are many studies used machine learning techniques for developing senti-wordnet [12]. In the study of Wei Zhang et al., they used Word2Vec model, cosine word vector similarity calculation and SO-PMI algorithm to build a Weibo emotional lexicon. Dataset is collected from Chinese emotional vocabulary ontology library. Authors used the Word2Vec tool to convert the original Weibo posts into word vectors and mapped them into the vector space, calculated the cosine similarity of two word vectors to obtain the correlation between the two words and then extended them to the benchmark lexicon. An other study from Pedro Miguel Dias Cardoso and Anindya Roy [7], they presented an approach for automatic creation of sentiment word lists. Firstly, words are mapped into a continuous latent space, then serves as input to a multilayer perceptron (MLP) trained using sentiment-annotated words.

3   Methodology of Vietnamese Sentiment Lexicon with Scoring

3.1   Support Measurement in Association Rule

Definition 1: Support Measurement
The number of transactions that include items in the {X} and {Y} parts of the rule as a percentage of the total number of transaction.It is a measure of how frequently the collection of items occur together as a percentage of all transactions.

Definition 2: Association Rule
An implication expression of the form X → Y, where X and Y are any 2 itemsets.

Given a database T is a set of transaction, T is expressed by $T = \{t_1, t_2, \ldots, t_n\}$. For each transaction t_i is included a set of objects I (called itemset). An itemset includes k items is called k-itemset $I = \{i_1, i_2, \ldots, i_k\}$. Support measurement of rule $X \Rightarrow Y$ the frequency of a transaction contains all the items in both sets X and Y. $\text{support}(X \rightarrow Y) = \frac{n(X \cup Y)}{N}$ with N is the number of all transactions.

Suppose that, $C = \{C_{pos}, C_{neg}\}$ is the two classess as positive and negative. For each class C_j have a set of reviews R and $R = \{r_1, r_2, \ldots, r_m\}$. Then, score of term are calculated by $supp(w_i \rightarrow C_j) = \frac{n(w_i)}{N_{C_j}}$. In which: N_{C_j} is the reviews belong to class C_j; $n(w_i)$: is the number sentences in class C_j that contain w_i.

3.2   Developing Vietnamese Sentiment Lexicon with Scoring

We develop a sentiment lexicon by extracting only nouns, adverbs, and adjectives. Because some nouns are expressed in positive meaning such as "the beauty" or negative as "the contempt". We performed four steps as follow (Fig. 1):

- Step 1: Data collecting: In this step, the domain of data is the online reviews from e-commerce sites. Let R is the set of reviews $R = \{r_1, r_2, \ldots, r_m\}$
- Step 2: Pre-processing: After that, The reviews are pre-processed by remove noises and segmented in to sentences. Let $S = \{s_1, s_2, \ldots, s_l\}$. These sentences are labeled into class C_j with $C_j \in C = \{C_{pos}, C_{neg}\}$.
- Step 3: Extracting potential words: Potential words are the kind of words as nouns, adverb and adjective. In this step, sentences are segmented into words. We used Vn-Tagger to process word segmentation and identify nouns, adverb and adjective. Set of potential words is expressed by $W = \{w_1, w_2, \ldots, w_k\}$.
- Step 4: Calculating weight of potential words: Assuming that, each sentence of class C_j is a transaction, the frequency of potential word w_i is considered an item, we calculate weight of w_i based on their frequency in transactions in class C_j. This weight is calculated by $supp(w_i \rightarrow C_j) = \frac{n(w_i)}{N_{C_j}}$.

```
Input:
      S: Set of labeled reviews
Output:
      W: Set of sentiword with scoring
Initialization
            L=Ø; M= Ø; i, j=0; F= Ø;
Begin
   1. Segmenting and labeling for words
          For each s_i in S do
          For i←1 to length (s_i)
          L←Tag(w_i); // Removing nouns, advebs and adjective into
          L
   2. Calculating number of sentences in each class C_j
          For each class C_j
          If s_i ∈ C_pos then
          N_pos ++
          Else
          N_neg ++
   3. Calculating number of transactions that contain word w_i
          For each s_i in each class C_j
          For i←1 to length (s_i)
          If w_i ∈ L then
          M ← num (s_i)++
   4. Calculating weight of words based on support measurement
          4.1 For each s_i in each C_pos
               For i←1 to length (s_i)
                  If w_i ∈ L then
                     F ← num (s_i)/ N_pos
          4.2 For each s_i in each C_neg
               For i←1 to length (s_i)
                  If w_i ∈ L then
                     F ← num (s_i)/ N_neg
End;
```

Fig. 1. Algorithm of developing Vietnamese Sentiment lexicon.

4 Experimental

4.1 Dataset

To build this dictionary, we use the data source are downloaded from some e-commerce website and then use HTML code analysis techniques as follows:

- Get the HTML tags structure of web page and saved.
- Get the content of web pages containing reviews
- Use regular expressions to separate the content of the reviews

We collected reviews for the developing Vietnamese sentiment lexicon. There are 1,128 reviews, after pre-processed, we obtained 5,052 sentences, in which, number of positive sentences: 1,963; number of negative sentences: 1,298 and number of neutral sentences: 1,796.

4.2 Developing System to Build Vietnamese Sentiment Lexicon with Scoring

For building automatic Sentiment lexicon, we developed a system to build and continuously supplement sentiment words. This system was developed by C# language.

4.3 Results

With this dataset, we have built 962 sentiment words with scoring of a total of over 1,800 words extracted from 1,128 reviews. Table 1 describes the detail of results.

Table 1. Vietnamese Sentiment Lexicon

Number of extracted words	1,825
Number of positive words	548
Number of negative words	234
Number of neutral words	180

To evaluate the method, we use a dataset of reviews. These reviews fist labeled positive or negative by manual, we used 249 sentences from 5,052 labeled sentences. At the same time, we also used Vietnamese sentiment lexicon to classify these reviews into positive and negative based on scoring of potential words. We considered classifying by manual as a baseline method, and then, accuracy is calculated by ratio of classified by lexicon to the classified by manual. We also compared our method (VSL-Vietnam Sentiment Lexicon) with an other that developed by Vu Xuan Son (called VSWN) [19]. VSL' accuracy is higher than VSWN because it was developed a specific – domain and used includes slang words that VSWN doesn't got (Table 2).

Table 2. Evaluating experiment result

Method	Results
VSL (ours)	0.92
VSWN	0.87

5 Conclusion

There are a number of SentimentwordNet for Vietnamese have been developed by research groups. Currently, for each Sentiwordnet has a specific training data set, corresponding to each specific task.

In this paper, we presented a method for building the Vietnamese sentiment lexicon based on association rule and social reviews dataset for electronic products domain. The characteristics of social reviews use slang words, hybrid words, so if using Vietnamese dictionary to develop will brings inadequate analysis accuracy reviews on social networks. We have built a Vietnamese sentiment lexicon by manually labeled dataset, so the results when analyzing are close to those performed by humans, giving a high accuracy of 92% with the social reviews dataset.

References

1. Ahmed, M., Chen, Q., Li, Z.: Constructing domain-dependent sentiment dictionary for sentiment analysis. Neural Comput. Appl. **32**, 14719–14732 (2020)
2. Alemneh, G.N., Rauber, A., Atnafu, S.: Dictionary Based Amharic Sentiment Lexicon Generation. Communications in Computer and Information Science, vol. 1026. Springer International Publishing (2019)
3. Almatarneh, S., Gamallo, P.: Automatic construction of domain-specific sentiment lexicons for polarity classification. Adv. Intell. Syst. Comput. **619**, 175–182 (2017)
4. Alshari, E.M., Azman, A., Doraisamy, S., Mustapha, N., Alkeshr, M.: Effective method for sentiment lexical dictionary enrichment based on Word2Vec for sentiment analysis. In: Proceedings - 2018 4th International Conference on Information Retrieval and Knowledge Management Diving into Data Science CAMP 2018, pp. 177–181 (2018). https://doi.org/10.1109/INFRKM.2018.8464775
5. Araque, O., Zhu, G., Iglesias, C.A.: A semantic similarity-based perspective of affect lexicons for sentiment analysis. Knowl.-Based Syst. **165**, 346–359 (2019)
6. Baccianella, S., Esuli, A., Sebastiani, F.: SENTIWORDNET 3.0: an enhanced lexical resource for sentiment analysis and opinion mining. In: Proceedings of the 7th International Conference on Language Resources and Evaluation, LR 2010, pp. 2200–2204 (2010)
7. Dias Cardoso, P., Roy, A.: Sentiment lexicon creation using continuous latent space and neural networks. In: Proceedings of the NAACL-HLT 2016, pp. 37–42 (2016). https://doi.org/10.18653/v1/w16-0409
8. Esuli, A., Sebastiani, F., Vetulani, Z.: Enhancing Opinion Extraction by Automatically Annotated Lexical Resources Human Language Technology. Challenges for Computer Science and Linguistics. LNAI, vol. 6562, pp. 500–511. Springer, Heidelberg (2011)

9. Nguyen, H.N., Van Le, T., Le, H.S., Pham, T.V.: Domain specific sentiment dictionary for opinion mining of Vietnamese text. Lecture Notes in Computer Science (LNCS) (Including Subseries Lecture Notes in Artificial Intelligence and Lecture Notes in Bioinformatics), vol. 8875 (2014)

10. Kaity, M., Balakrishnan, V.: An automatic non-English sentiment lexicon builder using unannotated corpus. J. Supercomput. **75**, 2243–2268 (2019)

11. Kaity, M., Balakrishnan, V.: Sentiment lexicons and non-English languages: a survey. Knowl. Inf. Syst. **62**, 4445–4480 (2020)

12. Li, W., Zhu, L., Guo, K., Shi, Y., Zheng, Y.: Build a tourism-specific sentiment lexicon via Word2vec. Ann. Data Sci. **5**, 1–7 (2018)

13. Netisopakul, P., Thong-Iad, K.: Thai sentiment resource using Thai wordnet. Adv. Intell. Syst. Comput. **772**, 329–340 (2019)

14. Nguyen-Thi, B.T., Duong, H.T.: A Vietnamese sentiment analysis system based on multiple classifiers with enhancing lexicon features. Lecture Notes of the Institute for Computer Sciences, Social-Informatics and Telecommunications Engineering, LNICST, vol. 293. Springer International Publishing (2019)

15. Peng, H., Cambria, E.: CSenticNet: a concept-level resource for sentiment analysis in Chinese language. Lecture Notes in Computer Science (Including Subseries Lecture Notes in Artificial Intelligence and Lecture Notes in Bioinformatics). LNCS, vol. 10762. Springer International Publishing (2018)

16. Rao, Y., Lei, J., Wenyin, L., Li, Q., Chen, M.: Building emotional dictionary for sentiment analysis of online news. World Wide Web **17**, 723–742 (2014)

17. Salah, Z., Coenen, F., Grossi, D.: Generating domain-specific sentiment lexicons for opinion mining. Lecture Notes in Computer Science (LNCS) (Including Subseries Lecture Notes in Artificial Intelligence and Lecture Notes in Bioinformatics), LNAI, vol. 8346, pp. 13–24 (2013)

18. Severyn, A., Moschitti, A.: On the automatic learning of sentiment lexicons. In: NAACL HLT 2015 - Proceedings of the Conference of the North American Chapter of the Association for Computational Linguistics: Human Language Technologies, pp. 1397–1402 (2015). https://doi.org/10.3115/v1/n15-1159

19. Vu, X.-S., Park, S.-B.: Construction of Vietnamese SentiWordNet by using Vietnamese Dictionary. Preprint 2–5 (2014)

20. Wang, L., Xia, R.: Sentiment lexicon construction with representation learning based on hierarchical sentiment supervision. In: EMNLP 2017 - Proceedings of the 2017 Conference on Empirical Methods in Natural Language Processing, pp. 502–510 (2017). https://doi.org/10.18653/v1/d17-1052

21. Wang, Y., Yin, F., Liu, J., Tosato, M.: Automatic construction of domain sentiment lexicon for semantic disambiguation. Multimed. Tools Appl. **79**, 22355–22373 (2020)

22. Zhang, W., Zhu, Y.C., Wang, J.P.: An intelligent textual corpus big data computing approach for lexicons construction and sentiment classification of public emergency events. Multimed. Tools Appl. **78**, 30159–30174 (2019)

23. Wang, Y., Zhang, Y., Liu, B.: Sentiment lexicon expansion based on neural PU learning, double dictionary lookup, and polarity association. In: EMNLP 2017 - Proceedings of the 2017 Conference on Empirical Methods in Natural Language Processing, pp. 553–563 (2017). https://doi.org/10.18653/v1/d17-1059

Designing Bidding Systems in Supply Chain Management Using Blockchain Technology

Nguyen Huu Hieu[1], Le Van Minh[2], Le Nhat Anh[2],
Nguyen Thanh Tai[3], Phan Hai Le[2], Phung Ngoc Lam[4],
Nguyen Thi Tam[5], and Le Anh Ngoc[6(✉)]

[1] Anh Son District Health Center, Nghe An, Vietnam
[2] School of Engineering and Technology, Vinh University,
Nghe An, Vietnam
[3] VNPT, Nghe An, Vietnam
[4] Vietnamese-German College, Nghe An, Vietnam
[5] Thai Hoa High School, Nghe An, Vietnam
[6] Faculty of Electronics and Telecommunications, Electric Power University,
Hanoi, Vietnam
anhngoc@epu.edu.vn

Abstract. Recently, problems related to the supply chain have attracted the attention of logistic companies and various researchers. The attractiveness of the supply chain lies not only in the optimization of conventional administration costs, but also in its legality, transparency and trust. The article describes how to implement a bidding system. This bidding system is used to select carriers for e-commerce sites. Blockchain's smart contract with its transparency and precision can help improving cost efficiency for existing supply chains.

Keywords: Bidding · Supply chain · Blockchain technology

1 Introduction

In 2020, the SARS-CoV 2 epidemic breaks out and leads to the bankruptcy of 35% of businesses worldwide [1]. However, many companies in the e-commerce sector benefit from the epidemic. Amazon increases its market capitalization by 570 billion USD in 2020 [2]. Alibaba posted revenue of 114.31 billion yuan, 7% higher than the average forecasted by analysts from a Bloomberg survey. This shows that the trend for online shopping is rising and becoming the main method to shop for daily needs.

When completing orders on e-commerce sites, customers sometimes have the option to choose shipper so that they can pick their preferred delivery service or the one with the cheapest fee. However, these options are quite limiting for the customers and lack fairness for the shippers. This leads to an increased cost for customers.

Smaller carriers in a small area face difficulty when competing nationally. But if an order meets their capacities, there may be a better price for that single delivery. For example, in remote provinces where major carriers have not built a complete freight chain, small carriers cannot connect with customers. This is due to the location of e-commerce sites. It is hard to cooperate when there is not enough trust. Blockchain

D.-T. Tran et al. (Eds.): ICISN 2021, LNNS 243, pp. 59–68, 2021.
https://doi.org/10.1007/978-981-16-2094-2_8

completely eliminates the issue with trust, providing opportunities for smaller shippers and reducing costs for customers' online purchases. This research paper develops a bidding system based on blockchain smart contract technology. It outlines how the system works on the Ethereum bidding program.

2 Supply Chain and Procurement

How to incorporate freight costs into production and delivery decisions is a major area of research for supply chain improvement. For example, Sharp [3] conducted a study on a manufacturer that owned multiple factories with non-linear production costs. This manufacturer supplied products to multiple locations that were far apart. When there is more than one production base and production costs are non-linear, the traditional approach of combining production and shipping costs per supply network and make production decisions based on a single cost function is no longer appropriate. In their research, Holmberg and Tuy [4] furthered the classical transport problem by incorporating fixed production costs at random supply-demand nodes and convex non-linear penalties due to shortage at bridge center. They simplified the problem and solved it using the branching and binding methods. Both the researchs discussed above provide a list of references to solve similar optimization problems. However, none of them considered setting the price with a bidding system.

Bidding behavior in auctions has been studied since 1950 using game theory and decision theory [5]. In the classical bidding theory and in a typical auction, there is a buyer who wants to buy one or more units of the product, and there are number of potential suppliers in the market. To participate in a bid, each supplier must submit a bid to the bidder. Based on these bids, the number of purchases and the payout per supplier are determined by the bidder through pre-specified rules known as the bidding mechanism. There is an incentive for suppliers to send their actual production costs as bids. This incentive is important because it is often the key to an effective auction.

There are three main types bidding with multiple units of products. They are Pay-as-You-Bid, Uniform-price and VCG bidding.

- Pay-as-You-Bid works as its name suggests.
- Uniform Bidding (uniform-price): a uniform price is paid for each unit purchased. The price can be the bid that is first to be declined or last to be accepted.
- As laid out by economics materials, neither Pay-as-You-Bid nor Uniform-price is compatible or efficient. On the other hand, the VCG bidding mechanism, named after the 1996 Nobel Prize Winner Vickrey [6], Clarke [7], Groves [8] and is the system Facebook is using for their advertising packages, can be both. In VCG bidding, the buyer's payment to a supplier is based not only on the bid submitted, but also on the suppliers' contribution to the system when they enter the bid. This payment structure motivates suppliers to improve the operational efficiency and lower product costs, thus increasing their contributions to the system (Refer to Nisan and Ronen [9] for the general definition of the VCG procurement mechanism). Despite its attractive properties, VCG bidding is not widely used in practice, largely because bidders may take advantage of honest the participants [10].

As mentioned above, VCG bidding is rarely used as bidders can cheat. However, in an online bidding, the bidder can be a website (a virtual bidder) that receives the bid and decides the winner and pays based on some predefined algorithm so that it can minimize or eliminate this problem. Furthermore, blockchain technology can make bids absolutely secure based on its properties. Bidders can come from all over the world and the algorithm can be designed not to disclose the personal information of buyers and suppliers. Rapid progress in computability make it possible to quickly execute complex allocation rules and at a low cost. With the recent advances of blockchain technology like Ethereum, VCG bidding has the potential to become an important mechanism for online bidding. This paper uses VCG bidding in a complex supply chain setup to achieve global optimization in both production and shipping costs. It also opens the door to solutions to a variety of supply chain optimization problems.

3   Introduce Problems and Solutions

From the supply side, a retailer wants a product. Knowing that the retailer is in need, the suppliers who have the product available notifies the retailer and offers individual quotes [11]. The total cost paid by the retailer is the product's price plus the estimated shipping cost. In the case that the supplier does not transport the goods themselves, retailers and suppliers need to negotiate with a third party delivery service for the best final price. Choosing a reliable third party to deliver good is a complex process that takes effort and time. In large contracts, this is absolutely needed. However, in smaller cases, too much spending on choosing a carrier will increase in the final price. This is obviously not a wise choice. So most suppliers and retailers tend to choose an intermediate logistics service. This service will connect retailers who need to transport goods based on certain volume and geographic location to decide the best option. In this case, the only thing these retailers and suppliers can do is to trust the service with finding a third party to ship the goods.

For example, for purchases on e-commerce platforms like Tiki, the delivery option is fixed and the supplier cannot choose another delivery service to reduce the cost of the product. This creates injustice for the delivery service providers who want to actively participate in Tiki's shipping system. To be chosen as a provider for TIKI, they need to build a good relationship and trust. However, TIKI's system might one day be strong and independent enough that it does not need a third party to deliver. TIKI will replace the third party and reduce the options for suppliers. This paper will not go into depth about the business model and strategy of e-commerce sites. Rather, it focuses on building a microservice that can be integrated into e-commerce sites like this.

The microservice will select freight carrier to deliver from the provider to the buyer in this order:

– After an order is confirmed between the two parties on the e-commerce site, the site will go to the selection of delivery service providers.
– All providers will receive a notification.
– All will begin to participate in VCG bidding for the shipment.
– The provider who offers the lowest price will win the bid.

4 Application Model

As discussed in Sects. 2 and 3, existing bidding solutions need to depend on the bidder. If the bidder cheats, other participants will suffer losses, which not only does not reduce fees for the supply chain but will also increase the final cost.

Given this risk, to ensure fairness for all bidders, the paper proposes to build a blockchain bidding system for the delivery service providers to participate in a bidding proves to select the final provider. It recommends a bidding microservice model associated with e-commerce sites. It also installs a blockchain bidding program on the Ethereum platform and performs trials of different scenarios in a bidding process.

4.1 Overview of Ethereum

In 2015, blockchain technology gained greater interest when Vitalik Buterin discovered a new use for blockchain beyond bitcoin and created the Ethereum blockchain. Tech enthusiasts have seen the potential in blockchain as an open software that enables a secure exchange of value by building decentralized applications, such as smart contracts. The ability to democratize ownership of transaction records without error allows the Ethereum blockchain to create great social and economic impacts. Smart contracts are "small computer programs built directly into the blockchain, allowing financial instruments, such as loans or bonds, to be represented rather than just tokens like bitcoin, thus giving greater value" [12]. Ethereum currently has a market value of around 42 billion USD.

Ethereum is an open software platform based on blockchain technology that allows developers to build and deploy decentralized applications. As a decentralized application running on blockchain, Ethereum has all the properties of the blockchain.

- Fixation - A third party cannot make any changes to the data.
- Proof of Corruption & Tampering - Applications based on a network are formed around consensus principles, making censorship impossible.
- Security – With central point of failure and being cryptographically secure, the apps are well protected against hacking attacks and phishing activities.
- Uninterrupted - The application never stops working and cannot be shut down.
- For all the reasons above, I decided to use this platform to do run my trial.

4.2 Introduction to Solidity and the Remix IDE

Ethereum is a system running on EVM virtual machine. In theory, we can program with machine code and put in EVM to run smart contracts. However, this is not the method in reality. Developers creates high-level programming languages for g smart contracts then the compiler will push into the EVM. There are many smart contract programming languages such as: Solidity, Vyper, Java, Python, Javascript…

Solidity is a high-level, object-oriented language for implementing smart contracts. Smart contracts are programs that govern the behavior of accounts in the Ethereum state. Solidity is influenced by C++, Python and JavaScript and is designed to target the

Ethereum Virtual Machine (EVM). Solidity is statically typed, supports inheritance, libraries, and complex user-defined types among other features.

In theory, any text editor can be used for programming, but Remix is a powerful open source tool for writing Solidity contracts directly from the browser. Written in JavaScript, Remix supports in-browser usage, thereby greatly simplifying the preparation needed for smart contract programming. Remix provides a complete stimulation of the Ethereum system and we can directly write code on it and run without installing anything.

4.3   Proposed Bidding Application Model

The parties involved in the application are laid out in Fig. 1.

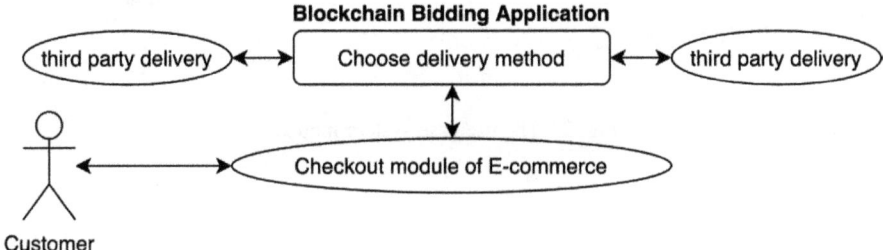

Fig. 1. The parties involved in the applications.

1. E-commerce site offers the option of "Use a shipper based on blockchain bidding" on the checkout page for sellers/ buyers to choose
2. When the shopper confirms the option of "Use shipper based on blockchain bidding", E-commerce sit will send a request including necessary information for selecting a carrier to "Blockchain Bidding Application"
3. "Blockchain Bidding Application" will announce this order to all registered 3rd party delivery service providers on the system.
4. The providers will set the minimum price to ensure the successful delivery of the order
5. Bidding algorithm VCG will select the winning provider and the shipping fee. Send it back to the e-commerce site
6. Buyer confirm to accept the shipping fee
7. The provider delivers to the buyer; the buyer and the shipper confirm that the goods have arrived at the right place
8. Money is transferred to the seller and the respective provider.

4.4   Smart Bidding Contract

The blockchain bidding program functions are depicted in the Fig. 2.

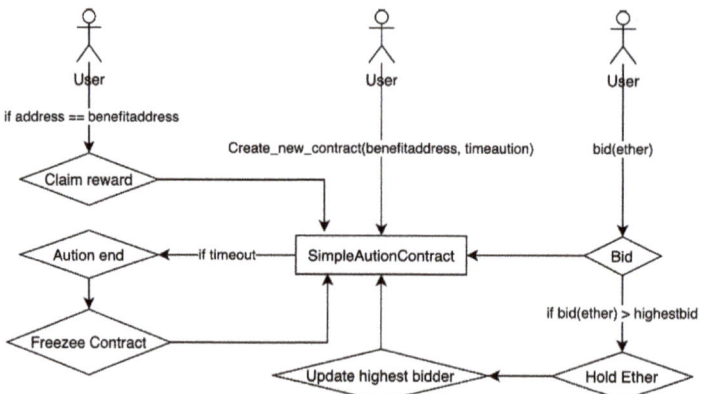

Fig. 2. The use case system model.

Functions in the program.
Steps to carry out the bidding process.

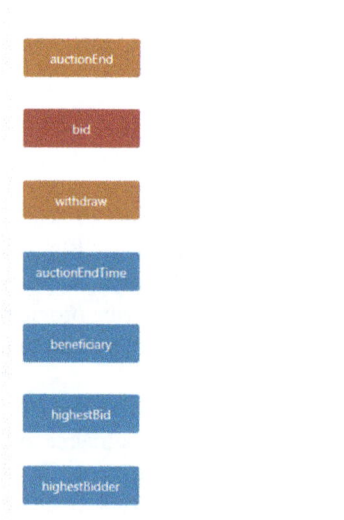

1. The bidder creates a bidding contract, with input: address of the payee and the time of the bidding. (This money is transferred to the e-commerce site; the shipper will receive the shipping money from the buyer).
2. Bidders execute bid () with the input of Ether for the bill of lading. If bid () is greater than the currently lowestbid () then it will be rejected.
3. The auctionendtime (), beneficiary (), lowestbid (), lowestbidder () functions can be called at any time without losing ether. They provide information on: bid time, ether receiving address, current lowest price, current lowest bidder.
4. The bidding time expires, the bidder in step 1 calls the auctionEnd () function to conclude the bid.
5. Winners will lose the deposit and get double the money back after the successful shipment. This amount is half of the shipping fee.

   Explanation of different scenarios.
   The program is deployed on a system that simulates the Ethereum virtual machine system. There are 10 nodes in this virtual system. A node has a balance of 100 ether.

In the Account section, Remix provides us with 10 accounts, each account has 100 ether. When the auction starts, we need to enter 2 input parameters that are the bid time and the address of the provider that receive the product at the end of the auction (Fig. 3).

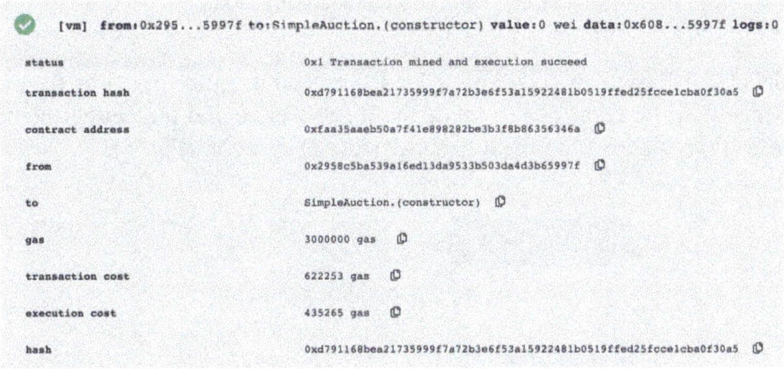

[vm] **from**:0x295...5997f **to**:SimpleAuction.(constructor) **value**:0 wei **data**:0x608...5997f **logs**:0	
status	0x1 Transaction mined and execution succeed
transaction hash	0xd791168bea21735999f7a72b3e6f53a15922481b0519ffed25fcce1cba0f30a5
contract address	0xfaa35aaeb50a7f41e898282be3b3f8b86356346a
from	0x2958c5ba539a16ed13da9533b503da4d3b65997f
to	SimpleAuction.(constructor)
gas	3000000 gas
transaction cost	622253 gas
execution cost	435265 gas
hash	0xd791168bea21735999f7a72b3e6f53a15922481b0519ffed25fcce1cba0f30a5

Fig. 3. Log screen.

After pressing deploy and successful, a few parameters are observed as below:

- Line contract address: This is the address of this smart contract, which means when other accounts want to work, they will send transactions or function call requests to this address (Fig. 4).
- Transaction cost stream: The fee to bring smart contracts into the ethereum system

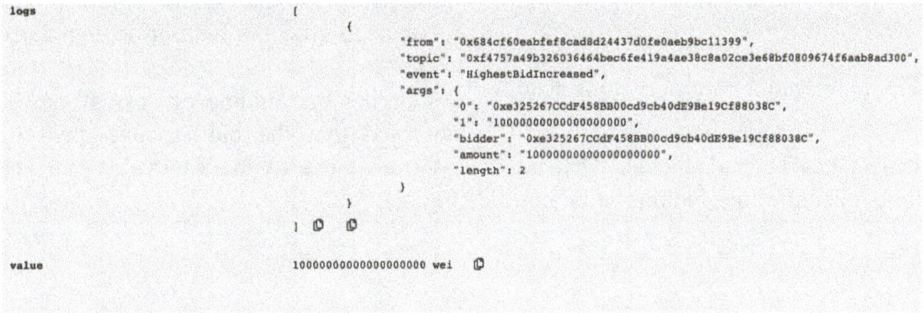

logs [
  {
    "from": "0x684cf60eabfef8cad8d24437d0fe0aeb9bc11399",
    "topic": "0xf4757a49b326036464bec6fe419a4ae38c8a02ce3e68bf0809674f6aab8ad300",
    "event": "HighestBidIncreased",
    "args": {
      "0": "0xe325267CCdF458BB00cd9cb40dE9Be19Cf88038C",
      "1": "10000000000000000000",
      "bidder": "0xe325267CCdF458BB00cd9cb40dE9Be19Cf88038C",
      "amount": "10000000000000000000",
      "length": 2
    }
  }
]

value   10000000000000000000 wei

Fig. 4. The log screen during the bidding process.

This is the log that records when an account successfully participates in the bid. We will see the bid account's address and bid price (Fig. 5).

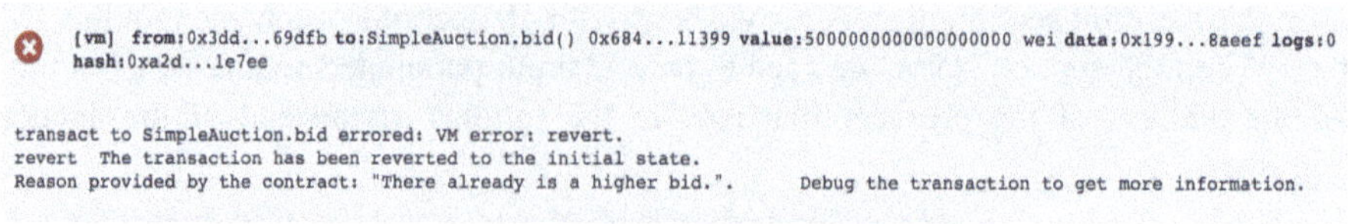

[vm] from:0x3dd...69dfb to:SimpleAuction.bid() 0x684...11399 value:5000000000000000000 wei data:0x199...8aeef logs:0
hash:0xa2d...1e7ee
transact to SimpleAuction.bid errored: VM error: revert.
revert The transaction has been reverted to the initial state.
Reason provided by the contract: "There already is a higher bid.".
Debug the transaction to get more information.

Fig. 5. Error in the bidding process.

In the case of providers offering a price lower than the price given by the previous person, there will be an error. To know the highest price and the current highest bid account, call the highestBid () and highestBidder () function (Fig. 6).

highestBid
0: uint256: 15000000000000000000
highestBidder
0: address: 0x50E55C2aea0FB8C121687
4c591829bb2801d26dB

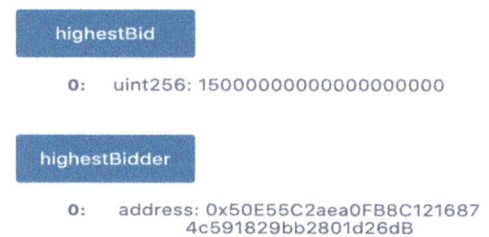

Fig. 6. Call the highestBid and highestBidder functions.

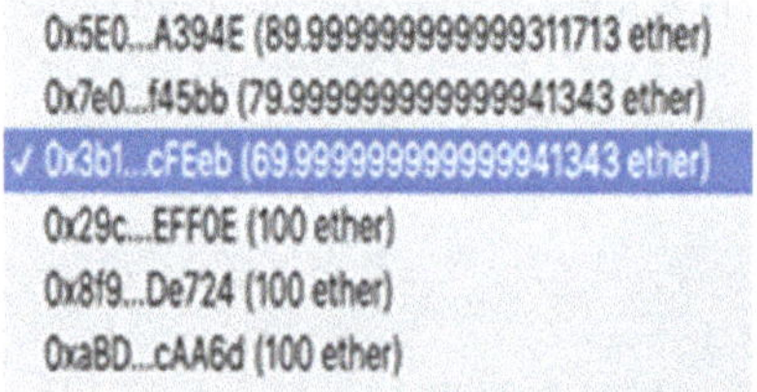

0x5E0...A394E (89.999999999999311713 ether)
0x7e0...f45bb (79.999999999999941343 ether)
✓ 0x3b1...cFEeb (69.999999999999941343 ether)
0x29c...EFF0E (100 ether)
0x8f9...De724 (100 ether)
0xaBD...cAA6d (100 ether)

Fig. 7. Balance of accounts in the node.

This ensures transparency and prevents fraud in the bidding processes.

During the bidding process, it can be observed that the bid amount spent by retailers has been deducted. This amount is transferred to the smart contract and reimbursed after the bidding is completed (Fig. 7).

During the bidding process, it can be observed that the bid amount spent by retailers has been deducted. This amount is transferred to the smart contract and reimbursed after the bidding is completed (Fig. 8).

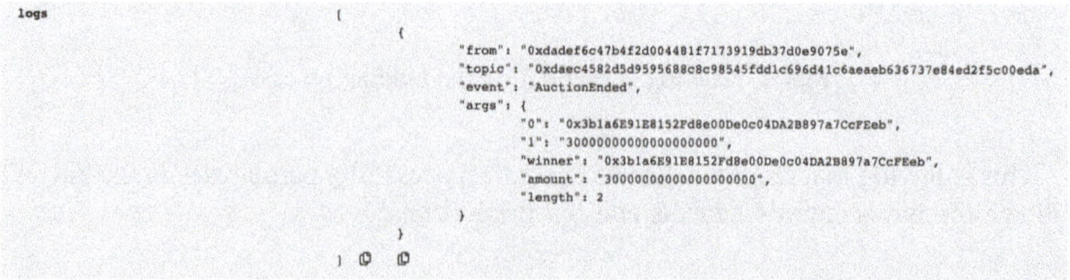

logs
[
{
"from": "0xdadef6c47b4f2d004481f7173919db37d0e9075e",
"topic": "0xdaec4582d5d9595688c8c98545fdd1c696d41c6aeaeb636737e84ed2f5c00eda",
"event": "AuctionEnded",
"args": {
"0": "0x3b1a6E91E8152Fd8e00De0c04DA2B897a7CcFEeb",
"1": "30000000000000000000",
"winner": "0x3b1a6E91E8152Fd8e00De0c04DA2B897a7CcFEeb",
"amount": "30000000000000000000",
"length": 2
}
}
]

Fig. 8. Log closing the auction.

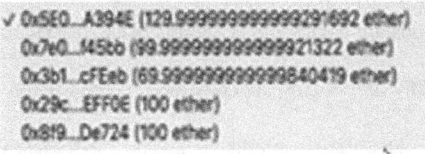

At the end of the bidding, the winning account can be seen in the "winner" line (Fig. 9).

Fig. 9. Balance of accounts in the node after the bid ends

5   Conclusion

Ethereum is currently heavily used for its rapid implementation. However, the EVM virtual machine system is Turing-completeness, so it is bound to have error. This was demonstrated by the DAO event. The solution when implementing this application into practice is to build a completely new blockchain system for private use. Hence, this increases the confidence of e-commerce companies to integrate this module into their applications.

Another difficulty is that cryptocurrencies are still not widely recognized especially in Vietnam. A system running on blockchain will inevitably have to use cryptocurrencies to ensure the elimination of third-party. If for legal reasons developers have to choose to use government coins instead of cryptocurrency, this would invalidate the original reasons for building this application. In general, the question of trust will no longer be guaranteed.

The last and most difficult barrier to overcome comes from e-commerce companies. As their companies grow, these companies will most likely build a seamless freight chain to maximize profits. And it is hard for them to share part of this profit with another company. The solution to this problem is to build a blockchain-based sales page where sellers and buyers can connect directly with each other. It would be possible with today's technology features if companies are really interested and recognize the opportunity.

References

1. Euler Hermes: www.eulerhermes.com. Accessed 16 July 2020
2. Hottovy, R.J.: Amazon's new opportunities in coronavirus; FVE to $3500. Stock analyst update, 31 July 2020
3. Sharp Logistics Pvt. Ltd. https://vrsharp.com/. Accessed 20 May 2020
4. Holmberg, K., Tuy, H.: A production-transportation problem with stochastic demands and concave production cost. J. Math. Program. **85**, 157–179 (1999)
5. Rothkopf, M.: A model of rational competitive bidding. J. Manag. Sci. **15**(7), 362–373 (1969)
6. Vickrey, W.S.: Counterspeculation, auctions, and competitive sealed tenders. J. Finan. **16**, 8–37 (1961)
7. Clarke, F.: The pinmen recall test as a measure of the individual. Britian J. Psychol. **62**(3), 381–394 (1971)

8. Groves, T.: Incentives in teams. Econ. Soc. **41**(4), 617–631 (1973)
9. Nisan, N., Ronen, A.: Algorithmic mechanism design, games and economic behavior, **35**(2), 166–196 (2001)
10. Rothkopf, M.H., Teisberg, T.J., Kahn, E.P.: Why are vickrey auctions rare? J. Polit. Econ. **98**(1), 94–109 (1990)
11. Satapathy, G., Kumara, S.R.T., Moore, L.M.: Distributed intelligent agents for logistics (DIAL). J. Expert Syst. Appl. Pract. **14**, 409–424 (1998)
12. Houben, R., Snyers, A.: Cryptocurrencies and blockchain. European Parliament's Special Committee on Financial Crimes (2018)

Feature Selection Using Genetic Algorithm and Bayesian Hyper-parameter Optimization for LSTM in Short-Term Load Forecasting

Nguyen Nhat Anh[1], Nguyen Hoang Quoc Anh[1], Nguyen Xuan Tung[2], and Nguyen Thi Ngoc Anh[1(✉)]

[1] SAMI, Hanoi University of Science and Technology, Hanoi, Vietnam
anh.nguyenthingoc@hust.edu.vn
[2] School of Electrical Engineering, Hanoi University of Science and Technology, Hanoi, Vietnam

Abstract. Electricity load forecasting at nationwide level is important in efficient energy management. Machine learning methods using big data multi-time series are widely applied to solve this problem. Data used in forecasting are collected from advanced SCADA system, smart sensors and other related sources. Therefore, feature selection should be carefully optimized for machine learning models. In this study, we propose a forecasting model using long short-term memory (LSTM) network with input features selected by genetic algorithm (GA). Then, we employ Bayesian optimization (BO) to fine-tune the hyper-parameters of LSTM network. The proposed model are utilized to forecast Vietnam electricity load for two days ahead. Test results have confirmed the model has better accuracy in comparison with currently used models.

Keywords: Genetic algorithm · LSTM · Time series · Forecasting · Bayes optimization

1 Introduction

Short term electricity load forecasting (STLF) is a vital task for national dispatching operation [4]. Load forecasting under uncertain conditions of weather, economy and consumer behavior is a challenging problem for researchers and industrial staff. To address the problem, three main approaches have been proposed and widely applied: (i) statistical approaches such as auto-regressive moving average (ARMA), auto-regressive integrated moving average (ARIMA); (ii) machine learning and deep learning approaches such as fuzzy learning, artificial neural network (ANN), recurrent neural network (RNN), long short-term memory (LSTM) [14]; (iii) hybrid approach that combine statistical and machine learning approaches [10]. Among the aforementioned approaches, LSTM is a popular and well-known method. It is suitable for forecasting non-linear time series.

D.-T. Tran et al. (Eds.): ICISN 2021, LNNS 243, pp. 69–79, 2021.
https://doi.org/10.1007/978-981-16-2094-2_9

Recently, LSTM is applied extensively in time series forecasting. LSTM is an upgraded variant of the vanilla recurrent neural network. It is capable of capturing the long-term effects of time series data. In [15], the authors employ LSTM and time correlation modification to forecast photovoltaic power. The predictions made by the LSTM network are adjusted by the time correlation method. Then an ensemble is constructed using individual models to obtain the final result. The ensemble model performs better than both individual models because it combines their strengths.

Electricity load forecasting has always been a challenging problem due to its dependency on various exogenous factors like temperature, humidity, holiday effects, etc. To address the problem, many researchers had applied metaheuristic approaches to determine important features, for instance, grasshopper optimization [12], ant colony optimization and genetic algorithm (GA) [13]. GA is a popular evolutionary algorithm and has been employed widely in many fields. In [6], a novel model is proposed to predict stock market measures.

The parameters of number of neurons and learning rate showed great influence on the prediction model [9] proposed by Anurag Kulshrestha et al. In addition, Anurag Kulshrestha et al. also pointed out that the dropout index and $L2$ regularization also greatly affect the results because they are the parameters to avoid overfitting. Because parameters have a great influence on the neural network models, especially the LSTM model, to choose right the optimal parameter, Bayesian Optimizing (BO) is an algorithm that can help us solve. BO is effectively algorithm that can process a large amount of input in an acceptable time [1].

This paper aims to achieve better performance of electricity load forecasting by combining the advantage of generic algorithm and Bayesian hyper-parameters optimization for LSTM. Concretely, generic algorithm is used for feature selection. Then, the long short term memory (LSTM) is used for short-term electricity load forecasting and Bayesian optimization is used to fine-tune the hyper-parameters of the LSTM.

The remainder of the paper is structured as follows: Sect. 2 explains the methodology of feature selection using genetic algorithm, Bayesian hyper-parameter optimization, LSTM and the proposed model. The experiments and results are presented in Sect. 3. Lastly, Sect. 4 provides the conclusion and discussions.

2   Methodology

Genetic Algorithm
Genetic algorithm (GA) is a branch of evolutionary algorithm based on Charles Darwin's theory of natural selection process. It was first proposed and developed by J. Holland in the 1970s. GA is a metaheuristic that use stochastic search technique to solve optimization problems. In GA context, an iteration is called a generation. In each generation, a set of candidates called a population is chosen based on the fittest individuals from the previous generation. To determine how

"good" a candidate is, a fitness function is constructed. This function should reflect the objective of the optimization problem to some degree. The evolution of a population is performed by genetic operators inspired by the evolution process in nature such as crossover and mutation. To apply these operators, each candidate is usually represented as a binary string, called a chromosome. First, in the selection phase, better individuals have higher chance of getting selected to produce offspring. The selected chromosomes are called parents. In crossover, two parents are chosen randomly to pair up and then they exchange segments of their chromosomes to form two new chromosomes. Mutation first selects a candidate, then its bit string representation is altered at random loci to create a new chromosome. The genetic operators encourage exploring the search space while the selection favors fitter individuals. The whole process is applied repeatedly until the stopping criteria is reached. The general flowchart of GA is described in Fig. 1.

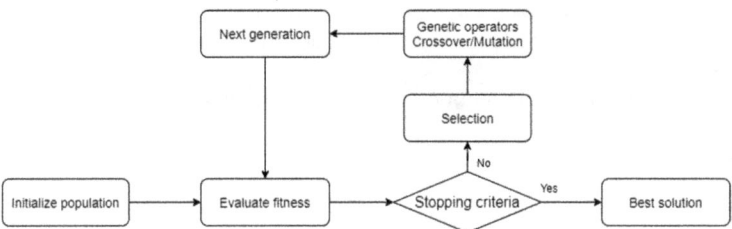

Fig. 1. General flowchart of genetic algorithm.

Long Short Term Memory
Recurrent neural network (RNN) is a special type of neural network, designed specifically to solve tasks that deals with sequence data such as time series forecasting, language modelling, speech recognition, video-based object tracking, etc. Despite its capability to handle long-term dependencies on paper, in practice RNN tends to suffer from gradient explosion and gradient shrinkage problems. Therefore, vanilla RNN's applications are limited. In 1997, a variant of RNN called long short-term memory (LSTM) was introduced by Hochreiter and Schmidhuber to address these problems [5]. In addition to the input and the hidden state of the previous time step, each LSTM cell also has a cell state to represent the long-term memory. Furthermore, the input gate and output gate are introduced to control the information flow within a cell. In 2000, Gers et al. proposed the addition of the forget gate which allowed the LSTM to filter out unwanted past information [3].

Bayesian Hyper-parameter Optimization
Given n pairs of input-observation $(x_i, y_i), i = 1, \ldots, n$, we want to pick the next value x_{n+1} that maximize y_{n+1}. Since f is a black box function, we can't employ gradient based methods to find the optimal value. This is where Bayesian

optimization (BO) comes into play. First, a surrogate function is constructed to approximate f. Our goal is to make this surrogate function as close as possible to f. Then, we find the optimal value for this function instead of the original f. The surrogate function is constructed in a way so that it is easier to optimize this function than the original f. Concretely, the construction makes use of the Gaussian process (GP) regression and the optimal value is chosen based on an objective function called acquisition function. The pseudo-code for BO is given in Algorithm 1 [2]. We discuss Gaussian process regression and acquisition function in details in the following sections.

Algorithm 1. Pseudo-code for Bayesian optimization

1: Initializing mean function and kernel function for Gaussian process f
2: Creating Gaussian process f according to previous mean function and kernel function. Set $n = n_0$
3: **while** $n \leq N$ **do**
4:     Calculate the new posterior probability distribution on Gaussian process f
5:     Update the acquisition function according to current posterior probability distribution
6:     Setting the global maximum of acquisition function to x_n
7:     Calculate $y_n = f(x_n)$
8:     **if** $f(x_n) > f(x_{max})$ **then**
9:         $x_{max} = x_n$
10:    **end if**
11:    n := n + 1
12: **end while**
13: Return: the point with the largest posterior mean x_{max}

Gaussian Process Regression

Consider the observation vector $(f(x_1), \ldots, f(x_n))$. In traditional statistics, we often suppose the observations come from a random variable. In GP regression, its distribution is assumed to be a multivariate normal distribution with fixed expectation vector and covariance matrix. This distribution is called prior probability distribution.

Using the inputs $x_i, i = 1, \ldots, n$, we construct the prior distribution function as follows. Initially, we choose a mean function μ_0 and the corresponding mean vector is $(\mu_0(x_1), \ldots, \mu_0(x_n))$. The mean function is usually a constant function $\mu_0(x) = \mu$. The covariance matrix is constructed based on a kernel function $k(x_i, x_j)$ and is in the following form

$$K = \begin{bmatrix} k(x_1, x_1) & \cdots & k(x_1, x_n) \\ \vdots & \ddots & \vdots \\ k(x_n, x_1) & \cdots & k(x_n, x_n) \end{bmatrix}$$

The kernel is chosen so that if x_i, x_j are close to each other, then they are strongly correlated. Furthermore, the covariance matrix must be positive semi-definite for every input. A commonly used kernel is the radial basis function kernel:

$$k(x, x') = \alpha_0^2 \exp\left(-\frac{1}{2l^2}\|x - x'\|^2\right)$$

where α_0^2 is the maximum allowed value of the covariance matrix, l is the length scale parameter which controls how quickly a function can change. The prior distribution is written as

$$f(x_{1:n}) \sim \mathcal{N}(\mu_0(x_{1:n}), K)$$

where $x_{1:n}$ denotes $x_1, \ldots, x_n$ and $f(x_{1:n}) = (f(x_1), \ldots, f(x_n)), \mu_0(x_{1:n}) = (\mu_0(x_1), \ldots, \mu_0(x_n)))$.

Now suppose x_{n+1} is given and we need to predict $f(x_{n+1})$. By Gaussian process assumption, $f(x_{n+1})$ also follows a normal distribution and concatenating $f(x_{n+1})$ and $f(x_{1:n})$ results in a multivariate normal distribution [16]:

$$(f(x_1), \ldots, f(x_{n+1})) \sim \mathcal{N}\left(\mu_0(x_{1:n+1}), \begin{bmatrix} K & K_{n+1}^T \\ K_{n+1} & K_{n+1,n+1} \end{bmatrix}\right)$$

where

$$K_{n+1} = (k(x_{n+1}, x_1), \ldots, k(x_{n+1}, x_{n+1}))$$
$$K_{n+1,n+1} = k(x_{n+1}, x_{n+1})$$

The conditional distribution function of $f(x_{n+1})$ is computed using Bayes' Theorem and we have the following results [11]:

$$f(x_{n+1})|f(x_{1:n}) \sim \mathcal{N}(\mu_{n+1}(x_{n+1}), \sigma_{n+1}^2(x_{n+1})) \tag{1}$$

where

$$\mu_n(x_{n+1}) = K_{n+1}K^{-1}f(x_{1:n})^T$$
$$\sigma_{n+1}^2 = K_{n+1,n+1} - K_{n+1}K^{-1}K_{n+1}^T$$

Acquisition Function. While GP predicts the distribution function of $f(x_{n+1})$, acquisition function is used to find candidates for global maximum. In Algorithm 1, let $f_n^* = \max_{m \leq n} f(x_m)$ be the best value we have found at iteration n and x^* is the corresponding input. At iteration $n + 1$, we have $f_{n+1}^* = \max(f_n^*, f(x))$. However, the value of $f(x)$ is unknown so we can't compare it with f_n^*. To address this problem, we can choose x based on the expected improvement function:

$$EI_n(x) = E_n[(f(x) - f_n^*)^+]$$

where $a^+ = \max(a, 0)$.

Note that the distribution function of $f(x)$ is given by (1). Hence we can choose $x_{n+1} = \arg \max EI_n(x)$. Finding maximum value of $EI_n(x)$ is feasible because according to [7], it is possible to calculate first and second order derivatives of $EI_n(x)$.

The expected improvement function is a reasonable choice because according to [8], it balances between exploration and exploitation at each iteration. Exploration means that we find candidates in uncertain region. Exploitation means that we find candidates in highly probable region.

There are alternative choices for acquisition function other than expected improvement, for instance, probability of improvement, upper confidence bound and lower confidence bound [16].

Proposed Model: We introduce the idea of incorporating GA and BO to optimize the LSTM network. The proposed model consists of several stages as follows:

- First, the time series is split into training, validation and test set in chronological order.
- **Feature selection using GA:** In this stage, the hyper-parameters of the LSTM are fixed to default values. The network's input consists of
  - calendar features: lunar calendar is also included due to its importance to Vietnamese culture. Many holidays, festivals and cultural activities are based on lunar calendar. The following features are considered: hour, day of week, month, day of year, day of month (by lunar calendar), lunar month, day of year, day of month (by lunar calendar), lunar month.
  - temperature features: measurements received from important weather monitoring stations in Northern Vietnam, Central Vietnam and Southern Vietnam are considered.
  - lag features: load from the previous two days, load from the last week and load from the last month are considered.

  To find the best feature subset using GA, it is necessary to represent each solution as a binary string. A natural way to represent a feature subset is encoding it as a binary string $a_1 a_2 \ldots a_n$ with length n, where n is the total number of features, a_i is 1 if the i-th feature is selected and 0 otherwise. For each candidate, we train an LSTM model with fixed hyper-parameters. The fitness function is chosen as the negative MAPE of the predictions on validation set.
- **Hyper-parameters tuning:** After obtaining the best feature subset from GA, proceed to the next stage. In this stage, the hyper-parameters of the LSTM network are optimized using BO. In this study, we consider the following hyper-parameters:
  - input width of the network (number of time steps used as input)
  - learning rate
  - number of units of the hidden state
- After fine-tuning the LSTM network using BO, train an LSTM network with these optimized settings (best features + fine-tuned hyper-parameters). This is the final model to be used for forecasting on the test set.

The whole process is summarized in Fig. 2.

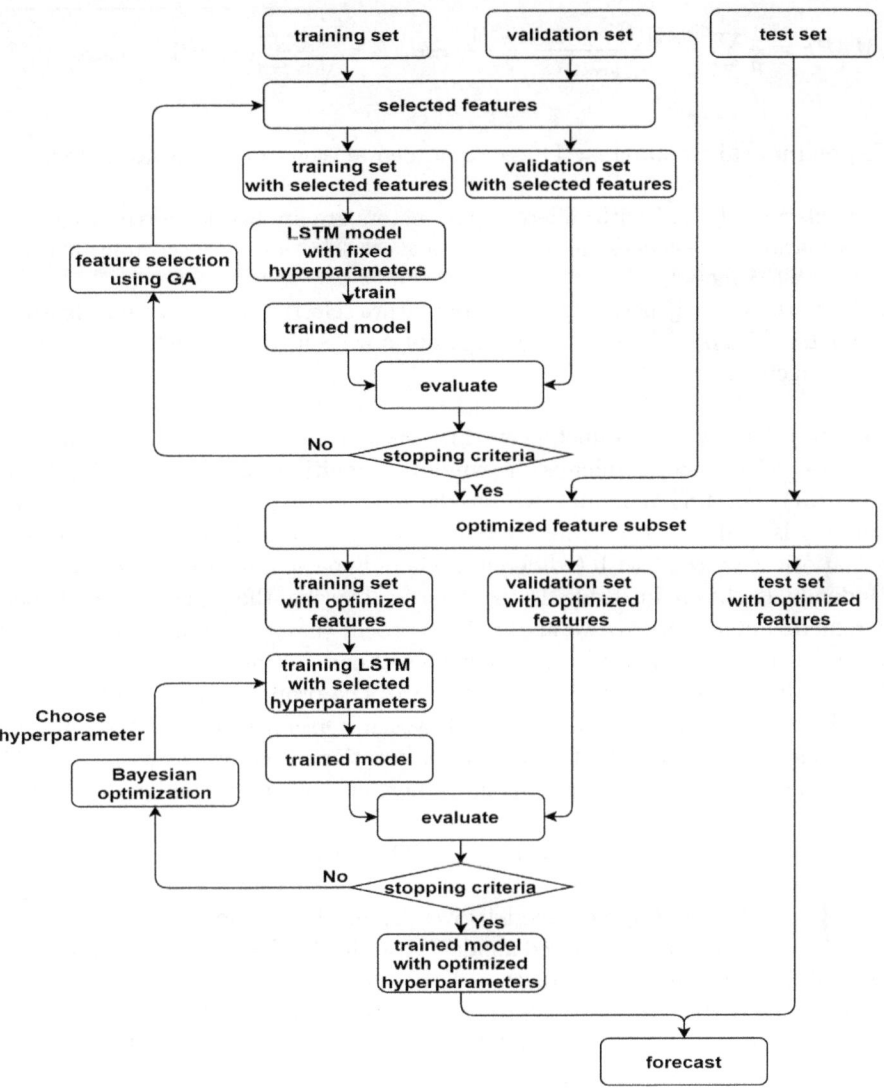

Fig. 2. Proposed model.

3   Experiments and Results

Data Set Description: The data used for experiment is the Vietnam electricity load dataset. The time series ranges from 1/1/2016 to 31/8/2020 and the resolution is one hour.

Evaluation Criteria: To evaluate the result of forecasting, mean absolute percentage error (MAPE) and root mean square error (RMSE) are considered. These criteria are defined as follows:

$$MAPE = \frac{1}{n}\sum_{t=1}^{n}\frac{|y_{real}(t) - y_{forecast}(t)|}{y_{real}(t)}; RMSE = \sqrt{\frac{1}{n}\sum_{t=1}^{n}(y_{real}(t) - y_{forecast}(t))^2}$$

$$(2)$$

Experimental Scenarios: Three experimental scenarios are investigated:

1. Scenario 1: LSTM with different parameters are applied for forecasting.
2. Scenario 2: Use genetic algorithm for feature selection to select input features for LSTM model.
3. Scenario 3: Use genetic algorithm for feature selection to select input features for LSTM model, then use Bayesian optimization to fine-tune LSTM hyper-parameters.

Results: The hyper-parameters we fine-tune in scenario 1 include number of neurons (NN), n-steps which is the window of multiple time series and learning rate (LR). For LSTM model, we set the number of layer is 1, the activation function is tanh function. Moreover, we use all features to train our model in scenario 1. The result with 6 difference sets of hyper-parameters which choosing randomly is shown in Table 1. In scenario 2, using the same sets of hyper-parameters in scenario 1, we apply GA to select the most appropriate features for each set of hyper-parameters. We will use Adam optimizer for GA algorithm. The result of scenario 2 is shown in Table 1. The proposed model not only use GA for feature selection but also use Bayesian Optimizer to fine the optimize hyper-parameters. For BO, the number of iterations (iter) we choose are 20, 30, 40, 50, 60. The sets of optimize hyper-parameters with each number of iteration are shown in Table 1.

For the proposed model LSTM-GA-BO, the average MAPE and average RMSE results are 4.31% and 1601.23 (MW), respectively. Compared to simple LSTM model, the proposed model gives us much superior results. However, when we compare the proposed model with the LSTM-GA model, there are two things that we need to discuss deeper. Firstly, in general, the proposed model gives better results than the LSTM-GA model. Although the average MAPE of the proposed model is 4.31% higher than the average MAPE of the LSTM-GA model, the average RMSE of the proposed model is 9.15% better. This can be explained that the loss function needed to optimize is RMSE, not MAPE. Secondly, among the parameters given for the LSTM-GA model, there are several sets of parameters that give a much better result than the model we propose, for example, parameter set NN: 300, n- steps = 25, LR = 0.001 yields a remarkable optimal result of 3.29% for MAPE and 1383.12 for RMSE. This raises the question of whether the BO algorithm is really effective for the parameter optimization problem. However, when we look at Table 1, we can see that when we increase the number of iterations of the BO algorithm, the RMSE value tends to decrease. This means that the BO algorithm is capable of converging to the optimal point with a large enough loop.

In summary, from the obtained results, we propose GA algorithm for features selection in short-term load forecasting problem. We also propose using Bayesian optimization to find the appropriate hyper-parameters for this problem. Note that if we want the Bayesian algorithm to be effective, we can increase the number of iterations. However, this means that the run time of algorithm will be longer and we should consider that.

Table 1. Results of all 3 scenarios

Model	NN	n-steps	LR	Iter	MAPE	RMSE (MW)
LSTM	128	24	0.001		7.14%	2634.31
	128	20	0.003		6.73%	2460.42
	200	12	0.005		6.71%	2492.68
	150	15	0.01		6.76%	2533.25
	300	25	0.001		7.25%	2599.50
	50	10	0.006		6.51%	2449.82
LSTM - GA	128	24	0.001		4.47%	1854.35
	128	20	0.003		4.37%	1634.23
	200	12	0.005		4.28%	1785.80
	150	15	0.01		4.19%	1982.02
	300	25	0.001		3.29%	1383.12
	50	10	0.006		4.68%	1775.60
LSTM - GA - BO	391	37	0.00069	20	4.26%	1624.67
	378	48	0.0012	30	4.21%	1590.17
	200	24	0.001	40	4.39%	1627.27
	362	26	0.015	50	4.28%	1589.77
	361	40	0.00082	60	4.29%	1563.21

4   Conclusion and Discussion

In this paper, a novel forecasting model has been proposed to obtain more accurate load prediction. By using genetic algorithm, feature selection is done automatically and is highly optimized rather than picking features manually. Additionally, the hyper-parameters of the LSTM network are optimized using Bayesian optimization. The test results have shown that the proposed model is suitable for Vietnam electricity load forecast at nationwide level. Actual data have been used to test short term load forecasting model for two days ahead.

In conclusion, in this paper some results were achieved as follows:

- Feature selection using genetic algorithm.
- Employ LSTM in multi time series in forecasting.
- Fine-tune the hyper-parameters for the LSTM using Bayesian optimization.

– Propose a forecasting model combining GA, LSTM and Bayesian optimization.
– Apply the proposed model in shot-term load forecasting in Vietnam.

With the results achieved by the proposed model, we believe it will contribute to energy consumption assessment and electricity market management.

In the future work, more time series forecasting applications are researched based on the proposed model. The stability of the model will be investigated.

References

1. Cheng, H., Ding, X., Zhou, W., Ding, R.: A hybrid electricity price forecasting model with Bayesian optimization for German energy exchange. Int. J. Electr. Power Energy Syst. **110**, 653–666 (2019)
2. Frazier, P.I.: A tutorial on Bayesian optimization (2018)
3. Gers, F.A., Schmidhuber, J.A., Cummins, F.A.: Learning to forget: continual prediction with LSTM. Neural Comput. **12**(10), 2451–2471 (2000)
4. Heydari, A., Majidi Nezhad, M., Pirshayan, E., Astiaso Garcia, D., Keynia, F., De Santoli, L.: Short-term electricity price and load forecasting in isolated power grids based on composite neural network and gravitational search optimization algorithm. Appl. Energy **277**, 115503 (2020)
5. Hochreiter, S., Schmidhuber, J.: Long short-term memory. Neural Comput. **9**(8), 1735–1780 (1997)
6. Huang, Y., Gao, Y., Gan, Y., Ye, M.: A new financial data forecasting model using genetic algorithm and long short-term memory network. Neurocomputing **425**, 207–218 (2020)
7. Jones, D., Schonlau, M., Welch, W.: Efficient global optimization of expensive black-box functions. J. Global Optim. **13**, 455–492 (1998)
8. Kaelbling, L.P., Littman, M.L., Moore, A.W.: Reinforcement learning: a survey. CoRR **cs.AI/9605103** (1996). https://arxiv.org/abs/cs/9605103
9. Kulshrestha, A., Krishnaswamy, V., Sharma, M.: Bayesian BILSTM approach for tourism demand forecasting. Ann. Tourism Res. **83**, 102,925 (2020)
10. Quang, D.N., Thi, N.A.N., Solanki, V.K., An, N.L.: Prediction of water level using time series, wavelet and neural network approaches. Int. J. Inf. Retriev. Res. (IJIRR) **10**, 1–19 (2020)
11. Rasmussen, C.E., Williams, C.K.I.: Gaussian Processes for Machine Learning (Adaptive Computation and Machine Learning). The MIT Press (2005)
12. Salami, M., Sobhani, F., Ghazizadeh, M.: A hybrid short-term load forecasting model developed by factor and feature selection algorithms using improved grasshopper optimization algorithm and principal component analysis. Electr. Eng. **102**, 437–460 (2019)
13. Sheikhan, M., Mohammadi, N.: Neural-based electricity load forecasting using hybrid of GA and ACO for feature selection. Neural Comput. Applications - NCA **21**, 1–10 (2011)
14. Sulandari, W., Subanar, Lee, M.H., Rodrigues, P.C.: Indonesian electricity load forecasting using singular spectrum analysis, fuzzy systems and neural networks. Energy **190**, 116,408 (2020)

15. Wang, F., Xuan, Z., Zhen, Z., Li, K., Wang, T., Shi, M.: A day-ahead PV power forecasting method based on LSTM-RNN model and time correlation modification under partial daily pattern prediction framework. Energy Convers. Manage. **212**, 112,766 (2020)
16. Zhang, Q., Hu, W., Liu, Z., Tan, J.: TBM performance prediction with Bayesian optimization and automated machine learning. Tunnelling Undergr. Space Technol. **103**, 103,493 (2020)

Improving Continuous Hand Gesture Detection and Recognition from Depth Using Convolutional Neural Networks

Thanh-Hai Tran[(✉)] and Van-Hieu Do

School of Electronics and Telecommunications,
Hanoi University of Science and Technology, Hanoi, Vietnam
`hai.tranthithanh1@hust.edu.vn`

Abstract. Hand gestures are becoming more and more efficient and intuitive means of communication between human and machine. While many proposed methods aim at increasing performance of recognition from spotted gestures segments, it lacks efficient solutions for both detection and recognition gesture from continuous video streams for practical application. In this paper, we approach by using a simple CNN detector to detect gesture candidates and a more precise and complicated CNN classifier to recognize gesture categories. We first deploy a method recently proposed in [5]. However, we improve that method by adjusting another condition for gesture decision making to avoid detection missing due to the detector performance. Our improved algorithm is compared with the original one, showing an improvement in term of overall accuracy (from 73.9% to 79.3%) on the same dataset.

Keywords: Hand gesture recognition · Deep learning · Human machine interaction

1 Introduction

Nowadays, human gestures are widely used as an efficient mean of communication between human and machine. They are also deployed for healthcare and AV-VR applications. To this end, hand gestures must be captured, pre-processed and recognized before being converted to commands for controlling machines. Among many sensors, camera is becoming more and more low-cost but provides rich information of the scene. In addition, with the development of technologies, cameras providing both RGB and depth data are popular than ever. Depth data is less sensitive to lighting condition and could easily extract hand from background.

This material is based upon work supported by the Air Force Office of Scientific Research under award number FA2386-20-1-4053.

To recognize hand gestures, most current approaches base on hand-crafted features or deep features. Currently, deep learning and convolutional neural networks (CNN) achieved very high performance on multi tasks in computer vision and action recognition in particular. CNNs try to extract hidden structures of objects of interest to generate coarse-to-fine representation. Recognition hand gestures relates several sub-problems such as hand segmentation [7] using Mask R-CNN on RGB images, [8] using different multimodalities such as skeleton, RGB and Depth and dynamic hand gesture classification [1,9] using C3D convolutional neural network. In [6], a two sub-networks model has been proposed for hand gesture recognition. In [5], different CNNs such as ResNext, C3D have been deployed for hand gesture recognition. Several methods have also mentioned to solve multiview problem [1].

Despite the fact that hand gestures recognition has achieved impressive performance on clip videos, their direct use in practical applications are not evident because in those applications, video streams come continuously that need to determine the starting and ending of a gesture before recognizing its category. To deal with this issue, recently in [5] the authors have proposed a two-model approach. Their proposed framework consists one binary classifier (detector) and one multi-class classifier (recognizer). The detector is a lightweight CNN (Resnet-10) to quickly detect the presence of a gesture and plays a switch role to activate the more complicated classifier (ResNext-101) once a gesture is detected. This system can run at 64fps when a gesture detected or at 460fps when only detector is called.

We then start to investigate this method and discover an important issue. The original method [5] relies upon a very strong assumption that is the detector must be nearly perfect that does not miss any gesture candidate. In case the detector fails to detect a candidate, the method in [5] will miss to classify the gestures. To overcome this issue, in this paper, we propose to make decision even the detector fails. Our proposition is integrated in the original algorithm and qualitatively evaluated on the same dataset according to the same evaluation protocol. The experiment shows that the improved algorithm outperforms the original one by 5.4% (the accuracy increased from 73.9% to 79.3%.

2 Detection and Recognition Methods

2.1 Gesture Detection and Classification

The design and training the detector must satisfy three main criteria. Firstly, the detector will be applied in a window whose size is smaller than the window size of classifier to speed up the detection. In our experiment, the window size for detector is 8 frames. Secondly, the detector has high true positive rate to do not miss any gestures. To this end, we will train the detector with cross-entropy loss function with three times more gesture samples than non-gesture samples. Finally, the detector should be light (small number of parameters). The Resnet-10 with only 862K parameters will be taken into account [4].

During the gesture implementation, the hand could be out of camera field of view that leads to detection missing. To overcome this, at the current time t, $k-1$ previous outputs of detector and the current one will be queued in a queue namely $Q_k = \{q_{t-k-1}, q_{t-k-2}, ..., q_t\}$. Then a median filter will be applied on Q_k to output the response of detector at the current time t. The size of queue is selected as 4 in the experiment.

Once the detector detects a gesture candidate, it will activate the gesture classifier. The classifier will be carefully designed so that can correctly classify multiple classes. In this work, ResNext-101, a state of the art method for video classification will be utilized [3]. ResNext-101 can extract both spatial and temporal features from video sequence. ResNext-101 will perform on a bigger window size (n = 32 frames in our experiment). It will output N probabilities corresponding to the N classes of gestures.

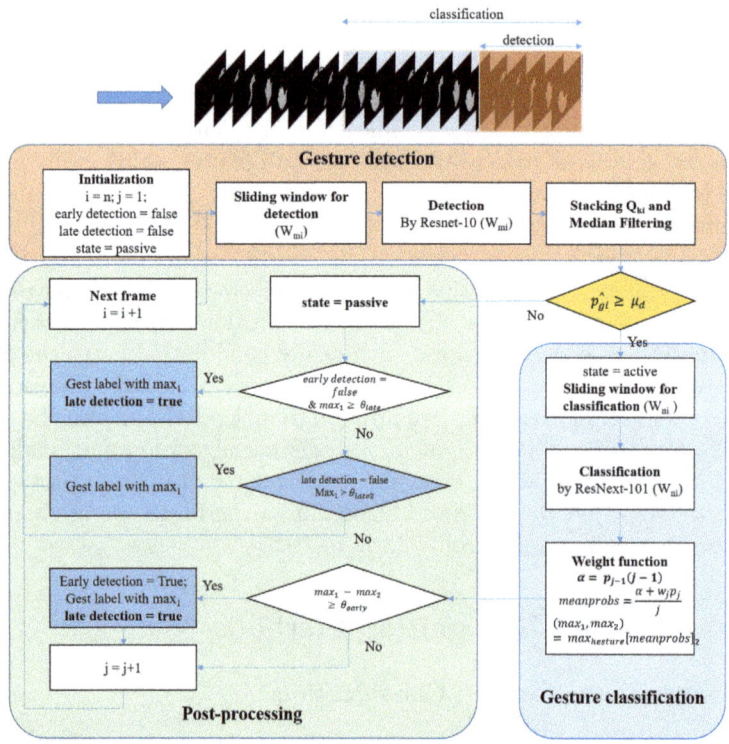

Fig. 1. Algorithm of hand gesture detection and classification

2.2 Combination of Gesture Detection and Classification

In continuous detection and recognition system, one criteria is to avoid multiple responses for one gesture. A gesture is usually composed of three main phases: preparation, implementation and retraction among which implementation phase is the most important to distinguish gestures while gestures could be similar each other at preparation or retraction phases. It's obvious that whenever a gesture is detected by the detector, the classifier will be activated. We consider two cases: early detection means a gesture will be classified before it finishes and late detection otherwise then the classifier will be activated.

Early Detection: As aforementioned, a gesture should be detected after the implementation phase to obtain the higher confidence because gestures at the preparation and retraction phase may be similar each other. One solution is to apply a weight function on classification scores. Then whenever the detector detects a gesture candidate, it will activate the classifier and the classification scores will be weighted at each iteration. Then the criteria on difference of two highest scores value will be checked.

Late Detection: If the criteria on difference of two highest scores is not satisfied, the system will wait until the classifier is switched off then the class with highest score above a late detection threshold θ_{late} will be prediction result.

2.3   Algorithm for Continuous Gesture Detection and Recognition

The Original Algorithm: Algorithm for continuous gesture detection and recognition is presented as follows. When starting the algorithm, state of classifier is set to "Passive", and an early detection is initialized to false. The algorithm will take input as a sequence of frames. At time i, it will take $w_i = m$ frames ($w_i - 1$ previous frames and the current one) to put into the detector. The detector will output two values p_{ni} and p_{gi}. The probability value for gesture prediction p_{gi} will be pushed into a queue Q_k that keeps $k - 1$ previous gesture probabilities and the current one p_{g_i}. We then apply a median filter to smooth the values of probabilities and generate the final probability for decision making $\hat{p}_{gi}$. A gesture is detected if a $\hat{p}_{gi} > \mu_d$ which is a detection threshold.

When a gesture is detected, the classifier will be activated. The classifier state is set to "Active". The classifier will back from the current frame to $(n-1)$ previous frames and perform the classification to output probabilities of N classes of gestures $p_1, p_2, ..., p_N$. These probabilities will be post-processed by applying a weighted function w_j to increase probabilities at the implementation phase of gesture. If the difference between two highest values are bigger than a threshold θ_{early}, then the class label having the best score will be taken. When the detector does not return a candidate, it could mean that the gesture has ended. The classifier state will be reset to "Passive". The algorithm will check if an early detection did not happen, then a gesture with the highest probability bigger than a threshold μ_{late} will be outputted.

Improved Algorithm for Continuous Gesture Detection and Recognition: As aforementioned, the original algorithm could not give any response if it is not early detection (the detection is wrong) or the late detection condition does not satisfy. To overcome this situation, we adjust a new boolean flag "late detection". When an "early detection" is set to True, then "late detection" flag will be also set to True. When the condition is satisfied, then "late detection" flag will be set to True. When both cases are not satisfied, the late detection is false (as initialized) then we check if $max_1 > \mu_{late_2}$ then we also consider the gesture with max_1 probability as the output of recognition. Our improved algorithm with difference compared to the original one is illustrated as Fig. 1 with adjustment in the three blue blocks.

3   Experiments

3.1   Dataset

In this paper, we utilize the dataset introduced in [2]. This dataset contains 25 gesture classes, which correspond to the commands for controlling devices in smart cars application. The dataset was recorded using multiple sensors at different viewpoints. Totally 1523 continuous video streams have been captured by 20 subjects. All of the subjects have performed with right hand. Each video length is about 5 s. In our work, we used depth data taken by SoftKinetic DS325. We also take 70% data for training and the 30% remaining data for testing as in the original paper [2]. Specifically, 1050 videos for training and 482 videos for testing. As the problem is to detect and recognize gestures from continuous video stream, the metric used for evaluation is Levenshtein distance.

3.2   Experimental Results

Quantitative Evaluation: In our experiment, we set early detection threshold $\theta_{early} = 0.3$ and the $\theta_{late} = 0.02$. Table 1 shows the experimental results obtained by our proposed method in comparison with the original method. We find that our proposed requires more deletion and replacement but don't require any insertion. That means that our proposed don't miss any gesture (0 insertion) while the original algorithm misses many (45 insertions). However, as our algorithm limits as less as possible missed gestures, it can give false alarms (3 false alarms more than the original algorithm (our algorithm requires 18 deletions while the original one requires only 15 deletions). In average, our proposed algorithm achieved better accuracy.

Table 1. Experimental results of the proposed algorithm for online detection and recognition of hand gestures

Method	#Transformations	Deletion	Insertion	Replacement	Accuracy(%)
Okan 2019 [5]	126	15	45	66	73.9
Our proposed method	100	18	0	82	79.3

Application of Gesture Recognition from Continuous Streams: Figure 2 shows a screenshot of GUI with an video stream as input (left), displays the detection score (middle) and classification score (right). The detection scores will activate the classifier which produces classification score. Relying upon the post reasoning, the final decision is made and the label of gesture is displayed on the original stream.

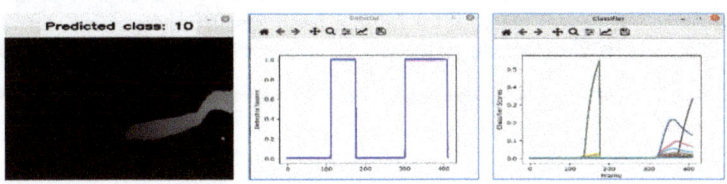

Fig. 2. GUI for continuous human gesture detection and recognition

4   Conclusions

In this paper, we have presented an improved algorithm for continuous gesture detection and recognition. We deployed the original method using a small and lightweight CNN (ResNet-10) for gesture detection and a more complicated CNN (ResNext-101) for gesture classification. The original method bases on the main assumption that the detector is absolutely accurate while it is not true in reality. We have improved this algorithm by adjusting a latest detection strategy to avoid missing detection and recognition. Our proposed algorithm increased accuracy of the whole system from 73.9% to 79.3%. In the future, we will try other CNN networks, add more streams (e.g. optical flows) or integrate the detector and classifier as well ass post-processor in an end-to-end framework for easier deployment and validation.

References

1. Doan, H.G., Tran, T.H., Vu, H., Le, T.L., Nguyen, V.T., Dinh, S.V., Nguyen, T.O., Nguyen, T.T., Nguyen, D.C.: Multi-view discriminant analysis for dynamic hand gesture recognition. In: ACPR, pp. 196–210. Springer (2019)
2. Gupta, P., Kautz, K., et al.: Online detection and classification of dynamic hand gestures with recurrent 3D convolutional neural networks. In: CVPR, vol. 1, p. 3 (2016)
3. Hara, K., Kataoka, H., Satoh, Y.: Can spatiotemporal 3D CNNs retrace the history of 2D CNNs and imagenet? In: CVPR, pp. 6546–6555 (2018)
4. He, K., Zhang, X., Ren, S., Sun, J.: Deep residual learning for image recognition. In: Proceedings of the IEEE Conference on Computer Vision and Pattern Recognition, pp. 770–778 (2016)
5. Köpüklü, O., Gunduz, A., Kose, N., Rigoll, G.: Real-time hand gesture detection and classification using convolutional neural networks. In: FG, pp. 1–8. IEEE (2019)
6. Molchanov, P., Gupta, S., Kim, K., Kautz, J.: Hand gesture recognition with 3D convolutional neural networks. In: Proceedings of the IEEE Conference on Computer Vision and Pattern Recognition Workshops, pp. 1–7 (2015)

7. Nguyen, D.H., Le, T.H., Tran, T.H., Vu, H., Le, T.L., Doan, H.G.: Hand segmentation under different viewpoints by combination of mask R-CNN with tracking. In: ACDT, pp. 14–20. IEEE (2018)
8. Pham, V.T., Le, T.L., Tran, T.H., Nguyen, T.P.: Hand detection and segmentation using multimodal information from Kinect. In: MAPR, pp. 1–6. IEEE (2020)
9. Truong, D.M., Doan, H.G., Tran, T.H., Vu, H., Le, T.L.: Robustness analysis of 3D convolutional neural network for human hand gesture recognition. Int. J. Mach. Learn. Comput. **9**(2), 135–142 (2019)

Research on Large-Scale Knowledge Base Management Frameworks for Open-Domain Question Answering Systems

Dat Tien Nguyen[1,2,3(✉)] and Hao Duc Do[3]

[1] University of Science, Ho Chi Minh City, Vietnam
[2] Vietnam National University, Ho Chi Minh City, Vietnam
[3] OLLI Technology JSC, Ho Chi Minh City, Vietnam
{ntdat,hao}@olli-ai.com

Abstract. In recent years, there are increasingly question answering systems based on large-scale knowledge bases that can answer natural questions. In this paper, we analyze the performance and efficiency of different knowledge base management frameworks when retrieving information from large-scale knowledge bases. The data model is built in the structure of a directed graph with vertices denoted the entities and edges denoted their relationships. With RDF (Resource Description Framework) model, Neo4j and Apache Jena built Graph Database Platform and Triple Store respectively to present the meaning networks. We analyzed, measured, and discussed how they store data from a knowledge base in the industry. We briefly showed the strengths and limits of each tool via experiments. Based on particular aims, researchers can choose the appropriate database management framework for their applications in large-scale open domain question answering systems.

Keywords: Neo4J · Apache Jena · Wikidata

1 Introduction

In fact, some natural questions are asked or searched by many people such as "What is the current population of Ho Chi Minh City?", "What is the current average salary in Ho Chi Minh City?", etc. are popular which are questions that appeared in many question answering systems which are generally based on knowledge bases. The knowledge bases need to be large, structured in a clear way so that the machine can replace people to read it. Some are provided free of charge, including Wikidata [1], YAGO [2], and DBpedia [3], of which Wikidata is an outstanding, multilingual, and regularly updated source.

This research is supported by OLLI Technology JSC, Ho Chi Minh City, Vietnamese.

D.-T. Tran et al. (Eds.): ICISN 2021, LNNS 243, pp. 87–92, 2021.
https://doi.org/10.1007/978-981-16-2094-2_11

Wikidata is a large-scale knowledge base, free, open. All of us including machines can read and edit data in it [4]. Wikidata serves as the central repository for Wikimedia sister projects, of which Wikipedia is the most important. Currently, some Wikipedia sites enumerate entities - e.g. Top-Earning Tech CEOs list - and information boxes appeared in the upper right of articles - e.g. a tech CEO's average annual income statistics - will be periodically updated manually. This may cause inconsistent with other information sources. The function of Wikidata is to keep unique and consistent factual information, facts only need to be edited once and information can be extracted from many different sites. Recently, Wikidata contains more than 88 million entities edited by larger than 45 thousand editors. Information can be retrieved legally online through the website[1]. In Wikidata, each entity is connected with other related entities or a data value through a named relationship, thus Wikidata is mapped into the RDF model. However, relationships in Wikidata are assigned as attribute-value pairs so as to further explain other contexts to the relationship that need definitions and references. This makes the representation of Wikidata content in RDF more complicated, so we need to use some form of transformation [5] in order to fully manipulate the knowledge base in RDF.

A database management framework is required to store large-scale knowledge bases like Wikidata and to support for question answering systems in an effective approach. We focus on studying NEO4J [6] and Apache Jena [7] which are popular Graph Database and Triple Store, respectively, how data is stored in both engines. From experimental results, we can find a effective solution for deploying large scale open domain question answering systems for Vietnamese in practice.

2   Wikidata Knowledge Base

The data model of Wikidata [8] is constructed based on a directed graph, entities are connected by edges labeled by attributes or relationships. An entity may be labeled in many languages but its actual identifier is unique: property using an identifier such as P569 illustrates the attribute "date of birth" whereas entity using an identifier such as Q254 represents "Wolfgang Amadeus Mozart" is Austrian composer of the Classical period. All properties and entities in Wikidata are freely generated by machines or users.

3   Apache Jena

3.1   Introduction to RDF

World Wide Web Consortium (W3C) has proposed a framework or a data model called Resource Description Framework (also known as RDF) [9, 10], to illustrate that data has been structured and linked on the web. RDF is generated with the purpose of describing a model for given statements that establish connections

---

[1] https://query.wikidata.org/.

between two resources or between resource and a value. Each RDF statement is formed by three components that are respectively subject-predicate-object (also known as RDF triple). There are many storage formats in RDF such as Turtle (.ttl), RDF/XML (.xml), N3 (.n3), etc. [11].

3.2 SPARQL

Recommended by W3C, SPARQL [12,13] (acronym for SPARQL Protocol and RDF Query Language) is a language for constructing queries over data in RDF format. A straightforward SPARQL query consists of a set of three elements: subject-predicate-object. There are many available libraries to formulate SPARQL queries and convert them into other query languages like SQL.

3.3 Apache Jena

Apache Jena [7,14] is a Semantic Web framework, open-source, free that supports functions to build semantic web, linked data applications that are developed by Java. It provides API to manipulate and retrieve data from the RDF graph as well as query engine ARQ [15,16] running queries over data in RDF format. SPARQL is executed on a model that is created by using RDF statements to retrieve RDF data. Figure 1 illustrates examples visualized on RDF data and the corresponding RDF Turtle serialization syntax.

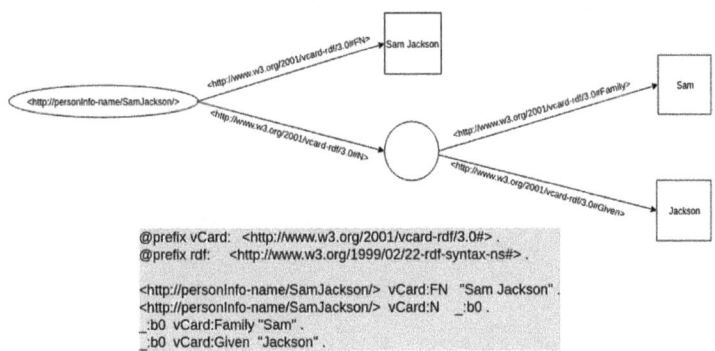

Fig. 1. The example is represented on RDF data and the corresponding RDF Turtle serialization syntax in APACHE JENA.

4   Neo4j

4.1   Neo4j and Its Plugin for RDF

Known as the most popular graph database, Neo4j was implemented by Java and first released in 2007 by Neo4j Inc [17,18]. One of the biggest advantages of Neo4j is having an ACID transaction concept - atomicity, consistency, isolation, durability - with native graph storage and processing. Each database transaction is made up of a chain of database operations which satisfy ACID properties, thus it is guaranteed of data validity despite any types of failures and errors. Neo4j has a plugin neosemantics (n10s) and authored by the Neo4j Labs Team, that allows to use RDF and its related vocabularies such as RDFS, OWL and others are used in Neo4j. Running as an extension of the Neo4j database, neosemantics includes some principal functionalities such as import/export RDF in multiple formats (Turtle, JSON-LD, N-Quads, N-Triples, N3, and RDF/XML), model mapping on import/export, graph validation, basic inferencing, etc.

4.2   Cypher

Cypher query language [19] allows users to store and retrieve data in Neo4j by constructing a query to add, delete and create new entities, relationships and properties. Furthermore, Cypher is also an open-source, hence it is possible to add more specifications.

5   Experimental Results and Evaluation

5.1   Experiment Installation

We focus on dumping Wikidata with Vietnamese label filtering only in June 2020 with N-Triples format. Data capacity is approximately 176 GB containing 1.1 billion statements with 87 million entities and more than 1.4 billion triples. We used Neo4j 4.1.1, Apache Jena 3.15.0 to import and store triples, Apache Jena Fuseki 3.9.0 as a server to access data. For effective evaluation, we use a single machine for all of our experiments with 1TB Samsung SSD 860 hard drive, Intel(R) Xeon(R) CPU E5-2620 v4 32 core @ 2.10 GHz and 64 GB of RAM.

5.2   Results

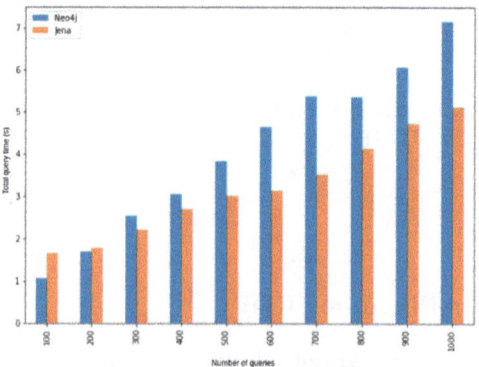

Fig. 2. Comparing result of processing time between Neo4J and Apache Jena.

Table 1. Comparison results in comparison between Neo4j and Apache Jena.

Criteria	Neo4J	Apache Jena
Secondary storage required	Need x1.2 the size of the original data to be imported	Need x2 the size of the original data to be imported
RAM required	More data costs more RAM, retrieving large data results often takes up RAM	Need enough RAM
Query performance	Best for deep or variable length traversals and path queries	Not good for deep queries, best for simple type queries
Data Integration	Allows integration from multiple sources, applying for highly dynamic datasets	Not yet supported
Flexibility when importing data	Be able to import and query concurrently	Unable to query when have not done imported
Database migration capabilities	Allows to move data to another machine	Also supported
Query language	Cypher	SPARQL

Results are shown in Table 1 in comparison between Neo4j and Apache Jena based on some criteria such as flexibility with data, processing speed, resource cost. In addition, we also measured the query performance of Neo4J and Apache Jena by randomly selecting from 100 queries to 1000 queries from Wikidata. As can be seen from the bar chart, as the number of query growth, Apache Jena results outperformed Neo4J which is clearly shown in the diagram in Fig. 2. In criteria of processing speed, Apache Jena outperforms Neo4J when the number of queries increases significantly. However, in criteria of flexibility with data, Neo4j is better support than Apache Jena when it is possible to integrate data from multiple sources and allows to query data while loading new data.

6   Conclusion

From our experimental results, it shows some advantages and disadvantages of each tool. For more details, this study achieves query performance assessments between Neo4J and Apache Jena on a large-scale knowledge base such as Wikidata. In criteria of processing speed, Apache Jena outperforms Neo4J. However, the graph database is capable of scaling storage with large numbers of entities and relationships. In fact, Neo4J can scale a network of several million to several billion entities and relationships and Neo4J can also combine data from a variety of sources to improve the quality and variety of the knowledge database. This research has a profound contribution for us to choose an efficient knowledge base data storage tool like Neo4J, from which for application to large scale open domain question answering systems for Vietnamese in practice.

References

1. Vrandečić, D., Krötzsch, M.: Wikidata: a free collaborative knowledgebase. Commun. ACM **57**, 78–85 (2014)
2. Suchanek, F.M., Kasneci, G., Weikum, G.: Yago: a core of semantic knowledge. In: Proceedings of the 16th International Conference on World Wide Web, WWW, pp. 697–706 (2007)
3. Auer, S., Bizer, C., Kobilarov, G., Lehmann, J., Cyganiak, R., Ives, Z.G.: DBpedia: a nucleus for a web of open data. In: 6th International Semantic Web Conference on the Semantic Web, ISWC, pp. 722–735 (2007)
4. Martín-Chozas, P., Ahmadi, S., Montiel-Ponsoda, E.: Defying wikidata: validation of terminological relations in the web of data. In: Proceedings of the 12th Language Resources and Evaluation Conference (2020)
5. Hayes, P., Patel-Schneider, P.F. (eds.): RDF 1.1 Semantics. W3C Recommendation, 25 February 2014
6. The Neo4j Team: The Neo4j Manual v2.3.1 (2015). http://neo4j.com/docs/
7. Wilkinson, K., Sayers, C., Kuno, H.A., Reynolds, D., Ding, L.: Supporting scalable, persistent semantic web applications. IEEE Data Eng. Bull. **26**(4), 33–39 (2003)
8. Piscopo, A., Vougiouklis, P., Kaffee, L., Phethean, C., Hare, J., Simperl, E.: What do Wikidata and Wikipedia Have in Common?: An Analysis of their Use of External References (2017). https://doi.org/10.1145/3125433.3125445
9. The Resource Description Framework (RDF) (2014). The W3C. http://www.w3.org/RDF/
10. Hussain, S.M., Kanakam, P.: SPARQL for semantic information retrieval from RDF knowledge base. Int. J. Eng. Trends Technol. **41**, 351–354 (2016). https://doi.org/10.14445/22315381/IJETT-V41P264
11. Optimized Index Structures for Querying RDF from the Web Andreas Harth, Stefan Decker, 3rd Latin American Web Congress, Buenos Aires, Argentina, 31 October to 2 November 2005, pp. 71–80
12. The SPARQL (2014). The Wikipedia. http://en.wikipedia.org/wiki/SPARQL
13. The SPARQL Query Language for RDF (2008). W3C
14. Bakkas, J., Bahaj, M.: Generating of RDF graph from a relational database using Jena API. Int. J. Eng. Technol. **5**, 1970–1975 (2013)
15. ARQ - A SPARQL Processor for Jena, version 1.3 March 2006, Hewlett-Packard Development Company. http://jena.sourceforge.net/ARQ
16. Pérez, J., Arenas, M., Gutierrez, C.: Semantics and complexity of SPARQL. In: Cruz, I., et al. (eds.) The Semantic Web, ISWC 2006. Lecture Notes in Computer Science, vol. 4273. Springer, Heidelberg (2006). https://doi.org/10.1007/11926078
17. Vukotic, A., Watt, N., Abedrabbo, T., Fox, D., Partner, J.: Neo4j in Action. Book Neo4j in Action (2014)
18. Faralli, S., Velardi, P., Yusifli, F.: Multiple knowledge GraphDB (MKGDB). In: Proceedings of the 12th Language Resources and Evaluation Conference (2020)
19. Ünal, Y., Oğuztüzün, H.: Migration of data from relational database to graph database, pp. 1–5 (2018). https://doi.org/10.1145/3200842.3200852

An Implementation of Firewall as a Service for OpenStack Virtualization Systems

Xuan Tung Hoang[1]([✉]) and Ngoc Dung Bui[2]

[1] VNU University of Engineering and Technology, Hanoi, Vietnam
tunghx@vnu.edu.vn
[2] University of Transport and Communications, Hanoi, Vietnam
dnbui@utc.edu.vn

Abstract. In this paper, we propose and implement a firewalling service for cloud system using OpenStack. The service, called FWaaS - Firewall as a Service, is offloaded from and loosely coupled with Openstack cloud system. It can be utilised to provision firewall functions and it supports a rich set of packet filtering capabilities, from link layer up to application layer. The service is lightweight but shows that it could prevent efficiently threats from outside of the networks with low level of resource consumption.

1 Introduction

OpenStack is a mature open-source cloud platform. It is supported by a large community of over 100,000 members worldwide. In OpenStack, firewall policies are managed by the concept of security group. A security group is a set of rules to firewall data traffics such as blocking or allowing packets that matched particular criteria related to port numbers, source/destination IP addresses or transport protocols. An administrator needs to build suitable security groups then, assign them to network interfaces of virtual machines that he/she manages.

The management of security policies using security group is efficient when system is small and simple. Once the number of virtual machines is larger and they scatter multiple connected network with a complex hierarchy, managing security group for each virtual network interface becomes cumbersome. Administrators have to spend a lot of time identifying and building security groups for their subnetworks and each virtual machine clusters. In this case, managing security policies at some strategic points on the network will be a much better approach.

In addition to the above weakness, the default security group implementation of OpenStack lacks of supports for advanced packet filtering capabilities including deep packet inspection and link-layer filtering. In this paper, we propose an off-the-shelf component called FWaaS, FireWall as a service, that is able to support administrators injecting firewall functions in Openstack virtual routers.

© The Author(s), under exclusive license to Springer Nature Singapore Pte Ltd. 2021
D.-T. Tran et al. (Eds.): ICISN 2021, LNNS 243, pp. 93–100, 2021.
https://doi.org/10.1007/978-981-16-2094-2_12

FWaaS supports all existing features of Openstack security group and advanced features including packet deep inspections and link-layer filtering. Our FWaaS uses iptables [8] as firewall engine and is compatible with Linux namespace-based virtual routers [9].

2   Related Work

The idea of a firewall as a service is not new. Openstack's Neutron project has a plan for the FWaaS service as its network plugins. However those projects is not finalized because of complexity in incorporating with Openstack and lack of human resources for maintaining them. The plugins are only available in OpenStack Ocata and older versions. Instead of deeply integrate firewall as a service into Openstack Networking core, for example Neutron, we implement our FWaaS as an off-the-shelf component, making it as loosely coupled with Openstack as possible. By that way, we can keep our FWaaS implementation simple but still provide powerful firewalling features to Openstack.

There have been several studies involved in building security services "as a Service" for cloud systems in general, and for OpenStack in particular. There exists a type of security service called NIDS - Network Intrusion Detection System. NIDS systems have the main function of detecting and monitoring attacks on computer networks. NIDS systems are usually deployed on a machine on the network and act as a network service. Packets flowing in the network will be mirrored into the NIDS system and the system will analyze the packet data, detect unauthorized access and make decisions to protect the network. A research group at Bakrie University, Jakarta, Indonesia has developed a NIDS system that uses signature-based method to identify packet data for an OpenStack private network [6]. Another team from the University of Tennessee, Chattanoog has also built a NIDS system for the OpenStack Cloud. They analyze the consumption and cost of network resources of the control mechanisms used by NIDS systems, thereby proposing a small NIDS as a Service by using fewer CPU resources [7]. However, these NIDS services often take up a large amount of network resources such as memory, CPU and network traffic. At the same time, building these solutions that are quite complex and effect intensively on the transmission performance of the network.

3   Proposed Method

Our proposed FWaaS is designed to be completely independent of OpenStack. Because of that, it can run on Linux operating system as a normal application for firewall provision and administration. The architecture of FWaaS is presented in Fig. 1. FWaaS, in fact, comprises a Webservice (*Firewall Web service*) *Firewall engines* which are agents running in OpenStack network servers. Firewall Engines takes responsibility to control iptables rules in virtual routers. Also, FWaaS includes a *Front-end Web Application*, which is a single page application

(SPA) that provides a graphical user interface for administrators to interact with firewall engines of the system.

Firewall Web Service provides a layer for user interface and other application features including user authentication and Openstack integration. It relays user configuration requests to Firewall Engine via Webservice APIs. It also retrieves virtual router information from integrated Openstack system via Openstack APIs.

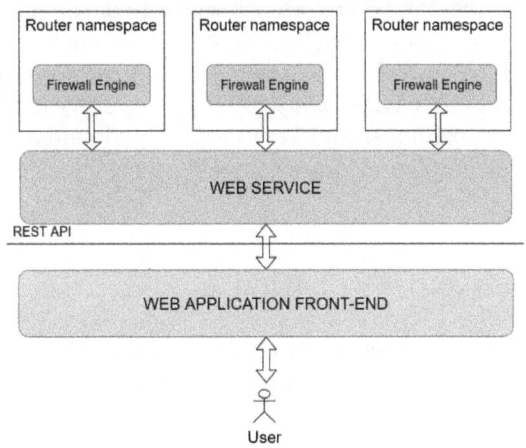

Fig. 1. High level architecture of service

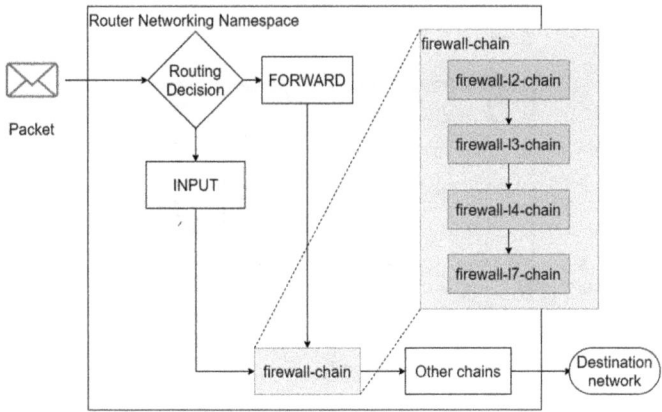

Fig. 2. Packet flow inside the Router namespace

The Front-end Web Application is generally an admin dashboard. It is implemented using React [10]. It provide an interactive GUI (Graphical User Interface) for user to work with FWaaS. It can help users in manage the Firewall rules and keep them away from errors due to human mistakes. Each operation of user over the GUI will be transformed into a respectively REST API request to the Fierwall Web Service by the application automatically.

Firewall Engine is the most important of the system. It is written in Python. In OpenStack, virtual routers are implemented by Linux networking namespaces. Upon receiving a configuration request from an administrator via the Front-end Web Application and Firewall Webservice, Firewall Engine jumps into the networking namespace, performs the corresponding manipulations on iptables's filtering table to provide desired firewall behaviors. In particular, Firewall Engine does not modify the existing rules in filter table created before. Instead, it creates new chains in filter table and inject them at the first position in the default iptables INPUT and FORWARD chains. By that way, every packet coming in the virtual router will be examined by FWaaS rules first before evaluated by other chains. How Firewall Engine modifies iptables packet filtering flow is depicted in Fig. 2.

4   Experiments and Evaluation

To evaluate the FWaaS service protection ability and the effectiveness of the service, we deployed a cloud project using OpenStack Stein. The cloud project consists of one provider network and two self-service networks. IP address range of the provider network is 192.168.0.0/24. And IP address range of the two self-service networks are 172.16.0.0/24 and 172.16.1.0/24 respectively. Within each network, several instances are created and assigned to the network. These instances are used to exchange information between networks to test functions and performance of FWaaS (Fig. 3).

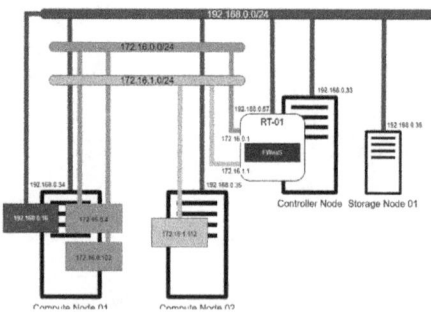

Fig. 3. Logical deployment of OpenStack system

Accuracy of packet filtering is an important aspect to evaluate. We perform tests on ability of filtering packets at each protocol layer that FWaaS can provide. In addition, impact of the service on data traffics should be carefully assessed. We examine the packet transmission while using a certain amount of protection rules within the FWaaS service. The result of this test will be compared with the case when FWaaS is not used. By that way, we can evaluate the impact of the service on the transmission performance of the network. Moreover, we further examine the level of resource usage by the service and draw conclusions about the impact of the service on the whole system.

Layer 2 Filtering: To evaluate how exactly the service filter the packets in Layer 2, we are going to ping 4 times between 2 instances called instance 1 and instance 2 with IP address 172.16.1.112, 172.16.0.4 and MAC address FA:16:3E:3F:D1:36, FA:16:3E:C6:83:81, respectively. The result of the ping test recorded below (Table 1).

Table 1. Result of L2 Filtering test

Source	Destination	Firewall		Result
		Source MAC	Action	
Instance 1	Instance 2	None	None	0% packet loss Round-trip avg: 2.270 ms
Instance 1	Instance 2	FA:16:3E:3F: D1:36	DROP	100% packet loss
Instance 1	Instance 2	FA:16:3E:C6: 83:81	DROP	100% packet loss
Instance 1	Instance 2	FA:16:3E:C6: 83:82	DROP	0% packet loss Round-trip avg: 2.782 ms

Layer 3 Filtering: In order to test L3 Filtering feature, we will test the ability of blocking traffic between networks by setting some of filtering rules inside the RT-01 Router. The test create ping traffic between 2 instance inside 2 network 172.16.1.0/24 and 172.16.0.0/24 with the IP address 172.16.1.112 (instance 1) and 172.16.0.4 (instance 2), respectively. The result is shown in the table below (Table 2).

Layer 4 Filtering: To test this feature of the FWaaS, we will send packets through 2 types of network transmission protocols are TCP and UDP using netcat tools over port 2222 in instance 1 (172.16.1.112). The result is shown in the table below (Table 3).

Layer 7 Filtering: This feature allows users to filter packets by matching data in application layer messages with a POSIX Regex string. Here are some tests on the L7 filtering feature for the 172.16.0.0/24 network corresponding to the machine with IP address 172.16.0.4 using services of layer 7 of the server with IP address 192.168.0.33 is Controller Node's address but not the OpenStack project's virtual network. On the Controller Node, there are a few active http servers to provide services on many ports such as: OpenStack Dashboard (port

Table 2. Result of L3 Filtering test

Firewall				Result
Protocol	Source IP	Destination IP	Action	
None	None	None	None	Instance 1 to instance 2: 0% packet loss Round-trip avg: 2.655 ms
icmp	Any	Any	DROP	Instance 1 to instance 2: 100% packet loss
icmp	Any	Any	DROP	Instance 2 to instance 1: 100% packet loss
icmp	192.168.0.0/24	Any	DROP	Instance 1 to instance 2: 0% packet loss Round-trip avg: 2.826 ms
icmp	192.168.0.0/24	Any	DROP	Instance 2 to instance 1: 0% packet loss Round-trip avg: 2.619 ms
Any	172.16.0.0/24	Any	DROP	Instance 1 to instance 2: 100% packet loss
Any	Any	172.16.0.0/24	DROP	Instance 1 to instance 2: 100% packet loss
icmp	172.16.0.0/24	Any	DROP	Send 512 x tcp 1 MB packets from instance 2 to instance 1: 536870912 bytes (537 MB, 512 MiB) copied, 12.524 s, 42.9 MB/s

Table 3. Result of L4 Filtering test

Firewall				Result
Protocol	Source port	Dest port	Action	
TCP	2222	Any	DROP	ping: 0% loss. RTT: 2.253 ms TCP: Connection timed out UDP: 76.0 MB/s
UDP	2222	Any	DROP	ping: 0% loss. RTT: 2.733 ms TCP: 54.6 MB/s UDP: not connected
UDP	Any	2222	REJECT	ping: 0% loss. RTT: 2.672 ms TCP: 89.5 MB/s UDP: can not send
TCP	Any	Any	DROP	ping: 0% loss. RTT: 2.650 ms TCP: Connection timed out UDP: 44.1 MB/s

80), FWaaS (port 3000), VNC console (port 6080). When only the L7 Filtering feature is applied, the underlying data flows between the two machines will function normally. The ping results always achieve 0% packet loss and the scan results of open TCP ports on server 192.168.0.33 with the netcat tool still results in ports 80, 3000 and 6080 still open and can connect to (Table 4).

Table 4. Result of L7 Filtering test

Firewall		Result
Pattern	Action	
None	None	✓ GET http://192.168.0.33:80
		✓ GET http://192.168.0.33:3000
		✓ GET http://192.168.0.33:6080
/http/i	DROP	× GET http://192.168.0.33:80
		× GET http://192.168.0.33:3000
		× GET http://192.168.0.33:6080
/firewall/i	DROP	✓ GET http://192.168.0.33:80
		× GET http://192.168.0.33:3000
		✓ GET http://192.168.0.33:6080
/:6080/i	DROP	✓ GET http://192.168.0.33:80
		✓ GET http://192.168.0.33:3000
		× GET http://192.168.0.33:6080

In order to test the effect of the service on the performance of the transmission performance, the service will measure the packet transmission speed between the two networks 172.16.0.0/24 and the network 172.16.1.0/24 corresponding to the two hosts with IP addresses are 172.16.0.4 and 172.16.1.112, respectively, when FWaaS service is not applied and when FWaaS service is applied with all filtering features in which each feature includes a certain amount of filtering rules. The filtering rules will be random rules but will not interrupt the connection to transmit and receive packets between the two test machines. The test will use the dd and netcat tools on the Linux operating system to create 5 packets sized 512 MB and send them over TCP protocol. The results of the test are given below (Table 5).

Table 5. Result of network transmission performance test

Number of Firewall rules				Result		
L2	L3	L4	L7	Data	Time (s)	Bandwidth (MB/s)
0	0	0	0	2.5 GB	18.5034	138
10	0	0	0	2.5 GB	21.9104	117
0	10	0	0	2.5 GB	21.9517	117
0	0	10	0	2.5 GB	25.2608	101
0	0	0	10	2.5 GB	254	10.1
10	10	10	10	2.5 GB	180	14.2

5   Conclusion

In this study, the Firewall as a Service for OpenStack Cloud was implemented to protect the private network from outside. The main goal is to help administrators of the Cloud System effortlessly monitor firewall rules in Neutron Routers. The service provide a central view to manage the traffic among the networks and also the Internet. On the other hand, the service also provide the extended functions of firewall for more ability of security.

The future work could include developing and enhancing the features of the firewall service, especially the Layer 7 filtering function. Moreover, we will apply other security approach that are suitable for any kinds of cloud environment.

References

1. OpenStack: OpenStack Docs: Firewall-as-a-Service (FWaaS), 04 March 2020. https://docs.openstack.org/neutron/pike/admin/fwaas.html. Accessed 05 Mar 2020
2. OpenStack: neutron-fwaas/README.rst, 01 June 2013. https://github.com/openstack/neutron-fwaas/blob/master/README.rst. Accessed 05 Mar 2020
3. OpenStack: OpenStack Docs: Firewall as a Service API 2.0, 10 April 2016. https://specs.openstack.org/openstack/neutron-specs/specs/newton/fwaas-api-2.0.html. Accessed 05 Mar 2020
4. OpenStack: OpenStack Docs: Manage project security, 14 February 2020. https://docs.openstack.org/nova/rocky/admin/security-groups.html. Accessed 06 Mar 2020
5. OpenStack: OpenStack Docs: Configure access and security for instances, 23 August 2019. https://docs.openstack.org/ocata/user-guide/cli-nova-configure-access-security-for-instances.html. Accessed 06 Mar 2020
6. Santoso, B.I., Idrus, M.R.S., Gunawan, I.P.: Designing network intrusion and detection system using signature-based method for protecting OpenStack private cloud. Informatics Department, Bakrie University Jakarta, Indonesia 12920 (2016)
7. Xu, C., Zhang, R., Xie, M., Yang, L.: Network intrusion detection system as a service. In: 2019 IEEE 27th International Conference on Network Protocols (ICNP), pp. 1–2 (2019)
8. Rash, M.: Linux Firewalls: Attack Detection and Response with iptables, psad, and fwsnort. No Starch Press, San Francisco (2007)
9. Schubert, D., Jaeger, B., Helm, M.: Network emulation using Linux network namespaces. Network **57** (2019)
10. Aggarwal, S.: Modern web-development using ReactJS. Int. J. Recent Res. Aspects **5**, 2349–7688 (2018)

A Hybrid Swarm Evolution Optimization for Solving Sensornet Node Localization

Trong-The Nguyen[1,2], Tien-Wen Sung[1], Duc-Tinh Pham[3,4],
Truong-Giang Ngo[5(✉)], and Van-Dinh Vu[6]

[1] School of Computer Science and Mathematics, Fujian University
of Technology, Fujian 350118, China
[2] University of Management and Technology, Haiphong 18000, Vietnam
[3] Center of Information Technology, Hanoi University of Industry,
Hanoi, Vietnam
[4] Graduate University of Science and Technology, Vietnam Academy of Science
and Technology, Hanoi, Vietnam
[5] Thuyloi-University, 175 Tay Son, Dong-Da, Hanoi, Vietnam
giangnt@tlu.edu.vn
[6] Information Technology Faculty, Electric Power University, Hanoi, Vietnam
dinhvv@epu.edu.vn

Abstract. A new approach to node localization is proposed by using the
combined localization model in Wireless sensor nets (WSN) with a hybridized
swarm evolution algorithm (HSEA). Mathematically, based on the distances of
the localizing model and Pareto, the fitness function is determined. To achieve
better efficiency, the HSEA is implemented by combining the particle swarm
optimizer (PSO) with the differential-evolution (DE) algorithms. Applied HSEA
deals with the node location situation in WSN. The simulation compares the
results of the proposed scheme with the others schemes to demonstrate that the
suggested task using HSEA achieves better performance than the competitors.

Keywords: Node position · Wireless sensor net · Swarm computing ·
Deferential evolution

1   Introduction

A wireless sensor nets (WSN) is composed by sensor nodes deployed in the monitoring
areas [1]. It can help to sense, collect, and process the information of monitoring
objects in real-time [2, 3]. Therefore, it is widely used in environmental detection and
disaster warning [4]. The nodes' localization will affect some applications' network
performance, such as monitoring, detecting WSN systems [5, 6]. Since it is related to
the NP-hard problem, determining the real coordinate of the sensor node for permu-
tation of distances [7]. The computational intelligence metaheuristic algorithm has
helped achieve optimal solutions to the given NP-hard problem [2, 8]. An intelligent
optimization algorithm that simulates the behavior of organisms on earth with self-
organization method for solving large-scale issues that have been applied to solving
positioning problems by these algorithms [9].

This paper considers a new localization approach using the combined localization model with a hybrid swarm evolution algorithm (HSEA) in WSN. As effective node localization remains improving in the real-time and precise location of nodes, the objective function is determined mathematically based on the localizing model's distances [10]. HSEA is implemented by combining the advantages of the swarm optimizer such as PSO [11] and Evolution algorithm e.g., DE [12] algorithms to achieve better performance. The WSN network node location scenario is put forward dealing with by applying the HSEA. The simulation compares the computational schemes' results to show that the proposed task using HSEA achieves better performance than the competitors.

2   Node Location Problem Modeling and Objective Function

Set the anchor $P_1(x_1, y_1), P_2(x_2, y_2), \cdots, P_n(x_n, y_n)$ and unknown node $P(x, y)$ is r_1, r_2, and $\cdots, r_n$, and the ranging error is $1, 2, \ldots, n, |ri - di| < \varepsilon_i;$ I, where $i = 1, 2, \ldots, n.$[9]. If the actual true length of $P_1(x_1, y_1), P_2(x_2, y_2), \cdots, P_n(x_n, y_n)$ and unknown node $P(x, y)$ is $r_1, r_2 \ldots r_n$ and the ranging error is $\varepsilon_1, \varepsilon_2, \ldots \varepsilon_n$ respectively, then $|r_i - d_i| < \varepsilon_i$ is satisfied, where $i = 1, 2, \ldots, n$ is, then the constraint condition of unknown node $P(x, y)$ is:

$$\begin{cases} d_1^2 - \varepsilon_1^2 \leq (x - x_1)^2 + (y - y_1)^2 \leq d_1^2 - \varepsilon_1^2 \\ d_2^2 - \varepsilon_2^2 \leq (x - x_2)^2 + (y - y_2)^2 \leq d_2^2 - \varepsilon_2^2 \\ \quad\quad\quad \vdots \\ d_n^2 - \varepsilon_n^2 \leq (x - x_n)^2 + (y - y_n)^2 \leq d_n^2 - \varepsilon_n^2 \end{cases} \quad (1)$$

Solution (x, y), so that:

$$f(x, y) = \sum_{i=1}^{n} \sqrt{(x_i - x)^2 + (y_i - y)^2 - d_i^2} \quad (2)$$

In this section, the improved intelligent optimization method is applied to the optimization problem of node position estimation to reduce the interference of error to positioning performance and improve positioning accuracy.

3   Hybrid Swarm Evolutionary Algorithm (HSEA)

3.1   Particles SWarms Optimizer (PSO)

The core idea of PSO algorithm is that particles fly randomly in the solution space, interact with other particles in each iteration, and then decide the next flight position according to personal experience and the influence of other particles [11]. Suppose a particle swarm of population size N searches for the obtained optimization in the searching space with a dimension D. The state properties of individual i at time t are set as follows:

Location: $x_i^t = \left[x_{i1}^t, x_{i2}^t, \ldots, x_{iD}^t\right], x_{id}^t \in [L_d, H_d]$, L_d, H_d are the min and max limits of the d-dimension of the searching problem space respectively; Speed: $V_i^t = [V_{i1}^t, V_{i2}^t \cdots V_{iD}^t]$, $V_{id}^t \in [V_{min,d}, V_{max,d}]$, $V_{min,d}$, $V_{max,d}$ are the minimum speed rate and maximum speed rate of an individual in the d-dimension successively. Individual optimal position: $p_i^t = [p_{i1}^t, p_{i2}^t \cdots p_{iD}^t]$; Global optimal position: $x_i^t = [x_{i1}^t, x_{i2}^t \cdots x_{iD}^t]$ where $1 \le e \le E$, $1 \le i \le N$. Then, in the process of t + 1 generation update, the individual speed and position update formula are respectively:

$$V_i^{t+1} = w * V_{id}^t + c_1 u_1 \left(p_{id}^t - x_{id}^t\right) + c_2 u_2 \left(p_{gd}^t - x_{id}^t\right) \tag{3}$$

$$x_i^{t+1} = x_i^t + V_i^{t+1} \tag{4}$$

In the formula, u_1 and u_2 are the Numbers in random normal distribution (0,1); c_1 and c_2 represent individual cognitive factors and social cognitive factors (both are collectively referred to as acceleration factors). Generally, $c_1 = c_2 = 2.0$.

3.2   Differential Evolutionary (dE)

Initialization: NP individuals is generated with dimension D. If xi is denoted as a vector, a species group can be represented as Pop = $\{x_1.x_2\ldots x_{NP}\}$ [12], with the upper and lower boundaries of the dimension of an individual are x_{max}^j and x_{min}^j respectively, then the individual should be initialized in the following way:

$$x_{j,i,0} = x_{min}^j + rand_j(0, 1) * \left(x_{max}^j - x_{min}^j\right) \tag{5}$$

In the expression, i = $1, 2, \ldots, NP$; j = $1, 2, \ldots, D$. $rand_j(0, 1)$ is a uniform distributed random number on (0,1).

Mutation operation: the vector is generated by mutation operation [12] with variation operations are DE/ rand /1, DE/best /1, DE/rand-to-best/1, DE/best/2, DE/rand/2, DE/current-to-best/1, DE/current-to-rand/1, etc. Where, the classic "DE/rand/1" mutation strategy is expressed:

$$v_{i,g} = x_{r1,g} + F * \left(x_{r2,g} - x_{r3,g}\right) \tag{6}$$

In the formula, $r_1, r_2, r_3 \in [$ 1, 2.. NP] are integers, and the scaling factor F is a constant in the interval (0,1) used to control the size of the difference vector.

Crossover operation: the target vector $x_{i,g}$ and the variation vector $v_{i,g}$ are required to cross binomial to generate the final experimental vector $u_{i,g} = [u_{i1,g}, u_{i2,g}, u_{iD,g}]$, and perform the crossover operation according to formula (7)

$$v_{ij,g} = \begin{cases} v_{ij,g}, rand(0, 1) \le CR \cup j = j_{rand} \\ x_{ij,g}, else \end{cases} \tag{7}$$

where, j_{rand} is from the set $\{1, 2, \cdots, D\}$ to ensure v and g have at least one dimension information retained. The crossover probability CR is a constant in the interval $(0,1)$.

Select operation: a one-to-one tournament selection is performed between the test vector and the corresponding distance vector. If the test vector has a better fitness function, replace the target vector with the test vector. Otherwise, the target vector remains the same. Objective function is optimized for the choosing operation can be expressed:

$$x_{i,g+1} = \begin{cases} u, f\left(u_{i,g}\right) \leq , f\left(x_{i,g}\right) \\ x_{i,g}, else \end{cases} \tag{8}$$

where, $f(\cdot)$ is the fitness function of the optimization issue.

4 HSEA Algorithm for Node Localization

This section presents the HSEA with a balance local optimization algorithm and global exploring ability-based on the adaptive inertia weight updating strategy. The hybrid evolution strategy create one has a strong searching ability and fast searching speed.

4.1 Adaptive Inertia Weight Updating Strategy

In PSO, W measures the speed's influence at the previous moment on the current rate that is used to balance the algorithm's local and global searching ability [11]. The population distribution in each iteration process is taken into account with calculating W. Mean distance d_i of the ith particle from other $N-1$ particles is defined as follows:

$$d_i = \frac{1}{N-1} \sum_{j=1, j\neq i}^{N} \sqrt{\sum_{k=1}^{D} \left(x_i^k - x_j^k\right)^2} \tag{9}$$

where, N and D respectively represent population size and coding dimension. Then define the evolutionary factor δ as follows.

$$\delta = \frac{d_g - d_{min}}{d_{max} - d_{min}} \tag{10}$$

where, d_g represents the mean distance of the global optimal individual, and d_{max} and d_{min} respectively represent the maximum and minimum mean distance in the population. The update W is set to $w(\delta) = \frac{1}{1+1.5e^{-2.6\delta}}$. The δ in changing range is $[0.4\ 0.9]$, in the process of early iterations, δ is larger, w is bigger also. The optimal solution is found in late iterations.

4.2   Hybrid Optimization Strategy

The hybrid algorithm can further develop the population with dividing the population into a superior group and an inferior group. That is, first set the threshold f_{avg}, and the calculation formula is as shown in Eq. (12):

$$f_{avg} = \frac{1}{N}f \tag{11}$$

Fitness value of each particle is compared with itself and selection mechanism is used to determine the optimal solution in the current population. The steps of the HSEA algorithm are summed up as follows:

(1) Initialize the velocity and position of particles, and set the number of iterations.
(2) Calculate the fitness value of each agent as Eq. (2), Calculate the threshold f_{avg} of the population Eq. (12), the population is divided into the superior group Su and the inferior group In.
(3) Mutate, cross and select operation are executed with Eqs. (6), (7) and (8), to further improve the fitness value of individuals in the inferior group.
(4) Combine superior group and inferior group, and update global optimal value and individual optimal value.
(5) If I is equal to the maximum number of iterations, it performs step (7); otherwise, step (2) is performed.
(6) Output the location of the optimal individual as the location estimation of unknown nodes.

5   Simulation Experiment and Analysis

The learning factor c_1 and c_0 are set to 1.4945 for the PSO; the search step radius X is set to 0.1, the constant crossover rate of probability c_r is set to 0.6 for DEOA and HSEA; the population size is set as 30, and the maximum number of iterations is set as 80 for all algorithms in comparison. A parameter is used to take average Location Error (AVE) as the evaluation standard that is formulated as follows.

$$AVE = \frac{1}{M}\sum_{i=1}^{M} \sqrt{(x - x_i) + (y - y_i)^2} \tag{12}$$

where M is the sum number of known nodes, (x, y) is the predicted position, and (x_i, y_i) is the actual position. Communication radius was set to be 10 m, 30 sensors were randomly placed in a 30 m $\times$ 30 m area, 10 nodes were randomly selected as anchor nodes, and the maxi number of generations of the algorithm is set to 200.

Table 1. Comparison of the predicted values of unknown nodes under algorithms

Unknown nodes	Actual positions/m	DEOA [15]		**HSEA**	
		Forecast/m	AVE/m	Forecast/m	AVE/m
1	(6.2044,4.4345)	(6.3968,4.5507)	0.2259	(6.2853,4.4439)	0.2824
2	(1.1491,0.9704)	(1.2873,1.0523)	0.1616	(1.1378,0.9555)	0.0197
3	(9.5997,18845)	(9.4456,2.0685)	0.2410	(9.5078,1.9039)	0.0230
4	(3.8650,2.8111)	(3.9913,3.0314)	0.2529	(3.9052,2.8391)	0.0480
5	(8.2767,5.0020)	(8.3135,5.1729)	0.3045	(8.2461,4.9332)	0.0763
6	(23.1429,23.127)	(23.02795,23.422)	0.3261	(23.1299,23.415)	0.2897
7	(26.7628,3.386)	(26.9157,3.5957)	0.2585	(26.7828,3.396)	0.0234
8	(26.1103,25.355)	(26.4185,25.454)	0.3248	(26.3579,25.387)	0.2488
9	(23.5800,15.924)	(23.3901,15.787)	0.2335	(23.5160,16.007)	0.1055
10	(28.8580,4.6746)	(28.6873,4.5079)	0.2396	(28.8525,4.6915)	0.0188

The obtained results from the HSEA compared with the other schemes, e.g., PSO [13], DE [14], DEOA [15] algorithms for node localization WSN. Table 1 shows the comparison of the selected forecast obtained results of the proposed scheme HSEA with the DFOA algorithm scheme. It can be seen that the results of the comparison show that the proposed approach proves the advantage of the proposed algorithm. Table 2 displays the comparison of the proposed HSEA method's time consumption with the PSO [13], DE [14], and DEOA approaches for node positioning problem with a variety of nodes numbers of deployed networks. It can be seen that most cases of the time running of the proposed method produces a shorter time than the competitors. Figure 1a shows the relationship between average positioning error and ranging error. The adaptive change of inertia weight value and the hybrid evolution strategy of HSEA proposed to make strong positioning ability and fast optimization speed over iteration. The ranging error of the proposed scheme increases from 25% to 30%. The slope of the curve of the proposed scheme increases significantly, indicating.

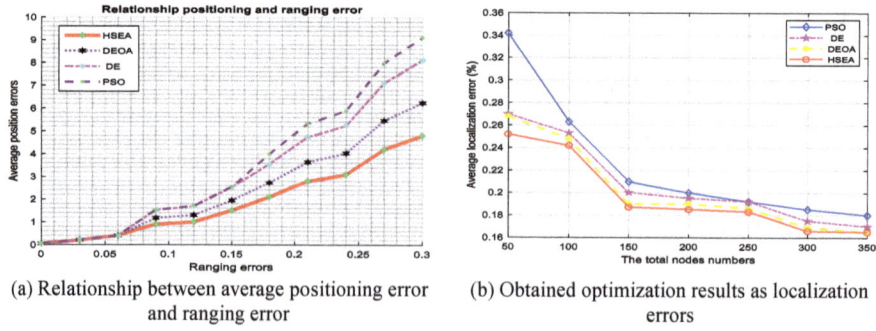

(a) Relationship between average positioning error and ranging error

(b) Obtained optimization results as localization errors

Fig. 1. Comparison of obtained optimization results as localization errors of the proposed HSEA scheme with PSO [13], DE [14], and DEOA [15] schemes

Figure 1b displays the comparison of the suggested scheme's error fluctuation curves with the algorithms for optimizing the node's localization. Obviously, the proposed method's error fluctuation curve ratio results in the smallest error in the error fluctuation curve.

Table 2. Comparison of the proposed HSEA scheme's time consumption with the PSO [13], DE [14], and DEOA [15] approaches for node positioning problem with different nodes numbers (N) of deployed networks.

Algorithms	Nodes numbers (N) of deployed networks					
	N = 20	N = 50	N = 80	N = 110	N = 130	N = 160
HSEA	2.38E+02	2.81E+02	4.19E+02	4.62E+02	4.02E+02	4.42E+02
DEOA [15]	2.42E+02	2.83E+02	4.25E+02	4.63E+02	4.07E+02	4.46E+02
DE [14]	4.97E+00	5.29E+01	2.33E+02	2.71E+02	2.20E+02	2.62E+02
PSO [13]	1.80E+03	5.40E+01	2.24E+02	2.92E+02	2.33E +02	2.56E+02

In general, position accuracy and measurement time comparison results indicate that the proposed algorithm can achieve better performance than competitors.

6   Conclusion

Focused on the localization model with a hybridizing swarm and evolution algorithm (HSEA), this paper suggested a node localization in wireless sensor nets (WSN). Based on the distances of the localizing model, the fitness function has been modeled mathematically. In order to achieve better efficiency, it was introduced to combine the benefits of particles swarms optimizer (PSO) and differential evolutions (DE) algorithms. The DE mutation strategy is used to boost local optimization. Applied HSEA defined in the location scenario for the WSN network node. The results of the computational schemes compared with the other techniques displays that the proposed task using HSEA achieves better performance in terms of the mean calculational cost of position error and the number of localized nodes than the competitors.

References

1. Yick, J., Mukherjee, B., Ghosal, D.: Wireless sensor network survey. Comput. Netw. **52**, 2292–2330 (2008). https://doi.org/10.1016/j.comnet.2008.04.002
2. Nguyen, T.T., Pan, J.S., Dao, T.K.: An improved flower pollination algorithm for optimizing layouts of nodes in wireless sensor network. IEEE Access. **7**, 75985–75998 (2019)
3. Nguyen, T.-T., Wang, H.-J., Dao, T.-K., Pan, J.-S., Ngo, T.-G., Yu, J.: A scheme of color image multithreshold segmentation based on improved moth-flame algorithm. IEEE Access. **8**, 174142–174159 (2020). https://doi.org/10.1109/ACCESS.2020.3025833
4. Nguyen, T.-T., Dao, T.-K., Horng, M.-F., Shieh, C.-S.: An energy-based cluster head selection algorithm to support long-lifetime in wireless sensor networks. J. Netw. Intell. **1**, 23–37 (2016)

5. Goyal, S., Patterh, M.S.: Wireless sensor network localization based on cuckoo search algorithm. Wirel. Pers. Commun. **79**, 223–234 (2014). https://doi.org/10.1007/s11277-014-1850-8

6. Nguyen, T.-T., Thom, H.T.H., Dao, T.-K.: Estimation localization in wireless sensor network based on multi-objective grey wolf optimizer (2017). https://doi.org/10.1007/978-3-319-49073-1_25

7. Pan, J.-S., Nguyen, T.-T., Chu, S.-C., Dao, T.-K., Ngo, T.-G.: Network, diversity enhanced ion motion optimization for localization in wireless sensor. J. Inf. Hiding Multimed. Signal Process. **10**, 221–229 (2019)

8. Dao, T., Nguyen, T., Pan, J., Qiao, Y., Lai, Q.: Identification failure data for cluster heads aggregation in WSN based on improving classification of SVM. IEEE Access. **8**, 61070–61084 (2020). https://doi.org/10.1109/ACCESS.2020.2983219

9. Nguyen, T.-T., Pan, J.-S., Chu, S.-C., Roddick, J.F., Dao, T.-K.: Optimization localization in wireless sensor network based on multi-objective firefly algorithm. J. Netw. Intell. **1**, 130–138 (2016)

10. Nguyen, T.-T., Pan, J.-S., Dao, T.-K., Sung, T.-W., Ngo, T.-G.: Pigeon-Inspired Optimization for Node Location in Wireless Sensor Network BT - Advances in Engineering Research and Application. Presented at the (2020)

11. Kennedy, J., Eberhart, R.B.T.-I.C. on N.N.: Particle swarm optimization. In: Proceedings of IEEE International Conference on Neural Networks. pp. 1942–1948. IEEE, Perth, WA (1995)

12. Storn, R., Price, K.: Differential evolution - a simple and efficient adaptive scheme for global optimization over continuous spaces. Science (80). 11, 1–15 (1995)

13. Monica, S., Ferrari, G.: Particle swarm optimization for auto-localization of nodes in wireless sensor networks BT. In: Adaptive and Natural Computing Algorithms (2013)

14. Harikrishnan, R., Kumar, V.J.S., Ponmalar, P.S.: Differential evolution approach for localization in wireless sensor networks. In: 2014 IEEE International Conference on Computational Intelligence and Computing Research, pp. 1–4 (2014)

15. Céspedes-Mota, A., Castañón, G., Martínez-Herrera, A.F., Cárdenas-Barrón, L.E.: Optimization of the distribution and localization of wireless sensor networks based on differential evolution approach. Math. Probl. Eng. **2016**, 7918581 (2016)

Deep Learning-Based Imbalanced Data Classification for Chest X-Ray Image Analysis

Dang Xuan Tho[1(✉)] and Dao Nam Anh[2]

[1] Hanoi National University of Education, Hanoi, Vietnam
thodx@hnue.edu.vn
[2] Electric Power University, Hanoi, Vietnam
anhdn@epu.edu.vn

Abstract. The aim of this research work is to improve the reliability of prediction of Virus Pneumonia for chest X-ray images by considering difference aspect of learning. Our method uses convolution neural networks to extract their final states, which became input features for further learning by the SVM. In training the distribution of classes over samples are checked and regulated for similarity of the distribution of classes. This is done by random under-sampling to decrease number of samples that belong to majority class. The collaborate nature of our method covering CNN, checking class distribution and SVM is demonstrated by remarkable results over a benchmark X-ray chest database with very large number of samples.

Keywords: Virus pneumonia · Chest X-ray image · Deep learning · Random under-sampling · Imbalance · ResNet-50 · SVM

1 Introduction

In recent years, many researchers use the convolutional neural networks (CNN) in classifying X-ray images to classify diseases with high efficiency. Pneumonia is an inflammation of the lungs that can be caused by a variety of factors such as bacteria, viruses, and fungi [1]. Our aim in this work is to propose deep learning to detect pneumonia by X-ray imaging in combining with a data re-balancing method for performance enhancement. In practice, the X-ray images of patients with pneumonia were collected much less than the X-ray images of other diseases and normal people. After applying the convolution neural network, the output layer can be checked with the imbalance issue to improve the efficiency and speed of computation. Our major contribution is demonstration of the method which combine CNN, class imbalance and SVM to improve effectiveness of medical diagnosis.

Among deep learning techniques, convolutional neural networks (CNN) is widely applied in image classification and has achieved encouraging results. For example, a ChestX-ray8 data set containing images of 32,717 different patients with 108,948 X-ray images were used in CNN by Wang et al. [2]. In another study, CNN could be used to detect lymph nodes in the case of poor imaging by Roth et al. [3]. Lymphopathy and lung disease were also addressed by Shin et al. [4] by applying different CNN. AlexNet and ResNet18 were used by Souza et al. [5] for predicting pneumonia in a two-stage combination to isolate and reconstruct the missing parts of the lung area. Standard network architectures (VGG16/19, Xception, Inception, and ResNet) were reported by Taylor et al. [6] for a conjunction to obtain fully connected layers. In addition, ResNet-50 was used to characterize and combine with SVM RBF machine learning to identify the malignancy of lung nodules by da Nobrega et al. [7]. On the other hand, we realize that the X-ray data of the lungs is often unbalanced, when the number of X-rays of pneumonia is much less than the X-ray of the normal person. Previous studies did not pay attention to this problem, and this affects the accuracy. Therefore, we propose to use data rebalancing to improve predictive efficiency.

2   The Method

We begin with the notation of s and c, which refer to the feature of sample and the class of the sample. For example, an X-ray chest image of a patient is a sample and the class of the image can be Normal or Virus Pneumonia. The features are generated by the ResNet-50 [7] which depends on the quality of the pre-processed tasks for image contrast enhancement and image size conversion. To express the learning process in our method, we use Bayes' Rule [8] that trace out conditional probability for the feature of sample s and class c.

$$p(c|s) = \frac{p(s|c)p(c)}{p(s)} \tag{1}$$

From any query image feature sample s, the maximum a posteriori (MAP) (most likely) class c, applicable for s can be identified by a Bayesian decision where C is the set of predetermined classes.

$$c_{MAP} = argmax_{c \in C} p(c|s) \tag{2}$$

Thus, Bayes' Rule (1) allows to represent the most likely class c. Then, the denominator $p(s)$ can be dropped.

$$c_{MAP} = argmax \frac{p(s|c)p(c)}{p(s)} \tag{3}$$

$$c_{MAP} = argmax\, p(s|c)p(c) \tag{4}$$

Potentially, distribution of classes over training samples are not homogeneous. We represent a minimum number of classes C by two and two corresponding sets of samples S as the follows.

$$C = \{c_1, c_2\} \tag{5}$$

$$S = S_1 \cup S_2, S_j = \{s, p(c_j|s) = 1)\}, j = 1, 2 \tag{6}$$

By (5) and 6), it is sufficient to express the inhomogeneous distribution by showing the large difference of number of samples for each class in training data:

$$numel(S_1) \ll numel(S_2) \tag{7}$$

Here, *numel* means number of elements and c_1 is minority class and c_2 is majority class. Once frequencies in the data are adequately estimated, with a count of samples of each class and knowing total number of sample (n), it is necessary to determine the maximum likelihood $p(c_j)$:

$$\hat{p}(c_j) = \frac{count(c = c_j)}{n} \tag{8}$$

That is, the Bayesian decision $p(s_i|c_j)$ from (4) now can be determined by the frequencies in the data with consideration of a concrete sample s_i and a specific class c_j:

$$\hat{p}(s_i|c_j) = \frac{count(s_i, c_j)}{\sum_{s \in S} count(s, c_j)} \tag{9}$$

As different distribution of classes in training data were discussed by (7), it became apparent that the real issue was the large distinction on the a prior belief $p(c)$ (as opposed to rare ideal equal distribution). Similarly, large divergence are expected to likelihood $p(s|c)$ applied for the minority c_1 and majority class c_2:

$$\hat{p}(c_1) \ll \hat{p}(c_2) \tag{10}$$

$$\hat{p}(s_i|c_1) \ll \hat{p}(s_i|c_2) \tag{11}$$

Note that starting from unequal distribution of class in training data, the Bayesian decision from (3) can be found also dissimilar for minority and majority classes:

$$\hat{p}(s_i|c_1)p(c_1) \ll \hat{p}(s_i|c_2)p(c_2) \tag{12}$$

Using an unbalanced train data can lead to incorrect classification in testing:

$$c_{MAP} \equiv c_2 \tag{13}$$

To resolve problem, two techniques can be used. First, the new samples for minority class S_1 can be generated. Second, some samples of majority class S_2 being selected randomly can be removed from the training test:

$$numel(S_1) \uparrow \tag{14}$$

$$numel(S_2) \downarrow \tag{15}$$

This random under-sampling approach can gain learning speed due to decreasing total number of training samples. From these, the problem of unbalance distribution can be cleared partially and we have equation:

$$numel(S_1) < numel(S_2) \qquad (16)$$

An important task is performance evaluation. Here, we use the Accuracy (ACC) [9] is to measure of how well a test with binary classification correctly identifies or excludes a test case and this is generally appropriate for the case when classes of virus pneumonia positive and negative have similar distribution on training data. In fact, the more general unequal distribution of samples for classes may need the area under the curve (AUC) [10]. The AUC between two points can be found by doing a definite integral between the two points. It is known ACC is not suitable for unbalanced classes but we propose to use it to compare with other general methods which do not consider if classes are balanced or not. Thus, performance of our experiments by AUC and ACC can be measured and compared.

3   Experiments

We performed a number of experiments in order to demonstrate the method described in this paper. The experiments consist of image analysis and virus pneumonia detection for a benchmark Chest X-ray image dataset [11]. In the moment we are writing, the database contains 5840 frontal images of patients classified into Bacterial Pneumonia, Normal and Virus Pneumonia. Table 1 shows the number of images available for each label in training set and test set expresses the structure of samples in training and test data. Here, given 1345 samples of Virus Pneumonia, the proportion of the class to whole training dataset is 26% and the proportion of Bacterial Pneumonia is 49%.

We combine the Normal class and Bacterial Pneumonia class into a negative class, and Virus Pneumonia is a positive class. Thus the class of Virus Pneumonia can be seen as a minority class. For cross validation we split randomly data into 10 folds, each of them has 1/10 of total available images for training and for test. Performance metrics are get by averaging from results of these 10 folds. Totally, we have 20 splits. Now, given a set of training images and a set of test image in one fold, our method can be performed in a structured ways. As can be observed, three different classification options are obtainable from the experiment:

1. Classification by ResNet-50.
2. Classification by SVM with features extracted by ResNet-50 without using random under-sampling.
3. Classification by SVM with features extracted by ResNet-50 with using random under-sampling.

The average AUC from experiments for SVM without using Random undersampling in all 20 splits are estimated as 89.98% while Accuracy is 86.50%.

Because we are using SVM with using Random under-sampling, the average AUC from experiments for all splits are estimated 90.99% while Accuracy is 89.87%. Table 2 reports the AUC for the database by ResNet-50. The classification by ResNet-50 (89.20%) was improved by SVM with without using any sampling methods (89.98%). While this is enhanced by using Random under-sampling, AUC is achievedat 90,99%. The Accuracy by ResNet-50 (86.20%) was improved by SVM with without using any sampling methods (86.50%). We see 89.87% as the result enhanced by using Random under-sampling. There has been a great deal of progress in making these chest X-ray image analysis suitable for the diagnostics of COVID-19 from chest X-ray images.

In particular, imbalanced distribution of the sizes of the infection regions between COVID-19 and pneumonia is focused in study of [13], where authors use a dual-sampling strategy to train the network, which is a 3D convolutional network ResNet34 combined with features of visual attention. Their dataset in experiments covers total 4,982 chest CT images. The method using version of attention ResNet34 with uniform sampling allowed to get AUC at 94.80% and Accuracy at 87.9%. While considering the DarkNet like a classifier for a deep learning based model, you only look once (YOLO) real time object detection system is proposed by authors of [14] to detect and classify COVID-19 cases from X-ray images. The COV database [12] and ChestX-ray8 database [2] were the material for the article and 87.02% was the Accuracy reported by the work. For automatic tuberculosis screening, the SVM in the work [15] used object detection inspired features which were extracted from appearance patterns of images. The experiments achieved AUC of 90.0%, and an accuracy of 84.10%. As can be seen from the Table 3, the mentioned above research results are resumed with note that different database were used for experiments.

Table 1. Structure of samples in training and test image data

Class	Train	Test	Train (%)	Test(%)
Bacterial Pneumonia	2530	242	49%	39%
Normal	1341	234	26%	38%
Virus Pneumonia	1345	148	**26%**	24%
Total	5216	624		

Table 2. AUC and Accuracy from experiments for SVM with/without using Random under-sampling

Row Labels	Average of AUC	Average of ACC
ResNet-50	89.20%	86.20%
SVM without using any sampling methods	89.98%	86.50%
SVM with using Random under-sampling	**90,99%**	**89.87%**

[a] The best results are printed in bold

Table 3. Comparison of the proposed method to other published results

Method	Database[a]	AUC (%)	Accuracy (%)
RUS with ResNet-50 (our)	Chest X-ray 5840 images [11]	90.99	89.87
ResNet34 [13]	4982 images	94.80	87.90
DarkNet-19 [14]	COV [12], Chest X-ray8 [2]	n/a	87.02
SVM [15]	3 sets with 1000 images	90.00	84.10

[a] Database are not the same

4   Conclusions

This paper proposes a method of Random Under-Sampling and ResNet-50 for Chest X-ray Analysis. The method uses ResNet-50 as a feature detector to extract image features. Imbalance of class distribution in training data set is analyzed in a Bayesian framework, which is used as an conceptual base to check imbalance problem and to solve this by random under-sampling.

The case study demonstrates that it is feasible for a collaborated method combining CNN, class imbalance and SVM to improve effectiveness of medical diagnosis. The proposed method is applied to chest X-ray images and can be studied further to other X-ray images.

References

1. Gilani, Z., Kwong, Y.D., Levine, O.S., Deloria-Knoll, M., Scott, J.A.G., O'Brien, K.L., Feikin, D.R.: A literature review, survey of childhood pneumonia etiology studies: 2000–2010. Clin. Infect. Dis. **54**(Suppl. 2), S102–S108 (2012)
2. Wang, X., Peng, Y., et al.: Chestx-ray8: hospital scale chest x-ray database, benchmarks on weakly-supervised classification, localization of common thorax diseases. In: IEEE Conference on Computer Vision, Pattern Recognition; 2097-2106 (2017)
3. Roth, H.R., Lu, L., et al.: A new 2.5 d representation for lymph node detection using random sets of deep convolutional neural network observations. In: International Conference on Medical Image Computing and Computer-Assisted Intervention, pp. 520–527. Springer (2014)
4. Shin, H.-C., Roth, H.R., Gao, M., Lu, L., et al.: Deep convolutional neural networks for computer-aided detection: CNN architectures, dataset characteristics, transfer learning. IEEE Trans. Med. Imaging **35**, 1285–1298 (2016)
5. Souza, J.C., Bandeira Diniz, J.O., Ferreira, J.L., et al.: An automatic method for lung segmentation, reconstruction in chest X-ray using deep neural networks. Comput. Methods Programs Biomed. **177**, 285–296 (2019)
6. Taylor, A.G., Mielke, C., Mongan, J.: Automated detection of moderate, large pneumothorax on frontal chest X-rays using deep convolutional neural networks: a retrospective study. PLoS Med e1002697 (2018)
7. da Nobrega, R.V.M., Reboucas Filho, P.P., Rodrigues, M.B., et al.: Lung nodule malignancy classification in chest computed tomography images using transfer learning, convolutional neural networks. Neural Comput. Appl. 1–18 (2018)
8. Barber, D.: Bayesian Reasoning, Machine Learning. Cambridge University Press, Cambridge (2012)

9. Taylor, J.R.: An Introduction to Error Analysis: The Study of Uncertainties in Physical Measurements. University Science Books, pp. 128–129 (1997)
10. Swets, J.A.: Signal Detection Theory, ROC Analysis in Psychology, Diagnostics: Collected Papers. Lawrence Erlbaum Associates (1996)
11. Kermany, D.S., Goldbaum, M., Cai, W.J., et al.: Identifying medical diagnoses, treatable diseases by image-based deep learning. CELL **172**(5), 1122 (2018)
12. Cohen, J.P., Morrison, P., et al.: COVID-19 Image Data Collection: Prospective Predictions Are the Future, arXiv:2006.11988 (2020)
13. Ouyang, X., Huo, J., et al.: Dual-Sampling Attention Network for Diagnosis of COVID-19 from Community Acquired Pneumonia, arXiv:2005.02690 [cs.CV] (2020)
14. Ozturk, T., Talo, M., et al.: Automated detection of COVID-19 cases using deep neural networks with X-ray images. Comput. Biol. Med. 121 (2020)
15. Jaeger, S., Karargyris, A., Candemir, S., Folio, L., et al.: Automatic tuberculosis screening using chest radiographs. IEEE Trans. Med. Imaging. **33**(2), 233–45 (2014)

Mathematical Modeling the Network's Non-linearity for Analyzing and Assessing the Small-Signal Stability in Multi-machine Power Systems

Luu Huu Vinh Quang[✉]

Ho Chi Minh City University of Technology, Ho Chi Minh City, Vietnam

Abstract. This paper proposes new mathematical modeling to assess the transient structures with the power network configuration varying in time caused by the small oscillations of angular velocity affect the rotating's variations of power generators operating under the small-disturbance condition. The typical results and examples presented to prove the new algorithm's features for analyzing the multiple contingent changes occurring under the daily operating condition in the multi-machine power system. The transient parameters, such as the synchronous, asynchronous, damping torques, angles of transient electromotive forces, are simulated with the impact of automatic voltage and frequency regulation. The transient voltage stability estimation values are determined to assess the transient structures influenced by the continuous small-disturbance of angular velocities of the generators operating in the multi-machine power system.

Keywords: Small-signal/oscillation stability · Multi-machine power system

Nomenclatures

$\Delta f_i(t)$	are the function vectors simulate the external forces varying in time and characterizing the operative changes in the power network.
a_{ij}, b_{ij}, c_{ij}	are the real coefficients consists of the partial derivatives of functions of $y(x_i)$ with respect to the state variables of x_i found at the point of the initial condition.
Δx_1, Δx_2, ΔG	are state, input and output variable vectors
A_1	is the state matrix of size nxn
A_2	is the input matrix of size nxr
A_3	is the output matrix of size mxn
A_4	is feedforward matrix of size mxr
Λ	is the eigenvector matrix
M_R	is the right eigenvector matrix
M_L	is the left eigenvector matrix

© The Author(s), under exclusive license to Springer Nature Singapore Pte Ltd. 2021
D.-T. Tran et al. (Eds.): ICISN 2021, LNNS 243, pp. 116–125, 2021.
https://doi.org/10.1007/978-981-16-2094-2_15

1 Introduction

The small-signal stability or small-oscillation or small-disturbance stability (SSS/SOS/SDS) is the ability of the power system to maintain synchronism under the condition of daily operative changes of the power system's configuration that cause the small state oscillations in the power system operation. Such daily operative changes continually occur in the power system relating to the small changes in bus power input/output amounts or the operative breaking/closing of the transmission line/transformer elements, and the automatic line reclosing implemented in the power network. In general, the small-disturbances are without the short-circuit fault occurring in power system operation in comparison distinguishing with the large disturbance always accompanying the short-circuit faults. The change in power system configuration, such as the power generator or substation outage accompanied by the shedding a part of power load, may cause forced oscillations. The small-disturbance transformed into a large disturbance. The small-signal stability problem may become the transient stability problem that needs to verify the power system's ability to maintain stability with the synchronism accompanying some operative change in the power system structure.

The existing methods for solving the small-signal stability problem are analytically using the first and second Lyapunov's methods to affirm the ability concerning the small-disturbance stability relating to the specified operative changes occurring in the power system. The power system's transient state simulated by a set of differential equations with the state variable vectors of x_j varying in time, and the first order linearization of this set of differential equations gives

$$\sum_{i,j=1}^{m} \left(a_{ij} \frac{d^2 \Delta x_j}{dt^2} + b_{ij} \frac{d \Delta x_j}{dt} + c_{ij} \Delta x_j \right) = \Delta f_i(t); \tag{1}$$

Using Laplace's transformation [1] to solve the above set of differential equations, the characteristic equation is in a high order polynomial form. In general, the state variable vector of $\Delta x_j(t)$ can be the real and complex conjugate roots of p_k of the characteristic equation and the real coefficients of C_{kj} as follows

$$\Delta x_j(t) = \sum_{k=1}^{n} C_{kj} \cdot e^{p_k t}; \tag{2}$$

There are two groups of methods using Lyapunov's theorem to analyze and affirm the power system's ability concerning the small-signal stability. The first group of methods expects to solve characteristic equation relating to (1) to find its roots. Based on the sighs and values of the real and complex conjugate characteristic roots to apply Lyapunov's theorem to analyze for affirming the power system's ability concerning the small-signal stability. The second group of methods does not expect to find the characteristic roots of the characteristic equation concerning (1) but to apply the mathematical theorem (Hurvitz, Raus) involving the mathematical ties relating to the coefficients of the characteristic equation and the characteristic roots to analyze for affirming the power system's ability concerning the small signal-stability in the power systems.

An alternative method [2] is used for solving a set of first-order differential equations that equivalently transformed from (1). In general, this method is called the state-space. The set of first-order differential equations simulating the small-signal stability in the power systems can be written as

$$\frac{dx_1}{dt} = f(x_1, x_2, t); \tag{3}$$

The output variable vector of G determined as a function of the state variable vector of x_1 and input variable vector of x_2:

$$G = F(x_1, x_2); \tag{4}$$

The first-order linearization of (3) and (4) gives

$$\begin{cases} \frac{d\Delta x_1}{dt} = A_1 \cdot \Delta x_1 + A_2 \cdot \Delta x_2; \\ \Delta G = A_3 \cdot \Delta x_1 + A_4 \cdot \Delta x_2; \end{cases} \tag{5}$$

The diagonalization of state matrix A_1 gives

$$A_1 = M_R \Lambda M_L; \tag{6}$$

Using the z-transformation to find the state variable vector varying in time:

$$\Delta x_1(t) = \sum_{i=1}^{n} M_i C_i e^{\lambda_i t}; \tag{7}$$

The eigenvalue matrix of Λ consists of the real and complex conjugate eigenvalue elements of λ_i that like the roots of the characteristic equation of p_k, allowing to affirm the power system's ability concerning the small-signal stability, to maintain the synchronism near the point of the initial state in the power system operation.

The small-signal stability methods suggest the small oscillation near the point of the initial steady-state. Solving the small-signal stability problem has some hard such as the power network size is large, the operative changes in the power network with the breaking/closing of a transmission line, transformer, or generator accompanying the shedding power load. And the small oscillations sometimes forcedly increase their amplitude and become some unexpected faults that lead to losing the synchronism with the angular frequency forced rising. The large oscillation amplitude of the state variable negatively affects the linearizing of the initial equations to obtain (1).

2 New Proposed Mathematical Modeling the Small- Oscillation Stability with the Angular Frequency Varying in Time

Referring to [3–6], the set of equations simulating the transient state of the power system written as

$$
\begin{cases}
\frac{d\Delta\omega_i}{dt} = \frac{\omega_o}{2H_i} T_{ai}; \\
\frac{d\delta_i}{dt} = \omega_o \Delta\omega_i; \quad i = 1, 2..M;
\end{cases}
\tag{8}
$$

Herein, ω_o is the rated angular frequency, and $\Delta\omega_i$ is the state variable vector simulating the relative angular frequencies of the rotating generators.

The step-by-step alternatively solving the (8) combined with the step-by-step solution of a system of algebraic Eqs. (9), which consists of the functions of output electromagnetic powers of the ith synchronous generator as

$$
\begin{cases}
P_i^t = \sum_{j}^{Ng} E_i^t E_j^t Y_{ij}^t \cos(\delta_i^t - \delta_j^t - \phi_{ij}^t); \\
Q_i^t = \sum_{j}^{Ng} E_i^t E_j^t Y_{ij}^t \sin(\delta_i^t - \delta_j^t - \phi_{ij}^t);
\end{cases}
\tag{9}
$$

Herein, $Y_{ij}^t \angle\Phi_{ij}^t$ are the elements of the driving-point-transfer admittance matrix of Y_{bus} determined in the time interval of t; $E_i^t \angle\delta_i^t$ is the electromotive force of the synchronous machine take into account the effects of actions of the automatic voltage and the automatic frequency regulations (AFR and AVR) in the time interval of t.

Using the state variable of bus voltage of V^t and the bus admittance matrix of y_{ij}^t, the system of algebraic Eqs. (9) can be equivalently expressed by a set of algebraic Eq. (10), as follows:

$$
\begin{cases}
P_i^t = \sum_{j=1}^{Nb} V_i^t V_j^t y_{ij}^t \cos(\delta_i^t - \delta_j^t - \varphi_{ij}^t); \\
Q_i^t = \sum_{j=1}^{Nb} V_i^t V_j^t y_{ij}^t \sin(\delta_i^t - \delta_j^t - \varphi_{ij}^t);
\end{cases}
\tag{10}
$$

Herein, $y_{ij}^t \angle\varphi_{ij}^t$ is the elements of the bus admittance matrix of y_{bus}; $V_i^t \angle\delta_i^t$ is the bus voltages of the power network in the time interval of t.

We simulate the non-linear transient structure of power network with the angular frequency varying in time is as follows: in majority, the main elements forming a power network are the ac transmission lines or the power transformers, which are simulated for a computation process by the quantities of branch impedances with the branch reactance elements of $X_{ij} = \omega L_{ij}$. In general, the bus admittance matrix of y_{bus}

relating to (10) is formed with the branch impedances of $Z_{ij} = R_{ij} + jX_{ij}$. And the elements of driving-point-transfer admittance matrix Y_{bus} are also equivalently determined with the branch reactance of $X_{ij} = \omega L_{ij}$.

In general, the bus admittance matrix of y_{bus} and Y_{bus} are determined and depending on the rated angular frequency of ω_o for the above states with their unchanged elements.

In this article, we propose a new algorithm for the step-by-step solving the (8), (9), and (10) to model the transient process involving the nonlinear power circuits of network configuration, the impedance and admittance matrix continually recalculated in time of t with the $Z_{ij}(t) = R_{ij} + j\omega^t L_{ij}$ or the $Y_{ij}(t) = 1/Z_{ij}(t)$. The elements of the bus admittance and the driving-point-transfer admittance matrix vary in time with the angular frequency of ω^t (state variables) that recomputed in the transient process. So, the transient structure of the power system is affected and assessed by the magnitude of transient voltage stability estimation (TVSE).

Referring to [5], the sets of nonlinear Eqs. (10) can be expressed as

$$\left(\frac{\partial F(x_1, x_2, z(\omega^t))}{\partial x}\right)^{(\Delta t, \omega^t)} \Delta x_1^{(\Delta t, \omega^t)} = \Delta F^{(\Delta t, \omega^t)}(x_1, u_2, z(\omega^t)); \tag{11}$$

Transforming the (11) in to a matrix form as follows:

$$L_{ev}^{(\Delta t, \omega^t)} \Delta x_1^{(\Delta t, \omega^t)} = \left[E_M^{(\Delta t, \omega^t)}\right]^{-1} L_{ev}^{(\Delta t, \omega^t)} \Delta F^{(\Delta t, \omega^t)}(x_1, x_2, z(\omega^t)); \tag{12}$$

Basing on the discussions in [5], the smallest transient eigenvalue among the elements of the transient eigenvalue matrix of $E_M^{(\Delta t)}$ is used to assess the small signal stability in the power system concerning the network's non-linearity caused by the angular frequency varying in time. Referring to [6], the transient voltage stability estimation (TVSE) quantity is modeled for estimating the transient voltage stability considering the network non-linearity with the angular frequency variation as

$$TVSE(\omega^t) = \int_0^T E_M^{min}(\omega^t, t) dt; \tag{13}$$

3   Examples

Let's investigate a 49-bus power system consists of 7 power plants (7 equivalent generators including 27 synchronous generators) supplying the total power load of 3272 MW.

The power system's schema is shown in Fig. 1, as follows:

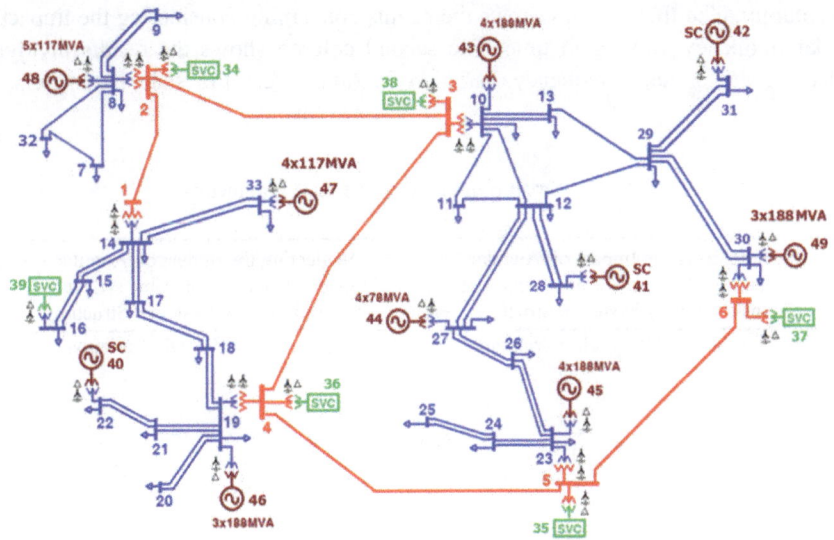

Fig. 1. Power system's schema.

The initial power generation is shown in Table 1, as follows:

Table 1. Initial power generation.

Bus Number	MVA Generation	Bus Number	MVA Generation
40	-0.000+j54.163	45	525+j143.743
41	-0.000+j52.516	46	485+j149.867
42	-0.000+j62.139	47	375+j175.544
43	585+j128.001	48	490+j146.570
44	287+ j69.546	49	575.090 +j108.532

The rated frequency is 50Hz. After the incident, the generators change their velocity with some accelerations. The angular frequency of ω_t of each generator begins to increase in time causes the changes in the reactance values of $X(t) = \omega_t.L$ concerning all of the inductive branch elements of the power network. So the power system configuration simulating with the inductive branch elements is varying in time, neglecting the magnetic saturation.

1st Example. Let's investigate the small-disturbance concerning the sudden line outage that two transmission lines of the numbers (10–12) and (12–29) simultaneously are cut-out, then, after 0.2 s, 10% of initial power load shedding (about of the 10+j4MVA) occurs in bus 11. The small oscillations of the state parameters are in Table 2 includes two columns. The first column shows the results concerning considering the impact of angular frequency varying in time. The second column shows the results involving neglecting the angular frequency variation in time affect the state parameters, as follows:

Table 2. Small oscillation of the state parameters

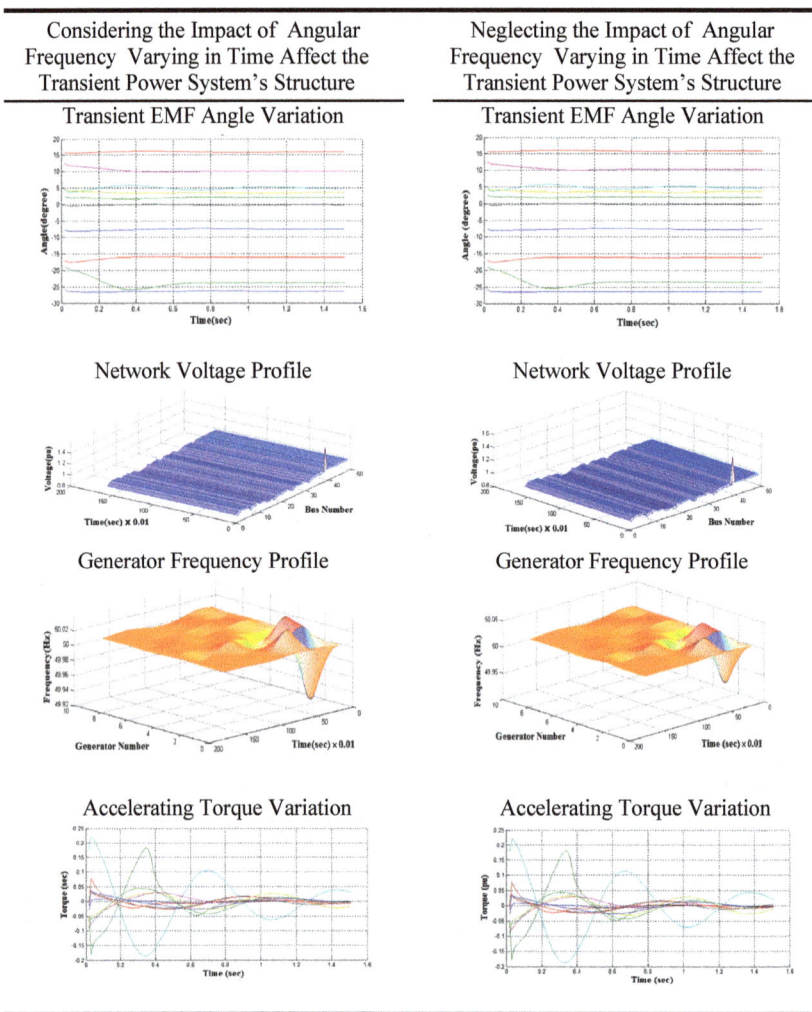

2ⁿᵈ Example. Let's investigate the sudden line outage (without any short-circuit fault) for ranking the transmission lines of numbers (1–2), (2–3), (3–4), and (4–5) for small oscillation stability by comparing the TVSE's values.

Similarly, the small oscillations of the state variable of transient EMF angles are in Table 3 with two columns. The first column shows the results concerning the case considering the impact of angular frequency varying in time. The second column shows the results involving the case neglecting the impact of angular frequency variation in time, as follows:

Table 3. Comparing the transient EMF angle's oscillations and the TVSE's values.

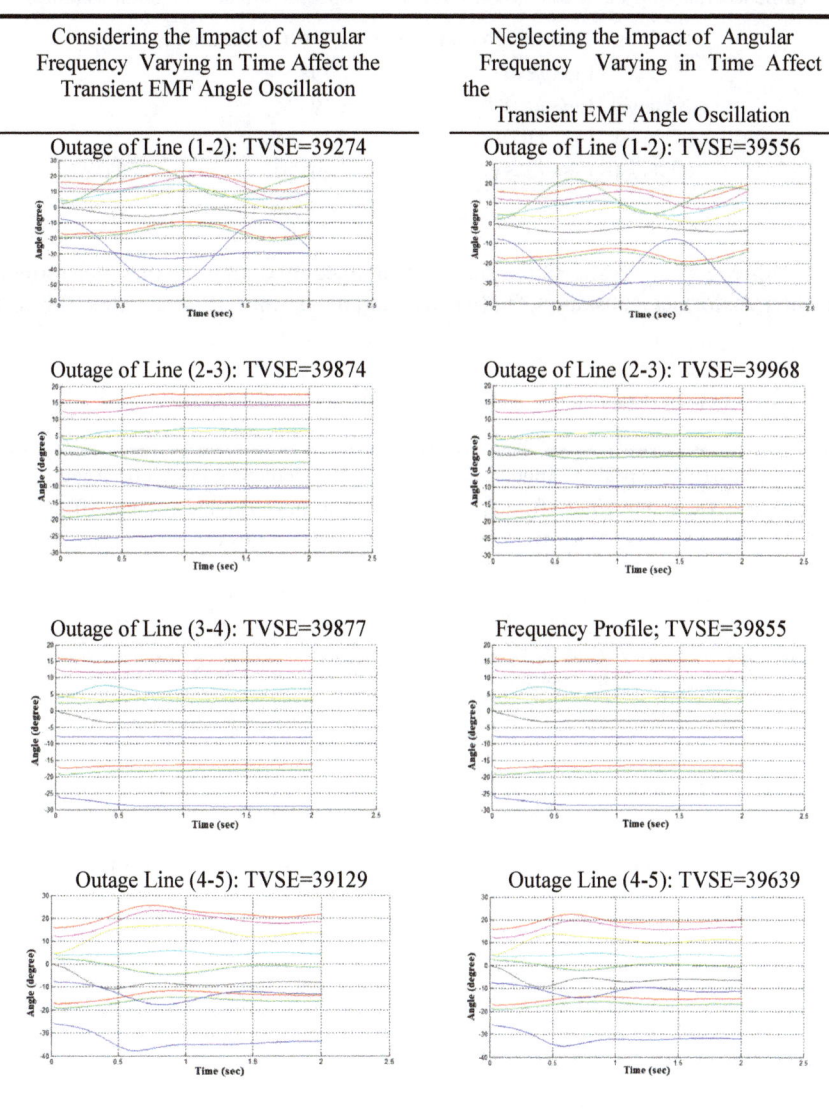

The TVSE's magnitude is proportional to the ability of the power system [5] to maintain synchronism take into account the operative changes of the power system's configuration, caused by the small oscillations of state variables in the power system operation. Using the TVSE's magnitude [6] involving the angular frequency varying impacts the transient power system's structure, to rank the transmission lines for the small-oscillation stability, the obtained results are shown in Table 4, as follows:

Table 4. Ranking the transmission lines with respect to small oscillation stability.

Ranking	Transmission line number	TVSE's magnitude
1st-level	Line (3–4)	39877
2nd-level	Line (2–3)	39874
3rd-level	Line (1–2)	39274
4th-level	Line (4–5)	39129

3rd Example. Let's investigate the small-disturbance involving the one transformer of 3-power transformer substation (1–14) incidentally is cut-out, then, after 0.2 s, the

Table 5. Oscillations of the synchronous and asynchronous torques.

Considering the Impact of Angular Frequency Varying in Time Affect the Transient Power System's Structure	Neglecting the Impact of Angular Frequency Varying in Time Affect the Transient Power System's Structure

transmission line (1–2) incidentally is an outage. The oscillations of the synchronous and asynchronous torques comparatively are shown in Table 5 includes two columns. The first column shows the results considering the impact of angular frequency varying in time affects the torques. The second column shows the results neglecting the angular frequency variation in time affect the torques, as follows:

4   Conclusion

The investigating results prove that the proposed algorithm can effectively assess the transient power system structure involving the network's nonlinearity caused by the continuous small disturbance of the angular frequency concerning the small-signal stability.

This new mathematical modeling proposes to solve the electrical transient problems with the changes in time of the coefficient matrix of the system of nonlinear algebraic equations, which step-by-step simulate the small-oscillation stability involves the nonlinearity of electrical network in the multi-machine power systems.

References

1. Venikov, V.: Transient Processes in Electrical Power Systems. Mir Publishers (1980)
2. Kundur, P.: Power System Stability and Control. Mc Graw Hill, Inc. (1993). ISBN 0-07035958-X
3. Quang, L.H.V.: A new algorithm assessing the transient stability for ranking the power network areas in large power system. In: 2018 South East Asian Technical University Consortium (SEATUC Symposium), Yogyakarta, Indonesia, pp. 1–6 (2018). https://ieeexplore.ieee.org/abstract/document/8788858
4. Quang Luu, H.V.: Modeling the transient energy margin for accessing the transient stability with double shot automatic line reclosing in power system. In: 2018 International Conference on Advanced Computing and Applications, Ho Chi Minh City, pp. 29–34 (2018). https://ieeexplore.ieee.org/document/8589485
5. Quang, L.H.V.: Assessing the transient structure with respect to the voltage stability in large power system. In: 2019 Intelligent Computing in Engineering. Advances in Intelligent Systems and Computing, vol. 1125, pp. 1017–1026. Springer, Singapore (2019). https://doi.org/10.1007/978-981-15-2780-7_106
6. Quang, L.H.V.: A new algorithm for ranking the transmission lines with respect to the transient voltage stability in multi-machine power system. In: 2019 International Conference on Advanced Computing and Applications, Nhatrang, Vietnam, pp. 135–140 (2019). https://ieeexplore.ieee.org/document/9044242

Hybrid Louvain-Clustering Model Using Knowledge Graph for Improvement of Clustering User's Behavior on Social Networks

Hai Van Pham[1(✉)] and Dong Nguyen Tien[1,2]

[1] Hanoi University of Science and Technology,
No. 1. Dai Co Viet Street, Hanoi, Vietnam
haipv@soict.hust.edu.vn
[2] CMC Institute of Science and Technology,
No. 11 Duy Tan Street, Hanoi, Vietnam

Abstract. Recently, social networks play significant roles in real-time content analysis. Many studies have investigated social networks by analysis of contents based on users. However, these studies have some limitations of social network analysis while clustering contents based on users' behavior. The paper has presented the Hybrid Louvain-Clustering model using a knowledge graph to cluster contents based on user behaviors in a social network. In the proposed model, all multi-dimensional user relationships represent in a knowledge graph while clustering contents based on user behaviors in real-time on social networks. The experiment demonstrated that the proposed method for a large-scale social network and reached efficient clustering content based on user's behaviors. Additionally, experimental results show that the proposed algorithm enhances the performance of clustering contents based on user's behavior of a social network analysis.

Keywords: Knowledge graph · Data analysis · Social network analysis · Clustering facebook page

1 Introduction

Recently, there are many studies related to social networks focusing on the investigations of the theory and applications of social networks [1]. Some studies have investigated approaches in social network analysis such as Attacking Similarity-Based Link Prediction in Social Networks [2], Clustering Social Networks [3], Mapping Users Across Social Networks integrated with Network Embedding [4] and Online profiling and clustering of Facebook users [5], and Grouping Facebook Users Based on the Similarity of Their Interests [6, 8]. All of the related result works have some limitations in clustering contents/users without weights among relationships of nodes so that the problems in clustering with multi-dimensional in social networks.

In Subreddit Recommendations within Reddit Communities [6] done by Vishnu Sundaresan, et al. in Stanford University, the researchers have proposed method to

D.-T. Tran et al. (Eds.): ICISN 2021, LNNS 243, pp. 126–133, 2021.
https://doi.org/10.1007/978-981-16-2094-2_16

assume the weight among shared users between subreddits, performed by Louvain algorithm to clusters of subreddit to cluster content-based users in the recommendation of social network analysis. Louvain method provides clustering contents/users in large social networks [6]. However, the method uses to clustering in fully connected nodes which is poor results. These results were limited to use for doing marketings, advertisements using Louvain algorithm to real-world problems on social networks.

The paper has presented the Hybrid Louvain-Clustering model using a knowledge graph to cluster contents based on user behavior. All multi-dimensional relationships based on knowledge graph to clustering Facebook Pages are based on user's behavior with these interactions in real-time on social networks. The proposed model presents two algorithms. The first algorithm called relationship user behaviors and content weights is a path directed relationships from user to pages by calculating the interactions of a users' behaviors and contents. The second algorithm called relationship content-page weights is calculated by the total numbers of interaction relationship of users' behavior that divided by the total the number of users. In the experiments, the proposed model has been validated tests in real-time on Facebook using real-time data collections and open data source [7].

2   Preliminaries

2.1   Knowledge Graph

A graph consists of two features: nodes and relationships. Each node has an entity (a person, thing, place, category and other attributes of data) as shown in Fig. 1. A relationship among nodes represent how nodes are connected. Unlike other databases, "Relationship" is a top priority in the graph database.

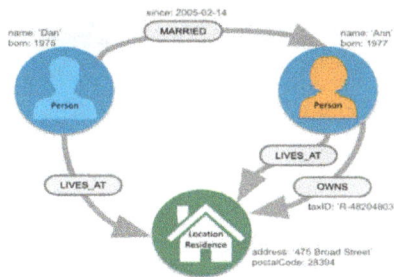

Fig. 1. The person as nodes in location, having relationship connected with its location node [6]

Knowledge graphs (KGs) provide an a structure information in information retrieval with question responses including Freebase, YAGO and DBpedia. A regular KG is usually represented as multi-relational data with large triple facts such as head entity, relation, and tail entity [6, 8]. The social network Facebook, where there are many objects and types of relationships, all interact with each other in a multi-dimensional relationship and show the interactions in real life.

2.2   Facebook Social Network

Social networking is mostly based on Internet-based social sites to connect with friends, colleagues, family, or clients, customers. Social networking as a social purpose, given in a business purpose [8]. Facebook represents a social network modeled as a graph, including typical types of nodes, groups and relationships. Each node represent an individual, page, group, or entity that contains various attributes such as name, gender, interest, location, image, content, and time [9].

The nodes contain user behaviors such as Page, Group, Place. Louvain Algorithm to social networks. The Louvain method is to apply detecting communities using networks [6]. The Louvain algorithm represents modularity-based algorithms dealing with large graphs. The Louvain algorithm has been recommended Reddit users in order to find optimal subreddits, based on the user behavior. The Louvain algorithm was applied to cluster topics from online social networks, such as Twitter and YouTube. While clustering of conventional Louvain algorithm, the results of clustering nodes of the algorithm are not high because of fully connected nodes.

3   The Proposed Model

The proposed model called Hybrid Louvain-Clustering model using a knowledge graph to cluster contents based on user behaviors is depicted in five steps, as shown in Fig. 2.

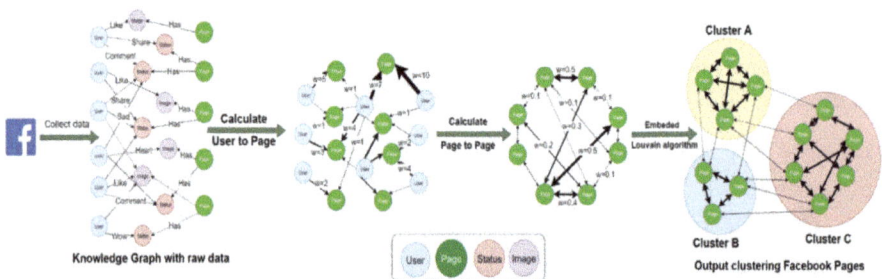

Fig. 2. The proposed model to clustering pages based on user

Step 1. Collect Data Sets from Facebook
Scrapy tool is used to get automatically data from Facebook.

Step 2: Present a Social Network to Knowledge Graph
This step is to send a request from Facebook data sets, then detect entities from Responses. Each entity in a sets of entities $V := \{U, P, G, S, I, A, C, R\}$.

Let $G = (V, E, \omega)$ be a social network where $V := \{U, P, G, S, I, A, C, R\}$ is the set type of nodes in Facebook knowledge graph and $E := \{(v_i, v_j)\}$ is the set of directed edges representing the relations.

From raw data obtained from Facebook, the raw data extracts field data from HTML to build Relationships that show the level of connection between nodes in the graph.

Step 3. Calculate Weights from User Behavior to a Page

ALGORITHM 1: Relationship User behavior and content weights

    Goals: Calculate weights from relationship user behavior and contents that is a path directed relationships from user to pages by calculating the interactions of a users' behavior and contents.

— **Input**: Knowledge graph with raw data.
— **Output**: Weighted relationships from User to Page

1	**Foreach** *status* **in** *list_status_or_images*
2	$P \leftarrow$ **Get** page of status
3	*List_R* $\leftarrow$ **Get** list reactions of status
4	**Foreach** *r* **in** *List_R*
5	$U \leftarrow$ **Get** User of reaction
6	*U_interact_P* $\leftarrow$ **Get** relationship interact from user to page
7	*If exist U_interact_P:*
8	*weight* $\leftarrow$ **Get** weight of relationship user to page
9	*Else:*
10	*is_like* $\leftarrow$ **Get** relationship user like page
11	**If exist** *is_like*:
12	*weight* $\leftarrow 1$
13	**Else**:
14	*weight* $\leftarrow 0$
15	*weight* += *reaction*→type→weight
16	**Return** *relationship_U_interact_P* $\leftarrow$ {weight: weight}

Step 4. Calculate Weights from Pages to Pages

Calculate relationship content-page weights is calculated by the total numbers of interaction relationship of users' behavior that divided by the total the number of users to get weighted relationships among pages.

Step 5. Transform Weights to Embedding Louvain Algorithm

The outputs of Algorithm 1 are weights that automatically embedded to input variable values of the proposed model using Louvain algorithm.

    The experiments of the model validate how the mode performance done successfully as shown in Sect. 4.

4   Experiments and Results

4.1   Dataset

The experiments of the proposed model has been tested in data sets, getting from Facebook stored at the link [7]: https://www.kaggle.com/dongnguyentien/facebook-neo4j-dataset.

4.2   Experimental Results

The experimental results in the proposed model show the best case for the highest accuracy (*threshold* = 0.148, *precision* = 0.86, *recall* = 0.72) also show five categories called clustering content nodes including **"E-commerce & Shop", "Sports, Entertainment, News, Music", "Tourism, cuisine", "Technology and entertainment", "Military and history"**.

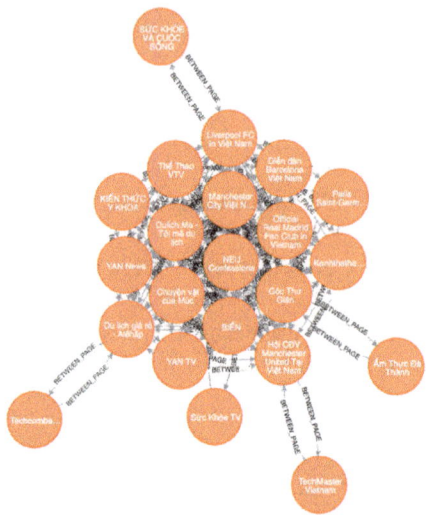

Fig. 3. Pages belong to clustering content node

In the experiments, the results of proposed model showed the relationships among Pages and Scored relationship among pages as shown in Figs. 3 and 4.

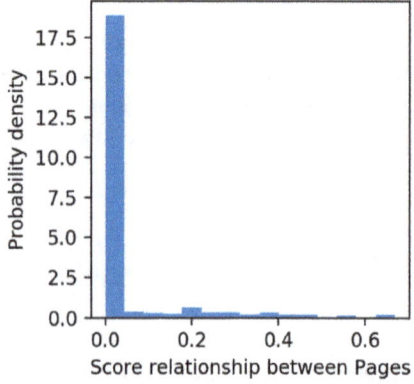

Fig. 4. Score of relationship between Pages, $w = (0, 0.669)$

Most of relationship are weaken linkage relations, with the centralized value under 0.01.

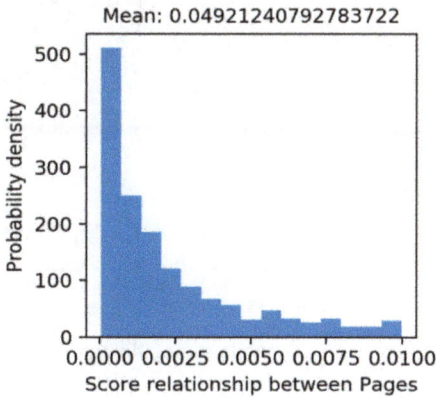

Fig. 5. Scored relationship among pages under 0.01

To control links based on nodes's weights, we adjusted the parameter as threshold when running the Louvain algorithm, calculating the relationships with its condition $w \geq threshold$. In order to evaluate the threshold, The dataset was sorted by contents of categories regarding to $tourism = [Page1, Page2, \ldots]$.

The evaluation formula is defined by True Positive (TP), False Positive (FP), True Negative (TN), False Negative (FN) of the following:

$$Presicion = \frac{TP}{TP + FP} \tag{1}$$

$$Recall\,(true\,positive\,rate) = \frac{TP}{TP + FN} \tag{2}$$

$$False\,positive\,rate = \frac{FP}{TP + FP} \tag{3}$$

To adjust parameters of the proposed models, the proposed model in experiments showed the best case when $threshold = 0.148$, $precision = 0.86$, $recall = 0.72$ in the average, as shown in Figs. 5 and 6, respectively.

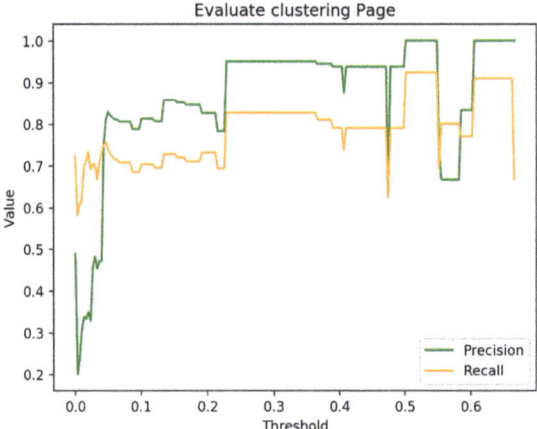

Fig. 6. Precision and recall rate while setting the threshold from 0 to 0.669

In further experiments, the experimental data consists of thousands of nodes in real-time Facebook and static data sets obtained from Scrappy tool in data source [7]. We carried out the experiments and compared the performance of the proposed approach and Louvain method through experimental results as summarized in Table 1. Average accuracy performance in evaluation of the proposed model has been calculated as shown in Table 1. The experimental results consistently show that the proposed approach achieved accuracy in terms of clustering nodes, precision, and recall.

Table 1. Comparison performance of the proposed model and Louvain method

Evaluation formula	The proposed model	Louvain
Clustering nodes	5	2
Precision	0.86	1
Recall	0.72	0.5

The performance average of all experiments with analysis expert surveys in experiments was calculated for both static data sets and real-time Facebook social network obtained at 120.000 nodes from the period of January to October 2019 [7]. The experimental results in clustering contents based on users' behavior demonstrate clearly that the proposed approach far outperforms the Louvain method.

5   Conclusion

The investigation in this paper presented the Hybrid Louvain-Clustering model using a knowledge graph to cluster contents based on user behaviors. The experiment demonstrated that the proposed method for a large-scale social network and reached efficient clustering content based on user's behavior. It is indicated that the proposed algorithm enhances the performance of clustering user-based contents for improvement of social network analysis. For further investigations, the proposed model will be applied in huge data sets of social networks, to dealing with dynamic environment of the Facebook.

Acknowledgements. This research is funded by Vietnam National Foundation for Science and Technology Development (NAFOSTED) under grant number 102.05-2019.316.

References

1. Marsden, P.V., Friedkin, N.E.: Network studies of social influence. Sociol. Methods Res. **22** (1), 127–151 (1993). https://doi.org/10.1177/0049124193022001006
2. Zhou, K.: Attacking similarity-based link prediction in social networks, pp. 1–9, 31 December 2018
3. Mishra, N.: Clustering social networks. In: Algorithms and Models for the Web-Graph, pp. 55–67, December 2017
4. Liu, L.: Aligning users across social networks using network embedding. In: IJCAI 2016, July 2016
5. Needham, M.: Neo4j (2019). https://neo4j.com/blog/graph-algorithms-neo4j-louvain-modu larity
6. Kido, G.S.: Topic modeling based on Louvain method in online social. In: Conference: XII Brazilian Symposium on Information Systems (2016)
7. Scrapy data source link. https://www.kaggle.com/dongnguyentien/facebook-neo4j-dataset
8. Pham, H., Tran, K., Moore, P.: Context matching with reasoning and decision support using hedge algebra with Kansei evaluation. In: Proceedings of the Fifth Symposium on Information and Communication Technology, pp. 202–210 (2014)
9. Nguyen, T., Van Pham, H.: A novel approach using context matching algorithm and knowledge inference for user identification in social networks. In: Proceedings of 4th International Conference on Machine Learning and Soft Computing (ICMLSC 2020), pp. 149–153 (2020)
10. Dinh, X.T., Pham, H.: A proposal of deep learning model for classifying user interests on social networks. In: Proceedings of the 4th International Conference on Machine Learning and Soft Computing (ICMLSC 2020), pp. 10–14 (2020)

Design of Air Bubble Detection System Used in Pleural Drainage Suction Systems

Kien Nguyen Phan$^{(\boxtimes)}$, Khoi Nguyen Doan, and Vu Tran Anh

Hanoi University of Science and Technology, Hanoi, Vietnam
{kien.nguyenphan, vu.trananh}@hust.edu.vn

Abstract. Drainage suction has been used in clinical for a long time and widely used in hospitals around the world. During pleural drainage suction process, being able to detect pneumothorax (the presence of air bubbles in the one-way-flow valve cylinder of the drainage suction system) is vital as it can directly affect the patient's life in a short time. This study suggests a condenser microphone device to detect the presence of air bubbles by measuring change of sound. The system has been developed on a platform using condenser microphones to detect pressure changes caused by the vibrations of the suction system in the presence of air bubbles. Experimental results on the 3-cylinder drain suction system show that the system is capable of detecting air bubbles with an accuracy level of up to 95%.

Keywords: Air bubble detection · Condenser microphone · Pleural drainage suction system

1 Introduction

Chest Drainage System is used to remove fluid and air from the pleural space, prevent drained air and fluid from returning to the pleural space, and restore negative pressure in the pleural space to re-expand the lung [1]. When draining the pleural fluid, if there is an accompanying air leaking, the lung will be compressed, unable to exchange gas as usual, endanger the patient [2]. Therefore, the detection of air spillage is very importance in the treatment and care of patients who have to use the pleural drain system. Clinicians and researchers have found that the present of air bubbles is quite common in the chest drainage systems [3].

Several medical companies have developed pleural drainage system with integrated electromagnetic components to monitor and detect air bubbles in the system [4]. Some techniques to detect air bubble in the systems such as optical sensors, pressure sensors, or capacitors [5–7]. The air bubble detection method using optical sensors [5] uses one transmitter and one receiver LED opposite each other, when there is a bubble appearing, this transmission will be interrupted and give a warning signal. With the method using pressure sensor [6], the pressure sensor will be connected directly to the suction system to receive continuous pressure value of the system, when air bubbles appear, the value of the pressure is changed, then through an amplifier which changes the output voltage of the system. The other method is to use a capacitor [7], this method places two plates of the capacitor on the fluid pipeline, if there air bubbles going through it will change the capacitance of the capacitor, the capacitor combined with a low pass filter circuit that will change the output voltage of the system. With all above methods, the method using optical sensors and the method using capacitors do not need a direct contact with the system as the method using pressure sensors. The Table 1 compares the pros and cons of those methods:

Table 1. Comparation of air bubbles detection methods.

Method	Advantages	Disadvantages
Optical sensor	- Low cost - Using in many system	- Not measured with some condensed liquids
Pressure sensor	- Can Monitor and save data on the system - Can use in many systems	- High cost - Use at fixed temperature
Capacitor	- Low cost - Easy to install	- Hard to work on Chest Drainage System

At that time in Vietnam, almost suction devices have not function to detect pneumothorax (the appearance of air bubble in the one-way valve bottle of the suction system) during the suction process of the pleural drain. To detect the pneumothorax incident, in some hospitals today, the medical staff will ask the patient's family members to regularly observe the draining system to see if there is a bubble appearance. If there are air bubbles, it means there is a pneumothorax incident (in the condition where the suction system is not leaked), they promptly notify the nurse or doctor so that it can be immediately treated. This follow-up by the patien's family members or the medical staff more or less creates difficulties and causes fatigue.

In this study, we aim to design a simple air bubble detection system with high accuracy and low cost. The system is developed on a platform that uses the capacitance change of the capacitor in condenser microphone [8] (the vibration of the condenser microphone's membrane) when detecting the pressure change generated from system vibrations of drainage system when air bubbles are present. It can be used for a variety of drain suction systems with three or more cylinders.

2 System Demonstration

Figure 1 presents a typical 3-cylinders drainage system. During the process of draining the pleural fluid, if there is pneumothorax, the system will show air bubbles at the outlet tube in the 1-way valve bottle as a red area marked in Fig. 2.

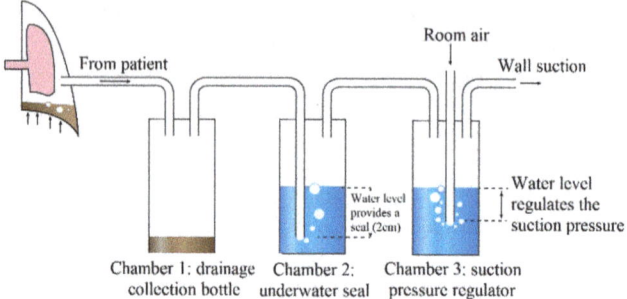

Fig. 1. A typical 3-cylinders drainage suction system

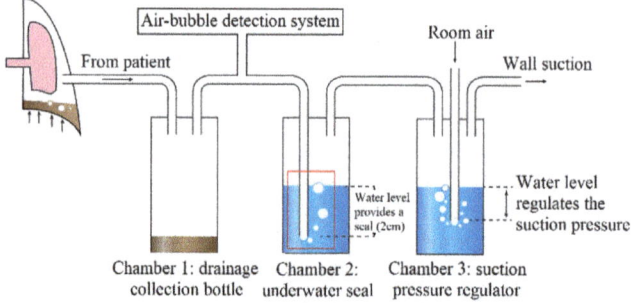

Fig. 2. Location of pneumothorax detection and air bubble detection system

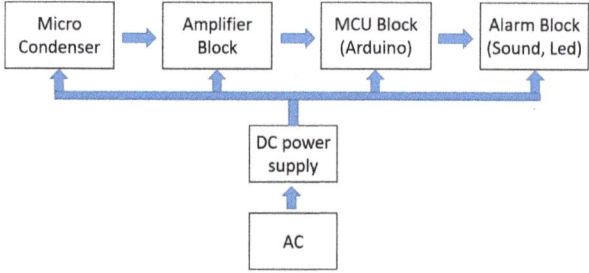

Fig. 3. System block diagram

As stated above, there are currently methods of using optical sensors, pressure sensors or capacitors, but no method of using condenser microphones in detecting air bubbles. From this idea, the system block diagram of the system that uses the condenser microphone to detect pneumothorax is shown in Fig. 3. In Fig. 3, the 220 V AC input source is converted to a 5 V DC source to supply to the amplifier unit, the central processing unit and the alarm block.

The air bubble detection system uses a condenser microphones to detect the change in pressure from the 3-cylinder suction system. When there is a pneumothorax, the pressure in the conductor changes and produces vibrations of the condenser microphone membrane, similar to vibrations. Condenser microphones are the most common type of probe used to detect or measure the loudness of an audio signal (Figs. 4 and 5).

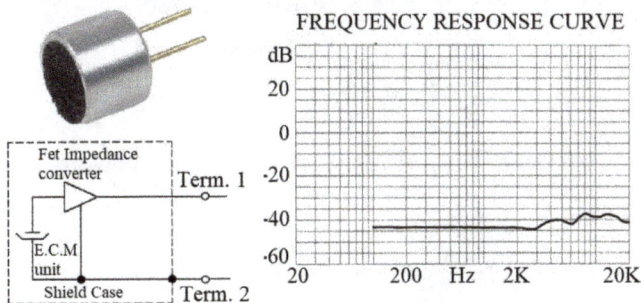

Fig. 4. Circuit diagram and frequency response of condenser microphones

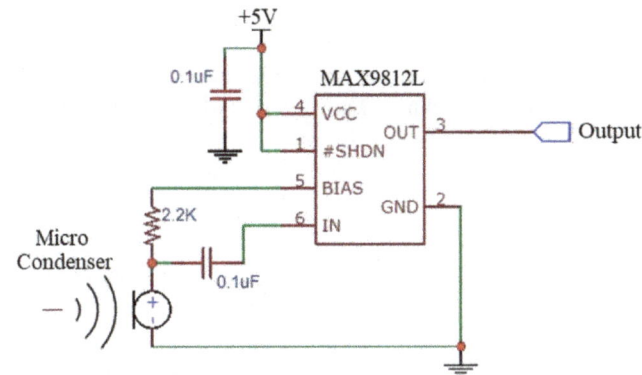

Fig. 5. Circuit diagram using MAX9812L chip to amplify audio signals

The signal is then sent to an audio amplifier circuit using the Maxim MAX9812L chip. This is a basic amplifier that has gain up to 10 times. It can be used as an audio/voice input device or to detect sounds. With the Maxim MAX9812L chip, the output reference voltage is kept always positive, which makes the processing unit easy to receive, analyze and process signals.

The amplified audio signal will be sent to the Arduino for processing. From the abnormal signal of the input signal will give an audible and light warning about the presence of air bubbles in the system. The designed system includes a switch so that when the alarm sounds from the system, the doctor/medical staff can press it to turn off the alarm sound, to avoid causing additional noise pollution. in the hospital room.

3   System Test

The finishing system has dimension of $70 \times 45 \times 30$ mm. The system was tested on a 3-cylinder suction system researched and produced by Bach Khoa Application Technology Company Limited (BKAT CO., LTD) (as showed in Fig. 6).

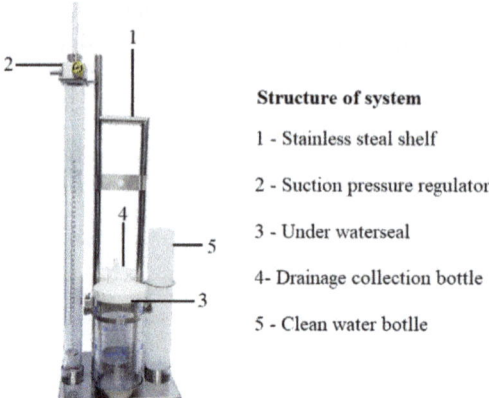

Structure of system

1 - Stainless steal shelf

2 - Suction pressure regulator

3 - Under waterseal

4- Drainage collection bottle

5 - Clean water botlle

Fig. 6. 3-cylinder drainage system product of BKAT company

Under practical test conditions, 2 cylinders, one-way valve cylinder (2) and fluid reservoir (3), are brought out for easier operation and performance. The air bubble detection system location as shown in Fig. 2 is directly connected to the connecting wire between the 1-way valve cylinders (2) and fluid reservoir (3). Figure 7 shows the complete system and the installation location of the gas detection system.

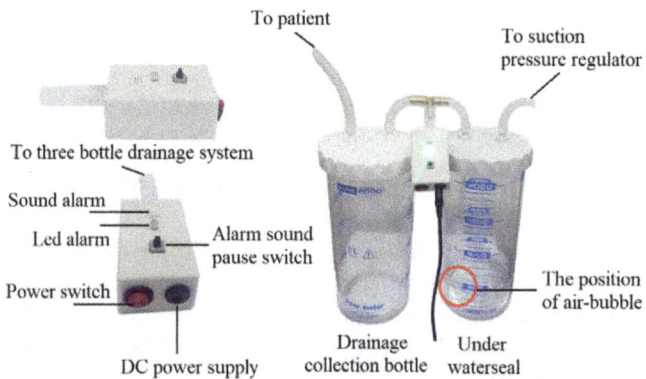

Fig. 7. Gas detection system and actual installation location

Signals obtained from the system are observed and simulated using MATLAB software. Figure 8(a) is the sound signal obtained from the system when the suction system is not operating, and Fig. 8(b) present the sound signal when the suction system is operating but no air bubbles appear. The noise presented in Fig. 8(b) is due to the influence of the sound from the water manometer when the system is operating.

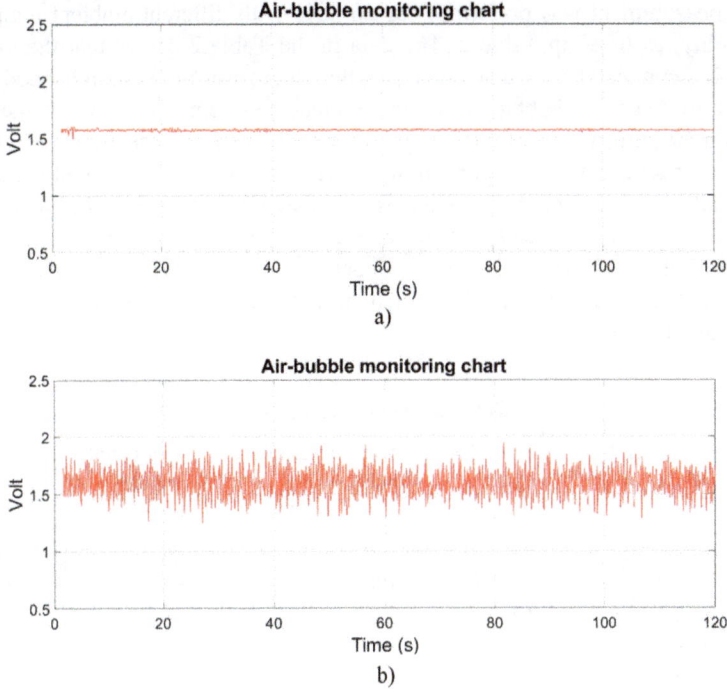

Fig. 8. Audio signal obtained when the drainage system is not operating (a) and operating in the absence of air bubbles (b)

This interference signal is very small compared to the signal when there is air bubbles appearance, so there is no need to process, as demonstrated in Fig. 9.

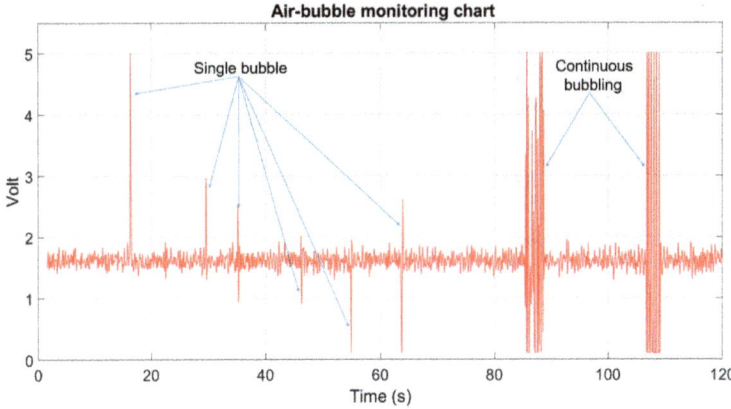

Fig. 9. Audio signal obtained when the drainage system is operating with air bubbles

Test measurement was performed for 20 days with different ambient temperature and humidity, as listed in Table 2. The data in the Table 2 shows that the system is about 95% accurate. This accuracy confirms that the measuring system is good enough to be able to detect air bubbles in draining suction systems. The size of the bubble depends on the surface area of the tube in the water seal. When the pressure in the tube increases excessively, the air will push the water in the water seal's tube to recede in a circular pattern, so the bubble volume is calculated by the height between the flat surface of the small tube end and the concave arc of the suction air. Usually, the smallest air bubble size that the system can detect is 0.3 cm^3. The data also shows that the detection of air bubbles in the system is not highly dependent on the ambient temperature and humidity.

Table 2. System test results.

Days	Temperature (°C)	Humidity (%)	# of appearance bubbles	# of counted bubbles	Error (%)
01	28.9	74	100	97	3
02	30	74	100	94	6
03	32.2	70	100	95	5
04	32.2	62	100	96	4
05	25	89	100	96	4
06	27.2	84	100	94	6
07	27.8	84	100	93	7
08	30	74	100	93	7
09	30	79	100	96	4
10	31.1	79	100	95	5
11	28.9	79	100	94	6
12	30	74	100	93	7
13	31.1	70	100	95	5
14	30	74	100	95	5
15	27.8	58	100	96	4
16	25	74	100	95	5
17	27.8	79	100	96	4
18	27.8	84	100	98	2
19	27.2	89	100	95	5
20	26.1	74	100	94	6
Average	28.8	76.2	100	95.1%	4.9%

4 Conclusion

In this research, a system to detect air bubbles in drainage system has been designed and tested. This system can detect and give a warning for pneumothorax applying in pulmonary drainage. The proposed system can partly reduce the care effort of nurses and family members for patients as well as enhance the quality of medical services.

Finished product is well looking design, high mobility, suitable for many current suction systems. The product has a power switch and jack, easy to handle and store. The shell of the box is made entirely of hard plastic, which increases safety in case of product breakdown. Experimental results on the 3-cylinder suction system show that the system can detect air bubbles with an accuracy level of up to 95%. In the future, we will develop an additional pressure monitoring module along with the fluid volume monitoring module in the fluid reservoir and display it on the display to complete a patient monitoring monitor in pleural drainage system.

Acknowledgments. The authors would like to thank Bach Khoa Applied Technology Company Limited (BKAT Co., LTD) for supporting and sponsoring the authors group to research and implement this research.

References

1. Zisis, C., et al.: Chest drainage systems in use. Ann. Transl. Med. **3**(3), 43 (2015)
2. Jany, B., Welte, T.: Pleural effusion in adults—etiology, diagnosis, and treatment. Deutsches Ärzteblatt Int. **116**(21), 377–386 (2019)
3. Barak, M., Katz, Y.: Microbubbles: pathophysiology and clinical implications. Chest **128**, 2918–2932 (2005)
4. Cerfolio, R.J., Varela, G., Brunelli, A.: Digital and smart chest drainage systems to monitor air leaks: the birth of a new era? Thorac. Surg. Clin. **20**, 413–420 (2010)
5. Benharash, P., Cameron, R., Zhu, A.: Automated Optical Detection of Air Leaks in Chest Tube Drainage System. University of California, California (2019)
6. Chen, C.-H., et al.: A chest drainage system with a real-time pressure monitoring device. J. Thorac. Dis. **7**(7), 1119–1124 (2015)
7. Gafare, M., Adam, A., Dennis, J.O.: Capacitor device for air bubbles monitoring. Int. J. Electr. Comput. Sci. **9**(10), 12–15(2009)
8. Devices, C.: Electret Condenser Microphone Datasheet, 27 04 2020. https://www.cuidevices.com/product/resource/cmc-5044pf-a.pdf

An Improved Structural Design for Elastic Beam of Six-Axis Force/Torque Sensor Based on Strain Gauges

Ho Quang Nguyen[1(✉)], Anh Tuan Nguyen[1], Huu Chuc Nguyen[1], and Trong Toai Truong[2]

[1] Thu Dau Mot University, Thu Dau Mot, Binh Duong, Vietnam
quangnh@tdmu.edu.vn
[2] 3C Machinery Co., Ltd, Ho Chi Minh City, Vietnam

Abstract. This paper presents a structural design for elastic beam of six-axis force/torque sensor which improves both static and dynamic performance of the sensor. Firstly, an elastic body structure of six-axis force/torque sensor is proposed based on structures of H-beam and cross-beam sensors. Secondly, both static and dynamic analyses of the elastomer structure have been studied by performing finite element analysis (FEA). Finally, the static and dynamic performances of the developed structure are compared to those of typical cross-beam structures which commonly used in commercial six-axis force/torque sensors. The results show that the proposed elastomer structure provides better sensitivity and natural frequency of the sensor than conventional sensor based on cross-beam.

Keywords: Elastic body design · Strain gauges · Finite element analysis · Six-axis force/torque sensors

1 Introduction

Six-axis force/torque sensors, which can measure forces and moments along three orthogonal directions based on strain gauges, have been developed for decades and widely used in robotics, especially in medical robot, collaborative robot and other intelligent robots.

The key mechanical component of six-axis force/torque sensors is the elastic body. The development of this component design to guarantee both high sensitivity in each axis due to deformation and dynamic performance of the sensors is still challenging.

Depending on mechanical structure, there are many various types of elastic beams, which are used in different situations. Their basic structures can be classified into different types including Stewart structure [1], composite [2–4], cylinder [5], cross beam [6–9] and so on. The comparison of main features of force/torque sensors is given in Table 1.

Table 1. The main features of various mechanical structures of elastic beams [10].

Mechanical structure	Advantages	Disadvantages
Cross-beam	High symmetry, compact structure, high stiffness, machinability	Existence of inter-dimensional coupling
Cylindrical beam	Good shock resistant ability, large load capacity	Serious inter-dimensional coupling
Composite beam	Near zero-coupling measurement, good sensitivity and large rigidity	Complicated structure, significantly affected by machining and assembly
Stewart beam	High load capacity, high stiffness	Low natural frequency, only use for the static force

Thanks to these advantages on the table above, cross beam is commonly applied in various tasks. Therefore, the elastic beam with cross-type structure is analyzed in this study. In this paper, a structure design of elastomer for force/torque sensor is developed to improve the sensitivity and dynamic performance of the sensor. The simplified mechanical model is analyzed by theory of mechanics. Analytical predictions are verified by FEM simulations with commercial software ANSYS. In the end, the comparison with an conventional elastic beam in [11] is employed to ensure the new design of force sensor to be better in both sensitivity and dynamic performance.

2   Structure Design

2.1   Typical Elastic Beam

The structure of the elastic beam includes cross beams connected to floating beams. The cross beams produce sensitive strain and the floating beams reduce inter-dimensional coupling. The sensor is subjected by three dimensional forces and moments. The strain gauges are used to determine the strain in cross beams. Figure 1 shows the structure of sensor with floating beam.

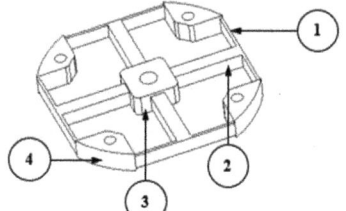

1. Floating beams
2. Cross beams
3. Central Platform
4. Rim

Fig. 1. The elastic beam with floating beams.

2.2   Proposed Design for Force/Torque Sensor

The proposed design of elastic beam is described in Fig. 2. The stiffness of the sensor can be increased by changing floating beams to H-beam in the proposed structure.

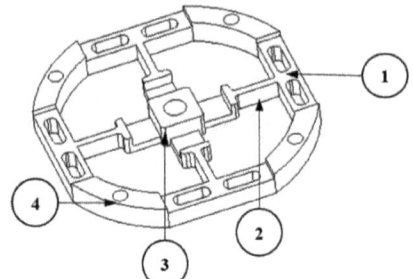

1. H beams
2. Cross beams
3. Central Platform
4. Rim

Fig. 2. Elastic beam of an improved structural design.

Moreover, shape of cross beams has been changed in order to improve the sensitivity. Besides, there are more spaces for the strain gauges to be glued on elastic beam.

In order to improve sensitivity, the cross beams need to get large deformation. In other words, the larger the strain obtained at the place where strain gauges are attached, the better the sensitivity. Accordingly, the size of cross beam and the thickness of floating beam should be reduced as much as possible. On the other hand, cross beam that is smaller in size can lead to the worse dynamic performance of the sensor. Therefore, there are many researchers [2–4, 6, 7] try to optimize the sensor that can satisfy both the sensitivity and dynamic performance requirement. In this study, we proposed the structure of elastic beam which improves both sensitivity and dynamic performance of the F/T sensor.

3   Modeling and Simulation

In order to show the preeminence of the improved structural elastic beam, the comparison with an conventional elastic beam in [11] is employed to ensure the new design of force/torque sensor is better in both sensitivity and dynamic performance.

The FEA method is used to evaluate the property of the elastic beams by commercial software ANSYS to get numerical solutions.

3.1   FEM Simulation

For the purpose of comparison, the outer diameter and mechanical properties of material are assumed to be the same. The mechanical property of material Aluminum Alloy 2024 [11] show in Table 2. The proposed geometry and dimension are showed in Fig. 3 and Table 3, respectively.

Table 2. Mechanical properties of structure material

Property of material	Density (kg/m³)	Young's modulus (GPa)	Poisson's ratio	Yield Strength (MPa)
	2,780	72	0.3	325

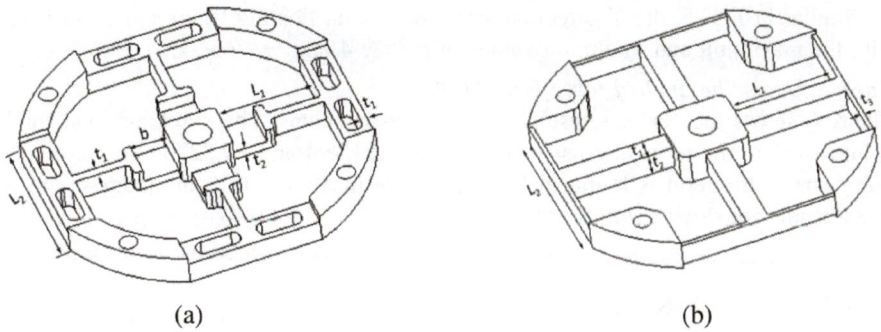

Fig. 3. (a) Proposed elastic beam, (b) Conventional elastic beam.

Table 3. Parameters of elastic beams (unit: mm).

Comparison of sensors	External diameter	Internal diameter	L_1	L_2	t_1	t_2	t_3	b
Conventional elastic beam [11]	76	–	25.0	36	5.0	5.0	1.0	–
Proposed elastic beam	76	64	25.7	30	2.0	2.0	1.5	8.0

Due to the symmetric structure of the elastic body, only four cases considering the effect of F_x, F_z, M_x and M_z to the elastic beam need to be analyzed. For the purpose of comparison, the material, boundary and loading conditions used for the two sensors are the same. In this work, loading $F_x = 100$ N, $F_z = 100$ N, $M_x = 3$ N.m, $M_z = 3$ N.m and a fixed support are applied to the elastic beam.

Case 1: Under the applied force F_x= 100 N
As show in Fig. 4, the Y direction strain occurs around four holes on H-beam due to bending caused by the force F_x. The maximum and minimum strains are 3.536e-4 and −3.537e-4 (Fig. 11).

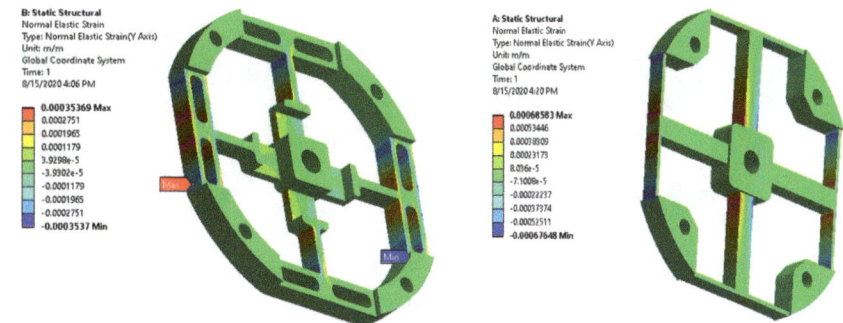

Fig. 4. Strain distribution of the proposed elastic beam under F_x= 100 N

Fig. 5. Strain distribution of conventional elastic beam under F_x= 100 N

Similar in Fig. 5, the Y direction strain occurs on floating beam and cross beam with the maximum and minimum values of 6.858e-4 and −6.765e-4.

Case 2: Under the applied force F_z= 100 N
Figure 6 shows the analysis results based on cross beams. The maximum and minimum strains at the roots at cross beams are of 6.815e-4 and −7.327e-4, respectively. The same distribution is found in Fig. 7, but the maximum and minimum strains at cross beams are slower with the values of 4.074e-4 and −4.091e-4, respectively.

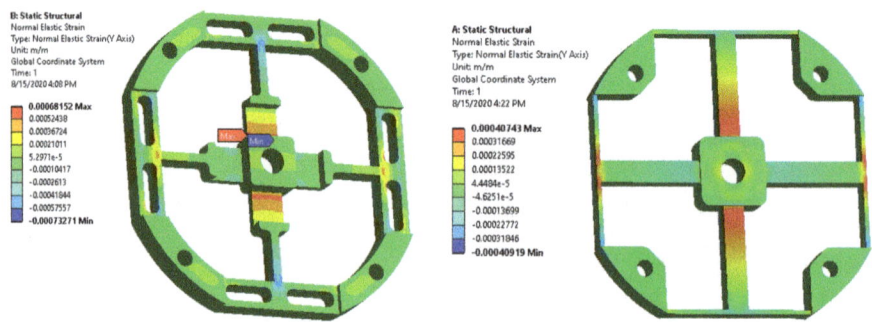

Fig. 6. Strain distribution of improved elastic beam under F_z= 100 N

Fig. 7. Strain distribution of conventional elastic beam under F_z = 100 N

Case 3: Under the apply moment M_x = 3 N.m
Figure 8 illustrates the results of Y direction strain on cross beam due to applied moment M_x. The maximum and the minimum strains at the root are 2.554e3 and −2.554e-3, respectively. Similarly, Fig. 9 shows the maximum and minimum strains with the values of 6.569e-4 and −6.506e-4

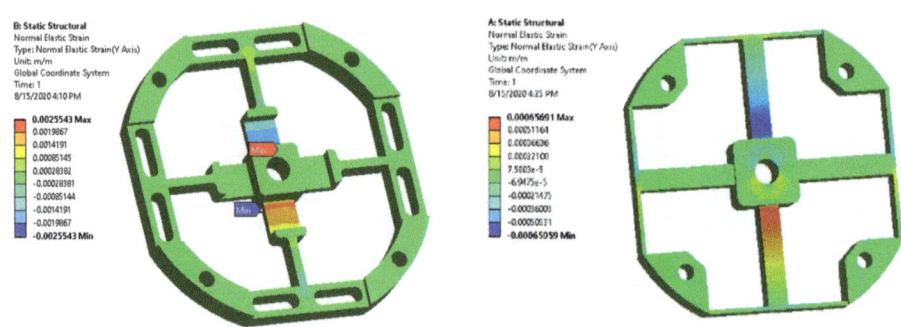

Fig. 8. Strain distribution of improved elastic beam under M_x= 3 N.m

Fig. 9. Strain distribution of conventional elastic beam under M_x = 3 N.m

Case 4: Under the apply moment M_z= 3 N.m

Y direction strains at vertical thin plate on cross beam get the maximum and minimum strains 7.919e-4 and −7.919e-4 on improved elastic beam in Fig. 9. Y direction strains at roots of cross beams in Fig. 10 got maximum and minimum strains around 3.773e-4 and −3.773e-4.

Y direction strains at vertical thin plate on cross beam get the maximum and minimum strains 7.919e-4 and −7.919e-4 on improved elastic beam in Fig. 9. Y direction strains at roots of cross beams in Fig. 10 got maximum and minimum strains around 3.773e-4 and −3.773e-4, respectively.

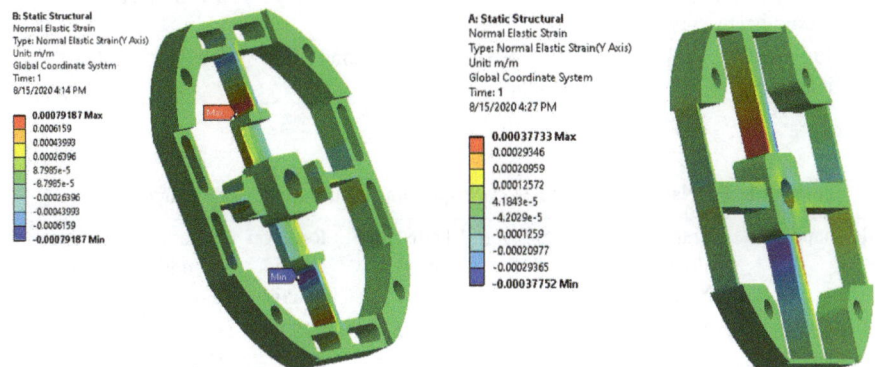

Fig. 10. Strain distribution of improved elastic beam under $M_z = 3$ N.m

Fig. 11. Strain distribution of conventional elastic beam under $M_z = 3$ N.m

4   Results and Discussion

In order to perform the static and the dynamic analyses. The finite element model of elastic body was built with boundary condition and different load case. The maximum strain values obtained from the static simulations of the two elastic beams under different loading modes are shown in Table 4. The natural frequencies corresponding to first six vibration mode shapes are given in Table 5. We also can see that the improvement of dynamic lead to reduction of static performance, so structure design for elastic body need to consider static and dynamic performance simultaneously. The results presented in Table 4 indicate that the developed elastic body mostly has higher sensitivity due to the size reduction and shape changing of cross beams to maximize the deformation in certain direction. The strain under the applied moment M_x, M_y and M_z are around twice higher than typical cross beam structure. Moreover, due to changing the shape of cross beams, the developed structure is able to provide more favorable positions to facilitate strain gauges attachment. However, the smaller the sizes of cross beam, the smaller the stiffness of elastic body which reduces the sensor dynamic performance. Additionally, as can be found in Table 5, natural frequencies of the elastic body are considerably higher than that of typical elastic beam in all cases thanks to integrating H-beam to the structure design to increase the stiffness of elastic beam, it can be seen that the stiffness of H-beam is higher that the stiffness of floating beam in all direction, especially the translational stiffness of X axis and Y axis. Therefore, the proposed structure is preeminent to the structure of the conventional sensor [11] in both static and dynamic performance, which can meet the requirements of high-speed and high-precision.

Table 4. Maximum strains under different loading modes

Model	$F_x = 100$ N	$F_z = 100$ N	$M_x = 3$ N. m	$M_z = 3$ N.m
Conventional elastic beam [11]	6.858×10^{-4}	4.074×10^{-4}	6.569×10^{-4}	3.773×10^{-4}
Proposed elastic beam	3.537×10^{-4}	6.815×10^{-4}	2.554×10^{-4}	7.919×10^{-4}

Table 5. Natural frequencies corresponding to different mode shapes

Mode shape	Translation along X-axis	Translation along Y-axis	Translation along Z-axis	Rotation around X-axis	Rotation around Y-axis	Rotation around Y-axis
Conventional elastic beam [11]	2016.6 (Hz)	2017.0 (Hz)	2334.5 (Hz)	6252.4 (Hz)	6253.0 (Hz)	9139.9 (Hz)
Proposed elastic beam	4945.5 (Hz)	4945.6 (Hz)	2642.3 (Hz)	6494.2 (Hz)	6494.2 (Hz)	9709.1 (Hz)

5 Conclusions

This article proposed an improved structure of cross beam in 6 axis force/torque sensor. Both sensitivity and dynamic analyses are carried out by adopting finite element analysis. The results show that the proposed structure is preeminent to the structure of the conventional sensor in both static and dynamic performance. In the future, the structure of elastic body will be optimized and fabricated and experimentally validated.

Acknowledgements. This research is supported by Vingroup Innovation Foudation (VINIF) in project code VINIF.2020.NCUD.DA059.

References

1. Hou, Y., et al.: Optimal design of a hyperstatic Stewart platform-based force/torque sensor with genetic algorithms. Mechatronics **19**(2), 199–204 (2009)
2. Yuan, C., et al.: Development and evaluation of a compact 6-axis force/moment sensor with a serial structure for the humanoid robot foot. Measurement **70**, 110–122 (2015)

3. Po Wu, P.C.: Decoupling analysis of a sliding structure six-axis force/torque sensor. Meas. Sci. **13**(4), 188 (2013)
4. Sun, Y., et al.: Design and optimization of a novel six-axis force/torque sensor with good isotropy and high sensitivity. In: 2013 IEEE International Conference on Robotics and Biomimetics (2013)
5. Aghili, F., Buehler, M., Hollerbach, J.M.: Design of a hollow hexaform torque sensor for robot joints. Int. J. Robot. Res. **20**(12), 967–976 (2016)
6. Chen, D., Song, A., Li, A.: Design and calibration of a six-axis force/torque sensor with large measurement range used for the space manipulator. Procedia Eng. **99**, 1164–1170 (2015)
7. Lou, Y., Wei, J., Song, S.: Design and optimization of a joint torque sensor for robot collision detection. IEEE Sens. J. **19**(16), 6618 (2019)
8. Kim, Y.-G., et al.: Miniaturized force-torque sensor built in a robot end-effector for delicate tool-tip gripping control. Elektronika ir Elektrotechnika **20**(6) (2014)
9. Wang, Y., et al.: Strain analysis of six-axis force/torque sensors based on analytical method. IEEE Sens. J. **17**, 4394–4404 (2017)
10. Aiguo, S., Liyue, F.U.: Multi-dimensional force sensor for haptic interaction: a review. Virtual Reality Intell. Hardware **1**(2), 121–135 (2019)
11. Ma, J., Song, A.: Fast estimation of strains for cross-beams six-axis force/torque sensors by mechanical modeling. Sensors **13**(5), 6669–6686 (2013)

QoS Evaluation of Multimedia Data over Multi-hop Ad Hoc Networks for Testbed System

Ngo Hai Anh[✉] and Pham Thanh Giang

Institute of Information Technology, Vietnam Academy of Science and Technology,
Hanoi, Vietnam
{ngohaianh,ptgiang}@ioit.ac.vn

Abstract. Wireless network technology has rapidly developed along with the development of the Internet and data transmission services, especially multimedia data which is becoming more and more popular with audio and video applications. Unlike an *Access Point* (AP)-based wireless network, an *ad hoc* network has the advantage that the network nodes can self-forming the network without a central AP, thus increases flexibility in different conditions of use. However, due to this ad hoc characteristic, ensuring network performance and *Quality of Service* (QoS) will be more difficult compared to networks with APs. This paper will introduce how to build an emulation system or *testbed* system to simulate a multi-hop ad hoc network and perform QoS evaluation of multimedia data including audio, video and background on this testbed system.

Keywords: Testbed · Network performance · Multimedia · Wireless · QoS

1 Introduction

Wireless network technology are becoming an important infrastructure for many environments includes industry, business, and home. But the network standards need to be analyzed, or checked, and evaluated before they are released for official use. Previously, research on wireless networking technologies was mainly evaluated based on modeling and simulating (based on mathematical models and simulation tools). These solutions have characteristics of not having to use hardwares because they mainly use software tools to write simulation scripts and analyze results, or demonstrate mathematical representations. However, their disadvantage is that they are limited by mathematical assumptions or software functions, so that methods such as simulation and modeling cannot reflect all the physical elements of the real devices in the network. The trend of using **testbed** to evaluate network parameters shows more and more advantages compared to modeling and simulation methods [1–3]. Our research focuses on using a network testbed system that we built ourselves to simulate a *multi-hop wireless ad hoc network* to evaluate the QoS of multimedia data service.

D.-T. Tran et al. (Eds.): ICISN 2021, LNNS 243, pp. 150–156, 2021.
https://doi.org/10.1007/978-981-16-2094-2_19

2   Related Works

Nowadays, advanced networking technologies and wireless networks has been developing rapidly. Along with the rapid development of network technology there are also limitations. It is difficult for researchers as well as industry manufacturers to design and release new technologies. These new network technologies should be evaluated and refined before they can be released as official products.

We can understand the simulation will allow for crude evaluation of the performance, behavior, and scalability of a new network technology. That's why simulation is a very helpful, and often inexpensive initial step in the evaluation of a new technology. However, with design and computational complexity, simulations often use simplified models, and that can easily lead to mistakes in real-world environment. The emulator is an alternative that combines the use of installed prototypes and entities to replicate real-world behaviors. Such replacements, however, often fail to capture the complexity of the actual environment. And the results were only experiments that used real-world prototype settings and used in controllable environments, e.g. users in real life, wireless network devices, vehicle movement, etc., will give results almost identical to reality. The platforms that allow such things are called *testbed systems.*

Therefore, network testbeds are major ingredients in the development of new wireless network technologies, and this will led to many cooperations at a global level aimed at things like design, build, share, operate, and share testbed system. Currently, many universities and researchers around the world have been building laboratory systems using their own testbeds, such as Planet [6] or ORBIT [7].

Testbeds can serve as infrastructures for the development and evaluation of new network technologies. Simulation-based evaluations provide inexpensive but valuable results in terms of network performance for a new approach or technology. Nevertheless, simulation tools such as NS3 [4], OMF [8] and OMNet++ [5], inherently simplify assumptions. By contrast, for network standards to be widely accepted in the computer networking industry and in the user community will require new proposed algorithms or techniques to be installed, tested and evaluated in real world with real users, and of course it would be costly and laborious. The testbeds are often seen as an efficient alternative, where new technologies are assessed in control, but with the same environment and scale as in real life.

From the above researches, we see the use of testbed to analyze and evaluate network performance is a cutting-edge trend in research on computer networking. However, the investment in the testbed system is expensive due to the need for a certain amount of hardware. Therefore, in our paper, we will propose a mechanism for evaluating multimedia data throughput based on a testbed system with the medium-sized laboratory.

3   Designing and Setting Up Multi-hop Wireless Ad Hoc Network Testbed

In this paper, information related to the streaming and playback of video files between wireless nodes in the scenario will be observed by Wireshark software, the result in trace files will be analyzed to give information about QoS of network such as throughput and packet loss rates. The experimental model we set up is to simulate a scenario of ad hoc wireless network as shown in Fig. 1.

In order to easily adjust wireless network parameters such as frequency, IP address, QoS parameters, we use a network node as an Access Point instead of an Access Point device, so we set hostapd software as the role of access manager, and using dnsmasq software to assign IP addresses to wireless nodes joining the network. We use static routing in this scenario instead of AODV [9] or DSR [10] routing protocols.

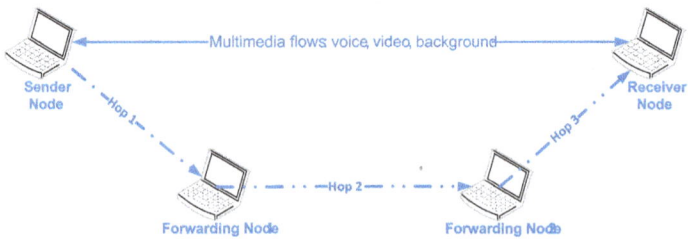

Fig. 1. Multi-hop testbed scenario.

The research method that we use in this paper is the experimental method, based on the workflow as shown in Fig. 2.

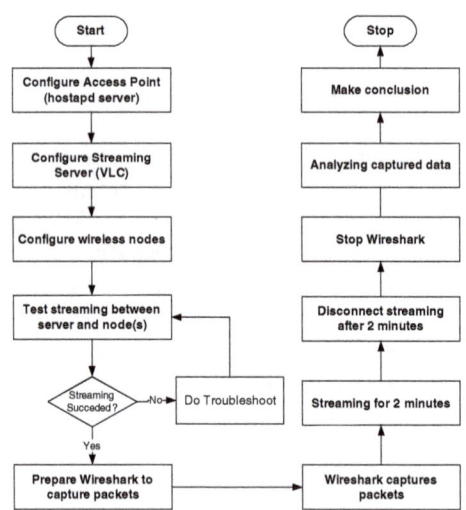

Fig. 2. Research methodology.

Starting the experiment is configuring the control node (Access Point node) based on hostapd software tool and the VLC server node for streaming multimedia data. The next step is to configure the wireless nodes to participate in the simulation scenario, for example between the nodes that will send and receive data, the wireless standard here can be either IEEE 802.11b or IEEE 802.11g. The next step is necessary to test the connection between the wireless nodes because if the connection is not correct, the simulation will be run wrong from the beginning. To observe the whole process of sending and receiving data between nodes in the simulation, the testbed needs to set up a *monitor node* to observe the send and receive information of all nodes, to do so wireless The monitoring node's card needs to be configured to run in monitor mode to capture all packets at the Data-Link layer, otherwise sending/receiving are required at all destination nodes (node receive) and it is not possible with many nodes participating in the simulation. The process of the next steps capturing the packets, and saving the simulation output. The final step is to analyze the data and evaluate the network performance information.

Based on the work flow of the experiment as above, the next step we build simulation scenarios and evaluate some test cases.

4   Evaluate the Impact of Multimedia Data on Network Performance

To evaluate the impact of multimedia data such as voice and video on wireless network performance, we use WMM (Wi-Fi MultiMedia) service. WMM is a specification based on IEEE 802.11 wireless network standards, with a descending priority Access Class (AC) including four data classes: AC_VO (voice), AC_VI (video), AC_BE (best-effort) and AC_BK (backround). VMM works well with wireless standards like IEEE 802.11a, b, g and n. In today's network devices, these ACs are mapped corresponding to the Differentiated Services Code Point (DSCP) tags, which is a packet header value of 6-bits length. The bit is intended to classify the traffic in network devices that correspond to ACs in the WMM. Table 1 shows the corresponding mappings between AC–DSCP.

Table 1. Mapping values of DSCP and AC

DSCP (decimal value)	WMM access category
8	Background
16	
0	Best effort
24	
32	Video
40	
48	Voice
56	

In the following simulation scenarios we use IEEE 802.11g with three types of AC: *voice*, *video* and *best-effort*. For each data type, two Quality of Service criteria are considered as *throughput*, which affects the user's service usage such as data downloads, television viewing, and *delay* (jitter), this parameter influences the online performance of the picture and sound, to gain insight into network performance.

4.1   Single-Hop Scenario

First, we evaluate network performance of a single-hop scenario, this network topology simply consists of two network nodes acting as the sender and receiver. The test data consists of three types: voice, video and background, and is sent in saturation to ensure a certain classification of data is carried out.

Table 2 shows simulation results with popular IEEE 802.11g Wi-Fi without WMM enabled (IEEE 802.11g DCF).

Table 2. Simulation result of single-hop scenario with IEEE 802.11g DCF

Data type	Throughput [Mbps]	Jitter [ms]
Voice	3.08	39.972
Video	3.07	34.141
Background	3.07	52.822

Looking at the results in Table 2 we can see that without using the WMM function, devices in the network will not have a distinction between different data types, and throughput and jitter parameters of all three types of voice, video and background are almost the same.

Next, we see simulation result of 802.11g Wi-Fi with WMM enabled (IEEE 802.11g EDCA).

Table 3. Simulation result of single-hop scenario with IEEE 802.11g EDCA

Data type	Throughput	Jitter
Voice	6.64 [Mbps]	39.972
Video	3.99 [Mbps]	34.141
Background	32.7 [Kbps]	52.822

Looking at the results in Table 3 we can see that when the WMM function is enabled, devices in the network will have differentiation between different data types, and throughput and jitter parameters of all three types of voice and video. and the background are both very different. The highest priority voice data accounts for more than 60% of the network bandwidth and has the best data stability (the lowest jitter), while the other two types of data have lower quality of service, data type. background material has the lowest priority so it takes up very little bandwidth and has very high jitter.

Thus, it can be said that multimedia function in network devices now has a significant effect in classifying multimedia data, the above single-hop scenario clearly shows that. However, this simulation scenario is almost the ideal case because in reality, the number of network devices will be more and more diverse (there is the influence of other devices such as television, microwave, mobile phone...) Therefore, the next part of this paper will present simulation of multimedia data transmission on a multi-hop network.

4.2 Multi-hop Scenario

In this section, we evaluate network performance of a multi-hop scenario, this network topology as shown in Fig. 1 includes of 4 (four) wireless nodes, with two end-end nodes serving as sender and receiver. The two middle nodes act as data forwarding to preserve end-to-end communication. The test data consists of three types: voice, video and background, and is sent in saturation to ensure a certain classification of data is carried out.

Table 4 shows the simulation results for multi-stage network with regular 802.11g Wi-Fi without WMM function enabled (IEEE 802.11g DCF), this is the result measured at destination node.

Table 4. Simulation result of multi-hop scenario with IEEE 802.11g DCF

Data type	Throughput [Mbps]	Jitter [ms]
Voice	2.30	38.474
Video	2.30	66.845
Background	2.30	63.203

Looking at the results in Table 4 we can see that without using the WMM function, devices in the network will not have a distinction between different data types, and throughput and jitter parameters of all three types of voice, The video and background are almost the same, similar to the single-stage scenario conducted, however, as the data transmission goes further, it has to pass through intermediate nodes to forward the data, so the network performance is degraded. The parameters of throughput and jitter are both worse than the simulation results in Table 2.

5   Conclusion

In our paper, we have proposed a solution to evaluate the Quality of Service for multimedia data on a testbed system that emulates multi-hop ad hoc network. The experiments on the real system has more realistic results than other modeling or software simulation methods, which is very meaningful for network performance evaluation.

Acknowledgement. This paper is completed under the sponsorship of the Institute of Information Technology (IOIT) for project number CS20.09.

References

1. Navarro-Ortiz, J., Cervelló-Pastor, C., Stea, G., Costa-Perez, X., Triay, J.: Testbeds for future wireless networks. Wirel. Commun. Mob. Comput. **2019**. https://doi.org/10.1155/2019/2382471
2. Güneş, M., Hahm, O., Schleiser, K.: G-mesh-lab wireless multi-hop network testbed for the G-lab. In: Lecture Notes of the Institute for Computer Sciences, Social Informatics and Telecommunications Engineering Book Series, LNICST, vol. 46, pp. 597–598 (2010)
3. Blywis, B., Güneş, M., Juraschek, F., Schiller, J.H.: Trends, advances, and challenges in testbed-based wireless mesh network research. Mob. Netw. Appl. **15**(3) (2010). https://doi.org/10.1007/s11036-010-0227-9
4. NS-3 Consortium: NS-3 Network Simulator (2019). https://www.nsnam.org/
5. OMNEST: OMNeT++: Discrete Event Simulator (2019). https://omnetpp.org/
6. PlanetLab Consortium: Planetlab: an open platform for developing, deploying, and accessing planetary-scale service (2015). http://www.planet-lab.org/
7. Raychaudhuri, D., Seskar, I., Ott, M., Ganu, S., Ramachandran, K., Kremo, H., Siracusa, R., Liu, H., Singh, M.: Overview of the ORBIT radio grid testbed for evaluation of next-generation wireless network protocols. In: Wireless Communications and Networking Conference, vol. 3, pp. 1664–1669, 13–17 March 2005 (2005)
8. Rakotoarivelo, T., Ott, M., Jourjon, G., Seskar, I.: OMF: a control and management framework for networking testbeds. ACM SIGOPS Oper. Syst. Rev. **4**, 54–59 (2011)
9. Chakeres, I.D., Belding-Royer, E.M.: AODV routing protocol implementation design. In: 2004 Proceedings of the 24th International Conference on Distributed Computing Systems Workshops, Hachioji, Tokyo, Japan, pp. 698–703 (2004). https://doi.org/10.1109/ICDCSW.2004.1284108
10. Khabir, K.M., Siraj, M.S., Habib, M.A., Hossain, T., Rahman Ahad, M.A.: A study on DSR routing protocol in Adhoc network for daily activities of elderly living. In: 2018 Joint 7th International Conference on Informatics, Electronics & Vision (ICIEV) and 2018 2nd International Conference on Imaging, Vision & Pattern Recognition (icIVPR), Kitakyushu, Japan, pp. 573–577 (2018). https://doi.org/10.1109/ICIEV.2018.8640994

Beamforming for Density-Based DBIM Scheme in Ultrasound Tomography

Tran Quang Huy[1], Tran Binh Duong[2], Phuc Thinh Doan[3,4],
and Duc-Tan Tran[5(✉)]

[1] Faculty of Physics, Hanoi Pedagogical University 2, Hanoi, Vietnam
tranquanghuy@hpu2.edu.vn
[2] School of Information Science and Engineering, Southeast University,
Nanjing, China
[3] NTT Hi-Tech Institute, Nguyen Tat Thanh University,
Ho Chi Minh City 700000, Vietnam
[4] Faculty of Mechanical, Electrical, Electronic and Automotive Engineering,
Nguyen Tat Thanh University, Ho Chi Minh City 700000, Vietnam
[5] Faculty of Electrical and Electronic Engineering, Phenikaa University,
Hanoi, Vietnam
tan.tranduc@phenikaa-uni.edu.vn

Abstract. The diagnostic ultrasound technique uses scattered information appearing limitations as the research models' motivation to create a new image to supplement quantitative ultrasound information in tomography ultrasound imaging. One promising solution is density imaging, which is capable of detecting diseased tissues. In tomography ultrasound imaging, the DBIM method is commonly used to restore the object to be imaged; this method's advantage is fast convergence, but easy to be affected by noise. Therefore, a beamforming technique using multiple probe elements transmitting simultaneously to produce a narrow beam capable of minimizing the effect of noise has been proposed for density imaging using DBIM. The numerical simulation results have shown that noise and normalization errors are significantly reduced when using beamforming DBIM.

Keywords: Ultrasound tomography · Inverse scattering · DBIM · Beamforming

1 Introduction

Ultrasonic imaging techniques and tomography techniques play an essential role in clinical diagnosis. Currently, the ultrasonic image acquisition process is mainly based on the acoustic feedback method, which means that a part of the energy will be reflected when the sound waves meet an obstacle [1]. The feedback signal will be picked up by the receivers and used for imaging. By expanding the number of angles around the subject, backscattering allows better quality image recovery in strong scattering fields [2].

Studies in ultrasound tomography (UT) often focus on imaging based on sound mechanical properties such as attenuation or sound contrast [3]. However, a recovery image based on sound contrast gives higher quality than a recovery image based on attenuation. Therefore, this paper focuses on restoration images based on sound contrast. The computed tomography technique based on the propagation of sound waves has been applied. The first research on computerized tomography uses sound contrast to detect cancer-causing cells in women [4]. Another method using the neural networks model has been applied to ultrasound tomography [5]. The ultrasonic imaging model consists of a circular cylinder (object of interest) containing the acoustic properties of tissues and the probes arranged on a circle around the object (including a transmitter and multiple receivers). The receiver is kept in a fixed position, and the transmitter will move around. Therefore, the time for collecting and processing is still long.

The technique of computed tomography based on the propagation of sound waves has been applied and is the first research on computerized tomography using sound contrast to detect cancer-causing cells in women. Most studies on UT are based on the Born approximation method [6, 10–14]. DBIM method was used to recover a two-dimensional image of the dielectric constant distribution [7]. The DBIM method was then applied to the ultrasound tomography technique with improved standard parameter selection [8]. In the DBIM method, the Tikhonov norm [9] is used in solving the inverse problem. In DBIM, the Green function is updated in each loop, so this method's advantage is a fast convergence speed, but its disadvantage is affected by noise. To solve the DBIM method's weaknesses, i.e., overcoming noise, in this study, we use the beamforming algorithm to create a narrow beam combined with the DBIM method to recover the target function's density information.

2 Working Principles

The establishing of the measurement is presented in Fig. 1. The pressure signals emitted from the transmitters will be spread, scattered, and measured by the receivers. DBIM is used to estimate the change in sound speed and density information of measured data. Any tissue will be recovered (if it exists) through changes in sound speed and density information. The object function O(r) is represented by:

$$O(\mathrm{r}) = \begin{cases} \omega^2 \left(\frac{1}{c^2} - \frac{1}{c_0^2} \right) - \rho^{\frac{1}{2}}(r)\nabla^2 \rho^{-\frac{1}{2}}(r) & \text{if } r \leq R \\ 0 & \text{if } r > R \end{cases} \tag{1}$$

The ambient density is denoted by ρ, and the radius of the heterogeneous object is R. As indicated by Eq. (1), O(r) depends on the frequency of the wave ($\omega = 2\pi f$), c_0 and c are the sound speeds in the background environment and the tumor, respectively, and the density of the investigated object (ρ). Density information of surveyed subjects is often overlooked in the previous explorative, causing the images to be reproduced with limited details. To contribute to improving the diagnosis and treatment by adding quantitative information about image reproduction information, both density and sound contrast of the imaging object should be considered. Unknown tumors in the women's breast are often detected by the widely used DBIM method; therefore, the circular DBIM configuration is arranged around the subject. On a measurement configuration, the receivers and transmitters can, in principle, be placed equally, random or pseudo-random. Depending on the scenarios, the number of receivers and transmitters will match the actual requirements. It were assumed that there are N_t transmitters and N_r receivers. The reception and transmission of ultrasonic signals are performed according to the following procedure: The first three transmitters contemporaneously emit ultrasonic signals (the remaining ones are inactive), all receivers (N_r) will receive ultrasonic scattering signals. The first positions of the transmitters are determined by $1 \times N_r$ measurements. Then, the next three transmitters begin the activity. All receivers will receive signals in the second positions of the next three transmitters, resulting in $2 \times N_r$ measurements. The same procedure repeated until the last three transmitters (N_t). We obtain $N_t/3$ sets of measured values (i.e., $N_t \times N_r/3$ measurements). Finally, $N_t/3$ sets of measured values are combined to obtain sufficient information for the estimation.

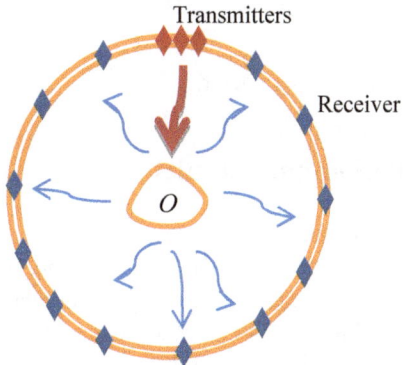

Fig. 1. Measuring principle of the ultrasonic tomography device

The simulation scenarios use the Bessel function as the transmitter signals or the incident wave whose frequency is f, and the wavelength is $\lambda = c_0/f$. When the incident waves propagate in the environment: a) The signal obtained is the incident pressure at the receivers if the environment is homogeneous; b) The following cases may occur if the environment is heterogeneous (with unknown tumors) when the incident wave reaches the target: When the wavelength of the incident wave is much smaller than the size of the object, the ultrasonic signal will be reflected; When the wavelength of the incident wave is greater than or equal to the size of the object, the ultrasonic signal will be scattered in all directions. In consideration of that, the frequency range from 20 kHz to 12 MHz belongs to the ultrasonic spectrum used in clinical diagnosis, if 1484 m/s is the sound speed in the background environment, λ ranges from 6.2 μm to 74.2 mm. The incident signal is

$$p^{inc} = J_0(k_0|r - r_k|) \tag{2}$$

where J_0 is a zero-order Bessel function, and $|r - r_k|$ is the distance from the generator to the kth pixel in the region of interest. Scattered signals will scatter in many directions around the object, provided that the incident waves have a wavelength equivalent to the size of the encountered object. Applying the Born approximation method, linear relationship can be found between the diffusion pressure difference Δp^{sc} and the target function $\Delta \overline{O}$ and is represented as follows:

$$\Delta p^{sc} = \overline{M}.\Delta \overline{O}, \tag{3}$$

where the matrix $\overline{M} = \overline{B}.D(\overline{p})$; total sound pressure field $\overline{p} = \left[\overline{I} - \overline{C}.D(\overline{O})\right]p^{inc}$ with $\overline{B}$ is a matrix whose coefficients are Green functions representing the interaction of the pixels to the receivers, $\overline{C}$ is a matrix whose coefficients are Green function denoting the interaction among pixels, $\overline{I}$ is the unit matrix, and $D(.)$ is the operator diagonalization. With a generator and a receiver, we have a matrix $\overline{M}$ and a vector quantity Δp^{sc}. $\Delta \overline{O}$ can be estimated by solving the Tikhonov regularization problem [9]:

$$\Delta \overline{O} = \arg \min_{\Delta \overline{O}} \left\| \Delta p^{sc} - \overline{M}\Delta \overline{O} \right\|_2^2 + \gamma \left\| \Delta \overline{O} \right\|_2^2 \tag{4}$$

Where γ is the regularization parameter.

3   Numerical Simulation and Results

Parameters: Transmitter frequency 0.6 kHz, object diameter 7.3 mm, transmitter/receiver-to-object distances 100 mm, sound speed in the background environment 1540 m/s, sound contrast 10%, number of iterations 5, number of transmitters 15, and of receivers 15, noise 10%. Figure 2 is an initial image containing sound contrast and density information (number of pixels is 15). It is the input of the DBIM method. The output is the recovered image when using the beamforming technique. The recovery results of the DBIM and beamforming DBIM methods are shown in Figs. 3 and 4. Visually, we see that, when using DBIM, noise occurs quite large, mostly background noise. As for the DBIM beamforming method, the noise is significantly reduced. Moreover, by comparing to the original DBIM, the object convergence using

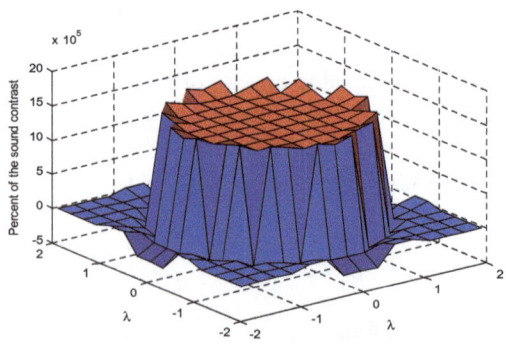

Fig. 2. Ideal initialization image (N = 15)

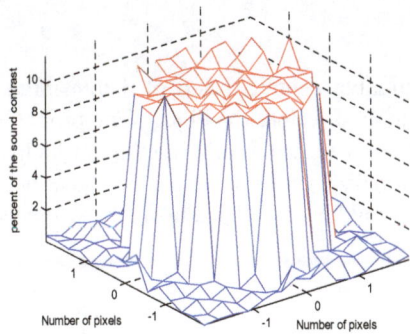 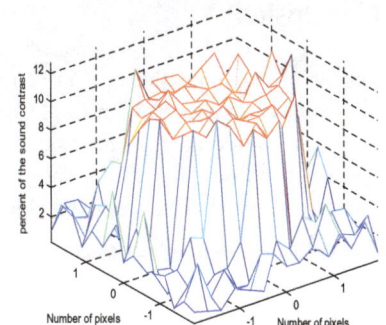

Fig. 3. Recovery image with beamforming after 5 loops (N = 15)

Fig. 4. Recovery image without beamforming after 5 loops (N = 15)

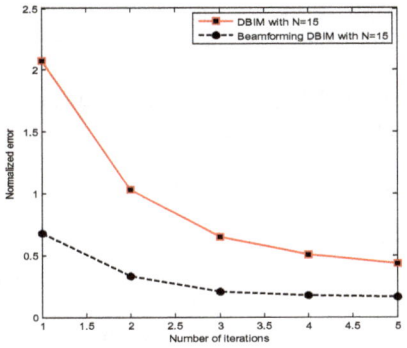 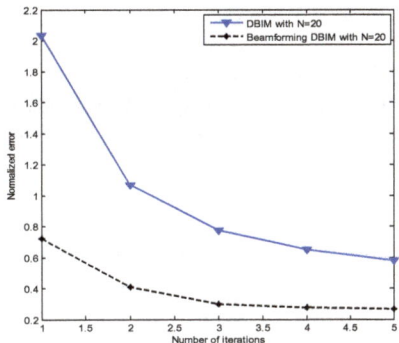

Fig. 5. Error comparison between DBIM and beamforming DBIM (N = 15)

Fig. 6. Error comparison between DBIM and beamforming DBIM (N = 20)

beamforming DBIM is quite good because the recoverable target function when using the proposed method is close to the initialization objective.

For a quantitative comparison of the imaging quality of DBIM and beamforming DBIM, normalization error was calculated and shown in Figs. 5, and 6 with a various number of pixels, corresponding to N = 15 and 20. All two figures show that the beamforming DBIM method's error is much better than the DBIM, especially right from the first iteration. It can be seen that the beamforming technique can overcome the noise quite well, and the estimation of object convergence has dramatically improved from the first loop.

4   Conclusions

This paper has applied the beamforming technique (using a beam of 3 transmitters simultaneously) combined with the DBIM method to restore the object. The proposed solution can overcome the intrinsic disadvantages of the DBIM method, which is susceptible to noise. The results show that normalization errors and noise have been significantly reduced. The proposed algorithm should be developed using experimental data before imaging in practice.

References

1. Osama, E.S., Haddadin, S., Sean, D.: Lucas: Solution to the inverse scattering problem using a modified distorted Born iterative algorithm. In: EEE Ultrasonics Symposium, pp. 1411–1414, IEEE (1995)
2. Wiskin, J., Borup, D.T., Johnson, S.A., Berggren, M., Abbott, T., Hanover, R.: Full-wave, non-linear, inverse scattering high resolution quantitative breast tissue tomography. Acoust. Imaging 28(3), 183–193 (2007)
3. Abubakar, A., Habashy, T.M., van den Berg, P.M., Gisolf, D.: The diagonalized contrast source approach: an inversion method beyond the Born approximation. Inverse Prob. 21(2), 685 (2005)
4. Grenleaf, F., Johnson, S.A., Bahn, R.C., Rajagopalan, B., Kenue, S.: Introduction to computed ultrasound tomography. In: IFIP TC-4 Working Conference, Computed Aided Tomography and Ultrasonics in Medicine, pp. 125–136, North-Holland (1970)
5. Conrath, B.C., Daft, M.W., O'Brien, W.: Applications of neural networks to ultrasound tomography. In: IEEE Ultrasonics Symposium, pp. 1007–1010 (1989)
6. Devaney, A.J.: Inversion formula for inverse scattering within the Born approximation. Opt. Lett. 7, 111–112 (1982)
7. Chew, W.C., Wang, Y.M.: Reconstruction of two-dimensional permittivity distribution using the distorted born iterative method. IEEE Trans. Med. Imag. 9, 218–225 (1990)
8. Lavarello, R., Oelze, M.: A study on the reconstruction of moderate contrast targets using the distorted Born iterative method. IEEE Trans. Ultrason. Ferroelectr. Freq. Control 55, 112–124 (2008)
9. Golub, G.H., Hansen, P.C., O'Leary, D.P.: Tikhonov regularization and total least squares. J. Acoust. Soc. Am. 21, 185–194 (1999)
10. Huy, T.Q., Dung, C.T.P., Ninh, B.T., Tan, T.D.: An Efficient Procedure of Multi-frequency Use for Image Reconstruction in Ultrasound Tomography. In: Intelligent Computing in Engineering. pp. 921–929, Springer, Singapore (2020)
11. Huy, T.Q., Duc-Tan, T.: The efficiency of applying compressed sampling and multi-resolution into ultrasound tomography. Ing. Solidaria 15(3), 1–16 (2019)
12. Huy, T.Q., Cuc, N.T., Long, T.T., Tan, T.D.: Tomographic density imaging using modified DF–DBIM approach. Biomed. Eng. Lett. 9(4), 449–465 (2019)
13. Tran, Q.H., Tran, D.T., Huynh, H.T., Ton-That, L., Nguyen, L.T.: Influence of dual-frequency combination on the quality improvement of ultrasound tomography. Simulation 92(3), 267–276 (2016)
14. Tran-Duc, T., Linh-Trung, N., Do, M.N.: Modified distorted Born iterative method for ultrasound tomography by random sampling. In: International Symposium on Communications and Information Technologies, pp. 1065–1068, IEEE (2012)

Improving Social Trend Detection Based on User Interaction and Combined with Keyphrase Extraction Using Text Features on Word Graph

XuanTruong Dinh[1,2][✉], TienDat Trinh[1,2], TuyenDo Ngoc[1,2], and VanHai Pham[2]

[1] Hanoi University of Science and Technology, and CMC Institute of Science and Technology, No. 11 Duy Tan Street, Hanoi, Vietnam
[2] Hanoi University of Science and Technology, No. 1 Dai Co Viet, Hanoi, Vietnam
`haipv@soict.hust.edu.vn`

Abstract. Recently, by the explosion of information technology, the valuable and available data exponentially increases in various social media platforms which allow us to exploit and attain convenient information and transform it into knowledge. This means that prominent topics are extracted on time in the social media community by leveraging the proper techniques and methods. Although there are various novel approaches in this area, they almost ignore the factors of the user interactions. Besides, since the enormous size of the textual dataset is distributed to any languages and the requirements for trending detection in a specific language, most of the proposed methods concentrating on English. In this paper, we proposed a graph-based method for the Vietnamese dataset in which graph nodes represent posts. More specifically, the approach combines the user interactions with a word graph representation which then is extracted to the topic trends by the RankClus Algorithm. By applying our proposal in several Facebook and Twitter datasets, we introduce dominantly dependable and coherent results.

Keywords: Social trending · Keyphrase extraction · Word graph

1 Introduction

Since the social media platforms have been continuing to grow exponentially, it's easy for users to express their opinions on emerging topics and engage in debates related to ongoing events of their lives. Trend is considered as user debates and interactions usually leading to massive content in favor of a specific topic [1]. It provides an overview of the topic and issue that are currently popular within an online community [2]. If real-time application developers and social media researchers could interpret early what emergent trending topics exchanged among people, those predicted topics' future popularity has an enormous impact on the academic and business value [3].

There are two main approaches for trend extraction: an embedding-based and a graph-based method. The first approach is based on representing the posts to corresponding vectors and detecting groups of similar posts. The other aims to find a graph representation and employ clustering algorithms to extract sub-graphs [1]. In social media, since user interactions ensure that any specific topics become a massive trend, so trending detection are required to include these interactions. However, they only focused on text-based directing to an underestimation of the credibility for detecting social trends [4].

In this research, posts are represented as a heterogeneous graph. Then, the clustering algorithm is used to extract posts into one cluster inspired by the RankClus algorithm to cluster graph-structured data [5]. Finally, using Keyphrase Extraction algorithm called Yet Another Keyword Extractor (YAKE) combining with social score to aggregate keyphrase score [6].

The proposed method is extensively evaluated on two corpora of Vietnamese and English posts. We have demonstrated the extracted trends using top posts and word in the Results section. Our contributions in this research are as follows:

- Representing posts as a heterogeneous graph.
- Utilizing RankClus algorithm to cluster nodes and ranking them.
- Employing YAKE method to extract and rank the important keyword phrases.
- Proposing keyphrase score including YAKE score and interaction scores and re-normalizing the YAKE score to suitable for trend extraction.
- Evaluating on Vietnamese and English posts on Facebook and Twitter

The paper remind is structured as follows: Sect. 2 considers related research on the social trend detection and Sect. 3 describes the proposed method. The results set out in Sect. 4, and Sect. 5. The paper closes in Sect. 6.

2   Related Work

In the field of keyword extraction, Swagata et al. proposed novel methods called Semantic Connectivity Aware Keyword Extraction (sCAKE) for constructing a graph presenting the related meaning and scoring connection between contextual words [8]. For the task of Keyword Extraction based on graph centrality measures, Didier presented the multi-centrality index. These different centrality measures are analyzed in the co-occurrence graph of the words in documents to find keywords [9].

To extract keywords from documents, a novel TextRank algorithm was introduced by Yang which is based on PMI weighting. By using PMI for proposing candidate words, the transition probabilities between them are calculated iteratively of the word graph model [10].

In order to overcome some limitations for Keyword Extraction like the Vector Space Model, Saroj K. B. proposed a novel unsupervised learning method call Keyword from Weighted Graph (KWG) which belongs to the class of graph-based approach. Since the graph uses Node Edge rank centrality measure, the ties among the nodes are broken due to the importance of nodes closeness [11]. The novel method by Leandro takes various advantages from an arbitrary number of classes with measurable strength and a better denoising process [12]. Shams presented a contribution to the LSTM model by applying with time-series feature and combining time series and applied in Gaming domain [13].

Keyword extraction is one of the simplest ways to leverage text mining for providing business value. They can use statistical features from the text itself and can be applied to large documents without re-training [14]. YAKE is a popular keyword extraction algorithm proposed in 2018 [15].

Another aspect of trend detection is keyword ranking which is not addressed before. More specifically, one needs to know the most important keyword in each trend to fully understand and analyze the extracted trends. In this research, we propose a method for trend detection as well as post and keyword using YAKE algorithm with social influence score.

3   Proposed Method

In our proposed method, we have two-step to generate keyphrase trending: post clustering and keyphrase extraction. In the post clustering step, we using RankClus for a word graph representation. After that, the graph will be clustered by a ranking based clustering algorithm to detect the relationship of nodes in this graph, and find the exact trend. Keyphrase trending step including keyphrase extraction and keyphrase ranking by social score.

3.1   Word Graph Representation

Firstly, the graph presentation is described in three sections:

- Word Graph construction: construct a graph from data.
- Graph clustering: clustering the constructed graph.
- Graph scoring: combined a keyword extraction with the user interaction weight score to extract the explicit trend.

Word graph construction with two types of nodes post and word. In our research, we focus on two types of edges between words and words. Weights between post and word are calculated by TF-IDF scores. Weights between two-word nodes are calculated by the PMI pairs of co-occurring words.

3.2   Rank Clustering

We choose RankClus algorithm to cluster nodes in the graph working on heterogeneous graphs and integrates ranking and clustering. There are two types of node are target node and attribute node based on bi-type information network given two types of object sets P - target node as Post and T - attribute node as Token and $W \in R_{(m+n)}$ is the adjacency matrix, graph $G = (P, T; W)$ is a bi-type information network.

Ranking function $f : G \longrightarrow (\boldsymbol{r}_P, \boldsymbol{r}_T)$ gives rank score for each object in post P and token T where $\boldsymbol{r}_P \geq 0, \sum_{p \in P} \overrightarrow{r}_P(p) = 1$ and $\boldsymbol{r}_T \geq 0, \sum_{t \in T} \overrightarrow{r}_T(t) = 1$. The cluster number K and the output is K clusters of P. Alogrithm in detail by Sun et al. to cluster nodes and ranking them of each cluster in [5].

3.3 Graph Scoring

To evaluate and eliminate clusters of trends, we use the matrix S [5] represents each word w appearing in a cluster c based on Entropy calculation. If a word has a high score in cluster i and low scores in other clusters, it is important for i, and words have high score in all clusters are not important.

3.4 Keyphrase Extraction

YAKE is a keyword extraction algorithms using five features to quantify word are a = casing, b = word position, c = word frequency, d = word relatedness to context, e = word different sentence. The weights are calculated very specifically using the formula in [6, 15]. These 5 features are combined into a score S(w):

$$score(w) = \frac{d * b}{a + (c/d) + (e/d)} \tag{1}$$

For each of our candidate keywords, a score is calculated using the following:

$$S(kw) = \frac{product(scores\ of\ words\ in\ keyword)}{1 + (sum\ of\ scores\ of\ words) * count(keyword)} \tag{2}$$

A keyword is more important if it has a lower score. $S(kw)$ is an inverse function, whose value from negative infinity to infinity so to scale the value back from 0 to 1, meaning that the keyphrase more valuable keyword, the score closer to 1 and less values are closer to 0 according to the equation below:

$$kScore(kw) = \frac{1}{\sqrt{1 + exp(S(kw))}} \tag{3}$$

3.5 Keyphrase Trending Detection

The social score is presented as the Klout Score, an influence scoring system across multiple social networks, the more you connect the higher the score [16]. Research by Octolabs - Research organization in Social Media and Trending Analysis, based on users' efforts to create social activities, they give the weights of them on social media [17]. Below for Facebook:

$$Likes:Comments:Shares = 1:20:30 \tag{4}$$

Facebook *Likes* do not require much effort from Facebook visitors, so they show the lowest. *Comments* and *Shares* are already a sign of engaging content.
The three user actions we account for Engagement across Twitter accounts:

$$Likes:Replies:Retweets = 1:20:40 \tag{5}$$

Social score on Youtube and Linkedin (Fig. 1):

$$Likes:Comments:Shares/Favs = 1:20:30 \tag{6}$$

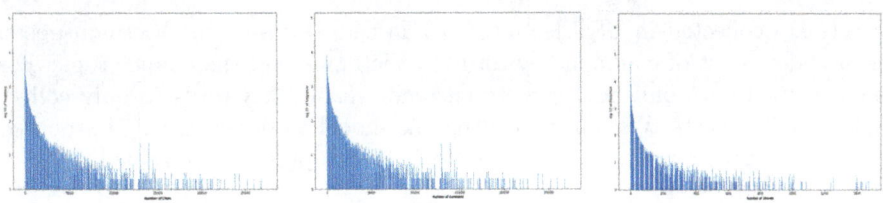

Fig. 1. The number of likes, comments and shares on Facebook posts in finance topics

In Facebook Finance Data [16], financial topics are collected with 25906 posts. By statistics, the average of Likes is 798, from 0 to 915300; Comment value from 0 to 259311; Share from 0 to 30073. We define the social score has a max value is 100 so the coefficient for the Likes is $\alpha_{FL} = 1.43 \times 10^{-5}$ and Comments is $\alpha_{FC} = 28.6 \times 10^{-5}$ and Shares is $\alpha_{FS} = 42.9 \times 10^{-5}$. The US Election 2020 Tweets Dataset with 1.727.000 tweets, the coefficient for the Likes is $\alpha_{TL} = 6.39 \times 10^{-5}$ and Replies is $\alpha_{TRp} = 127.8 \times 10^{-5}$ and Retweets is $\alpha_{TRw} = 255.6 \times 10^{-5}$.

$$SocialScore(i) = \begin{cases} \alpha_{FL}.\#Likes + \alpha_{FC}.\#Comments + \alpha_{FS}.\#Shares & i \text{ is a post} \\ \alpha_{TL}.\#Likes + \alpha_{TRp}.\#Replies + \alpha_{TRw}.\#Retweets & i \text{ is a tweet} \end{cases} \quad (7)$$

From there we have the trending score of a keyword phrase Kw in cluster c belong to post i calculated by the following formula:

$$TrendScore(Kw) = \sum_{i=1}^{\#c} kScore(Kw) \times SocialScore(i) \quad (8)$$

A list of keyword phrases along with their trending scores, it is more trending if it has a higher score and take the top N keyword phrase as the output of each cluster post.

4 Experiments

In order to evaluate the proposed approach, we considered daily posts as trends usually change each day. We have collected posts in English as well as Vietnam.

Table 1. Data statistics: tweet in English (En) and post in Vietnam (Vie).

Language	Date	#Post	#Trend
En	2020-11-08	129000	2
	2020-11-02	52000	3
	2020-10-24	31000	3
Vie	2020-10-14	1727	2
	2020-10-28	2491	2

Tweets are collected in US Election 2020 Tweets Dataset. For Vietnam posts, we provided a list of common keywords to Vietnamese of the finance topic. For English tweets, we only used crypto-currency related keywords to only collect topic-specific tweets. We aim at testing the model to detect trends in focused datasets in this scenario. The dataset statistics are provided in Table 1.

5   Results and Discussion

Table 2. Clusters' scores calculated for two days

	C1	C2	C3	C4	C5	C6	C7	C8	C9
2020-11-08	3.17	0.0	0.0	0.0	0.0	2.01	–	–	–
2020-10-28	0.289	4.321	0.311	0.897	0.221	3.302	0.692	0.135	0.428

Table 2 presents cluster scores on two datasets US Election Dataset and Facebook Finance Dataset. As a result, for the first one, the clustering algorithm emitted 6 clusters and our score assigned 0 scores to four clusters. Hence only two trends are extracted: C1, and C6. For each trend, top post, keyphrases and trend score are presented. Extracted posts and sample posts from each trend for the first date in Table 2 are illustrated in Table 3.

Table 3. Extracted trend samples on English tweets

2020-11-08		
C1	Tweets/Posts	If you're a #Trump supporter, you should be thankful and optimistic
		I wanted #Trump to win the election. If he loses fairly, let's accept the
	Keyphrases	**#trump, #election2020, #trumpout, #electionresults2020, donald trump's vote count**
	Score	**708.42, 418.51,257.41, 177.61, 85.25**
C6	Tweets/Posts	Wonderful speeches tonight by President and Vice-President elect
		I watched that speech and I'm crying, not bc I think #JoeBiden
	Keyphrases	**#joebiden, #bidenharris2020, # election2020, joe biden as president, #uspresident**
	Score	**355.86, 210.25, 154.18, 84.09, 50.46**

The proposed algorithm is capable of providing a ranking for posts of a trend and keyword phrases of it. Also in Table 4, two extracted trends for the second date in the Vietnam dataset are demonstrated. Table 5 provides a comparison between our proposed algorithm and baseline models such as K-Means [20], LDA [21] based on accuracy in finding the trends.

Table 4. Extracted trend samples on Vietnam posts.

2020-10-28

C2	Tweets /Posts	ƯU ĐÃI ĐẶC QUYỀN KHI MUA SIÊU PHẨM GALAXY NOTE 10
		Mới sắm cái túi xách 6 chiệu .Cầm hông nổi. ->Nhanh chân đến FPT
	Keyphrases	galaxy note 10, đặt cọc galaxy note 10, trả góp lãi suất 0%
	Score	118.24, 81.45,57.32, 27.64, 12.51
C6	Tweets /Posts	Vừa giảm giá gần 300 triệu, Vinfast lại chơi lớn khi áp dụng mua mới
		Dành cho các bác quan tâm nè:THAY ĐỔI ĐỂ ĐÓN NHỮNG ĐIỀU
	Keyphrases	vinfast, doiculaymoi, đổi cũ lấy mới, trả góp, voucher vinfast
	Score	135.62, 91.22, 58.42, 14.94, 10.26

Table 5. Arrcuracy of our model

Lang	Date	K-Means	LDA	Graph-based
En	2020-11-08	50%	100%	**100%**
	2020-11-02	66%	100%	**100%**
	2020-10-24	66%	100%	**100%**
Vie	2020-10-14	50%	50%	**100%**
	2020-10-28	100%	0%	**100%**

We have evaluated performance of the graph-base and compare it to the base-line methods on the cluster problem using accuracy the number of correct clusters to total number of input samples. Average validation accuracy values was calculated in Eq. 9.

$$Accuracy = \frac{Number\ of\ Correct\ clusters}{Total\ number\ of\ clusters\ made} \qquad (9)$$

As the results suggest, our proposed model outperforms the baselines in most cases.

6   Conclusion

By utilizing a huge of user-posted data on social media, trending detection could provide an overview of the topic in online community. This paper introduced a novel approach for social trend extraction and ranking posts and keyphrases in each trend. We employed the method to extract and ranking the most important keyphrases and aggregated keyphrase score to find the trending keyword phrases. Despite of the promising results of our proposed method, there are various open questions to invest in down the road. By leveraging many types of reactions in order to add more weights to improve the final result and lead to new approaches.

Acknowledgment. This research is funded by CMC Institute of Science and Technology (CIST), CMC Corporation, Vietnam.

References

1. Majdabadi, Z., Sabeti, B., Golazizian, P., Asli, S.A.A., Momenzadeh, O.: A graph-based approach for tweet and hashtag ranking, utilizing no-hashtag tweets. In: 12th Language Resources and Evaluation Conference (2020)
2. Benhardus, J., Kalita, J.: Streaming trend detection in Twitter. Int. J. Web Based Communities **9**(1), 122–139 (2013)
3. Madani, A., Boussaid, O., Zegour, D.E.: Real-time trending topics detection and description from Twitter content. Soc. Netw. Anal. Min. **5**(1) (2015)
4. Dinh, X.T., Van Pham, H.: A proposal of deep learning model for classifying user interests on social networks. In: Proceedings of the 4th International Conference on Machine Learning and Soft Computing, pp. 10–14 (2020)
5. Sun, Y., Han, J., Zhao, P., Yin, Z., Cheng, H., Wu, T.: RankClus: integrating clustering with ranking for heterogeneous information network analysis. In: Proceedings of the 12th International Conference on Extending Database Technology: Advances in Database Technology, pp. 565–576 (2009)
6. Campos, R., Mangaravite, V., Pasquali, A., Jorge, A.M., Nunes, C., Jatowt, A.: YAKE! Collection-independent automatic keyword extractor. In: European Conference on Information Retrieval, pp. 806–810. Springer, Cham (2018)
7. Kim, S., Jeon, S., Kim, J., Park, Y.-H., Yu, H.: Finding core topics: topic extraction with clustering on tweet. In: 2012 Second International Conference on Cloud and Green Computing, pp. 777–782. IEEE (2012)
8. Ameri, M.R., Stauffer, M., Riesen, K., Bui, T.D., Fischer, A.: Graph-based keyword spotting in historical manuscripts using Hausdorff edit distance. Pattern Recogn. Lett. **121**, 61–67 (2018)
9. Vega, D.A., Gomes, P.S., Milios, E., Berton, L.: A multi-centrality index for graph-based keyword extraction. In: Information Processing and Management (2019)
10. Tao, Y., Cui, Z., Jiazhe, Z.: Research on keyword extraction algorithm using PMI and TextRank. In: 2019 IEEE 2nd International Conference on Information and Computer Technologies (ICICT), pp. 5–9 (2019)
11. Biswas, S.K.: Keyword extraction from tweets using weighted graph. In: Cognitive Informatics and Soft Computing, pp. 475–483. Springer, Singapore (2019)
12. Anghinoni, L., Zhao, L., Ji, D., Pan, H.: Time series trend detection and forecasting using complex network topology analysis. Neural Netw. **117**, 295–306 (2019)
13. Shams, M.B., Hossain, M.: Trend analysis with Twitter hashtags (2012)
14. Siddiqi, S., Sharan, A.: Keyword and keyphrase extraction techniques: a literature review. Int. J. Comput. Appl. **109**(2) (2015)
15. Rose, S., Engel, D., Cramer, N., Cowley, W.: Automatic keyword extraction from individual documents. In: Text Mining: Applications and Theory (2010)
16. Rao, A., Spasojevic, N., Li, Z., Dsouza, T.: Klout score: measuring influence across multiple social networks. In: International Conference on Big Data (2015)
17. Octolabs, measuring social score: the what, the how and the why. https://www.octoboard.com/blog/measuring-social-score
18. Facebook Finance Data. https://www.kaggle.com/tdtrinh1198/finance-data
19. US Election 2020 Tweets Dataset. https://www.kaggle.com/manchunhui/us-election-2020-tweets
20. Hartigan, J.A., Wong, M.A.: Algorithm AS 136: a k-means clustering algorithm. J. Roy. Stat. Soc. Ser. C **28**(1), 100–108 (1979)
21. Blei, D.M., Ng, A.Y., Jordan, M.I.: Latent dirichlet allocation. J. Mach. Learn. Res. **3**, 993–1022 (2003)

Synthesis of Remote Control Law When Taking into Dynamics and Nonlinear of the Missile Stage

Nguyen Ngoc Tuan[1], Nguyen Duc Thi[2], Nguyen Van Bang[3(✉)], and Tran Van Tuyen[1]

[1] Military Technical Academy of Vietnam, Hanoi, Vietnam
[2] General Department for Defence Industry of the Vietnam People's Army, Hanoi, Vietnam
[3] Air Defence - Air Force Academy of Viet Nam, Hanoi, Vietnam

Abstract. The paper presents the results of research and propose synthesis methods of the remote control law when taking into the dynamics and nonlinear of the missile stage. Simple control law, capable of realization in practice. The simulation results show the advantages of control law in cases of target movement, small guidance error, ensuring good response to changing dynamic parameters of the missile during flight. Contribute to improving the quality of the missile control system, at the same time serving as a basis in the upgrade, improvement and new design of the flight device control system.

Keywords: Missile stage · Remote control law · Dynamic · Nonlinear

1 Introduction

When studying remote missile control systems, three major issues need to be addressed [10–14]:

- Missile stabilization: Improved dynamic properties for missile.
- Guidance method and missile control laws: Guidance methods are intended to determine the desired trajectory of the missile and control laws ensure that the missile is flying in that desired trajectory.
- Stabilize the remote control loop: Improved dynamics for control loop when taking into account the dynamics of all stages in the control loop.

Currently, remote missile control systems are widely used in practice [12, 14]. However, due to various objective factors and subjective, the document was published just mention the main contents are as follows [10, 11, 13]: Remote control loop structure, mathematical modeling of stages in remote control loop, meaning of dynamic error compensation…

In fact, the missile is essentially a nonlinear system, not a linear [1, 4, 10–14]. Which has given the way to synthesize the law of remote control, but only gives the general way in which the formation of the control law is done without taking into account the dynamics of the missile itself, the dynamics of systems in the control loop

D.-T. Tran et al. (Eds.): ICISN 2021, LNNS 243, pp. 171–180, 2021.
https://doi.org/10.1007/978-981-16-2094-2_22

and the relation to the missile's motion geometry is considered to be approximate [10] or take into account missile stage dynamics, where the missile is still assumed to be linear [1, 3–9]. That does not fully reflect the nature of a control system. Currently, there is not any scientific article published on the synthesis of missile control laws, which takes into account the dynamics and nonlinear properties of the missile stage [1].

Therefore, when synthesizing the law of remote missile control, it is necessary to fully consider the dynamics and nonlinear nature of the missile itself. Therefore, in the scope of this article, the authors would like to present how to synthesize the law of remote missile control, with the assumption that the missile has an on-board automatic stabilization system, the command setting system using the command according to the 3-point guidance method.

2   The Dynamic Equation of the Missile

Missile system with symmetrical wings around the axis (cruciform-type missile systems), the missile is stabilized throwing angle so that the control channels (nod and direction) are not related to each other, so just consider in a control plane is enough. The dynamic equation of the missile in the nodal plane is given by [3, 10, 13] (Fig. 1):

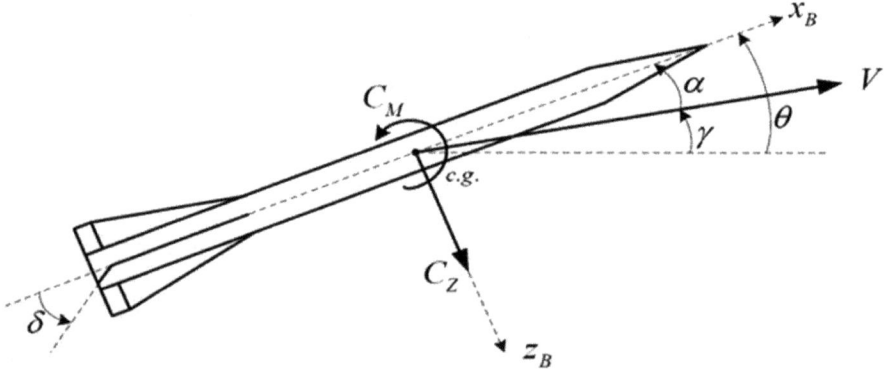

Fig. 1. Missile have symmetrical wings around the axis of movement in space

$$
\begin{cases}
\dfrac{dV}{dt} = \dfrac{P\cos\alpha}{m} - \dfrac{X}{m} - g\sin\theta \\[2mm]
\dfrac{d\theta}{dt} = \dfrac{P\sin\alpha + Y^{\alpha} + Y^{\delta}}{mV} - g\cos\theta \\[2mm]
\dfrac{d\vartheta}{dt} = \omega_z \\[2mm]
\dfrac{d\omega_z}{dt} = \dfrac{M_z}{I_z} \\[2mm]
\alpha = \vartheta - \theta
\end{cases}
\tag{1}
$$

Where,

m - Mass of the missile; V - Velocity of the missile; P - Axial motor thrust; G - Gravity of the missile; X - Main drag; θ - Angle of inclination of the trajectory; α - Angle-of-attack; g - Gravity acceleration; I_z - The torque of inertia of the missile on the axis of the linked coordinate system; ω_z - The projection of the angular velocity vector on the axis of the link coordinate system; M_z - Torque projection of external torque on the axis of the link coordinate system; ϑ - Missile nod angle.

3 Synthesis of Control Law

To transform Eq. (1), we have:

$$
\begin{cases}
\dot{\alpha} = -\dfrac{P\sin\alpha + qSC_{Z_0}(M,\alpha)}{mV} + g\cos\theta - \dfrac{qSC_{Z_\delta}(M,\alpha)\delta}{mV} + \omega_z \\[3mm]
\dot{\omega}_z = \dfrac{qSL}{I_z}\left[C_{M_0}(M,\alpha) + C_{M_q}(M,\alpha)\dfrac{qL}{2V} + C_{M_\delta}(M,\alpha)\delta\right] \\[3mm]
a_y = \dfrac{P\sin\alpha + qSC_{Z_0}(M,\alpha)}{m} - Vg\cos\theta + \dfrac{qSC_{Z_\delta}(M,\alpha)\delta}{m}
\end{cases}
\tag{2}
$$

With, L - Length of the missile; δ - Angle rotation of wing, is the control input; q - Aerodynamic pressure; S - Equivalent surface area.

The aerodynamic coefficients described by C_{Z_0}, C_{Z_δ}, C_{M_0}, C_{M_q} and C_{M_δ} are dependent on Mach number and angle-of-attack [3, 4, 7, 12]. These parameters, and their derivatives have a small value, exist as time continuous functions.

Because the lift on the drive wing is much less than that on the lift wing ($Y^\delta \ll Y^\alpha$ or $|C_{Y_\delta}(M,\alpha)\delta| \ll |C_{M_\delta}(M,\alpha)\delta|$) and $g\cos\theta$ can be compensated in the command setting system [3–5, 9–11, 14]. Equation (2) then, has the form:

$$
\begin{cases}
\dot{\alpha} = -\dfrac{P\sin\alpha + qSC_{Z_0}(M,\alpha)}{mV} + \omega_z \\[3mm]
\dot{\omega}_z = \dfrac{qSL}{I_z}\left[C_{M_0}(M,\alpha) + C_{M_q}(M,\alpha)\dfrac{qL}{2V} + C_{M_\delta}(M,\alpha)\delta\right] \\[3mm]
a_y = \dfrac{P\sin\alpha + qSC_{Z_0}(M,\alpha)}{m}
\end{cases}
\tag{3}
$$

Set a new variables: $x_1 = \alpha; x_2 = \omega_z; u = \delta; y = a_y$

Where, x_1 and x_2 represent the state variables;

u - The control input; y - the output of system.

The control system is then represented in the following state space:

$$
\begin{cases}
\dot{x}_1 = f_1(x_1) + x_2 \\
\dot{x}_2 = f_2(x_1, x_2) + g_2(x_1)u \\
y = h(x_1)
\end{cases}
\tag{4}
$$

$$
\text{Where,} \quad
\begin{cases}
f_1(x_1) = -\dfrac{P sin\alpha + qSC_{Z_0}(M,\alpha)}{mV} \\[2mm]
f_2(x_1,x_2) = \dfrac{qSL}{I_z}\left[C_{M_0}(M,\alpha) + C_{M_q}(M,\alpha)\dfrac{qL}{2V}\right] \\[2mm]
g_2(x_1) = \dfrac{qSL}{I_z}C_{M_\delta}(M,\alpha) \\[2mm]
h(x_1) = \dfrac{P sin\alpha + qSC_{Z_0}(M,\alpha)}{m}
\end{cases}
\tag{5}
$$

From Eq. (4) and (5), We have:

$$
\dot{y} = \frac{\partial h(x_1)}{\partial x_1}\dot{x}_1 = \frac{\partial h(x_1)}{\partial x_1}[f_1(x_1) + x_2]
\tag{6}
$$

$$
\ddot{y} = \left(\frac{\partial h(x_1)}{\partial x_1}\dot{x}_1\right)' = \left(\frac{\partial h(x_1)}{\partial x_1}\right)'\dot{x}_1 + \frac{\partial h(x_1)}{\partial x_1}\ddot{x}_1
\tag{7}
$$

$$
\ddot{y} = \frac{\partial^2 h(x_1)}{\partial x_1^2}[f_1(x_1) + x_2]^2 + \frac{\partial h(x_1)}{\partial x_1}\left[\frac{\partial f_1(x_1)}{\partial x_1}[f_1(x_1) + x_2] + f_2(x_1,x_2) + g_2(x_1)u\right]
\tag{8}
$$

Equation (8) can be rewritten using shorthand notation as follows:

$$
\ddot{y} = f_3 + g_3 u
\tag{9}
$$

Set; $u = \frac{v - f_3}{g_3}$ (purpose to find $\ddot{y}$).

When that, $\ddot{y} = f_3 + g_3 \frac{v - f_3}{g_3} = v$

The notations of K_1 and K_2 denote the stage amplification of the inner loop, the outer loop. It is selected in the simulation.

Calculate the transfer function with the diagram in Fig. 2, we get:

$$
K_1(p) = \frac{\frac{1}{p}}{1 + K_1\frac{1}{p}} = \frac{1}{K_1}\frac{1}{T_1 p + 1} \text{ with } T_1 = \frac{1}{K_1}
$$

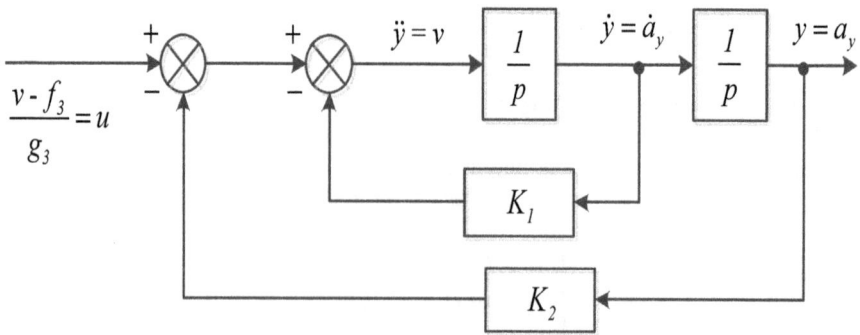

Fig. 2. Structure diagram of missile stage model

$$K(p) = \frac{\frac{1}{K_1}\frac{1}{T_1p+1}\frac{1}{p}}{1+K_2\frac{1}{K_1}\frac{1}{T_1p+1}\frac{1}{p}} = \frac{1}{K_1(T_1p+1)p+K_2}$$

$$K(p) = \frac{K}{T^2p^2 + 2\xi Tp + 1}$$

Where, $K = \frac{1}{K_2}$; $T = \sqrt{\frac{K_1T_1}{K_2}} = \sqrt{\frac{1}{K_2}}$; $\frac{K_1}{K_2} = 2\xi\sqrt{\frac{1}{K_2}}$; $\xi = \frac{K_1}{2\sqrt{K_2}}$

$$v = K_1\dot{y} + K_2y = K_1\dot{a}_z + K_2a_z \tag{10}$$

Next, we need to find f_3, g_3. From formula 3 of Eq. (5), We have:

$$\frac{\partial h(x_1)}{\partial x_1} = \frac{P}{m}cos\alpha + \frac{qSC_Y^\alpha}{m} \tag{11}$$

$$\frac{\partial^2 h(x_1)}{\partial x_1^2} = -\frac{P}{m}sin\alpha + \frac{qS\partial C_Y^\alpha/\partial\alpha}{m} \tag{12}$$

From formula 1 of Eq. (5), We have:

$$\frac{\partial f_1(x_1)}{\partial x_1} = -\frac{Pcos\alpha + qSC_Y^\alpha}{mV} \tag{13}$$

From Eq. (11), (12) and Eq. (13), (5), combined with Eq. (1), then:

$$f_3(x_1,x_2) = \frac{\partial^2 h(x_1)}{\partial x_1^2}[f_1(x_1)+x_2]^2 + \frac{\partial h(x_1)}{\partial x_1}\left[\frac{\partial f_1(x_1)}{\partial x_1}[f_1(x_1)+x_2]+f_2(x_1,x_2)\right] \tag{14}$$

$$f_3(x_1,x_2) = \left(-\frac{P}{m}sin\alpha + \frac{qS\partial C_Y^\alpha/\partial\alpha}{m}\right)\left(-\dot{\theta}+\omega_z\right)^2$$
$$+ \left(\frac{P}{m}cos\alpha + \frac{qSC_Y^\alpha}{m}\right)\left[\left(-\frac{Pcos\alpha + qSC_Y^\alpha}{mV}\right)\left(-\dot{\theta}+\omega_z\right) + \frac{qSL}{I_z}\left(C_{M_0}(M,\alpha) + C_{M_q}(M,\alpha)\frac{qL}{2V}\right)\right] \tag{15}$$

From Eq. (8), We have:

$$g_3 = \frac{\partial h(x_1)}{\partial x_1}g_2(x_1) \tag{16}$$

$$g_3 = \left[\frac{P}{m}cos\alpha + \frac{qSC_Y^\alpha}{m}\right]\frac{qSL}{I_z}C_{M_\delta}(M,\alpha) \tag{17}$$

Substituting Eq. (10), (15), (17) into Eq. (13), get the law was a format:

$$u = \delta = \frac{K_1 \dot{a}_z + K_2 a_z}{\left(\frac{P}{m}cos\alpha + \frac{qSC_Y^\alpha}{m}\right)\frac{qSL}{I_z}C_{M_\delta}(M,\alpha)} - \frac{\left[-\frac{P}{m}sin\alpha + \frac{qS\partial C_Y^\alpha/\partial\alpha}{m}\right]\left[-\dot{\theta} + \omega_z\right]^2}{\left(\frac{P}{m}cos\alpha + \frac{qSC_Y^\alpha}{m}\right)\frac{qSL}{I_z}C_{M_\delta}(M,\alpha)}$$

$$- \frac{\left(\frac{P}{m}cos\alpha + \frac{qSC_Y^\alpha}{m}\right)\left[\left(-\frac{Pcos\alpha + qSC_Y^\alpha}{mV}\right)\left(-\dot{\theta} + \omega_z\right) + \frac{qSL}{I_z}\left(C_{M_0}(M,\alpha) + C_{M_q}(M,\alpha)\frac{qL}{2V}\right)\right]}{\left(\frac{P}{m}cos\alpha + \frac{qSC_Y^\alpha}{m}\right)\frac{qSL}{I_z}C_{M_\delta}(M,\alpha)}$$

$$(18)$$

4 Simulation and Evaluate the Results

Perform simulation control loops with control law (18) have the structure diagram is shown in Fig. 3, with the following assumptions.

Where, the target's motion pattern is determined by:

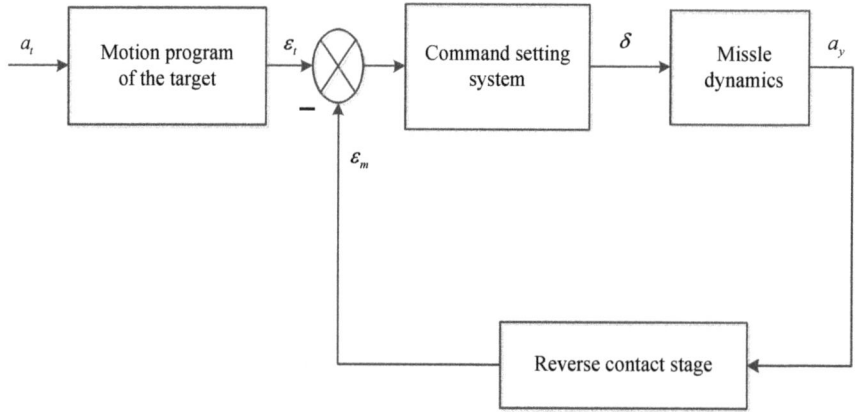

Fig. 3. Simulation organization diagram

$$\dot{r}_t = V_t cos(\theta_t - \varepsilon_t); \ \dot{\varepsilon}_t = \frac{V_t}{r_t}sin(\theta_t - \varepsilon_t); \ \dot{\theta}_t = \frac{a_t}{V_{mt}}; \ \dot{x}_t = V_t cos\theta_t; \ \dot{y}_t = V_t sin\theta_t$$

Reverse contact stage using the model;

$$r_m \ddot{\varepsilon}_m + 2\dot{r}_m \dot{\varepsilon}_m = \dot{V}sin(\theta - \varepsilon_m) + V\dot{\theta}cos(\theta - \varepsilon_m)$$

Case 1: Velocity of the target $V_t = 500$ (m/s), start 50 m/s^2 maneuvering from 10th second to 15th. Fly at high altitude $H_t = 6$ (km) (Figs. 4, 5, 6, 7, 8, 9).

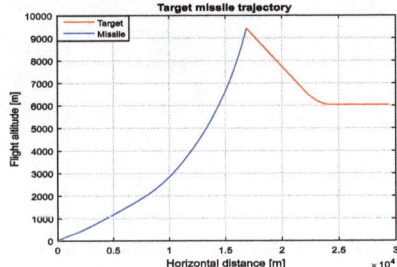

Fig. 4. Target missile trajectory

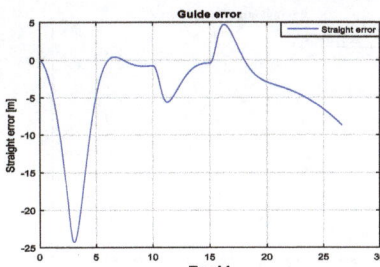

Fig. 5. Straight error

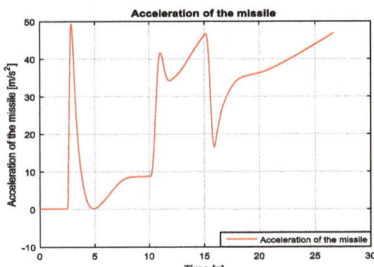

Fig. 6. The missile's acceleration

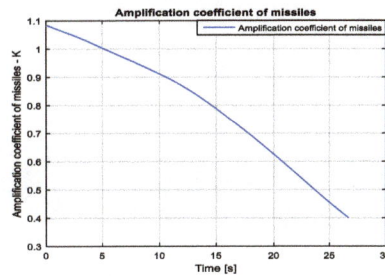

Fig. 7. The change of the missile's amplification coefficient

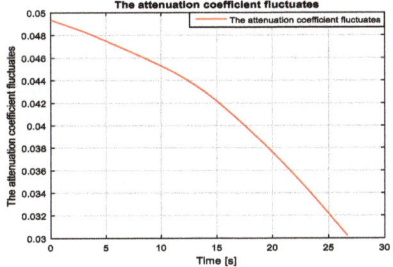

Fig. 8. The change of the missile's oscillation attenuation coefficient

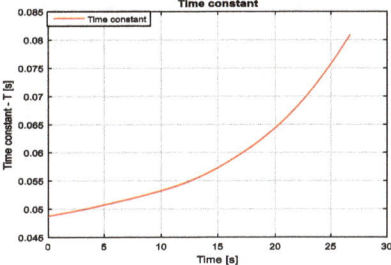

Fig. 9. The change of the missile's time constant

Comment: During the flight of the missile, the dynamic parameters of the missile itself are not known exactly and change, due to the influence of flight altitude, velocity, angle-of-attack limitation… The survey shows the accuracy of the built algorithm. Small guidance error. The control law responds well to changes in missile dynamic parameters. It is possible to apply a built algorithm in practice to control the missile.

Case 2: Velocity of the target $V_t = 500$ (m/s), start 30 m/s^2 maneuvering from 12th second to 18th. Fly at high altitude $H_t = 8$ (km) (Figs. 10, 11, 12, 13, 14, 15).

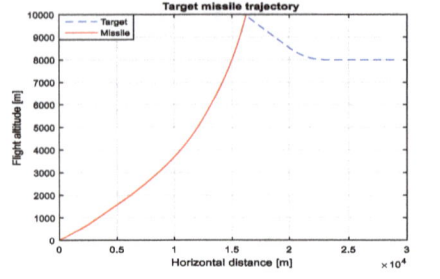

Fig. 10. Target missile trajectory

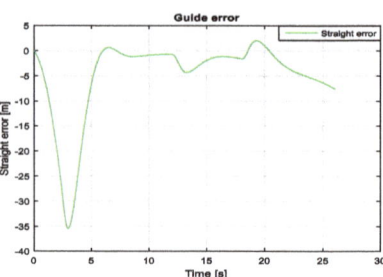

Fig. 11. Straight error

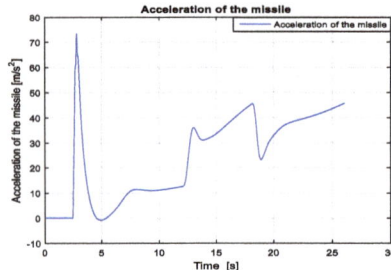

Fig. 12. The missile's acceleration

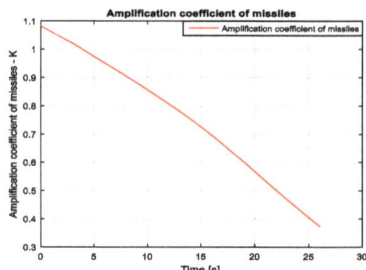

Fig. 13. The change of the missile's amplification coefficient

Comment: In all cases of the target's maneuverability, the integrated control law is always well met. Ensure the error at the meeting point is small.

Comment: From Fig. 16, the simulation results show that the trajectory curvature and guidance error when applying the control law (18) are greater than the traditional

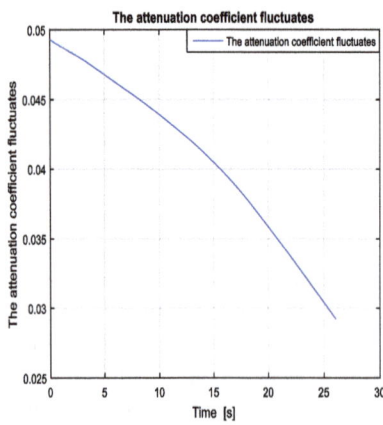

Fig. 14. The change of the missile's oscillation attenuation coefficient

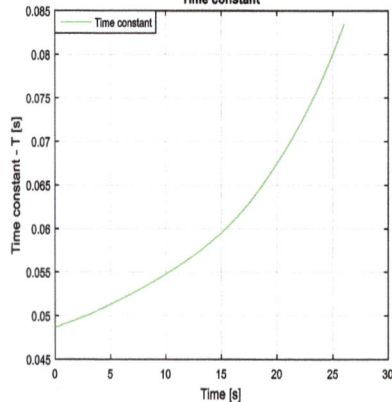

Fig. 15. The change of the missile's time constant

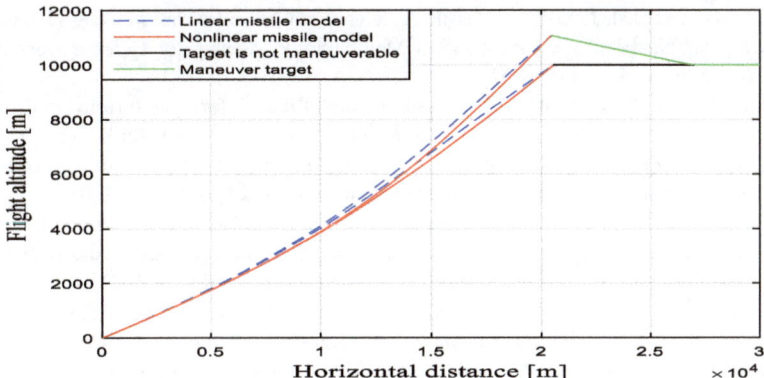

Fig. 16. Target missile trajectory

control laws [10–14]. However, it is still acceptable. Because of the traditional control laws (missile have a linear form) the $\alpha, \theta, \vartheta$ angles are assumed to be very small and can be ignored.

5 Conclusion

The paper proposed a method to synthesize new control law for missile, remote control according to the 3-point guidance method when taking into account the dynamics and nonlinear factors of the missile stage. Simulate the proposed algorithm to verify and evaluate. Also compare the control system quality in cases where the missile model is linear or nonlinear, confirming the soundness of the research results.

Proposing a method to synthesize the remote control loop into account of missile stage dynamics and nonlinear in the remote control loop. The result is a generalized command formula on the basis of quality criteria. From the general commanding expression, build the actual command expression and set out the requirements with the target coordinate determination system and flight device.

The proposed algorithms are applicable to synthesize remote control loop and missile stability in current technological and technical conditions, as well as in upgrading, improving and new design of air defense missile complexes.

References

1. Van Bang, N.: Research and synthesize remote control loop of flying device using modern control technology. Technical doctoral thesis, Military Technical Academy, Viet Nam (2020)
2. Das, A., Das, R., Mukhopadhyay, S., Patra, A.: Robust nonlinear design of three axes missile autopilot via feedback linearization. In: Proceedings of the 27th IEEE Conference on Decision and Control, pp. 730–743 (2014)

3. Lee, C.-H., Lee, J.-I, Jun, B.-E.: Missile Acceleration Controller Design using PI and Time - Delay Adaptive Feedback Linearization Methodology. Agency for Defense Development (ADD), Daejeon, Korea (2001)

4. Lee, C.H., Kim, T.H., Tahk, M.J.: Missile autopilot design for agile turn using time delay control with nonlinear observer. Int. J. Aeronaut. Space Sci. Technol. **12**(3), 266–273 (2011)

5. Devaud, E., Harcaut, J.P., Siguerdidjane, H.: Three-axes missile autopilot design: from linear to nonlinear control strategies. J. Guidance Control Dyn. **24**(1), 64–71 (2001)

6. Bruyere, L., Tsourdos, A., White, B.A.: Robust augmented lateral acceleration flight control design for a quasi-linear parameter-varying missile. In: Proceedings of the Institution of Mechanical Engineers, Part G: Journal of Aerospace Engineering, vol. 219 (2005)

7. Menon, P.K., Yousefpor, M.: Design of Nonlinear Autopilot for High Angle of Attack Missiles. In: AIAA Guidance, Navigation and Control Conference, pp. 3896–3913 (1996)

8. Menon, P.K., Iragavarapu, V.R., Ohlmeyer, E.J.: Nonlinear Missile Autopilot Using Time Scale Separation. AIAA Paper 96–3765 (1997)

9. Kim, S.H., Tahk, M.J.: Missile acceleration controller design using proportional-integral and non-linear dynamic control design method. In: Proceedings of Institution of Mechanical Engineers, Part G: Journal of Aerospace Engineering, vol. 225 (2011)

10. Анатолий Борисович Скорик. "Системы управления современными зенитными ракетами. Математическое моделирование контура наведения", Сучасци информащйя технологи сфере безпеки та оборони (том 1) (2013)

11. Федосов, Е.А., Бобронников, В.Т., Красилыщиков, М.Н., Кухтенко, В.И., Лебедев, А. А., Малыщев, В.В., Орлов, Е.В., Пучков, Б.В., Силаев, А.И., Стефанов, В.А.: Динамическое проектирование систем управления автоматических маневренных летательных аппаратов, Машиностроение, Москва, стр. 49–269 (1997)

12. Голубев И.С. и Светлов В.Г.: Проектирование зенитных управляемых ракет, Издательство МАИ, Москва (2001)

13. Толпегин, О.А, Новиков, В.Г.: Математические моделисистем наведения летательных аппаратов, Коломна, Издательство КИ (ф) МГОУ (2011)

14. Мизрохи, В.Я.: Проектирование управления зенитных ракет, Учебно-научное издание - М.: Изд., Экслибрис-Пресс (2010)

Determining for Launcher Distance and Intersection Point of Missile and Target Using Simulation Method

Thi Nguyen Duc[1], Tuan Nguyen Ngoc[2], Giang Le Ngoc[3(✉)], and Tinh Tran Xuan[3]

[1] General Department for Defence Industry of the Vietnam People's Army, Ha Noi, Vietnam
[2] Military Technical Academy, Ha Noi, Vietnam
[3] AD-AF Academy of Viet Nam, Ha Noi, Vietnam

Abstract. The launcher distance and intersection point of missile and target is the key indicator for measuring surface-to-air missile weapon systems. The paper has built a simulation program for the trajectory of a missile to reach its target, thereby determining launcher distance and intersection point, serve in the process of learning, research and teaching. At the same time, as a basis to do a good job of preparation, to obtain reasonable performance in all expected situations. The simulation results have shown the effectiveness of this method.

Keywords: Missile · Launch area · Trajectory · Simulation

1 Introduction

Missile attack zone research is an indispensable technical basis for missile design and performance research. Attack zone refers to an area around the target under certain operational conditions. When launches a missile in this specific area, it can hit the target with a certain probability [1–3].

Anti-aircraft missiles are one of the modern weapons, it is capable of destroying aerial targets such as aircraft, unmanned aerial vehicles, cruise missiles, ground and water targets if necessary.In combat with air defense missiles, the trajectory of the missile must be accurately controlled to approach the target trajectory, when the distance to the target is close enough to be able to damage then detonate the missile to destroy the target completely. To improve the probability of destroying the target, need to determine the exact for launcher distance and intersection point of missile and target [4–6].

In the course of the development of precision-guided missiles, a large number of experiments were used to assess the performance of the missile meets requirements whether or not [1]. Cost of range test experiment is high and cycle is long; therefore, method of simulation is adopted, thereby determining launcher distance and intersection point, as a basis to do a good job of preparation, to obtain reasonable performance in all expected situations [7–9].

2   Kinematics Equation System of Missiles and Targets

Assume that a missile is launched from surface of the earth while an aerial target is
located at a horizontal distance of d km and at a height of H km above the surface.
A schematic representation of engagement is shown in Fig. 1. The target is moving in
horizontal direction receding away from missile at speed of V_T m/s. Assume that the
missile is launched at a speed of V_M m/s at an angle of γ_M with the horizontal reference.

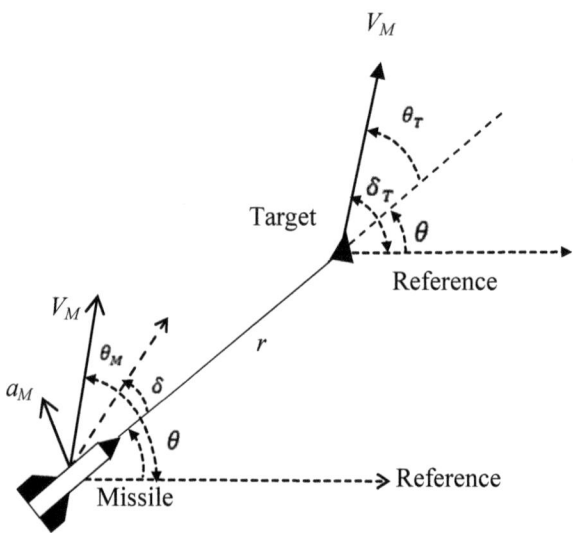

Fig. 1. Planar engagement geometry

For simulation purpose, assume the missile and the target to be the point mass
vehicles. Also, use the cartesian coordinate system for propagation of missile and target
states. Assume that the position of missile and target are denoted by pairs (X_M, Y_M) and
(X_T, Y_T), and their flight path angles are denoted by γ_M, γ_T, respectively. The lateral
acceleration of missile and target are denoted as a_M and a_T, respectively. The missile
and target kinematics in Cartesian coordinate system is given by:

$$\begin{cases} \dot{X}_M = V_M . \cos \gamma_M \\ \dot{Y}_M = V_M . \sin \gamma_M \\ \dot{X}_T = V_T . \cos \gamma_T \\ \dot{Y}_T = V_T . \sin \gamma_T \\ \dot{\gamma}_M = a_M / V_M \\ \dot{\gamma}_T = a_T / V_T \end{cases} \tag{1}$$

$$\dot{\theta} = \frac{V_T \sin(\gamma_T - \theta) - V_M \sin(\gamma_M - \theta)}{\gamma} \tag{2}$$

3   Launch Area Calculation

The launch zone is a space around the control tower that contains the target at the time of launch, ensuring the missile meets the target in the damage area. To determine the launch area limits when firing targets with limited maneuverability, from the meeting points above the limit of the damage area, taken in the opposite direction to the target movement by a distance r_L, is the target flying distance within the time period of missile launch t_P and flying missile to the meeting point t_B, that is:

$$r_L = V_T.(t_P + t_B) \qquad (3)$$

The launch area calculation of missiles is actually solving the maximum and minimum launch distance under the given launch conditions. For this, it is necessary to integrate the equations of motion of a missile. The initial conditions of the integral and the assumption of the target's motion judgment (direction maneuver, maneuver overload) should be input as known parameters, and the limiting factors such as the stable flight time of the missile, the maximum off-axis angle of the missile and the ability of the missile to withstand overload... are included in the integration. As for the determination of the boundary of the launch zone, the main considerations are the approach speed of the missile and the target, the missile's control flight time. Usually, proceed to launch the missile when the target is at the optimal launch range R_{opt}. The optimal launch range is the range of the target at that time to conduct missile launch, make sure the missile hits its target at the far limit of the damage area, to be able to launch additional missiles when needed.

The steps to find the boundary of the launch area are: input a target's entry angle q value, pre-estimate an initial shooting distance r, and then perform numerical integration, and judge whether the missile hits the target according to the hit limit conditions specified by the missile characteristics. If it does not hit, make corrections and re-calculate until the boundary that satisfies the constraints is found. The initial shooting distance r, can be estimated according to the method in [2], or a constant value can be given.

In order to improve the calculation speed, a variable step size method is used in the calculation process. Considering the factor of accuracy, a larger step size is selected at the beginning of the calculation, and a smaller step size is used when the missile approaches the target. In addition, the calculation of pneumatic parameters is simplified in programming, and piecewise linear interpolation is used instead of quadratic interpolation.

4   Intersection Point Calculation

The distance of incline to the intersection point of the missile and the target can be determined by the formula:

$$r = V_M.t_B \qquad (4)$$

where, t_B - the missile's flight time to the intersection point;

V_M - Average speed of the missile for moving to the intersection point.
The flight time of the missile is calculated by the following formula:

$$t = \frac{r_0[V_T \cos(\gamma_T - \theta_0) + V_M] - r[V_t \cos(\gamma_T - \theta) + V_M]}{-V_T^2 + V_M^2} \tag{5}$$

At intersection point: $t = t_B$ and $r = 0$
We are given the expression:

$$t_B = \frac{r_0[V_T \cos(\gamma_T - \theta_0) + V_M]}{-V_T^2 + V_M^2} \tag{6}$$

5   Simulation and Result Analysis

Taking the performance data of a certain type of missile on behalf of the calculation model, the launch area under different attack conditions is shown in Fig. 2. Figure 2a is obtained when the target is not maneuvering, Fig. 2b reflects the missile's launch area when the target is maneuvering to the left.

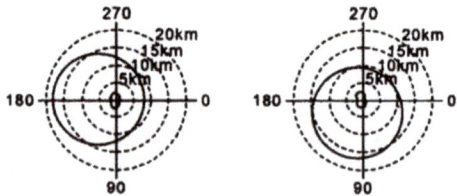

Fig. 2. Launch area: (a) when the target is not maneuvering (b) when the target moves to the left

It can be seen from the figure that when the target is not maneuvering, the launch area is axisymmetrical about the angle of entry 0^0 and 180^0; when the target is maneuvering, the launch area deflects to the side where the target is maneuvering.
Conducting general simulation with the following initial parameters:
The parameter of the target: $\gamma_T = 200°$; $V_T = 250$ m/s; $\theta_0 = 20°$; $H_T = 20$ km
The parameter of the missile: $\gamma_M = 40°$; $V_M = 750$ m/s; $t_p = 2$ s;

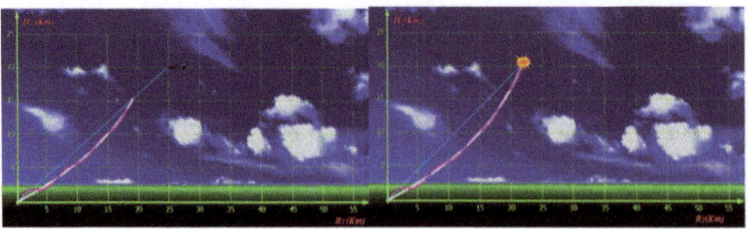

Fig. 3. The trajectory of the missile

In Fig. 3, simulation for the calculated trajectory and actual trajectory of the missile when it flies to the target in the vertical plane. In which Fig. 3a simulates the missile guidance process; Fig. 3b simulates the intersection point of the missile and the target.

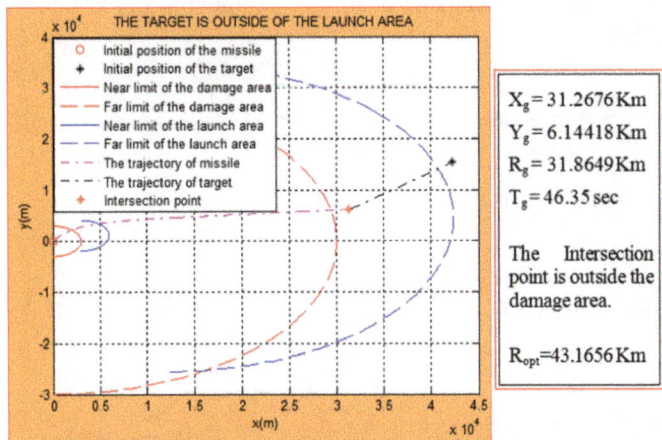

Fig. 4. Simulation results for the case: launching the missile when the target is outside the launch zone $r_0 = 45$ km

When the target is outside the launch zone $r_0 = 45$ km, proceed to launch the missile. The software automatically calculates the results as follows: after 46.35 s, the missile will meet the target at the coordinates: Xg = 31.2676 km, Yg = 6.14418 km, corresponding the intersection range Rg = 31.8649 km. The software warns that: The Intersection point is outside the damage area, can choose the optimal launch range is $R_{opt} = 43.1656$ km.

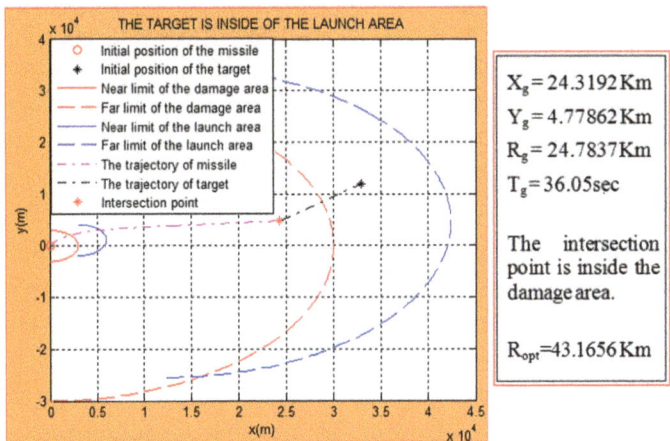

Fig. 5. Simulation results for the case: launching the missile when the target is inside the launch zone $r_0 = 35$ km

When the target is inside the launch zone $r_0 = 35$ km, proceed to launch the missile. The software automatically calculates the results as follows: after 36.05 s, the missile will meet the target at the coordinates: Xg = 24.3 km, Yg = 4.77 km, corresponding the intersection range Rg = 24.78 km. The software warns that: The Intersection point is inside the damage area, can choose the optimal launch range is $R_{opt} = 43.1656$ km.

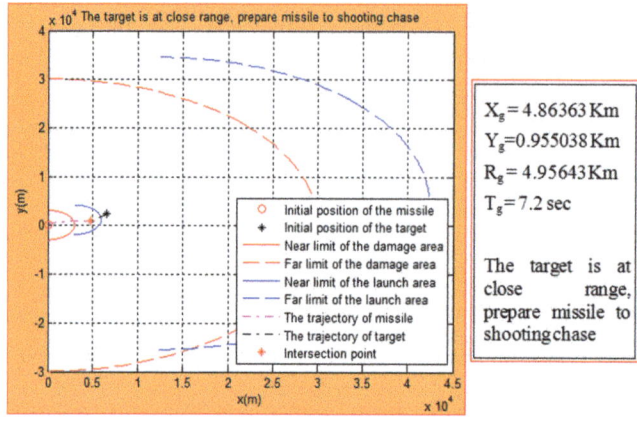

Fig. 6. Simulation results for the case: launching the missile when the target is at close range $r_0 = 7$ km

When the target is at close range $r_0 = 7$ km, proceed to launch the missile. The software automatically calculates the results as follows: after 7.2 s, the missile will meet the target at the coordinates: Xg = 4.8 km, Yg = 0.95 km, corresponding the intersection range Rg = 4.95 km. The software warns that: The target is at close range, prepare missile to shooting chase.

Usually, proceed to launch the missile when the target is at the far limit of the launch area, then the intersection point of the missile and the target is at the far limit of the damage area, to be able to launch additional missiles when needed. Choose the optimal launch range is $R_{opt} = 43.1656$ km (Fig. 7).

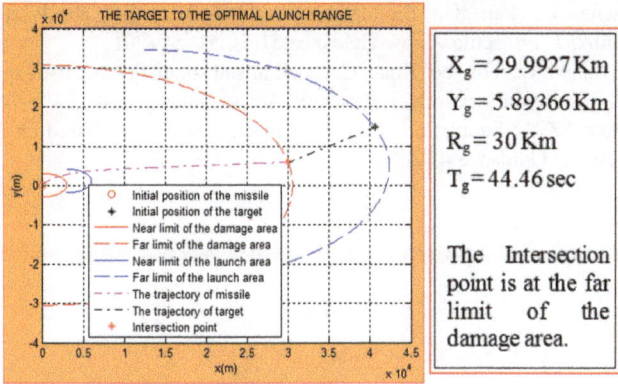

Fig. 7. Simulation results for the case: launching the missile when the target is at the far limit of the launch area $r_0 = 43.1656$ km

6 Conclusion

In the paper, the research team proposed a program that simulates the trajectory of the missile approaching the target and the program to determining for optimal launcher distance and intersection point for surface-to-air missile. Through calculations and verifications, it is shown that calculation model of the established launch area, the intersection point of the missile and the target has a fast calculation speed and can perform real-time calculations during air combat. The simulation of a certain model example shows that the characteristics of the launch area, the intersection point of the missile and the target can be quickly generated by this method, are the same as the real launch area, is the basis for good preparation and to obtain reasonable performance in all expected situations.

References

1. Li, A., Meng, Y., He, Z.: Simulation research on new model of air-to-air missile attack zone. IEEE Xplore (2020)
2. Wu, W.H., Zhou, S.Y., Gao, Li: Improvements of situation assessment for beyond-visual-range air combat based on missile launching envelope analysis. J. Syst. Eng. Electron. **33**(1), 2679–2685 (2011)
3. Kou, Y., Fu, Z., Feng, G.: Air to air missile allows launch area simulation and analyze under networked targeting environment. J. Air Force Eng. Univ.: Nat. Sci. Edn. **13**(2), 75–78 (2012)
4. Meng, G., Pan, H., Liang, X., Tian, F.: Allowable missile launch zone calculation for multi-fighter coordination attack under network targeting environment. IEEE Xplore (2016)
5. Xu, G., Liang, X., Zhang, J.: Simulation study on cooperative attack zone of dual-aircraft air-to-air missile. Fire Command Control **44**(1), 36–41 (2019)
6. Ping, Zhang, Fang, Y., Jin, C.: A new method for real-time calculation of air-to-air missile attack zone. J. Ballistic Eng. **22**(4), 11–14 (2010)

7. Qu, X., Zhang, L., Fan, G.: Simulation of air-to-air missile attack zone based on Matlab/Simulink. J. Projectile Arrow Guidance **31**(5), 51–54 (2011)
8. Kai, Wang, Wangxi, Li, Yongbo, Xuan: Calculation and simulation of air-to-air missile attack zone based on BP neural network. J. Projectile Arrow Guidance **30**(1), 75–77 (2010)
9. Deng, J., Uranus, Z.Y.: Simulation of air-to-air missile attack zone based on data modeling. J. Projectile Arrow Guidance **4** (2016)

Integrating Multi-threading into Query-Subquery Nets

Son Thanh Cao[✉], Phan Anh Phong, and Le Quoc Anh

School of Engineering and Technology, Vinh University, 182 Lc Duan Street,
Vinh, Nghe An, Vietnam
{sonct,phongpa,anhlq}@vinhuni.edu.vn

Abstract. In this paper, we propose a new method, named QSQN-MT, for the evaluation of queries to Horn knowledge bases. Particularly, we integrate multi-threading into query-subquery nets to reduce the execution time for evaluating a query over a logic program regardless of the order of clauses. The usefulness of the proposed method is indicated by the experimental results.

Keywords: Horn knowledge bases · Query processing · Deductive databases · QSQN · QSQN-MT · Multi-threading

1 Introduction

In first-order logic (FOL), the Horn fragment has received much attention from researchers because of its important roles in the logic programming and deductive database communities. Horn knowledge bases (Horn KBs) are an extension of Datalog deductive databases [1]. Various methods have been arised for Datalog or Horn KBs such as (i) the top-down methods including QSQ [14], QSQR [9], QSQN [5,11] and (ii) the bottom-up method including Magic-Set [2].

Normally, clauses in a logic program are processed in order. This means that a clause is executed when the processing of preceding clauses has finished. Particularly, there is always only one process being executed at a specific time. Multi-threading has been adopted in logic program implementations in order to improve the execution time [10,12,13]. By using multi-threading, we can perform multiple operations at once (simultaneously) in a program. This integration allows a single processor to share multiple and concurrent threads. Each thread executes its own sequence of instructions or clauses.

Example 1. When executing a logic program, the order of clauses in this program may be important and can affect the execution time to find solutions. For instance, consider the logic program P including (i) intensional predicates: $reachable$, $reachable_1$, $reachable_2$ and $reachable_3$; (ii) extensional predicates: $link_1$, $link_2$, and $link_3$; (iii) variables: x, y and z; (iv) constant symbols: a_i, b_i and c_i; and (v) a natural number: n.

D.-T. Tran et al. (Eds.): ICISN 2021, LNNS 243, pp. 189–196, 2021.
https://doi.org/10.1007/978-981-16-2094-2_24

– the logic program P (for defining $reachable$, $reachable_1$, $reachable_2$ and $reachable_3$):

$$reachable(x,y) \leftarrow reachable_1(x,y) \tag{1}$$
$$reachable(x,y) \leftarrow reachable_2(x,y) \tag{2}$$
$$reachable(x,y) \leftarrow reachable_3(x,y) \tag{3}$$
$$reachable_1(x,y) \leftarrow link_1(x,y) \tag{4}$$
$$reachable_1(x,y) \leftarrow link_1(x,z), reachable_1(z,y) \tag{5}$$
$$reachable_2(x,y) \leftarrow link_2(x,y) \tag{6}$$
$$reachable_2(x,y) \leftarrow link_2(x,z), reachable_2(z,y) \tag{7}$$
$$reachable_3(x,y) \leftarrow link_3(x,y) \tag{8}$$
$$reachable_3(x,y) \leftarrow link_3(x,z), reachable_3(z,y). \tag{9}$$

– the extensional instance I (for specifying $link_1$, $link_2$ and $link_3$) is demonstrated in Fig. 1),
– the query: $reachable(a_0, a_n)$.   ◁

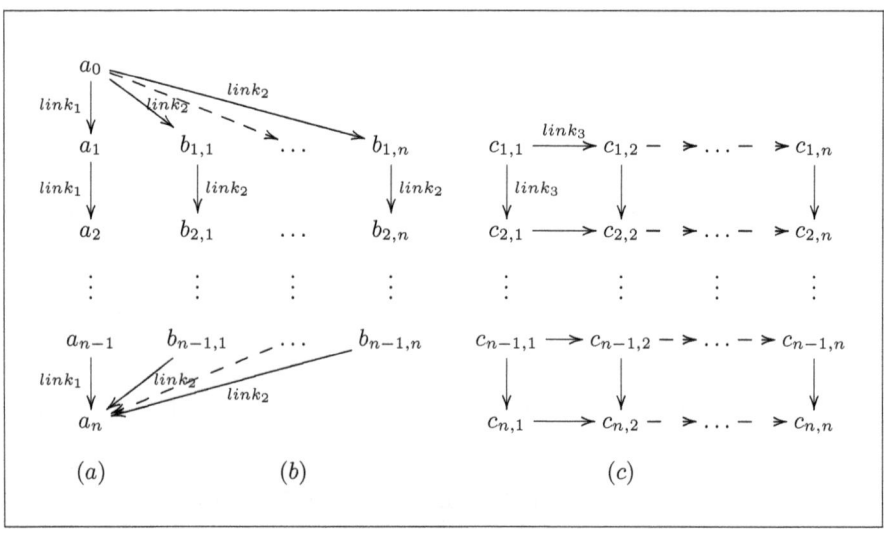

Fig. 1. The extensional instance I: (a) $I(link_1)$, (b) $I(link_2)$, and (c) $I(link_3)$.

As can easily be seen in Fig. 1, the relation $link_1$ (reps. $link_2$ and $link_3$) contains n (reps. n^2 and $2(n^2 - n)$) instances. For instance, if $n = 100$ then the relations $link_1$, $link_2$ and $link_3$ include 100, 10 000 and 19 800 instances, respectively. The relations $link_1$ and $link_2$ contain instances related to answer the query $reachable(a_0, a_n)$ but the relation $link_3$ does not.

Normally, a logic program is executed sequentially until getting results (e.g., the order of executing the program P is a sequence of the clauses (1), (2) and (3), together with the related ones). The question is, what will happen if we swap the order of program clauses in P so that the clause (2) is executed first? Clearly, this takes a long time to get the answer for the mentioned query since the size (i.e., the number of instances) of relation $link_2$ (which is defined by the clauses (6) and (7)) is much bigger than $link_1$ (which is defined by the clauses (4) and (5)). It is worth studying how to compute the answers to a query over a logic program regardless of the order of program clauses.

In this paper, we integrate multi-threading into QSQN framework to develop a new method for evaluating queries to Horn KBs, named QSQN-MT. Our intention is to reduce the execution time for evaluating a query over a logic program regardless of the order of clauses in this program. The experimental results indicate the outperformance of the QSQN-MT method. Due to space limitations, the reader could refer to [5,8] for the basic notions and definitions such as *term*, *atom*, *substitution*, *predicate*, *unification*, *Horn KBs*, *query* and other related ones. The rest of the paper is structured as follows. Sect. 2 outlines an overview of QSQN[1] and presents a new method called QSQN-MT. The tested results are provided in Sect. 3. Section 4 gives conclusions of the paper.

2   Query-Subquery Nets with Multi-threading

In this section, we first give an overview of the QSQN method and then present a new method for evaluating queries to Horn KBs, named QSQN-MT.

2.1   An Overview of Query-Subquery Nets

In [5,11], Nguyen and Cao formulated a framework *query-subquery net*, which is used to develop methods for evaluating queries to Horn KBs with the intension of improving the efficiency of query processing by (i) decreasing redundant computation, (ii) increasing flexibility, and (iii) minimizing the number of read/write operations to disk. Using this framework, we proposed an evaluation method named QSQN. The method is goal-directed, set-at-a-time, and has been developed to allow dividing the query processing into smaller steps to maximize adjustability (i.e., we can apply various flow-of-control strategies in QSQN, which are similar to search strategies in a graph and called *control strategies* for short). In particular, the given logic program is transformed into a corresponding net structure, which is used to specify set of tuples/subqueries in each node should be processed at each step. The proofs given in [3] showed that the QSQN evaluation method (as well as its extensions) is sound, complete and has PTIME data complexity with a condition of fixing the term-depth bound. For a more explanation of running example and relating QSQN to SLD-Resolution with tabulation, the reader could refer to [6, Section 3] for further reading. The other definitions related to QSQN structure, QSQN and a *subquery* are provided

---

[1] A demonstration in the PowerPoint-like mode to help the readers figure out the gist of QSQN is provided in [4].

in [11]. Also, to help the readers figure out the gist of QSQN, a detailed demonstration in the PowerPoint-like mode is provided in [4]. The experimental results shown in [3,6] indicate the usefulness of the QSQN evaluation method and its extensions.

2.2   Integrating Multi-threading into Query-Subquery Nets

In this subsection, we present an extension of the QSQN method proposed in [11] by integrating multi-threading into QSQN forming a new evaluation method called QSQN-MT.

The definition of a QSQN-MT structure (reps. QSQN-MT) is analogous to the definition of QSQN structure (reps. QSQN). Due to space limitations, we omit to present the details for brevity. We refer the reader to [5,11] for further understanding. In [5], we proposed an algorithm for evaluating queries to Horn KBs. We now present an extension of this algorithm to deal with multi-threading. From now on in this section, a logic program is denoted by P.

Algorithm 1: evaluating the query $(P, q(\overline{x}))$ on EDB instance I.

---

1 initialize a QSQN-MT of P and related ones;
2 let n be the number of threads detected;
3 **for** $i = 0$ **to** $n - 1$ **do**
4 ⌊ thread[i] = new Thread();   *// initialize the thread i*
5 **for** $i = 0$ **to** $n - 1$ **do**
6 ⌊ thread[i].start();   *// excute the procedure $run(i)$ for thread i*
7 **for** $i = 0$ **to** $n - 1$ **do**
8 ⌊ thread[i].join();   *// wait until sub-threads finish*

9 **return the results.**

Procedure run(i)

Purpose: running the thread i.
1 **while** *there exists* $(u, v) \in E$ *w.r.t. thread i s.t.* **active-edge**(u, v) *returns true*
   do
2 │   select $(u, v) \in E$ w.r.t. thread i s.t. **active-edge**(u, v) return *true*;
   │   *// arbitrary control strategies for the selection of (u, v)*
3 ⌊ fire(u, v);

Algorithm 1 describes steps of the QSQN-MT evaluation method for Horn KBs, which is a modified version of the one given in [5] for QSQN by integrating multi-threading into QSQN. The algorithm first automatically detects the number of threads and then concurrently executes these threads until getting results (the method start() in the step 6 of Algorithm 1 calls the procedure run(i) w.r.t. the thread i). Each thread represents a flow-of-control strategies in QSQN-MT.

The procedure run (on page 4) uses the function active-edge(u, v) (specified in [5]). For an edge (u, v), the function active-edge(u, v) returns *true* if there

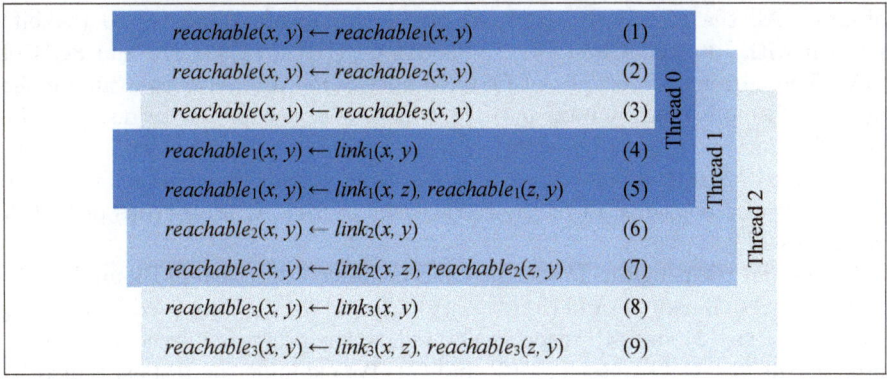

Fig. 2. An example of splitting the program P stated in Example 1 into 3 threads.

are some collected data in u (i.e., tuples or subqueries) that can be evaluated to generate data and transfer through (u, v), otherwise, this function returns *false*. If `active-edge`(u, v) returns *true*, the procedure `fire`(u, v) (specified in [5]) will evaluate unprocessed data collected in u and then transfer relevant data through (u, v). The procedure `fire`(u, v) calls the procedure `transfer`(D, u, v) (stated in [5]), which is used to determine the effectiveness of transferring data D through (u, v). These functions and procedures are also used for QSQN-MT.

An example of splitting the program P given in Example 1 into 3 threads is illustrated in Fig. 2. The first thread (thread 0) includes clauses (1), (4) and (5) of P. The second thread (thread 1) consists of clauses (2), (6) and (7) of P. The last one (thread 2) includes clauses (3), (8) and (9) of P. These threads run concurrently and independently of each other. Hence, some threads may run faster than others w.r.t. the execution time according to the size of EDB relations.

3   Preliminary Experiments

In [6], we have made a comparison between QSQN (together with its extensions) and DES-DBMS[2] as well as SWI-Prolog w.r.t the execution time. The experimental results described in [6] indicate the usefulness of QSQN as well as its extensions.

3.1   Experimental Settings

We have implemented prototypes of QSQN and QSQN-MT in Java, using a control strategy name IDFS proposed in [3], which is intuitively demonstrated in [4]. These prototypes use extensional relations which are stored in a MySQL

---

[2] The Datalog Education System (DES), a deductive database system with a DBMS via ODBC, available at http://des.sourceforge.net.

database. All the tests were executed on the Microsoft Windows 10 (64 bit) platform with Intel(R) Core(TM) i3-2350M CPU @ 2×2.30 GHz and 8GB of RAM. The current prototypes of QSQN and QSQN-MT allow to evaluate the query of the following syntax: $q(\bar{t})$, in which, $\bar{t}$ is a tuple of terms. For the query that has one answer being either *true* or *false* (e.g., $reachable(a_0, a_n)$), we use a flag to break the computation at the time getting the *true* answer. The package [4] also contains all of the below tests as well as prototypes of QSQN and QSQN-MT.

Reconsider the program P and the EDB instance I specified in Example 1. As mentioned, the clauses (4) and (5) (reps. (6), (7) and (8), (9)) are used for defining clause (1) (reps. (2) and (3), respectively). Thus, the program is executed mainly based on the order of clauses (1), (2) and (3). We examine the tests specified by changing the order of the first three rules of program clauses in P as follows: Test 1 ((1), (2), (3)); Test 2 ((2), (1), (3)) and Test 3 ((3), (2), (1)). Each test is performed with the query $reachable(a_0, a_n)$ using the following values of n: 20, 40, 60, 80 and 100, respectively.

3.2   Experimental Results

Figure 3 illustrates a comparison between our prototype of QSQN-MT and QSQN w.r.t. the execution time for Tests 1–3. We have executed each test case ten times to measure the execution time in milliseconds and averaged the results. To provide better data visualization, the average execution time of each method reported in Fig. 3 was converted to log_{10}.

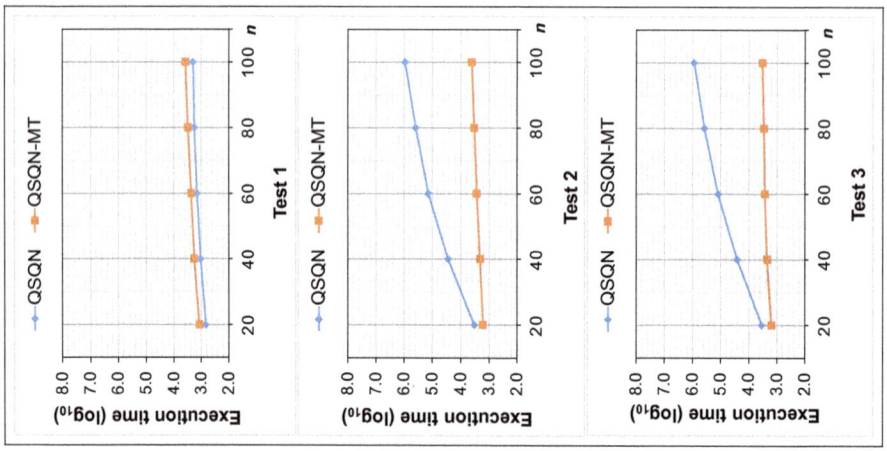

Fig. 3. Experimental results for Tests 1, 2 and 3.

As can be seen in Fig. 3 for Tests 1, 2 and 3, the execution time for the QSQN method is different when the order of program clauses are changed. In Test 1, the

QSQN-MT method takes a little bit more time than the QSQN method since it has to wait until all threads terminate. In all mentioned tests, when integrating multi-threading in QSQN, the execution time of the QSQN-MT is almost the same regardless of the order of program clauses.

4   Conclusions

A method, named QSQN-MT, for evaluating queries over a logic program has been proposed. With multi-threading, the logic program does multiple tasks concurrently in order to increase the performance. The results of experiments indicated that QSQN together with multi-threading can make the proposed method performs better than it would be with a single thread. The flow of answering a query using multi-threading in QSQN can be treated as a control strategy, thus QSQN-MT inherits all good properties of QSQN such that: goal-directed, set-at-a-time, sound, complete and having PTIME data complexity. Of course, having multiple threads is not always efficient and there are many other issues related to multi-threading that we need to be concerned about. As future work, we will make a comparison between the proposed method and other related applications as well as apply our method in parallel computation with multi-processors [7].

Acknowledgments. We are extremely grateful to dr hab. L.A. Nguyen from the Institute of Informatics, University of Warsaw, Poland for his helpful comments.

References

1. Abiteboul, S., Hull, R., Vianu, V.: Foundations of Databases. Addison Wesley (1995)
2. Beeri, C., Ramakrishnan, R.: On the power of magic. J. Log. Program. **10**, 255–299 (1991)
3. Cao, S.T.: Methods for evaluating queries to Horn knowledge bases in first-order logic. Ph.D. dissertation. University of Warsaw (2016). http://mimuw.edu.pl/~sonct/stc-thesis.pdf
4. Cao, S.T.: A prototype implemented in Java of the QSQN and QSQN-MT evaluation methods (2020). http://mimuw.edu.pl/~sonct/QSQN-MT.zip
5. Cao, S.T., Nguyen, L.A.: Query-subquery nets for Horn knowledge bases in first-order logic. J. Inf. Telecommun. **1**(1), 77–99 (2017)
6. Cao, S.T., Nguyen, L.A.: Incorporating stratified negation into query-subquery nets for evaluating queries to stratified deductive databases. Comput. Inform. **38**, 19–56 (2019)
7. Fidjeland, A.K., Luk, W., Muggleton, S.H.: Customisable multi-processor acceleration of inductive logic programming. In: Latest Advances in Inductive Logic Programming, pp. 123–141 (2014)
8. Lloyd, J.W.: Foundations of Logic Programming, 2nd edn. Springer (1987)
9. Madalińska-Bugaj, E., Nguyen, L.A.: A generalized QSQR evaluation method for Horn knowledge bases. ACM Trans. Comput. Log. **13**(4), 32 (2012)

10. Marques, R., Swift, T., Cunha, J.: Extending tabled logic programming with multi-threading : a systems perspective. In: Proceedings of CICLOPS 2008, pp. 91–106 (2008)
11. Nguyen, L.A., Cao, S.T.: Query-subquery nets. In: Proceedings of ICCCI 2012. LNCS, vol. 7635, pp. 239–248. Springer (2012)
12. Taokok, S., Pongpanich, P., Kerdprasop, N., Kerdprasop, K.: A multi-threading in prolog to implement K-mean clustering. In: Latest Advances in Systems Science and Computational Intelligence, pp. 120–126. WSEAS Press (2012)
13. Umeda, M., Katamine, K., Nagasawa, I., Hashimoto, M., Takata, O.: Multi-threading inside prolog for knowledge-based enterprise applications. In: Proceedings of INAP 2005, pp. 200–214. Springer (2005)
14. Vieille, L.: Recursive axioms in deductive databases: the query/subquery approach. In: Proceedings of Expert Database Systems, pp. 179–193 (1986)

The Next Generation Multifunctional Window System

Julker Nien Akib[1], Shakik Mahmud[1,2(✉)],
and Mohammad Farhan Ferdous[2]

[1] Department of Computer Science and Engineering,
United International University, Dhaka, Bangladesh
smahmud172174@bscse.uiu.ac.bd
[2] Japan Bangladesh Robotics and Advanced Technology Research Center,
Dhaka, Bangladesh

Abstract. In the era of digitalization and automation, individuals endlessly need to switch previous manual operating systems. During this paper, Authors tend to describe the Window automation system framework, and that they demonstrate here however it works and additionally describes all functions. This window is one type of distinctive good window that may gather its nearest condition info and end its activity like open, close, and send knowledge to the man of affairs, using IoT. It tends to be controlled by physical and autonomous (automated and internet applications). It will quantify the thickness of residue, dampness, temperature, and it additionally will determine gas, water (rain or snow), smoke, light, fire, dust, once distinguishing it took immediate actions supported matters. Here, the actions are; open and close the window receive commands from the user, and send collected knowledge to the user. At long last, authors have directed some real tests of assorted kinds of the sensing element for the shrewd window framework.

Keywords: Internet of Things (IoT) · Multifunctional window system · Cloud computing · Health-care · Room monitoring

1 Introduction

People are thoroughly dependent on innovation these days. People need to decrease their physical work in each progression of life also, they need the additional component for staying up with the latest likewise they need to feel free from traditional works. IoT is currently an emerging technology that is related to the typical features of a conventional system that can trade data with another. Recently IoT has stepped around all sectors like industries, technology, academia, and government. This method can include any device, programming, or sensors for this reason; however, various research is still carried in this domain [1, 2]. In remote locations, the internet used to access the process and order ongoing parameters [1, 3]. Moreover, IoT creates a platform that allows us to control remotely across a network architecture based on device connection. A large number of sensors used to control different devices for a long time for domestic automation. However, it is not appropriately implemented, then cost-effectiveness and efficiency do not improve [7]. That time we require innovation that spares their time and lessens physical work. They additionally need to dispose of the little family works; the yield is they can concentrate on a significant issue. The 'Next generation multifunctional window' is these sorts of advancement which can work automatically, can diminish physical exertion, guarantee safeness in-home, hospital, office, industry and it's anything but easy to utilize, and the setup cost is low. We should endeavor to comprehend a precedent: assume today is a radiant day, when you went out for going to the office you neglected to close your window yet few time later, tragically, the rain came, and the house got wet however if you utilize the window it won't occur, Or the spillage of gas in the house or hospital is excessively risky it can end your life yet shrewd window can ensure it, envision you room temperature turn out to be high from the ordinary condition than the window will consequently open and alter the temperature with the outside temperature. Just because of gas leakage in the kitchen, a vast number of people are facing death every year [8–10]. This multifunctional window can reduce these problems by sending and receiving commands from the user. It truly can make human life easy and relief from fewer physical work, somewhat quicker and smidgen strain-free. We installed the device in a hospital and got the expected results. It can make a significant contribution to health care.

The authors [1] proposed a project idea in their article named 'smart window' which is responded to by the sets of sensor data and able to send data to clients, using IoT but they are not appropriately demonstrated. Moreover this is the primary idea of our work. In our method the client can remotely control the window (open or close) by the help of IoT, and we used NodeMCU for Wi-Fi communication also, for the mainboard, we used Arduino Mega where authors [1] proposed to use Bluetooth or Wi-Fi modules. As a mainboard, they suggest Arduino or Raspberry Pi.

Here in the paper [4] carried out a framework, simulation, and parametric analysis of an automated window system for ventilate cooling, reducing the discomfort risk in the summer. Authors attempt to mimic there a serious ventilate cooling calculation for a window framework on building execution reproduction apparatuses. This window can only respond by the difference rate between ambient air temperature and indoor operative zone temperature. Here, authors design a window that can respond in various situations and send and receive data from users.

In [5, 6], authors discuss here some IoT based smart home commercial devices, each of the devices discussed here can perform specific functions like smoke detectors, humidity detectors thermostat. Authors in this paper have brought multiple devices into one device and also add more features.

2 Proposed Window Automation System

2.1 Proposed System Feature

Through this paper, authors have developed an ideal plan that meets all kinds of unique needs that will come in handy for the growing population in today's world. The most significant advantage of this model is installing all kinds of sensors to close and open the window automatically or manually. By the standards of the author, the window system can be manually controlled remotely, which saves much time. Another feature of our proposed model is that if gas, smoke, and high humidity are detected, the system opens the window and keeps it open until the gas or smoke leaves the room, which reduces the risk of fire. This model has three more sensors that take steps to close the window, such as if the user forgets to close the window if it is raining, dusting, or the light is low (dark), and the window closes automatically. In automatic mode, the sensors are arranged on a priority basis such as gas and smoke are given higher priority; for example, if the window is closed, but it is raining outside the window will remain open. Savvy gadgets should be synchronized with the principal worker to control these gadgets distantly. Clients may utilize login IDs and passwords to change the status of any applications to spare time, energy, and cash. Moreover, the creators suggested that the model gives wonderful insurance. Nonetheless, the creator's multifunctional window framework strategy gives 100% effectiveness, for example, sparing the client time if there should arise an occurrence of a mishap and giving the best insurance to a protected and up-to-date place.

2.2   Network Architecture

A vital component of any IoT-based operation is a server. The centralized server serves as the heart of all IoT rooted operations. For this paper, authors have created their own server using PC, and PHPMyAdmin has been used to store it. A virtual connection needs to be created between the server and the IoT device. There are many ways to create an internet connection. In this case, authors have used point-to-point web sockets. The PHP programming language is applied to design a web application for interacting with point-to-point web sockets and IoT devices for logical decision making. Point-to-point web sockets are used to create Internet connections between models and servers. Since a point-to-point attachment has been created, any IoT device can now connect to server and send data to the cloud server. On the other hand, there is a database attached to the server which will store the data. All the smart devices connected to the server will now be able to access the database connected to the server. To connect the web form application with the central server, the user needs to use the local host URL. During the installation, the user will be given a user ID and password, which they can login to the web application. Any logged-in user will now be able to access IoT devices connected to the central server and databases connected to the cloud. May show errors if the entire system is not connected to the Internet. The Internet will play a vital role in the whole system. The authors are hopeful that this proposed model can be handled accurately and securely if there are no internet connection errors (Figs. 1 and 2).

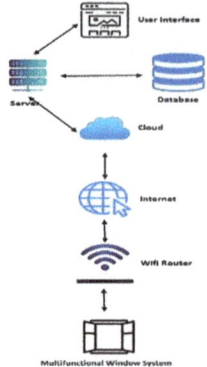

Fig. 1. Network architecture

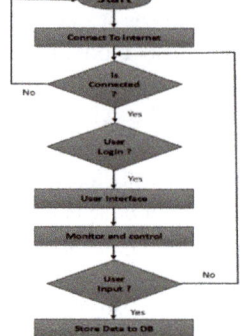

Fig. 2. Software design

3   Implementation

3.1   Software Design

PHP programming language has been used to create web applications and point-to-point web sockets. After creating this kind of socket, it is connected to our server with the help of an internet connection. The user will be given a user ID and Password at the time of installation to login. Now the user can monitor and control the data from the database. Data will be refreshed from the database each time. The data will be updated from the database, and these updated data will be sent to the user interface. The user will then be able to control the IoT device with those data. The data obtained from the control user interface is displayed on the control screen. The data obtained from the cloud is displayed on the monitor or control screen. Now the user can see which mode his/her device has selected, and he/she can choose any mode if he/she wants. If the user selects automatic mode, then he/she will have two options on and off button. Clicking on and off buttons will input data from the user, and that input data will be saved in the database. This way, the user can use this model through the web application.

3.2   Implementation Setup

We have divided the implementation of this ideal into three parts. First, the author will talk about the complete hardware installation, secondly, how the hardware will fetch the data from the database, and thirdly the author will describe how the sensors work. This model uses seven sensors (Ultrasonic, MQ 2 Smoke, MQ 5 Gas, DHT11 Humidity, Water, Dust, and LDR) and two microcontrollers, as shown in Fig. 3 by a flowchart. The author used Arduino as the central controller unit, and NodeMCU used to bring data from the cloud and send it to Arduino. Serial communication methods have been used to send data from NodeMCU to Arduino. These sensors capture data in the form of analog signals using a processing module microcontroller. Signals are converted to digital format by analog to digital converters (ADC).A motor used as the output is attached to the window, which will open and close the window through that data. Two modes are used to open the window automatically or manually. To select the mode the user will be sent data from the user interface to the cloud that data will be fetched via NodeMCU and sent to Arduino.

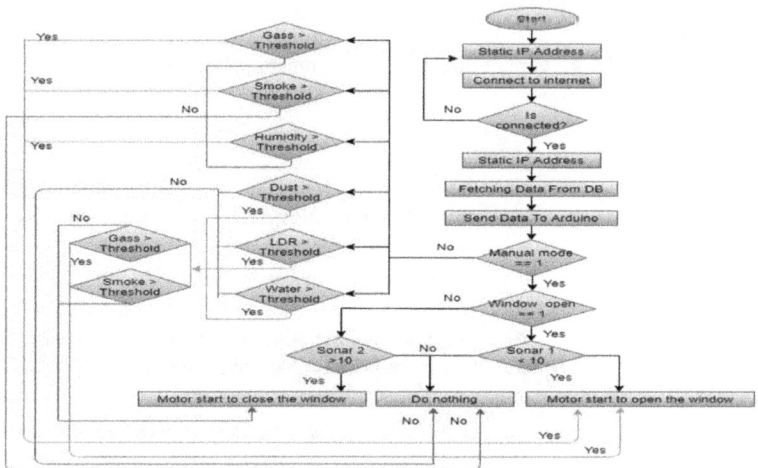

Fig. 3. Flowchart of multifunctional window system

Then the mode will be selected by checking the condition for the main controller unit. If the user selects manual mode, the user can manually open or close the window. If the user selects automatic mode, the data from the sensors will automatically open or close the window. The authors mentioned in the previous paragraph that NodeMCU would fetch data from the cloud and send that data to Arduino. NodeMCU needs to be connected to the internet via Wi-Fi to fetch the data. An IP will then be assigned to the NodeMCU. For fetching data, an API has been created that will be fetched via the HTTP protocol.

A sensor is a device that can detect changes in the environment and then send an electrical signal. Sensitivity, linearity, range, accuracy, and response time are the essential features of a sensor. Many times there are errors in the sensors which are eliminated by calibration. Reducing this error increases the performance of the sensor. If the performance is low, then the sensor does not get proper output, then the accuracy is low. So it is better to check the faulty sensor and then work.

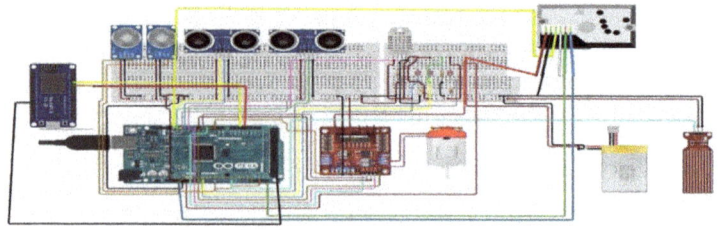

Fig. 4. Circuit diagram of multifunctional window

Number	Sensor name	Using purpose
01	Ultrasonic Sonar sensor	Here the author will determine the position of the window using this ultrasonic sensor. They have used two sensors to know the position of the window
02	MQ 2 smoke sensor	It can detect gases in densities in the range of 200 to 10000 ppm. A condition check by setting a threshold of 200 ppm will open the window if it is closed. Smoke sensors are given more priority than other sensors because smoke is only when there is a fire somewhere
03	MQ 5 gas sensor	This is an Analog output sensor. This needs to be connected to any Analog pin. Like the smoke sensor, a threshold value has been taken here. If the detection value in the sensor is higher than that threshold value, the window will open automatically. Like smoke sensors, these sensors are given higher priority because there is a possibility of fire if the gas line leaks
04	Humidity sensor	This DHT11 sensor detects high humidity and opens the window automatically so that fresh air from outside enters the room and keeps the room cool
05	Dust sensor	An infrared transmitting diode and a phototransistor are slantingly orchestrated into this gadget to recognize the mirrored light of residue noticeable all around. It is particularly helpful in recognizing the best particles like tobacco smoke and is usually utilized in air purifier frameworks. So If the level of dust is high the window will close
06	Water sensor	It can detect the water, suppose if rain comes then the window will close
07	LDR sensor	When an LDR is kept in the dark place, its resistance is higher, and its resistance will decrease when the LDR is kept in the light. The LDR sensor will also close the window automatically if it is an evening or when the light is low

4 Result

The IoT system we have designed is tested by installing the smart sensors and setting up a server for one home.

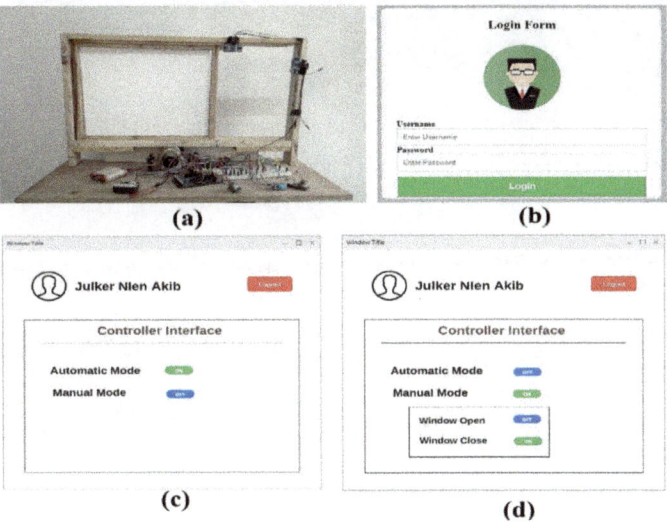

Fig. 5. User interface and device picture.

After installing the smart sensors, the user has a user ID and password by which the user can login through the web applications shown in Fig. 5(b) After successfully logging in, the controller dashboard will be displayed in front of the user through the web application, as shown in Fig. 5(c). Now the user can select any mode from this controller page. The user can select a single mode at a time. If the user selects automatic mode, the manual mode will be off, and if the user selects manual mode, then two more options will be enabled, which is Fig. 5(d). The user can then manually turn the window on or off from that option. By our standard, the user will be able to control this process from any web browser if the system is installed at his/her home or office. Using these automated windows designed by us will reduce user time and risk. We installed the device in a local hospital of Bangladesh for 10 month and got the expected results. It can make a significant contribution to health care.

Next Generation Multifunctional Window Data: After installing this device in a local hospital in Bangladesh we got some amazing data based on the sensor value and it worked amazingly.

Sensor data are shown in the Fig. 6 these sensors are LED, water, gas and smoke. For water and LDR sensors both working logic are the same. Here we attach the both sensor data from our device observation. It's a data scenario of a rainy evening. For gas and smoke sensors observation we leakage a LPG gas tube and found these data and the experiment duration was 60 s. For the dust sensor we got these (Fig. 7) data. When the dust sensor value is higher than 165 then the window took action. We attach here 60 min data. Usually, in our home, hospital or residence the environment is almost conventional. That's why we present here individual sensor data for better an understanding.

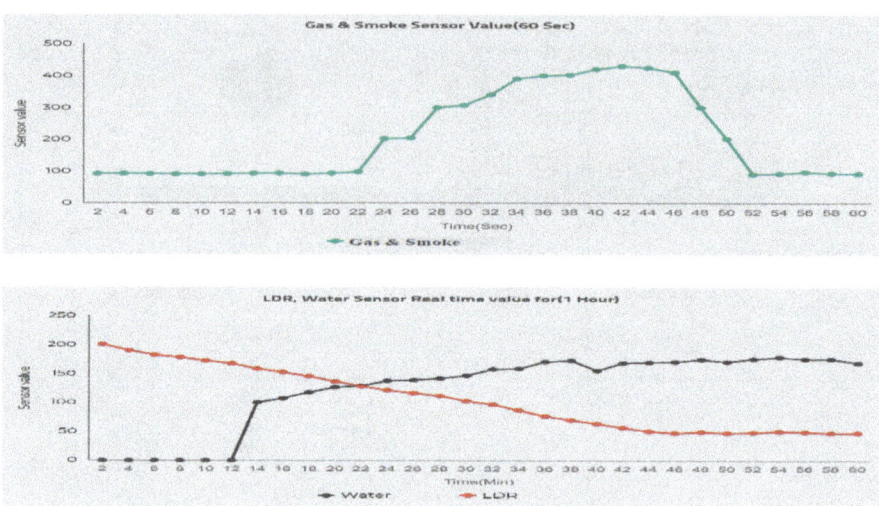

Fig. 6. Sensor data-1

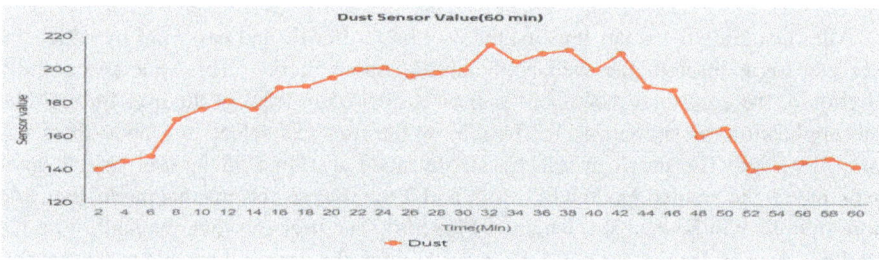

Fig. 7. Sensor data-2

5 Conclusion

People need an additional component for staying up with the latest. Likewise, they need to feel free from traditional works. Thus, the multi functionary window thought is ideal. By utilizing this window, the client can guarantee that his/her home is protected from the flame, dust, rain, gas, and smoke. Additionally, the proprietor can gather any single circumstance data from this window. It can make its proprietor live more brilliantly. Later on, some new features will be included in this system, and the improvement work is going on the expectation that it will be more helpful in the future. We are currently working on running the household equipment like AC, fan, Gas line Fire extinguisher, based on room temperature, Humidity and other sensor data. Also, controlling the window and monitoring the room using a camera another one is predicting the emergency situation by analyzing the previous data. Now the security issue has come, it will be an issue when we give one sort window and the same working framework and the same programming. At that point, numerous groups will come for the same issue, such as protection, information mining, and security. Yet, these things won't occur because window classification isn't the same; it relies entirely upon the client like size, shape, and look. In this way, it is best to give individual working programming based on the client.

References

1. Mahmud, S., Ferdous, M., Chakraborty, D., Tasnim, L.: Domestic mechanization system with IoT and robotics. IJMLNCE **2**(04), 151–162 (2018)
2. Dey, S., Roy, A., Das, S.: Home automation using Internet of Thing. In: 2016 IEEE 7th Annual Ubiquitous Computing, Electronics Mobile Communication Conference (UEM-CON), New York, NY, 2016, pp. 1–6. https://www.doi.org/10.1109/UEMCON.2016.7777826
3. Zhang, W.: Study about IOT's application in digital agriculture construction. In: 2011 International Conference on Electrical and Control Engineering (ICECE), Yichang. IEEE, pp. 2578–2581 (2011)
4. Psomas, T., Fioventini, M., Kokogia, G., Heiselber, P.: Ventilative cooling through automated window opening control systems to address thermal discomfort risk during the summer period: framework, simulation and parametric analysis. Energy Build. **153**, 18–30 (2017)
5. Nag, A., Alahi, M., Afsarimanesh, N., Prabhu, S., Mukhapadhay, S.: IoT for smart homes. In: Sensors in the Age of the Internet of Things: Technologies and Applications (2019)
6. Pavithra, D., Balakrishnan, R.: IoT based monitoring and control system for home automation. In: 2015 Global Conference on Communication Technologies. IEEE, pp. 169–173 (2015)
7. Altahrawi, M.A., Ismail, M., Mahdi, H., Ramli, N.: Routing protocol in a hybrid sensor and vehicular network for different mobility scenario. In: 2017 IEEE 13th Malaysia International Conference on Communications (MICC), Johor Bahru, pp. 113–118 (2017)
8. Saver couple burnt death after gas leak. https://www.thedailystar.net/backpage/savar-couple-burnt-death-after-gas-leak-kitchen-1342015. Accessed 28 May 2020
9. 1 killed 6 hurt fire gas cylinder leak, web address:https://www.thedailystar.net/city/news/1-killed-6-hurt-fire-gas-cylinder-leak-1661410. Accessed 28 May 2020
10. Danger from gas line and cylinders. https://thefinancialexpress.com.bd/editorial/danger-from-gas-line-and-cylinders-1574260130

An Improved Automatic Lung Segmentation Algorithm for Thoracic CT Image Based on Features Selection

Tran Anh Vu[1], Pham Duy Khanh[1], Hoang Quang Huy[1],
Han Trong Thanh[1], Nguyen Tuan Dung[2],
and Pham Thi Viet Huong[3(✉)]

[1] School of Electronics and Telecommunications, Hanoi University of Science
and Technology, Hanoi, Vietnam
[2] Radiology Center - Bach Mai Hospital, Hanoi, Vietnam
[3] International School, Vietnam National University, Hanoi, Vietnam
huongptv@isvnu.vn

Abstract. In order to evaluate thoracic disease, lung segmentation is a basic and necessary step. In computed tomography, different methods have been developed for automatic lung segmentation. Deep learning models are considered as one of the most important approaches in this field because it can process a lot of images in a timely manner. However, in order for deep learning work efficiently, it needs significant large dataset with precise manually lung segmentation. In this article, we concentrate on approaches and algorithms for image processing to work with pixels value of images. We are able to remove lung mask with high precision from initial computed tomography images by managing the pixel value array. By evaluating the pixel value of its binary image output with 2-D ground truth images, automatic segmentation is assessed. The result shows that for most sections of thoracic pictures, this procedure worked well and in case if there is any fading in the image, such as bone or soft tissue, the algorithm needed to be enhanced.

Keywords: Lung segmentation · Deep learning · Thoracic disease

1 Introduction

Lungs are the one of the main components of the respiratory system. Around 65 million people over the world tolerated from chronic obstructive pulmonary disease (COPD), according to The Global Effect of Respiratory Disease [1], making COPD the world's third leading cause of death. Lung cancer strikes 225,000 people annually in the United States, which accounts for health care expenses of $12 billion [2]. In emergency situations, our planet is also faced with the epidemic triggered by Coronavirus Disease 2019 (COVID-19) distributed worldwide in the year 2020. For patients, early diagnosis is also important, and tends to improve the probability of healing and longevity.

On order to test lung disorders, helical computed tomography (CT) of the thorax is commonly utilized [3]. Accurate estimation of lung volumes firmly facilitates the detection and therapy of ventilatory defects. Using the CT scan with spirometry gating,

D.-T. Tran et al. (Eds.): ICISN 2021, LNNS 243, pp. 206–214, 2021.
https://doi.org/10.1007/978-981-16-2094-2_26

the calculation of lung volumes in patients with respiratory disorders given identical lung volume values using plethysmography methods [4]. In terms of higher resolution of nodules, the gain of CT scan of the chest is better than that of a simple chest X-ray [5]. Consequently, CT images can be used for early diagnosis and treatment to diagnose pulmonary nodules [6] and lung cancer [7]. Automated CT image monitoring of lung infections provides great ability to strengthen the conventional healthcare approach for COVID-19 tackling [8]. In order to provide valuable and accurate knowledge about the recognition of lung boundaries within the CT images, all these quantitative analysis applications involve a preprocessing stage known as "lung segmentation". On most CT control systems, a manual outline of regions using an interactive graphics input system is available. While this approach is clearly based on the operator's expertise and experience, it suffers from some downside. This technique appears to be inefficient and repetitive because it requires technical expertise to execute the operator's roles. Therefore, the exponential rise in the number of thoracic CT research and the CT images have motivated many researchers to improve their computer-aided diagnostic (CAD) techniques to support in CT image research.

Several approaches for segmentation, using computer aided method for pulmonary CT images have been tested [9–15]. In [11], three-dimensional (3-D) CT scan of the lung are formed as the combination of 2-D slices. According to [10, 11, 14], in some situations, gray-scale thresholds do not distinguish the left and right lungs near the anterior and posterior junctions, since they are quite small and weakly contrasting. 2-D morphological erosion is utilized to find the junction lines and distinguish the right and left lungs entirely. Otherwise, for estimating regional air and tissue volumes in normal people's lungs, manually traced boundaries were used [13]. Various machine learning models have been used, such as U-net [16], ResU-net [17], Dilated Residual Network-D-22 [18]. Although these models work well in a number of circumstances, they require a significant time of preparation and running to maximize the precision.

In our research, we create a complete automated algorithm for extracting the lungs in CT images saved in Tag Image File (TIF) format. We obtain highly reliable outputs in most thoracic CT slices by applying various image processing algorithms. Several steps are used: K-means clustering, gray-level thresholding, scikit-image algorithms. In addition, the paper suggests a new method of selecting features to improve the precision of automated filtering. After each examination, segmented results are analyzed and explained to incorporate conditions in the select area of the lung and exceeding the blurred spectrum.

Our paper is structured as below. Section 2 presents a comprehensive description of the algorithm. Section 3 presents our setup of experiment and corresponding results. Section 4 summaries the paper.

2   Materials and Methods

2.1   Databases

In literature, there are competitions like Lung nodule analysis 2016 (LUNA) and the Kaggle Data Science Bowl 2017, which includes processing and attempting to locate lesions in lung CT images. It is important to define the lung component specifically in order to find disorders in these images. The dataset includes 267 original TIF images of 512 × 512 pixels and their corresponding manually segmented 2D binary images. In our paper, we use this dataset to create lung segmentation. In the preprocessing phase, we found four images with redundant regions and missing half of the lung, so we removed these images from our testing phase (Fig. 1).

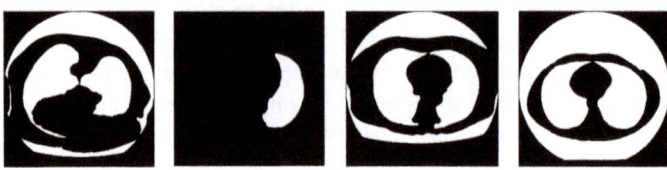

Fig. 1. Four removed images.

Moreover, this dataset contained a CSV statistical file containing specific information on the lungs in each image, such as the lung area in both pixel and mm2. We also calculate the fraction of the volume of the lung and the mean value of the Hounsfield region of the lung unit for further investigations. Our algorithm will use the initial CT images and 2D images.

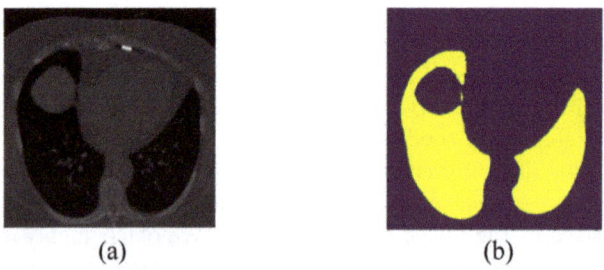

(a)   (b)

Fig. 2. (a) The original CT tomography image. (b) Ground truth

2.2   The Flowchart of Automated Lung Segmentation Algorithm

In order to get completely segmented lungs, the CT images of the lung will be examined using the flowchart shown in Fig. 3. The complete dataset is initially loaded and saved as NumPy arrays. Our task is to process these arrays to isolate the lung mask from the original images and to establish boundaries with manual ground truth to determine our segmented result. In this step, to separate the lung region from the CT image, we use the gray-level thresholding and K-means algorithm. Secondly, using the sci-kit-image module in Python, the CT images are then segmented into corresponding regions. Next, by setting a series of criteria, we are able to remove the blurred areas and retain only lung regions. Eventually, we eliminate all dark structures within the lung masks as well as smoothen the distorted border around the mediastinum to achieve more clear results. We present findings that compare our automated process with the

Kaggle 2D and 3D image dataset with manually segmented lungs. We plan to refine our segmentation method with the highest possible accuracy statistics as the first step to establish lung volume calculation and reconstruction of lung 3D images corresponding to thoracic CT.

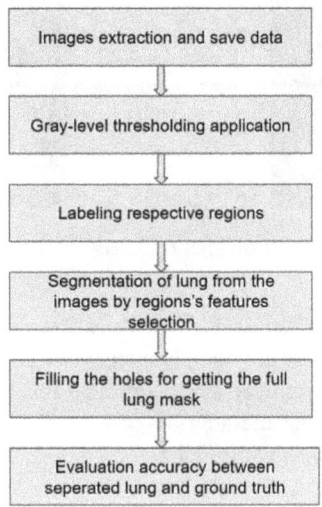

Fig. 3. The flowchart of automated lung segmentation algorithm

2.3 Description of the Algorithm

Images Preprocessing
In this step, each image and its corresponding 2D segment are loaded and transfered to a 512×512 NumPy arrays, which display HU (Hounsfield Units) values in the image for each pixel. Based on these HU values, we can get a general view of the structure. Inside the lungs, there is not only a lot of air, but also soft tissue, mainly muscle, etc. There is some fat and just a little amount of bone. In addition, some pixels have a HU of about -2000, whereas the HU of air is only -1000, which leads us to see that some type of artifact could exist.

Gray-Level Thresholding for Converting CT Images into Binary Images
Firstly, we normalized the pixel value. Then, by constructing two K-means clusters that concentrate on soft tissue/bone and lung/air areas, we find the appropriate threshold. By varying the maximum and minimum value within the array into the mean value of the pixel near the lungs, we transfer the overflow and underflow to the pixel spectrum. The washed-out regions or images can be renormalized by doing this, and threshold results can be reinforced. When the pixel's values of images are set, K-means is applied distinguish foreground (soft tissue/bone) and backdrop (lung/air). To divide the pixels in the image into two groups, we use K-means and the algorithm can locate these two group center points (pixels). By having the mean pixel value of these two points, the

threshold is calculated. After that, in binary images of lower threshold area, two regions are explicitly differentiated to equate to 1 with white, while the foreground region gets 0 value and darker color.

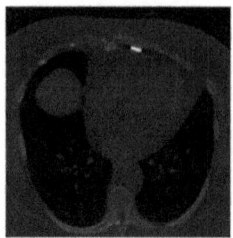 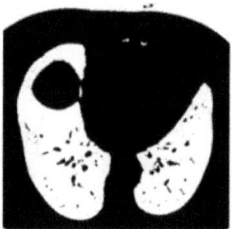

Fig. 4. The binary image of original CT image before and after thresholding

In conjunction with the picture in Fig. 4, we can see that while the lung has been removed from the thoracic, there are already many issues, including tiny holes, nodules and blood vessels. The most critical issue after splitting the image into two different black and white sections is that the area beyond the thoracic (supposed to be black) along with the lung is marked as white. Therefore, before the segmentation process can proceed, the region beyond the thoracic portion must be removed.

Labeling the Images

We use the skimage package to segment the lung region after converting the image into a gray scale 1. The background (labelled as 0) is kept constant in this process, the foreground is labelled as 1. The pixels are scanned from left to right, up to down, following a raster scan: where they are neighbours, two pixels are related, and the same value is marked. The same index is assigned to all related regions. The number of regions will be recorded until the procedure is completed.

As labeling is performed in Fig. 5(a), we have details about each image region, such as: the number of regions (marked from index 0), the area of each region (represented as the number of pixels belonging to each region), the size of each region (defined by the number of rows and columns and the boundary boxes of each region).

Getting the Segmented Lung

We are able to remove the blurred regions and retain only the section that includes the lung by setting criteria for all regions. When looking at the labelled image in Fig. 5a, we set the criteria and evaluate the precision in the next steps. To get the segmented lung as accurate as possible, we continue to repeat this process. Next, regions connected to the image border are excluded, indicating that the area beyond the thoracic portion and foreground region is set to 0. In addition, in order to eliminate the tiny nodules impacting the precision of the segmentation, we eliminate regions that are less than 100 pixels. Finally, the regions with conditions are set to 1, while the remaining regions are set to 0, as seen in the binary image in Fig. 5b.

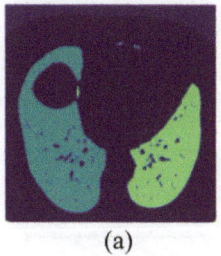 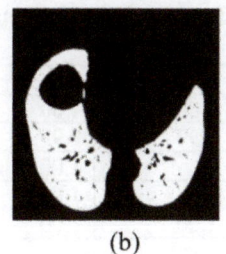 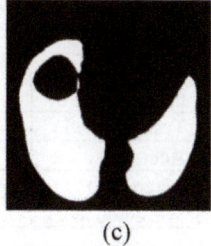

(a)   (b)   (c)

Fig. 5. (a) Labeled image. (b) Retained regions. (c) Final result of segmentation.

Finish the Lung Mask
To minimize small gaps, we use a morphology closing algorithm with a radius disk of 10 (pixels). We eliminate the tiny dark spots inside the lung part and connect the bright cracks by closing. The lung nodules are isolated from the blood vessels first by using this algorithm and then kept fixed to the lung, which represents in Fig. 5(c).

3 Experimental Findings

In this section, we present the metrics for evaluating the accuracy: precision, recall and f1-score. We compare the pixels value of two segmented lung images, one is done with our segmentation algorithm, one is done with manually segmentation. Table 1 represents the confusion matrix of algorithms for images in Fig. 2.

Table 1. Confusion matrix of our proposed method

		True value	
		True (Class 1)	False (Class 0)
Prediction	Positive	75324	423
algorithm	Negative	428	185969

The true positive value (75324) is labeled as 1 for all the number of pixels of both the ground truth and our segmented image. The false positive (423) is the number of pixels labelled with a value 0 in ground truth, but our algorithm identified as 1. The true negative (428) is the number of pixels labelled as 1 and our algorithm identified as 0. The false negative (185969) is the number of image pixels labelled as 0 and also be identified as 0.

Table 2 provides the precision of automated lung segmentation for the picture in Fig. 2. Accuracy and recall were evaluated by true positive (TP), false positive (FP), true negative (TN), false negative (FN) in Table 1. F1-score is the harmonic mean of the precision and recall. The accuracy is determined by the number of correctly predicted pixels (TP, TN) over a total of 262144 pixels in images of almost 1,00 pixels.

Table 2. Classification metrics of testing image in Fig. 2

	precision	recall	F1-score	ground truth	predict
False (Class 0)	1.00	1.00	1.00	186392	186397
True (Class 1)	0.99	0.99	0.99	75752	75747
Accuracy			1.00	262144	
Macro avg	1.00	1.00	1.00	262144	
Weighted avg	1.00	1.00	1.00	262144	

In addition, the borders of the segmented lung mask are drawn in both images on the original thoracic CT to provide a complete understanding, the boundary generated by our algorithm is red, while the boundary of the segmented manual lung is green (Fig. 6a). By leaving them in a binary picture and making a contrast, we can see the difference between these two masks. The pixels with the same value are labelled as 0 for two masks and 1 for others, so that the difference between the two images is clearly shown in white (Fig. 6b).

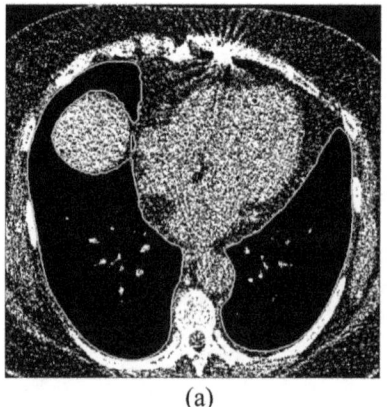   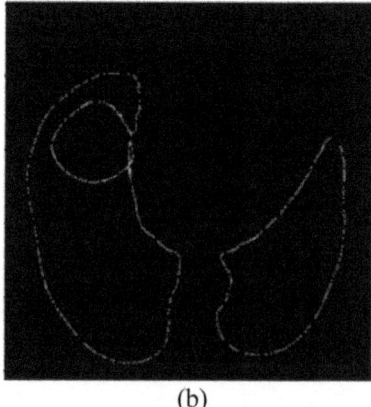

(a)   (b)

Fig. 6. (a) The lung borders of our algorithm (red) and ground truth (green). (b) Marking the difference between two segmented lung masks.

After analyzing a particular image, for all 267 images in the data, we begin to apply the same method and record the distribution of f1-score in these images. In order to evaluate and improve algorithms, the best and worst images are presented.

In Fig. 7, there are 232 images of a total of 263 with an average f1-score of approximately 0.986, and 21 images with an average f1-score of 0.939. There are 2 images evaluating the worst score of about 0.70 that keep us trying to update the algorithm in more various situations. Automatic segmentation can be enhanced to be carried out on three-dimensional volumes after working well on two-dimensional slices. The sum of 263 images resulted in an average value of f1-scores of 0.97887

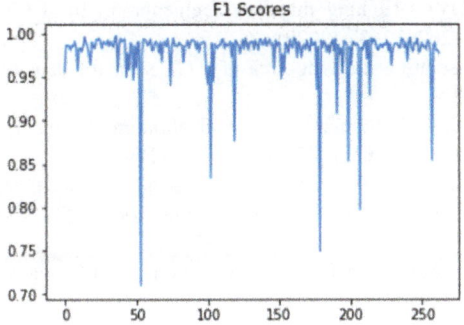
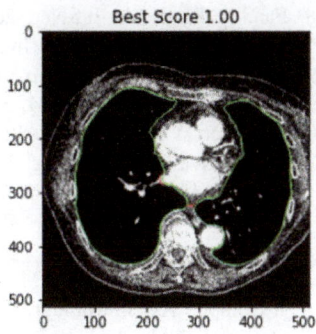

Fig. 7. The distribution of F1-score of the whole dataset and the best F1-score image.

compared to the result of a mean value of f1-scores of 0.96291 when analyzing using the automatic Kaggle notebook segmentation method [19].

4   Conclusion

Our automated segmentation algorithm performs remarkably well, and it can manage most CT images in the dataset. As the smoothing surface processes that make the nodules linked to the lung mask, the precision of the lung's edge could be lost. If the thoracic portion is not distinctly presented or faded when applied to the CT scans, some of the soft tissue and bond part of the lung mask are attached in the same area and allow the accuracy to decrease considerably. In order to tailor our CAD approaches to particular tasks, more functionality of choosing features should be introduced. Our algorithm can be updated by identifying lung nodules, calculating lung volumes, creating three-dimensional lung images from multiple two-dimensional slices.

References

1. The Global Impact of Respiratory Disease, 2nd edn. European Respiratory Society, Sheffield (2017)
2. Data Science Bowl 2017, 1 February 2017. https://www.technology.org/2017/02/01/data-science-bowl-2017/
3. Remy-Jardin, M., Remy, J.: Spiral CT of the Chest. Springer, Berlin (1996)
4. Tantucci, C., et al.: Methods for measuring lung volumes: is there a better one? Respiration **91**(4), 273–280 (2016)
5. Cardinale, L., et al.: The pulmonary nodule: clinical and radiological characteristics affecting a diagnosis of malignancy. La radiologia medica **114**, 871–889 (2009)
6. Dehmeshki, J., et al.: Shape based region growing using derivatives of 3D medical images: application to semiautomated detection of pulmonary nodules. In: Proceedings 2003 International Conference on Image Processing (2003)
7. Makaju, S., et al.: Lung cancer detection using CT scan images. Proc. Comput. Sci. **125**, 107–114 (2018)

8. Fan, D.-P., et al.: Inf-Net: automatic COVID-19 lung infection segmentation from CT images. IEEE Trans. Med. Imaging **39**(8), 2626–2637 (2020)

9. Hedlund, L.W., et al.: Two methods for isolating the lung area of a CT scan for density information. Radiology **144**, 353–357 (1982)

10. Kalender, W.A., Fichte, H., Bautz, W., Skalej, M.: Semiautomatic evaluation procedures for quantitative CT of the lung. J. Comput. Assist. Tomogr. **15**(2), 248–255 (1991)

11. Hu, S., Hoffman, E.A., Reinhardt, J.M.: Automatic lung segmentation for accurate quantitation of volumetric X-Ray CT images. IEEE Trans. Med. Imaging **20**(6), 490–498 (2001)

12. Keller, J.M., Edwards, F.M., Rundle, R.: Automatic outlining of regions on CT scans. J. Comput. Assist. Tomogr. **5**(2), 240–245 (1981)

13. Denison, D.M., Morgan, M.D.L., Millar, A.B.: Estimation of regional gas and tissue volumes of the lung in supine man using computed tomography. Thorax **41**, 620–628 (1986)

14. Brown, M.S., et al.: Method for segmenting chest CT image data using an anatomical model: preliminary results. IEEE Trans. Med Imaging **16**, 828–839 (Dec. 1997)

15. Memon, N.A., Mirza, A.M., Gilani, S.A.M.: Segmentation of Lungs from CT Scan Images for Early Diagnosis of Lung Cancer. ISSN (2006)

16. Ronneberger, O., Fischer, P., Brox, T.: U-net: convolutional networks for biomedical image segmentation. In: International Conference on Medical image computing and computer-assisted intervention, vol. 9351, pp. 234–241 (2015)

17. Srivastava, R.K., Greff, K., Schmidhuber, J.: Training very deep networks. In: Advances in Neural Information Processing Systems, pp. 2377–2385 (2015)

18. Yu, F., Koltun, V.: Multi-scale context aggregation by dilated convolutions. Published as a Conference Paper at ICLR 2016

19. Kaggle expert: 2D&3D Lung segmentation. https://www.kaggle.com/azaemon/2d-3d-lung-segmentation

Lung Sounds Classification Using Wavelet Reconstructed Sub-bands Signal and Machine Learning

Vu Tran[1], Tung Trinh[1], Huyen Nguyen[1], Hieu Tran[1], Kien Nguyen[1], Huy Hoang[1], and Huong Pham[2(✉)]

[1] School of Electronics and Telecommunications, Hanoi University of Science and Technology, Hanoi, Vietnam
[2] International School, Vietnam National University, Hanoi, Vietnam
huongptv@isvnu.vn

Abstract. The incidence of respiratory diseases is increasing rapidly due to environmental pollution affecting everyone around the world. Diagnosis from cardiopulmonary hearing has been made for hundreds of years. However, this method is influenced by noise and subjectivity by the doctor, which creates uncertainty and inconsistency in screening for lung disease. Numerous studies have tried to solve these problems by recording lung sounds digitally and processing them. In this article, we use a discrete Wavelet transformation to classify wheeze, crackle and normal lung sounds to reduce computation time and cost. The dataset is taken from a published database initiated by the Internal Biomedical Health Informatics Conference (ICBHI). As characteristics and machine learning models are used to learn between respiratory traits, reconstructed sub-band energies are extracted. The feasibility and efficiency of our proposed approach have been verified by our findings.

Keywords: Wheeze · Crackle · Lung sounds · Discrete wavelet transform · Machine learning · Classification

1 Introduction

Respiratory related diseases are among the world's most dangerous diseases. Approximately 334 million individuals suffer from asthma and more than 60 million individuals suffered from extreme chronic obstructive pulmonary disease (COPD), in which around 3 million die each year, making COPD the 3rd leading cause of death after heart disease and stroke worldwide [1]. Pulmonologists used the technique of chest auscultation as a simple non-invasive way of monitoring suspicious traces in the pulmonary system. However, depending primarily on the skill and expertise of the doctor, the limitation of this method is noise and lack of lung sound storage equipment [2]. Hence, the systematic investigation of lung sounds can be highlighted as a positive approach to screening for lung disorders.

Lung sounds are used to identify lung diseases. Lung sound refers to the sound of natural breathing heard from below 100 Hz to 1000 Hz, with a rapid decrease of 100 to

200 Hz [3] on the surface of the lung. Abnormal or unintended lung sounds contain several various types of sounds. Wheeze and crackle are two kinds of unintended lung sounds that are most important in lung sound studies [4–7]. Wheezes is a musical tone that is long beyond 100ms that can often be perceived by the human ear. Wheezes are produced in the range of 100 to 1 000 Hz as sinusoidal oscillations with harmonics that sometimes exceed 1000 Hz. Crackles are brief, explosive, non-musical sounds that are heard by inspiration, as opposed to wheezes. Two main types of crackles in the lung sound were noticed: coarse crack-les (found at 350 Hz and 15 ms) and fine crackles (typical 650 Hz frequency, and the typical 5 ms duration) [3].

In order to characterize lung sound, various techniques are used, primarily divided into two major approaches: lung sound analysis in the time domain and in the frequency domain. In the time domain analysis, some features of lung sound such as kurtosis (χ), skewness (ξ), lacunarity (ζ), Teager energy and sample entropy have been long used in researching lung sounds [4]. The processing of frequency domains is a popular approach. The short time Fourier transform [5], Mel-frequency spectral [6], Welch spectral [7] were implemented as the algorithm that recognize the features in the frequency spectrum.

In 2017, the Internal Conference on Biomedical Health Informatics (ICBHI) [8] introduced a broad dataset of lung sounds. Since then, several publications have used the dataset to solve the classification task [9, 10] in order to identify pulmonary abnormalities. The spectrogram and deep convolutional neural network were used by most of them to identify various lung sounds or different disease types. In the field of image and signal processing, a convolutional neural network has received great attention [11]. The research in [9] presented the accuracy of 86% and 80% ICBHI score on 2-class and 4-class classification, respectively using spectrogram and CNN-MoE model. The tremendous feature-fitting potential of deep neural networks, however, comes with several trade-offs. One of which is the time and computational cost to analyze the signal because the complexity of neural networks when processing data. In addition, the neural networks have long been considered as a "black box" with the user because the method of processing data is under control.

The discrete wavelet transforms, and sub-band energy estimate can be used to extract features for the classification process, which will help solve the above problems while still ensuring a high level of accuracy [12–14]. Sample normalization is not mentioned in the current studies, so we add the processes of normalization to increase accuracy, which is the highlight of the study. Finally, to evaluate the efficiency of such algorithms in lung sounds classification, the use of machine learning models such as support vector machines (SVM), k-nearest neighbor (K-NN), decision tree and bagged tree ensemble learning is implemented.

2   Methodology

2.1   Database

In 2017, the Int. Conf. on Biomedical Health Informatics (ICBHI) was held with the aim to publicize the meta-database of lung sounds recordings [8]. Kaggle is a site for data scientists to share and access large database, including the lung sounds meta-database from ICBHI that we used in our paper [15]. There were a total of 920 annotated recordings of lung sounds. These sound clips vary in length, extending from 10s to 90 s, and recorded on 126 patients, which are comprised of healthy and unhealthy individuals. Each audio clip is supplemented with a text in which there are four columns: start and end of a respiration cycle, presence of wheeze or crackle in the cycle (which is decoded as 0 for non-existence and 1 for existence of the sound). After a careful examination, we were aware that the subset of a total 1511 sound cycle were the most prominent of respiratory traits (596 normal, 344 wheezes and 571 crackle, lengths swinging from 0.1 s to 2 s). These recordings are measured from chest anterior, posterior and lateral positions.

2.2 Preprocessing

The preprocessing are composed of the Pre-processing stage (Denoise with wavelet denoising, resample data, and normalize them) and the Features Extraction (Decomposition and Features Extraction) (Fig. 1).

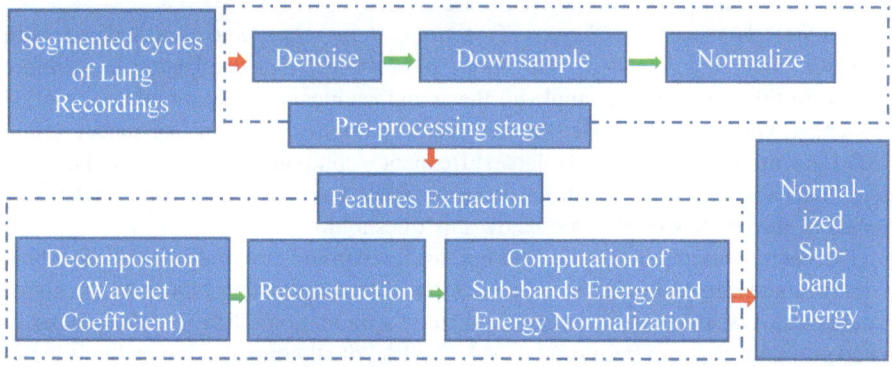

Fig. 1. Block diagram of pre-processing step

2.2.1 Denoising

In [12, 13], researchers contemplated sounds denoising as a pre-processing stage of data cleaning. In [12], Yan Shi and colleagues introduced a method that utilized thresholded wavelet reconstructed signals as a mean to eliminate the interference of the heart sounds imposed on the lung audios. Similarly, in [13], Kandaswamy and colleagues eradicated the noise with the method of wavelet shrinkage denoising. We applied the similar method in our work to denoise the data. Sym4 wavelet, minimax and soft threshold were set in the algorithm parameters. Illustration of a segment of sound signal with and without denoise are shown in Fig. 2.

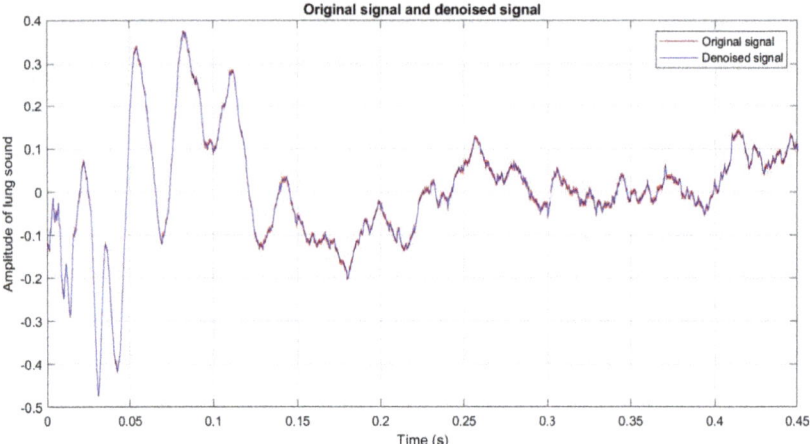

Fig. 2. Original signal and denoised signal comparison

2.2.2 Down-Sampling and Normalizing

In the ICBHI database, several types of electronic stethoscope were used, and each with a different sampling frequency, maximum at 44100 Hz. Standardization is therefore needed. In this paper, we resample all the sound segments to the minimum sampling frequency, which was 4000 Hz. As the Nyquist sampling theorem had stated, with a 4000 Hz sampling frequency, the largest frequency component of signals in the audio recordings was 2000 Hz, which ensure a complete traits extraction since most lung sounds of the paper's interest are below this threshold.

The instrument inhomogeneity and non-ideal recording environments all contribute to the unstandardized loudness of audio sounds, leading to uncertainty in the signal feature extraction if we use these audios directly without normalization. Many types of sound normalization had been prosed [16]. The trial of all normalization methods led us to realize that with the normalization of signal amplitude, namely the amplitude normalized are in the range of [−1, 1], a generally higher accuracy was obtained:

$$X_{nor} = \frac{X_i - ((X\_max + X_{min})/2)}{(X_{max} - X_{min})/2}$$

X_i represents sample's time domain amplitude. X_{max} and X_{min} are the maximum and minimum. X_{nor} is the normalization value of X_i. The normalization is computed with every point value X_i in the sample (i.e., i = 1,2 ,...,N; N sample length).

2.2.3 Lung Sound Signals Decomposition with Multiresolution

Continuous Wavelet transform (CWT) is increasingly common practice in signal processing applications. However, the drawback of this method is its redundancy, which leads to the construction of discrete wavelet transform (DWT) algorithm. DWT is seen as an alternative for CWT but with a much lighter computational cost in signal decomposition, subsequently breaks down the signal into smaller frequency bands on an octave scale [17].

The resulting coefficients of the decomposition are composed of details (the high frequency component) and approximation (low frequency component that can be further decomposed) coefficients. Each sub-band in the decomposition lies on subsequent but non-overlapping frequency spectrum. For example, a signal of 4000 Hz can be decomposed on the 4 level DWT as below (Fig. 3).

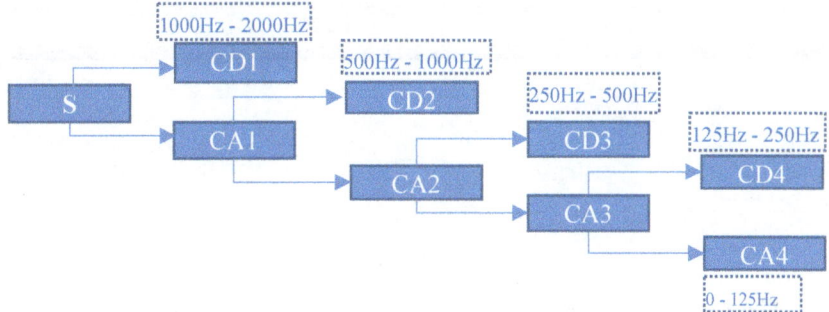

Fig. 3. Approximations of decomposed signal S at 4th level

In this paper, a 10 level DWT is used for lung sound decomposition. Each sound cycles decomposition resulted in 10 reconstructed detail signals (detail 1 to detail 10, D1–D10), and the approximation signal (A10). Each sub-band represent a spectrum in the range from 0–2000 Hz. The decomposition in [12, 13] were 5 and 7 respectively. A 10 level was used in this paper, which results in more features extracted. Also, we can visualize more clearly sub-bands energy. The energy distribution of higher level decomposition was also noticed. Figures 4, 5 and 6 demonstrate three decomposition sound types, namely normal vesicular lung sound, wheeze sounds and crackle sounds. The coif2 wavelet was utilized for the transform.

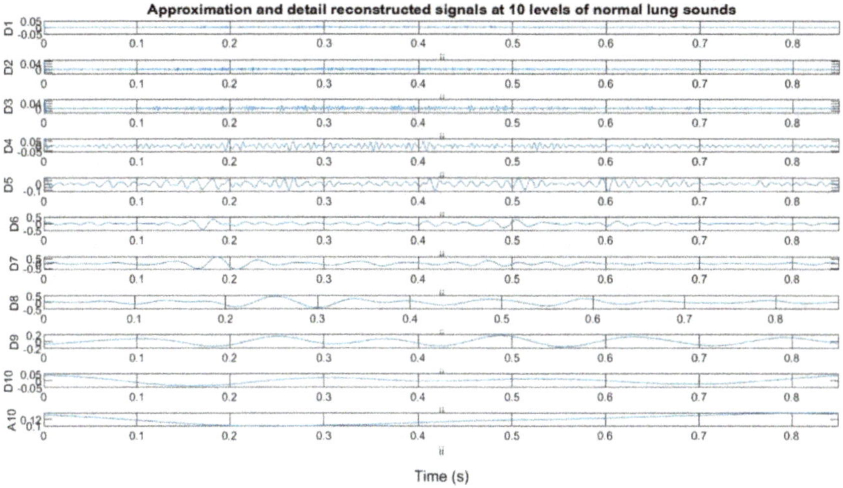

Approximation and detail reconstructed signals at 10 levels of normal lung sounds

Time (s)

Fig. 4. 10 levels detail and approximation reconstructed signals of normal vesicular sounds.

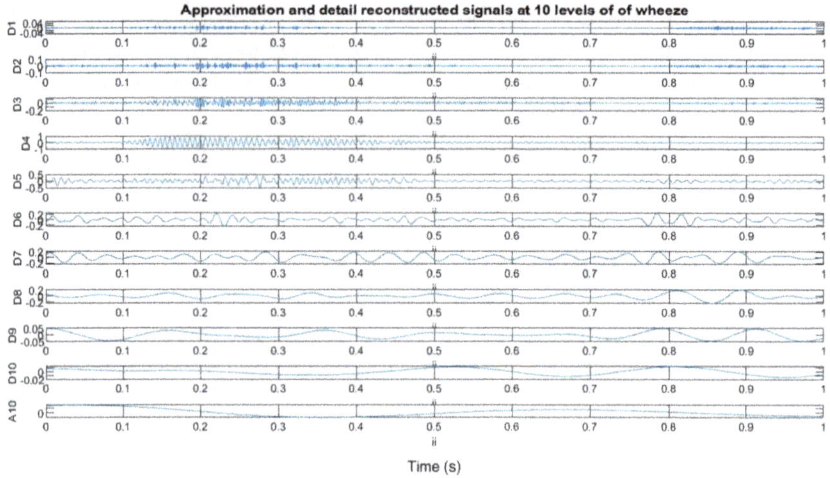

Fig. 5. 10 levels detail and approximation reconstructed signals of wheeze.

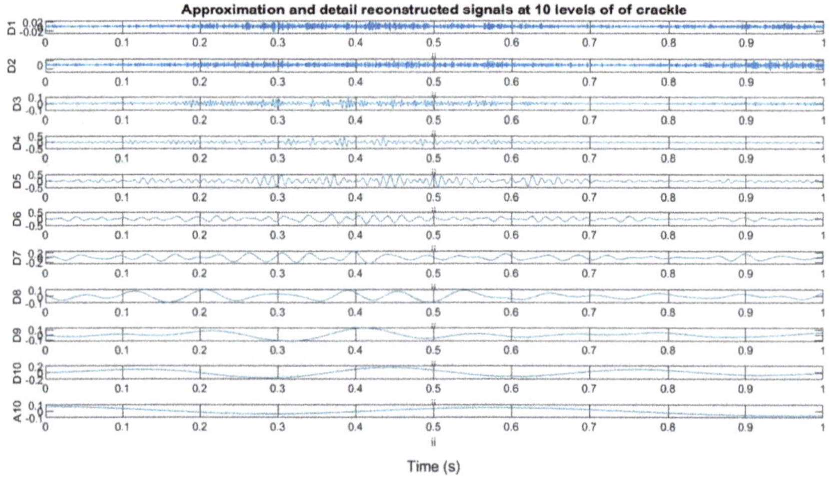

Fig. 6. 10 levels detail and approximation reconstructed signals of crackle.

The three above figures show the patterns observed from three lung sound types. As can be seen, wheeze energy is higher around the frequency of 400 Hz, crackle's frequency range is more widespread with the distribution in both high and low up to 300 Hz, and the normal sound's energy focus in the lower part of the spectrum, at below 200 Hz.

2.2.4 Features Extraction

The author of [13] considered some sub-bands as irrelevant to the spectrum range of interested signals and eliminated them. Similarly, the signal component A10 is removed in this paper due to the fact that it lied outside the interested frequency range ad was the subject to great noise and trend effects. The feature set are collected from the remaining sub-band other than A10, which resulted in a more informative, intuitive signal representation.

The features used are the energy contained in each sub-band. From D1 to D10 components, the energies are calculated and then normalized:

$$E_i = \|D_i\|^2 = \sum_{j=0}^{n-1} |D_{ij}|^2 \quad E_{N_i} = \frac{E_i}{\sum_{i=0}^{m-1} E_i}$$

D_i the detail reconstructed signal at i^{th} level; E_i is the computed energy of D_i; m the level of decomposition; n the sample length (the original sound length); and E_{N_i} is normalized E_i.

Figure 7 illustrates the difference among the lung sound types normal, crackle and wheeze in terms of their energy distribution over various sub-bands. The patterns of different respiratory sounds can be observed: normal sound energy focus at level higher than D5; wheeze energy concentrates around D3, D4 bands and crackle's energy can be seen disperse on a wider range.

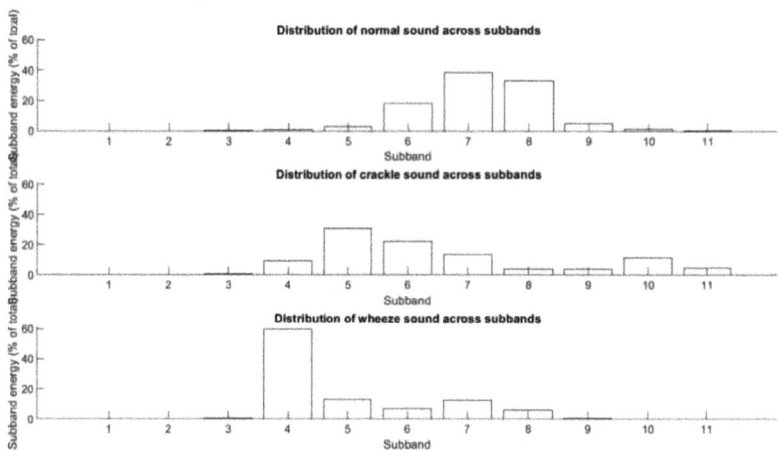

Fig. 7. Distribution of lung sound energy across sub-bands

3 Result and Discussion

In this research, the authors carried out classification experiments of different types of lung sound. The dataset used was the audio recordings from the ICBHI 2017 respiratory sound database. A subset of total 1511 respiratory cycles were used in this scope.

With the use of DWT and machine learning, a faster implementation of lung signal decomposition and classification can be attained, which solves the accuracy-computational cost trade-offs that associate with training deep neural networks. The decomposed sub-bands signal is used to extract energy distribution of the sound samples. The sub-band energies are used as input features for machine learning models to learn the traits of the different lung sounds: normal vesicular, wheeze and crackle sound. Normalization and denoise are incorporated in this paper to assess the influence of these steps to final accuracy. The learning models used are support vector machine (SVM), K-nearest neighbor (K-NN), decision tree and ensemble learning, which are used widely in bio-signal and biomedical image processing applications [18]. In this paper, a train test ratio of 70–30 is used with a 5-fold cross validation, all performed with MatLab R2019a.

Table 1 illustrates the results associated with each training models on the four metrics: sensitivity, specificity, accuracy, and F-1 score. The results of training with respect to normalized and not normalized sound samples are also shown.

Table 1. Scores using different learning models, with and without normalization.

Models	Cubic SVM (%)			Weighted KNN (%)			Fine Decision Tree (%)			Bagged Ensemble Tree (%)		
	Normal	Wheezes	Crackles	Normal	Wheezes	Crackles	Normal	Wheezes	Crackles	Normal	Wheezes	Crackles
Accuracy												
Unnormalized	83,26	88,55	85,90	83,04	89,21	82,82	82,38	87,44	80,84	85,46	90,09	85,68
Normalized	83,26	88,55	85,90	83,04	89,21	82,82	82,38	87,44	80,84	87,22	92,07	87,67
Sensitivity												
Unnormalized	82,76	78,64	75,14	78,16	65,05	84,18	79,31	73,79	72,32	81,61	81,55	79,10
Normalized	82,76	78,64	75,14	78,16	65,05	84,18	79,31	73,79	72,32	83,33	84,47	83,05
Specificity												
Unnormalized	83,57	91,45	92,78	86,07	96,30	81,95	84,29	91,45	86,28	87,86	92,59	89,89
Normalized	83,57	91,45	92,78	86,07	96,30	81,95	84,29	91,45	86,28	89,64	94,30	90,61
F1-score												
Unnormalized	79,12	75,70	80,61	77,94	73,22	79,26	77,53	72,73	74,64	81,14	78,87	81,16
Normalized	79,12	75,70	80,61	77,94	73,22	79,26	77,53	72,73	74,64	83,33	82,86	84,00

From the table, we can observe that the highest crackle sensitivity and wheeze specificity are obtained with the weighted K-NN model, which suggests that this model has high classification capacity with respect to this dataset. Overall, Ensemble tree model was the model that proved to be most effective with F1-score of 0.83, 0.82, and 0,84 for the three types of normal vesicular, wheezes, and crackles sound, respectively. These number shows that classification with feature extraction by wavelet transform and machine learning models have a potential in differentiating various respiratory traits. The scores of the normalized and not normalized dataset indicates the influence of this step. For SVM, K-NN, and decision tree, the changes are too small but for ensemble learning model, the accuracy boost is prominent with scores increased from 2%–4%. The results can be compared with the work in [12, 13]. In [12], the best accuracy obtained with the test set was 94.56%, but this work used deep neural network and the training data was small at only 126 samples. The authors in [13] achieved accuracy of 92.5% with training data of 58 samples. In our paper, we used of a much larger dataset, which includes 1511 samples with simpler machine learning models. Using our method, we can have a clear understanding of the signal processing at every detail step. Unlike deep neural network, the algorithm processing is a black box that we cannot control. The average accuracy of 89% obtained by our method indicates the high capability of this approach in lung sounds analysis and realization of lung diseases screening tools.

4   Conclusion

In this paper, discrete wavelet transform was used to classify the different types of lung sounds, namely wheeze, crackle and normal vesicular sounds. The energies of reconstructed sub-band signals are used as features for machine learning models, achieving an average accuracy of 89%. Compared to other works with the same approach, the accuracy is acceptable in the sense that it deals with a larger dataset and simple computation. The use of this approach in lung sound classification cuts down on the computational time and resources significantly, inferring that real-time screening applications are within grasp. Normalization of all audio samples were performed, and experimental results confirmed the efficiency of the method.

References

1. https://www.who.int/gard/publications/The_Global_Impact_of_Respiratory_Disease.pdf. Accessed 25 Oct 2020
2. Palaniappan, R., Sundaraj, K., Ahamed, N.U.: Machine learning in lung sound analysis: a systematic review. In: Biocybernetics and Biomedical Engineering 2013, vol. 33, pp. 129–135 (2013)
3. Bohadana, A., Izbicki, G., Kraman, S.: Fundamentals of lung auscultation. New Engl. J. Med. **370**, 744–751 (2014)
4. Mendes, L., et al.: Detection of wheezes using their signature in the spectrogram space and musical features. In: 37th Annual International Conference of the IEEE Engineering in Medicine and Biology Society (EMBC), Milan, pp. 5581–5584 (2015)
5. Serbes, G., Sakar, C.O., Kahya, Y.P., Aydin, N.: Pulmonary crackle detection using time–frequency and time–scale analysis. In: Digital Signal Processing 2013, vol. 23(3), pp. 1012–1021 (2013)
6. Mayorga, P., Druzgalski, C., Morelos, R.L., Gonzalez, O.H., Vidales, J.: Acoustics based assessment of respiratory diseases using GMM classification. In: Annual International Conference of the IEEE Engineering in Medicine and Biology Society 2010, pp. 6312–6316 (2010)

7. Oud, M., Dooijes, E.H., van der Zee, J.S.: Asthmatic airways obstruction assessment based on detailed analysis of respiratory sound spectra. IEEE Trans. Biomed. Eng. **47**(11), 1450–1455 (2000)
8. Rocha, B.M., et al.: α respiratory sound database for the development of automated classification. In: International Conference on Biomedical and Health Informatics 2017, pp. 33–37 (2017)
9. Pham, L., McLoughlin, I., Phan, H., Tran, M., Nguyen, T., Palaniappan, R.: Robust deep learning framework for predicting respiratory anomalies and diseases. In: 42nd Annual International Conference of the IEEE Engineering in Medicine & Biology Society (EMBC), Montreal, QC, Canada, pp. 164–167 (2020)
10. Ma, Y., Xu, X., Li, Y.: LungRN+NL: An improved adventitious lung sound classification using non-local block ResNet neural network with Mixup data augmentation. Interspeech **2020**, 2902–2906 (2020)
11. Yamashita, R., Nishio, M., Do, R., Togashi, K.: Convolutional neural networks: an overview and application in radiology. Insights Imaging **9**(4), 611–629 (2018)
12. Shi, Y., Li, Y., Cai, M., Zhang, X.D.: a lung sound category recognition method based on wavelet decomposition and BP neural network. Int. J. Biol. Sci. **15**(1), 195–207 (2019)
13. Kandaswamy, A., Kumar, C.S., Ramanathan, R.P., Jayaraman, S., Malmurugan, N.: Neural classification of lung sounds using wavelet coefficients. Comput. Biol. Med. **34**(6), 523–537 (2004)
14. Ulukaya, S., Serbes, G., Kahya, Y.: Overcomplete discrete wavelet transform based respiratory sound discrimination with feature and decision level fusion. Biomed. Signal Process. Control **38**, 322–336 (2017)
15. https://www.kaggle.com/vbookshelf/respiratory-sound-database. Accessed 25 Oct 2020
16. Bigand, E., Delbé, C., Gérard, Y., Tillmann, B.: Categorization of extremely brief auditory stimuli: domain-specific or domain-general processes? PloS One **6**(10), e27024 (2011)
17. Goodman, R.W.: Discrete Fourier and Wavelet Transforms: an Introduction Through Linear Algebra with Applications to Signal Processing. World Scientific, Singapore (2016)
18. Ray, S.: A quick review of machine learning algorithms. In: International Conference on Machine Learning, Big Data, Cloud and Parallel Computing (COMITCon) 2019, Faridabad, India, pp. 35–39 (2019)

Portfolio Selection with Risk Aversion Index by Optimizing over Pareto Set

Tran Ngoc Thang[(✉)] and Nguyen Duc Vuong

School of Applied Mathematics and Informatics,
Hanoi University of Science and Technology, Hanoi, Vietnam
thang.tranngoc@hust.edu.vn

Abstract. In decision making theory, portfolio selection problems have an important role, which suggest the best choices among many investing alternatives. Within our article, we take into consideration the portfolio selection problem with risk aversion index as an optimization problem over the efficient set of a convex biobjective programming problem based on Markowitz mean-variance model. By using the outcome space approach, we alter the considered problem to a convex programming problem, which is solved efficiently by some computational tools. The proposed algorithm is applied to optimize security portfolios and some experiments are reported.

Keywords: Portfolio selection · Markowitz model · Risk aversion index · Bi-objective programming · Pareto set

1 Introduction

In economy and finance, portfolio optimization helps investors research, understand and make right decisions about their investments and strategies. Let's consider a set of potential portfolios, the process of selecting the best portfolio according to some constraints is portfolio optimization.

Harry Markowitz was the first one to give a clear definition of modern portfolio theory (see [1]) in which an investor wants to maximize a portfolio's expected return on any given amount of risk. Our research approaches the portfolio optimization problem by the Mean-Variance model as a special case of bio-objective convex programming (BOP). To find an optimal solution in the Pareto set, we consider an auxiliary factor. In other words, we solve the optimization problem over the Pareto set. This problem belongs to NP-hard class even in the case we have linear objective functions and a polyhedral feasible set. Therefore, we alter the main problem to a convex programming problem.

Portfolio selection has attracted special attention of researchers. A lot of results have been released in the fields of economics, finance, agriculture and so on (see [2,4,5] and references therein). Some popular methods can be listed as follows: Weighted Sum of Deviations (see [7]), Chebyshev Goal Programming (see [7]) or Lagrangian Multiplier (see [8]). Most of previous methods apply scarlarization to transfrom Markowitz model to single-objective problem. However, whether scarlarization is applied to non-linear functions or not is still a controversial question. We approach problem (PS) as an optimization problem over the efficient set $(P_{\mathcal{X}})$. Based on approaching by the outcome space, we propose to transform problem $(P_{\mathcal{X}})$ to a single-objective convex programming problem.

In Sect. 2, the theoretical preliminaries are presented to analyze the portfolio selection problem with risk aversion index and propose the equivalent convex programming problem. Section 3 introduces the algorithm to solve the problem including some computational experiments as illustrations. A few conclusions are given within the final section.

2 Theoretical Preliminaries

2.1 Portfolio Selection by Markowitz Model

Consider the portfolio determination problem as a bi-objective programming problem based on Markowitz mean-variance model. Recall $x = (x_1, x_2, ..., x_n)^T$, in which x_j represents the relative amount invested in asset number j. Since, vector x is called a portfolio. Note that x_j is positive for all $j = \overline{1,n}$ and $\sum_{j=1}^{n} x_j = 1$. Return L_j of asset number $j, j = \overline{1,n}$ is a random variable with expected value $l_j = E(L_j)$. Notice $L = (L_1, L_2, ..., L_n)^T, l = (l_1, l_2, ..., l_n)^T$. Then, the whole portfolio has a return given as $L^T x = \sum_{j=1}^{n} L_j x_j$ and expected return given as $\mathcal{E}(x) = E(L^T x) = \sum_{j=1}^{n} l_j x_j$. Calling co-variance matrix of random vector R is $Q = (\sigma_{ij})_{n \times n}$. Then, variance of return is $\mathcal{V}(x) = Var(L^T x) = \sum_{i=1}^{n} \sum_{j=1}^{n} \sigma_{ij} x_j x_i$ where σ_j^2 is variance and ρ_{ij} is correlation coefficient between L_j and L_i, $i, j = 1, 2, ..., n$. Easy to see, matrix Q is symmetric and positive semi-definite.

Thus, the profit is calculated based on the expected return or the average value of the returns, while the risk is displayed by the variance of the returns on the whole portfolio. Investors perform optimally on two objectives that are maximally profitable and minimally risky functions with conditions bound on the value of the portfolio. Then, the problem of optimizing the portfolio is given as the following:

$$\max \mathcal{E}(x) = \sum_{j=1}^{n} l_j x_j$$

$$\min \mathcal{V}(x) = \sum_{i=1}^{n} \sum_{j=1}^{n} \sigma_{ij} x_j x_i \qquad (PS)$$

$$\text{s.t. } x_1 + x_2 + ... + x_n = 1, \ x_j \geq 0, j = \overline{1,n}.$$

Here, the risk function can be written in quadratic form $\mathcal{V}(x) = x^T Q x$. Because Q is positive semi-definite so $\mathcal{V}(x)$ is a convex function. Also easy to see, return function $\mathcal{E}(x)$ is linear. So problem (PS) is a special case of convex biobjective programming which can be given as the following

$$\text{Min}\{f(x)|x \in \mathcal{X}\}, \qquad (BOP)$$

where $\mathcal{X}$ is a nonempty compact convex set and the objective function $f = (f_1, f_2)^T$ is convex on $\mathcal{X}$. Recall that a vector function $f(x)$ is called *convex* on $\mathcal{X}$ if its component functions $f_i(x), i = \overline{1,2}$, are convex on $\mathcal{X}$. The problem of selecting a portfolio is known as finding the efficient solution set of problem (BOP).

2.2   Portfolio Selection Problem with Risk Aversion Index

As mentioned in the previous subsection, it is stated by Markowitz portfolio theory that an speculator should select a portfolio from the efficient solution set of problem (BOP), depending on his risk aversion. The investor can select a portfolio by maximizing an utility function, such as $U(\mathcal{E}(x), \mathcal{V}(x)) = \mathcal{E}(x) - A * \mathcal{V}(x)$, where A is the risk aversion coefficient, which is a non-negative real number. We can replace A by $1/\lambda - 1$ with a real number $\lambda \in [0, 1]$. Then, we consider the function $\bar{U}(\mathcal{E}(x), \mathcal{V}(x)) = -\lambda * U(\mathcal{E}(x), \mathcal{V}(x)) = \lambda * (-\mathcal{E}(x)) + (1 - \lambda) * \mathcal{V}(x)$. We can see that function $\bar{U}(\mathcal{E}(x), \mathcal{V}(x))$ is a weighted composition of return and risk. Selecting a portfolio is solved by minimizing this function over the Pareto set of (BOP). In general, the formulation of the main problem is given as

$$\min\{\Phi(x)|x \in \mathcal{X}_E\}, \qquad (P_{\mathcal{X}})$$

where $\Phi(x) = \langle \lambda, f(x) \rangle$ and $\mathcal{X}_E$ is the efficient solution set for problem (BOP), i.e. Recall that $\mathcal{X}_E$ contains every $x^0 \in \mathcal{X}$ which satisfies that there is no $x \in \mathcal{X}$ where $f(x^0) \geq f(x), \; f(x^0) \neq f(x)$.

As usual, the notation $y^1 \geq y^2$, where $y^1, y^2 \in \mathbb{R}^m$, is used to indicate $y_i^1 \geq y_i^2$ for all $i = 1, \ldots, m$.

It is widely acclaim that, in general, the set $\mathcal{X}_E$ is nonconvex and given implicitly as the form of a standard mathematical programming problem, even in the case $m = 2$ and we have linear objective functions f_1, f_2 as well as the polyhedral feasible set $\mathcal{X}$. Hence, problem $(P_{\mathcal{X}})$ is a global optimization problem and belongs to NP-hard problem class. In this article, we propose to transform the main problem to a convex programming problem.

2.3   The Equivalent Convex Programming Problem

Let $\mathcal{Y} = \{y \in \mathbb{R}^m | y = f(x) \text{ for some } x \in \mathcal{X}\}$. Normally, the set $\mathcal{Y}$ represents the *outcome set* for problem (BOP).

Let $Q \subset \mathbb{R}^m$ be a nonempty set. $\text{Min}Q$ is denoted as the efficient set of Q. Note that $\text{Min}Q = \{q^0 \in Q | \nexists q \in Q : q^0 \geq q, q^0 \neq q\}$. It is clear that a point $q^0 \in \text{Min}Q$ if $Q \cap (q^0 - \mathbb{R}_+^m) = \{q^0\}$, where $\mathbb{R}_+^m = \{y \in \mathbb{R}^m | y_i \geq 0, \; i = 1, \ldots, m\}$.

Since the functions $f_i, i = 1, \ldots, m$, are continuous and $\mathcal{X} \subset \mathbb{R}^n$ is a nonempty compact set, the outcome set $\mathcal{Y}$ is also compact set in $\mathbb{R}^m$. Therefore, the efficient set $\text{Min}\mathcal{Y}$ is nonempty [3]. Let $\mathcal{Y}_E$ be denoted as the *efficient outcome set* for problem (BOP), where $\mathcal{Y}_E = \{f(x) | x \in \mathcal{X}_E\}$. We can infer by definition that $\mathcal{Y}_E = \text{Min}\mathcal{Y}$. The relationship between the efficient solution set $\mathcal{X}_E$ and the efficient set $\text{Min}\mathcal{Y}$ is described as follows.

Proposition 1.

i) For any $y^0 \in \text{Min}\mathcal{Y}$, if $x^0 \in \mathcal{X}$ satisfies $f(x^0) \leq y^0$, then $x^0 \in \mathcal{X}_E$.
ii) For any $x^0 \in \mathcal{X}_E$, if $y^0 = f(x^0)$, then $y^0 \in \text{Min}\mathcal{Y}$.

Let $\mathcal{Z} = \mathcal{Y} + \mathbb{R}^m_+ = \{z \in \mathbb{R}^m \mid z \geq y \text{ for some } y \in \mathcal{Y}\}$. Apparently, $\mathcal{Z}$ is not empty. Also, it is a full-dimension closed set but it is nonconvex in general (see Fig. 1) The following interesting property of $\mathcal{Z}$ will be used in the sequel.

Proposition 2. $\text{Min}\mathcal{Z} = \text{Min}\mathcal{Y}$.

The problem $(P_\mathcal{X})$ can be reformulated by outcome-space approach as

$$\min \varphi(y) \text{ s.t. } y \in \mathcal{Y}_E. \tag{$P_\mathcal{Y}$}$$

Combining $\mathcal{Y}_E = \text{Min}\mathcal{Y}$ and Proposition 2, problem $(P_\mathcal{Y})$ can be rewritten as

$$\min \varphi(y) \text{ s.t. } y \in \text{Min}\mathcal{Z}. \tag{$P_\mathcal{Z}$}$$

Thus, we solve problem $(P_\mathcal{Z})$ instead of problem $(P_\mathcal{Y})$.

Proposition 3. *If f is convex, the set $\mathcal{Z} = f(X) + \mathbb{R}^2_+$ is convex in $\mathbb{R}^2$.*

As $\mathcal{Z}$ is a nonempty convex subset in $\mathbb{R}^2$ by Proposition 3, the fact that $\text{Min}\mathcal{Z}$ is homeomorphic to a line segment $[a, b] \subset \mathbb{R}$ is proven in [4]. By geometry, it is easily seen that the optimal solution set of the following problem

$$\min\{y_2 \mid y \in \mathcal{Z}, \ y_1 = y_1^I\} \tag{P_S}$$

contains only one solution y^S and similarly, the problem

$$\min\{y_1 \mid y \in \mathcal{Z}, \ y_2 = y_2^I\} \tag{P_E}$$

has a unique optimal solution y^E. Since $\mathcal{Z}$ is convex, problems (P_S) and (P_E) are convex programming problems. If $y^S = y^E$ then $\mathcal{X}_E = \{y^S\}$ and y^S is the only optimal solution of problem $(P_\mathcal{X})$, else if $y^S \neq y^E$ then $\text{Min}\mathcal{Z}$ is a curved line on the boundary of $\mathcal{Z}$ with starting point y^S and the end point y^E such that

$$y_1^E > y_1^S \text{ and } y_2^S > y_2^E. \tag{1}$$

Note that we also get the efficient solutions $x^S, x^E \in \mathcal{X}_E$ such that $y^S = f(x^S)$ and $y^E = f(x^E)$ while solving problems (P_S) and (P_E). For the convenience, x^S, x^E is called to be *the efficient solutions respect to y^S, y^E*, respectively. In the second case, we consider problem (P_Z) where $\varphi(y) = \langle \lambda, y \rangle$. Direct computation reveals the line through y^S and y^E equation is $\langle c, y \rangle = \alpha$, in which

$$\begin{cases} c = \left(\frac{1}{y_1^E - y_1^S}, \frac{1}{y_2^S - y_2^E} \right), \\ \alpha = \frac{y_1^E}{y_1^E - y_1^S} + \frac{y_2^E}{y_2^S - y_2^E}. \end{cases} \tag{2}$$

From (1), it is easily seen that the vector c is strictly positive. Now, let $\tilde{Z} = \{ y \in Z \mid \langle c, y \rangle \le \alpha \}$ and $\Gamma = \partial \tilde{Z} \backslash (y^S, y^E)$, where $(y^S, y^E) = \{ y = ty^S + (1-t)y^E \mid 0 < t < 1 \}$ and $\partial \tilde{Z}$ is the boundary of the set $\tilde{Z}$.

It is clear that $\tilde{Z}$ is a compact convex set because Z is convex. By the definition and geometry, we can see that Γ contains every extreme points of $\tilde{Z}$, moreover, $\mathrm{Min} Z = \Gamma$. Consider the following convex problem

$$\min \langle \lambda, y \rangle \text{ s.t. } y \in \tilde{Z}. \tag{CP^0}$$

that has the explicit reformulation as follows

$$\begin{array}{ll} \min & \langle \lambda, y \rangle \\ \text{s.t.} & f(x) - y \le 0 \\ & x \in \mathcal{X}, \langle c, y \rangle \le \alpha, \end{array} \tag{CP^1}$$

where the vector $c \in \mathbb{R}^2$ and the real number α is determined by (2). Because of the convexity of the objective function f and the feasible set $\mathcal{X}$, problem (CP^1) is a convex programming problem with $n + 2$ real variables.

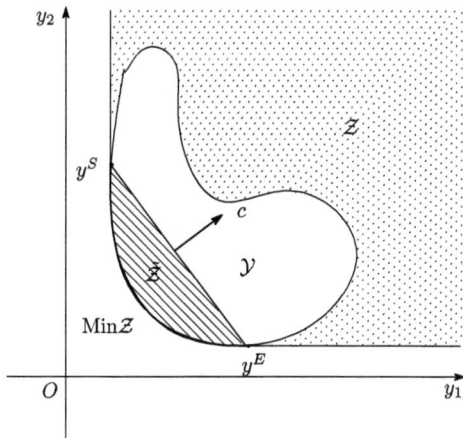

Fig. 1. The convex set $\tilde{Z}$

Proposition 4. *If* $(x^*, y^*) \in \arg\min_x (CP^1)$, *then* $x^* \in \arg\min_x (P_\mathcal{X})$.

Proof. It is well known that a convex programming problem with the linear objective function has an optimal solution which belongs to the extreme point set of the feasible solution set [6]. Therefore, problem (CP^0) has an optimal solution $y^* \in \Gamma$. This fact and $\mathcal{Y}_E = \text{Min}\mathcal{Y}$ implies that $y^* \in \text{Min}\mathcal{Z}$. Since $\text{Min}\mathcal{Z} \subset \tilde{\mathcal{Z}}$, it implies that $y^* \in \arg\min_x (P_{\mathcal{Z}})$.

Since $\text{Min}\mathcal{Z} = \mathcal{Y}_E = \text{Min}\mathcal{Y}$, by definition, we have $\langle \lambda, y^* \rangle \leq \langle \lambda, y \rangle$ for all $y \in \mathcal{Y}_E$ and $y^* \in \text{Min}\mathcal{Y}$. Then

$$\langle \lambda, y^* \rangle \leq \langle \lambda, f(x) \rangle, \ \forall x \in \mathcal{X}_E. \tag{3}$$

Because $(x^*, y^*) \in \arg\min_x (CP^1)$, $f(x^*) \leq y^*$. By Proposition 1, $x^* \in \mathcal{X}_E$. Furthermore, $f(x^*) \in \mathcal{Y}$ and $y^* \in \text{Min}\mathcal{Y}$. The definition of efficient points infers that $y^* = f(x^*)$. Combining this fact and (3), we get $\Phi(x^*) \leq \Phi(x) \ \forall \ x \in \mathcal{X}_E$ which means $x^* \in \arg\min_x (P_{\mathcal{X}})$.   ∎

3   Procedures and Computing Experiments

The procedure for solving problem $(P_{\mathcal{X}})$ is established by Proposition 4.

Procedure 1.

Step 1. Solve problem (P_S) and problem (P_E) to find the efficient points y^S, y^E and the efficient solutions x^S, x^E respect to y^S, y^E, respectively.

Step 2. **If** $y^S \neq y^E$ **Then** Go to Step 3. **Else** The procedure is terminated ($y^S \in \arg\min_x (P_{\mathcal{X}})$).

Step 3. Solve problem (CP^1) to find an optimal solution (x^*, y^*). The procedure is terminated ($x^* \in \arg\min_x (P_{\mathcal{X}})$).

Below are the example to illustrate Procedure 1.

Example 1. Consider problem $(P_{\mathcal{X}})$, where $\mathcal{X}_E$ is the efficient solution set to problem (BOP) with $f_1(x) = x_1^2 + 1$, $f_2(x) = (x_2 - 3)^2 + 1$,

$$\mathcal{X} = \left\{ x \in \mathbb{R}^2 \mid (x_1 - 1)^2 + (x_2 - 2)^2 \leq 1, 2x_1 - x_2 \leq 1 \right\},$$

and $\Phi(x) = \lambda_1 f_1(x) + \lambda_2 f_2(x)$. The computational results are shown in Table 1.

Table 1. Results of Example 1 by computation

λ	x^*	y^*	$\Phi(x^*)$
$(0.0, 1.0)$	$(0.9612, 2.9991)$	$(1.9412, 1.0000)$	1.0000
$(0.2, 0.8)$	$(0.4460, 2.8325)$	$(1.1989, 1.0281)$	1.0622
$(0.5, 0.5)$	$(0.2929, 2.7071)$	$(1.0858, 1.0858)$	1.0858
$(0.8, 0.2)$	$(0.1675, 2.5540)$	$(1.0208, 1.1989)$	1.0622
$(1.0, 0.0)$	$(0.0011, 2.0435)$	$(1.0000, 1.9326)$	1.0000
$(-0.2, 0.8)$	$(0.9654, 2.9992)$	$(1.9412, 1.0000)$	0.4118
$(0.8, -0.2)$	$(0.0011, 2.0435)$	$(1.0000, 1.9326)$	0.4146

Example 2. (see [8], pp. 7) There are five stocks with the list of ticker symbols as DGX, AAPL, BAC, IBM, MSFT. Chronicled stock cost and profit installment from 2002 to 2007 are used to calculate the anticipated return and covariance matrix of five stocks (Table 2).

Table 2. Expected Returns and Covariance Matrix of chosen stocks

Stock	Expected Profit	Covariance Matrix				
		DGX	AAPL	BAC	IBM	MSFT
DGX	1.006%	0.0056	0.0014	0.0004	0.0024	−0.0002
AAPL	4.085%	0.0014	0.0127	0.0013	0.0024	0.0010
BAC	1.236%	0.0004	0.0013	0.0016	0.0010	0.0006
IBM	0.400%	0.0024	0.0024	0.0010	0.0065	0.0030
MSFT	0.513%	−0.0002	0.0010	0.0006	0.0030	0.0039

We use Matlab to resolve the problem, then the speculator can get an insight into the Expected Return and Risk for any risk aversion index A in Fig. 2. From the figure, it can be revealed that when A varies from 20 to 50 or greater, there is not a significant change in Expected Return. The same trend is also applied to Risk when A ranges between 5 and 50 or higher.

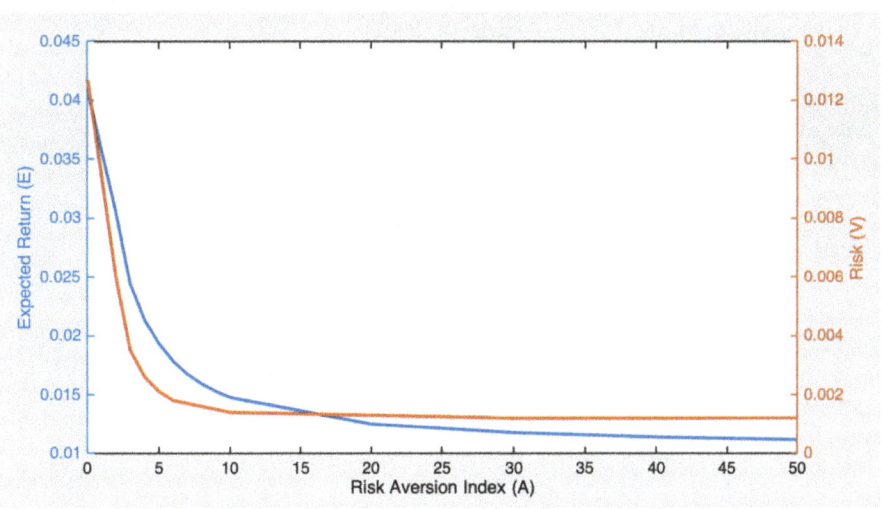

Fig. 2. Expected Return and Risk vs. Risk Aversion Index

4   Conclusion

Within this research, a method has been developed to solve the issue of portfolio selection with risk aversion index by proposing the equivalent single-objective optimization problem over Pareto Set. Practical experiments reveal that this method required little computational effort in comparison to other methods because only some convex programming need solving. Therefore, it can be widely applied in most fields where exist portfolio optimization.

Acknowledgment. This research is funded by Hanoi University of Science and Technology (HUST) under grant number T2018-TT-007.

References

1. Markowitz, H.: Portfolio selection. J. Fin. **7**(1), 77–91 (1952)
2. Kim, N.T.B., Thang, T.N.: Optimization over the efficient set of a bicriteria convex programming problem. Pac. J. Optim. **9**, 103–115 (2013)
3. Luc, D.T.: Theory of Vector Optimization. Springer, Berlin (1989)
4. Phu, H.X.: On efficient sets in $\mathbb{R}^2$. Vietnam J. Math. **33**, 463–468 (2005)
5. Thoai, N.V.: Reverse convex programming approach in the space of extreme criteria for optimization over efficient sets. J. Optim. Theory Appl. **147**(2), 263–277 (2010). https://doi.org/10.1007/s10957-010-9721-2
6. Tuy, H.: Convex Analysis and Global Optimization. Kluwer (1998)
7. Roberts, M.C., Dizier, A.S., Vaughan, J.: Multiobjective Optimization: Portfolio Optimization Base on Goal Programming Methods (2012)
8. Duan, Y.C.: A multi-objective approach to portfolio optimization. Rose Hulman Undergraduate Math. J. **8**(1), Article 12 (2007)

Control Design for Two-Channel Missile Systems

Ba Thanh To[1], Duc Thuan Tran[1], and Phuong Nam Dao[2(✉)] 🔾

[1] Ministry of Defense, Hanoi, Vietnam
[2] School of Electrical Engineering, Hanoi University of Science
and Technology, Hanoi, Vietnam
nam.daophuong@hust.edu.vn

Abstract. This paper presents the algorithm for determining the control law of two steering channels of the air-to-air missile rotating around the vertical axis by using a relay steering mechanism. The approach of proposed control law is to meet the requirements of steering channels seperation and ensures that the generated normal force tracks the guidance problem requirements. Moreover, several related control results are given out based on the linearization signals to be utilized the standard signal in the form of harmonic function. The theoretical analysis validated the effectiveness of proposed control in separating the control channels of steering mechanism and identifying the linear dependence between the control pulse parameters and the normal force of missile.

Keywords: Two-channel missile · Normal force · Steering mechanism

1 Introduction

The control design techniques play an important role in industrial automation, such as robotic systems, electrical drive systems, power systems. Several basic approaches in robust adaptive control, optimal strategies have been mentioned in robotic systems [1–9]. Firstly, the control systems of holonomic manipulators can be designed by only one control loop in the presence of external disturbances, unknown dynamic and variable time-delays [1]. In [1], the sliding variable is introduced to allow the order reduction of Bilateral Teleoperations (BTs). Based on the particular properties of classical robots, it is able to implement the adaptive controller with an appropriate adaptation law [1]. Unlike the traditional method for BTs using scattering theory and wave variables, authors in [1] deal with the challenges of variable time-delays by energy based Lyapunov candidate function [1]. Secondly, the cascade control scheme are always utilized in mobile robotic systems with Backstepping technique being the remarkable method for stabilization of whole control system [2–9]. The key idea of Backstepping technique enables to obtain the Lyaunov candidate function of whole control system from the outer control. Moreover, due to the difficulty in obtaining the state variables, the output feedback control is established in the cascade control system [7]. On the other hand, the stability of model predictive control design is also determined by appropriate Lyapunov function [9]. Additionally, for the missile system, several mathematical

D.-T. Tran et al. (Eds.): ICISN 2021, LNNS 243, pp. 233–238, 2021.
https://doi.org/10.1007/978-981-16-2094-2_29

models have been given with the purpose of finding the completed control schemes [10–12].

It should be noted that almost previous control design techniques for dynamical systems are based on Lyapunov stability theory. This article presents a different approach for control system of missiles using the linearization signals and separation technique of steering mechanism in the absence of Lyapunov theory. The rest of paper is organized as follows. We describe the physical principle and mathematical model of Missiles in Sect. 2. The proposed algorithm without Lyapunov stability theory is represented in Sect. 3. Finally, the conclusions are provided in Sect. 4.

2   The Physical Principal and Mathematical Model of Missiles

Considering the flying vehicles (FV) with the structure as shown in Fig. 1. Like other flying vehicles, in order to control the flight trajectory, the steering wings (4 wings) must be impacted with the aim of creating the angle of attack and the angle of slideslip to have normal power. The normal force here is essentially aerodynamic forces.

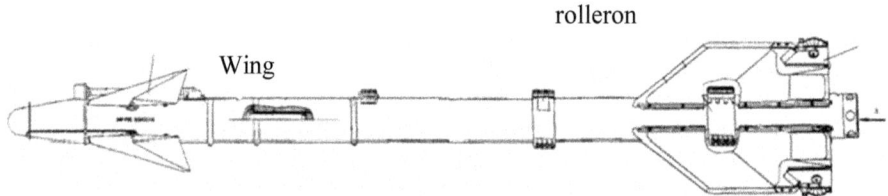

Fig. 1. Two-channel flying vehicle with rolleron mechanism

Figure 2 shows the convention for the coordinate system of two-channel missiles: the OX axis coincides with the vertical axis, the OXZ plane contains the B-B steering wing to create the angle of attack, the OXY plane contains the A-A steering wing to create the angle of slideslip.

The mathematical Model of Missiles can be given as described in [11, 12]:

$$mV \cdot \frac{d\theta}{dt} = (PK_\alpha + C_\alpha)\delta_B \cos \omega_X t + (PK_\beta - C_\beta)\delta_H \sin \omega_X t - G \qquad (1)$$

$$-mV \cdot \frac{d\Psi}{dt} = (PK_\alpha + C_\alpha)\delta_B \sin \omega_X t + (-PK_\beta + C_\beta)\delta_H \cos \omega_X t \qquad (2)$$

Two Eqs. (1) and (2) are linear differential equations, thus they satisfying the superposition principle. We now consider the force components created by the effect of the angle of attack and the pair wings δ_B, δ_H and the effect of the angle of slideslip and the pair wings δ_B, δ_H:

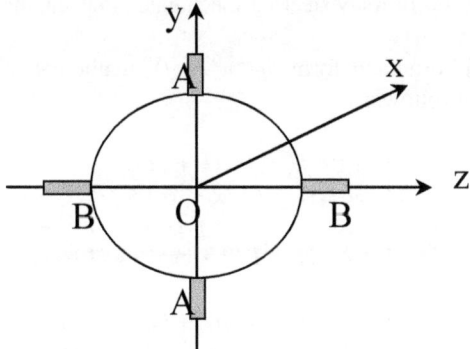

Fig. 2. Coordinate system on two-channel missiles

$$mV \cdot \frac{d\theta}{dt} = (PK_\alpha + C_\alpha)\delta_B \cos \omega_X t + (PK_\beta - C_\beta)\delta_H \sin \omega_X t \qquad (3)$$

$$-mV \cdot \frac{d\Psi}{dt} = (PK_\alpha + C_\alpha)\delta_B \sin \omega_X t + (-PK_\beta + C_\beta)\delta_H \cos \omega_X t \qquad (4)$$

3   Proposed Algorithm

The left sides of (3) and (4) are parts of the normal force. Suppose that we need to change the rotation angle of the wings so that the lift force reaches the desired value d_1 and the drag force reaches the desired value d_2 (the values d_1, d_2 are determined by the guidance problem and depended on the specified guided method), it means:

$$(PK_\alpha + C_\alpha)\delta_B \cos \omega_X t + (PK_\beta - C_\beta)\delta_H \sin \omega_X t = d_1 \qquad (5)$$

$$(PK_\alpha + C_\alpha)\delta_B \sin \omega_X t + (-PK_\beta + C_\beta)\delta_H \cos \omega_X t = d_2 \qquad (6)$$

For the steering mechanisms that produces continuous values δ_B, δ_H, the creation of the change rule δ_B, δ_H to meet the requirements in the expressions (5) and (6) is shown in [3]. However, the continuous mechanisms often have a complicated structure, large size and weight, so they are unsuitable for small sized missiles. For small sized missiles, relay steering mechanism is often used, therefore, δ_B, δ_H only receive one of the two threshold values (one with negative threshold, one with positive threshold and the magnitude of these two thresholds is usually equal to $\delta*$). However, the laws of creating them to satisfy the two Eqs. (5) and (6) will be more complicated. Thanks to the development of IC technology, it is possible to create the command generator to change δ_B, δ_H according to that complex algorithm. The following is a suggested solution to change δ_B, δ_H of relay steering mechanism that satisfies the expressions (5) and (6).

Using the control command form $\delta_B(t)$, $\delta_H(t)$, mathematical description of that command is shown as follows:

$$\delta_B(t) = T \begin{cases} +\delta*; iT \le t < iT + (T/2 - \tau_B), \\ -\delta*; iT + (T/2 - \tau_B) \le t < iT + T/2, \\ +\delta*; iT + T/2 \le t < iT + T/2 + \tau_B, \\ -\delta*; iT + T/2 + \tau_B \le t < iT + T. \end{cases} \tag{7}$$

$$\delta_H(t) = T \begin{cases} +\delta*; iT \le t < iT + (T/2 - \tau_H), \\ -\delta*; iT + (T/2 - \tau_H) \le t < iT + T/2, \\ +\delta*; iT + T/2 \le t < iT + T/2 + \tau_H, \\ -\delta*; iT + T/2 + \tau_H \le t < iT + T. \end{cases} \tag{8}$$

Combining (5) and (6) with commands (7) and (8), it is shown that the normal force (lift and drag force) also has impulsive form. Because missile rotates around the vertical axis, we need to determine average lift force and drag force in a rotation period as follows:

$$\begin{aligned} F_{hTB} &= \frac{1}{T}(PK_\alpha + C_\alpha) \int_0^T (\delta_B \cos \omega_X t) dt + \frac{1}{T}(PK_\beta - C_\beta) \int_0^T (\delta_H \sin \omega_X t) dt \\ &= \frac{1}{T}(PK_\beta + C_\beta) 4 \frac{\delta*}{\omega_X} \cos(\omega_X \tau_H) = \frac{4\delta*}{T\omega_X}(PK_\beta - C_\beta) \cos(\omega_X \tau_H) \\ &= \frac{2\delta*}{\pi}(PK_\beta - C_\beta) \cos(\omega_X \tau_H) \end{aligned} \tag{9}$$

$$\begin{aligned} F_{ngTB} &= \frac{2\delta*}{\pi}(PK_\alpha + C_\alpha) \cos(\omega_X \tau_B) = \frac{2\delta*}{\pi}(PK_\alpha + C_\alpha) \cos \omega_X(t_3 - t_2) \\ &= \frac{2\delta*}{\pi}(PK_\alpha + C_\alpha) \cos(\omega_X t_3 - \omega_X t_2) = \frac{2\delta*}{\pi}(PK_\alpha + C_\alpha) \cos[\pi - \arccos(-a/2b)] \\ &= \frac{2\delta*}{\pi}(PK_\alpha + C_\alpha)\{\cos \pi \cos[\arccos(-a/2b)] - \sin \pi \sin[\arccos(-a/2b)]\} \end{aligned}$$

Or:

$$F_{ngTB} = \frac{2\delta*}{\pi}(PK_\alpha + C_\alpha)\frac{a}{2b} = K_{ng}a \tag{10}$$

Where K_{ng} is the scale factor:

$$K_{ng} = \frac{\delta*}{b \cdot \pi}(PK_\alpha + C_\alpha) \tag{11}$$

If b is kept at a constant value (b = const) and we change the value of a, the average control force affecting on the center of the missile is proportional to the amplitude a of the harmonic signal $y_1(t)$. Mathematically, the linear ratio property of the expression (10) is explained as follows: The value of average force which has impact on the center of the missile relates to pulse width according to the law of cos function (Eq. 5 and 6), but the relationship between that pulse width and the amplitude a of the control signal is according to the law of Arcos function (expression 39). So the relationship between the average force and the amplitude a will be linear (Cos(Arcos (D) = D).

From the above results, it is possible to state the important property as follows:

Theorem: *If the information on the control value is the amplitude of the harmonic function, the average control force is proportional to it. Thus, by utilizing the standard signal in the form of a harmonic function which with twice angular velocity when compared to the angular velocity rotating around the missile's longitudinal axis, the relationship between the average control force and the control command has been linearized. That is the reason for calling the above standard signal as a linearized signal.*

The comparison between (10) and (5) can be shown:

$$K_{ng}a = d_2, \ a = \frac{d_2}{K_{ng}} \tag{12}$$

The control command for steering mechanism δ_B can be obtained:

$$\begin{aligned} \delta_B(t) &= \delta^* Sing(f(t)) = \delta^* Sing\{a.sin(\omega_x t) + b.sin(2\omega_x t)\} \\ &= \delta^* Sing\{\frac{d_2}{K_{ng}}.sin(\omega_x t) + b.sin(2\omega_x t)\} \end{aligned} \tag{13}$$

Similarly, the control command for steering mechanism δ_H will be:

$$\delta_H(t) = \delta^* Sing(f(t)) = \delta^* Sing\{\frac{d_1}{K_h}.sin(\omega_x t) + b.sin(2\omega_x t)\} \tag{14}$$

Where K_h is similar to K_{ng} and they are defined as follows:

$$K_{ng} = \frac{\delta^*}{b \cdot \pi}(PK_\beta - C_\beta) \tag{15}$$

4   Conclusions

The use of a relay steering mechanism helps to reduce the weight and size of the its installation location, which simplifies the process of designing and manufacturing a two-channel missile (the missile that rotates around the vertical axis). However, the control algorithm of steering mechanism algorithm which meets the requirements of guidance problem for missile, it will be more complicated at this time. Thanks to the control solution proposed in this paper, the control channels for the steering mechanisms have been separated and the linear dependence between the control pulse parameters and the requirements of normal force required for the missile has also been identified.

References

1. Liu, Y., Dao, P.N., Zhao, K.Y.: On robust control of nonlinear teleoperators under dynamic uncertainties with variable time delays and without relative velocity. IEEE Trans. Ind. Inform. **16**(2), 1272–1280 (2020). https://doi.org/10.1109/TII.2019.2950617
2. Nam, D.P., et al.: A cascade controller for tracking and stabilization of wheeled mobile robotic systems. In: The 5th IEEE International Conference on System Science and Engineering (ICSSE), Ho Chi Minh City, pp. 305–309, July 2017
3. Van Hau, P., et al.: Asymptotic stability of the whole tractor-trailer control system. In: Proceedings of The 5thIEEE International Conference on System Science and Engineering (ICSSE), Ho Chi Minh City, pp. 444–449, July 2017
4. Binh, N.T., et al.: An adaptive backstepping trajectory tracking control of a tractor trailer wheeled mobile robot. Int. J. Control Autom. Syst. **17**(2), 465–473 (2019)
5. Van Tinh, N., et al.: A Gaussian wavelet network-based robust adaptive tracking controller for a wheeled mobile robot with unknown wheel slips. Int. J. Control (2019). https://doi.org/10.1080/00207179.2018.1458156
6. Asif, M., Khan, M.J., Cai, N.: Adaptive sliding mode dynamic controller with integrator in the loop for nonholonomic wheeled mobile robot trajectory tracking. Int. J. Control **87**(5), 964-975 (2014)
7. Huang, J., Wen, C., Wang, W., Jiang, Z.-P.: Adaptive output feedback tracking control of a nonholonomic mobile robot. Automatica **50**, 821–831 (2014)
8. Sun, W., Tang, S., Gao, H., Zhao, J.: Two time – scale tracking control of nonholonomic wheeled mobile robots. IEEE Trans. Control Syst. Technol. **24**(6), 2059-2019 (2016)
9. Sun, Z., Dai, L., Liu, K., Xia, Y., Johansson, K.H.: Robust MPC for tracking constrained unicycle robots with additive disturbances. Automatica **90**, 172–184 (2018)
10. Казаков, ИЕ., Мишаков, А.Ф: Авиациольные управляемые ракеты. ВВИА им проф. Н. Е, Жуковского (1988)
11. Лебедев А.А., Чернобровкин Л.С: Динамика полёта беспилотных летательных аппаратов, "Машиностроение" Москва (1973)
12. Thuan, T.D., Tue, P.V., Quan, N.H., Hung, T.M.: Building a model to describe two-channel missiles rotating around the vertical axis. J. Mil. Sci. Technol. Res. 179–184 (2009). Special version 3-2009. Ha Noi

Detecting Leftover Food and the Shrimp for Estimating of the Shrimp Body Length Based on CNN

Van Quy Hoang and Dinh Cong Nguyen[✉]

Hong Duc University, Thanh Hoa, Vietnam
{hoangvanquy,nguyendinhcong}@hdu.edu.vn

Abstract. Controlling the body length of the shrimp and estimating leftover food are the major indicators for feeding strategies in shrimp farms. Normally, the works are done by naked-eye observation and mostly counted on the experiences. Nevertheless, it could sometimes be subjective, especially with leftover food. The miscalculation could affect to the feeding plans of farmer, and hence, commercial losses in shrimp aquaculture. In this paper, a new approach to automatically calculate the shrimp body length and to estimate leftover food in pond is given by using a underwater cameras and a convolutional neural network (CNN) model. The proposed method obtained a mAP of 87.3% in the shrimp and leftover food detection and localization. In addition, just 7% of a MSE in the body length of shrimp is produced.

Keywords: Leftover food · Shrimp body length · Underwater camera

1 Introduction

Recently, aquaculture industry has been affirmed as a key profession of the economic development in Vietnam. As claimed by Global Aquaculture Alliance, the number of global shrimp production achieved to approximately 3.3 million tons. In Vietnam, according to Vietnam Association of Seafood Exporters and Producers, the production of shrimp reached to 877.2 thousand tons. Among all of the shrimp species, white leg shrimp (Penaeus vannamei) holds extremely a high economic value and be favored by Vietnamese customers. In addition, the production of this shrimp accounted for 547.6 thousand tons and the export value was 2.6 billion USD in 2019[1].

One of the significant costs for the shrimp aquaculture is known as the feed. The dietary of shrimps could be based on their body length and hence to optimize their growth. Basically, this estimation is done by naked-eye observation, it therefore is quite subjective Fig. 1(a). This can affect to the feeding strategies of farmers. In addition, leftover food Fig. 1(b) existed in grow-out pond in a long time is one of the main components to contaminate the water. For the reasons,

---

[1] http://seafood.vasep.com.vn.

D.-T. Tran et al. (Eds.): ICISN 2021, LNNS 243, pp. 239–247, 2021.
https://doi.org/10.1007/978-981-16-2094-2_30

an alternative to estimate the body length of shrimp and to detect leftover food in grow-out pond is required.

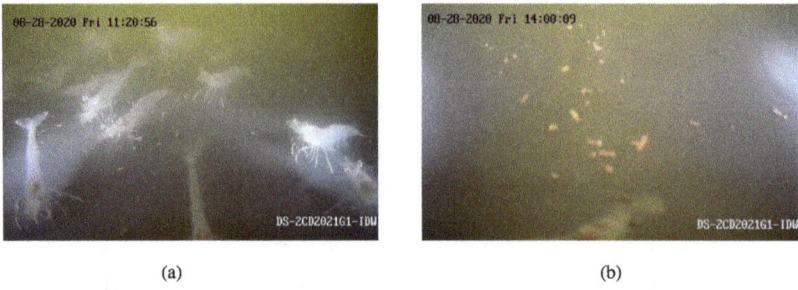

(a)   (b)

Fig. 1. Image acquisition from camera with (a) the shrimps (b) leftover food.

Several published methods to compute the sizes of objects have been proposed over past few years [1–3] in the agriculture domain. For examples, [1] detected and measured fruits based on histogram of oriented gradient in 2017. In this year, [2] localized and detected potato by applied watershed segmentation algorithm, then measured the length of each potato. These methods required some special rules to image acquisition, and less be possible to adapt with the large scales of the input data.

Recently, most of works on object detection and localization are mainly dominated by CNN (convolutional neural network) with their efficient performances on accuracy and time-processing [4]. Along with this trend, a large number of recent publications is driven to the field of agriculture by employing CNN with a huge amount data and uncontrolled conditions. Fast R-CNN [5] was applied to detect fish species [6]. For detecting obstacles and anomalies in field, [7] applied Faster R-CNN. Besides, [8] used SSD model [9] to detect on-tree mango. For the shrimp detection, localization and classification, some works in [10, 11] attempted to use the CNN on their purposes. However, at the best of our knowledge, the methods to detect the shrimps and leftover food in grow-out pond and then to estimate the body length of shrimp are little discussed in literature.

In this paper, several key contributions are given as follows.

- We present a new pipeline for image acquisition in grow-out pond.
- We provide a new approach to detect and localize shrimps and leftover food in grow-out pond, then to measure the body lengths of the shrimps.

The rest of paper is organized as follows. Section 2 gives the description of our protocols to set up materials in pond. Next, our approach is discussed in Sect. 3. Then, Sect. 4 presents our results. At last, Sect. 5 will summarize and give out some perspectives.

2   Materials

Data Collection

In this research, the shrimps have been raised at a stocking density of 150–200$/m^2$ with a square of the grow-out pond about $3575\,m^2$ (65 m × 55 m) rounded by plastic tarps in Tinh Gia district, Thanh Hoa province. The shrimps are fed four times per day at 6 am, 10 am, 2 pm and 6 pm. The feed is based on the weight of the shrimps. However, this ration could be changed by the observation of leftover food. Other parameters such as temperature, dissolved oxygen, pH and salinity are controlled and modified as needs.

An underwater camera, one IP HIKvision with sensor of 1/2.8 in., is used to acquire videos of the shrimps. Video resolution is 1920 × 1280 pixels; frame rate per second is 30; focal length of camera system is 2.8/4 mm; color model is Lux @ (F1.2, AGC ON), 0.028 lx @ (F2.0, AGC ON). Other techniques could help to reduce noise, improve the video quality in the low-contrast environment. We developed a new protocol to collect videos in real-time through wifi connection based on Kurento[2]. Within the Kurento system, there are two main components given as PlayerEndPoint and WebRtcEndpoint. For the our content, we embedded the RecoderEndPoint component where data is compressed and saved. It allows a fast connection and could access to scout randomly.

In grow-out pond, we use the fan systems, which create cyclically a moving flow of water. Therefore, the shrimps will change position based on the direction of this flow. Food is distributed equally around the pond. In addition, videos are recorded in a long time, which guarantee that one camera can represent the whole shrimp pond. During the testing phase, we changed several locations, it then gave the same results.

Different videos were recorded and then extracted into a large number of images for training and testing processes. Two thousand images are collected randomly for training and testing phases at the different illumination changes as shown in Fig. 2(a, b). For the training phase, the data labels are manually cropped by LabelImg toolkit[3] into two classes: Shrimp Fig. 2(a, b) and food Fig. 2(c). There are approximately 3960 shrimp labels and 2300 food labels were given. Then, we separated into rations of 60%, 10%, 30% for train, val, test sets, respectively. The shrimp images were cropped at different resolutions due to the distance from the shrimps to camera. In addition, we also notice here that the ground truth of length of shrimp is estimated by the linear distance of pixels in image from head to tail.

3   Proposed Approach

The general architecture is presented in Fig. 3. At the first step, we employ the CNN model with pre-trained weights named You Only Look Once version 4

---

[2] http://www.kurento.org/.

[3] https://github.com/tzutalin/labelImg.

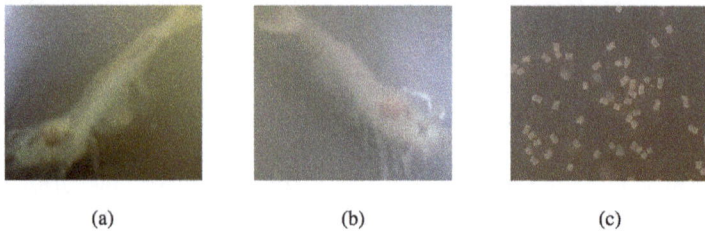

(a)   (b)   (c)

Fig. 2. Image acquisition from camera at different illumination changes. It results from the recorded time such as in the morning, noon, or afternoon.

(YOLOv4) [12] in order to detect and localize the shrimps and leftover food from images. For convenience, we remind some main points of this model. YOLOv4 includes of three main parts as: backbone - CSPDarnet53, neck - SPP (Spatial pyramid pooling), and head - YOLOv3. YOLOv4 uses BoF (Bag of Freebies), BoS (Bag of Specials) for backbone and detector. At our problem, we only apply YOLO-TinyV4[4] to adapt with the low-level of hardware architecture.

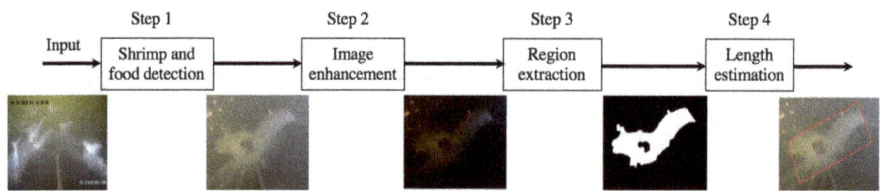

Fig. 3. Our framework description composes of four main steps.

At the second step, we take care only of the images with shrimp label. The main purpose of this step is to improve the quality of images. This can help to reduce the burden for the next step. To do this, the noise from images should be extenuated, and then to remove the influence of illumination changes. Moreover, contrast adjustment is also included to highly distinguish between background and foreground.

First of all, a median filter will be applied with size of (5×5) to blur images. Next, we follow the multi-scale retinex (MSR) algorithm [13] to reduce the influence of illumination changes. Finally, gamma correction could be applied for the contrast adjustment. The output of this step is shown in Fig. 4.

At the third step, the exact region covered the shrimp should be obtained. The OTSU method [14] is applied to convert into binary images Fig. 5(a). Within the binarized image, a connected-component labeling algorithm [15] is employed to extract candidate regions. The largest region is filtered and chosen as the shrimp region Fig. 5(b).

---

[4] https://github.com/AlexeyAB/darknet.

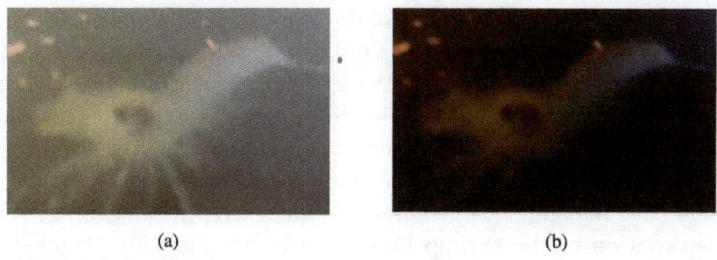

(a)   (b)

Fig. 4. Applying a median filter with size of (5×5): (a) input image (b) output of the image enhancement step.

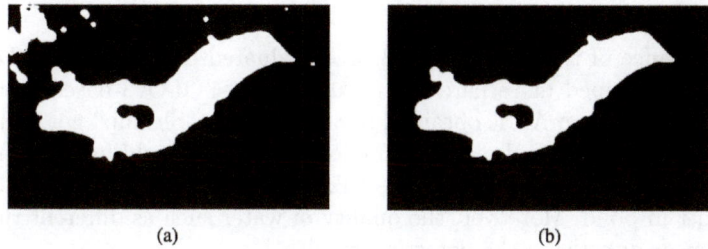

(a)   (b)

Fig. 5. Applying the OTSU method on: (a) binarized image (b) binarized image with the largest region.

At last, the minimum-area enclosing rectangle [16] to provide consistent bounding box is applied to cover the shrimps. Then, the body length of the shrimp is measured as the long side of the bounding box. Some visual examples are presented in Fig. 6.

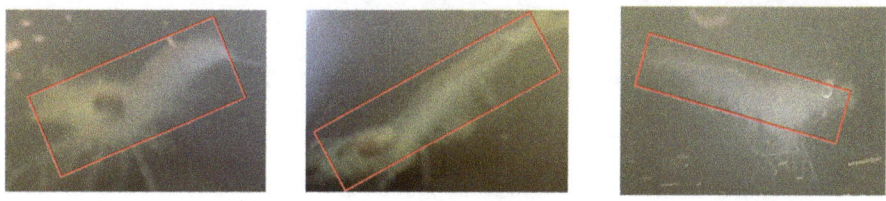

Fig. 6. Results of the shrimp images after our framework are presented here with some visual examples of the bounding boxes covered the shrimp body lengths.

4   Experiments

Experimental Setups
At the training step, the pre-trained weights of YOLOv4 was cloned from Darknet. Then, we modified the config file as follows: batch $= 64$, subdivisions $= 16$, max-batches $= 6000$, step $= 4800, 5400$, classes $= 2$, filters $= 21$.

In order to improve the robustness of the training model, image augmentation was deployed as hue variation (in range 0.9 to 1.1), exposure variation (in range 1 to 1.5), and rotation (in range $-60°$ to $-60°$). In addition, we developed in dilation and erosion of training samples to enrich the dataset. The model was obviously trained by an open-source python environment using Darknet library.

For the characterization metrics, we employed to compute the regular metrics as: Precision (%), Recall (%), F1-scores (%) and mAP (mean average precision) (%). For estimation of the shrimp body length, we manually cropped 350 the shrimp images from images. Then, a MSE score (%) and a normalized quadratic error were used to evaluate for the our method.

Results

The performance of the detection task was evaluated on the test set. For comparison, we developed more here the results of the YOLOv3-based system. As shown in Table 1, the mAP is obtained of 87.3%. Class "Shrimp" got higher Precision, Recall, F1-score and mAP than class "Food". It could be explained that the features of leftover food are simple (like blobs) and quite similar with some factors exist in pond. Moreover, the quality of water such as different rubbishes also influences directly to the detection results.

As discussed in Sect. 2, videos were recorded in the long time and the shrimps and food are moved continuously with the water flow. The model for detection still works well when the density of shrimp is large and food is less. It is worth noting here that the overlapping objects do not affect to the detection results with YOLO-based object detection system. However, it could influence on the measurement of the shrimps. We discuss this aspect in the following paragraph. Some visual examples of detection results of our systems are shown in Fig. 7.

Table 1. Performance of detection with metrics Precision (%), Recall (%), F1-scores (%), mAP (%) with our system and YOLOv3-based system.

Class	Precision (%)	Recall (%)	F1-Score (%)	mAP (%)
Our system				
Shrimp	96.2	90.1	93.05	**87.3**
Food	81.6	79	80.2	
YOLOv3-based system				
Shrimp	83.6	81.7	82.6	79.3
Food	76	78.2	77.1	

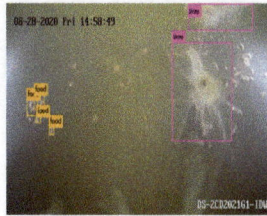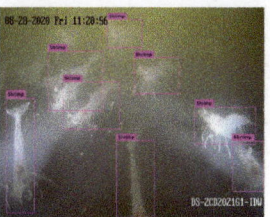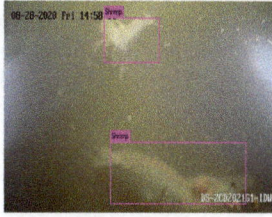

Fig. 7. Visual examples of the our system at different videos and slot times.

For the body length measurement, the our proposed method achieved the results as expectation. With a standard deviation 50 pixels, only about 7% of a MSE is obtained. For the further analysis, Fig. 8 gives the distribution and accumulative distribution of the normalized quadratic error between ground truth and prediction. It can be seen that more than 95% of the body length of the shrimps can be approximated with less than 10% of error. This error could result from the complicated background and illumination changes when we extracted the shrimp features. In addition, with overlapping parts of the shrimps, the system cannot measure exactly the body length. Therefore, we suggest the owner of the pond give their decisions on the body lengths in the best view.

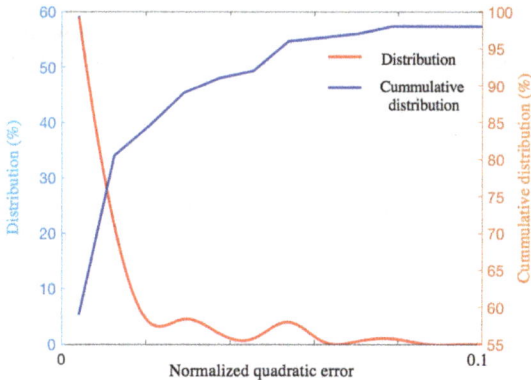

Fig. 8. Normalized quadratic error distributions.

5   Conclusions and Perspectives

In this paper, we proposed an approach to detect and localize the shrimp and leftover food collecting directly from grow-out pond. A mAP of 87.3% was obtained for detection and localization of the shrimp and leftover food by employing YOLOv4 model. . Then, the body length of the shrimp was computed based on the delicate image processing algorithms. a MSE of 7% was exposed.

As perspectives, a dedicated segmentation approach could be competitive with our system. Furthermore, we could pay more attention on detecting some kinds of diseases of the shrimps, which could explore through the color of head or vein of the shrimps. Moreover, waste materials from grow-out pond could be monitored through the detecting process.

Acknowledgments. The authors would like to thank Mr. Van Duong Nguyen, the owner of the shrimp farm in Nghi Son, Tinh Gia, Thanh Hoa who allowed us to implement our proposed system in his grow-out ponds. This work is supported by the Department of Science and Technology of Thanh Hoa.

References

1. Wang, Z., Walsh, K.B., Verma, B.: Sensors. On-tree mango fruit size estimation using RGB-D images **17**(12), 2738 (2017)
2. Si, Y., Sankaran, S., Knowles, N.R.: Potato tuber length-width ratio assessment using image analysis. Am. J. Potato Res. **94**(1), 88–93 (2017)
3. Zhou, C., Yang, X., Zhang, B., Lin, K., Xu, D., Guo, Q.: An adaptive image enhancement method for a recirculating aquaculture system. Sci. Rep. **7**(1), 6243 (2017)
4. Liu, W., Wang, Z., Liu, X., Zeng, N., Liu, Y., Alsaadi, F.E.: A survey of deep neural network architectures and their applications. Neurocomputing **234**, 11–26 (2017)
5. Ren, S., He, K., Girshick, R., Sun, J.: Faster R-CNN: towards real-time object detection with region proposal networks. In: Advances in Neural Information Processing Systems, pp. 91–99 (2015)
6. Li, X., Shang, M., Qin, H., Chen, L.: Fast accurate fish detection and recognition of underwater images with fast R-CNN. In: OCEANS 2015-MTS/IEEE, Washington (2015)
7. Christiansen, P., Nielsen, L.N., Steen, K.A., Jørgensen, R.N.: DeepAnomaly: combining background subtraction and deep learning for detecting obstacles and anomalies in an agricultural field. Sensors **16**(11), 1904 (2016)
8. Liang, Q., Zhu, W., Long, J., Wang, Y., Sun, W.: A real-time detection framework for on-tree mango based on SSD network. In: International Conference on Intelligent Robotics and Applications, pp. 423–436 (2018)
9. Anguelov, D., Erhan, D., Szegedy, C., Reed, S.: SSD: single shot multibox detector. In: European Conference on Computer Vision, pp. 21–37 (2016)
10. Lin, H.Y., Lee, H.C., Ng, W.L., Pai, J.N., Chu, Y.N.: Estimating shrimp body length using deep convolutional neural network. In: ASABE Annual International Meeting (2019)
11. Hu, W.C., Wu, H.T., Zhang, Y.F., Zhang, S.H.: Shrimp recognition using Shrimp-Net based on convolutional neural network. J. Ambient Intell. Humaniz. Comput. (2020)
12. Bochkovskiy, A., Wang, C.Y., Liao, H.Y.M.: YOLOv4: optimal speed and accuracy of object detection. arXiv preprint arXiv:2004.10934 (2020)
13. Lin, H., Shi, Z.: Multi-scale retinex improvement for nighttime image enhancement. Optik **125**(24), 7143–7148 (2014)

14. Otsu, N.: A threshold selection method from gray-level histograms. Trans. Syst. Man Cybern. **9**(1), 62–66 (1979)
15. Wu, K., Otoo, E., Suzuki, K.: Optimizing two-pass connected-component labeling algorithms. Pattern Anal. Appl. **12**(2), 117–135 (2009)
16. Toussaint, G.T.: Solving geometric problems with the rotating calipers. In: Proceedings of the IEEE Melecon, vol. 83, p. A10 (1983)

Real-Time End-to-End 3D Human Pose Prediction on AI Edge Devices

Hai-Thien To, Trung-Kien Le, and Chi-Luan Le[✉]

University of Transport and Technology, Hanoi, Vietnam
{thienth,kienlt,luanlc}@utt.edu.vn

Abstract. This paper tackles the problem of predicting 3D Human Pose on AI Edge Devices. An AI edge device is an AI-enabled Internet of Things (IoT) device that can run Deep Learning (DL) models right on the device without connecting to the Internet. In recent years, researchers have proposed numerous DL-based models for 3D human pose estimation (HPE), but no work focuses on solving this task in edge devices. Building a DL model for an edge device has unique challenges such as limited computation capacity, small memory, and low power. This paper investigates how to optimize a big, heavy-computing 3D pose estimation DL model into a light-weight, small-size model that can run efficiently in a device. Specially, we propose an End-to-End pipeline to run 3D human pose prediction (HPP) in Real-time on AI edge devices. Furthermore, our proposed end-to-end pipeline is general and can be employed by other AI edge device based real-world applications.

Keywords: AI Edge · 3D human pose · Real-time · End-to-end

1 Introduction

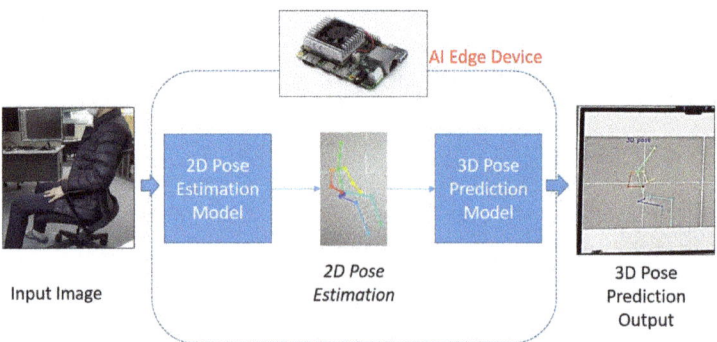

Fig. 1. 3D HPP on an AI Edge device. The input is images from a camera, and output is the predicted 3D human poses displayed on a screen. All models are running in the AI edge device.

© The Author(s), under exclusive license to Springer Nature Singapore Pte Ltd. 2021
D.-T. Tran et al. (Eds.): ICISN 2021, LNNS 243, pp. 248–255, 2021.
https://doi.org/10.1007/978-981-16-2094-2_31

IoT Edge devices are becoming more and more popular and essential for all aspects of our lives, from personal devices, home appliances to industrial machines. AI-enabled edge devices or AI edge devices are intelligent devices that can run AI algorithms locally on a hardware device, using the data born on the device without requiring any connection to a server or the Internet. This allows us to process data in a few milliseconds, which gives us real-time information. Moreover, with AI edge devices, AI processing is now moving to a device, and keeping data locally on the device, preserves data privacy.

Recently, many companies have designed and manufactured edge devices that can deploy AI algorithms. Nvidia Jetson Nano Developer Kit [1] is a small, powerful edge device that can run multiple DL models for applications like image classification, object detection, segmentation, and speech processing. Google Coral Board [2] is a complete toolkit that helps developers can build products with local AI. The above devices are capable of inference allow us to build efficient, fast, private, and offline products. Table 1 shows the characteristics of the Nvidia Jetson Nano and the Google Coral Board.

Table 1. Characteristics of the Google Coral Board and Nvidia Jetson Nano.

Features	Google Coral Board	Nvidia Jetson Nano
DL accelerator	Google EdgeTPU (4 TOPs, Int8 only)	128-core Maxwell GPU (472 GFLOPs, Float16)
Memory	8 MB SRAM + 1 GB RAM	4 GB RAM
DL frameworks	Tensorflow Lite	Tensorflow, Pytorch, Keras
Power	5 W	10 W
Price	$144.99	$99

HPE is an important problem in AI in general and Computer Vision (CV) in particular. It is a critical step to understand the people in images and videos. DL-based models have recently been popularly researched and successfully applied for 2D, and 3D HPE like in [3–6]. However, little effort focuses on running this HPE task on an edge device. This paper introduces recent approaches for optimizing a DL model to fit on an edge device. In particular, we use Google Coral Board [2] as our real use-case to demonstrate the transformation from a server-based 3D HPE DL model to an edge-based model (Sect. 3.1).

One promising research in 3D HPP is to estimate the 3D coordinates of each human key-point from its 2D estimations. This research proposes an End-to-End 2-stage pipeline for 3D HPE in AI edge devices. Figure 1 shows our pipeline and the real demonstration with image input from a camera and the 3D pose prediction results displayed on a screen. Our proposed pipeline is general and can be applied to other cases. For example, in a Real-time Face-Masked detection problem, we can apply a 2-stage end-to-end pipeline by using a pre-trained face detection model at the first stage and then train a mask recognition model from the face output of the first step. A mask recognition model from a face input is much easier to train than a Face-Masked detection model from a raw image.

To summarize, in this paper, we have the following contributions.

- First, we investigate and demonstrate how to optimize a server-based, heavy-computing DL model to an edge-based, light-weight model that can run efficiently on AI edge devices.
- Second, we propose an end-to-end pipeline to real-time predict 3D human pose on an edge device. The pipeline is general and can be employed for AI edge based applications. To the best of our knowledge, we are the first to introduce an end-to-end pipeline for real-time 3D HPE on AI edge devices.
- Third, we demonstrate our real-time 3D HPP on a live video using the Google Coral Board [2], a popular AI edge device.

2   Related Work

2.1   Human Pose Estimation

HPE is the localization of human joints (also known as key-points such as neck, elbows, wrists, knees) in images or videos. HPE has two directions of research, including 2D and 3D HPE. In 2D HPE, we need to estimate a 2D pose (x, y) coordinates for each human joint from an image [3,4].

3D HPE is the task of producing a 3D pose that matches the person's spatial position in an image. One common method is to reason 3D pose through intermediate 2D pose predictions. The inputs are set of paired (2D, 3D) data and predictions from a 2D pose estimation algorithm, the depths for the 3D pose are returned [5,6].

2.2   Deep Learning on AI Edge Devices

AI Edge Devices can run Deep Learning models directly, without any connection to the Internet. Some popular DL models running on AI edge devices include Image classification, Object detection, Face recognition, 2D Pose estimation. Building DL models for an AI edge device needs specific approaches because of its limited resources. Google Coral [2] is an AI edge toolkit from Google that uses EdgeTPU accelerator [9] to run AI at the edge. A DL model running on a Coral device must be converted to use only 8-bit weights. Similarly, Nvidia Jetson Nano [1] uses a TensorRT SDK [10] that includes a deep learning inference optimizer and run-time that delivers low latency and high-throughput for deep learning inference applications. Jetson Nano attains real-time performance with many DL models such as ResNet-50, SSD MobileNet, and TinyYolo [11].

3   Methodologies

This section presents our approaches to building an End-to-End Pipeline for real-time 3D HPP on AI Edge Devices.

3.1   Optimize DL Models for AI Edge Devices

An AI edge device has limited memory, computing resources, and power. It means that a DL network must be optimized for embedded deployment. Reducing network size is the most common approach. Quantization, which is the process of transforming DL model parameters from a higher floating-point precision, like 32-bit floats, to a lower bit representation, like 8-bit integers, is prevalent in edge-based neural network optimization. There are two methods for training quantized networks: post-training quantization and quantize-aware training [12]. Post-training quantization is a common approach that takes a pre-trained model in floating-point and then quantizes the weights by applying post-training dynamic range quantization, to 16-bit floats or 8-bit full integers. Quantize-aware training approach simulates quantization effects in the forward pass of training. It ensures that the forward pass matches the precision for both training and inference time. In this paper, we choose the post-training quantization method for our 3D HPE model conversion because of its ease of implementation without re-training the DL model. A pre-trained 3D HPP model in the Tensorflow framework is quantized to a Tensorflow Lite model with the weights in 8-bit full integers. In the next section, we will introduce the details of our utilized 3D HPE model.

3.2   The GAN Based 3D Human Pose Prediction Model

In this part, we present the 3D HPP model used in this research. We use the method in [6] that follows the 3D pose estimation approach without any 3D labels. It is based on the generative adversarial networks (GANs) [7], and the networks are trained in an unsupervised process. The framework of the GAN based model for 3D HPE from 2D pose input is shown in Fig. 2.

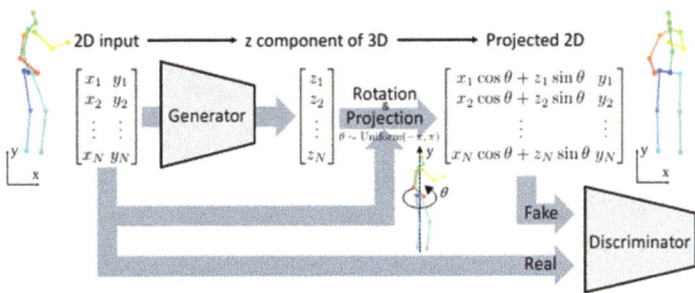

Fig. 2. The framework of GAN based 3D HPE. Taken from [6]

The framework above works as the following pipeline. Given the 2D pose estimations in a single image, the generator network estimates the z-component for each input joint. Then, we get 3D pose by rotating around the y-axis. The rotated pose is later projected onto the 2D-plane. Next step, the discriminator network is used. It is trained to classify clearly between a projected 3D pose and a real 2D pose. We adopt this method for our 3D HPP model, do not require any 3D labels, and can be trained in an unsupervised manner. To reduce the training time, we employ a pre-trained model from [13] and follow the steps in Sect. 3.1 to convert it into an edge device based 3D HPE model. The next section will explain how we use the converted edge-based model into a real-time 3d HPP task.

3.3   The End-to-End Pipeline for Real-Time 3D HPP

An end-to-end pipeline is defined as a process producing the result as a complete functional solution, without needing new input from outside. Regarding this research, we define our pipeline starting from a camera connecting to an edge device. The input video is split into multiple frames; each frame represents an image input to the pipeline. The 3D HPP process stays in the AI edge device and includes two models: (1) a 2D pose estimation model to estimate 2D poses of the person existing in each image; (2) the converted 3D pose prediction model learns a 3D pose from the human 2D joints. The output is the predicted 3D pose of the person in each input frame. The output frames are then concatenated to make a video as the final output of the pipeline. The whole process is end-to-end because we do not need any other input, and all models are run on the edge device. The Demonstration section will show this process can run in real-time with a popular AI edge device like Google Coral [2]. Moreover, in the Introduction part, we have revealed that this proposed pipeline is general and can be applied for other real-world edge-based applications like Face-Masked detection. Figure 3 depicts the details of our proposed pipeline.

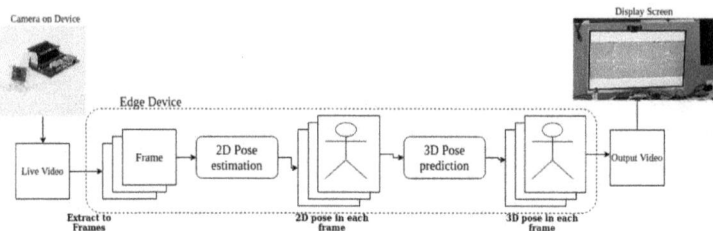

Fig. 3. The end-to-end pipeline for 3D HPP on an edge device.

4   Demonstration and Discussion

4.1   Demonstration

This section presents our real-world demonstration of the 3D HPP from a live camera connecting to an AI edge device. Currently, there is no previous work for this real-time 3D human pose problem on edge devices. Thus, we only show our results and leave them as baseline for future research. In this demo, we use Google Coral Board [2], a popular AI edge device from Google. The pre-trained 2D pose estimation model in Google Coral is implementing the PoseNet model [8]. The 17 pose key-points returned by the PoseNet are divided to the three main parts which are the head (nose, left eye, right eye, left ear, right ear), the trunk (left shoulder, right shoulder, left hip, right hip) and the limbs - extremities(left wrist, right wrist, left knee, right knee, left ankle, and right ankle). Figure 4 shows 2D pose results with a person sitting and standing in front of a camera on the Google Coral device. The 3D pose model outputs the following 3D body joints: Head, Neck/Nose, Thorax, Left Shoulder, Right Shoulder, Spine, Left Hip, Hip, Right Hip, Left Knee, Right Knee, Left Elbow, Right Elbow, Left Wrist, Right Wrist, Left Foot, and Right Foot. Table 2 presents the Inference time (running speed) of the pre-trained 2D pose estimation model and the whole end-to-end 3D pose prediction process on a Google Coral device.

Table 2. Running speed of our testing pose estimation models in Google Coral.

Model name	Inference time	Frames per second (FPS)
2D pose estimation model	12 ms	84 FPS
End-to-end 3D pose prediction	25 ms	39 FPS

In our real-world demo, a person will sit in front of the camera. After executing the pipeline, the output as his 3D key body joints such as head, neck, elbow, shoulder, or foot showed up on the screen. The person's action, such as raise the hand, is also captured in the 3D pose. Besides the 3D body joints, we implement a 3D rotation from the predicted 3D pose by some degrees (0° to 360°) in the demo. It shows the 3D pose in the spatial space, which proves the predicted 3D pose's accuracy. Figure 5 illustrates our demo's outputs with two cases, a person sitting, and a person raising hand. The speed of the whole process is presented in Table 2.

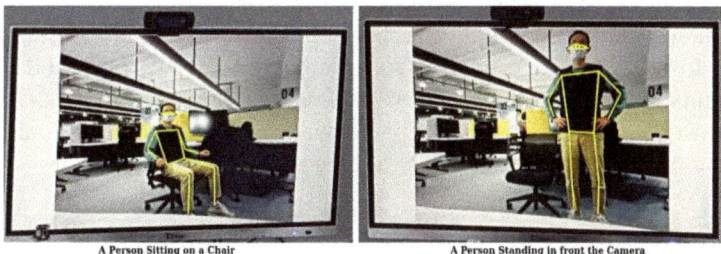

A Person Sitting on a Chair   A Person Standing in front the Camera

Fig. 4. 2D pose outputs with a person sitting and standing in front of a camera. The demo is running on a Google Coral Device.

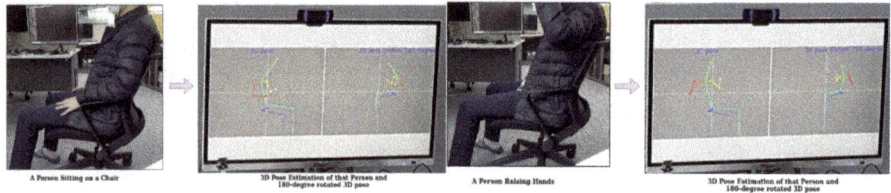

A chair-sitting person.   A hand-raising person.

Fig. 5. The outputs of our demo 3D human pose prediction. (a) A chair-sitting person, (b) a hand-raising person.

4.2   Discussion

From Table 2 we can realize the running time of the whole 3D pose prediction process is **39 FPS**, that can be considered running in real-time. The running speed of the end-to-end 3D pose estimation pipeline is half of a 2D pose model. It means a 3D pose prediction model alone has a similar inference speed as a 2D pose estimation model. Although a 3D pose estimation algorithm is more complicated than a 2D pose method, with the 2-stage pipeline in our paper, we can have a 3D pose estimation model as fast as a 2D model.

5   Conclusions and Future Work

In this paper, we address a DL problem like 3D HPE on AI edge devices. We show that a server-based DL model should not directly run in an edge device but needs edge-based optimization like the quantization process. We show the detailed process of converting a DL model into an edge-based model and using Google Coral devices as our real use-case. We suggest adopting a GAN based 3D pose estimation model that predicts 3D body joints from the 2D pose inputs without using any 3D labels. After successfully converting the 3D pose model into a device-size model, we use it in a 2-stage pipeline. The proposed pipeline is demonstrated to run in real-time on an edge device such as Google Coral and can be generalized to other edge-based applications.

Regarding the future work, in the short-term, we will evaluate our introduced approach to other AI edge devices such as Nvidia Jetson Nano and compare the performance. In the long-term, we will extend this research of 3D HPE to other real-world applications in AI edge devices such as real-time human action recognition or real-time video understanding.

References

1. Nvidia Jetson Nano Developer Kit (2020). https://developer.nvidia.com/embedded/jetson-nano-developer-kit. Accessed 26 Nov 2020
2. Google Coral Dev Board (2020). https://coral.ai/products/dev-board/. Accessed 26 Nov 2020

3. Alexander, T., Christian, S.: DeepPose: human pose estimation via deep neural networks. In: IEEE Conference on Computer Vision and Pattern Recognition, pp. 1653–1660 (2014)
4. Alejandro, N., Kaiyu, Y., Jia, D.: Stacked hourglass networks for human pose estimation. In: European Conference on Computer Vision, pp. 483–499 (2016)
5. Julieta, M., Rayat, H., Javier, R., James, J.L.: A simple yet effective baseline for 3D human pose estimation (2017)
6. Wei, Y., Wanli, O., Xiaolong, W., Jimmy, R., Hongsheng, L., Xiaogang, W.: 3D human pose estimation in the wild by adversarial learning. In: IEEE Conference on Computer Vision and Pattern Recognition (2018)
7. Goodfellow, I., Pouget-Abadie, J., Mirza, M., Xu, B., Warde-Farley, D., Ozair, S., Courville, A., Bengio, Y.: Generative adversarial networks. In: International Conference on Neural Information Processing Systems, pp. 2672–2680 (2014)
8. George, P., Tyler, Z., Nori, K., Alexander, T., Jonathan, T., Chris, B., Kevin, M.: Towards accurate multi-person pose estimation in the wild. In: IEEE Conference on Computer Vision and Pattern Recognition (2017)
9. EdgeTPU - Run Inference at the Edge (2020). https://cloud.google.com/edge-tpu. Accessed 26 Nov 2020
10. NVIDIA TensorRT (2020). https://developer.nvidia.com/tensorrt. Accessed 26 Nov 2020
11. Jetson Nano: Deep Learning Inference Benchmarks (2020). https://developer.nvidia.com/embedded/jetson-nano-dl-inference-benchmarks. Accessed 26 Nov 2020
12. Benoit, J., Skirmantas, K., Bo, C., Menglong, Z., Matthew, T., Andrew, H., Hartwig, A., Dmitry, K.: Quantization and training of neural networks for efficient integer-arithmetic-only inference. In: IEEE Conference on Computer Vision and Pattern Recognition (2018)
13. Unsupervised Adversarial Learning of 3D Human Pose from 2D Joint Locations, Author's implementation (2020). https://github.com/DwangoMediaVillage/3dpose$_$gan. Accessed 26 Nov 2020

Nonlinear Adaptive Filter Based on Pipelined Bilinear Function Link Neural Networks Architecture

Dinh Cong Le[(⌀)], Van Minh Le, Thai Son Dang, The Anh Mai, and Manh Cuong Nguyen

School of Engineering and Technology, Vinh University, Vinh, Vietnam
ldcong@vinhuni.edu.vn

Abstract. In order to further enhance the computational efficiency and application scope of the bilinear functional links neural networks (BFLNN) filter, a pipelined BFLNN (PBFLNN) filter has been developed in this paper. The idea of the method is to divide the complex BFLNN structure into multiple simple BFLNN modules (with a smaller memory-length) and cascade connection in a pipelined fashion. Thanks to the simultaneous processing and the nested nonlinearity of the modules, the PBFLNN achieves a significant improvement in computation without degrading its performance. The simulation results have demonstrated the effectiveness of the proposed method and the potentials of the PBFLNN filter in many different applications.

Keywords: Pipelined · Generalized FLNN · Nonlinear adaptive filtering

1 Introduction

Many practical systems (such as system identification, signal prediction, channel equalization, and echo and noise cancelation,…) may contain nonlinearity. The linear adaptation technique cannot model well enough because of the nonlinear nature of these systems [1]. To overcome this problem, many new classes of nonlinear filters are based on neural networks (NNs) and truncated Volterra series (VFs) has been developed [1, 2]. However, they also reveal many disadvantages such as complex architecture and heavy computing burden in their implementation [3, 4].

It is well known that the FLNN has been proposed to replace the multilayer artificial neural network (MLANN) in some simple nonlinear applications because it has a single layer structure, low computational complexity, and the simple learning rule [4]. It has been successfully applied in other areas of nonlinear filtering including nonlinear dynamic systems identification, channel equalization, active noise control, nonlinear acoustic echo cancellation [4–8]. However, the performance of the FLNN-based model may be significantly impaired when faced with systems containing strong nonlinear distortion. As pointed out in [9], the main reason may be that the basic functions of FLNN lack the cross-terms (for example $x(n) * x(n-1)$, $x(n-1) * x(n-2)$,…. To mitigate this disadvantage, some studies have added appropriate cross-terms into the conventional FLNN structure [9, 10]. Research results in [9, 10] indicate that these new models outperform the Volterra-based model in the noise control application.

On the other hand, in order to increase the computational efficiency for recurrent neural networks (RNN), a pipelined RNN (RNN) structure was developed in a speech predictor [11]. The benefits of pipelined architecture make the nonlinear predictor significantly reducing the total computational of recurrent neural networks (RNN). Following this study, many computational efficiency systems using pipeline architecture have been developed and successfully applied for speech signal prediction [12], channel equalization [13].

In order to be able to model certain nonlinear systems well enough, the BFLNN filter needs to be designed with a sufficiently large memory length. However, this also leads to the computational complexity of BFLNN becoming quite heavy. Inspired by efficient pipelined architecture, a pipelined BFLNN (PBFLNN) filter is proposed in this paper. In this method, a complex BFLNN structure (contains many cross-terms) is divided into several simple small-scale BFLNN modules (smaller memory-length, contains less than cross-terms) and cascaded in a pipelined parallel fashion. Thanks to the parallel processing of small-scale modules, its total computational efficiency is significantly improved.

2   Nonlinear Adaptive PBFLNN Filter

The PBFLNN Structure: The proposed PBFLNN structure consists of simple small-scale BFLNN modules connected in a pipelined fashion. In addition, to ensure the overall output is a global estimate, the outputs of each module are filtered through a conventional transversal filter. The design of the PBFLNN is illustrated in Fig. 1.

The simple BFLNN modules are identical in design (i.e. the parameters selected for the external signal input and the cross-terms are the same). Thus, the modules have the same synaptic weight matrix. As designed, it is easy to see that the input to each module consists of two types of signals: one of them is the external signal and the other is the output signal of the previous module.

Suppose we define N external input signals of the ith module as

$$X_{Ei}(\xi) = [x(\xi - i), x(\xi - i - 1), ..., x(\xi - i - N + 1)]^T \tag{1}$$

Hence, the input signal of the ith module is

$$X_i(\xi) = \left[X_{Ei}^T(\xi), U_i(\xi)\right]^T = [x(\xi - i), x(\xi - i - 1), ..., x(\xi - i - N + 1), U_i(\xi)]^T, \qquad i = 1, ..., M \tag{2}$$

where $U_i(\xi) = y_{i+1}(\xi)$ when the module differs from the Mth module; $U_i(\xi) = y_M(\xi - 1)$, when the module is the Mth module.

Since each module is a BFLNN structure, the input signal is expanded to

$$Xf_i(\xi) = \left[Xf1_i^T(\xi), Xf2_i^T(\xi), \cdots, XfV_i^T(\xi)\right]^T \tag{3}$$

where $Vi = (6 + 4k)$ is the number of channels of the ith module, and k is the cross-term selection parameter.

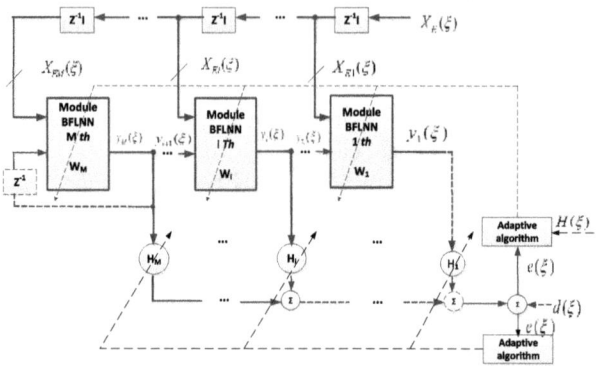

Fig. 1. The proposed nonlinear adaptive PBFLNN filter

Based on the ideal of the BFLNN filter (refer to [10]), the signals for each channel $(Xf1_i(\xi), Xf2_i(\xi), ..., XfV_i(\xi))$ of the ith module can be expressed as follows

$$Xf1_i(\xi) = [x(\xi - i), x(\xi - i - 1), ...x(\xi - i - N + 1), U_i(\xi)]^T \tag{4}$$

$$Xf2_i(\xi) = [cos(\pi x(\xi - i)), ...cos(\pi x(\xi - i - N + 1)), cos(\pi U_i(\xi))]^T \tag{5}$$

$$Xf3_i(\xi) = [sin(\pi x(\xi - i)), ...sin(\pi x(\xi - i - N + 1)), sin(\pi U_i(\xi))]^T \tag{6}$$

$$Xf4_i(\xi) = \left[y(\xi - i - 1), y(\xi - i - 2), ...y(\xi - i - N_{py})\right]^T \tag{7}$$

$$Xf5_i(n) = [y(\xi - i - 1)cos(\pi x(\xi - i)), y(\xi - i - 2)cos(\pi x(\xi - i - 1)), ..., y(\xi - i - k)cos(\pi x(\xi - k + 1))]^T \tag{8}$$

$$Xf6_i(\xi) = [y(\xi - i - 1)sin(\pi x(\xi - i)), y(\xi - i - 2)sin(\pi x(\xi - i - 1)), ..., y(\xi - i - k)sin(\pi x(\xi - k + 1))]^T \tag{9}$$

$$\vdots \qquad = \qquad \vdots$$

$$Xf(2k+5)_i(n) = [x(\xi - i - 1)cos(\pi x(\xi - i)), ..., x(\xi - i - k)cos(\pi x(\xi - i - k + 1)), U_i(\xi)\, cos(\pi x(\xi - i - k))]^T \tag{10}$$

$$Xf(2k+6)_i(\xi) = [x(\xi - i - 1)sin(\pi x(\xi - i)), ..., x(\xi - i - k)sin(\pi x(\xi - i - k + 1)), U_i(\xi)sin(\pi x(\xi - i - k))]^T \tag{11}$$

$$\vdots \qquad\qquad = \qquad\qquad \vdots \tag{12}$$
$$XfV - 1_i(\xi) = [U_i(\xi)cos(\pi x(\xi - i))]^T$$

$$XfV_i(\xi) = [U_i(\xi)sin(\pi x(\xi - i))]^T \tag{13}$$

As analyzed above, the synaptic weight vector of each module is designed similarly. We, therefore, define the weight vectors for all modules as

$$W(\xi) = \left[w_1(\xi), w_2(\xi), ..., w_{Lf}(\xi)\right]^T \tag{14}$$

where the length L_f of expanded input signal $Xf_i(\xi)$ is defined by $L_f = 3(N+1) + N_{py} + 2(k+1)(k+2)$. and N_{py} is the feedback coefficient selection parameter.

Therefore, the output of the ith modules is a linear combination of the weights and the extended signals of the ith module as follows

$$y_i(\xi) = W^T(\xi)Xf_i(\xi) \tag{15}$$

The outputs of each module are then filtered through an adaptive finite impulse response (FIR) filter to obtain a global estimate. It is easy to see that the output of this FIR filter is also the output of the PBFLNN nonlinear filter and is defined as follows.

$$\hat{y}(\xi) = H^T(\xi)Y(\xi) \tag{16}$$

where $H(\xi) = [h_1(\xi), h_2(\xi), ..., h_M(\xi)]^T$ is the weight vector of the FIR filter, and $Y(\xi) = [y_1(\xi), y_2(\xi), ..., y_M(\xi)]^T$ is the input vector of the FIR filter.

Adaptive Algorithm: In this section, the weight vectors of the modules $W(\xi)$ and the FIR filter $H(\xi)$ are updated to minimize $J(\xi)$ (instantaneous square error). In this way, we define the cost function as follows

$$J(\xi) = e^2(\xi) \tag{17}$$

where the $e(\xi) = d(\xi) - \hat{y}(\xi) = d(\xi) - H^T(\xi)Y(\xi)$ is the instantaneous output error at time ξ.

Thus, weight vectors of the modules $W(\xi)$ and the FIR filter $H(\xi)$ are updated in accordance with the rule as follows

$$H(\xi+1) = H(\xi) - \frac{1}{2}\mu \nabla_{H(\xi)} J(\xi) \tag{18}$$

$$W(\xi+1) = W(\xi) - \frac{1}{2}\eta \nabla_{W(\xi)} J(\xi) \tag{19}$$

where $\nabla_{W(\xi)}J(\xi)$ and $\nabla_{H(\xi)}J(\xi)$ are the gradient of cost function $J(\xi)$ with respect to the $W(\xi)$ and the $H_i(\xi)$, respectively; μ and η are the learning rate. The gradients can be calculated as follows

$$\nabla_{W(\xi)}J(\xi) = \frac{\partial J(\xi)}{\partial W(\xi)} = 2e(\xi)H^T(\xi)\begin{bmatrix} \frac{\partial(y_1(\xi))}{\partial W(\xi)} \\ \vdots \\ \frac{\partial(y_M(\xi))}{\partial W(\xi)} \end{bmatrix} = -2e(\xi)\sum_{i=1}^{M} h_i(\xi)Xf_i(\xi) \quad (20)$$

$$\nabla_{H(\xi)}J(\xi) = \frac{\partial J(\xi)}{\partial H(\xi)} = 2e(\xi)\frac{\partial d(\xi) - H^T(\xi)Y(\xi)}{\partial H(\xi)} = -2e(\xi)Y(\xi) \quad (21)$$

Substituting (21) in (18) we yield the update equation of the weigh vector $H(\xi)$ as

$$H(\xi+1) = H(\xi) + \eta e(\xi)Y(\xi) \quad (22)$$

Similarly, substituting (20) in (19) we obtain the update equation of the weigh vector $W(\xi)$ as

$$W(\xi+1) = W(\xi) + \mu e(\xi)\sum_{i=1}^{M} h_i(\xi)Xf_i(\xi) \quad (23)$$

Table 1. Computational complexity of FLNN, BFLNN and PBFLNN filters

Type of filter	Multiplications	Additions
FLNN	$2(2B + 1)N_f + 1$	$2(2B + 1)N_f1$
BFLNN	$2[3N_b + N_y + N_{db}(N_{db} + 1)] + 2N_{db} + 2$	$2[3N_b + N_y + N_{db}(N_{db} + 1)] - 1$
PBFLNN	$2[3(N + 1) + N_{py} + 2(k + 1)](k + 2)] + 2k + 2M + 4$	$2[3(N + 1) + N_{py} + 2(k + 1)](k + 2)] + 2M$

Computational Complexity: To evaluate the effectiveness of the proposed method, a comparison of the computational complexity of the 3 filters (FLNN, BFLNN and proposed PBFLNN) is summarized in Table 1. Assuming N_f, N_b, and N, is external input signals of FLNN, BFLNN and proposed PBFLNN, respectively. N_{db} and N_y, are the cross-term and feedback selection parameter of BFLNN; M is the number of modules.

3   Simulation

To demonstrate the effectiveness of the proposed method, several experiments were conducted to compare the PBFLNN filter with the FLNN and BFLNN filters in terms of performance and computational complexity. The parameters of the FLNN, BFLNN filters and the PBFLNN were selected the same for all experiments. Specifically, the parameter of FLNN is $Nf = 10$; $B = 3$, that of BFLNN is $N_b = 10$; $N_y = 10$; $N_{db} = 9$ and that of PBFLNN is $N = 4$; $N_{py} = 4$; $k = 3$. The function expansion is first-order type for PBFLNN and BFLNN. The experimental results are all taken by averages on 100 independent runs.

Experiment 1: In this experiment, we conducted the identification of a nonlinear dynamic model as described below

$$d(n) = \frac{d(n-1)}{1+d^2(n-1)} + x^2(n)x(n-1) \tag{24}$$

where $d(n)$ and $x(n)$ are the observed signal and the input signal of the system. The performance of the adaptive filters is evaluated based on the mean square error $MSE = 10 \log 10(e^2(n))$. Assuming that the $x(n)$ is the random sequence, and its range is chosen as $(0,1)$.

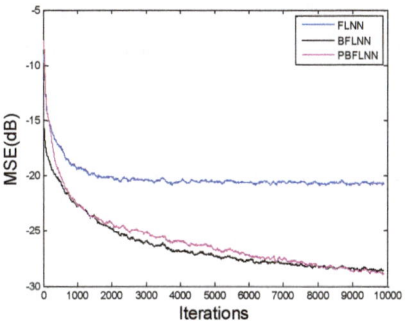

Fig. 2. Comparison of MSE for random input signal.

The step-size of the synaptic weigh vector $W(n)$ (the expanded by BFLNN function) includes linear part $\mu_1 = 0.03$; sin (.) cos (.) part $\mu_2 = 0.05$; feedback part $\mu_3 = 0.03$ and cross-terms part $\mu_4 = 0.04$. The step-size of the FIR filter of the PBFLNN is $\eta = 0.04$. Figure 2 shows the averaged MSE performance curves for the random input signal. It is clear that the performance of the proposed PBFLNN filter is equivalent to that of BFLNN.

In addition, the computational requirements of the filters are summarized in Table 2. It is obvious that the computational requirements of the PBFLNN are about 51% less than that of the BFLNN.

Table. 2 Computational complexity of FLNN, BFLNN and PBFLNN filters

Type of filter	Parameter	Multiplications	Additions
FLNN	$(N_f = 10, B = 3)$	141	139
BFLNN	$(B = 1, N_{db} = 9, N_b = 10, N_y = 10)$	280	259
PBFLNN	$N = 4, k = 3; N_{py} = 4; B = 1, M = 5$	138	128

Experiment 2: In this experiment, we carried out to compare identification of a nonlinear dynamic system as described in [15].

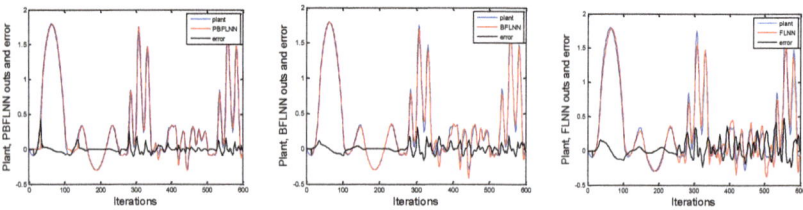

Fig. 3. The identification results of the nonlinear dynamic system are based on the PBFLNN, BFLNN, and FLNN filters

The step-size of the PBFLNN filter are set to $\eta = 0.87$, $\mu_1 = 0.79$, $\mu_2 = 0.68$, and $\mu_3 = 0.83$. The Fig. 3 show the identification results with the corresponding BFLNN, FLNN and PBFLNN filters. It is clear that the PBFLNN achieves equivalent performance to the BFLNN with a lower computational complexity.

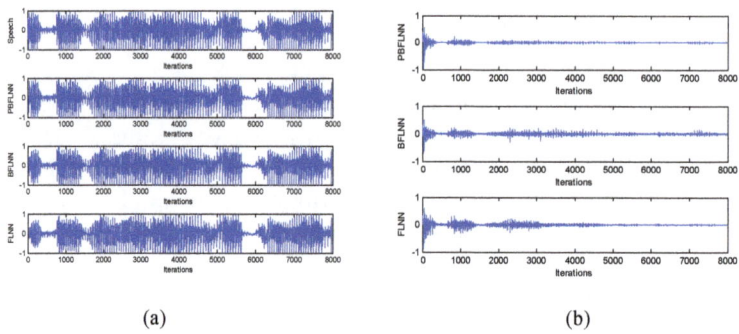

(a)   (b)

Fig. 4. (a) Original signal and the corresponding prediction signals, (b) corresponding prediction error

Experiment 3: To demonstrate speech signal predictive performance of the proposed PBFLNN, we use the experiment as described in [14]. The one-step forward prediction is employed to measure the predicting capability and defined as in [14].

Figure 4a illustrates the results of speech prediction using the PBFLNN, BFLNN, FLNN filters respectively, and the original speech signal. Figure 4b depicts the correspoding predicting errors. The one-step prediction gain of the PBFLNN, BFLNN, and FLNN are 18.912 dB, 19.324 dB and 17.160dB, respectively. From Fig. 4 and the value of the one-step prediction, we find that the speech signal prediction ability of the PBFLNN is equivalent to that of the BFLNN.

4   Conclusion

This paper has proposed a PBFLNN filter, aiming to reduce computation cost and extend the application scope for the BFLNN. The architecture of the proposed filter is simpler with a shorter memory length. Computational analysis and simulation results have shown that the PBFLNN filter significantly reduces computation cost without degrading performance compared to BFLNN. Furthermore, the simulations have also demonstrated the potential of the PBFLNN filter for nonlinear dynamic identification and speech signal prediction.

Acknowledgements. This work was supported by Ministry of Education and Training, Vietnam fund (project title: Research to reduce computational complexity and impulsive noise impact for nonlinear active noise control (ANC) system, grant B2021-TDV-03).

References

1. Ogunfunmi, T.: Adaptive Nonlinear System Identification. Springer, New York (2007)
2. Diniz, P.: Adaptive Filtering Algorithms and Practical Implementation, 3rd edn. Springer, New York (2008)
3. Li Tan, Jiang, J.: Adaptive Volterra filters for active control of nonlinear noise processes. IEEE Trans. Sig. Process. 49(8), 1667–1676 (2001)
4. Patra, J.C., Pal, R.N.: A functional link artificial neural network for adaptive channel equalization. Sig. Process. 43(2), 181–195 (1995)
5. Comminiello, D., Scarpiniti, M., Azpicueta-Ruiz, L.A., Arenas-García, J., Uncini, A.: Functional link adaptive filters for nonlinear acoustic echo cancellation. IEEE Trans. Audio Speech Lang. Process 21(7), 1502–1512 (2013)
6. Le, D.C., Zhang, J., Li, D.F., Zhang, S.: A generalized exponential functional link artificial neural networks filter with channel-reduced diagonal structure for nonlinear active noise control. Appl. Acoust. 139, 174–81 (2018)
7. Chakravorti, T., Satyanarayana, P.: Nonlinear system identification using kernel based eponentially extended random vector functional link network. Appl. Soft Comput. 89, 1–14 (2020)
8. Le, D.C., Zhang, J., Li, D.: Hierarchical partial update generalized functional link artificial neural network filter for nonlinear active noise control. Digit. Sig. Process 93, 160–171 (2019)
9. Sicuranza, G.L., Carini, A.: A generalized FLANN filter for nonlinear active noise control. IEEE Trans. Audio Speech Lang. Process 19(8), 2412–2417 (2011)
10. Le, D.C., Zhang, J., Pang, Y.: A bilinear functional link artificial neural network filter for nonlinear active noise control and its stability condition. Appl. Acoust. 132, 19–25 (2018)

11. Haykin, S., Li, L.: Nonlinear adaptive prediction of nonstationary signals. IEEE Trans. Sig. Process **43**(2), 526–535 (1995)
12. Baltersee, J., Chambers, J.A.: Nonlinear adaptive prediction of speech with a pipelined recurrent neural network. IEEE Trans. Sig. Process **46**(8), 2207–2216 (1998)
13. Goh, S.L., Mandic, D.P.: Nonlinear adaptive prediction of complex-valued signals by complex-valued PRNN. IEEE Trans. Sig. Process **53**(5), 1827–1836 (2005)
14. Zhang, S., Zhang, J., Pang, Y.: Pipelined set-membership approach to adaptive Volterra filtering. Sig. Process **129**, 195–203 (2016)
15. Patra, J.C., Pal, R.N., Chatterji, B.N., Panda, G.: Identification of nonlinear dynamic systems using functional link artificial neural networks. IEEE Trans. Syst. Man. Cybern. B **29**(2), 254–262 (1999)

Adaptive Reinforcement Learning Motion/Force Control of Multiple Uncertain Manipulators

Phuong Nam Dao$^{(\boxtimes)}$, Dinh Duong Pham, Xuan Khai Nguyen, and Tat Chung Nguyen

School of Electrical Engineering,
Hanoi University of Science and Technology, Hanoi, Vietnam
nam.daophuong@hust.edu.vn

Abstract. Cooperating Mobile Manipulators (CMMs) have been successfully employed as a powerful equipment in industrial automation. The classical nonlinear controllers have been implemented in presence of dynamic uncertainties, external disturbances. However, they are not appropriate in the situation of actuator saturation, input constraint as well as optimality requirements. This paper presents the adaptive reinforcement learning (ARL) based optimal control for CMMs. This approach enables us to overcome the disadvantage of solving Hamilton Jacobi Bellman (HJB) equation to obtain the optimal controller after obtaining the modified motion dynamic model. The simulation results illustrate the performance of the proposed control algorithm.

Keywords: Cooperating mobile manipulators · Adaptive reinforcement learning · Robust adaptive control

1 Introduction

Last decades have witnessed nonlinear control as a promising and powerful technique in designing control scheme for robotic systems. In [1–3], the Lyapunov stability theory are utilized for considering the selection of appropriate Lyapunov function candidate and backstepping technique in Bilateral Teleoperation systems and mobile robots. For Cooperating Mobile Manipulator (CMM) Systems, the control objective was classified into several situations depending on the control design requirement [4–12]. Firstly, multiple mobile robot manipulators in cooperation carrying a common object in presence of dynamic uncertainties, disturbances [11]. Secondly, one of them tightly holds object by end-effector and the end effector of remaining mobile manipulator follows a trajectory on the surface of the object [7]. Due to these control objectives, one realized the methods including centralized or decentralized, cooperation control methods with the tracking requirement of object as well as each mobile manipulators. The scheme in [10] implemented the centralized coordination control to ensure the tracking of each mobile manipulators using approximated Neural Networks by considering the actuators. However, almost existing control strategies for CMMs are implemented by conventional nonlinear control technique. As a result, it is difficult to

© The Author(s), under exclusive license to Springer Nature Singapore Pte Ltd. 2021
D.-T. Tran et al. (Eds.): ICISN 2021, LNNS 243, pp. 265–271, 2021.
https://doi.org/10.1007/978-981-16-2094-2_33

handle the situation of actuator saturations, full-state constraint, etc. This paper considers the novel direction of employing the optimal control for CMMs with the adaptive reinforcement learning (ARL) technique. As we have known that it is impossible to analytically solve Hamilton-Jacobi-Bellman (HJB) equation being an intermediate step of finding the optimal controller. To overcome this challenges, the approach using ARL technique is able to solve online the HJB equation to obtain the optimal scheme.

2   Preliminaries and Problem Statement

We investigate a n two-link manipulators, as shown in Fig. 1, which includes n robots interacted by each pair of manipulators, the unknown environment with the original Cartesian coordinate system $O(X, Y, Z)$ and the corresponding coordinate E_i of end-effector of each robot. The interaction between each pair of manipulators leads to constraint condition as described in the following equation:

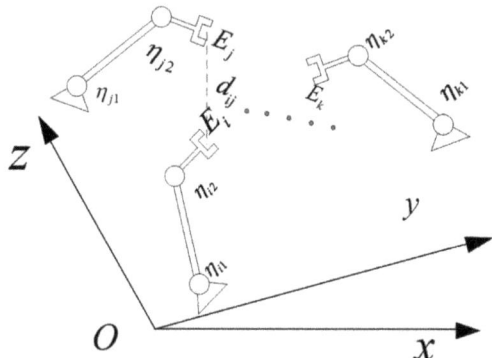

Fig. 1. The cooperating manipulator systems

$$\Xi_{ij} = |\mathrm{OE}_i - \mathrm{OE}_j| - d_{ij} = 0 (\forall 1 \leq i < j \leq n) \qquad (1)$$

where OE_i is a vector with the first point being the original point and the end-effector E_i under the Cartesian coordinate system, d_{ij} is the distance between end-effectors of the i th robot manipulator and the j th robot manipulator. According to (1), we can be obtain the constraint vector as follows:

$$\Xi(\eta) = [\ldots, \quad \Xi_{ij}(\eta_i, \eta_j), \quad \ldots]^T = 0 \qquad (2)$$

where $\Xi(\eta) \in R^{m \times 1}$, η is the joint vector of n two-link manipulators systems.

Assumption 1. The number of joints is more than the number of constraint conditions, i.e., $2n > m$.

In [8], the dynamic equation of n two-link robot manipulators with constraint force by interaction between each couples of manipulators can be described as:

$$M(\eta)\ddot{\eta} + C(\eta,\dot{\eta})\dot{\eta} + F(\eta,\dot{\eta},t) = \tau - J^T(\eta)\lambda \qquad (3)$$

where $J(\eta) = \frac{\partial\Xi}{\partial\eta} \in R^{m \times 2n}$, $\lambda \in R^{m \times 1}$ is the vector of Lagrange multiplier.

The control objective is to obtain not only the tracking of joints to desired trajectories but also guaranteeing the constant distance between end-effectors of each pair of manipulators. Moreover, the Motion/Force control objective is also mentioned in the Largarangian coefficient (3). It is worth emphasizing that the consideration of ARL technique in control design is a significant difference between this work and [11].

3 Motion/Force Control for Multiple Uncertain Manipulators

In this section, we focus on the motion/force control scheme for multiple uncertain manipulators as described in Fig. 2. It will be seen that the robust control algorithm is the starting point of ARL based control design. Furthermore, the stability will be mentioned in these controllers.

3.1 Robust Controller for Multiple Uncertain Manipulators

In this section, the motion/force control design can be established from motion dynamic model and constraint force model. In order to obtain the motion dynamic model, the constraint force $J^T(\eta)\lambda$ in (3) needs to be eliminated by multiplying the matrix $Q = I - A^T(AA^T)^{-1}A$ on both sides of (3) to achive the following motion dynamic equation:

$$QM(\eta)\ddot{\eta} + QC(\eta,\dot{\eta}) + QF(\eta,\dot{\eta},t) = \tau_{||} \qquad (4)$$

It sholud be noted that due to the multiple manipulators system is considered in the case of $2n > m$ (assumption 1), it leads to that the completed motion dynamic model (5) is the differential equation of independent joint $\eta_{ind} \in R^{(2n-m) \times 1}$ with the detailed transformation being seen in [11]:

$$K_1\overline{M}\ddot{\eta}_{ind} + K_1\overline{C}(\eta,\dot{\eta})\dot{\eta}_{ind} + K_1\overline{F} = \tau_{||ind} \qquad (5)$$

Similar to the work in [11], the constraint force can be obtained by the following equation:

$$F_c = \tau_\perp - h_\perp - F_\perp - (I-Q)MM_c^{-1}\left(\tau_{||} - Q(C\dot{\eta} + F(\eta,\dot{\eta},t)) + M(\eta)S\dot{\eta}\right) \qquad (6)$$

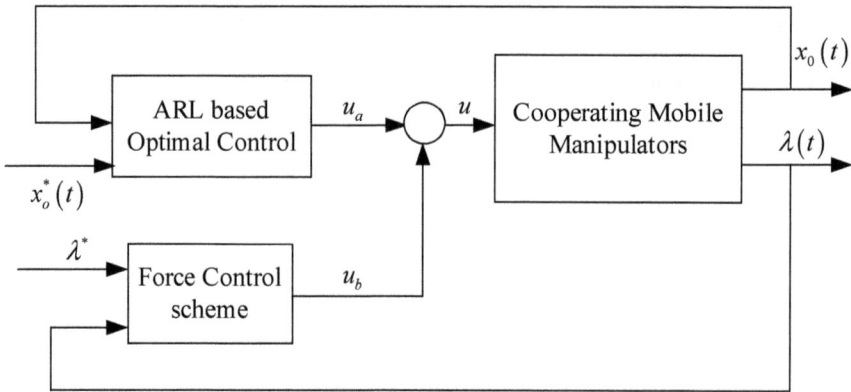

Fig. 2. The motion/force controller of cooperating mobile manipulators.

Based on the certainty equivalence principle, appropriate assumptions as described in [8], the robust adaptive controller can be designed as follows:

$$\tau_{\|ind} = Y\hat{\theta} - e_1 - \overline{\overline{M}}e_2 - k_2e_2 - k_3\|e_1\|^2e_2$$
$$\tau_{\|idep} = f\left(\tau_{\|ind}, \dot{\eta}_{ind}\right)$$

$$(7)$$

3.2 Adaptive Reinforcement Learning for Multiple Uncertain Manipulators

In this section, we further extend the results in the previous section to ARL based control design with the consideration of motion dynamic model (5) and the appropriate tracking error model. It can be seen that the tracking error model is a time-varying system. As a result, it should be transformed into the autonomous systems by using the additional terms as:

$$\tau_d = K_1\overline{M}\ddot{\eta}_{ind}^{ref} + K_1\overline{C}\dot{\eta}_{ind}^{ref} + K_1\overline{F};$$

$$X = \begin{pmatrix} z_{dq} \\ z_{qind} \\ q_{ind}^d \end{pmatrix}; u = \tau - \tau_d; l(x) = K_1(x)\overline{C}(x).x; h_1 = \dot{\eta}_{ind}^d; h_2 = \ddot{\eta}_{ind}^d$$

$$(8)$$

According to (5) and (8), the affine system can be given as:

$$\dot{X} = \begin{bmatrix} -(K_1\overline{M})^{-1}.l(z_{dq} + \dot{\eta}_{ind}^{ref}) + (K_1\overline{M})^{-1}.l(\dot{\eta}_{ind}^{ref}) \\ (z_{qind} + \eta_{ind}^d)z_{dq} - \beta z_{qind} \\ h_1 \end{bmatrix} + \begin{bmatrix} (K_1\overline{M})^{-1} \\ 0 \\ 0 \end{bmatrix} u$$

$$(9)$$

Therefore, the ARL based controller can be proposed with performance index $V = \int\limits_0^\infty (X^T Q_T X + u^T R u) ds$ as follows:

$$\hat{V} = \hat{W}_C^T \phi(X) \tag{10}$$

$$\hat{u}(X) = -\frac{1}{2} R^{-1} H^T(X) \left(\frac{\partial \phi}{\partial X}\right)^T \hat{W}_a \tag{11}$$

$$\dot{\hat{W}}_C = -\eta_C \Gamma \frac{\gamma}{1 + v\gamma^T \Gamma \gamma} \sigma_{hjb} \tag{12}$$

$$\dot{\hat{W}}_a = -\frac{\eta_{a1}}{\sqrt{1 + \gamma^T \gamma}} \cdot \frac{\partial \phi}{\partial X} . H(X) . R^{-1} H^T(X) . \left(\frac{\partial \phi}{\partial X}\right)^T . (\hat{W}_a - \hat{W}_c) \sigma_{hjb} - \eta_{a2}(\hat{W}_a - \hat{W}_c) \tag{13}$$

4   Simulation Results

In this section, the system contains three two-link manipulators is considered to simulate with the purpose of validating the proposed controller. The parameters of three manipulators used in this simulation are as follows: $D_{12} = 7$ m, $D_{13} = 3.5$ m, $l_1 = l_2 = 1$ m, $m_1 = m_2 = 1$ kg, $r_1 = r_2 = 0.9$ m, $l_3 = l_4 = 1.5$ m, $m_3 = m_4 = 2$ kg, $r_3 = r_4 = 1.2$ m, $l_5 = l_6 = 1.2$ m, $m_5 = m_6 = 1.5$ kg, $r_5 = r_6 = 1$ m. Matrix M_1, M_2, and M_3 of three manipulators can be rewritten as

$$M_1 = \left[\hat{\theta}_1 + 2\hat{\theta}_2 cos\eta_2 \hat{\theta}_3 + \hat{\theta}_2 cos\eta_2; \hat{\theta}_3 + \hat{\theta}_2 cos\eta_2 \hat{\theta}_3\right]$$

$$M_2 = \left[\hat{\theta}_4 + 2\hat{\theta}_5 cos\eta_4 \hat{\theta}_6 + \hat{\theta}_5 cos\eta_4; \hat{\theta}_6 + \hat{\theta}_5 cos\eta_5 \hat{\theta}_6\right]$$

$$M_3 = \left[\hat{\theta}_7 + 2\hat{\theta}_8 cos\eta_6 \hat{\theta}_9 + \hat{\theta}_8 cos\eta_6; \hat{\theta}_9 + \hat{\theta}_8 cos\eta_6 \hat{\theta}_9\right]$$

The desired trajectories for four independent joints are previously established as $\eta_{1d} = \frac{\pi}{12}; \eta_{2d} = 1,91\pi + 0,2\sin(t); \eta_{3d} = 0,51\pi; \eta_{5d} = 0,191\pi$. The tracking effectiveness of join variables responses in three manipulators system and adaptation Law of uncertain parameters in (7) are shown in Fig. 3.

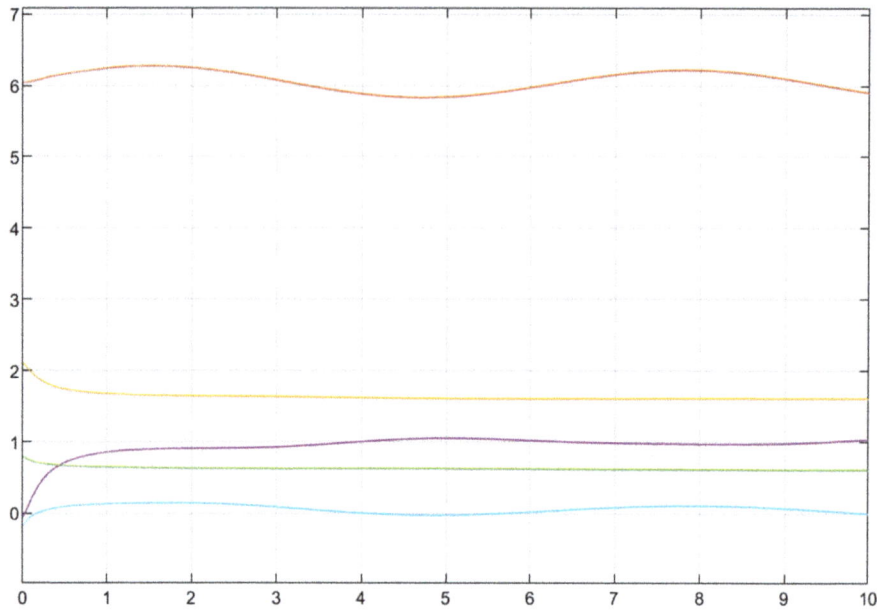

Fig. 3. The response of trajectories of η_i

5 Conclusions

In this paper, ARL based motion/force control have been proposed for solving the optimal control problem of multiple uncertain manipulators. The motion/force control structure is given to develop the ARL technique based on motion dynamic model after the separation technique of cooperating manipulators was implemented. However, the control design requires the ARL based solution for time-varying systems. For this reason, the additional terms in (8) are proposed to handle this disadvantage. Finally, the online Actor/Critic ARL is developed with Neural Network and optimality principle for modified motion dynamic model.

References

1. Liu, Y., Dao, P.N., Zhao, K.Y.: On robust control of nonlinear teleoperators under dynamic uncertainties with variable time delays and without relative velocity. IEEE Trans. Ind. Inform. **16**(2), 1272–1280 (2020). https://doi.org/10.1109/TII.2019.2950617
2. Nguyen, T.B., et al.: An adaptive backstepping trajectory tracking control of a tractor trailer wheeled mobile robot. International J. Control Autom. Syst. **17**(2), 465–473 (2019)
3. Van Tinh, N., et al.: A Gaussian wavelet network-based robust adaptive tracking controller for a wheeled mobile robot with unknown wheel slips. Int. J. Control (2019). https://doi.org/10.1080/00207179.2018.1458156
4. Baek, J., Kwon, W., Kim, B., Han, S.: A widely adaptive time delayed control and its application to robot manipulators. IEEE Trans. Ind. Electron. **66**, 5332–5342 (2018)

5. Bian, T., Jiang, Y., Jiang, Z.P.: Adaptive dynamic programming and optimal control of nonlinear nonaffine systems. Automatica **50**, 2624–2632 (2014)
6. Cao, F., Liu, J.: Three-dimensional modeling and input saturation control for a two-link flexible manipulator based on infinite dimensional model. J. Franklin Inst. **357**, 1026–1042 (2020)
7. Chen, C., Liu, Z., Zhang, Y., Xie, S.: Coordinated motion/force control of multiarm robot with unknown sensor nonlinearity and manipulated object's uncertainty. IEEE Trans. Syst. Man Cybern.: Syst. **47**, 1123–1134 (2016)
8. Deng, M., Li, Z., Kang, Y., Chen, C.P., Chu, X.: A learning based hierarchical control scheme for an exoskeleton robot in human–robot cooperative manipulation. IEEE Trans. Cybern. **50**, 112–125 (2018)
9. Dohmann, P.B.G., Hirche, S.: Distributed control for cooperative manipulation with event-triggered communication. IEEE Trans. Robot. **36**(4), 1038–1052 (2020)
10. Dong, Y., Ren, B.: UDE-based variable impedance control of uncertain robot systems. IEEE Trans. Syst. Man Cybern.: Syst. **49**(12), 2487–2498 (2017)
11. Dou, H., Wang, S.: Robust adaptive motion/force control for motion synchronization of multiple uncertain two-link manipulators. Mech. Mach. Theory **67**, 77–93 (2013)
12. Guo, X., Yan, W., CUI, R.: Integral reinforcement learning based adaptive NN control for continuous-time nonlinear MiMo systems with unknown control directions. IEEE Trans. Syst. Man Cybern.: Syst. **50**(11), 4068–4077 (2019)

Road Sign Segmentation for Smart Bicycles

Blake Hament[1] and Vikram Puri[2(✉)]

[1] University of Nevada, Las Vegas, USA
hament@unlv.nevada.edu
[2] Duy Tan University, Da Nang, Vietnam
purivikram@duytan.edu.vn

Abstract. Bicycles are low-carbon emitting and inexpensive means of transportation that are popular worldwide. In densely populated areas, bicycle sharing programs are helping to expand access to these vehicles. As bicycle sharing programs grow and more people use bicycles on the road, there is a need to improve the performance and safety of bicycle operation. One opportunity for improving the safety of bicycle operation is to add automated road sign detection and identification. Once a road sign is detected and identified, this data can be used to improve autonomous navigation of the vehicle, to share data between bicycles on the road for better traffic control, and to remind or prompt the operator about current road status. The Mask-RCNN deep learning architecture is trained to detect, identify, and segment several common Vietnamese road signs. Training images were captured from a camera mounted on a bicycle traveling on the road. Results show that the trained neural net successfully recognizes signs, with varying success rates across trials correlating heavily with the number and quality of training annotations.

Keywords: Deep learning · Mask RCNN · Internet of things · Bicycle · Transportation · Artificial intelligence

1 Introduction

Consumption of the natural resources is rapidly increasing as compared to their replenishment as per the report of the Worldwide Footprint Network [1]. If natural resources consumption increases as usual, the requirement for replenishment of these resources would require more than 3 earth-sized planets by 2050 [2]. Sharing economy [3] is one possible solution to sky-rocketing consumption, providing unique opportunities for entrepreneurs and government to jointly reshape shareable public transportation systems. Marcus and Joel [4] presented the concept of sharing economy in 1978 called "collaborative consumption" that based on mutual consumption of goods and fuel temporary or permanent basis. Public bicycle sharing system (PBSS) acts as a rooter for the sharing economy and enlarge the usage of public transportation system. In this era, many public initiatives are already involved to promote the bicycle usage in the urban areas,

D.-T. Tran et al. (Eds.): ICISN 2021, LNNS 243, pp. 272–279, 2021.
https://doi.org/10.1007/978-981-16-2094-2_34

PBSS is one of the initiatives that have achieved most attractive results within a shorter period of time [5]. PBSS [6] defines as "according to the need anyone can rent bicycle without take any permission from the third party as well as mutual cost setting". The main features of the BSS [7–9] are to reduce the consumption of fuel in the transportation, reduce traffic congestions and with the environmental point of view, it depletes the gases such as carbon monoxide gas.

To see wide-spread adoption of bicycle sharing systems, stakeholders need to ensure that these systems are as safe, convenient, and efficient as possible. One way to increase safety, convenience, and efficiency is to incorporate automation that detects, tracks, and responds to other vehicles, changes in the road, and other environmental factors. Road sign detection and identification is a first step towards such a system. For wide-spread adoption of bicycle sharing systems (see in Fig. 1).

Fig. 1. The Mask RCNN architecture was retrained for detection and identification of common road signs in Vietnam. Above images show "Road junction with priority" or "207b" signs identified with the neural net.

2   Literature Review

Various models and techniques has been explored for detection and recognition of road signs. Benallal [10] proposed real time system that based on color segmentation to recognize signs in the road and also monitor the RGB variation in the road signs from the day to night time. Their proposed system methodology is limited to a single camera. Lai [11] proposed traffic sign recognition system based on the integration of in-vehicle computing device and smartphone. This system is worked on four different stages: 1) real time video capturing 2) image processing 3) road signs detection 4) recognition of road signs. The drawback of the system is low computing performance. In [12], the Authors presented an algorithmic program for the detection of traffic signs through the aid of color centroid matching technique. The YCbCr color space is employed to create the detection process for the color segmentation. Bahlman [13] presented real-time road sign recognition system based on visual feature recognition. This system is categorized into two parts: 1) Propagation framework is used for to detect the objects 2) Bayesian generative modelling for the traffic signs classification. Timofte [14] illustrated a combination of two dimensional and three dimensional techniques for the detection and recognition of the traffic signs. In this study, 8 cameras are mounted on the roof of a van to capture the traffic road signs. In addition, multiple detection of 2D are combined to generate the 3D hypotheses. De la Escalera et al. [15] use neural networks to detect traffic sign and create a genetic algorithm that helps to detect the step changes in position, variation in the localization position, rotations, and weather condition. Neural Networks [16–18] are one of the best techniques to detect the traffic road sign and recognize them. In [19–21], researchers applied neural network to detect traffic signs. A huge benefit of modern deep learning methods are that they can be applied across different alphabets, such that similar methods can be used for signs in English, Chinese, Vietnamese, etc. In terms of real time implementation, neural networks outperform all other options. The method of image segmentation using Mask R-CNN(Region-based Convolutional Network) is an extension of previous object detection algorithms. One of the algorithms is the R-CNN which is used for accurate object detection and segmentation [22]. This algorithm generates several independent region proposals with a high probability of containing an object using a selective search from the input image. Then a feature vector helps to extract each region using a CNN for each proposal which generates high computing. Several drawbacks arose with the R-CNN network such as slow object detection (47s/image) as well as space and time expensive during training which led to Fast R-CNN. This method applies the CNN first, then obtains the region proposals from the image [23]. Switching this order increased the testing time per image to 2 s compared to the 45 s of the R-CNN. Chen proposes using an R-CNN framework to recognize the minor objects in a scene [24]. This work augments the R-CNN algorithm that boosts performance by 29.8% on the benchmark dataset made up of the existing COCO and SUN datasets. The change included using object proposals instead of a sliding window approach which leads to fewer and accurate proposals, along with good R-CNN performance. Most recently, Mask R-CNN has emerged as the premier architecture for classifying pixel-by-pixel, we use the Keras implementation by Waleed Abdulla [25].

3   Methodology

3.1   Annotation

Two rounds of annotation were performed. In the first round of annotation, approximately 1000 images were annotated with bounding boxes around any visible signs. These signs were then labeled according to colloquial references, such as "Right Turn" or "No Parking". After training results were less than satisfactory with this dataset, the first annotations were discarded, and a second round of annotation was performed. In the second round of annotation, a finite number of common signs were identified from the national Vietnamese catalogue of road signs using their official numbering as labels, i.e. "207b" and "408" [26]. In the second round, annotations consisted of drawing masks point-by-point around each observed sign, then classifying according to available labels.

Table 1. Class counts after first round of annotation.

Class	Training	Validation
Bus Stop	120	13
Direction Sign	244	27
Instruction Sign	192	22
No Stopping	163	18
Parking	119	13
Road Junction Priority	132	15
"Other" Sign	261	29
Speed Limit	130	15

Table 2. Class counts after second round of annotation.

Class	Training	Validation
207b	28	7
408	20	5
407a	35	9
131a	50	13

3.2   Transfer Learning

The Mask RCNN architecture is trained via transfer learning. First, the full net including resnet101 backbone, region, and feature proposal networks are initiated. Next, the model is pre-loaded with weights trained on the COCO dataset from [25]. The final layers for predicting features, bounding boxes, masks, and classes specific to the COCO trained model are wiped for retraining with new data, but most of the net is left untouched. Because the original model is trained on over 200,000 images from the diverse COCO dataset, it has encoded activations for a multitude of salient visual features. In levels close to the input, these activations might pick up features like edges and corners, while deeper layers combine many higher-level features into complex, but unique and efficient visual patterns useful for pixel classification. The pre-trained model is next retrained on a much smaller dataset of road signs, this time with 10's of images rather than 10^6's. Only network heads, or fully connected layers, are adjusted through backpropagation, leaving most feature activation and region proposal layers untouched. The layers that see most changes in weight are layers that combine region and feature actions for final estimations.

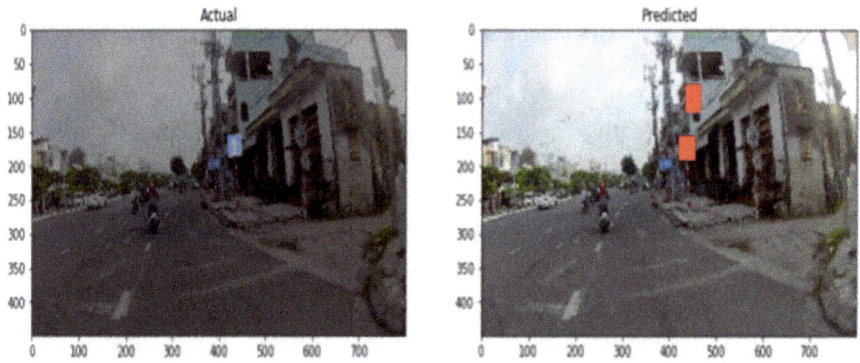

Fig. 2. Left image shows target sign, "No Parking", highlighted with white mask. Right shows net guess after several epochs of training. The net begins to learn visual features that identify 'sign-like' pixel neighborhoods, but needs more information to refine estimations and determine class.

4   Results

Results from the first round of annotations were decidedly unsatisfactory, as seen in Fig. 2 and Fig. 3 After training the net, sign-like pixel regions were often identified, but the correct class of sign was virtually never labeled correctly. This prompted re-annotation. On the second round of annotation, much more care was given to labeling of the sign classes. To avoid redundant, confusing, and inaccurate labeling, classes were taken from the official, numbered catalogue of Vietnamese road signs. Additionally, images in which the target sign had a very small area and noisy images containing many signs were pruned from the dataset. Table 1 and 2 compares class counts of the two rounds of annotations.

After the second round of annotations, training experiments began with a subset of our fully annotated dataset containing only the most common sign. Training loss reaches an asymptote after roughly 30 epochs as seen in Fig. 3 It can be seen in Fig. 4 that average precision and average recall also stabilize after roughly 30 epochs, this validation testing considers all predictions with confidence ranging from 50% to 95%. Performance statistics are slightly higher when predictions with greater than 95% confidence are considered, as seen in Table 3.

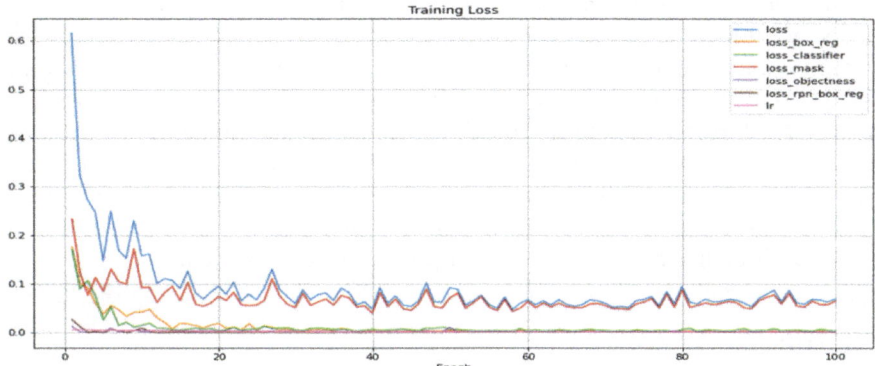

Fig. 3. Losses during multi-sign training with final round of annotations.

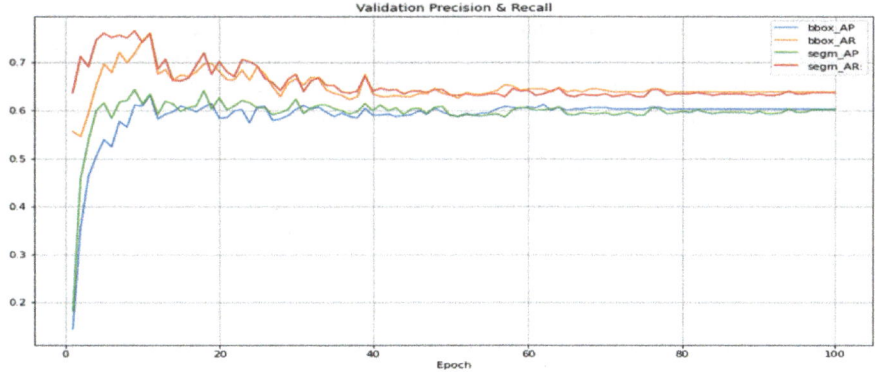

Fig. 4. Average precision and average recall for both bounding box and segmentation metrics during validation testing. 20% of data was held back for use in validation testing. Confidence scores between 50 and 95% are included.

278 B. Hament and V. Puri

Table 3. Final statistics from best model after validation testing. Only predictions with confidence greater than 95% are included.

Accuracy	Precision	Recall	Specificity	F1 Score
89.23%	83.33%	66.67%	96.00%	74.07%

5   Conclusion

In conclusion, these experiments demonstrate the possibility of using the Mask RCNN or similar neural network architecture to recognize and classify road signs from a moving bicycle. By using transfer learning methods, the network is trained using an order of 10^3 images rather than training from scratch with an order of 10^6 images. Annotation quality is of the utmost importance! Training and validation results improved significantly after reannotating the dataset and pruning data. By discarding images with very low information content (small sign areas) and images with high amounts of noise (spurious, unlabeled road signs), much higher mean accuracy and precision was observed.

References

1. Worldwide   Footprint   report.   https://www.footprintnetwork.org/our-work/ecological-footprint. Accessed 20 Oct 2020
2. UN fuel consumption. https://www.un.org/sustainabledevelopment/wp-content/uploads/2016/08/16-00055L_Why-it-Matters_Goal-12_Consumption_2p.pdf. Accessed 20 Oct 2020
3. Winslow, J., Mont, O.: Bicycle sharing: sustainable value creation and institutionalisation strategies in Barcelona. Sustainability **11**(3), 728 (2019)
4. Felson, M., Spaeth, J.L.: Community structure and collaborative consumption: a routine activity approach. Am. Behav. Sci. **21**(4), 614–624 (1978)
5. Castillo-Manzano, J.I., Sánchez-Braza, A.: Managing a smart bicycle system when demand outstrips supply: the case of the university community in Seville. Transportation **40**(2), 459–477 (2013)
6. Shaheen, S.A., Guzman, S., Zhang, H.: Bikesharing in Europe, the Americas, and Asia: past, present, and future. Transp. Res. Rec. **2143**(1), 159–167 (2010)
7. Midgley, P.: The role of smart bike-sharing systems in urban mobility. Journeys **2**(1), 23–31 (2009)
8. O'brien, O., Cheshire, J., Batty, M.: Mining bicycle sharing data for generating insights into sustainable transport systems. J. Transp. Geogr. **34**, 262–273 (2014)
9. Peng, C., OuYang, Z., Liu, Y.: Understanding bike sharing use over time by employing extended technology continuance theory. Transp. Res. Part A: Policy Pract. **124**, 433–443 (2019)
10. Benallal, M., Meunier, J.: Real-time color segmentation of road signs. In: CCECE 2003-Canadian Conference on Electrical and Computer Engineering. Toward a Caring and Humane Technology (Cat. No. 03CH37436), vol. 3, pp. 1823–1826. IEEE (2003)

11. Lai, C.H., Yu, C.C.: An efficient real-time traffic sign recognition system for intelligent vehicles with smart phones. In: 2010 International Conference on Technologies and Applications of Artificial Intelligence, pp. 195–202. IEEE (2010)

12. Chourasia, J.N., Bajaj, P.: Centroid based detection algorithm for hybrid traffic sign recognition system. In: 2010 3rd International Conference on Emerging Trends in Engineering and Technology, pp. 96–100. IEEE (2010)

13. Bahlmann, C., Zhu, Y., Ramesh, V., Pellkofer, M., Koehler, T.: A system for traffic sign detection, tracking, and recognition using color, shape, and motion information. In: IEEE Proceedings. Intelligent Vehicles Symposium, pp. 255–260. IEEE (2005)

14. Timofte, R., Zimmermann, K., Van Gool, L.: Multi-view traffic sign detection, recognition, and 3D localisation. Mach. Vis. Appl. 25(3), 633–647 (2014)

15. De la Escalera, A., Armingol, J.M., Mata, M.: Traffic sign recognition and analysis for intelligent vehicles. Image Vis. Comput. 21(3), 247–258 (2003)

16. Puri, V., Jha, S., Kumar, R., Priyadarshini, I., Abdel-Basset, M., Elhoseny, M., Long, H.V.: A hybrid artificial intelligence and internet of things model for generation of renewable resource of energy. IEEE Access 7, 111181–111191 (2019)

17. Jha, S., Kumar, R., Abdel-Basset, M., Priyadarshini, I., Sharma, R., Long, H.V.: Deep learning approach for software maintainability metrics prediction. IEEE Access 7, 61840–61855 (2019)

18. Priyadarshini, I., Cotton, C.: Intelligence in cyberspace: the road to cyber singularity. J. Exp. Theoret. Artif. Intell. 1–35 (2020). https://www.tandfonline.com/doi/figure/10.1080/0952813X.2020.1784296?scroll=top&needAccess=true

19. Qian, R., Zhang, B., Yue, Y., Wang, Z., Coenen, F.: Robust Chinese traffic sign detection and recognition with deep convolutional neural network. In: 2015 11th International Conference on Natural Computation (ICNC), pp. 791–796. IEEE (2015)

20. Lee, H.S., Kim, K.: Simultaneous traffic sign detection and boundary estimation using convolutional neural network. IEEE Trans. Intell. Transp. Syst. 19(5), 1652–1663 (2018)

21. Zhu, Y., Zhang, C., Zhou, D., Wang, X., Bai, X., Liu, W.: Traffic sign detection and recognition using fully convolutional network guided proposals. Neurocomputing 214, 758–766 (2016)

22. Girshick, R., Donahue, J., Darrell, T., Malik, J.: Region-based convolutional networks for accurate object detection and segmentation. IEEE Trans. Pattern Anal. Mach. Intell. 38(1), 142–158 (2016). https://doi.org/10.1109/TPAMI.2015.2437384

23. Girshick, R.: Fast R-CNN. In: Proceedings of the IEEE International Conference on Computer Vision, pp. 1440–1448 (2015)

24. Chen, C., Liu, M.Y., Tuzel, O., Xiao, J.: R-CNN for Small Object Detection. In: Lai, S.H., Lepetit, V., Nishino, K., Sato, Y. (eds.) Computer Vision - ACCV. ACCV 2006. Lecture Notes in Computer Science, vol. 10115. Springer, Cham (2016)

25. Abdulla, W.: Mask R-CNN for object detection and instance segmentation on Keras and TensorFlow. Github (2017). https://github.com/matterport/Mask_RCNN

26. Road signs in Vietnam. https://en.wikipedia.org/wiki/Road_signs_in_Vietnam. Accessed 26 Oct 2020

Building an Automobile Accident Detection and Messaging System Using Arduino

Ninh Bui Trung, Huy Luong Duc[✉], and Hung Nguyen Viet[✉]

VNU University of Engineering and Technology (VNU-UET),
Vietnam National University, Hanoi, Vietnam
ninhbt@vnu.edu.vn

Abstract. In Vietnam, the number of traffic vehicles and traffic accidents is increasing. Currently, in the mid-tier model and below, there is no warning and rescue system that is equipped to deal with vehicle crashes, especially when it occurs at night. It is difficult for people in distress and can even lead to death in some emergencies. In this paper, we propose to build a model of automobile accident detection and messaging system using Automatic utilizing Accelerometer sensor, GPS module, and GSM module. The experimental test results show that the system works well, the response time is fast and the price is cheap. It is suitable for installation on vehicles in Vietnam in the future.

Keywords: Messaging system · Global positioning system · Global system for mobile communications

1 Introduction

Every day, hundreds of road vehicle accidents occur around the world, a large percentage of death is because the incidents happened at remote locations, and help couldn't reach the victims within the Golden time following a traumatic injury. Analytical reports show that reducing the response time for accidents by one minute will increase life-saving chances by six percent [1]. Complete accident prevention is unavoidable but the repercussions can be reduced. Viet Nam's population density is very high and traffic here is also non-lane based and chaotic. Infrastructure growth is slow as compared to the growth in the number of population and vehicles. Existing systems are almost manual controlled and gradually, the number of people working on the roads and efficiency will be out of control [2].

Currently, on a few expensive luxury car models of BMW, Mercedes, some safety features such as anti-lock brakes, stability control systems and the automatic accident detection and notification systems are equipped with the most recent manufactures vehicles, which depend on the vehicle onboard sensors to detect the accident and utilize the built-in radio cellular to notify the emergency responders [3]. These systems will become useless after having an accident. Especially in mid-tier model and below, the lack of safety features is extremely serious. Moreover, the installation cost of these systems inside the vehicles is very expensive. Also, they are not considered as a standard option for all vehicles in Vietnam and other developing countries.

A variety of project have been designed to protect drivers from accident, but most of them are for smartphone's app. The most prominent one is S-Bike, which SamSung installed in Galaxy J 2017 series to support for motorbike riders during riding. However, S-Bike is useless in detecting an accident, and it does not have the function to neither call rescue center nor relatives. On the other hand, riders do not always take smartphones with them, so it will be a big problem when they forget their phones [6].

Our project's strength compared to those apps is a fully automatic system, that has the ability to detect when the vehicle had crashed and sends messages to rescue centers and the client's contact list. We have developed a cheap, small, low power and especially fixed installation in car or motorcylces. It is an Arduino based system, using ADXL 335 Accelerometer sensor, Neo 6M GPS module, SIM800A GSM module. This system is small and compact, and can be easily installed on the vehicle, user do not have to bring anything to protect themself in accidental senarios.

2   Proposed System Architecture

An accelerometer output the changes in the vehicle's condition. This signal is regarded as the input of the Control unit [4]. GPS simultaneously detects the location through satellite signal and in case of accidents, data are sent to the Rescue messages of GSM, from here an alert message with the location coordinates is sent to both the contact list of the driver and rescue centers. Based on information from the message, location tracking will become easier, drastically reducing the time required by the ambulance to reach the crash site, increasing the survival chances of the casualties [2]. Figure 1 and 2 show block diagram and modules used in the system (Fig. 3).

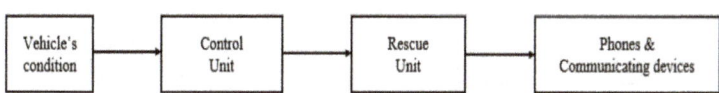

Fig. 1. Block diagram of the system

ADXL335   GPS NEO 6M   GSM SIM 800A

Fig. 2. Main modules used in the system

The system consists of two main units, which link together and make sure that the rescue arrives on time: The Control Unit and the Rescue Unit.

Fig. 3. Completed system

a) **Control unit**

Regarded as the center of this project, the control unit (Fig. 4) should be implemented in every vehicle to optimize the system's efficiency. An accelerometer sensor will be the key to detecting an accident. It can be used as a crash or rollover detector of the vehicle during or after a crash [5]. The signal from the accelerometer will be sent to the main controller to determine whether an accident has occurred. GPS works consistently to find out the current location of the vehicle, such as longitude, latitude, speed, time and date,... These pieces of information are handled by the main controller [1]. In case of accidents, these data will become the input of the Rescue unit, and LEDs, buzzers are turned on to warn nearby vehicles.

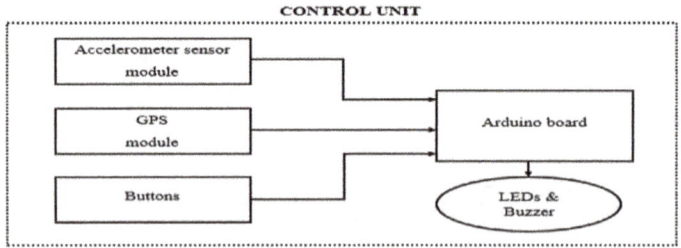

Fig. 4. Control unit

b) **Rescue unit**

Recognizing a real accident is the first step to reaching a safe state. GSM module collects data from GPS and sends a distress message including the vehicle's location and situation to the rescue center and the client's contact list. So the rescue team and police can immediately trace the latest known location through GPS [2].

In this project, we have added two more functionalities, SOS and cancel messages. In many cases, when there is just a minor accident, or when there is nothing hazardous to people's lives, they can stop the messaging process in about twenty seconds. After this time, if the system doesn't receive any responses from the clients, a distress message will be sent automatically. Therefore, this can save much precious time for the medical rescue center. Besides, we continue to develop an SOS function, especially effective when you encounter situations not related to accidents such as kidnap, robbery, … After press and holding the SOS button for 3 s, a distress message will be sent, including the call for help and current location. Rescue unit is shown in Fig. 5.

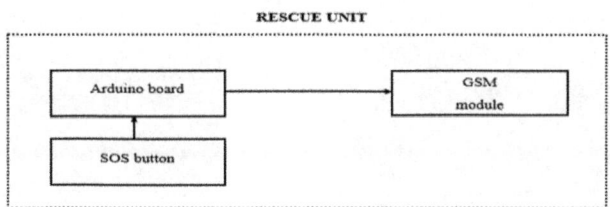

Fig. 5. Rescue unit

3   Experiment Results

The model has been carried out to understand the behaviors of Collision Detect and Messaging Systems. System responses to a crash with nearly zero delays, turning on LED and buzzer, user will have about 60 s to confirm everything is under control or not. Getting through that time without any responses, a message will be sent automatically. In the best-case scenario, the message will take approximately 10 s to reach the destination. Figure 6 represents the flowchart diagram and prototype of the system.

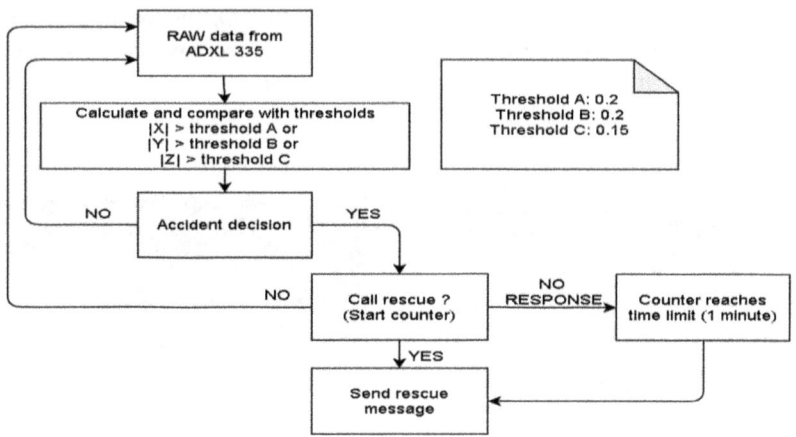

Fig. 6. System's flowchart

For the first time running the system, it will take 2–3 min to receive GPS signal. Figure 7 shows the LCD displaying vehicle's current location. System responding in case of an accident, waiting for users to confirm to send or not, via YES/NO button. After 1 min, if there aren't any responses, the system assumed that users are severely injured and unconscious. The message will be sent automatically (Fig. 8). Figure 9 shows the LCD screen after sending the message. It is continuosly flashing to warn nearby residents.

Fig. 7. Lattitude & Longitude displayed on the LCD screen

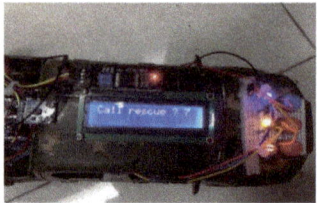

Fig. 8. LCD screen after accident occur

Fig. 9. The content displayed on the LCD screen after the message is sent

In an emergency case, users can press and holding SOS button in 3 s, an SOS message will be sent immediately to the rescue center and the driver's family (Figs. 10 and 11). Contents in messages sent to rescue centers and user's contact list, including GPS coordinates and Google Maps link to the accident site (Fig. 12).

Fig. 10. SOS count before sending the emergency message

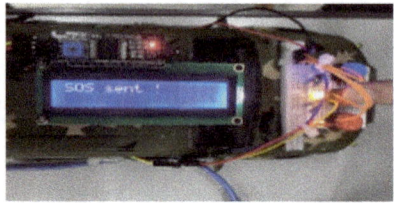

Fig. 11. SOS sent when pressing SOS button.

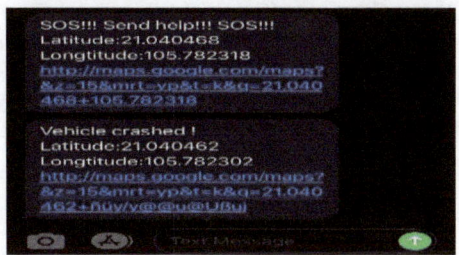

Fig. 12. Distress messages transmitted by the system

4 Conclusion

In this paper we have discussed existing problems, requirements from which to build an automatic accident detection and notification system on the vehicles. The system is particularly useful in detecting collisions precisely, utilizing both accelerometer sensor and micro electro Mechanical system (MEMS). The system basically detects collisions and sends out warning messages in a short time to both the rescue centers and the driver's relatives. In addition, the system is also equipped with a button to confirm the danger level of an accident and an SOS button to use in the situation of kidnapping or robbery… In order for the system to be deployed in practice, economical and with no loss of time, it is necessary to reduce the latency of the system. Therefore, the challenge is to reduce delay of GSM messages since it is a queue based technique, which can be resolved by giving more priority to messages communicated through the controller.

In the future we plan to have some upgrades in both hardware and software. The vibration sensor can be added to the system, combined with the accelerometer sensor and vehicle's speed from GPS to further increase the reliability in detecting accidents. Also, we have planned an Android application version of the system to provide one more option for drivers. The software version will be much cheaper and more user friendly.

References

1. Evanco, W.E.: The Impact of Rapid Incident Detection on Freeway Accident Fatalities, technical report available from Mitretek (center for information system), McLean, Virginia, USA, report No. WN 96W0000071, June 1996
2. Niranjana, V.M.I., Manikandan, R., Suganthan. S., Malayandisamy, P.: Automatic accident detection, ambulance rescue and traffic signal controlle. IJSTE – Int. J. Sci. Technol. Eng. **3**(09), 244–250 (2017)
3. Chris, T., White, J., Dougherty, B., Albright, A., Schmidt, D.C.: WreckWatch: automatic traffic accident detection and notification with smartphones. Int. J. Mob. Netw. Appl. **16**(3), 285–303 (2011)
4. Wang, S.: Automatic vehicle accident detection and messaging system using GSM and GPS modem. J. Ind. Electron. Appl. **4**(2), 252–254 (2020)

5. Dinesh Kumar, H.S.D.K., Gupta, S., Kumar, S., Srivastava, S.: Accident detection and reporting system using GPS and GSM module. J. Emerg. Technol. Innov. Res. (JETIR) **2**(5), 1433–1436 (2015)
6. Van Nguyen, T., Nguyen, V.D., Nguyen, T.-T., Khanh, P.C.P., Nguyen, T.-A., Tran, D.-T.: Motorsafe: an android application for motorbike drivers using decision tree algorithm. Int. J. Interact. Mobile Technol. (iJIM) **14**(02), 119–129 (2020)

Upgrade and Complete the Electrical Impedance Measuring System Applying for Meat Quality Analysis and Evaluation

Kien Nguyen Phan, Vu Tran Anh$^{(\boxtimes)}$, Trung Dang Thanh$^{(\boxtimes)}$, and Thang Phung Xuan$^{(\boxtimes)}$

School of Electronics and Telecommunications, Hanoi University of Science and Technology, Hanoi, Vietnam
kien.nguyenphan@hust.edu.vn

Abstract. Impedance spectrum analysis for the study of biological tissue is a useful method. In previous studies, impedance spectroscopy has been used to analyze the changes in pork after slaughter over time while also assessing meat quality. However, the old system is still limited in collecting data with the accuracy that has not been verified and performed only on one sample at a time. In this study, a new impedance measuring system was developed to measure two parallel samples with the accuracy of the measured data up to 97%. The system is tested with Fricke model and Cole-Cole model circuit, and pork meat samples. The measured data are identical to those obtained in the previous study but with higher accuracy and faster. In the future, the system will continue to be developed with more simultaneous samples to collect data to perform further analysis of the variation in meat quality corresponding to the variation in measured impedance.

Keywords: Impedance spectrum · Meat quality · Fricke model · Cole-Cole model

1 Introduction

Currently there are many methods of determining the quality of pork. Colorimetry and near infrared spectroscopy (NIRS) to predict meat freshness are presented in researches [1, 2]. However, those method of measurements use a very expensive spectrum analyzer and complex processing algorithm. Hyperspectral imaging and computer vision are used to evaluate pork, beef, and chicken quality [3–5]. The freshness of meat is analyzed through the meat's color, texture and shape. The advantage of the two methods above is to accurately assess the quality of meat in terms of freshness but cannot distinguish whether the meat is contaminated with harmful agents or not. Several researches [6–9] have used mechanical analysis to assess meat hardness, based on which to distinguish raw and cooked meat. Among the methods of meat quality analysis, the meat biological impedance analysis method is a quick and effective approach. These methods often use specialized measuring devices (impedance analyzers, LCR meters and tissue spectroscopy) and a computer for data acquisition and processing [10–12]. Such a system offers high precision and efficiency in meat quality analysis, but the cost of such a

D.-T. Tran et al. (Eds.): ICISN 2021, LNNS 243, pp. 287–297, 2021.
https://doi.org/10.1007/978-981-16-2094-2_36

system is very high. Therefore, it is necessary to build a system with low cost and still meet the requirement for accurate analysis of meat quality.

That is the premise of previous researches that have been developed [12–15]. In the [12, 15] researches, the first system was designed with manual frequency control and had satisfactory results. Next, based on the previous researches, the automatic impedance measuring system was developed [13, 14]. The results of the research for the new measuring system have many advantages over the old system by measuring the sample automatically. But the limitation of this automated system is that it only measures one sample per day, which does not meet the requirement for measuring large numbers of samples. But the limitation of this automated system is that it only measures one sample per day, which does not meet the requirement for measuring large numbers of samples.

2   System Consideration

2.1   Tissue Biological Impedance Model

The physicochemical properties of biological tissues depend on their constituents, of which the three most concerned are intracellular fluids (ICF), extracellular fluids (ECF) and cell membranes (Cm). Fricke et al. demonstrated the electrical equivalent of a biological tissue [16]. This model is shown in Fig. 1.

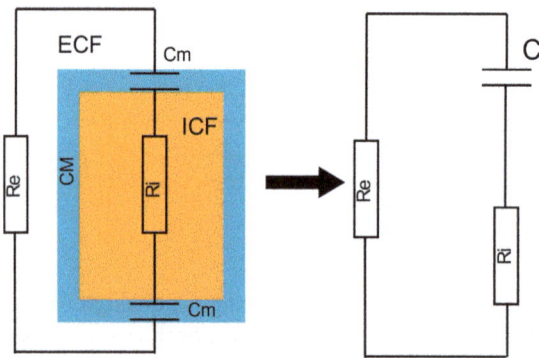

Fig. 1. Fricke equivalent impedance model

Based on Fig. 1, the complex impedance of a tissue is calculated using Eq. (1).

$$Z = Re(Z) + jIm(Z) = |Z|e^{j\theta} \qquad (1)$$

In which, Re(Z) and Im(Z) are real and imaginary parts, |Z| is the amplitude of the impedance, θ is the phase displacement angle. Parameters Re, Ri and Cm are the typical parameters to evaluate the tissue structure electrically.

In addition to the Fricke model, Cole-Cole is a simple model and is widely used for the description of the electrochemical properties of biological [17, 18]. The model consists of three components: impedance at very low frequency (characteristic of the extracellular medium), impedance at high frequency (representing both intracellular and extracellular media) and a constant component phase (CPE - Constant Phase Element); is shown in Fig. 2.

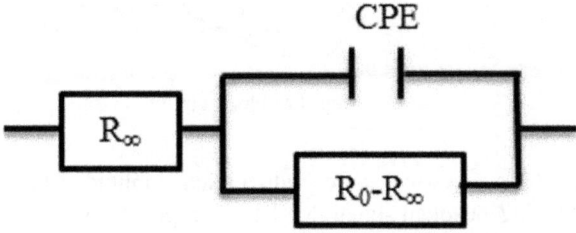

Fig. 2. Cole-Cole impedance model

Equations of the Cole-Cole model are shown in the Eq. (2).

$$Z* = R_\infty + \frac{R_0 - R_\infty}{1 + (iw\tau)^\alpha} = Z' + i.Z'' \tag{2}$$

In which:

- R_0 and R_∞ are the impedances at low frequencies and high frequencies in the Cole-Cole model.
- τ is a characteristic time constant
- α is within $(0,1)$, related to the heterogeneity of cell size and morphology of living tissue.
- $Z*$ is the complex impedance, Z' is the real part and Z'' is the virtual part.

The characteristic parameters for the research object $(R_0, R_\infty, \tau$ and $\alpha)$ can be found when "matching" experimental data to Eq. 2. Based on these parameters, it is possible to monitor, evaluate any changes in biological tissue.

2.2   Meat Impedance Measuring System

2.2.1   Measurement Principle
The inverting amplifier circuit is used to measure the impedance of meat, where the test sample replaces the OpAmp feedback resistance as in Fig. 3.

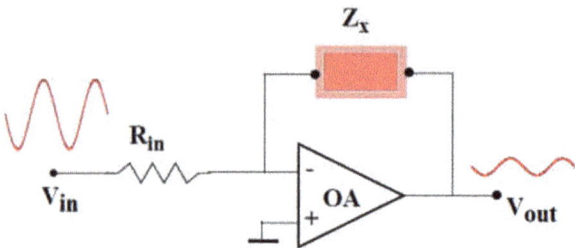

Fig. 3. Amplifier block circuit.

The input signal is a sine wave source with a fixed amplitude and a fixed frequency range. The output is a sinusoidal signal that has been phase inverted in comparison to the input signal. This principle was described in detail in the study [13].

2.2.2 System Design
The block diagram of the measuring system is shown in the Fig. 4.

The computer sends commands to the control unit to control the oscillator that produces signals with a fixed amplitude and frequency ranging from 10 Hz to 1 MHz.

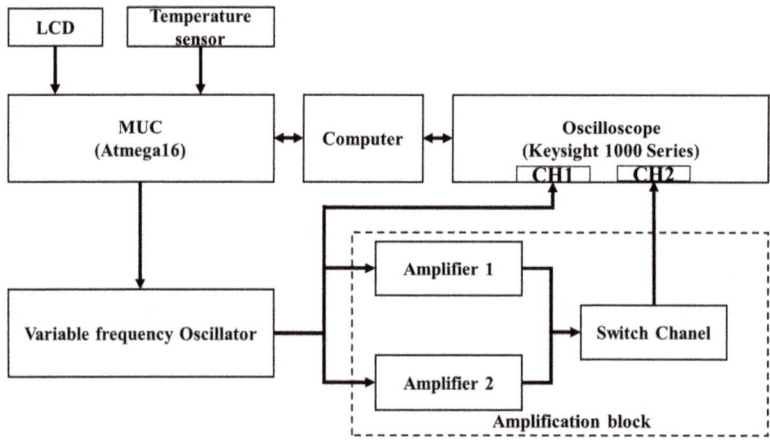

Fig. 4. Measurement system block

The signal from the oscillator will be received at channel 1 of the Oscilloscope, the controller generates the control signal, providing the power for amplifier 1 to operate, along with the controller to transmit the signal to connect the output to the channel 2 of the Oscilloscope. After the frequency sweep from 10 Hz to 1 MHz is terminated at amplifier 1, the controller sends a signal to amplifier 2 to begin measuring. The data acquisition process was similar to that in study [14]. Every hour the measurement is repeated, this is done for 24 h.

Figure 5 shows the complete measuring circuit after design.

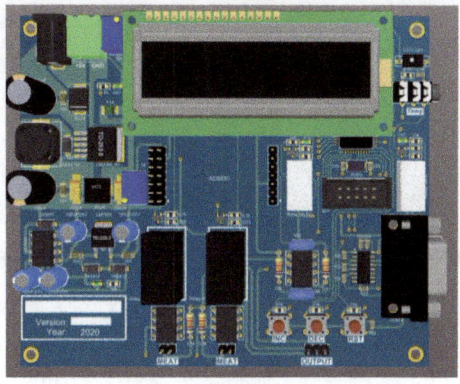

Fig. 5. Measurement circuit

2.3   Data Analysis

Raw data after the measurement is processed and obtained results include phase and impedance components. After that, data were matched into models by CNLS Fitting [19] method on EIS Spectrum Analyzer software. The data matching process is described in the study [14].

3   Experimental Results and Discussion

The resistor and capacitor values are measured with the KEYSIGHT U1233A multimeter. The input resistor has R_{in} value, and the feedback resistor has Z_x value in the amplifier unit.

3.1   Fricke Model Circuit Under the Test

a) *Test with Fricke simulation circuit*
   In this test, the resistor Rin is choosen by 463 Ω, Zx is replaced by the circuit connected as shown in Fig. 1, where, Re = 2751 Ω, Ri = 216.6 Ω, C = 102 nF.

   Figure 6(a) shows the actual impedance (from measured values) and theoretical impedance (from component values). The actual value is relatively matched to the theoretical value. Comparing between two values, the maximum error is 10.35% at the frequency of 1000 Hz and the RMSE value is 63.22 Ω.

   Figure 6(b) compares actual and theoretical phase when changing the frequency in the range from 10 Hz to 1 MHz. The actual line quite matches the theoretical line with the maximum error of 0.09 radians and the RMSE value of 0.03 radians.

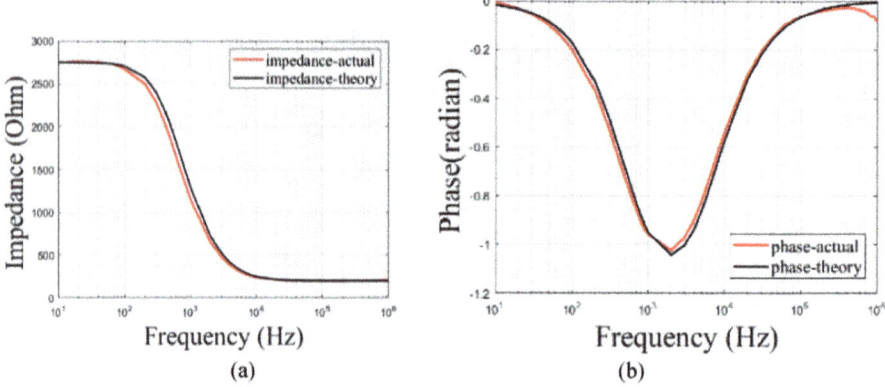

Fig. 6. Actual and theory value of impedence Z (a) and phase θ (b)

b) *Matching data with Fricke model by EIS Spectrum Analyzer and CNLS Fitting method*

With the obtained data, we compute Re(Z) and Im(Z) and use the internal Fricke model to estimate the parameters contained in the simulation circuit. Those estimated parameters are compared with real parameters including Re, Ri and Cm (Fig. 7).

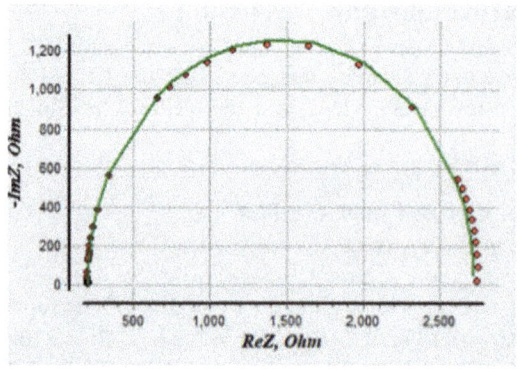

Fig. 7. Impedance spectrum of real part and imaginary part with fitting process

The parameters of Fricke model (C, R_e, R_i) are estimated from measured data. The comparison between estimated and real values are showed in Table 1.

Table 1. Estimated parameters under Fricke model

Parameter	Estimated value (nF)	Real value (nF)	Relative error
C	114.43	102	1.5884|%
R_e	2724.6	2751	1.276%
R_i	219.28	216.6	0.33287%

From estimated parameters, the relationship between impedance ($|Z|$) and frequency, and Phase (θ) and frequency are presented in Fig. 8.

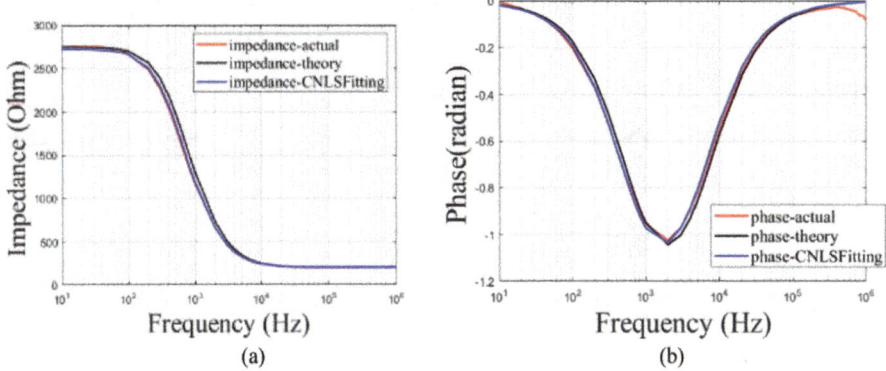

Fig. 8. Impedance Z (a) and Phase θ (b) of actual values, theory values and estimated by CNLS-Fitting method values

Figure 8(a) shows the actual impedance (from measured values), theoretical impedance (from component values) and the estimated impedance (from CNLS Fitting method) when changing the frequency in the frequency range from 10 Hz to 1 MHz. Comparing between the theoretical value and the value from the CNLS-Fitting method, the maximum error is 1.07% and the RMSE value is 13.35 Ω. Figure 8(b) shows the actual phase (from measured values), theoretical phase (from component values) and the estimated phase (from CNLS Fitting method) when changing frequency in the frequency range from 10 Hz to 1 MHz. Through calculation, the maximum error is calculated by 0.0068 radians and the RMSE value is 0.0036 radians.

3.2   Cole-Cole Model Circuit Under the Test

a) *Test with Cole-Cole simulation circuit*
   The resistor Rin is chosen by 459 Ω, Zx is replaced by the circuit connected as shown in Fig. 2, where, $R_\infty = 219.8$ Ω, $R_0-R_\infty = 2677$ Ω, and $C_{PE} = 40$ nF.
   Figure 9(a) shows the actual impedance (from measured values) and theoretical impedance (from component values). The actual value is relatively matched to the theoretical value. Comparing between two values, the maximum error is 6.24% at the frequency of 1000Hz and the RMSE value is 18,35 Ω. Figure 9(b) compares actual and theoretical phase when changing the frequency in the range from 10 Hz to 1 MHz. The actual line quite matches the theoretical line with the maximum error of 0,078 radians and the RMSE value of 0.03 radians.

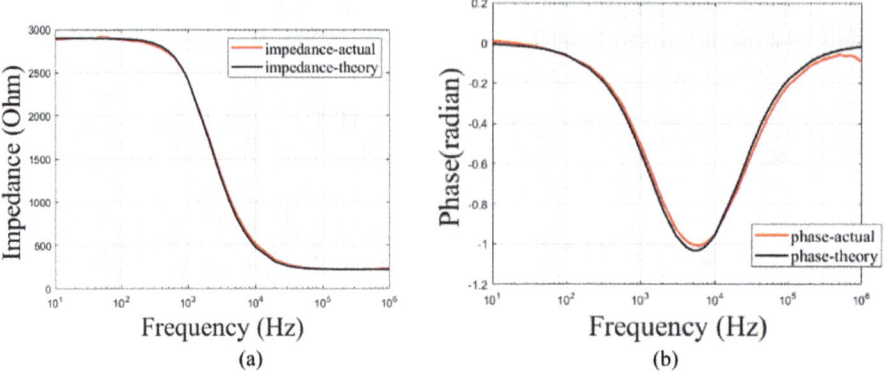

Fig. 9. Actual and theory value of impedence Z (a) and phase θ (b)

b) *Matching data with Cole-Cole model by EIS Spectrum Analyzer and CNLS Fitting method*

With the obtained data, we compute Re(Z) and Im(Z) and use the Cole-Cole model as in Fig. 2 to estimate the parameters contained in the simulation circuit. Those estimated parameters are compared with real parameters including R_∞, R_0 và C_{PE}. The fitting process is performed by EIS Spectrum Analyser sofware (Fig. 10).

The parameters of Cole-Cole model (C_{PE}, R_e, R_i) are estimated from measured data. The comparison between estimated and real values are showed in Table 2.

From estimated parameters, the relationship between impedance (|Z|) and frequency, and Phase (θ) and frequency are presented in Fig. 11.

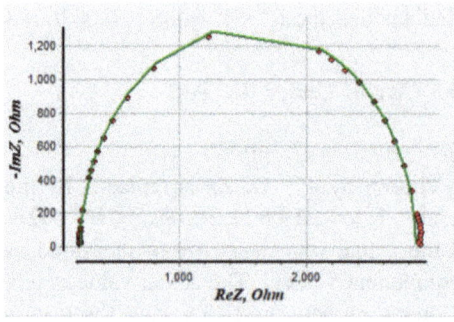

Fig. 10. Impedance spectrum of real part and imaginary part with fitting process

Table 2. Estimated parameters under Cole-Cole model

Parameter	Estimated value (nF)	Real value (nF)	Relative error
C_{PE}	37.323	40	2.2684
R_∞	222.96	219.8	0.55138
$R_0 - R_\infty$	2634.4	2677	1.5717

Figure 11(a) shows the actual impedance (from measured values), theoretical impedance (from component values) and the estimated impedance (from CNLS Fitting method) when changing the frequency in the frequency range from 10 Hz to 1 MHz. Comparing between the theoretical value and the value from the CNLS-Fitting method, the maximum error is 6.58% and the RMSE value is 31.28 Ω.

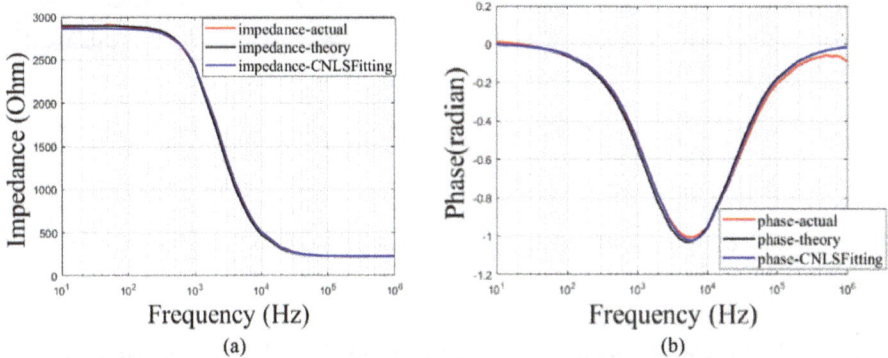

Fig. 11. Impedance Z (a) and Phase θ (b) of actual values, theory values and estimated by CNLS-Fitting method values.

Figure 11(b) shows the actual phase (measured values), theoretical phase (component values) and the estimated phase (from CNLS Fitting method) when changing frequency in the frequency range from 10 Hz to 1 MHz. Through calculation, the maximum error is calculated by 0.037 radians and the RMSE value is 0,0164 radians.

The results after matching data by using CNLS Fitting method have approximate value compared with given theory. The errors are small and within acceptable range. Therefore, this method has high applicability in the biological impedance analysis of meat samples that we will implement in the upcoming studies.

3.3   Pork Meat Impedance Measurement

The purpose of the system is to analyze the biological impedance of a pork sample. We perform impedance measurements with Zx replaced with a pork sample. The measured sample is cut into box-shaped pieces, with dimensions of length, width and height equal to 5×3, 5×3 cm, then is placed in the box for keeping the shape. The electrode used for the measurement is a stainless-steel needle electrode. Measured samples are measured continuously for 24 h.

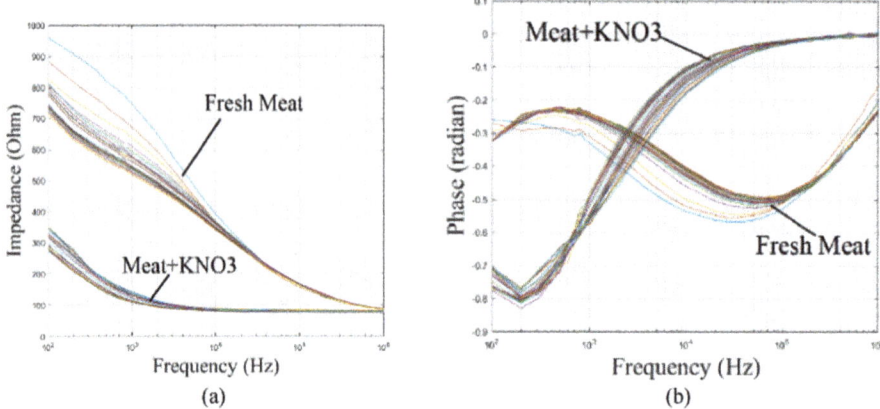

Fig. 12. Values of impedance Z (a) and phase θ(b) of the same pork sample before and after washed by KNO3 measured during 24 h.

Figure 12 presents the value of impedance |Z| and phase θ in the range of frequency of two samples before and after washed by KNO3. Notice that the relationship of impedance and frequency has a shape similar to that of the models tested above.

The pork meat sample, after measuring 24 h, noticed a bad smell, the meat became ruined, not as fresh as it was at first. We then washed the meat with KNO3, and found that the meat after washing became "fresher" (looked like fresh one), the odor of the meat was also completely eliminated. Phase and impedance values between fresh meat samples and washed with KNO3 show clear differences. The impedance of the KNO3 washing sample tended to decrease gradually from 450 Ω while the clean meat tended to decrease from 1000 Ω. The phase of the measured sample also has this difference, which is most clearly shown in Fig. 12. The measurement result from sample 2 is the same as that from sample 1, this shows the stability of the measurement. This can confirm that the measuring system is able to distinguish clean meat from processed meat with KNO3.

4   Conclusion

As the obtained results, the system has a small error, the error value is 1.35% with pure resistance circuit, 10.35% with simulated Fricke circuit, and 6.24% with simulated Cole-Cole circuit. Together with the simultaneous measurement of 02 samples at the same time, there is no difference in measured values between same type meat. The measurement data obtained during the data matching gives high accuracy results. In the future, the system will be used to measure pork meat to distinguish between fresh meat and spoiled meat that has been treated with KNO3 that cannot distinguished by the naked eye.

References

1. Stoyanchev, T., Ribarski, S., Atanassova, S.: Evaluation of pork meat quality and freshness using colorimetric and spectral methods. Agric. Sci. Technol./AST/ **5**, 1 (2013)
2. Candek-Potokar, M., Maja Prevolnik, Š.: Ability of NIR spectroscopy to predict meat chemical composition and quality - a review. Czech J. Anim. Sci. **500**(515), 49(11) (2004)
3. Sun, D.-W., Zheng, L., Tan, J., Valous, N.: Quality evaluation of meat cuts. In: Computer Vision Technology for Food Quality Evaluation, pp. 175–193. Elsevier, Amsterdam, May 2016
4. AminTaheri-Garavand, S.M.Y.: Meat quality evaluation based on computer vision technique: a review. Meat Sci. **156**, 183–195 (2019)
5. MiroljubMladenov, S.M.: Complex assessment of the quality of foodstuffs through the analysis of visual images, spectrophotometric and hyperspectral characteristics. IFAC-PapersOnLine **48**(24), 60–65 (2015)
6. Lepetit, J.J.: Mechanical properties of meat. Meat Sci. **36**(1–2), 203–237 (1994)
7. Tornberg, E.: Biophysical aspects of meat tenderness. Meat Sci. **43**, 175–191 (1996)
8. Stephens, J.W., Unruh, J.A., Dikeman, M.E., Hunt, M.C., Lawrence, T.E., Loughin, T.M.: Mechanical probes can predict tenderness of cooked beef longissimus using uncooked measurements J. Anim. Sci. **82**(7) 86 (2004)
9. Wheeler, T.L., Shackelford, S.D., Koohmaraie, M.:Beef longissimus slice shear force measurement among steak locations and institutions. Jo. Anim. Sci. **85**(9), 9 (2007)
10. Damez, J.L., Clerjon, S., Abouelkaram, S., Lepetit, J.: Dielectric behavior of beef meat in the 1–1500 kHz range: simulation with the Fricke/Cole–Cole model. Meat Sci. **77**(4), 512–519 (2007)
11. Yin, C.K., Khamil, K.N.: An analysis of lard by using dielectric sensing method. J. Telecommun. **8**(9), 131–137 (2016)
12. Trung, D.T., Kien, N.P., Hung, T.D., Vu, T.A.: Electrical impedance measurement for assessment of the pork aging: a preliminary study. In: Third International conference on Biomedical Engineering, Hanoi, October 2016
13. Kien, N.P., Vu, T.A.: Cost Effective System Using for Bioimpedance Measurement, Hanoi, December 2019
14. Phan, K.N., Tran, V.A., Dang, T.T.: Modified biological model of meat in the frequency range from 50 Hz to 1 MHz. In: Intelligent Computing in Engineering, Hanoi, pp. 1055–1068, April 2020
15. Kien, N.P., Vu, T.A., Trung, D.T., Duong, T.A.: A novel method to determine the bio-impedance. Int. J. Sci. Res. **6**(10), 649–654 (2017)
16. Fricke, H.: A mathematical treatment of the electrical conductivity of colloids and cell suspensions. J. General Physiol. **6**(4), 375–384 (1924)
17. Kenneth, R.H.C., Cole, S.: Dispersion and absorption in dielectrics I. Alternating current characteristics. J. Chem. Phys. **9**(4), 341–351 (2004)
18. Kenneth, R.H.C., Cole, S.: Dispersion and absorption in dielectrics II. Direct current characteristics. J. Chem. Phys. **10**(2), 98–105 (2004)
19. Macdonald, J.R., Garber, J.A.: Analysis of impedance and admittance data for solids and liquids.J. Electrochem. Soc. **124**(7), 1022–1030 (1977)

Attentive RNN for HS Code Hierarchy Classification on Vietnamese Goods Declaration

Nguyen Thanh Binh[1], Huy Anh Nguyen[2], Pham Ngoc Linh[1],
Nguyen Linh Giang[3], and Tran Ngoc Thang[1(✉)]

[1] School of Applied Mathematics and Informatics,
Hanoi University of Science and Technology, Hanoi, Vietnam
thang.tranngoc@hust.edu.vn
[2] Computer Science Department, Stony Brook University,
Stony Brook, NY 11794, USA
anh.h.nguyen@stonybrook.edu
[3] School of Information and Communication Technology,
Hanoi University of Science and Technology, Hanoi, Vietnam
giang.nguyenlinh@hust.edu.vn

Abstract. The harmonized commodity description and corresponding coding system (HS Code System) created by the World Customs Organization (WCO) are internationally used to classify standard transaction goods from their descriptions. The system uses the four-level hierarchical structure to arrange thousands of different codes. However, in practice, the traditional and manual methods for classifying a large number of items is a labor-consuming work and also prone to error. In order to assist the customs officers as well as many companies, we proposed a deep learning model with self-attention mechanism along side hierarchical classifying layers to improve the accuracy of classification of Vietnamese short text from goods declarations. Experimental results indicated the potential of these approaches with high accuracy.

Keywords: Hierarchical classification · Self-attention mechanism · HS code · Vietnamese text · Recurrent neural network

1 Introduction

When the international trade between countries drastically expanded in terms of both volume and value, reducing the clearance time, especially those related to goods declaration process, was the most important goal that every nation desired to achieve. However, the former system was regional and inconsistent that took a long time, maybe up to a week to complete the declaration procedure. Thus, in 1998, World Customs Organization (WCO) introduced the Harmonized Commodity Description and Coding System (or simply, Harmonized System - HS), a complex hierarchical system based on the economic activity or component

D.-T. Tran et al. (Eds.): ICISN 2021, LNNS 243, pp. 298–304, 2021.
https://doi.org/10.1007/978-981-16-2094-2_37

material, in order to standardize the name and number of traded products. The system has been adopted widely as over 200 countries and territories, and 98% of merchandises in international trade are used according to [1].

However, choosing the right HS code is not an easy task. In particular, a noticeable amount of merchandises has been miss-classified even though companies have dedicated experts or experienced agencies for this kind of work. To facilitate these difficulties, Vietnam customs provided a website that users can fill their goods description and then receive a list of potential HS codes. However, this website as well as many third party services mostly use searching keywords which has many hindrances to find the codes that matches the semantic content, not only the words itself. For instance, "83099099, Dây buộc PVC lõi kẽm" and "56090000, Dây buộc giày Polyester" and "39269099, Dây buộc cáp bằng nhựa while they certainly belongs to different codes, the tools or websites return the same result. For this reason, building an automated classification HS code system using knowledge based techniques to capture the meaning of descriptions is a highly promised approach to solve the aforementioned problems.

Generally, HS Code classification task deals with short-texts (one to two sentences) and hierarchical labels. There are some approaches to tackle the similar task such as [4] proposed a method using Background Net to classify the chapter 22 and chapter 90 goods declaration. However, the description by Vietnamese has some unique characteristics. Hence, the main purpose of this paper is introducing an efficient system to classify the declaration using Vietnamese.

This paper contains four sections: the introduction represents the importance of HS Code classification problem and some related works. Section 2 introduces the data and problem formulation. Section 3 states the model, evaluation metrics and the experiment results. The last section is the conclusion and discussion.

2 Problem Formulation

Harmonized System, as mentioned above, is a hierarchical system in which divided into 99 chapters, 1224 headings and 5224 subheadings.

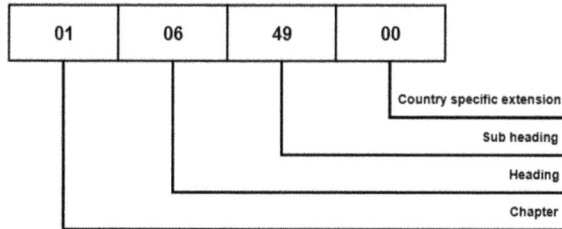

The first 6-digits code is used internationally across countries in the organization. WCO also allows members to add numbers after the first 6-digits depending on the purpose of the countries (normally for statistics reason). Vietnam uses the 8-digits system to classify commodities, which means there are totally 11371 different HS codes (in 2017). These large number of different codes, or in other words, distinct labels in a classification task, is a high complexity challenge to solve. These labels also have hierarchical attribute. For example, an 40.16.10.10 can be interpreted as four levels.

In our model, we propose using four fully-connected (FC) layer in which each layer outputs one level of HS code. The first layer determines which chapter the product belongs to, the second and third layer generate the heading and subheading, respectively. The last layer decides the HS code by combining the outputs or three previous layers and its own output.

The second problem relies on differences between the description and HS nomenclature in terms of semantic content. Vietnamese has many vagueness when combining words. To deal with this problem, we improve the performance of embedding layer by using a bidirectional LSTM [3]. A layer of attention is added after the embedding layer to accurately represent the importance of particular words in each sentence. More detailed explanation will expressed in the following section.

The last question is how to cope with HS amendments. WCO makes HS revision every 5–6 years, where they add, remove or change a small number of codes to reflect the international trade's variation. The last amendments was in 2017, where they accepted 233 changes.

3   Methodology

3.1   Preprocessing Data

Input data of our proposed model is the description of the goods in text form. For raw data, which is records in Vietnamese, having trouble writing with difficult codes, with strange characters or without accents.

In addition, the data also has a lot of descriptions of the same HS code and a type of item, but only the difference in quantity or volume. We proceeded to clean raw data by removing redundant characters, false characters that lose the meaning of words in sentences, and accent remark for Vietnamese missing accent descriptions.

Then with the duplicate data that were mentioned above, we used the LCS (the longest common substring) method [2] with a similarity of them is 0.78 or more. The description of the goods could be short text without many features, so we kept a longer and more detailed description of the goods with the same HS Code for the tokenizing process word, and remove sketchy descriptions.

3.2   Building Model

As mentioned above, we have built 2 deep learning models: a model has hierarchical classification and self-attention mechanism (Fig. 1), and the second without. We propose the first model in this section. It consists of three parts: Word representation layer, Bidirectional LSTM Layer and Self-attention, 4 Fully Connected Layers and Output Layer.

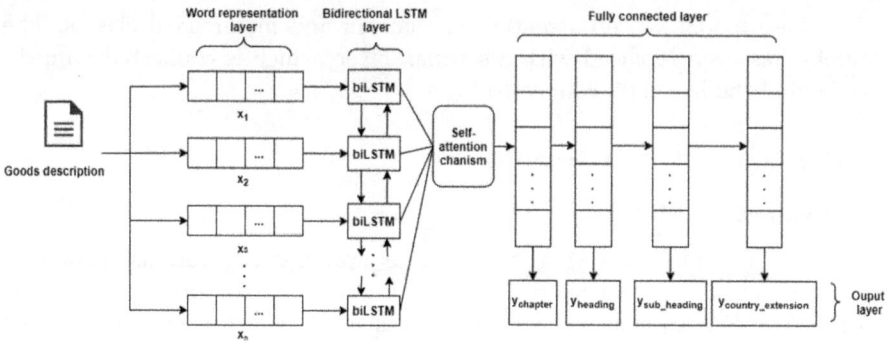

Fig. 1. The architecture of the model we build, the first is word representation, after that sentence embedding is made up of two parts. The first part is a bidirectional LSTM, the second part is the self-attention mechanism. The rest of the model is the fully connected layer and the output layer.

3.2.1   Word Representation Layer

With the commodity description preprocessed above, the input of the model is tokenized. We denote n be the maximum number of words of a description, with each word represented by a vector x_t whose word dimension is d. The output of this layer is $X = (x_1, x_2, ..., x_n)$ with $x_t \in \mathbb{R}^d, t = 1, ..., n$.

3.2.2   Bidirectional LSTM Layer and Self-attention

We decided to build a sentence embedding model based on [5] with two components: a bidirectional LSTM layer and the following part is the self-attention mechanism. Hence, we will summarize this process. We feed each word $x_t \in X = (x_1, x_2, ..., x_n)$ into a Bi-directional LSTM to learn some mutual information between neighboring words in one sentence.

$$\overrightarrow{h_t} = \overrightarrow{LSTM}(x_t, \overrightarrow{h_{t-1}}) \tag{1}$$

$$\overleftarrow{h_t} = \overleftarrow{LSTM}(x_t, \overleftarrow{h_{t-1}}) \tag{2}$$

Each h_t vectors, are concatenated from $\overrightarrow{h_t}$ and $\overleftarrow{h_t}$, is included in section "Self-attention mechanism" to form a structure of matrix sentence embedding. The purpose is to represent many aspects of the sentence semantics. Hence, the output of this section is a structure of matrix sentence embedding. It is discussed in more detail in [5].

3.2.3   Four Fully Connected Layers and Output Layer

Our model uses 4 Fully connected layer connected with each output respectively $y_{chapter}$, $y_{heading}$, $y_{sub\_heading}$, $y_{country\_extension}$ respectively is the prediction probability of y_i with corresponding label i.

The base model has no attention mechanism and hierarchical classes. The hierarchy has been replaced with a softmax layer, which is connected immediately behind the last fully connected layer.

4   Experiment and Results

4.1   Evaluation Metrics

We present the measures that are used to evaluate the various models in our experiments. These include Precision, Recall and F1-score. Assume that $B(TP, TN, FP, FN)$ is a binary evaluation measure, which is computed based on the true positives (TP), true negatives(TN), false positives (FP) and false negatives (FN). We have that

$$Precision = \frac{TP}{TP + FP}, Recall = \frac{TP}{TP + FN}$$

In order to combine these two metrics, a harmonic mean called F1-Score is proposed as:

$$F1 - score = \frac{2 * Precision * Recall}{Precision + Recall}$$

The average result over all classes is determined using micro-averaging. In Micro-average method, it sums up the individual true positives (TP), false positives (TP), and false negatives (FN) of the different classes and apply to get statistics. The average precision and recall are calculated:

$$Micro - average\ of\ precision = \sum_{i=1}^{n} \frac{TP_i}{TP_i + FP_i}$$

$$Micro - average\ of\ recall = \sum_{i=1}^{n} \frac{TP_i}{TP_i + FN_i}$$

Micro F1-Score is simply the harmonic mean of precision and recall, respectively.

4.2   Data Description and Preparation

We gathered the dataset from real transactions which recorded by Vietnam Customs Department. In addition, we purposely chose the good descriptions from chapter 03, chapter 62, chapter 85, chapter 38 and chapter 40 since they are heavily misclassified and also have large impact on detecting tax evasion. Normally, good descriptions have no more than a sentence and fall into only one HS code. Hence, we feed a pairs of sentences and HS code into our model.

However, before using those data, we performed data preprocessing tasks which consist of taking away numbers and punctuations, lowering all the characters, and removing any duplicated records for each category.

We used the data of the customs declarations with descriptions of Vietnamese in chapters 3, 62, 85, 30, 38 and 40 available from 2017 to the present with 1,167 different labels along with 751,328 samples (Table 1).

Table 1. The description data

Chapter	Number of labels	Number of samples
3	201	84306
62	168	187266
85	459	159890
30	88	197308
38	104	56240
40	147	66318

4.3 Results

We tested and compared the performance of two models: a hierarchical model and the baseline model. For medium dataset of 3 chapters 3, 62, and 85 and large dataset of chapters 3, 62, 85, 30, 38 and 40, we used 200 hidden units in LSTM and four fully connected layer (with 100 units each layer) and 1,167 units each layer in baseline with large dataset but still keep 100 unit each layer in the hierarchical model. Our hierarchical model and baseline model were trained by Adam optimizer with the learning rate of 0.001 and the batch size was 1024. We ran 20 and 50 epochs for mediumscale and large-scale datasets, respectively. We splited 80% for training and 20% to testing. We used an early stopping strategy to avoid overfitting and accelerate training.

Table 2 gives the micro F1-score for each part in an HS Code. Table 3 gives the Micro F1-score for each model. We used the same train and test datasets for both models, but the baseline model was not predictable with larger data.

Table 2. Micro F1-Score of hierarchical model for 2 datasets

Dataset	Medium dataset			Large dataset		
	Precision	Recall	Micro F1	Precision	Recall	Micro F1
Chapter	0.9981	0.9981	0.9981	0.9587	0.9587	0.9587
Heading	0.9015	0.9015	0.9015	0.8507	0.8507	0.8507
Sub heading	0.8004	0.8004	0.8004	0.7556	0.7556	0.7556
Country extension	0.8207	0.8207	0.8207	0.7687	0.7687	0.7687
First 6 digits	0.7778	0.7778	0.7778	0.7279	0.7279	0.7279
Full HS Code	0.6982	0.6982	0.6982	0.6443	0.6443	0.6443

Table 3. Performance micro F1-score comparison of 2 models

Dataset	Baseline model	Hierarchical model
Medium dataset	0.7500	0.6982
Large dataset	0.0436	0.6443

5   Conclusion and Discussion

We demonstrated a novel architecture for classifying HS codes automatically. Although this result has some limitations but also introduced an idea of an HS code hierarchy classification model, which could help solve the HS code classification when the number of codes becomes larger. We realized that as the amount of data and the number of labels got larger, the baseline model seemed to be ineffective in the classification process. For the baseline model, the error rate of all 8 digits is higher compared to the hierarchical model, each code is separate, not hierarchical. Hence, the code is wrong and the code is correct can not be related to each other. For the hierarchical model, the code that HS predicts incorrectly can be the same chapter, heading, subheading, and only the different country extension code. We will improve it in the future by incorporating Vietnamese word processing techniques or pretrained Vietnamese embedding. In addition, we will combine the use of deep learning models combined to achieve better results with a larger dataset.

Acknowledgement. This paper is supported by CMC Institute of Science and Technology (CIST), CMC Corporation; and Hanoi University of Science and Technology, Vietnam. This research is funded by Vietnam National Foundation for Science and Technology Development (NAFOSTED) grant number [102.05-2019.316].

References

1. World Customs Organization. http://www.wcoomd.org/
2. Gusfield, D.: Algorithms on Strings, Trees and Sequences: Computer Science and Computational Biology. Cambridge University Press, Cambridge (1997)
3. Schuster, M., Paliwal, K.K.: Bidirectional recurrent neural networks. IEEE Trans. Sig. Process. **45**(11), 2673–2681 (1997). https://doi.org/10.1109/78.650093
4. Liya, D., Zhen, Z.F., Dong, L.C.: Auto-categorization of HS code using background net approach. Procedia Comput. Sci. Proc. **60**, 1462–1471 (2015). https://doi.org/10.1016/j.procs.2015.08.224
5. Lin, Z., Feng, M., Santos, C.N.D., Yu, M., Xiang, B., Zhou, B., Bengio, Y.: A structured self-attentive sentence embedding. In: ICLR Conference (2017). arXiv preprint arXiv:1703.03130

A Solution to UAV Route Planning Problem Based on Chaotic Genetic Algorithm

Trong-The Nguyen[1,2], Jun-Feng Guo[1,3], Jin-Yang Lin[1,3],
Tien-Wen Sung[1], and Truong-Giang Ngo[4(✉)]

[1] Fujian Provincial Key Laboratory of Big Data Mining and Applications,
Fujian University of Technology, Fuzhou 350118, Fujian, China
[2] Haiphong University of Management and Technology, Haiphong, Vietnam
[3] Research Center for Microelectronics Technology in Fujian University
of Technology, Fujian University of Technology, Fuzhou 350118, Fujian, China
[4] Thuyloi-University, 175 Tay Son, Dong-Da, Hanoi, Vietnam
giangnt@tlu.edu.vn

Abstract. The paper addresses un-crewed aerial vehicles (UAV) route planning with its tracking trajectory of preventing obstacles by implementing a chaotic genetic algorithm (CGA). In complex environments, the dynamic 2D-grid model of low-altitude aircraft is discretized according to the constraints of 3D-space, divided into several 2D spaces. The improved CGA code is used to optimize the UAV's route planning phase for obstacles in space. The experimental results depict that the proposed scheme can effectively design an estimated optimal path and overcome route planning in the 3D space by converting engineering practicability spaces.

Keywords: Three-dimensional route planning · Un-crewed aerial vehicle · Chaotic genetic algorithm

1 Introduction

Un-crewed aerial vehicles (UAV) trajectory planning refers to careful consideration of maneuverability under such constraints as the probability of penetration [1], the probability of hitting the ground [2], and the flight time [3]. Its solution is to look for an optimal or feasible flight trajectory from the starting spot to the target spot [4]. Due to the robustness of the genetic algorithm (GA) [5], many scholars have used evolution and genetic algorithms [6] for dealing with practical problems in the fields of engineering and financial sections [7]. The path planning research is one of the exciting problems [8]. The UAV is a complex multidimensional nonlinear engineering optimization problem in three-dimensional (3D) path planning [9]. Sparse A* algorithm is a heuristic search algorithm, which combined with constraints to narrow the search space greatly, greatly reduce the search time, but in three-dimensional space structure objective cost function and heuristic function, because the function solving process need to be calculated in the 3D space, greatly increased the difficulty of the algorithm [10]. In sparse A* algorithm combining evolutionary theory, based on the introduction of the cultural algorithm, based on sparse A* algorithm and the cultural algorithm

D.-T. Tran et al. (Eds.): ICISN 2021, LNNS 243, pp. 305–312, 2021.
https://doi.org/10.1007/978-981-16-2094-2_38

hybrid algorithm has realized the dynamic target flight path planning, to solve complex multidimensional nonlinear engineering optimization problem effectively provides an analysis method, has great reference value [11]. The chaotic genetic algorithm (CGA) [12] is a robust and powerful algorithm inspired by chaos theory. In the CGA, the chaotic is used mapping to modify the genetic algorithm (GA) [5].

This paper makes flight path planning comprehensively considering the maneuver performance, terrain elevation obstacle threat, flight mission, and UAV factors to improve the disturbance factor by applying CGA.

2   Majority Steps for Trajectory Planning

The described constraints of optimization problem of UAV is under different conditions [13], e.g., the min cost, maxsteering, maxclimb/dive angle, track distance constraint, and fixed direction. Let S_{min}, and (θ_{max}), be a minimum cost and maximum climb/dive angle. Tracking distance constraint and fixed direction are factor entering for approaching the target position, and flight height limit (H_{max}). The constraints, the following cost function is adopted:

$$g(n) = \sum_{i=1}^{n} (l_i + h_i) \tag{1}$$

where l_i is set to $\sqrt{(x_i - x_{i-1})^2 + (y_i - y_{i-1})^2 + (z_i - z_{i-1})^2}$; l_i is the length of the path in segment i; h_i represents the deflection cost of the i node. The deflection of horizontal plane: $h_i = \sqrt{2} S_{min}$; The deflection on the vertical surface, that is, when climbing and diving, $h_i = \frac{2\sqrt{3}}{3} S_{min}$, where S_{min} represents the minimum step size. The linear distance from the extension node to the target node is taken as the heuristic *function* is expressed as.

$$f(n) = \sqrt{(x_n - x_{good})^2 + (y_n - y_{good})^2 + (z_n - z_{good})^2} \tag{2}$$

Actual cost from the current node to the target node satisfies both the accessibility and the consistency. The majority of steps for trajectory planning are expressed as follows:

Step 1. Rough filtering of obstacles. Input the obstacle's coordinates and determine whether the obstacle is within the cuboid range composed of (S_x, S_y, S_z) and (E_x, E_y, E_z). The region with n times of S_{min}, set an upper limit for N.

Step 2. On the line from the starting spot S to the target position E, judge whether the obstacle intersects the line according to the cuboid range set by Step1. If it crosses, calculate the intersection point between the line segment and the obstacle; otherwise, go to Step 5.

Step 3. A plane parallel and perpendicular is generated to the top at the point of intersection with two-dimensional path planning to find the optimal path. For the vertical plane, the extension factor is used to carry out three-dimensional expansion in the plane. The expanded plane also adopts path optimization, and then the path segment with the shortest distance is selected as the final trajectory.

Step 4. Intersection point 1, intersection point 2, the coordinates of all intersection points of trajectory lines are added to the Vega path to realize obstacle avoidance path planning;

Step 5. Move point S to the current position, while point E remains unchanged. Continue the search in the SE direction and exit if you reach the target location. If you come across an obstacle, go to Step 2.

3   Route Planning for UAV by CGA

A raster modeling in two-dimensional space is calculated then real coding for CGA is used to optimize the path. CGA is the integration of chaos and GA. The chaos disturbance is introduced to solve the premature and local convergence of a single GA. CGA's design idea follows the teps: the initial population generation, selection, crossover, and mutation generation paths. Due to space limitation, the solution set after GA mutation is directly substituted into the chaotic optimization formula, and the detailed calculation process described. Random disturbances determination is expressed as follows.

$$\delta'_k = \alpha\delta_k + (1 - \alpha)\delta^* \tag{3}$$

where δ^* is the vector formed after the current optimal solution $(x_1^*, \cdots, x_r^*)$ is mapped to the interval [0,1], called the optimal chaotic vector; δ_k is the chaotic vector after k iterations, and δ'_k is the random disturbance $(x_1, \cdots, x_r)$ The corresponding chaotic vector [6]. Among them, $0 < \alpha < 1$, the adaptive selection is adopted, because the initial search hope $(x_1, \cdots, x_r)$ changes considerably, and a larger α is required. As the search progresses, $(x_1, \cdots, x_r)$ gradually approaches the best point, so it is necessary to select a smaller α to search within the short range where $(x_1^*, \cdots, x_r^*)$ is *located* [7]. This article applies formula (4) to determine α.

$$\alpha = 1 - \left[\frac{k-1}{k}\right]^m \tag{4}$$

where m is an integer, which depends on the optimization fitness function; k is the number of iterations. The steps of optimizing the route planning with the CGA algorithm are described as follows.

Step 1. The parameters are set, e.g., the value range $[a_i, b_i]$, population size m, attractor μ_i in chaotic operator, and the interchange rate P_{c1}, P_{c2} and the variation rate P_m of progeny.

Step 2. The optional logistic mapping is shown in Eq. (5) with the relationship is as follows:

$$\beta_i(u+1) = \mu_i \beta_i(u)(1 - \beta_i(u)) \tag{5}$$

Among them, i represents the sequence number of the chaotic variable, $i = 1, \cdots, r$; u represents the population sequence number, u $= 0, 1, \cdots, m$; β_i represents the chaotic variable, $0 \leq \beta_i \leq 1$; μ_i represents the attractor. Variable u $= 0$ and $\mu_i = 4$; r initial values of small differences to obtain r chaotic variables $\beta_i(1)$, $(i = 1, \cdots, r)$. Then u $= 1, 2, \cdots, m$ in turn to get m initial solution groups.

Step 3. The carrier of r chaotic variable $\beta_i(u+1)$ is selected to r optimization variable of Eq. (6), which is converted into chaotic variable x_i'. The value range of the chaotic variable would be correspondingly switched to the value range of the corresponding optimization variable.

$$x_i' = c_i + d_i \beta_i^{(u+1)} \tag{6}$$

where, c_i, d_i is the transformation constant, $i = 1, 2, \cdots, r$.

$$X = (x_1, x_2, \cdots, x_r) \tag{7}$$

$$X' = \left(x_1', x_2', \cdots, x_r'\right) \tag{8}$$

Step 4. The criteria of the fitness function, calculation Eq. (8) generated by the fitness value as descending order, because of $f(X')$ is less than 0, and even nonnegative $f(X')$, otherwise if $f(X')$ relative changes in a population scope is too small, the equivalent of two generations of value would be close to or similar, will cause the algorithm convergence rate is slow; so also need to $f(X')$ press type to make small changes: $f_k'(X')$

$$f_k'(X') = f_k(X') - f(X')_{min} + \frac{1}{m}\left(f(X')_{max} - f(X')_{min}\right) \tag{9}$$

where $f_k'(X')$ is for small changes after the fitness value, $f_k(X')$ is for small changes in front of the fitness value, $f(X')_{min}$ is for small changes before the minimum fitness value, $f(X')_{max}$ to small changes in front of the largest fitness values, size m for groups, according to Eq. (9) after adjustment, fitness values are greater than zero, and the relative changes of fitness value increased range, easy to increase the convergence.

Step 5. Variables are coded according to the decimal system: e.g., if fitness value ranking top 10% in the previous generation population do not participate in three genetic operations (replication, crossover, variation) and directly enter the next-generation population; these three operations will generate the remaining 90%, and the offspring population will be decoded according to rules.

Step 6. Next generation, groups according to the rules of the fitness value is calculated following conditions, and then execute Eq. (11) and the rules of the adjustments, adjust the size of finished according to the adjusted fitness value, the sort of groups, and the average fitness, and the average and maximum fitness according to Eq. (10) compare, if established Eq. (10), argues that end of the optimization process, the output as the optimal value; If Eq. (10) is not established, Step7 shall be executed for next process.

$$\left|\overline{f'(X')} - f'(X')_{max}\right| < \varepsilon_1 \tag{10}$$

$$\overline{f'(X')} = \frac{1}{m}\sum_{j=1}^{m} f_j'(X') \tag{11}$$

$$f'(X')_{max} = max\left\{f_j'(X')\right\}, (j = 1, 2, \cdots m) \tag{12}$$

ε_1 is some small positive number given up front.

Step 7. Current generation, select fitness smaller 90% of individuals, to find the corresponding optimal solution, according to Eq. (5), a chaos disturbance is added, and then according to Eq. (6), the mapping as chaos optimization variables, finally the chaos and chaos optimization variables are substituted Eq. (3) over iterative calculation. As the number of iterations increases, α value calculated in Eq. (4) keeps changing. The optimal solution is obtained until the difference between the fitness averages calculated twice is less than a small positive; ε_2 is given in advance as operation.

$$\left|\overline{f_k'(X')} - \overline{f_{k+1}'(X')}\right| < \varepsilon_2 \tag{13}$$

where k is the number of iterations, $k = 0, 1, \cdots$;

Step 8. According to the fitness value to the group again, calculate the fitness average, according to Eq. (10) will rule will compare it with the maximum, if the formula establishes, optimization end and output the optimal value, otherwise, go to Step 5.

A chaos disturbance added to the mutated genes according to the above eight steps with 90% genes with small fitness in a certain generation of the population is equivalent to carrying out heuristic mutation operation on these genes, which can reduce the evolutionary algebra of the algorithm and find the optimal solution as early as possible. The disturbance may produce higher than the 10% and the corresponding genes' fitness better, effectively avoid the simple genetic algorithm to local convergence and prematurity. Due to the high genetic fitness that has generated 10% of genes, the chaos disturbance, for only the remaining 90% of the genes to narrow the GA's searching space, can accelerate the optimization rate.

4 Simulation Results

Matlab is used as a tool software for a simulation computing system. The hardware environment is in a laptop with CPU Intel(R) Core(TM) i5-7600, 8G memory, in Windows 10. First simulated in a two-dimensional environment for a path optimization simulation. Figure 1 shows a GUI setting UAV's workspace model environments with (a) center ID detection of generated random obstacles and (b) configuration of 3D workspace parsing solutions.

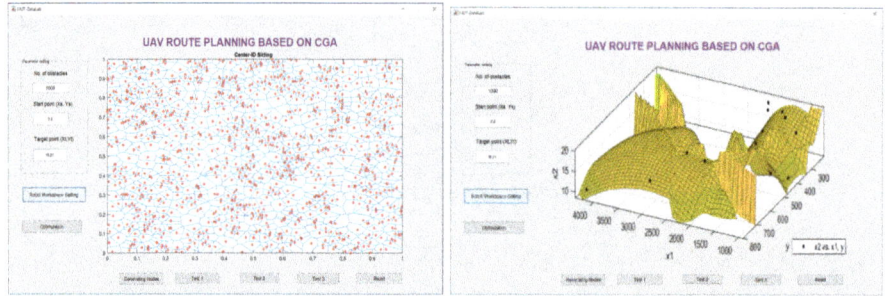

| (a). Setting UAV's workspace model environments with center ID detection of generated random obstacles | (b). Simulation of route planning optimization with the configuration of 3D workspace parsing solutions. |

Fig. 1. A GUI setting UAV's workspace model environments with (a) center ID detection of generated random obstacles and (b) configuration of 3D workspace parsing solutions.

The obtained outcomes of the proposed task are compared with the other methods, e.g., A* algorithm [11], GA algorithm [14] for the route planning problem. The algorithm parameters are set as follows. Then, a 10 km × 10 km 3D geographical environment model is established, including deserts, marshes, lakes, towns, villages, roads, tall plants, and so on. Urban building sand t all plants were used as obstacles to demonstrate the optimal track of low-altitude UAV when it hit the target, avoiding these obstacles. The algorithms parameters are set as follows: Population size $N = 100$, crossover rate c_r is set to 0.75, Maximum emigration rate $E = 0.86$, mutation rate $p_r = 0.005$. The number of iterations is set to 2000 (Fig. 2).

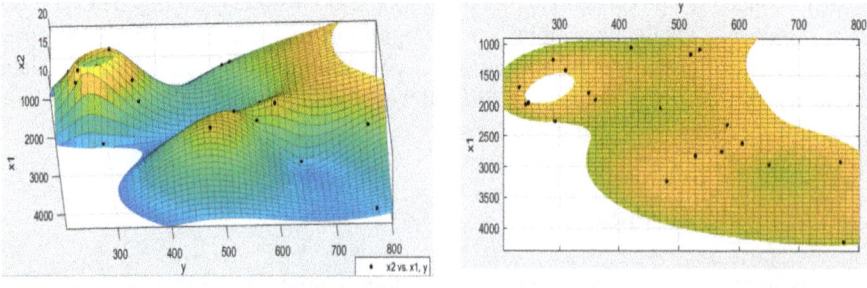

| (a) complex environment of constraints of 3D-space | (b) environment of constraints 2D divided space |

Fig. 2. The complex environments, the dynamic 2D-grid model of low-altitude aircraft, is discretized according to the constraints of 3D-space, divided into several 2D spaces.

2D environment simulation, the grid method is used to divide the environment. The premise of satisfying the constraints, a path close to the optimal solution is found. Because of the genetic algorithm's randomness, many experiments have obtained the mean value of the genetic algorithm.

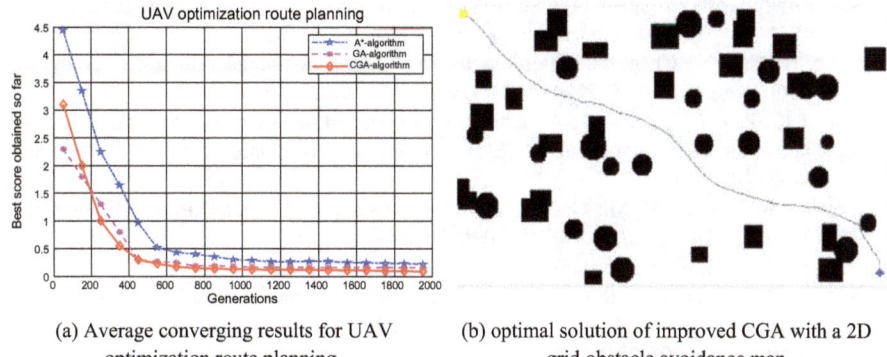

(a) Average converging results for UAV optimization route planning

(b) optimal solution of improved CGA with a 2D grid obstacle avoidance map

Fig. 3. The optimal solution of improved CGA under different algorithms for the UVA route planning

Comparing applying CGA and other algorithms, e.g., the GA [14] and A* algorithm for the route planning problem under the same conditions, is displayed in Fig. 3a. It can be seen from the figure that in terms of convergence speed, the quality output observation results suggest that the optimization approach used is superior to other methods. Figure 3b shows the route planning results from enough to consider only a two-dimensional environment's path planning. It can be seen that the environmental path planning using the method designed runs smoothly, avoid obstacles, and basically meets the requirements of authenticity and real-time performance.

5   Conclusions

This paper proposed a solution to addresses un-crewed aerial vehicles (UAV) route planning with its tracking trajectory of preventing obstacles based on a chaotic genetic algorithm (CGA). In complex environments, the dynamic 2D-grid model of low-altitude aircraft is discretized according to the constraints of 3D-space, divided into several 2D spaces—the UAV's route planning phase for obstacles in space. The UAV's route planning has been modeled mathematically under considering the minimum cost, obstacles, and track distance constraint as an objective function. This function has been optimized by applying CGA. The simulation results show that the proposed approach can effectively design an estimated optimal path and overcome route planning in the 3D space by converting engineering practicability spaces.

References

1. Xie, T., Zheng, J.: A joint attitude control method for small unmanned aerial vehicles based on prediction incremental backstepping. J. Inf. Hiding Multimed. Signal Process. **7**, 277–285 (2016)
2. Nguyen, T.-T., Wang, H.-J., Dao, T.-K., Pan, J.-S., Ngo, T.-G., Yu, J.: A scheme of color image multithreshold segmentation based on improved moth-flame algorithm. IEEE Access. **8**, 174142–174159 (2020). https://doi.org/10.1109/ACCESS.2020.3025833
3. Chao, H., Cao, Y., Chen, Y.: Autopilots for small unmanned aerial vehicles: a survey. Int. J. Control. Autom. Syst. **8**, 36–44 (2010). https://doi.org/10.1007/s12555-010-0105-z
4. Dao, T.K., Pan, T.S., Pan, J.S.: A multi-objective optimal mobile robot path planning based on whale optimization algorithm. In: IEEE 13th International Conference on Signal Process, pp. 337–342 (2016)
5. Srinivas, M., Patnaik, L.M.: Genetic algorithms: a survey. Computer (Long. Beach. Calif) **27**, 17–26 (1994). https://doi.org/10.1109/2.294849
6. Holland, J.H.: Genetic algorithms. Sci. Am. **267**, 66–73 (1992)
7. Stewart, T.J., Janssen, R., van Herwijnen, M.: A genetic algorithm approach to multiobjective land use planning. Comput. Oper. Res. **31**(14), 2293-2313 (2004). https://doi.org/10.1016/S0305-0548(03)00188-6
8. Pan, J.-S., Nguyen, T.-T., Chu, S.-C., Dao, T.-K., Ngo, T.-G.: A multi-objective ions motion optimization for robot path planning (2019). https://doi.org/10.1007/978-3-030-04792-4_8
9. Goerzen, C., Kong, Z., et al.: A survey of motion planning algorithms from the perspective of autonomous UAV guidance. J. Intell. Robot. Syst. Theory Appl. **57**, 65–100 (2010)
10. Duchoň, F., Babinec, A., Kajan, M., Beňo, P., Florek, M., Fico, T., Jurišica, L.: Path planning with modified a star algorithm for a mobile robot. Proc. Eng. **96**, 59–69 (2014)
11. Tseng, F.H., Liang, T.T., Lee, C.H., Der Chou, L., Chao, H.C.: A star search algorithm for civil UAV path planning with 3G communication. In: 2014 Tenth International Conference on Intelligent Information Hiding and Multimedia Signal Processing, pp. 942–945. IEEE (2014)
12. Snaselova, P., Zboril, F.: Genetic algorithm using theory of chaos. Proc. Comput. Sci. **51**, 316–325 (2015). https://doi.org/10.1016/j.procs.2015.05.248.
13. Roberge, V., Tarbouchi, M., Labonté, G.: Comparison of parallel genetic algorithm and particle swarm optimization for real-time UAV path planning. IEEE Trans. Ind. Inform. **9**, 132–141 (2012)
14. Roberge, V., Tarbouchi, M., Labonté, G.: Fast genetic algorithm path planner for fixed-wing military UAV using GPU. IEEE Trans. Aerosp. Electron. Syst. **54**, 2105–2117 (2018)

Incorporation of Panoramic View in Fall Detection Using Omnidirectional Camera

Viet Dung Nguyen$^{(\boxtimes)}$, Phuc Ngoc Pham, Xuan Bach Nguyen,
Thi Men Tran, and Minh Quan Nguyen

School of Electronics and Telecommunications, Hanoi University
of Science and Technology, Hanoi, Vietnam
dung.nguyenviet1@hust.edu.vn

Abstract. Falling is one of the major problems that threaten the health of the elderly and particularly dangerous for people that live alone. Recently, surveillance systems using omnidirectional cameras in general and fisheye cameras particularly have become an attractive choice, as they provide a wide Field of Vision without the need for multiple cameras. However, objects captured by fisheye cameras are highly distorted, therefore computer vision approaches that are developed for conventional cameras require modification to work with such systems. The aim of this work is to incorporate the de-warping of fisheye image using polar to Cartesian transformation to generate a panoramic view. Objects are detected by background subtraction method. Depending on the objects lay inside or outside of a center circle of the omnidirectional frame, features based on contour and rotated bounding box of the objects, are extracted correspondingly on original omnidirectional view or panoramic view. Experiments show that by incorporating both panoramic and omnidirectional view, we can achieve significant improvement in fall detection, particularly in peripheral areas. This result could be a useful reference for further studies.

Keywords: Falling · Omnidirectional · Panoramic · Detection

1 Introduction

The world's population is getting increasingly older. The proportion of elderly people (aged 60 and over) increased from 7.15% in 1989 to 8.93% in 2009 [1]. About 8% of all elderly people live alone and 13% elderly couples live alone. Living alone poses a great danger as the elderly are easily affected by emergency situations: natural disasters, health problems such as stroke, heart attack, and unusual actions in daily life, especially falls, strokes…[2]. These situations need to be discovered, dealt with promptly and quickly and accurately notified to relatives or careers. Therefore, having a video surveillance system is extremely necessary. Over the past decades, there has been various researchers on computer-vision based surveillance system. Most of these researchers focus on regular pin-hole cameras. However, while pin-hole camera has an advantage of simplicity as well as giving an accurate representation of the human's visual system, monitoring a wide area with conventional pinhole cameras can be troublesome, as it requires a Pan-Tilt-Zoom (PTZ) or multiple conventional cameras.

D.-T. Tran et al. (Eds.): ICISN 2021, LNNS 243, pp. 313–318, 2021.
https://doi.org/10.1007/978-981-16-2094-2_39

Using multiple camera is not optimal, as it costs more to maintain and install, as well as requires more bandwidth and synchronization among cameras. PTZ cameras on the other hand, while more flexible than conventional, can still only capture image from only one direction at any given time, and mechanical parts also are prone to failure. To extend the FOV of a surveillance system, omnidirectional cameras can also be used.

Omnidirectional camera, which consists of fisheye-lens cameras and catadioptric lens cameras [3], allows a larger field of vision compared to traditional pin-hole camera by sacrificing resolution in peripheral areas and image's structural integrity, which makes it difficult for conventional action recognition and object detection algorithms to work on fisheye camera. Some techniques have been proposed to mitigate the drawbacks of fisheye cameras. Chiang and Wang [4] applied HOG+SVM detector and rotate people in fisheye images so as they appear upright. This practice can be computationally expensive, as in full-frame fisheye images, the image must be rotated many times. Other works modify the features used for conventional camera to work on omnidirectional camera [5], however, this might get complex and the results fall behind compared to modern standards. Some papers propose to use Convolutional Neural Network (CNN) directly on fisheye dataset [6]. However, the variety of human poses captured by the fisheye camera can lead to problems, as there are not enough fisheye databases currently. Other authors use fisheye calibration to undistort the fisheye image [7]. However, to get a full straight equiangular quadrilateral shaped image, cutting out a considerable number of pixels from the edges is required. Meanwhile this method is applicable in some situations, it is not optimal to do so in surveillance applications. Another projection is fisheye de-warping, in which a fisheye image is unwrapped into 360-degree equirectangular image [8]. This projection is also flawed, as the nearer the object is to the center of the fisheye image, the more deformed it becomes when unwrapped.

In this work, we propose an approach to reduce the effect of omnidirectional camera distortion by incorporating de-warped panorama view for a better fall detection result. Section 2 describes our method of incorporating panoramic view in fall detection using omnidirectional video. Experimental results of the proposed method are illustrated in Sect. 3. Finally, conclusions are given in Sect. 4.

2 Method for Incorporating Panoramic View in Fall Detection on Omnidirectional Video

In original fisheye images, the further away an object is from the image's center, the more distorted it becomes, so they become unsuitable for image processing. On the contrary, in de-warped panoramic view, the closer an object is to the center of the fisheye image, the more deformed it becomes. From these analyses, instead of falling detection solely on omnidirectional video, panoramic view is incorporated in the process.

First, panoramic view (or equirectangular view) is created by de-warping the omnidirectional frame using polar to cartesian transformation. Next, positions of background subtracted objects is checked to determine as if objects should be further processed in panoramic or omnidirectional view. Objects near the center of the original omnidirectional image will be processed in omnidirectional view, while objects near edges of the original image will be processed in the panoramic view. Then, number of features of objects are extracted. It should be noted that they are extracted in corresponding view of each object. Finally, classification methods are applied to recognize the object as in falling or not fall state.

2.1 Background Subtraction

Background subtraction methods are utilized to detect moving objects in omnidirectional videos. The background after being converted to grayscale has a lot of background noise, therefore Gaussian Blur is applied to smooth and to reduce noise. Then, shadow removal and opening morphology are applied to further separate the main moving object from the background [9].

2.2 Panoramic De-warping by Polar to Cartesian Transformation

Mapping each pixel in omnidirectional frame to its corresponding location in panoramic view is shown in Fig. 1.

Every omnidirectional pixel (in polar coordinate) is represented by four parameters (θ, r, X_f, Y_f) while position of its corresponding pixel (in cartesian coordinate) is represented by (X_c, Y_c). Suppose that height and width of the panoramic image are H_p and W_p; radius of omnidirectional frame is R; center position of the omnidirectional frame is (C_x, C_y). Then we have

$$r = \frac{Y_c \cdot R}{H_p} p \qquad \theta = \frac{X_c}{W_p} \qquad X_f = C_x + r \cdot \sin\theta \quad Y_f = C_y + r \cdot \sin\theta$$

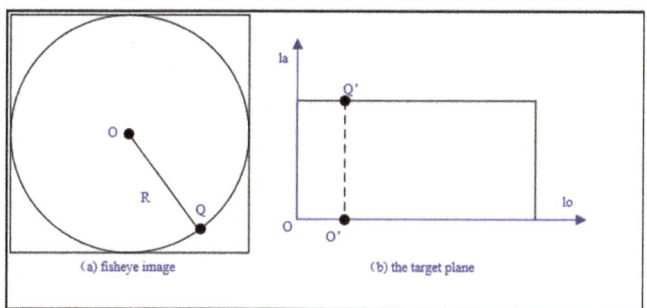

Fig. 1. Pixel mapping diagram

2.3 Feature Extraction

Despite of detected objects are lying on omnidirectional view or panoramic view; 6 following features are extracted based on object contour and rotated rectangle that encompasses the object including area of contour (A1), angle of line fitted to contour, ratio of width and height of rotated bounding box, area of bounding box (A2), ratio of area (A1/A2), angle of bounding box.

2.4   Classification

Several classification methods are implemented to recognize each detected object as in falling or not falling state including Support Vector Machine [10], K-Nearest Neighbor [11], Naïve Bayes [12] and Decision Tree [13, 14].

3   Results and Discussions

The proposed method is implemented in an I5–6300 2.3Ghz processor laptop using Python programing language with OpenCV library.

Omnidirectional video used in the experiments are taken from BOMNI database [15]. Resolution of the video is 640 × 480 pixels. For omnidirectional de-warping, we achieved a frame rate of 10 FPS and good results, as shown in Fig. 2.

Fig. 2. De-wrapping

As previously proven in [9], K-Nearest Neighbors (KNN) gave the best object detection performance in comparison with Mixture-of-Gaussian (MOG), MOG2 and Geometric MultiGrip (GMG). Therefore, in this paper, we use KNN as background subtraction method.

Described in Sect. 2, if the detected objects lay outside a fixed-radius circle at center of the original omnidirectional frame, it will be further processed at the panoramic view of the outer part where it is appearing. By examining different value of radius of centered circle, the radius of 0.145xR with R is the radius of the omnidirectional frame, is chosen. In this work, it is approximately 70 pixels. Fig. 3 shows the detected objects and lines fitted to their contours in each case. In Fig. 4 are enlarged illustrations of object contour, fitted line, and rotated bounding box.

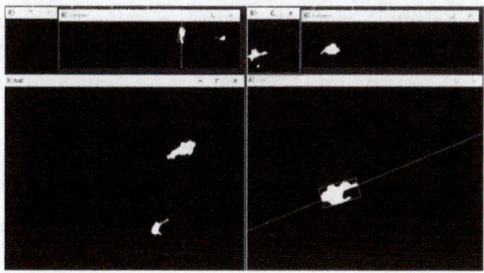

Fig. 3. Detected object in outer part (a) and inner part (right) of omnidirectional frame.

Fig. 4. Contour (left). Line fitting to contour (middle). Rotated bounding box (right)

Falling or not falling classification accuracies for 10-fold cross-validation are depicted in Table 1. It is obvious that DT gives the highest accuracy and standard deviation. It also proves our idea of combining panoramic view in fall detection using omnidirectional video.

Table 1. Classification accuracy

		DT	KNN	SVM	Naive
Omnidirectional video		78.03	75.51	73.30	75.99
Proposed method	R = 50	91.15 ± 2.10	85.34 ± 1.85	87.38 ± 1.96	33.32 ± 0.30
	R = 70	**92.49 ± 1.50**	83.78 ± 2.62	85.89 ± 2.71	35.21 ± 0.20
	R = 100	78.75 ± 6.00	73.55 ± 6.31	75.76 ± 6.55	18.73 ± 5.76

Although the average result is quite high, the proposed method does not perform well when the objects stand nearby to bright regions or when the objects open or close the door. That will be our future research.

4 Conclusions

In this paper, we propose an approach to detect falling in omnidirectional video by incorporating panoramic and omnidirectional view. Compared to previous research, our method could reliably deal with the drawbacks of omnidirectional distortion when applying conventional detection methods as well as solve the problem of loss of vital information compared to conventional omnidirectional calibration methods. We achieve satisfactory classification results on both panoramic as well as omnidirectional view. However, our approach is still sensitive to multiple object detection as well as occlusion or sudden change in illumination. The result of this work can be used for further research on omnidirectional camera, and with further optimization, it can be reliably used for real-time omnidirectional surveillance system.

References

1. https://www.un.org/en/sections/issues-depth/ageing. Accessed 20 Oct 2020
2. https://www.who.int/ageing/publications/EmergenciesEnglish13August.pdf. Accessed 20 Oct 2020
3. Scaramuzza, D.: Omnidirectional camera. In: Ikeuchi, K. (eds.) Computer Vision. Springer, Boston (2016)
4. Chiang, A.-T., Wang, Y.: Human detection in fish-eye images using HOG-based detectors over rotated window. In: Proceedings of 2014 IEEE International Conference on Multimedia and Expo Workshops (ICMEW) (2014)
5. Cinaroglu, I., Bastanlar, Y.: A direct approach for human detection with catadioptric omnidirectional cameras. In: Proceedings of 2014 22nd Signal Processing and Communications Applications Conference (SIU) (2014)
6. Dupuis, Y., et al.: A direct approach for face detection on omnidirectional images. In: Proceedings of 2011 IEEE International Symposium on Robotic and Sensors Environments (ROSE) (2011)
7. Scaramuzza, D., Martinelli, A., Siegwart, R.: A flexible technique for accurate omnidirectional camera calibration and structure from motion. In: Proceedings of 4th IEEE International Conference on Computer Vision Systems (ICVS 2006) (2006)
8. Kim, H., Jung, J., Paik, J.: Fisheye lens camera-based surveillance system for wide field of view monitoring. Optik **127**(14), 5636–5646 (2016)
9. Nguyen, V.D., et. al.: Evaluation of background subtraction methods in omnidirectional video. In: Proceeding of 2019 KICS Korea-Vietnam International Joint Workshop on Communications and Information Sciences, Hanoi, Vietnam (2019)
10. Rohith, G.: Support Vector Machine - Introduction to machine learning algorithms. https://towardsdatascience.com/support-vector-machine-introduction-to-machine-learning-algorithms-934a444fca47. Accessed 20 Oct 2020
11. Ling, W., Dong-Mei, F.: Estimation of missing values using a weighted k-Nearest Neighbors algorithm. In: Proceedings of 2009 International Conference on Environmental Science and Information Application Technology (ESIAT) (2009)
12. Huang, Y., Li, L.: Naïve Bayes classification algorithm based on small sample set. International J. Comput. Appl. 99(16), pp.14–18 (2011)
13. Priyam, A., et. al.: Comparative analysis of decision tree classification algorithms. Int. J. Curr. Eng. Technol. 3(2), 334–337 (2013)
14. https://www.datacamp.com/community/tutorials/decision-tree-classification-python. Accessed 20 Oct 2020
15. Demiröz, B.E., et. al.: Feature-based tracking on a multi-omnidirectional camera dataset. In: Proceedings of 2012 5th International Symposium on Communications, Control, and Signal Processing (ISCCSP12) (2012)

Development of an Autonomous Intelligent Mobile Robot Based on AI and SLAM Technology

Ha Nguyen-Xuan[1]([⊠]), Tung Ngo-Thanh[1], and Huy Nguyen-Van[2]

[1] Hanoi University of Science and Technology, 1 Dai Co Viet, Hanoi, Vietnam
ha.nguyenxuan@hust.edu.vn
[2] Meiko Automation Company, EMS 2, CN9,
Thach That Industry Zone, Hanoi, Vietnam

Abstract. This paper describes the development of an autonomous intelligent mobile robot. By using a multi-layer sensor fusion, an algorithm which is the combination of the SLAM algorithm and the bubble rebound algorithm is presented. The algorithm allows the robot to perform navigation tasks more efficient. Furthermore, the robot is integrated with artificial intelligence applications including the object detection and the voice processing. All of algorithms and models are so efficient that they can be deployed on a limited-resource embedded computer offering many potential applications.

Keywords: Autonomous intelligent robot · Artificial intelligence · Robot operating system · Slam

1 Introduction

Autonomous intelligent mobile robots have shown much of attention in recent years due to their high potential of applications in many fields, for example, autonomous vehicles [1], smart logistics [2], service robots [3]. For these applications, mobile robots are required to have manifold abilities including the instant automatic navigation and the interaction with working environments using artificial intelligence technologies. Therefore, there have been many studies focusing on these topics.

The navigation function is very complicated. It requires robots to have abilities of simultaneous localization and mapping (SLAM), obstacle avoidance and path planning. Key technologies for this function are the sensor fusion and optimized-mathematic algorithms. Several approaches, which use different types of sensors and processing algorithms, have been proposed in [4–10]. These approaches show advantages of the localization, the mapping as well as the obstacle avoidance. However, there has been still lack of an efficient method which exploits the multi-sensor fusion for combining both the SLAM and the object avoidance in an algorithm with a low computational cost.

Artificial intelligence (AI) has been applied in robotics for a long period. Recently, deep learning, a branch of AI, has facilitated solving a huge number of robotic problems in a more successful way than traditional methods [11]. The object detection and

D.-T. Tran et al. (Eds.): ICISN 2021, LNNS 243, pp. 319–326, 2021.
https://doi.org/10.1007/978-981-16-2094-2_40

recognition with deep learning models are premises for the robot to gain perception of the world [12]. A clear use case is the ARI robot [13]. The application of AI technologies for intelligent functions of robots is therefore a promising approach.

In this paper, the development of an autonomous intelligent mobile robot, namely the AIR-HUST, is introduced. A navigation system, which is the combination of the SLAM and the so-called bubble rebound algorithm for simultaneous localization, mapping and obstacle avoidance, was developed. A noticeable feature of the robot is that it is equipped with three layers of sensors arranging in three height-different plans of the robot's housing. The fusion of the three sensor layers allows the robot to avoid obstacles with different shapes and sizes, which could not be achieved by using sensors independently. In addition, an object detection model using deep learning was developed. A module for speech processing from Respeaker [14], was used. Experimental implementations on an embedded computer (Raspberry Pi) show that the navigation system of the robot works efficiently. Also, manifold features of an intelligent service robot, such as human-robot voice interaction and object/fire detection were obtained.

2   System Description

2.1   Hardware System

There are three main components of the hardware system including a navigation system, human interaction devices and an embedded computer, as shown in Fig. 1.a.

The core part of the navigation system is an EAI Dashgo D1 mobile platform, which is presented in Fig. 1.b. The robot is driven by two driving wheels controlled by two 'Dynamixel' motors. Two auxiliary wheels at the bottom of the robot assist the stability of the robot. The robot is also equipped with 4 ultrasonic sensors. All motors and sensors of the robot base are managed by an Arduino circuit. A YDLidar G4, the input data for SLAM algorithm, is set up at the middle level of the robot. A system of 10 IR sensors was integrated at the top level of the robot in order to collect data for our algorithm. The navigation system is described more detail in [15].

Furthermore, the robot is equipped with multiple devices for intelligent applications. A 7-inch touchscreen is set up at the highest part of the robot. A microphone

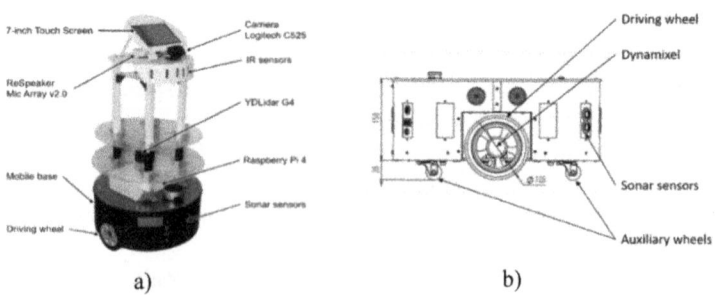

a)   b)

Fig. 1. Hardware system of AIR-HUST: a) hardware device, b) structure of the robot's base
Software System

array for speech recognition is arranged behind the touchscreen. A Logitech C525 camera for computer vision tasks is attached in front of the screen. All the hardware components are controlled by a Raspberry Pi 4 embedded computer based on ROS.

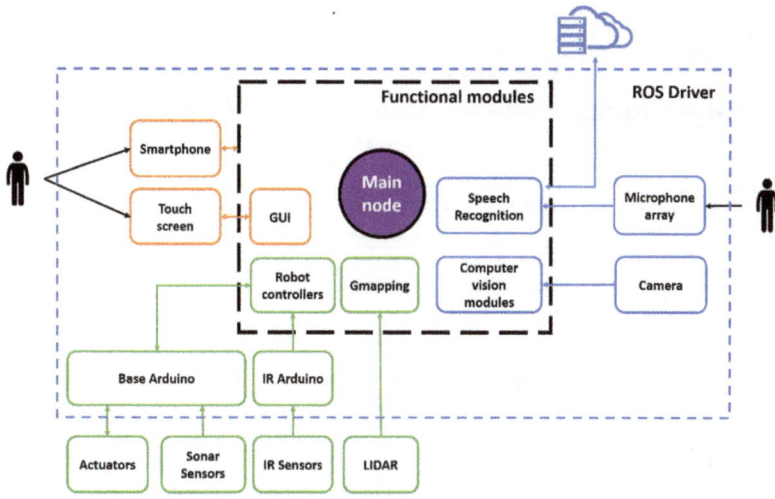

Fig. 2. The software architecture of AIR-HUST robot

The software system of the robot is shown in Fig. 2. It is divided into 3 levels:

- *ROS driver*: Software programs which interface directly to the robot hardware devices. The objective of this level is to convert raw data from devices to ROS topics and transfer commands from the embedded computer to the devices correctly.
- *Functional Modules*: Software programs which receive data from the ROS topics (created from *ROS driver*), process the data and send commands to the *ROS driver* to perform functional tasks.
- *Main Node*: The ROS node which manages all other functional modules.

3   Robot Control

The navigation system is in line with a variation of that used in [16] with some modifications. The relationship between navigation components is presented in Fig. 3. The 'move_base' package takes responsibility for the navigation tasks. This package connects all parts of the navigation system depending on the 'transform frame' information between components of the system. The package creates the 'global_-costmap' based on the data from the 'map' topic. Then, the 'global_planner' estimates the path for the robot to reach the goal. By making use of the data from the sensor system, the package also creates a local 'costmap'. Depending on the 'local_costmap', the 'local_planner' estimates a local path to handle the unpredicted situation while still follows the global path. After all, the 'local_planner' sends control commands to the base controller.

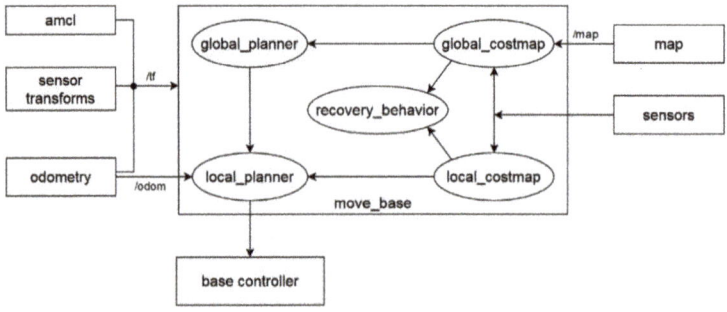

Fig. 3. Navigation components

Hierarchical Control System

There is a hierarchy of robot controllers which generally includes navigation stack, manual controller and safety controller. Starting bottom-up, the safety controller receives data from the sensor system and avoids obstacles instantly. On top of that, the manual controller allows human to control the robot manually. On the highest level, the navigation stack lets the robot follow the path created at the path planning step. A safety controller related to the IR sensors was added to the hierarchical control system.

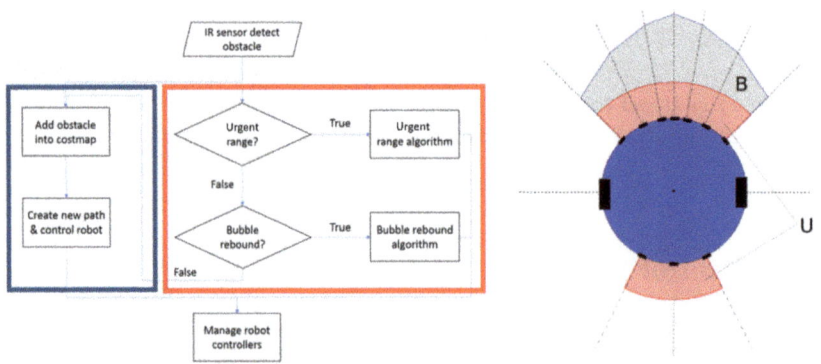

a) Obstacle avoidance workflow

b) Urgent range U and Bubble rebound range B

Fig. 4. Urgent range U and Bubble rebound range B.

To avoid obstacles, two algorithms corresponding to two ranges around the robot are used, including *urgent range* U and *bubble rebound range* B, as shown in Fig. 4. These two ranges are defined as follow:

- *Urgent range* U: A circle range whose center is the center of the robot. The radius of the circle R_U can be set manually. If an obstacle appears in this range, the robot will move away from the obstacle until the obstacle is out of the range.
- *Bubble rebound range* B: A range whose contour is related to the state of the robot. The radius of the bubble at each sensor's direction is calculated by the formula: $bb[i] = K_i.V_t.\Delta_t$. While K_i is the weight of the ith sensor; V_t is the velocity of the robot; Δ_t is the time step between two states. The bubble range allows the robot to move without collision.
- More details of our algorithms are given in our previous paper [15].

Experiments: The robot has to avoid an object which can be detected as only a small point by LIDAR in the map as shown in Fig. 5, for example, a spinning chair. The global path goes through the object, so the robot will be involved in a collision without our method. The obstacle is detected in the Bubble rebound range B as presented in Fig. 5a. After that, the local path of the robot is changed to avoid the obstacle as shown in Fig. 5b. The robot was tested to avoid the obstacle in two cases: a static obstacle and a dynamic one. In the static test, the robot can move back and forth 20 times without any collision. It also passed the dynamic test in 20 moves. However, the relative velocity between the robot and the obstacle has not been measured. As expected, the experiments show that the multi-layer sensor fusion approach could avoid complex-shape obstacles efficiently. More details of the experiments can be found in [15].

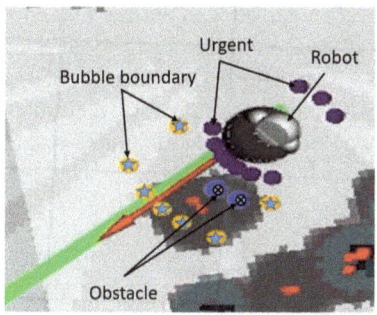 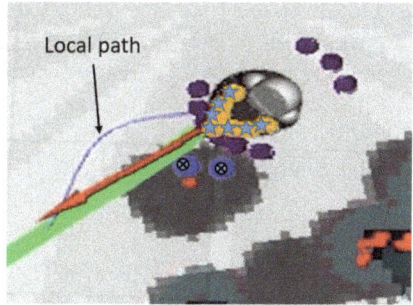

a) The robot detected an obstacle   b) The robot changes the local path to avoid the obstacle

Fig. 5. The obstacle reaction of the robot

4   Image and Speech Processing

4.1   Object Detection

By exploiting deep learning models, the object detection tasks could reach exceedingly higher accuracy, compared to traditional methods. In this work, a pre-trained YOLOv3 [17] model was transferred learning to perform object detection tasks including detecting normal objects and fire alert. This model was chosen because it is the most suitable model depending on its precision, speed and ease of deployment on the embedded computer Pi 4. Figure 6 presents the transfer learning process of a deep neural network.

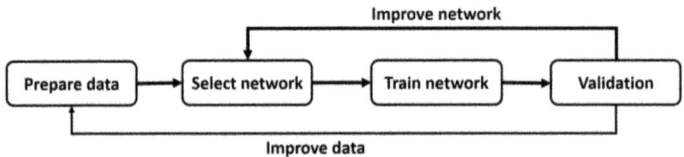

Fig. 6. Transfer learning process of a deep learning network

There are a huge number of datasets of normal objects, the 2014 COCO dataset [18], which is a large-scale object detection dataset with 120K images and 80 object categories, was selected. After transfer learning, the precision of the model on the valid set was about 31 mAP.In contrast, there is very little labelled data of fire while the fire alert using a camera problem was considered. 4300 images of authentic flame and 4500 images of inauthentic flame were selected and labelled manually to create our own dataset for this issue. Then, the YOLOv3 model was finetuned with our fire dataset and got a precision of 42,84 mAP. This approach enables the robot to learn to detect any new types of objects. All parameters when performed transfer learning two models are shown in Table 1. Figure 7 illustrated two frames when the two models were run. The speed of the model was nearly 10 fps which was acceptable on a limited-resource embedded computer Raspberry Pi 4.

Table 1.   Transfer learning parameters in two problems

Parameters	Normal objects	Fire
Batch size	64	64
Subdivision	16	2
Learning rate	0.001	0.001
Optimizer	Adam	Adam
Image size	416×416	416×416
No. of image	120000	8800
Iteration	160000	10000
No. of class	80	2

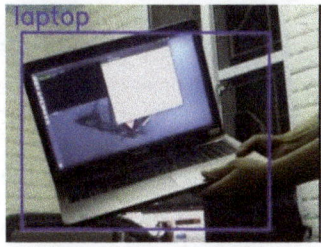
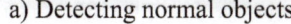

a) Detecting normal objects   b) Fire alert by a camera

Fig. 7. Examples of object detection models

4.2 Speech Recognition

A microphone array Respeaker v2.0 was used for voice control tasks because of its useful features for human-robot interaction. The device could detect the voice direction and filter the noise sound. Then, the voice is sent to Google Cloud via "respeaker_ros" ROS package from [14]. Depending on the recognized words from the cloud, the robot would perform relevant tasks to support human. The feedback rate is less than 1 s which was suitable for real-time human interaction applications.

5 Conclusion

In this work, an autonomous intelligent mobile robot (AIR-HUST) was developed successfully. An IR sensor layer was integrated with a sonar sensor layer and a LIDAR to create a multi-layer sensor fusion system. The sensor system complements strengths and weaknesses of each type of sensors and enhances the efficiency of the navigation system of the robot, as proved in the experiments. Compared to other methods, our algorithm can be used to construct a 'virtual' 2.5D SLAM and control the robot to avoid different types of obstacles, which is not solved so far. Besides, A deep learning model for object detection can be deployed in a very limited-resource embedded computer, which shows very high potential practical applications. Moreover, a number of intelligent applications such as speech recognition, object detection were combined to create a fully functional service robot. Thus, the robot has a wide range of human-friendly applications, for example, reception, nursing, rescue, to name but a few. In the future, more experiments should be performed to evaluate our sensor fusion system in more complex situations. In addition, developing more AI applications also should be taken into account.

Acknowledgement. The second author (Tung Ngo-Thanh) was supported by Vingroup Joint Stock Company and supported by the Domestic Master Scholarship Programme of Vingroup Innovation Foundation (VINIF), Vingroup Big Data Institute (VINBIGDATA).

References

1. Tokody, D., Mezei, I.J., Schuster, G.: An overview of autonomous intelligent vehicle systems. In: Jármai, K., Bolló, B. (eds.) Vehicle and Automotive Engineering. Lecture Notes in Mechanical Engineering, pp. 287–307. Springer, Cham (2017)
2. Wurman, P., D'Andrea, R., Mountz, M.: Coordinating hundreds of cooperative, autonomous vehicles in warehouses. AI Mag. (29), 9–20 (2008)
3. Chung, M.J.Y., Cakmak, M.: 'How was your stay?': exploring the use of robots for gathering customer feedback in the hospitality industry. In: 27th IEEE International Symposium on Robot Human Interactive Communication, pp. 947–954. IEEE, China (2018)
4. Wu, P., Xie, S., Liu, H., Luo, J., Li, Q.: A novel algorithm of autonomous obstacle-avoidance for mobile robot based on LIDAR data. In: 2015 IEEE Conference on Robotics and Biomimetics (ROBIO), pp. 2377–2383. IEEE, China (2015)
5. Gao, M., Tang, J., Yang, Y., He, Z., Zeng, Y.: An obstacle detection and avoidance system for mobile robot with a laser radar. In: 16th IEEE International Conference on Networking, Sensing and Control (ICNSC), pp. 63–68. IEEE, Canada (2019)
6. Baras, N., Nantzios, G., Ziouzios, D., Dasygenis, M.: Autonomous obstacle avoidance vehicle using LIDAR and an embedded system. In: 8th International Conference on Modern Circuits and Systems Technologies (MOCAST), pp. 1–4. IEEE, Greece (2019)
7. Susnea, I., Minzu, V., Vasiliu, G.: Simple, real-time obstacle avoidance algorithm for mobile robots. In: 8th WSEAS International Conference on Computational Intelligence Man-Machine Systems and Cybernetics (CIMMACS 2009), Spain, pp. 24–29 (2009)
8. Susnea, I., Filipescu, A., Vasiliu, G., Coman, G., Radaschin, A.: The bubble rebound obstacle avoidance algorithm for mobile robots. In: 8th IEEE International Conference on Control and Automation, pp. 540–545. IEEE, China (2010)
9. Quinlan, S., Khatib, O.: Elastic bands: connecting path planning and control. In: IEEE International Conference on Robotics and Automation, vol. 2, pp. 802–807. IEEE, USA (1993)
10. Fox, D., Burgard, W., Thrun, S., Cremers, A.B.: A hybrid collision avoidance method for mobile robots. In: IEEE International Conference on Robotics and Automation, vol. 2, pp. 1238–1243. IEEE, Belgium (1998)
11. Ruiz-del-Solar, J., Loncomilla, P., Soto, N.: A survey on deep learning methods for robot vision. arXiv:1803.10862 (2018)
12. Zhang, Y., Wang, H., Xu, F.: Object detection and recognition of intelligent service robot based on deep learning. In: 8th IEEE International Conference on CIS & RAM, pp. 171–176. IEEE, China (2017)
13. Ferro, F., Nardi, F., Cooper, S., Marchionni, L.: Robot control and navigation: ARI' s autonomous system. 29th IEEE International Conference on Robot & Human Interactive Communication (ROMAN-2020), pp. 4–5. IEEE, China (2018).
14. Respeaker_ros. https://github.com/furushchev/respeaker_ros. Accessed 10 Nov 2020
15. Ha, N.X., Huy, N.V., Tung, N.T., Anh, N.D.: Improvement of control algorithm for mobile robot using multi-layer sensor fusion. Vietnam J. Sci. Tech. **9**(1), 110–119 (2021). https://doi.org/10.15625/2525-2518/59/1/15301
16. ROS Wiki – navigation/Tutorials/RobotSetup. https://wiki.ros.org/navigation/Tutorials/RobotSetup. Accessed 18 Dec 2020
17. Redmon, J., Farhadi, A.: YOLOv3: an incremental improvement. arXiv:1804.02767 (2018)
18. Lin, T.Y., Maire, M., Belongie, S., Bourdev, L., Girshick, R., Hays, J., Perona, P., Ramanan, D., Zitnick, C.L., Dollar, P.: Microsoft COCO: common objects in context. arXiv:1405.0312 (2015)

Solving Resource Forecasting in Wifi Networks by Hybrid AR-LSTM Model

Ta Anh Son[1(✉)], Nguyen Thi Thuy Linh[1], and Nguyen Ngoc Dang[2]

[1] Hanoi University of Science and Technology, Hanoi, Vietnam
`son.taanh@hust.edu.vn`
[2] RD Group, AWING Company, Hanoi, Vietnam

Abstract. Time series forecasting is a field that is fairly complex and requires a lot of processing techniques. Identifying trends, detecting anomaly data points, are the first steps to improve forecast results. A point is called an anomaly when it is far away from the mean of the data series. In this paper, we apply an automatic anomaly detection method that combines the decomposition of data into linear and nonlinear components. After that, Autoregressive (AR) and Long Short Term Memory (LSTM) models are used separately, forecasting linear and nonlinear components respectively. The method is tested on a dataset of the times that public wifi was accessed per day, each day is a data point, and the goal is to forecast data in the next 30 days. The forecasted outcome is compared with that of Autoregressive Integrated Moving Average (ARIMA) and LSTM, which indicates that the proposed model achieves the best forecast results.

Keywords: Time series forecasting · Anomaly detection · ARIMA model · LSTM model · Hybrid AR-LSTM model

1 Introduction

A great number of predictive models have been built and developed. Among them, the ARIMA model [1] and some Neural Networks variants such as Artificial Neural Network (ANN), Convolutional Neural Network (CNN), Recurrent Neural Network (RNN), . . . are widely used. The ARIMA model is easy to understand, and it is applied to many different time series problems. However, it has limitations when processing with the nonlinear components of the data series. Besides, the LSTM model [2], a special form of RNN can be applied to time series forecasting. Actual data is usually the sum of linear and nonlinear components. It can be seen from both theories and experiments that ARIMA and LSTM can not solve the data series independently. That is the reason why hybrid models have been introduced with a view to getting better predictive results.

In this paper, we predict the number of times that public wifi was accessed over the next 30 days. From now on, we will use 'views' instead of the times that public wifi was accessed. The abnormal fluctuation of data on holidays and the

D.-T. Tran et al. (Eds.): ICISN 2021, LNNS 243, pp. 327–336, 2021.
https://doi.org/10.1007/978-981-16-2094-2_41

inaccurate data cause forecasts to be more difficult. Currently, there are several automatic anomaly detection approaches such as Luminol [3], Median Absolute Deviation (MAD) [4], Spectral Residual (SR) [5],.... Through experiments with this data set, SR achieved the optimal results compared to the other two methods (details in Sect. 3).

We propose a hybrid model that has combined AR with LSTM to process each component separately after using STL[1] to decompose time series. The proposed technique is inspired by a comparable idea, spearheaded by Zhang [6] who brought up that the actual time series generally consists of linear and nonlinear components.

The rest of the article includes the following sections: The next section describes model building and the steps during the forecasting process; Sect. 3 describes the dataset and how it is processed; Sect. 4 presents the ARIMA, LSTM, and hybrid AR- LSTM model; Sect. 5 is about the forecast error; Sect. 6 shows the numerical results, and finally, conclusions are presented in Sect. 7.

2   Model Building

The most important thing is that we can choose a model which is useful and suitable for our problems and goals. In our project, the dataset is about the number of public wifi views per day in public locations. Our goal is to forecast the number of views in these locations over the next 30 days.

The method consists of 6 steps described in detail below. (visualized in Fig. 1):

Step 1: Detect anomalies
There are two cases:
- With holidays, define them as anomalies;
- With other days, use the Spectral Residual approach to detect them;
Step 2: Separate train/test dataset;
Step 3: Use Augmented Dickey-Fuller Test [7] to check the stationarity
- If time series is stationary, go to step 5;
- If time series is not stationary, go to step 4;
Step 4: Calculate the differencing of time series and return to step 3;
Step 5: Put training data into the training model;
Step 6: Evaluate the model.

3   Data Preparation

Time series data often requires cleaning. Perhaps, it has corrupt or extreme anomaly values that need to be identified and handled. Sometimes, there are gaps or missing data that need to be interpolated or imputed.

---

[1] https://www.statsmodels.org/devel/generated/statsmodels.tsa.seasonal.STL.html.

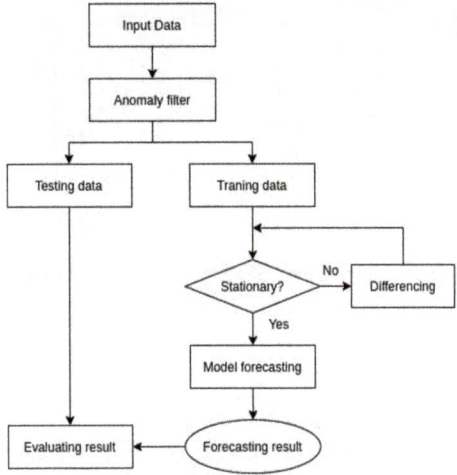

Fig. 1. The diagram of model

Generally, actual views are usually higher at the weekend (Fig. 2) and fluctuate abnormally on holidays. Figure 3 is the result of anomaly detection using three approaches Luminol, MAD, and SR. Luminol and MAD filter out anomalies more than the actual anomalies. The SR approach has the best accuracy.

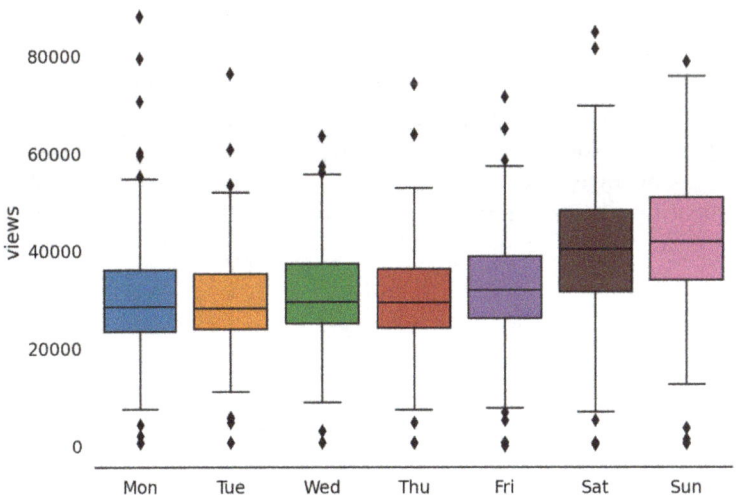

Fig. 2. Views grouped by days of the week distribution (Distribution of total views in Highlands Coffee Diamond Plaza - Ho Chi Minh City series.)

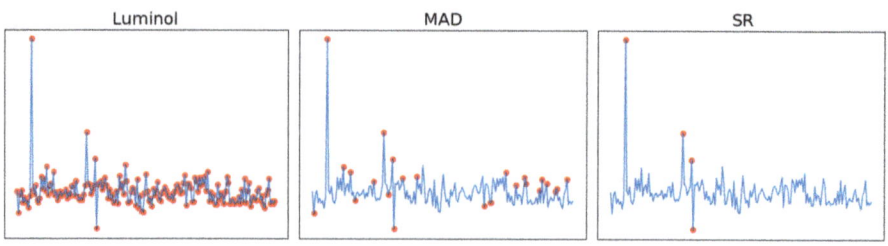

Fig. 3. Anomaly Detection with a) Luminol approach, b) MAD approach, c) SR approach

3.1   Anomaly Detection

Giving a sequence **y**, three major steps of the SR algorithm are: (1) Get the logarithm of the amplitude spectrum by using the Fourier Transform (FT); (2) Calculate the spectral residual; (3) Transform the sequence back to the spatial domain by using the Inverse Fourier Transform (IFT).[2]

$$A(g) = Amplitude(\delta(\mathbf{y})) \tag{1}$$
$$P(g) = Phrase(\delta(\mathbf{y})) \tag{2}$$
$$L(g) = log(A(g)) \tag{3}$$
$$AL(g) = h_m(g) \cdot L(g) \tag{4}$$
$$R(g) = L(g) - AL(g) \tag{5}$$
$$S(\mathbf{y}) = ||\delta^{-1}(exp(R(g) + iP(g)))|| \tag{6}$$

In (1), (2), and (6), δ and δ^{-1} represent FT and IFT, respectively. The input sequence **y** has shape $n \times 1$; $A(g)$ and $L(g)$ are the amplitude spectrum of input sequence **y** and the log representation of $A(g)$; Sequence **y** has phrase spectrum $P(g)$, and the average spectrum $AL(g)$ of $L(g)$ can be closed to convoluting the input sequence by $h_m(g)$ where matrix $h_m(g)$ with shape $m \times m$ is defined as:

$$h_m(g) = \frac{1}{m^2} \begin{bmatrix} 1 & 1 & 1 & \cdots & 1 \\ 1 & 1 & 1 & \cdots & 1 \\ \vdots & \vdots & \vdots & \ddots & \vdots \\ 1 & 1 & 1 & \cdots & 1 \end{bmatrix}$$

We substract the averaged log spectrum $AL(g)$ from the log spectrum $L(g)$ to get the spectral residual $R(g)$. In the last step, we convert the sequence into the spatial domain via IFT. We call the outcome sequence $S(\mathbf{y})$ is *saliency map* (Fig. 4).

---

[2] Daily views of Highlands Coffee Diamond Plaza - Ho Chi Minh City.

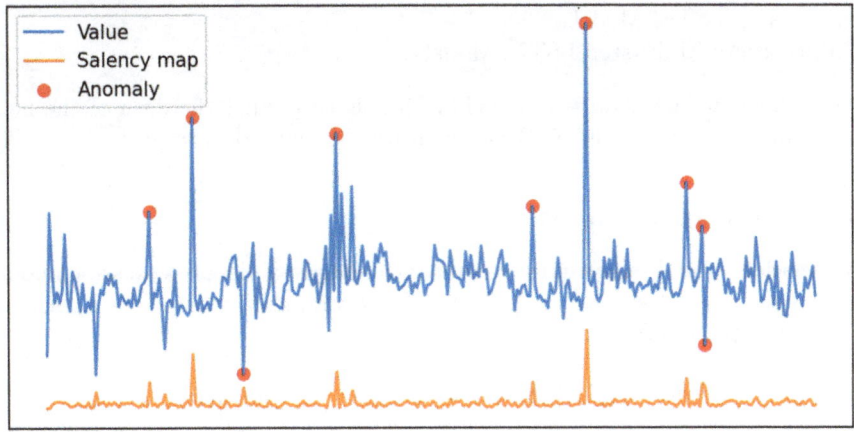

Fig. 4. Example of saliency map

3.2 Data Processing

We divide the dataset which is detected anomalies into two sets: a training set and a testing set. Because the goal is to forecast value of the next 30 days of data series, so we get 30 last values of dataset to use as testing data. For anomaly points, we process them by replacing it with the average of two adjacent weeks. In 2020, due to the influence of Covid-19, Vietnam implemented social distancing for several weeks in the first quarter. So that, we have a special sequence anomalies at this time. (See footnote 2)

4 Model Forecasting

4.1 ARIMA Model

Box and Jenkins [8] have pioneered an ARIMA model. The model is a kind of statistical model applied in time series analysis and prediction.

An ARIMA model identified by three terms: p, d, q where p and q are the order of AR and MA terms respectively. The time series becomes stationary at the order d of differencing.

4.2 LSTM Model

LSTMs is a special kind of RNN, capable of learning long-term dependencies. For each specific type of time series forecasting problem, we need a specific type of model that can be useful and suitable for this problem. Four effective LSTM models for forecasting problem are:

1. Univariate LSTM Models
2. Multivariate LSTM Models

3. Multi-step LSTM Models
4. Multivariate Multi-step LSTM Models

In this paper, we use Multi-step LSTM Models to predict the data of the next 30 days of time series problem. Details on LSTM networks are presented in [9].

4.3   Hybrid AR-LSTM Model

AR Models for Forecasting. An AR model is a linear regression model using lagged variables as input variables. For example, given a value x_t mean x measured in time period t:

$$xhat = b_0 + b_1 x_{t-1} + b_2 x_{t-2} + \cdots + b_p x_{t-p} + \epsilon_t$$

where $xhat$ is the prediction, coefficient set obtained by the optimizing model on training data is $\{b_0, b_1, \ldots, b_p\}$, $\{x_{t-1}, x_{t-2}, \ldots, x_{t-p}\}$ is the input value and ϵ_t is white noise. We refer to this as an $AR(p)$ model of order p.

Model Forecasting of Hybrid AR-LSTM Model. On the concept of the hybrid model ARIMA-ANN of Zhang and by Khashei and Bijari [10], we build a hybrid model combining the AR model with the LSTM model.

Figure 5 describes the specific structure of model forecasting part in the hybrid model.

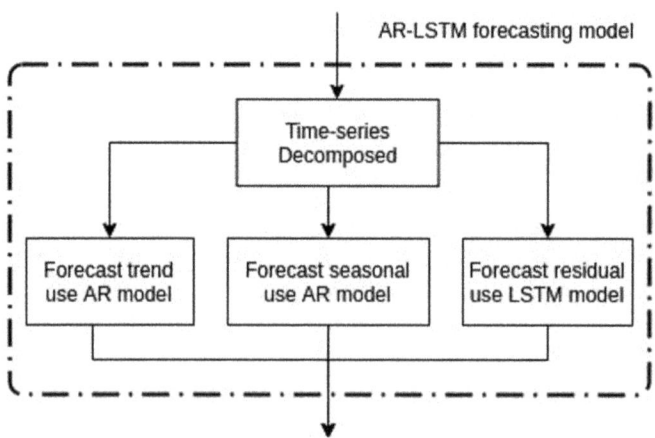

Fig. 5. AR-LSTM forecasting model

After checking stationarity, we use the STL method to decompose the time series into three constituent parts which are the trend, the seasonal component and the residual component. Trend and Seasonal components are forecasted by using AR model while Residual component is forecasted by the LSTM model. The final forecasting outcome is obtained by combining all forecasted components.

5   Evaluating Model

In our project, we evaluate the model by using Mean Absolute Percentage Error (MAPE). The MAPE calculation is as follows:

$$\text{MAPE} = \frac{\sum_{i=1}^{n} \frac{|x_t - \bar{x}_t|}{x_t}}{n}$$

where x_t and $\bar{x}_t$ represent the actual value and the prediction value in time period t respectively, n is the number of observations.

6   Experiments

We have conducted experiments with daily views of Highlands Coffee Diamond Plaza - Ho Chi Minh City from 4/10/2017 to 5/10/2020 (Fig. 6). The dataset includes 1129 records. We use 1099 records for training and 30 records for testing.

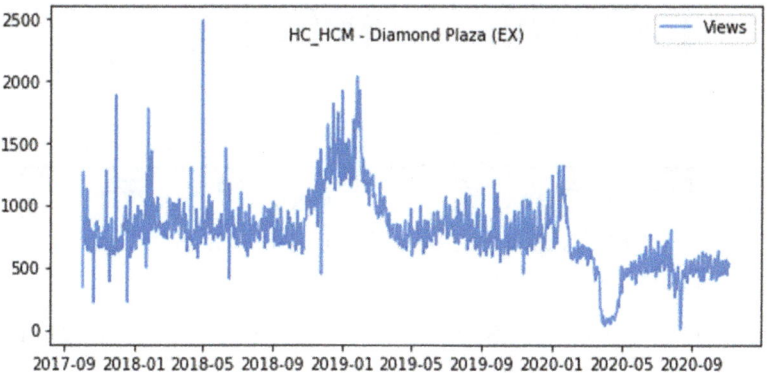

Fig. 6. Daily views of Highlands Coffee Diamond Plaza

After filtering and processing anomalies, we got the results shown in Fig. 7 and Fig. 8.

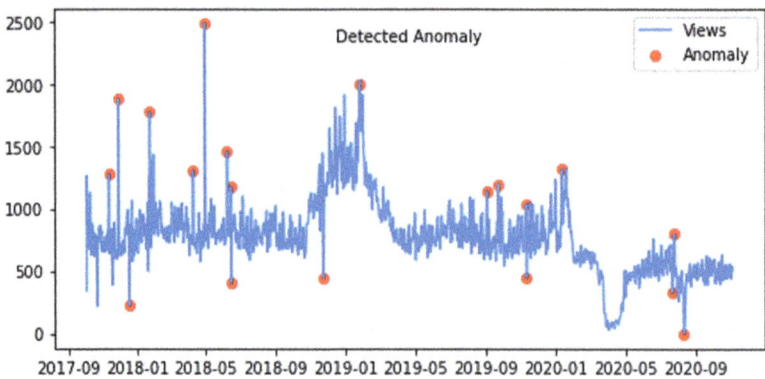

Fig. 7. Anomaly detection

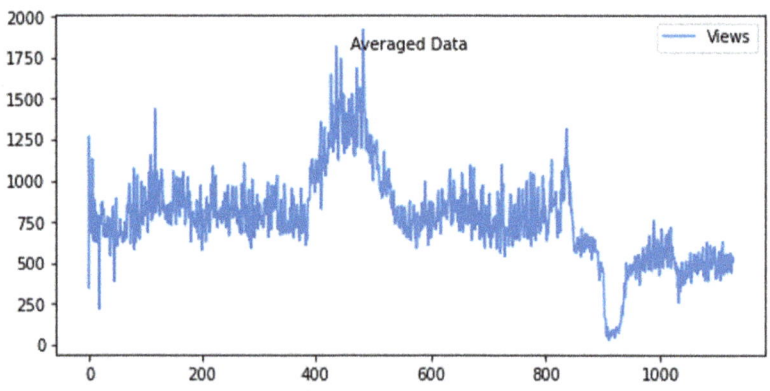

Fig. 8. Replacing anomalies by averaged data

Applying the hybrid AR-LSTM model, we decompose the series and forecast them by AR and LSTM model, the results are as follows (Fig. 9):

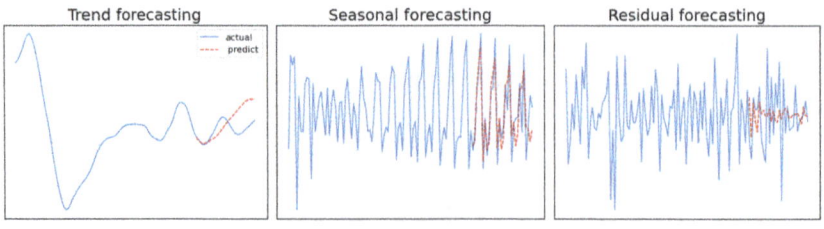

Fig. 9. Components forecasting

The forecasting results of three models ARIMA, LSTM, and AR-LSTM are as follows (Fig. 10):

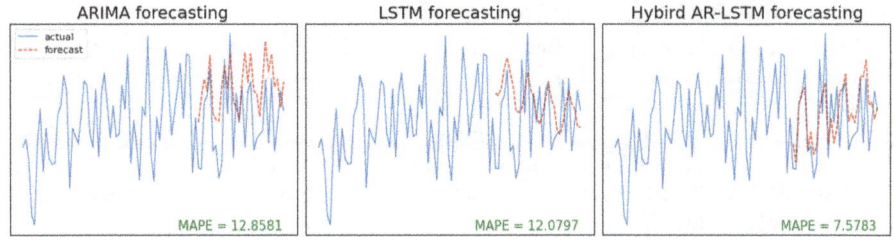

Fig. 10. Forecasting results a) ARIMA model, b) LSTM model, c) AR-LSTM model

Furthermore, we conducted experiments on three datasets at Circle K 74-76 Nguyen Khang - HN, Hutong Pho Duc Chinh - HN, and Vincom Dong Khoi - HCMC.

The empirical outcomes of three time series problems clearly show the hybrid model yielded notably better predicts than ARIMA and LSTM model (Table 1).

Table 1. Compare MAPE of three models following three datasets

Place	Min	Max	No.days	ARIMA	LSTM	AR-LSTM
Circle K 74-76 Nguyen Khang	1	204	386	10.1006	22.5995	8.9373
Hutong Pho Duc Chinh	1	353	774	28.473	16.4041	7.5546
Vincom Dong Khoi, HCMC	2	14810	1770	14.4627	17.9615	7.8809

7   Conclusions

Achieving good results in time series forecasting is challenging. ARIMA and LSTM also encounter many difficulties in solving the data series obtained by linear and nonlinear parts. ARIMA is suitable for linear data series and LSTM is suitable for nonlinear data series. Therefore, in this article, we propose a hybrid model between AR and LSTM based on the hybrid model concept of Zhang and by Khashei and Bijari but instead of decomposing time series by MA filter, we use STL. For better results, before inputting data into the forecasting model, we process the anomaly data using the SR approach. Comparing the results between the three models proves that the hybrid model is better than ARIMA and LSTM.

In the future, we also want to apply the most modern methods to improve the predictive results and apply other types of data series.

References

1. Brownlee, J.: Introduction to Time Series Forecasting with Python (2020)
2. Brownlee, J.: Deep Learning for Time Series (2019)
3. [n.d]. https://github.com/linkedin/luminol
4. Leys, C., Ley, C., Klein, O., Bernard, P., Licata, L.: Detecting outliers: do not use standard deviation around the mean, use absolute deviation around the median. J. Exp. Soc. Psychol. **49**(4), 764–766 (2013)
5. Ren, H., Xu, B., Wang, Y., Yi, C., Huang, C., Kou, X., Xing, T., Yang, M., Tong, J., Zhang, Q.: Time-series anomaly detection service at microsoft. In: Proceedings of the 25th ACM SIGKDD International Conference on Knowledge Discovery & Data Mining (2019)
6. Zhang, G.P.: Time series forecasting using a hybrid ARIMA and neural network model. Neurocomputing **50**, 159–175 (2003)
7. Mushtaq, R.: Augmented dickey fuller test. SSRN Electron. J. (2011)
8. Jenkins, G.M., Box, G.E.P.: Time Series Analysis, Forecasting and Control, 3rd edn. Holden-Day, California (1970)
9. Son, T.A., Phuong, N.T.: Long-short term memory networks for resource allocation forecasting in Wifi networks, pp. 348–351 (2019)
10. Khashei, M., Bijari, M.: A novel hybridization of artificial neural networks and ARIMA models for time series forecasting. Appl. Soft Comput. **11**(2), 2664–2675 (2011). The Impact of Soft Computing for the Progress of Artificial Intelligence

A New Approach for Large-Scale Face-Feature Matching Based on LSH and FPGA for Edge Processing

Hoang-Nhu Dong[2(✉)], Nguyen-Xuan Ha[1,2], and Dang-Minh Tuan[2,3]

[1] Hanoi University of Science and Technology, 1 Dai Co Viet, Hanoi, Vietnam
ha.nguyenxuan@hust.edu.vn
[2] CMC Institute of Science and Technology, 11 Duy Tan, Hanoi, Vietnam
hndong@cmc.com.vn, tuandm@ptit.edu.vn
[3] Posts and Telecommunication Institute of Technology, Hanoi, Vietnam

Abstract. In this work a new approach for the large-scale face-feature matching is introduced. Based on this approach, the LSH was used to hash the database to a smaller set of face-feature vectors have the most similarities with the searched vector. This hashed database is then further processed by linear search on FPGA. Experimental implementations show that the searching results of our method have higher accuracy of 99,6% and less computational time of 43ms in comparison to other methods such as pure-LSH, or LSH combined with CPU. This approach is very promising for the problem of large-scale data matching on edge devices having very limited hardware resources.

Keywords: Large-scale data matching · LSH · FPGA · Artificial intelligent · Edge processing

1 Introduction

The rapid development and advancement of the deep learning have shown many advantages for the image processing problem. The face recognition has received much of attention in recent year due to its very high potential of practical applications in, for example, access/check-in-out control systems, public security surveillance systems, and electronic-commercial transactions [1]. Like any typical image processing problem, the face recognition has also three main modules including: i) The preprocessing module is responsible for face detection, anti-spoofing, alignment, and quality check; ii) The face feature extraction module exploits a deep neural network which typically results to face-feature vectors (FFVs) with a size of 512 elements; iii) The face feature matching module calculates Euclidean or Cosine distances between FFVs to get the final identification of a person of interest. For high-dimensional large-scale datasets, the face feature matching is extremely challenging in term of the accuracy and the computational time, especially with the edge processing having very limited computational resources.

There have been many investigations on the topic of the high-dimensional large-scale data matching which exploit both hardware and software approaches [2–9].

© The Author(s), under exclusive license to Springer Nature Singapore Pte Ltd. 2021
D.-T. Tran et al. (Eds.): ICISN 2021, LNNS 243, pp. 337–344, 2021.
https://doi.org/10.1007/978-981-16-2094-2_42

A very well-known software approach is the so-called Locality-Sensitive Hashing (LSH) [2, 3]. LSH is an algorithm for solving the nearest neighbor search problem in high-dimensional large-scale datasets which shows many advantages in term of computational efficiency. A practical deployment of the LSH is the library FALCONN which provides very efficient and well-tested implementations of LSH based datasets [4]. For the hardware approach, the FPGA (Field Programmable Gate Array) is the best candidate for the computational acceleration due to its noticeable features of the parallel computing and the very low power consumption [5–9].

Although the two above-mentioned approaches have many advantages, several bottlenecks remain. The LSH is an approximation search which would never result in 100% rate of accuracy in comparison to the linear search. The FPGA only shows efficiencies with datasets which fit on chips. For large-scale datasets with limited hardware resources of edge computing devices, the shuffling data on or off chips is bottlenecked by costly merging operations or the data transfer time [6].

In this work, a new hybrid-approach for large-scale face feature matching which combines the LSH algorithm for the data hashing and the FPGA for the computational accelerator, is proposed. Our approach takes advantages of both the LSH and the FPGA and overcomes their limitations. Based on this approach, the large-scale face feature database is hashed to a smaller one which has a high probability of similarity with an interested FFV. The size of the hashed database is fitted to the memory of the FPGA. This smaller database is then transferred to the FPGA's chip for the linear search which is computationally accelerated. This allows for a very reasonable accuracy, computational time, accompanying with limited hardware resources of FPGA-based edge devices.

2   Methodology

2.1   Database

The database has thousands of identification (ID) containing personal information, for example, the name, the date of birth, etc. Each ID has at least ten face images having different viewpoints, light and quality conditions. A trained deep learning model was used to convert this database to a new database containing feature vectors of size 512 corresponding to each image.

Each FFV is a 512-dimensional space. The ID searching is the matching of the FFV of an interested ID with the FFV of the source database to find out the nearest neighbor point having the best similarity. The metric for the matching is typically the calculation of Euclidean or Cosine distances between interested FFVs and FFVs of the database (Fig. 1).

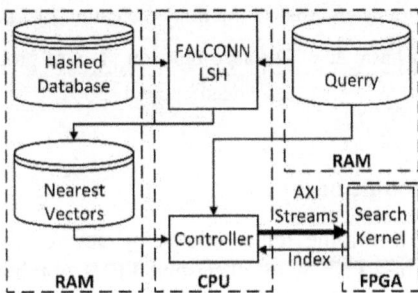

Fig. 1. Processing pipeline of the proposed approach

2.2   Processing Pipeline

The system was combined with two searching methods including the LSH data hashing and the FPGA for linear search. The LSH algorithm is used to quickly determine a set of FFVs having a high similarity (nearest vectors) with the searched FFV. This hashed database is transferred to the FPGA, where the linear search is performed to exactly find out an FFV having the best similarity with the searched FFV. This combination allows for an optimization of two factors including the searching time and the searching accuracy.

The FFV database is saved in the RAM as sequence records. The FALCONN library with random hashing functions is used to generate hash table in which FFVs having high similarities will be isolated to a smaller hashed database. When a search is required, the searched FFV is evaluated by hashing functions to find out a set of FFVs having high similarity with the searched vector.

After obtaining the set of high similarity FFVs, the searched FFV is computed with each FFV of this set to determine a vector having a maximum Cosine calculation (linear search). To accelerate the computation, the Cosine calculation is implemented on the FPGA. On the CPU, the controller continuously streams the searched FFV and the set of high similarity hashed FFVs to the Cosine computing kernel. The Cosine kernel will return the index of FFV having a maximum Cosine value with the searched vector. To accelerate the data transfer, FFVs are streamed to the FPGA via AXI interfaces using data bus width of 512 bits and a frequency of 300 MHz.

2.3   LSH Algorithm

The first step of the searching process is the quick determination of FFVs having high similarity with the searched FFV. In this work, the FALCONN library is used. More details of this library can be found in [4]. Parameters for hashing the whole database are set according to Table 1. The number of hash function is automatically calculated based on the parameter 'number of hash bits'. The parameter 'number of hash table' is used to config the FALCONN perform a loop to generate different hash tables on the same source dataset. This loop ensures to eliminate errors that two similar vectors are divided into two different hashed sets of the hash table. The more the number of the hash table is, the longer the hashed time is, but the accuracy is higher.

Table 1. Parameters used in the implementation of Falconn LSH

Lshfamily	No. of hash table	Distance function	No. of probes	No. of hash bits
Cross-polytope	100	EuclideanSquared	200	18

2.4   Linear Search Module on FPGA

The FPGA is responsible for the linear search by accelerating the computation of similarity between searched FFV with all hashed FFVs obtaining from LSH module. A Cosine kernel is set up on the FPGA and controlled by the processing controller implementing on CPU. FFVs are streamed from the RAM to the FPGA with burst type via AXI interface supported by a complier of the Vivado HLS tool. In this work, the tool named 'Vitis Unified Software Platform 2020.1' from Xilinx was used to program and compile kernels for edge MPSoC platforms [9]. The design of the kernel is described in Fig. 2.

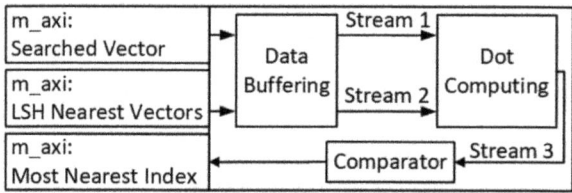

Fig. 2. Design of the linear search kernel on FPGA

FFVs are streamed to FPGA via the 'm_axi' interface. The kernel has three calculation blocks running in dataflow manner. The first block receives the data of the burst type from the CPUand constructs the data to pairs of vectors and then transfer them to the 'dot computing' block. The 'dot computing' uses multiply matrix function of the library 'xf::blast' of Vitis allowing for an optimization of the computation. The computed resultsare consequently transferred to the last block where the comparison is performed to find out a pair of vectors having a maximal Consine value. The index of this found pair of vectors is sent to CPU via the 'm_axi' connection. The data is transferred between blocks via interfaces of the 'hls_stream'. All blocks operate parallelly following the pipeline regime helping optimize data throughput. The clock of the kernel is set to 300 MHz.

3   Experimental Evaluation

To evaluate our approach, five search methods were performed and compared where the accuracy and the computational time are considered:

- Only LSH was used (LSH_only);
- LSH for hashing data and CPU for linear search of the hashed data (LSH_Linear_CPU);

- LSH for hashing data and FPGA for linear search of the hashed data (LSH_Linear_FPGA);
- Linear search on CPU (Linear_CPU);
- Linear search on FPGA (Linear_FPGA).

The dataset CelebA [10] is chosen to evaluate the accuracy and the computation time of the search methods. The dataset has 10000 IDs containing 128007 images. Each image is converted to an FFV size of 512 by a deep learning model. The circuit ZCU 104 [11] using the chip ZU7EV of MPSoC with 4 cores ARM® Cortex™-A53, 2 GB RAM, combining with an array of FPGA FinFET+ 16 nm is used to accelerate the computation.

Parameters for the LSH are configured as follows:

- Constant parameters including the 'Lsh_family', 'Distance_function' and 'Number of hash bits', are set according to Table 1.
- Changing parameters including the 'Number of hash table' and 'Number of probe' are varied to generate different test scenarios.

3.1   Testcase 1

In this experiment, the 'hash_table' is set to 50 and the 'probe_number' is varied following the Table 2. The accuracy and the computational time are evaluated. A set of 500 FFVs is extracted from the source database of 128007 FFVs to be a searched set. Search results of the linear search on CPU (Linear_CPU) are the reference for other search methods.

Table 2. Parameters for testcase 1. The 'hash_table' is set to 50.

Case	1	2	3	4	5	6
Probe number	50	100	200	300	400	600

Search methods using 'Linear_FPGA' or 'Linear_CPU' will have an unchanged computational time and a maximal accuracy for all tests since the methods are based on the sequence matching between the searched FFV and the FFVs in the database. Experimental results show that the linear search using FPGA has 3.75 times quicker than the one using CPU (320 ms vs. 1200 ms). In general, the computational time of these two methods is still high in comparison to the other methods.

Evaluation results of the three left methods are shown in Fig. 3. In Fig. 3a the relation between number of FFVs, which is most similar to the searched FFV, and the 'Probe_number' is represented.Each Probe will determine a set of similarity FFVs. When the 'Probe_number' is increased, the number of similarities is also expanded. This allows for the probability to find out the most similarities increase.

Figure 3b shows that when the 'Probe_number' is increased the accuracy of both 'LSH_CPU' and 'LSH_FPGA' is higher than the 'LSH_only'. It is explained that the similarities hashed by LSH increase if the 'Probe_number rises leading to an increase of searching radius resulting to a higher accuracy.

Figure 3c shows differences on the computational time of the three methods when the 'Probe_number' is increased. The 'LSH_only' always has best performances with lowest computation time. When the 'Probe_number' is small, the 'LSH_CPU' requires less computational time than the 'LSH_FPGA' and inversely. This can be explained that the 'LSH_FPGA' is only efficient if the number of the similarities is large enough and especially best fits with the memory of the FPGA. The 'LSH_FPGA needs time to transfer the data to the search kernel of the FPGA whereas the CPU do not need. The computational time includes the processing and the data transfer.

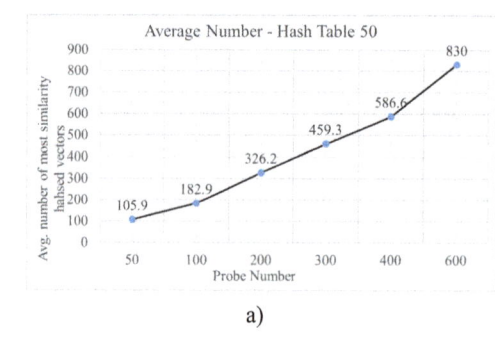

a)

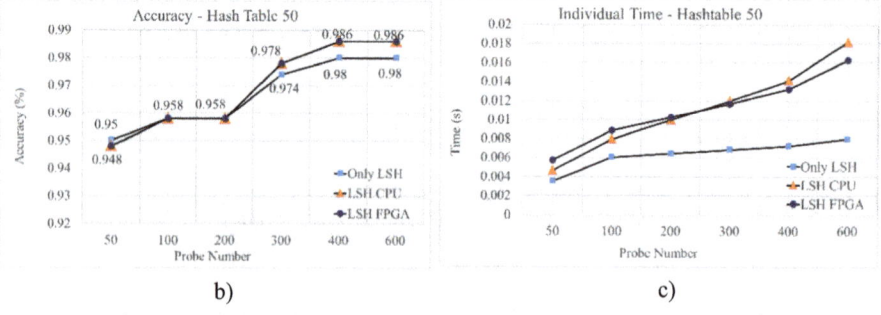

b)   c)

Fig. 3. Experimental results for testcase 1: a) LSH hashing, b) Comparison of the accuracy, c) Comparison of the computational time.

3.2   Testcase 2

In the testcase 2, the 'hash_table' is set to 100. The 'Probe_number' is varied in a larger scale as shown in Table 3.

Table 3. Parameters for testcase 2. The 'hash_table' is set to 100.

Case	1	2	3	4	5	6	7	8	9	10
Probe_number	100	200	300	400	600	800	1000	1200	1400	1600

Figure 4b shows that when the number of hash table is set to 200, the accuracy of "LSH only" is saturated at the value of 0.986. Meanwhile, when the number of probes increase, the accuracy of both "LSH CPU" and "LSH FPGA" are improved and reach 0.996 at 1600 probes. It is explained by Fig. 4a, the number of most similar FFVs reach nearly 2030 when the number of probes is 1600. This provides higher probability to find out the target FFV for the searched FFV. However, the processing time is therefore much higher as a trade-off for higher accuracy.

Figure 4c shows that the computational time of "LSH Only" increases slowly, in comparison to "LSH CPU" and "LSH FPGA". This is because "LSH Only" can directly predict the most similar FFV without linearly comparing through a temporally similar set of FFVs. The only way to improve accuracy for "LSH Only" is to increase the number of hash table, which requires the increase of memory resources and computational time. This should be avoided on edge processing devices with limited hardware capability.

Figure 4c also confirms the obvious difference of the computational time between "LSH CPU" and "LSH FPGA". The higher number of similar FFVs, the better computational time of "LSH FPGA" comparing to "LSH CPU".

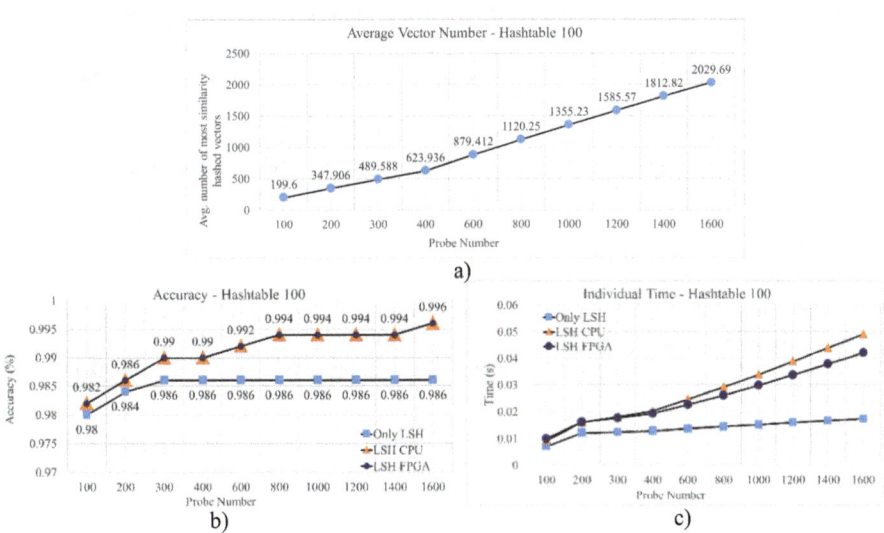

Fig. 4. Experimental results for testcase 2: a) LSH hashing, b) Comparison of the accuracy, c) Comparison of the computational time.

4   Conclusion and Outlook

In this paper a new approach for large-scale face-feature matching has been presented. The proposed method 'LSH_FPGA' can satisfy both key factors of the search on large-scale datasets including the accuracy and the computational time. Using LSH combining with the linear search on FPGA helps not only improve the computational speed and the accuracy but also release the CPU out of linear tasks. This help optimize the performance of edge devices having very limited hardware resources. In the future, we will further evaluate the approach on different edge hardware platforms where devices having high bandwidth memory are focused.

Acknowledgement. This work was funded by CMC Institute of Science and Technology, CMC Corporation, Hanoi, Vietnam.

References

1. Wang, M., Deng, W.: Deep Face Recognition: a Survey. arXiv:1804.06655 (2018)
2. Locality sensitive hashing. https://www.mit.edu/~andoni/LSH/. Accessed 20 Nov 2020
3. Andoni, A., Indyk, P., Laarhoven, T., Razenshteyn, I., Schmidt, L.: Practical and Optimal LSH for Angular Distance. arXiv:1509.02897 (2015)
4. FALCONN github repository. https://github.com/FALCONN-LIB/FALCONN. Accessed 24 Nov 2020
5. Melikoglu, O., Ergin, O., Salami, B., Pavon, J., Unsal, O., Cristal, A.: A Novel FPGA-Based High Throughput Accelerator for Binary Search Trees. arXiv:1912.01556 (2019)
6. Chen, H., Madaminov, S., Ferdman, M., Milder, P.: FPGA-accelerated samplesort for large data sets. In: 2020 ACM/SIGDA International Symposium on Field-Programmable Gate Arrays (FPGA 2020), pp. 222–232. Association for Computing Machinery, New York (2020)
7. Jun, S., Chung, C., Arvind: Large-scale high-dimensional nearest neighbor search using flash memory with in-store processing. In: 2015 International Conference on ReConFigurable Computing and FPGAs (ReConFig), pp. 1–8. IEEE, Mexico (2015)
8. Zhang, J., Li, J., Khoram, S.: Efficient large-scale approximate nearest neighbor search on OpenCL FPGA. In: 2018 IEEE/CVF Conference on Computer Vision and Pattern Recognition (CVPR), pp. 4924–4932. IEEE, Mexico (2018)
9. Cosine Similarity Using Xilinx Alveo. https://developer.xilinx.com/en/articles/cosine-similarity-using-xilinx-alveo.html. Accessed 20 Nov 2020
10. Large-scale CelebFaces Attributes (CelebA) Dataset. https://mmlab.ie.cuhk.edu.hk/projects/CelebA.html. Accessed 29 Nov 2020
11. Zynq UltraScale+ MPSoC ZCU104 Evaluation Kit. https://www.xilinx.com/products/boards-and-kits/zcu104.html. Accessed 29 Nov 2020

An Integer Programming Model for Course Timetabling Problem at HUST

Ta Anh Son$^{(\boxtimes)}$ and Nguyen Thi Ngan

School of Applied Mathematics and Informatics (SAMI), Hanoi University of Science and Technology (HUST), Hanoi, Vietnam
son.taanh@hust.edu.vn, ngan.nt162890@sis.hust.edu.vn

Abstract. Creating a course timetable for university is a complex problem. It is difficult and time-consuming to solve due to many constraints, including 'hard' constraints and 'soft' constraints. The proposed model is a 0-1 integer programming model that provides constraints for a host of operational rules and requirements found in Hanoi University of Science and Technology (HUST). The objective is to minimize a linear cost function related to the cost of infrastructure (classrooms) and teaching staff. The problem was solved successfully for medium-sized data sets by CPLEX 12.8. The case of a school (SAMI) in HUST with a considerable number of courses and lecturers is presented along with its solution as resulted from the proposed model.

Keywords: Timetabling · University timetabling · Integer programming

1 Introduction

In the world, most schools and universities have been unable to create course timetables automatically, and Hanoi University of Science and Technology is no exception. They almost duplicate the timetables of previous years and adopt some changes to satisfy newly occurred situations. However, when changes increase in the number, this approach is not always the best solution. These circumstances have encouraged the scientific community to continually work on the timetabling problem to develop an automated system for building efficient timetables [2].

   i. *Timetabling in general*
     Timetabling is the process of assigning university courses to specific time slots for five working days of the week and to specific classrooms suitable for the respective courses and the registered number of students.
   ii. *Timetabling: Hard constraints*
     Hard constraints are important constraints that can not be violated at all costs. In the course timetabling problem, hard constraints are:

D.-T. Tran et al. (Eds.): ICISN 2021, LNNS 243, pp. 345–354, 2021.
https://doi.org/10.1007/978-981-16-2094-2_43

+ At a time, a teacher or a student is in one most place and one most course,

+ At a time, each classroom is assigned to most one course and most one group of students...

iii. *Timetabling: Soft constraints*

Soft constraints are additional constraints and not as important as hard constraints. They are added into the model for better solutions only if all hard constraints are satisfied. Soft constraints for course scheduling can be the preferences and choices among teaching staff, the reasonable arrangement of some courses, or the minimization of classroom changeovers.

iv. *Timetabling as a NP problem*

Manual scheduling is a complicated and time-consuming task for people in charge. That is the reason why many efforts have been made to adopt computer methods to facilitate this task. However, the timetabling problem is known to be an NP-complete problem [3], given the fact that when the number of students increases and the teaching programs gets more complicated, the computation time to find reasonable solutions would rise exponentially.

v. *Some approaches to automated timetabling*

+ Using database management systems,

+ Genetic algorithm: This is a heuristic search method which has been used to solve many scheduling problems [6],

+ Graph coloring algorithms: Welsh and Power [4] had some discussions about the application of this technique, especially on timetabling problem,

+ Mathematical programming: Many mathematical models have been presented to solve various scheduling problems at universities [5]...

vi. *Course timetabling at Hanoi University of Science and Technology*

Most previous approaches for course timetabling problem presented a comprehensive model to solve the problem by assigning courses to students, lecturers, classrooms, and time slots at the same time. At Hanoi University of Science and Technology, the scheduling problem is different and complicated due to the large data sets and many requirements. Therefore, the timetabling task is divided into two smaller tasks. The first task is assigning courses to students, classrooms, and time slots to form classes, while the second task is assigning lecturers to these classes. When the data sets are large and complex, this division can ease the task of scheduling. Building the model based on this feature, we can considerably reduce the size of the problem as well as the computation time.

At HUST, every student studies certain courses every semester, including lecture courses and tutorial courses (some groups of students will have lab work courses). For lecture courses, students are included in large-sized groups to minimize the cost of infrastructure and teaching staff. For tutorial courses, students stay in small-sized groups for easier management, better interactions, and instructions.

In terms of infrastructure, the university has different classrooms suitable for different groups of students and different courses. However, in this model, we do not consider the function of each classroom for the sake of a simpler model. We will consider it in future work. There are two sub-problems in course timetabling at HUST, namely 'the first problem' and 'the second' problem.

   a. *The first problem* (course-student-classroom-time problem): Assigning courses to specific periods in five working days of a week and to specific classrooms suitable for groups of registered students.

   b. *The second problem* (class-lecturer problem): After having the timetable of classes with the arrangement of courses, students, classrooms, and periods, the university will entrust each department with the task of assigning lecturers to classes.

vii. *Integer programming problem*

   (Linear) Integer Program written as [1]:

$$\min \ c^T x$$
$$\text{(IP)} \qquad Ax \leq b$$
$$x \geq 0 \quad \text{and integer}$$

where x denotes the vector of variables which need to be determined, c and b are vectors of (known) coefficients, A is a (known) matrix of coefficients, $(.)^T$ is the matrix transpose, $c^T x$ is called the objective function (minimized or maximized), $Ax \leq b$ and $x \geq 0$ are the constraints specifying a convex polytope over which the objective function is to be optimized.

and if all variables are restricted to 0-1 values, we have a
0-1 or Binary Integer Program

$$\min \ c^T x$$
$$\text{(BIP)} \qquad Ax \leq b$$
$$x \in \{0, 1\}^n$$

A wide variety of practical problems were modeled as 0-1 integer programming problems and got good results. Therefore, we chose 0-1 integer programming model to solve course scheduling problem at HUST.

viii. *CPLEX Optimizer*

IBM CPLEX Optimizer is a solver developed by International Business Machines (IBM). It provides flexible and high-performance mathematical programming solvers for linear programming, mixed-integer programming, quadratic programming, and quadratically constrained programming problems. CPLEX Optimizer can solve large-sized optimization problems in reality with pretty fast speed and produce precise results. It has been reported to successfully solve optimization models with millions of constraints and variables [7].

2 Modeling the Course Timetabling Problem at HUST

2.1 The First Problem: Course-Student-Classroom-Time Problem

2.1.1 Data Sets

In this problem, the following data sets are required

1. Sets of groups of students

+ G: An available set which contains all groups of students $G = \{g_0, g_1, g_2, \ldots, g_{|G|}\}$.

    $G = \{\text{'G0'}: [\text{'Co khi 01-K62'}], \text{'G1'}: [\text{'Co khi 01-K62'}, \text{'Co khi 02-K62'}], \ldots\}$.

+ G_{big}: A set containing large-sized groups of students is generated from G

    $G_{big} = \{b_1, b_2, \ldots, b_i, \ldots, b_{|B|}\}$.

    ex: $G_{big} = \{\text{'G1'}: [\text{'Co khi 01-K62'}, \text{'Co khi 02-K62'}], \ldots\}$.

+ G_{small}: A set containing small-sized groups of students is generated from G

    $G_{small} = \{s_1, s_2, \ldots, s_j \ldots, s_{|S|}\}$.

    $G_{small} = \{\text{'G0'}: [\text{'Co khi 01-K62'}], \text{'G2'}: [\text{'Co khi 02-K62'}], \ldots\}$.

    $s_j \subset b_i$, $G_{big} \cap G_{small} = \emptyset$, $G = G_{big} \cup G_{small}$.

2. C: An available set which contains all the courses in the semester $C = \{c_0, c_1, c_2, \ldots, c_{|C|}\}$.

    ex: $C = \{\text{C0}: \{\text{Sub: 'Math I', Type: 'LT', Duration:'0:2,1:3'}\}, \ldots\}$.

    + C0: Code of course, Sub: Name of subject,

    + Type: In this model, there are 2 types of courses, the lecture course (denoted by 'LT') and the tutorial course (denoted by 'BT'). In reality, we have one more type which is the lab work course.

    + Duration: '0:a,1:b': the number of periods assigned for 2 sessions in a week, the first session lasts 'a' periods, the second one lasts 'b' periods.

    + $H_c = \{a, b\}$, $U_c = \{0, 1\}$.

3. Sets of classrooms

    + R_{all}: An available set which contains all the classrooms: $R_{all} = \{r_1, r_2, \ldots, r_{|R|}\}$.

    ex: $R_{all} = \{\text{'D3-101', 'D6-201', 'TC 210'}, \ldots\}$.

    + R: An available set which contains classrooms grouped by their capacity

    ex: $R = \{200 : \{\text{'D3-101', 'D3-201', 'D3-301'}\}, 370 : \{\text{'GD-B1'}\} \ldots\}$.

Here, there are three classrooms that can accommodate 200 students: 'D3-101', 'D3-201', 'D3-301'. And, there is only one classroom that can accommodate 370 students: 'GD-B1'.

4. R_g: An available set which contains groups of students accompanied by the classroom's capacity that fits them

    ex: $R_g = \{\text{'G0': '84', 'G1': '200', 'G2': 84}, \ldots\}$.

5. C_g: A set which contains groups of students accompanied by the courses that they need to register in the semester

    ex: $C_g = \{\text{'G0': ['C0', 'C1], 'G1': ['C2'], 'G2': ['C0', 'C1']}, \ldots\}$.

6. C_{gbig}: A set which contains big-sized groups of students accompanied by the courses that they need to register in the semester

    $C_{gbig} = \{[\text{'G0': ['C0', 'C1],'G2': ['C0', 'C1']}, \ldots\}$.

7. C_{gsmall}: A set which contains small-sized groups of students accompanied by the courses that they need to register in the semester

$C_{gsmall} = \{['G1': ['C2'],...\}, C_{gbig} \cap C_{gsmall} = \emptyset, C_g = C_{gsmall} \cup C_{gbig}.$
8. D: A set which contains weekdays (Mon, Tue, Wed, Thur, Fri), $D = \{0,1,2,3,4\}.$
9. P: A set which contains 6 periods in a session of a day (morning or afternoon), each period lasts 60 min, $P = \{0,1,2,3,4,5\}.$

2.1.2 Sets of Variables

Three different sets of variables are adopted in the model
1. Basic set of variables: binary variables denoted by $x_{d,p,g,c,r}$ where $d \in D, p \in P, g \in G, c \in C, r \in R_{all}$
   $+ x_{d,p,g,c,r} = 1$ when course c taught to the group of student g, is scheduled for the period p of day d in classroom r.
   $+ x_{d,p,g,c,r} = 0$ otherwise.
2. Set of auxiliary variables:
   $+$ Binary variables denoted by $y_{d,u,g,h,c,r}$ where $d \in D, g \in G, r \in R_{all}, u$ and h are natural numbers
   $+ y_{d,u,g,h,c,r} = 1$ when course c, which requires a session of h consecutive periods, is scheduled for day d for the group of student g in classroom r. Index u takes the value of '0' or '1'
   $+ y_{d,u,g,h,c,r} = 0$, otherwise.
For example, if course c requires totally 6 periods per week, it can be split into one session of 2 periods and one session of 4 periods ($h = 2, h = 4$), or into 2 sessions of 3 periods each ($h = 3$).
   $+$ Binary variables denoted by w_{dr}

$$w_{dr} = \begin{cases} 1 \text{ if } \sum_{p,g,c} x_{d,p,g,c,r} > 0 & (1a) \\ 0 \text{ if } \sum_{p,g,c} x_{d,p,g,c,r} = 0 & (1b) \end{cases}$$

where $\sum_{p,g,c} x_{d,p,g,c,r}$ is the total number of times that classroom r is used in day d, then (1a) refers to $w_{dr} = 1$ when classroom r is used in day d and (1b) refers to $w_{dr} = 0$ when classroom r is not used in day d.

2.1.3 Constraints for the Model

1. For every group of student at most one course, one classroom shall be assigned to every teaching period

$$\forall g \in G, \forall d \in D, \forall p \in P, \sum_{r \in R_g} \sum_{c \in C_g} x_{d,p,g,c,r} \leq 1. \qquad (2)$$

2. At a time, each classroom shall be assigned to at most one course and at most one group of students

$$\forall r \in R, \forall d \in D, \forall p \in P, \sum_{g \in G} \sum_{c \in C_g} x_{d,p,g,c,r} \leq 1. \qquad (3)$$

3. The constraint makes sure all the periods course c requires are assigned

$$\forall g \in G, \forall c \in C_g, \sum_{r \in R_g} \sum_{d \in D} \sum_{p \in P} x_{d,p,g,c,r} = p_c. \tag{4}$$

where $p_c = a_c + b_c$ is the total number periods of course c in a week.

4. The constraint makes sure every period in a session of a course is consecutive
$\forall d \in D, \forall g \in G, \forall c \in C, \forall r \in R_g, \forall t \in \{0, \ldots, h-2\}$,

$$x_{d,0,g,c,r} - x_{d,t+1,g,c,r} \leq 0. \tag{5}$$

$\forall d \in D, \forall g \in G, \forall c \in C_g, \forall r \in R_g, \forall p \in P, \forall t \in \{2, \ldots, h\}, t+j \leq 6$,

$$-x_{d,p,g,c,r} + x_{d,p+1,g,c,r} - x_{d,p+t,g,c,r} \leq 0. \tag{6}$$

5. Every teaching period of course c is taught in the same classroom (optional):
$\forall d \in D, \forall c \in C, \forall r \in R_g$,

$$p_c * \sum_{d \in D} \sum_{g \in G_c} x_{d,p,g,c,r} \leq \sum_{d \in D} \sum_{p \in P} \sum_{g \in G_c} x_{d,p,g,c,r}. \tag{7}$$

where p_c is the total number of periods of course c.

6. The constraint that makes sure 2 sessions of a course is not on the same day
$\forall g \in G, \forall h \in H_c, \forall r \in R_g, \forall d \in D$,

$$\sum_{h \in H_c} \sum_{u \in U_c} y_{d,u,g,h,c,r} \leq 1. \tag{8}$$

7. The constraint makes sure h consecutive periods of a session of a course are assigned fully on a given day
$\forall d \in D, \forall g \in G, \forall c \in C_g, \forall r \in R_g$,

$$\sum_{p \in P} x_{d,p,g,c,r} - \sum_{h \in H_c} \sum_{u \in U_c} (y_{d,u,g,h,c,r} * h) = 0. \tag{9}$$

8. The constraint makes sure a student study most one course at a time (because a student can be included in many large-sized groups)
We need to generate a data set named G_{in} from G_{big} and G_{small}. G_{in} contains all small-sized groups accompanied by the groups containing them.
$G_{big} = \{b_1, b_2, \ldots, b_i, \ldots, b_{|B|}\}$,
$G_{small} = \{s_1, s_2, \ldots, s_j \ldots, s_{|S|}\}$,
$G_{in} = \{s_j : b_i | s_j \subset b_i\}$.

$\forall d \in D, \forall p \in P, \forall s_j \in G_{small}$

$$\sum_{c_{s_j} \in C_{s_j}} \sum_{r_{s_j} \in R_{s_j}} x_{d,p,s_j,c_{s_j},r_{s_j}} + \sum_{b_i \in G_{in_{s_j}}} \sum_{c_{b_i} \in C_{b_i}} \sum_{r_{b_i} \in R_{b_i}} x_{d,p,b_i,c_{b_i},r_{b_i}} \leq 1 \tag{10}$$

9. The constraint which relates to variables w_{dr}:

$$w_{dr} \leq \sum_{p,g,c} x_{d,p,g,c,r} \leq P * w_{dr}, \tag{11}$$

where $P =$ the total number of periods in a week (ex: $P = 30$).

d. Objective Function for the Model:
Minimize the total number of classrooms used

$$\min \sum_{d,r} w_{dr}.$$

2.2   The Second Problem: Class-Lecturer Problem

After solving *the first problem*, we got the data set E containing all classes that were scheduled.

2.2.1   Data Sets

In this problem, the following data sets are required
1. T: A set of all periods that classes are assigned to. This set is generated from the output of the first problem
$T = \{dp\}$ where dp is the pth period on day d, ex: $T = \{00, 10, 20, 30, \ldots\}$.
2. E: set of all classes got from the output of the first problem
$E = \{e_1, e_2, \ldots, e_{|E|}\}$, ex: E = {'E0' : ['G1', 'C2', {'10','11'}], ... }
Class 'E0': Course C2 is taught to group G1 at the first and the second period on Thursday.
3. L: An available set of all lecturers in a department, $L = \{l_1, l_2, \ldots, l_{|L|}\}$.
ex: L = {L1: 'Ta Anh Son', L2: 'Le Chi Ngoc', ... }
4. C_l: An available set of all courses that lecturer l can teach
ex: C_l = {'L0': ['C0', 'C1'], 'L1': ['C2'],...}.
5. E_l: A set of all classes that can be taught by lecturer l. This set is generated from set C_l and set E
ex: E_l = {'L0': ['E1', 'E2', 'E3'], 'L1': ['E0'],...}.
6. T_e: A set of all periods in T accompanied by all the classes that are assigned to each period
ex: T_e = {'00': ['E1', 'E2', 'E3'], '10': ['E0', 'E4'],...}.
7. E_t: A set of all periods of every class $e \in E$
ex: E_t = {'E0' : ['10', '11'], 'E1': ['21', '22', '23', '32', '33'],...} .

2.2.2   Sets of Variables

There are two sets of variables.
+ Basic set of variables: Binary variables denoted by $x_{l,t,e}$ where $l \in L, t \in T, e \in E$,

$x_{l,t,e} = 1$ when lecturer l is assigned to teach class e at time t.
$x_{l,t,e} = 0$, otherwise.
+ Set of auxiliary variables: Binary variables denoted by w_l

$$w_l = \begin{cases} 1 \text{ if } \sum_{t,e} x_{l,t,e} > 0 & \text{(12a)} \\ \\ 0 \text{ if } \sum_{t,e} x_{l,t,e} = 0 & \text{(12b)} \end{cases}$$

where $\sum_{t,e} x_{l,t,e}$ is the total number of periods that lecturer l is assigned to teach
in a week, then (12a) refers to $w_l = 1$ when lecturer l is assigned to classes in a
week and (12b) refers to $w_l = 0$ when lecturer l is not assigned to any classes in
a week.

2.2.3　Constraints for the Model

1. At a time, a teacher is assigned to most one class
$\forall t \in T, \forall l \in L$,

$$\sum_{\substack{e \in T_e[e] \\ e \in E_l[l]}} x_{l,t,e} \leq 1 \quad \text{and} \quad \sum_{\substack{e \in T_e[e] \\ e \notin E_l[l]}} x_{l,t,e} = 0 \tag{13}$$

2. At a time, a class is assigned to exactly one lecturer

$$\forall t \in T, \forall e \in T_e[t], \sum_{l \in L_e[e]} x_{l,t,e} = 1 \tag{14}$$

3. All periods of a class are taught by exactly one teacher

$$\forall e \in E, \forall l \in L_e[e], \quad x_{[l,t_1,e]} = x_{[l,t_2,e]} = \ldots = x_{[l,t_i,e]} \tag{15}$$

where $t_1, t_2, \ldots, t_i \in E_t[e]$
4. The constraint which relates to variables w_l

$$A * w_l \leq \sum_{t,e} x_{l,t,e} \leq B * w_l \tag{16}$$

where A is the minimum number of periods that every lecturer is required,
and B is the maximum number of periods that every lecturer is required.

2.2.4　Objective Function for the Model
Minimize the number of lecturers

$$\min \sum_{l \in L} w_l$$

3   Model Implementation

The code is written in Python, using CPLEX 12.8 solver and running on Core i5-5200U 2.2 GHz, 16G RAM computer. We first implemented the model with small-sized data sets and got the result presented in Table 1 and Table 2.

Following this, we solved problems with medium-sized data sets of *School of Applied Mathematics and Informatics*, the sizes of problems are presented in Table 3 and Table 4.

Computational Results: Computing the first problem and the second problem with small-sized data sets requires only 3 min and 2 min respectively, while computing the first problem and the second problem with medium-sized data sets requires 15 min and 10 min respectively.

Table 1. Example result of the first problem

Code	Group	Sub	Type	Room	Per	Day
C0	TT01	Giai tich I	BT	D6-301	1, 2, 3	Mon
C1	TT02	Giai tich I	BT	D6-304	1, 2, 3	Mon
C2	TT01	Giai tich II	BT	D6-301	4, 5, 6	Mon
C3	TT02	Giai tich II	BT	D6-304	4, 5, 6	Mon
C4	TT01, TT02	Giai tich II	LT	D9-204	4, 5	Tue
C4	TT01, TT02	Giai tich II	LT	D9-204	1, 2	Wed
C5	TT01, TT02	Giai tich I	LT	D9-204	3, 4, 5, 6	Wed
C6	TT01, TT02, HTTTQL	Toan roi rac	LT+BT	D5-103	1, 2, 3	Thur
C6	TT01, TT02, HTTTQL	Toan roi rac	LT+BT	D5-103	5, 6	Fri

Table 2. Example result of the second problem.

Code	Group	Sub	Type	Lecturer	Per	Day
C0	TT01	Giai tich I	BT	V.M.Phuong	1, 2, 3	Mon
C1	TT02	Giai tich I	BT	N.X.Truong	1, 2, 3	Mon
C2	TT01	Giai tich II	BT	N.T.T Huong	4, 5, 6	Mon
C3	TT02	Giai tich II	BT	N.X.Truong	4, 5, 6	Mon
C4	TT01, TT02	Giai tich II	LT	N.T.T.Huong	4, 5	Tue
C4	TT01, TT02	Giai tich II	LT	N.T.T.Huong	1, 2	Wed
C5	TT01, TT02	Giai tich I	LT	N.X.Truong	3, 4, 5, 6	Wed
C6	TT01, TT02, HTTTQL	Toan roi rac	LT+BT	N.T.T.Huong	1, 2, 3	Thur
C6	TT01, TT02, HTTTQL	Toan roi rac	LT+BT	N.T.T.Huong	5, 6	Fri

Table 3. Size of the first problem

	Number of courses	Number of groups	Number of rooms
Problem size	32	103	53
	Number of rows	Number of columns	Number of non-zeros
Model size	36250	32815	222722

Table 4. Size of the second problem

	Number of lecturers	Number of classes	
Problem size	53	140	
	Number of rows	Number of columns	Number of non-zeros
Model size	1058	4873	19281

4   Conclusion

After implementing the model on real data sets of *School of Applied Mathematics and Informatics* we got a timetable that satisfies many requirements and rules. The model can be further developed with the addition of some soft constraints such as constraints regarding classroom changeovers of students in a day or constraints regarding pre-assignments. We also plan to apply genetic algorithms or other heuristic approaches to this model for the sake of efficiency. In the future, after improvements, this model can be used to solve the course timetabling problem at HUST and other universities.

References

1. Wolsey, L.A.: Integer Programming (1998)
2. Daskalaki, S., Birbas, T., Housos, E.: An integer programming for a case study in university timetabling. Eur. J. Oper. Res. **153**, 117–135 (2004)
3. Bruke, E.K., Petrovic, S.: Recent research directions in automated timetabling. Eur. J. Oper. Res. **140**, 266–280 (2002)
4. Welsh, D.J.A., Powell, M.B.: An upper bound to the chromatic number of a graph and its application to timetabling problem. Comput. J. **10**, 85–86 (1967)
5. Tripathy, A.: School timetabling - a case in large binary integer linear-programming. Manag. Sci. **30**, 1473–1489 (1984)
6. Man, K.F., Tang, K.S., Kwong, S.: Genetic algorithms: concepts and applications. IEEE Trans. Ind. Electron. **43**, 519–534 (1996)
7. IBM CPLEX Optimizer. https://www.ibm.com/analytics/cplex-optimizer

Real-Time Inference Approach Based on Gateway-Centric Edge Computing for Intelligent Services

Wenquan Jin[1], Vijender Kumar Solanki[2], Anh Ngoc Le[3], and Dohyeun Kim[4(✉)]

[1] Big Data Research Center, Jeju National University, Jeju 63243, Republic of Korea
wenquan.jin@jejunu.ac.kr
[2] CMR Institute of Technology, Hyderabad, TS, India
spesinfo@yahoo.com
[3] Electric Power University, Hanoi, Vietnam
anhngoc@epu.edu.vn
[4] Department of Computer Engineering, Jeju National University, Jeju 63243, Republic of Korea
kimdh@jejunu.ac.kr

Abstract. For providing intelligent services in the network edge, in this paper, we propose a real-time inference approach base on gateway-centric edge computing. The edge computing is deployed on the edge gateway that operates intelligent services in the network entry. The intelligent services are provided based on the inference approaches that are included in the edge gateway. The key component is an intelligent control algorithm based on deep learning. The training of the model is offloaded to the high-performance computing machine, and the resulting inference model is deployed on the edge gateway, which closes the control loop with the IoT device. The inference model is derived by user data and learning model through training the user data on the learning model. Based on deploying the inference model to the edge gateway, the intelligent approach is enabled close to the environment where the data is generated. Therefore, the edge computing architecture reduces the latency to get the control factor for updating the environment based on the real-time sensing data.

Keywords: Edge computing · Computation offloading · Deep learning · Artificial intelligence · Gateway

1 Introduction

According to the paradigm of edge computing, the computing ability including computation and storage is enabled to be deployed in the edge of networks to provide various functions to devices for collecting data and updating environmental parameters based on sensors and actuators. Edge computing brings heterogeneous solutions such as management, intelligent approaches, proxies, and autonomous control mechanisms to the network edge for providing rich services using sufficient computing ability [1–4]. In the constrained environment, the limited power supply, processor, storage, and

D.-T. Tran et al. (Eds.): ICISN 2021, LNNS 243, pp. 355–361, 2021.
https://doi.org/10.1007/978-981-16-2094-2_44

network capability are required for developing most Internet of Things (IoT) devices [5–7]. Therefore, edge computing overcomes the limitation of IoT devices to support sufficient computing ability close to the environment. Based on edge computing, intelligent services are enabled to be deployed in the environment to control the actuators based on sensing data in real-time. Using a volume of user data, deep learning derives the inference model for providing an intelligent approach. However, the training process takes time in deep learning [8]. Therefore, offloading the training model with user data to the cloud server and deploying the inference model to the network edge enables the operation of intelligent services in real-time.

This paper proposes a real-time inference approach base on gateway-centric edge computing. The edge gateway provides intelligent services based on the inference model that is derived from the learning model. The learning model is offloaded to the cloud server that provides the inference model based on the user data. Based on deploying the inference model to the edge gateway, the intelligent approach is enabled close to the environment where the data is generated. Therefore, the edge computing architecture reduces the latency to get the control factor for updating the environment based on the real-time sensing data.

The remainder of this paper is organized as follows. The proposed intelligent edge computing architecture is introduced in Sect. 2. The implementation details of the proposed intelligent edge computing is presented in Sect. 3. Finally, we conclude this paper in Sect. 4.

2 Proposed Intelligent Edge Computing Architecture

We propose intelligent edge computing for providing real-time intelligent services based on the edge gateway.

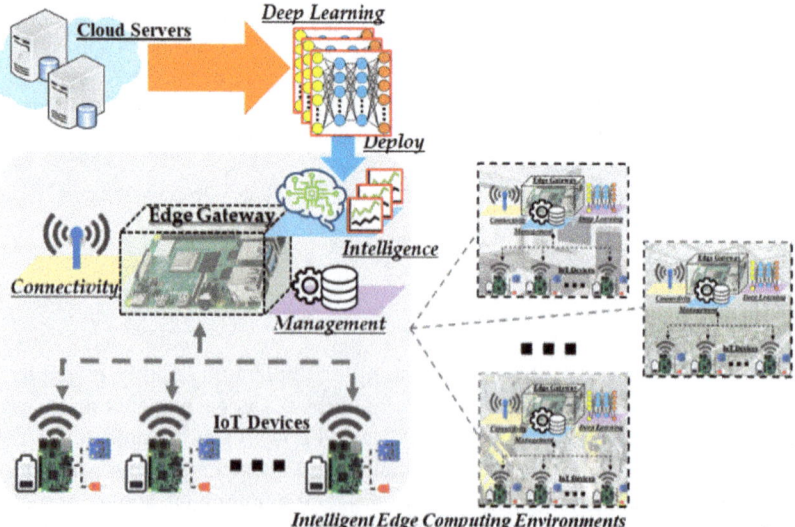

Fig. 1. Proposed intelligent edge computing architecture.

As shown in Fig. 1, the proposed edge gateway is comprised of functionalities of connectivity, management, and intelligence. For deploying the intelligent edge computing in the environment, the edge gateway provides intelligent approaches through offloading the inference model to the network edge.

The functions of the proposed edge computing are separated into three layers including client, edge and device layers. In the device layer, IoT devices are included for collecting data from the environment and update the environmental parameters through controlling actuators. The IoT device equips with sensors and actuators that are represented in the IoT device by resources. The edge layer includes edge server and edge gateway to distribute computing operating IoT devices based on intelligent approaches in the edge of networks. The deep learning models are designed to generate a network that outputs the results by passing the inputs between nodes. The client layer includes the web client that is used for providing content to the users and sending the command to the system by interacting with users and the edge gateway. Based on the web client, the client layer involves functions of User Interface (UI), management interface, control interface, and data visualization.

3   Implementation of Intelligent Edge Computing

As shown in Fig. 2, the proposed edge computing architecture is implemented through the web client, IoT device, and edge gateway.

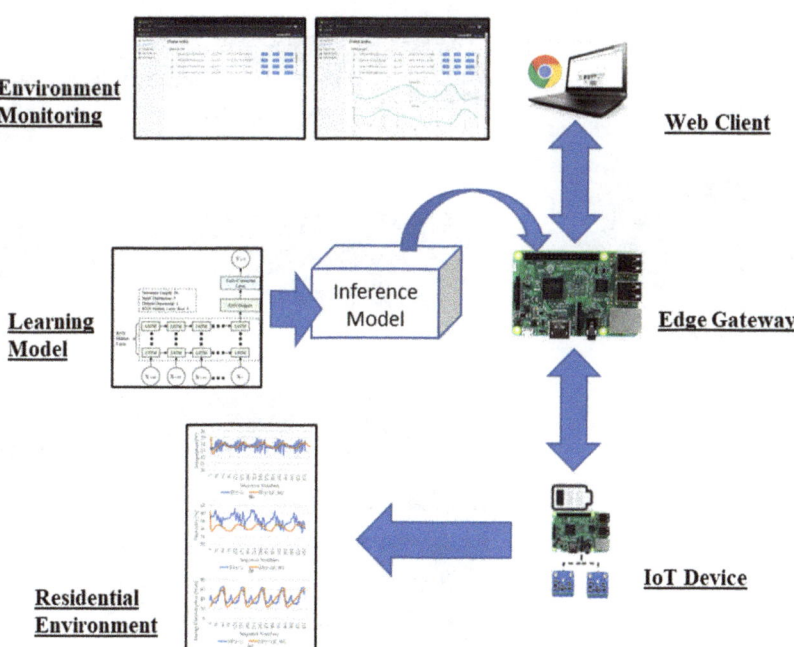

Fig. 2. Implementation details.

The web client provides the information through UIs to users as well as receiving the information from users and delivering it to the edge gateway. Based on the EdgeX, the edge gateway is implemented to include multiple microservices modules to provide management of devices, data, etc. Also, the intelligent services are provided by microservices modules that enable the intelligent approaches are operated and interacted with other functions in the edge gateway. The IoT device is an Android application that runs on the Android Thins platform based on Raspberry Pi 3. Jetty is a Java library for implementing the HTTP server on the IoT device to provide web services. For deriving the inference models, TensorFlow 2.0 is used for training the user data in the cloud server.

Figure 3 shows the implementation result of the proposed edge gateway that is comprised of the client service provider, intelligent service provider, rules engine, EdgeX core, and device proxy. The edge gateway runs the Ubuntu on the Raspberry Pi 4 that provides 4 GB memory for operating the microservices. The inference model is deployed in the intelligent service provider that provides the intelligent function through the microservices. The microservices are consumed by the rules engine when the deployed rules are triggered.

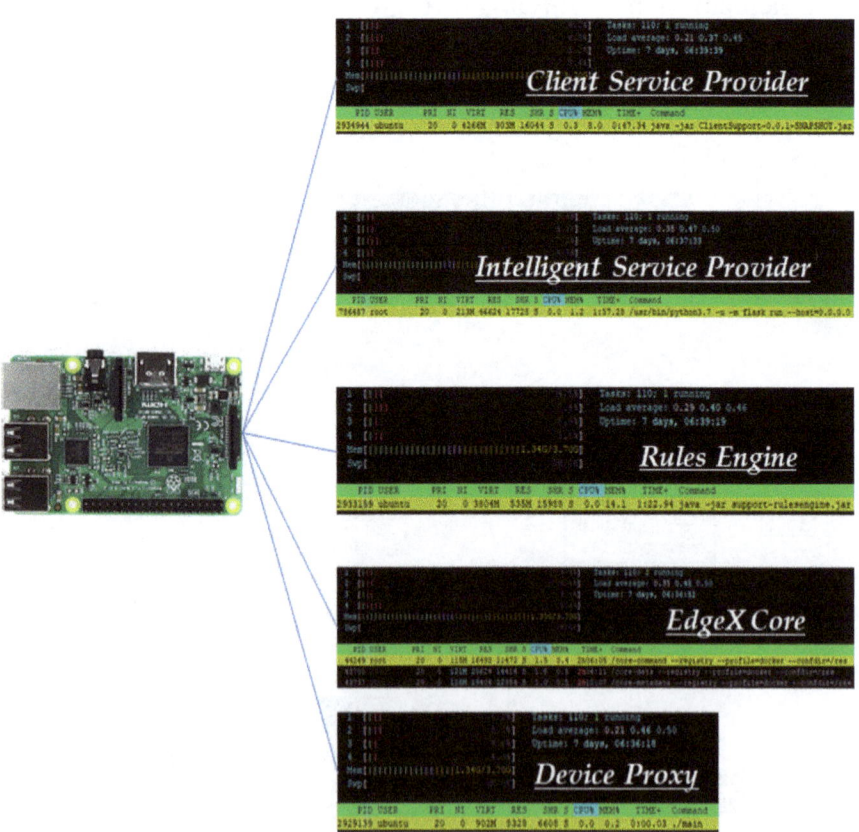

Fig. 3. Implementation result of edge gateway.

Figure 4 shows the implementation inference model that is derived by the learning model using the building user data for a building environment. The user data is collected by the Oak Ridge National Laboratory (ORNL) and the location in Campbell, USA. The size of the data is 35040 rows including indoor and outdoor temperature and power that affects the temperature and indoor temperature and humidity. Data is collected at 15 min intervals from 00:00 on October 1, 2013 to 23:45 on December 31, 2013, and from 00:00 on January 1, 2014 to 23:45 on September 30, 2014. The collected indoor temperature data is an environment created by operating various home appliances to meet the indoor environment desired by the user. Therefore, this indoor temperature does not decrease much even if the value of the outdoor temperature drops to 0 degrees. Indoor humidity data is also an environment created through user appliances. The heater is installed in the room where the temperature and humidity are collected, so this heater affects this indoor environment. The temperature and humidity in this environment change according to the power consumption of the heater, and the user adjusts the heater to create the desired environment.

TS, IT, IH, OT and OH stand for time sequence, indoor temperature, indoor humidity, outdoor temperature and outdoor humidity that are used for input of the neural network and EC is the heater energy consumption that is used for the output of the neural network. The learning model is implemented based on TensorFlow 2.0 for the smart heater learning model that derives the inference model. The inference model is formed by the file ai_model.tflite that is used for providing the intelligent service in the edge gateway.

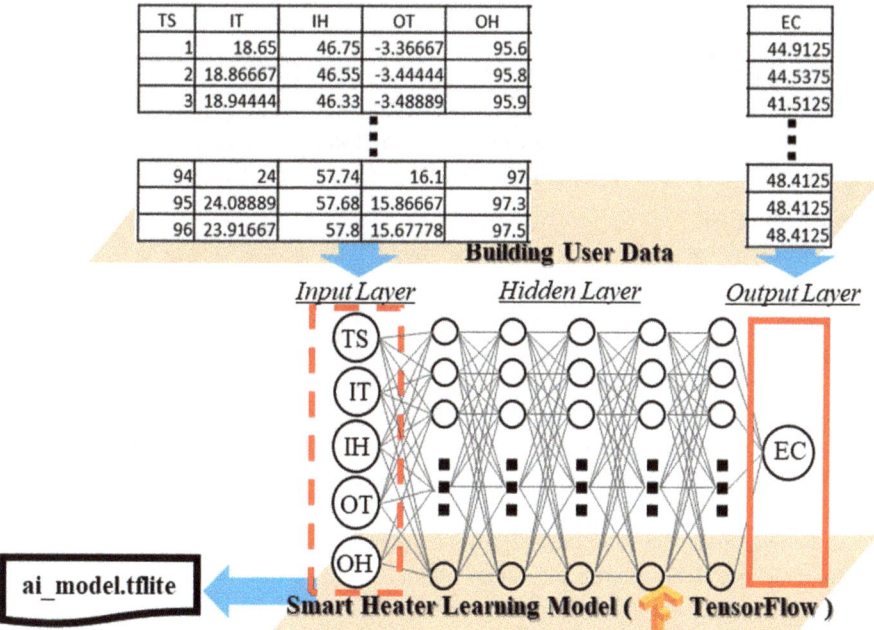

Fig. 4. Implementation of inference model based on learning model using user data.

4   Experiment Results

For evaluating the proposed edge computing, we collected the memory usage of the edge gateway. The edge gateway runs an Ubuntu OS on a Raspberry Pi that is equipped with 4 GB memory. The edge gateway includes client support provider, intelligent service provider, rules engine, ExgeX core and device proxy to provide edge computing services. The detected memory size of the device is 3,884,376 kb, in which, the running processes take 1,499,948 kb. The experiment results show that the edge gateway is sufficient to operate the proposed intelligent approach in the network edge for providing intelligent services (Fig. 5).

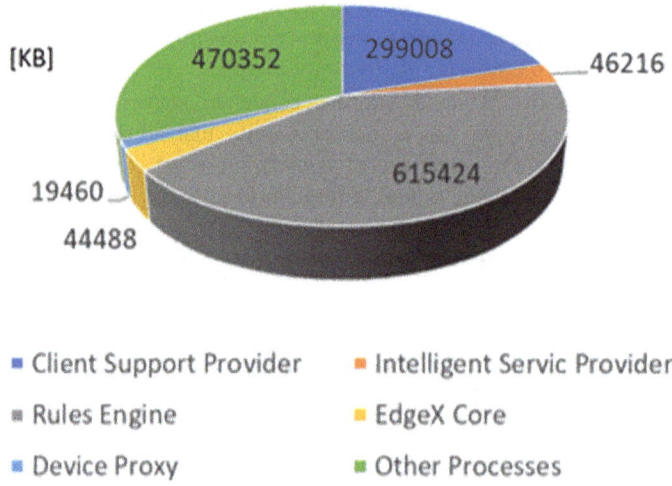

Fig. 5.   Proposed edge gateway memory usage.

5   Conclusions

We proposed the real-time inference approach which is based on deploying the edge gateway in the entry of the network edge for providing intelligent services in the environment where the data is generated. Deep learning is trained in the cloud server using the user data and derives the inference model that is deployed in the edge gateway to provide intelligent services. Multiple inference models are enabled to be deployed to the edge gateway based on intelligent service providers. The intelligent service provider is implemented based on the microservices module that enables the deployment of intelligent services to be independent from other functions of the edge gateway. Therefore, the proposed edge gateway brings intelligent services close to the environment based on inference models.

Acknowledgement. This work was supported in part by the National Research Foundation of Korea (NRF) Grant funded by the Korean Government under Grant NRF-2019R1I1A1A01 062456, and in part by Basic Science Research Program through the National Research Foundation of Korea (NRF) funded by the Ministry of Education (2018R1D1A1A09082919).

References

1. Shi, W., Cao, J., Zhang, Q., Li, Y., Xu, L.: Edge computing: vision and challenges. IEEE Internet Things J. **3**(5), 637–646 (2016)
2. Shi, W., Dustdar, S.: The promise of edge computing. Computer **49**(5), 78–81 (2016)
3. Hong, C.-H., Varghese, B.: Resource management in fog/edge computing: a survey on architectures, infrastructure, and algorithms. ACM Comput. Surv. (CSUR) **52**(5), 1–37 (2019)
4. Khan, W.Z., Ahmed, E., Hakak, S., Yaqoob, I., Ahmed, A.: Edge computing: a survey. Fut. Gener. Comput. Syst. **97**, 219–235 (2019)
5. Want, R., Schilit, B.N., Jenson, S.: Enabling the Internet of Things. Computer **48**(1), 28–35 (2015)
6. Jin, W., Kim, D.: A sleep scheme based on MQ broker using subscribe/publish in IoT network. Int. J. Adv. Sci. Eng. Inf. Technol. **8**, 539–545 (2018)
7. Al-Fuqaha, A., Guizani, M., Mohammadi, M., Aledhari, M., Ayyash, M.: Internet of things: a survey on enabling technologies, protocols, and applications. IEEE Commun. Surv. Tutor. **17**(4), 2347–2376 (2015)
8. Hoi, S.C., Wang, J., Zhao, P.: Libol: a library for online learning algorithms. J. Mach. Learn. Res. **15**(1), 495–499 (2014)

Mannequin Modeling from 3D Scanning Data

Do Thi Thuy[1,2(✉)], Nguyen Thi Thuy Ngoc[2], Ngo Chi Trung[2],
and Nguyen Thi Le[1(✉)]

[1] Faculty of Garment Technology and Fashion Design,
Hanoi University of Industry, Hanoi, Vietnam
{thuy.dothi, le.nguyenthi}@haui.edu.vn
[2] School of Textile – Leather and Fashion, Hanoi University of Science
and Technology, Hanoi, Vietnam

Abstract. 3D apparel design on computers is largely based on the human body model. To create the apparel fitting, the human body model needs to be the same size and shape as the objects scanned. Many researchers have proposed a method to create a human body model from 3D scanning data. However, the verification, evaluation, and modeling methods of the human body to achieve consistency between the model and the scanned data have not been fully mentioned. A mannequin modeling method of the female torso from 3D scan data is presented in this paper. Rapidform software is used to model and measure the mannequin's horizontal slices. The verification and evaluation of generated models are analyzed by R software. As the results, the female torso mannequin model with errors less than ±0.5 mm of the points located on the model's surface is set up from horizontal, parallel, and apart 5 mm slices. Each slice needs to have 180 points on the contour.

Keywords: Human body model · 3D scanning · Mannequin model

1 Introduction

Modeling is to replace the original object with a model to receive information about the object by conducting experiments and calculations on the model [1]. A model redesign is a complete re-modeling of objects from 3D scanning data. When studying the human body, Chalie found that although the dimensions of the chest, waist, and hips of the person are the same, the shape of the chest, the distance between the top of the chest and the chest root line, length from shoulder to chest, distance between chest and waist varies significantly [2].

To digitize the human body for the modelling, the methods and technical of digitalization are constantly being improved, the 3D scanning systems are more effective, powerful and faster and cost less. However, when scanning the human body still need to consider the following factors: movement, the limiting on stature, size and weight; comfort during scanning; acceptable level of scanning technology; personal data protection.

To be able to model a 3D human body, Jun showed that model methods often follow directions: small polygons and patches, implicit surfaces, curves and B-spline surfaces [3]. Zhang et al. [4] modeled a 3D human body using a triangular mesh.

D.-T. Tran et al. (Eds.): ICISN 2021, LNNS 243, pp. 362–372, 2021.
https://doi.org/10.1007/978-981-16-2094-2_45

Information about the feature points is needed for automatically creating 3D clothes. To automatically identify feature points, this study uses a cross-sectional plane to cut the body from head to legs to obtain a series of cross-sections. Implicit surfaces are used in the design and animation of 3D objects, because of their ability to smooth and deform 3D model objects. Douros et al. [5] combined implicit surface techniques and spline techniques, to enhance the ability to model implicit surfaces in fact, increasing display speed and effective control.

The B-spline curve and surface method, Hsiao and Chen reconstructed the surface of the 3D mannequin based on characteristic curves [6] and B-spline surfaces. The continuity between connected B-spline surfaces is governed by tangent vector adjustment methods.

Seo et al. [7] studied the method of using size tables to allow users to create virtual models by input some dimension parameters. This is especially useful for remote design through the web interface. Xuyuan et al. collected data of 3D human body and anthropometric analysis of human body model to construct models from these data and analysis [8]. To describe and represent the complex smooth human body surface, the support of anthropometric points and anthropometric lines on the body is needed.

In short, there are many different methods to create a model of the human body. The researchers strive to have the model created to meet the aim and reflect the characteristics of the human body. However, the verification, evaluation, and method of modeling the human body to achieve consistency between the model and the scanned data have not been fully mentioned by the researchers. Therefore, in order to ensure the size, shape, and posture of the human body, overcome a number of factors related to 3D scanning, a method of redesigning the female body model to be consistent with the 3D scanning data is presented in this paper. Rapidform software is used to model and measure the mannequin's horizontal slices. The generated model is validated and evaluated.

2 Methods

The medium-sized female body mannequin (Fig. 1) was selected according to the national standard TCVN 5782:2009 for this experimental study. Use the Artec Eva scanner to scan the mannequin. Rapidform XOR3 software is used to process scanned data, create models from 3D scanning data, and measure the dimensions of mannequin slices. The verification and evaluation of generated models are analyzed by R software.

2.1 3D Mannequin Scanning

Determine the Anthropometric Points on the Mannequin

The determination of anthropometric markers and human body landmarks is important, affecting the accuracy and consistency of data collected between scans. The landmarks are taken according to the international standard ISO 8559-1: 2017 (E) and positioned using a marker. The landmarks to support and identify in the process of data processing are the 7th vertebrae, pharynx, 2 points of shoulder, 2 points of the breast, 2 points of chest root line, 2 points of underarm, 2 points of the waist at the side, 2 points of protruding hip.

3D Scanning of Female Torso Mannequin

Artec Eva is used to creating a fast and precise 3D model. The device is suitable for scanning subjects with the size corresponding to the human body. Fast scanning device with high resolution, vivid colors allows many applications. Artec Eva's handheld 3D scanner products have fast sample scanning time, provide accurate data from 0.1 to 0.5 mm and sharp, flexible 360° in all angles and space, quality stable machine. Scanning data is collected in normal environmental conditions. The scanned data is processed using Rapidform XOR3 software. Objects after 3D scanning need to handle error data, smooth errors, point errors, errors in the combined scan area and fill the gap. Select the origin and adjust the coordinate axes for 3D data.

2.2   3D Mannequin Model of the Female Torso

Determine Main Cross-Sections on the Mannequin

The main cross-sections are horizontal slices, parallel to each other, and pass through the anthropological points on the mannequin. The anthropometric points on the mannequin determine in this study include the 7th vertebrae, pharynx, 2 points of shoulder, 2 points of the breast, 2 points of the waist at the side, 2 points of protruding hip.

Determine Sub-cross-Sections on the Mannequin

Sub-sections are horizontal slices, interspersed between and parallel to the main sections. In places where the body's curvature is complicated, such as the chest or the shoulder, more slices are needed to ensure the size and shape, the contour of the body after the model. At places with simple curvature, fewer slices be required. If fewer slices are used, the complex curved sections of the body will not be fully and accurately reproduced; if too many slices are used, it will take more time to process and the capacity to store data. To model the female torso mannequin from the neck to the hip, six main cross-sections through the anthropological points are used in this study. The sub-cross sections be started to take on both sides of the main cross-sections through the two breasts and the sub-cross-Sections 5 mm apart. There are a total of 115 sub-cross-sections interspersed between main cross-sections. Each slice is divided into 72 sections, each of which is equal to 5° around the center of the slice. Therefore, on the contour of each slice, there will be 72 points. If the sub-cross-sections overlap with the main cross-sections, the sub-cross-sections will be omitted in subsequent processing.

Construction a Surface that Passes Through the Contour of Cross-Sections

The intersection of each cross-section with the surface of the scanned data is a boundary curve. To model the mannequin, it is necessary to design a spline curve similar to the boundary of the scanned data. When the number of points increases, the spline will have a shape that is closer to the scanned data contour (Figs. 2 and 3). Depending on the location of the cross-section with the scanned data there will be the different shapes of boundary curves.

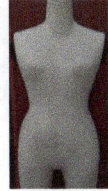

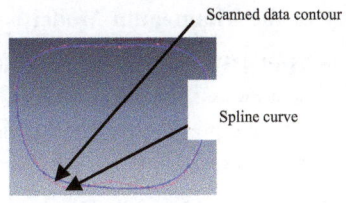

Scanned data contour

Spline curve

Fig. 1. Real mannequin   **Fig. 2.** The spline through 10 points   **Fig. 3.** The spline through 15 points

To construct the mannequin model surface, we can use the method of interpolation of boundary curves [9]. Suppose, there are two parameter curves $u_0(u)$ and $u_1(u)$ with $0 \leq u \leq 1$. The straight lines connect the two curves to form the piece of the face. Equation with a face has the form (Fig. 4):

$$r(u, v) = r_0(u) + v\{r_1(u) - r_0(u)\} \qquad 0 \leq u, v \leq 1$$

a)   b)

Fig. 4. The surface (a) and straight interpolated Taylo (b) [9]

Vectos in u direction, $r_1(u) - r_0(u)$ is vectos. Using vectos $t_0(u)$ of the boundary curve $r_0(u)$ to determine the equation of the plane:

$$r(u, v) = r_0(u) + vt_0(u)$$

Because the pairs of boundary curves are different and tangent to the diagonal boundary, it is possible to construct the Loft surface as follows:

$$r(u, v) = F_{0,3}(v)\, r_0(u) + F_{1,3}(v)\, t_0(u) + F_{2,3}(v)\, t_1(u) + F_{3,3}(v)\, r_1(u)$$

Therein:
$$\begin{array}{ll} F_{0,3}(v) = 1 - 3v^2 + 2v^2 & F_{1,3}(v) = v - 2v^2 + v^3 \\ F_{2,3}(v) = -v^2 + v^3 & F_{3,3}(v) = 3v^2 - 2v^3 \end{array}$$

$r_i(u)$ is the boundary curve with i = 0;1; $t_i(u)$ is diagonal tangent.

Assuming the mannequin model surface is symmetrically right and left. Construct a smooth surface passing the left half cross sections. Then, symmetrically to the right half and attach the two half surface of the mannequin model to form the body mannequin model.

2.3   Evaluate the Accuracy of the Mannequin Model

Evaluate the Accuracy of the Model Through the Cross-Section Contour Length
Measure the contour length of each cross-section of 3D scanning data and the contour length of each cross-section of a mannequin model corresponding. Compare the length difference between these two contours and t-test by software R.

Evaluate the Accuracy of the Model Through Dimensions from the Center to Points Belong to the Contour of a Cross-Section
Mannequin model results from 3D scanning data need to achieve consistency in the sizes from the center to the points located on the slices profile of the scanning data and the 3D model with the differences are not more than 0.5 mm.

To check the model's consistency with the scanned data, each slice of the scanned data and the model is divided into 72 sections, each with an angle of 5° around the center of the cross-section. Thus, the contour of each cross-section will have 72 points.

The distances from the center of the cross-section to each point on the contour of the 3D scan data and the mannequin model are measured. These two distances to evaluate the accuracy of the model redesign are compared. Using the t-test function in the R software to test the statistical significance of the distance difference from the center to each point on the contour of 3D scanning data and mannequin model and t test by using R software. If the test results show that the mannequin model has not achieved the required accuracy, it is necessary to divide each slice into smaller parts from the center, which means an increase in the number of points on the contour of each cross-section.

3   Results and Discussion

3.1   3D Scan Results for Female Torso Mannequin and Scan Data Processing

The result identifies the anthropometric signs and mannequin landmarks according to ISO 8559-1:2017(E) and marks the location of landmarks on the mannequin. 3D scanning results of female body mannequin with Artex scanner are shown in Fig. 5.

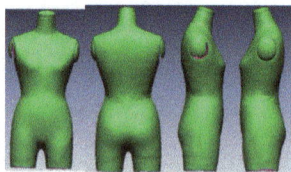

Fig. 5. Front, back and side images of the mannequin from scanned data

The data received after scanning need smooth handling of errors, spot errors, errors combining the scanning areas, and fill the scanning gap to have sufficient data for the next experimental steps. The coordinate origin chosen in this study is the intersection of the cross-section of the chest, the center vertical section and the vertical profile passing through the two axillary points.

3.2   Result of a 3D Female Torso Mannequin Model

The result of identifying main cross-sections through anthropological points: the 7th vertebrae, pharynx, 2 points of shoulder, 2 points of the breast, 2 points of the waist at the side, 2 points of the buttocks (Fig. 6).

   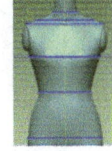   

Fig. 6. Main cross-sections   **Fig. 7.** Main and sub-cross-sections   **Fig. 8.** Main and sub- cross-section contours

Result of Determining Sub-Cross-Section: To model the female torso mannequin based on the cross-section contour, the distance between the cross-sections was empirically 5 mm. Then, the model will more fully reflect the shape of the body at each position, especially at places with complex curvature. There are 41 sub-cross-sections for the mannequin model of the body from the bust to the 7th vertebrae, 69 sub-cross-sections for the body model mannequin from the bust to buttocks, 9 sub-cross-sections were added to recreate the neck. The resulting of the main and sub-cross-sections (Figs. 6, 7 and 8). The result of creating a 3D mannequin model surface through main and sub-cross-sections is shown in Fig. 9.

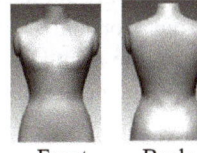

Front   Back   Side

Fig. 9. Mannequin surface extracted

3.3   Evaluation of the Accuracy of the Constructed Mannequin Model

Each cross-section of 3D scan data and mannequin model is divided into 72 sections, each with a 5° angle around the center of the cross-sections. Thus, each of the cross-sections contours has 72 points. The difference between the distance from the center of the cross-section to each point on the surface of the scanned mannequin and the distance from the center of the cross-section to the surface of the construct mannequin model is identified.

In this experiment, these distance are measured and analyzed in the left half body of scanned data and model. The result of the division of 3-dimensional scan data cross-section into 72 sections, each having a 5° angle around the center of the cross-section is shown in Fig. 10a. The result of cutting the 3D mannequin model into 72 parts, each part 5° has an angle equal to the rotation around the center of the cross-section is shown in Figs. 10b, c, d and e.

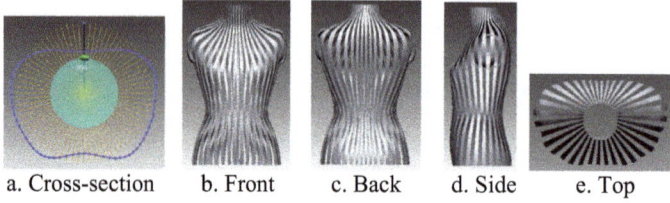

a. Cross-section   b. Front   c. Back   d. Side   e. Top

Fig. 10. Divide the 3D mannequin model into 72 parts

There are 10 cross-sections with the distance from the center to the contour less than -0.5 mm and 14 cross-sections with a difference greater than 0.5 mm. However, these cross-sections have p-values <0.05. There are 32 cross-sections with p-values >0.05, but the difference is less than ±0.5 mm.

Suppose, the distance from the center to the contour of the scanning data is x_1 and the 3D model is x_2. The difference between the distance from the center to the contour of the scanning data and the 3D model is $d_i = x_{1i} - x_{2i}$, iis the point on the contour, i = 1,2,3,…, n. Results of evaluating the accuracy of the mannequin model compared with 3-dimensional scan data (n = 72 points):$\overline{d}$ = −0.0559 mm; var(d) = 0.3142; standard errors(d) = 0.5605 mm; confidence interval 95%is −1.1545 mm to 1.0428 mm. This showed that the consistency between the scanned data and the 3D model does not meet the requirements and not uniform at the slices. To clarify the degree of inconsistency at points on each slice, determine the distance from the center to the contour of the 3D scan data and the mannequin model. The body is divided into vertical zones for the body half: center front, front side, rear side, and center back zone. The center front zone includes points 1 to 8, the front side zone includes points 9 to 19, the backside zone includes points 20 to 28, the center back zone includes points 29 to 37 of all cross-sections.

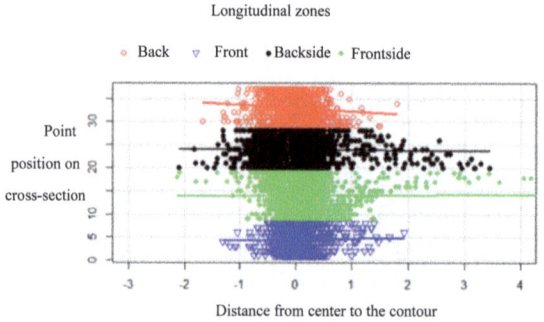

Fig. 11. Longitudinal zones chart (72 points)

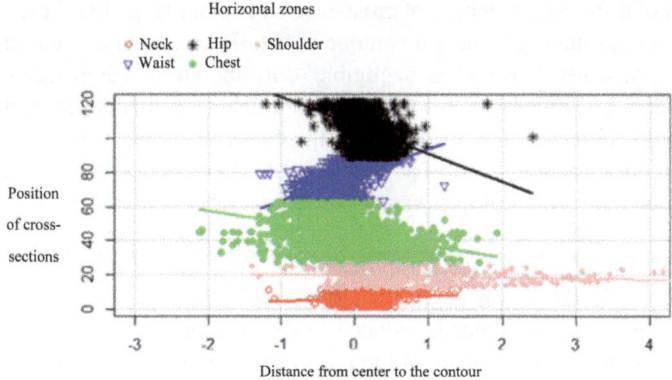

Fig. 12. Horizontal zones chart (72 points)

The longitudinal zones chart (Fig. 11) shows that there are many points of cross-section located in the zone between the center front, front side and backside, there are differences in dimensions from the center of the cross-section to the contour of the 3D scan data and the mannequin model is out of the range [−0.5, 0.5] mm.

To determine this difference in each region from neck to buttock, the body is divided into regions including neck, shoulders, chest, waist, and hip: The neck consists of cross-sections: 6 and 47 to 56; The shoulders include cross-sections: 5 and 7 and from 33 to 46; The chest consists of cross-sections: 4 and 10 to 32, from 57 to 68; The waist consists of cross-sections: 8 and 69 to 93; The hip consists of slices: 9 and 94 to 125.

The horizontal zones chart (Fig. 12) shows, at most points of the cross-sections, the distance difference is from the center of the cross-sections to the contour of the scan data and this mannequin model is in the range [−0.5, 0.5] mm. However, there are many positions at the shoulder and chest that have a difference of greater than 0.5 mm. Results show the difference in the length of the contours:

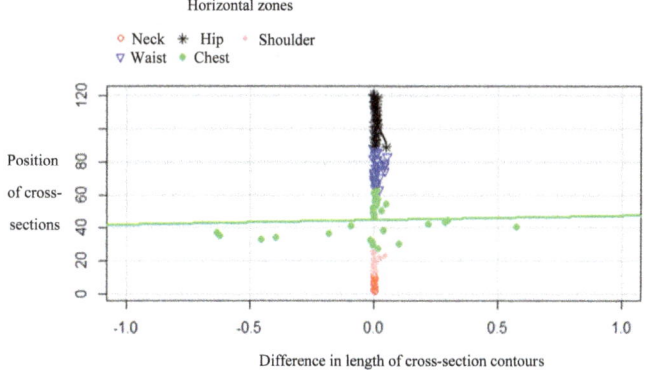

Fig. 13. Diagram of difference in length of cross-section contour (72 points)

Diagram of difference in length of cross-section contour (Fig. 13) shows, the length difference between the cross-section contour of the 3D scan data and the model at the neck, shoulders, waist, buttocks is negligible, only the chest region has some cross-section that is more different. Therefore, when combining model test results with p values and the differences, it was found that the consistency between the scanned data and the 3D model was not uniform at the chest and shoulder sections, so the accuracy of the designed mannequin model was not as satisfactory. Therefore, it is necessary to re-model the shoulders and chest, especially in the region between the center front side and backside .During the modeling process, the study experimented with a cross-section through the chest, the angle from the center was reduced to 4°, 3°, and 2 °. When the angle from the center is reduced to 2° corresponding to 180 points on the cross-section contour, the accuracy of the points on the cross-section contour when the model is achieved as required. Since then, modeling experiments and evaluation for the remaining slices continued to be performed. The result of a mannequin model with a central angle of 2° corresponding to 180 points on the contour of each cross-section contour.

The body is divided into vertical zones for the body half: center front, front side, rear side, and center back zone: The center front zone includes points 1 to 21, the front side zone includes points 22 to 46, the backside zone includes points 47 to 67, the center back zone includes points 68 to 91 of all cross-sections; The body is divided into regions including shoulders and chest: The shoulders include cross-sections: 5 and 7 and from 33 to 46; The chest consists of cross-sections: 4 and 10 to 32, from 57 to 68.

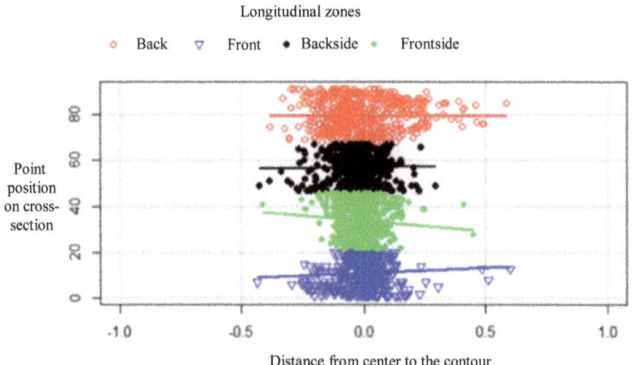

Fig.14. Longitudinal zones chart (180 points)

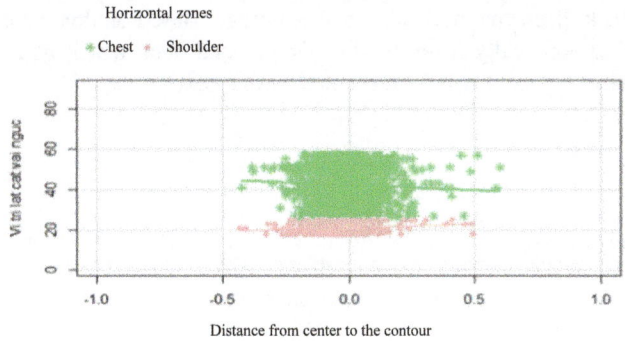

Fig. 15. Horizontal zones chart (180 points)

The longitudinal zones chart (Fig. 14) shows, most points of the cross-sections are in the center front, front side, rear side, and center back zone, with the difference in dimensions from the center of the cross-sections to the contour of the 3D scanning data and the mannequin model, is within [−0.5, 0.5] mm.

The horizontal zones chart (Fig. 15) shows, mostly at the points of the cross-sections, the difference in dimensions from the center of the cross-sections to the contour of the 3D scanning data and the mannequin model is within [−0.5, 0.5] mm.

Results of evaluating the accuracy of the mannequin model compared with 3-dimensional scan data (n = 180 points) using R software:

$\bar{d}$ = 0.0151 mm; var(x) = 0.0113; standard error s = 0.1061 mm; confidence interval 95% is −0.1929 mm to 0.2231 mm. This showed that the consistency between the scanned data and the 3D model meets the requirements.

Thus, the female torso mannequin model achieved an accuracy of ±0.5 mm compared to the 3D scan data, the contour of each cross-section is the shoulder and chest to go through 180 points. Two adjacent points create with the center of the cross-section at an angle of 2°.

4   Conclusions

This study created a 3D model with the accuracy required from the scan data. The method of creating the model has high reliability. Specifically, to obtain the female body model mannequin model from 3D scan data, the model has errors within ±0.5 mm of the points located on the model surface, it is necessary to divide the model into a horizontal cross-section parallel and 5 mm apart. Shoulder and chest cross-section should have at least 180 points on the contour of each cross-section. The neck, waist, and buttocks should have at least 72 points on each contour. The model results were evaluated for the consistency between the data using R software.

The determination of the number of points on each contour and the distance between the cross-sections is the basis for the body model and the shirt model to achieve consistency with the scanned data. This is also an important key and another method for data analysis, thereby developing a 3D shirt construction.

The drawback of this method is large data storage. Based on this method, when the model is built automatically from the directly scanned data, it will be faster and data stored less. These contents will be presented in the next articles.

References

1. NguyễnCôngHiền, NguyễnPhạmThục Anh: Môhìnhhóahệthốngvàmôphỏng. Nhàxuấtbản-KhoahọcvàkỹthuậtHàNội (2006)
2. Charlie C.L., Wang: Parameterization and parametric design of mannequins. Comput.-Aided Des. **37**, 83–98 (2005)
3. Jun, Z.: Study on 3D modeling and pattern making for upper garment. Doctoral Dissertation, Shinshu University (2017)
4. Zhang, D., Liu, Y., Wang, J., Li, J.: An integrated method of 3D garment design. J. Text. Inst. **109**(12), 1595–1605 (2018)
5. Douros, I., Dekker, L., Buxton, B.F.: An Improved Algorithm for Reconstruction of the Surface of the Human Body from 3D Scanner Data Using Local B-spline patches. Dept. of Computer Science University College London London WC1E 6BT, UK (1999)
6. Hsiao, S..-W.., Chen, R..-Q..: A study of surface reconstruction for 3D mannequins based on feature curves. Comput.-Aided Des. **45**(2013), 1426–1441 (2013)
7. Seo, H., Magnenat-Thalmann, N.: An automatic modeling of human bodies from sizing parameters. In: ACM SIGGRAPH 2003 Symposium on Interactive 3D Graphics 2003, pp. 19–26 (2003)
8. Tao, X., Bruniaux, P.: Toward advanced three-dimensional modeling of garment prototype from draping technique. Int. J. Clothing Sci. Technol. **25**(4), 266–283 (2013)
9. BùiQuýLực: Phươngphápxâydựngbềmặtcho cad/cam, nhàxuấtbảnKhoahọcvàKỹthuật (2006)

Development of 3D Breast Measurement System Using Structured Light for Bra Design

Luu Thi Hong Nhung[1,2(✉)], Nguyen Ngoc Tu[2], Nguyen Nhat Trinh[2],
Pham Minh Hieu[3], Nguyen Thi Dieu Linh[3], and Nguyen Thi Le[3(✉)]

[1] Hung Yen University of Technology and Education, Khoai Chau, Viet Nam
[2] Ha Noi University of Science and Technology, Hanoi, Viet Nam
[3] Hanoi University of Industry, Hanoi, Viet Nam
le.nguyenthi@haui.edu.vn

Abstract. Female breast size is an essential parameter to the designing, production, and selection of bra. To determine thoracic dimensions, designers use
manual measurement with a ruler or 3D measurement with measuring devices.
In Vietnam, the research and development of 3D measuring devices to serve the
design and manufacture of clothing is still new and limited. This article presents
the results of the study on designing and manufacturing the Scan3D MB2019
female breast measurement system. This device operates based on using the
structured light method, white light source (creating from 3 combined R-G-B
LED sources), which is not harmful to human eyes and skin, with a measurement accuracy of ± 0.4 mm, and measuring-scanning range 800×1000
600mm (length $\times$ width $\times$ depth).

The Scan3D MB2019 measuring device is designed to measure the shape and
size of the breast to meet following goals: low cost, fast measurement speed,
high accuracy, and convenience. The 3D breast measurement results are
equivalent to those of the manual measurement results. The study shows that the
Scan3D MB2019 system can be used to measure breast for design and manufacturing in the Garment and Fashion industry.

Keywords: Manual measurement · The structured light method · 3D
measurement

1 Introduction

Determining the dimension of the breast plays a crucial part in the design and manufacture of the bra. Currently, there are two methods of measuring breast size: Manual
measurement and 3D measurement. Zheng et al. [1] in the size of the 3D measurement
method compared to the manual measurement method is within the accepted range. As
a result, the application of Scan3D MB2019 scanning device and Geomagic Design X
measuring software in collecting breast size shape for the design process is suitable and
meets the requirements of the garment industry.

In recent years, the evolution of 3D scanning technology has made significant
changes, contributing to the design and sizing of bras. However, the scanning of
women's breasts still faces difficulties due to the limited equipment and inconsistent

© The Author(s), under exclusive license to Springer Nature Singapore Pte Ltd. 2021
D.-T. Tran et al. (Eds.): ICISN 2021, LNNS 243, pp. 373–385, 2021.
https://doi.org/10.1007/978-981-16-2094-2_46

measurement procedures. Pei et al. [2] investigated CAESAR 3D body scan data to classify breast shape. Seolyoung and Chun [3] gave the necessary parameters and several factors that influence the bra design. Zheng et al. [4] measured breast circumference, depth, and breast width. Studies were conducted to scan the breast when subjects wear thin, soft bras. However, measuring nude breasts is required for research about breast anthropology.

Breast measurement with a 3D scanning method has many advantages over manual measurement and can extract anthropometric data in a short time, without inconvenience for the tester. Several human body scanners have been developed as Artec TM Eva, 3D Scanner ImageTwin, Cyberware và SYMCAD, Neck-Midneck, CaMega 3D CF-1200. However, the 3D measurement method of the breast still meets difficulties due to complicated breast shape and unclear breast boundaries. Meanwhile, some handheld scanners are of sharpness, but make it difficult for the scanner. For handheld scanners, the scanner must have good scanning skills and move around the subjects, causing a lot of inconvenience and effort.

At the same time, the 3D scanning using structured light is stable and of great accuracy. Notwithstanding, 3D scanning equipment using the structured light method for the sewing industry is cumbersome and expensive. In the low-cost scanner system, the camera and the projector must move forward rather cumbersome, which takes up space [5]. The goal of this research is to design and manufacture the Scan3D MB2019 scanning system is a simplea, compact design that can be disassembled and folded, easy to move, highly accurate, and non-impacting human breast. This scanning system based on the 3D scanning method using Gray code and LineShifting. The study results contribute to the objective measurement of the bust size to serve the design and production of suitable bras.

2 Methods

2.1 Setup the 3D Scanning System for Measuring Female Breasts

2.1.1 System Building Based on Gray Code and LineShifting Methods

The 3D profile measurement method bases on the trigonometric principle [6]. Equipment for projecting 2D image samples designed to ensure every pixel encodes in color or intensity. When showing the 2D image samples on the surface of the object, the 3D profile of the object surface distorts projected sample images and is captured by the camera system.

Gray code image was created by using patterned and ordered gray code images. Suppose that you need to project a pattern that has dimensions W × H, and the projection creates for rows and columns.

First, it is necessary to calculate the number of N bits required to encode the pixels per row-column. Whereby, N value calculates by the formula:

$$N \geq \log 2(W), \text{ where W is the striated width.} \tag{1}$$

Assuming an image with a width of 16 pixels will need N = log2 (16) = 4 bits to encode those 16 pixels to project four images corresponding to 4 bits. Thus, each column pixel is encoded by a Gray code corresponding to the position of that column.

To create a LineShifting line, the method is to use a projection line with the width of 1 pixel, the distance between the lines is 6 or 8 pixels in x and y directions respectively.

Determining the Lighting Components from the Projector

In the 2D camera image obtained by the pattern projection, hypothesizes that the gray-level component analysis includes the magnitude of the light from the projector and the external environmental components [7]. In this study, to create a LineShifting line, the authors use a line projection with 1-pixel width, while the distance between the lines is 6 or 8 pixels in x and y directions respectively.

Suppose we have S = {I1, I2,..., Ik} is a set of images with a pattern, the illumination values at points p in the working area of the projector are determined as follows:

$$L_p^+ = max_{0 < 15k} I_i(p), \cdot L_p^- = min_{0 < 15k} I_i(p) \tag{2}$$

$$L_d(p) = \frac{L_p^+ - L_p^-}{1 - b}, \quad L_g(p) = 2\frac{L_p^- - bL_p^+}{1 - b^2} \tag{3}$$

In which: p is a pixel value of which the "direct" or "global" components are determined; L_p^+ is the maximum luminous value at the point p considering in a set of consecutive gray code projected images; L_p^- is the minimum illumination value at the point p considered in a set of successive gray code projected images; $L_d(p)$ is the illumination value determined to be a manual projector component; $L_g(p)$ is the relative illumination value of the medium and scattering light from other regions; $b \in [0, 1]$ is the value modeling the light information transmitted by turning on or off the pixels of the projector.

ON-OFF Bit Classification

The light composition of the image after pixel classification consists of three differentiation levels: ON is the pixel of the light fringes, OFF is the pixel of the dark texture, and Uncertain is the uncertain component of whether the pixel is ON or OFF, to this could be a shadow of a darken object [8]. For unexposed pixels, its gray density p contains only the indirect light component, and p is represented by the function below:

$$P = \begin{cases} d + i_{on} \text{ if pixel on} \\ i_{off} \text{ if pixel off} \end{cases} \tag{4}$$

Inside: Synthetic indirect light components: $i_{total} = i_{on} + i_{off}$

Light composition d and i_{total} calculated using Nayar's separation method.

To divide pixels [9] gives the following definition: d < m pixel as kind UNCERTAIN.

p < min(d, i_{total}) - > pixel as kind OFF; p > max(d, i_{total}) - > pixel as kind ON.

The opposite pixel as kind UNCERTAIN. With m is the minimum threshold defined to distinguish the shadow composition pixels of the object.

Determining the Vertices of the Light Lines

The LineShifting method is introduced to improve the Gray code by replacing the equal integer value of the coordinates of the camera points in pixels to sub-pixels so that the top of the projection fringes is not on the position of the integer pixel it has. The value is a real number so that the point cloud accuracy is better obtained. When using a detector irrespective of the illumination from the environment, group 4 or group 8 can be applied:

$$g_4(i) = f(i-2) + f(i-1) - f(i+1) - f(i+2) \tag{5}$$

$$g_8(i) = f(i-4) + f(i-3) + f(i-2) + f(i-1) - f(i+1) - f(i+2) - f(i+3) - f(i+4) \tag{6}$$

Thus, the position of the center sub-pixel vertex is determined by the linear interpolation that intersects the property line 0.

Determining the Distance of Pixels Using the Approximate Method

After obtaining the corresponding coordinates between the pixels on the camera and projector (the decoded image and the camera column coordinate image from Line Shift) together with the system parameters, then perform the final calculation of the acquisition. 3D is a determination of the pixel's z value, assuming the form of finding the intersection point of two lines when knowing the origin points and direction of the vector. In this calculation, the authors follow the approximation method for two lines intersecting each other.

Suppose p_{12} is the point to find with the coordinates x, y, and z where z is the distance from the pixel to be calculated to the acquisition system (Fig. 5). In which the points q_1 and q_2 are the origin pixels on the projector and camera respectively. The points p_1 and p_2 are decoded, with p1 already multiplied by the rotation matrix R and p2 are the result of the solution code after subtraction with translational matrix T and multiplication with rotation matrix R. The p_{12} value is calculated by the formula:

$$p_{12} = p_1 + \frac{1}{2}(p_2 - p_1),$$
$$p_1 = q_1 + \lambda_2 v_1, \quad p_2 = q_2 + \lambda_2 v_2 \tag{7}$$

In which, parameters λ_1, λ_2 are determined as below:

$$\begin{pmatrix} \lambda_1 \\ \lambda_2 \end{pmatrix} = \begin{pmatrix} \|v_1\|^2 & -v_1^0 v_2 \\ -v_2^1 v_1 & \|v_2\|^2 \end{pmatrix}^{-1} \begin{pmatrix} v_1^0(q_2-q_1) \\ v_2^0(q_1-q_2) \end{pmatrix} \tag{8}$$

2.1.2   Operation Principles of the System 3D Scanning

After the system has been calibrated with an error within the permissible range, then scan the sample. The computer creates a Gray projection pattern and shifts the phase, and then the projector sequentially projects these light patterns onto the surface within the subject's measuring area. The camera records the projected pattern image. The detailed structure of the Scan3D MB2019 device: Projector InFocus LightPro IN1146, Camera: Sca1400 - 30gm resolution 1392 × 1040 pixels, Lens 16 mm Computar Japan AOV, Laptop Dell Inspiron 15, Ram 8 GB, HDD: 500 GB, Turning tables, Measuring chamber (Fig. 1).

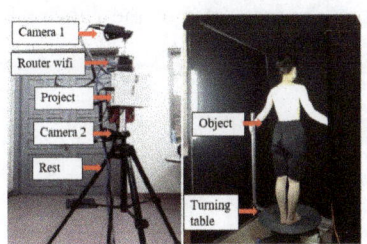

Fig. 1. Measuring system Scan3D MB2019

Measurement equipment parameters are as follows: Distance from camera center: 40 mm; Height above the ground of camera 1: 1540 mm; Height above the ground of camera 2: 940 mm; Maximum rack height: 1435 mm; Rack height when neatly folded: 570 mm; rack weight: 1.2 kg; Support load: 2 kg.

Measuring Software
The "NTU 3D Body Scanner" software for breast scanning control is programmed in C ++ and installed on a computer connected to the measuring device via USB, allowing scanning and turntable adjustments to follow automatic and semi-automatic modes. The camera's light capture modes can be customized to suit the measuring subject. 3D scanning allows collecting point cloud data in the form of a ".ply" file that includes X, Y, Z coordinate parameters, and monochrome color values in grayscale.

In particular, this software performs several outstanding features such as noise removal, automatic and manual coupling, smoothing, color change (see Fig. 2).

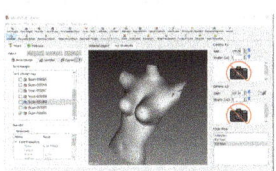

Fig. 2. 3D measurement control software interface

When scanning the pixel clouds, interference is required so that the software can receive and merge the clustered clouds. After testing, the overlap of 50% between the images, which corresponds to 10 rotation angles, gives the best results. 30% overlap leads to less accurate grafting results. Besides, it is impossible to merge the point clouds when the overlap is 0% because there is no commonality between pixel clouds.

Equipment Calibration

To calibrate the instrument for the $800 \times 1000 \times 600$ (mm) measuring area using a black and white chessboard, calculate the chessboard square size of 20 (mm) square, with a dimension of 12×15 cm. Calibrate at least 3 images in 3 different dimensions, each with a difference of about $30°–45°$ (see Fig. 3). The system is calibrated on the optical table to find the rotation matrix R and the T-shift vector between the two cameras, the internal calibration parameter of the camera Kc. The camera system is calibrated according to the OpenCV method.

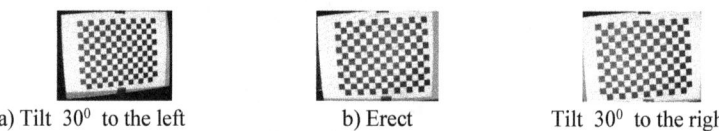

a) Tilt 30^0 to the left   b) Erect   Tilt 30^0 to the right

Fig. 3. Image of a calibrated chessboard at three different angles

2.2   3D Measurement of the Breast

Measuring Object: Model and Mannequin

With an aim to examining the measuring ability of the device, the research was carried out on the mannequin and 12 unmarried female students aged 18–24 years old. The subjects wore pants and no shirts to ensure measuring the actual bust size.

***Standing Posture*:** The mannequin was placed at rest when measured by two methods and did not change the state or position. The female subject is measured upright according to ISO 20685 in a vertical position, the head orients in the Frankfurt plane, while the long axes of the feet parallel to each other and 20 cm apart, arms outstretched to form an angle 15–20° with the side face of the body, elbows straight. To standardize body orientation, subjects would keep their heels in line with the footprints marked on the turntable. Besides, the women hold their breath for 2 s in each scan so as not to affect the errors.

Determining the Location of the Landmark: Use a circle of decals with a thickness of 0.1 mm and a diameter of 5 mm and stick to the marked positions on the chest area. The thoracic anthropometric measurement points, which describes in Table 2, are marked manually on the human body, while the virtual image shows in Fig. 4.

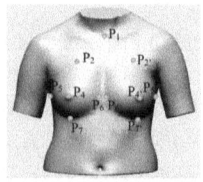

Fig. 4. Location of the markers on female breasts

Table 1. Determination of the location of measuring landmarks

No.	Location	Symbol	Description
1	Mid-neck point	P_1	The point is between the two collarbones, at the center of the anterior neck diameter
2	Point on breast	P_2	The highest point on the vertical breast line
3	The highest protrusion point of the breast	P_3	The highest protrusion of the breast that may be the nipple point or not the nipple point (if the breast is saggy)
4	Right nipple point	P_4	The point where the nipple protrudes outside the right breast
5	Left nipple point	P_4'	The point of the nipple protruding from the outside of the left breast
6	The outermost point of the right breast	P_5	The outermost point of the right breast which lies on the boundary line between the right breast and the body. At the same time is the intersection of the bust line and the lower edge of the breast
7	The outermost point of the left breast	P_5'	The outermost point of the left breast lies on the boundary line between the left breast and the body. At the same time is the intersection of the bust line and the lower edge of the breast
8	The innermost point of the right breast	P_6	The innermost point of the right breast and lies on the boundaries of the right breast and torso
9	The innermost point of the left breast	P_6'	The innermost point of the left breast and lies on the boundary of the left breast and torso
10	The midpoint of the right breast arch	P_7	The lowest midpoint of the bow below the right breast
11	The midpoint of the left breast arch	P_7'	The lowest midpoint of the bow below the left breast

Each dimension was measured three times. These values compare with those of 3D measurements on the designed Scan3D MB2019 system. Based on the markers determined on the body, the authors measure the bust size (Table 1).

3D Scanning Method: First of all, use the Scan3D MB2019 system to scan the breast. The subject stands on a turntable placed in a standard posture chamber, the distance from the projector lens to the center of the measuring object is 850 ± 50 mm. The height of the measuring device from the projector to the ground is 820 mm to ensure a full scan of the breast. The tester stands on an automatically computer-controlled turntable, with the speed of the turntable is 6 rpm. The total time to complete a complete scan sample is 2 min.

The Preparation of Measurement: Before breast measurements, 3D data on the upper body will remove from the noise, and unnecessary parts are removed to reduce the processing load. Scanned data obtained in the form of pixel clouds is brought back to Cartesian coordinates in perpendicular projection.

Measuring Procedure: 3D markers were selected with the landmarks marked in Table 1 to determine the size of the breast using Geomagic Design X software.

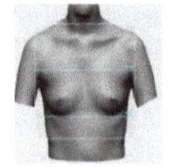 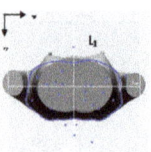

Fig. 5. Upper-bust girth

Upper-Bust Girth: On 3D body scan data of subjects, the upper bust girth (L_1) determines by creating a plane parallel to the body axis cut through the body at a horizontal position close to the armpits (Fig. 5). Further, the circumference around the outer points of the body and lying on the cutting plane is the circumference of the upper bust of the subject's body.

Bust Girth: Bust girth (L_2) is determined by the circumference of the cross-section by the horizontal plane perpendicular to the axis of the body, cutting through the highest protrusion point of the breast (P_3) (Fig. 6). In case the subject has the highest protrusion point coinciding with the nipple point, the cutting plane will pass through the nipple point. In the event of sagging breasts, the highest protrusion point of the breast does not coincide with the nipple point. Then, the bust size is calculated when the section passes through the highest protrusion point of the breast.

 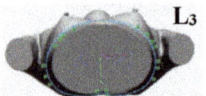

Fig. 6. Bust girth   **Fig. 7.** Under-bust girth

Under-Bust Girth: Under-bust girth (L_3) is determined by the circumference of the perpendicular plane to the body axis through the lowest point of the breast leg (P_7) (Fig. 7).

Breast Pain: One of the ideal breast evaluation parameters is to consider the ratio of the distance from the plane of the breast to the nipple point and from one nipple point to another (Fig. 8). Upper breast prolapse is calculated as the distance from the nipple point (P_4) to the plane cut through the dot on the thorax (P_1). Lower breast prolapse D_2 is the distance from the nipple point (P_4) to the plane cut through the point below the thorax (P_4).

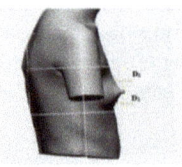 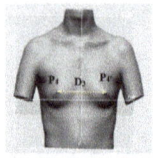

Fig. 8. Breast pain   **Fig. 9.** Nipple to nipple

Nipple to Nipple: The distance of the nipples made by measuring the distance from the left nipple point (P_4) to the right nipple point (P_4') (Fig. 9). In case the nipple points are not on the same horizontal plane, the distance from the breast calculates by the total distance from the nipple point (P_4) and (P_4') to the longitudinal line between the front body.

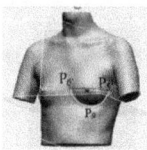

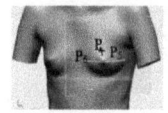

Fig. 10. Under breast arc length   **Fig. 11.** Breast arc measurement

Under Breast Arc Length: The curvature of the left pectoralis calculated by a curve from the innermost point of the left breast (P_6') to the outer dot of the left breast (P_5'). The arch of the thorax should be curvature from the innermost point of the right breast (P_6) to the outermost dot of the right breast (P_5) (see Fig. 10).

Breast Arc Measurement: The outer breast arch was calculated from the midpoint of the breast (P_5) to the nipple point (P_4), measured by the breast curve. The inner breast arch calculates from the score (P_6) to the nipple point, whereas breast supply calculates from scores P_5, P_4, P_6 (see Fig. 11).

   Results of measurement of 20 times directly and 20 times indirectly for the breast size of the mannequin were compared by ANOVA analysis to assess the difference. Dimensions for the thoracic part of 12 subjects, by manual method (3 times) and 3D scan method (3 times), were also analyzed. Next, the authors conducted a post-final analysis of R software to assess differences.

3   Results and Discussion

3.1   Measurement Equipment Calibration Results

After calibration, the system automatically adjusts the measuring range to the smallest error possible. The chessboard size obtained on the software, with an error of 0.01mm, is checked and compared with the actual chessboard size. Meanwhile, the device has calibrated with the chessboard. The results of the calibration parameters of the 3D scanning system shown in Fig. 6 with the following calibration errors:

$$
\begin{array}{cc}
\text{Camera 1} & \text{Camera 2} \\
K_c = \begin{bmatrix} 2688 & 0 & 6955 \\ 0 & 2688 & 5195 \\ 0 & 0 & 1 \end{bmatrix} & K_c = \begin{bmatrix} 2688 & 0 & 6955 \\ 0 & 2688 & 5195 \\ 0 & 0 & 1 \end{bmatrix} \\
K = \begin{bmatrix} -0.082078 & 0 & 0 & 0 & 0 \end{bmatrix} & K = \begin{bmatrix} 0.02038 & 0 & 0 & 0 & 0 \end{bmatrix}
\end{array}
$$

Rotation matrix between camera 1 and camera 2 R

$$R = \begin{bmatrix} 0.8723 & 0.0294 & 0.4879 \\ -0.0289 & 0.9995 & -0.0084 \\ -0.4880 & -0.0067 & 0.8728 \end{bmatrix}$$

The displacement matrix between camera 1 and camera 2 T:

$$T = \begin{bmatrix} -52.3 \\ -4.6246 \\ 14.6 \end{bmatrix}$$

Parameter calibration error: Camera $1 = 0.5507$; Camera $2 = 0.6238$; Stereo $= 0.6191$.

This calibration result allows the instrument to make accurate measurements within the permissible range for the Garment and Fashion industries.

3.2   Measured Results on the Mannequin

Results of 20 measurements of 8 dimensions on the breast part of the mannequin by the manual and 3D scan method were analyzed by ANOVA (Table 3).

Table 2. Comparison of the measurement results of the two methods

No.	Dimensions measured on the mannequin	Measure manually		Measure 3D	
		Medium (cm)	Standard deviation (cm)	Medium (cm)	Standard deviation (cm)
1	Upper-bust girth	84.55	0.32	85.56	0.28
2	Bust girth	86.55	0.25	86.56	0.33
3	Under-bust girth	73.63	0.27	73.38	0.31
4	Upper breast pain	11.58	0.15	11.47	0.27
5	Nipple to nipple	19.34	0.43	18.76	0.26
6	Sternum-nipple	22.73	0.11	22.74	0.14
7	Outer breast arc length	12.59	0.25	12.32	0.24
8	Under left breast arc length	18.73	0.21	19.55	0.11

Table 3. Results of analyzing variance comparing the measured dimensions on the mannequin

	Df	Sum Sq	Mean Sq	F value	Pr(>F)
Group	39	12	0.3	0	1
Residuals	280	318294	1136.8		

The results showed no statistically significant difference between twenty manual and twenty 3D measurements with eight dimensions measured on the mannequin's breast. The reason is that mannequin is a stationary object, so the error is small, and the results of the direct measurement and the 3D measurement are similar (p-value > 0.05).

3.3   Results Measured on the Human Body

ANOVA analysis results when comparing six groups (three groups of manual measurement and three groups of 3D measurement) on 12 subjects with 12 sizes on the breast areas are presented in Table 4.

Table 4. Results of analysis of variance when comparing the size measured on the human body

No.	Measurement size	Df	Sum Sq	Mean Sq	F value	Pr(>F)
1	Upper-bust girth	5	8	1.56	0.026	1
2	Bust girth	5	48.5	9.71	0.355	0.877
3	Under-bust girth	5	13.3	2.67	0.07	0.996
4	Nipple to nipple	5	0.24	0.049	0.015	1
5	Sternum - nipple	5	13.3	2.67	0.07	0.959
6	Upper breast pain	5	0.24	0.049	0.015	1
7	Lower breast prolapse	5	7.32	1.4631	2.377	0.0481*
8	Deep breast	5	16	3.199	0.551	0.737
9	Under right breast arc length	5	6.9	1.380	0.331	0.892
10	Under left breast arc length	5	13.55	2.709	0.576	0.718
11	Outer breast arc length	5	1.7	0.340	0.099	0.992
12	Inner breast arc length	5	1.21	0.2427	0.099	0.992

With P > 0.05, the above results show that the difference was not statistically significant between 6 value groups of the measured dimensions.

To ensure the equivalent results between groups, the values of the dimensions in the two groups of manual measurements and 3D measurements were analyzed by the TuykeyHSD post-deterministic test method.

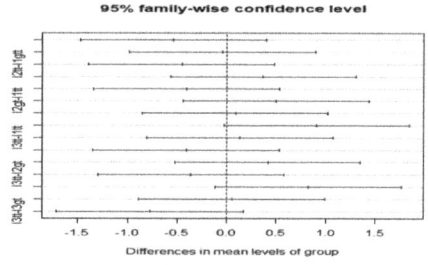

95% family-wise confidence level

Differences in mean levels of group

Fig. 12. Difference between 3 groups of manual measurements and 3 groups of 3D measurements with 95% confidence intervals

The difference between groups with a confidence interval of 95% is plotted using the plot function shown in Fig. 12. Thus, the values of the dimensions measured in six measurements are not of significant differences. However, several dimensions measured 3D through soft tissue tend to give higher values than manual measurements such as upper – bust girth or bust girth, especially with large bust size. This result, with the 95% confidence interval, shows on the graph of the difference between the manual and 3D measured value groups. When measuring the circumference using manual measurement, the tape measure against the body causes the soft tissues of the body to compress results in a smaller measured size compared to indirect measurement. In contrast, measuring thoracic dimensions of the plane circumference by a 3D measurement method, more accurate dimensions will be obtained while not being affected by soft tissue deformation.

The Scan3D MB2019 scanning system is a simple, compact design that can be disassembled and folded, easy to move. In particular, this 3D scanning system allows automatic transplantation of images obtained in a short time. Besides, it ensures the error of the body chest is 0.5 cm.

3.4   Conclusion

The virtual image of the female breast is similar to the simulated reality thanks to the application of 3D scanning technology. The measurement of the size of sensitive parts such as the breast is of great possibility without affecting the psychology of the measuring object. Scan3D MB2019 measuring system uses compact structured light, fast data acquisition, the measurement error of the system ±0.4 mm. The 3D breast size gives equivalent results and degree of accuracy to manual measurement. Further, the measurement time is fast and less affected by deformation when measuring. Therefore, 3D measurement technology allows measuring a wide range of breast sizes in large quantities, simple data storage.

However, it is necessary to have great scanning techniques as well as ensuring standard posture. The size difference between the manual measurement method and the 3D one appeared in some sizes such as upper-bust girth or under-bust girth, especially for subjects with large breasts. The difference in the size of the 3D measurement method compared to the manual measurement method is within the accepted range. As a result, the application of Scan3D MB2019 scanning device and Geomagic Design X measuring software in collecting breast size shape for the design process is suitable and meets the requirements of the garment industry.

References

1. Zheng, R., Yu, W., Fan, J.: Quantitative Analysis of Breast Shapes, pp. 113–119 (2014). https://doi.org/10.15221/10.113
2. Pei, J., Park, H., Ashdown, S.P.: Female breast shape categorization based on analysis of CAESAR 3D body scan data. Text. Res. J. **89**, 590–611 (2019)
3. Seolyoung, O., Chun, J.: New breast measurement technique and bra sizing system based on 3D body scan data. J. Ergon. Soc. Korea **34**, 377–399 (2015)

4. Zheng, R., Yu, W., Fan, J.: Development of a new Chinese bra sizing system based on breast anthropometric measurements. Int. J. Ind. Ergon. **37**, 697–705 (2007)
5. Rocchini, C., Cignoni, P., Montani, C., Pingi, P., Scopigno, R.: Istituto. a Low Cost 3D Scanner Based on Structured Light, vol. 20, no. 3. The Eurographics Association and Blackwell Publishers (2001)
6. Geng, J.: Structured-light 3D surface imaging: a tutorial. Adv. Opt. Photonics **3**, 128 (2011)
7. Nayar, S.K., Krishnan, G., Grossberg, M.D., Raskar, R.: Fast separation of manual and global components of a scene using high frequency illumination. In: ACM SIGGRAPH 2006 Papers, SIGGRAPH 2006, vol. 1, no. 212, pp. 935–44 (2006). https://doi.org/10.1145/1179352. 1141977
8. Xu, Y., Aliaga, D.G.: Robust pixel classification for 3D modeling with structured light. In: Proceedings - Graph. Interface, pp. 233–240 (2007). https://doi.org/10.1145/1268517.1268 556
9. Moreno, D., Taubin, G.: Simple, accurate, and robust projector-camera calibration. In: Proceedings - 2nd Jt. 3DIM/3DPVT Conference 3D Imaging, Modeling Processing Visuali zation & Transmission 3DIMPVT 2012, pp. 464–471 (2012). https://doi.org/10.1109/ 3DIMPVT.2012.77

Design Driver Sleep Warning System Through Image Recognition and Processing in Python, Dlib, and OpenCV

Tran Thi Hien[1,2(✉)], Qiaokang Liang[2], and Nguyen Thi Dieu Linh[3]

[1] Nam Dinh University of Technology Education, Nam Dinh 07100, Vietnam
tranthihien@hnu.edu.cn
[2] Hunan University, Changsha 410082, China
qiaokang@hnu.edu.cn
[3] Hanoi University of Industry, Hanoi, Vietnam
nguyen.linh@haui.edu.vn

Abstract. A large number of road accidents today are related to the drowsiness of the driver of a vehicle due to many causes such as fatigue, drunkenness. However, detecting driver drowsiness and warning is completely possible. In this paper, the authors propose a design pattern for driver drowsiness warning systems using Python, Dlib and OpenCV. Experimental results show that the system works in many different environmental conditions, meeting the requirements such as: Flexibility and convenience, high accuracy, processing time and warning fast enough for a real time request application.

Keywords: Drowsiness warning · OpenCV · Eyes closeness detection

1 Introduction

Today, along with the rapid development of the number of traffic vehicles, the number of traffic accidents is also increasing. One of the main causes of traffic accidents is the lack of driver concentration, due to fatigue or sleepiness. The installation of a warning system, reminders when driving is the top concern of the driver in particular as well as the transport industry in general.

Proposed solutions to detect drowsiness are usually based on the image of the driver's eyes closed or open, when it detects the driver's drowsiness, then the system will use the sound to alert driver. There have been many proposed methods, such as Dasgupta [1] that have used the number of closed eyes and the interval with each close eyes as a sign of sleepy. The circular Hough transform to detect the iris in an eye image to classify the eye as being open or closed is used in [2]. In [3, 4] using template matching method, in [5, 6] using local image features, and Hidden Markov Model in [7]. Teyeb uses Head Posture Estimation, which measures head tilt and specific time intervals to determine the driver's level of alertness [8].

In the article, the authors give a plan to identify, then study in depth the method of monitoring and warning the driver's sleepiness in the human face recognition application, by monitoring the state of closed/open eyes and yawning mouth of the driver. Since then, researching and building equipment that can be applied into practice in

Vietnam with the majority of means of passenger transport, long-distance goods transported easily by traffic accidents due to fatigue and driver's sleepiness.

2 Design System

The requirement for the system is to monitor and detect the driver's sleepiness during driving with high accuracy, real-time response, and compact.

The system is built with four main parts: Hardware, operating system, algorithms and display functions of the system (Fig. 1).

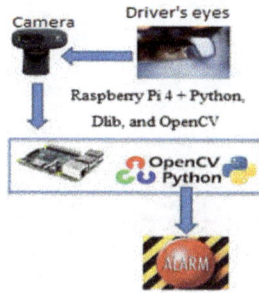

Fig. 1. System hardware structure

a) System Hardware Structure

- Central processing unit: Raspberry Pi 4.
- Webcam captures the image of the driver: Use the BlueOver T3200 with compact dimensions: 80 × 43 × 65 mm; shooting angle up to 60°; 30 fps shooting speed. Use a USB connection. No need to install drivers, compatible with all types of operating systems of PCs, laptops such as Windows, Ubuntu, MacOS... Using 5-layer HD lens, eliminating optical phenomenon causing image blur. Video resolution 640 × 480 pixels.
- The screen displays the results of the system: It is necessary to select a monitor with HDMI port to connect to the Raspberry.
- Speaker: alarms when the system detects the driver's sleepy state.
- Mouse and keyboard: Help the driver to control the system.

b) Ubuntu Operating System
The Ubuntu operating system is small in size, suitable for devices with limited memory of the Raspberry Pi 4, as well as possessing a huge application store. In addition, it also has the advantages of: Hardware requirements are not too high, high security, diverse and free application store, high compatibility, built-in necessary tools, and community massive support.

c) Drowsiness Detection Algorithm
In the study, the authors built algorithms based on the Python programming language to help identify and detect driver drowsiness. The main tools used include (Fig. 2):

- Dlib open source library: The algorithm that Dlib uses to extract image properties is HOG (Histogram of Oriented Gradients) and SVM (Support Vector Machine).
- Facial Landmark Recognition Algorithm:
  The mission of Facial Landmark is to locate the face in the photo and identify the points that make up the structure of the face. To perform these two tasks, the facial landmark recognition algorithm will determine 68 main points according to the coordinates (x, y) that make up the human face. Thereby distinguishing: mouth, right eyebrow, left eyebrow, right eye, left eye, nose, jaw.

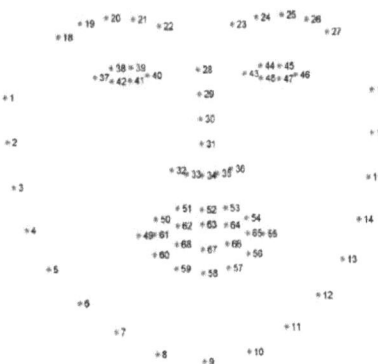

Fig. 2. 68 points on face

- Euclidean distance:

The distance between two points A and B is calculated according to the Pythagorean formula used by scipy library to calculate the Eucliden distance between the landmarks on the face.

d) Display Function of the System
This function displays the results of the OpenCV computer vision image processing on an open source basis...

3   An Algorithm to Detect Driver's Drowsiness

This section will present algorithm to detect driver's drowsiness, as follows:

- Process video input data: Collecting image data while driving by Webcam. Frame conversion using OpenCV's built-in function to convert input video from color image to multi-level grayscale to meet the processing speed of the system.
- Face detection: Use results from Dlib's algorithms combine with OpenCV to detect the driver's face.
- Extract the parts of the face: Using Dlib and OpenCV, identify 68 facial points from which extract facial parts, namely: 2 eyes and mouth

– Detection of drowsiness: Using Euclidean distance, calculate the distance of the coordinates of the points on the face to calculate the eye aspect ratio (EAR), mouth aspect rate (MAR).

Of the 68 facial points, each eye is denoted by 6 points with separate coordinates (Fig. 3)

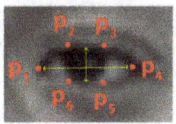

Fig. 3. Eye point coordinates

The formula for calculating EAR from point p_1 to p_6:

$$EAR = \frac{\|p_2 - p_6\| + \|p_3 - p_5\|}{2\|p_1 - p_4\|}$$

The results of the EAR calculation give us an approximate value that remains constant at the time of the eye opening but will rapidly decrease to 0 when the eye closing (Fig. 4).

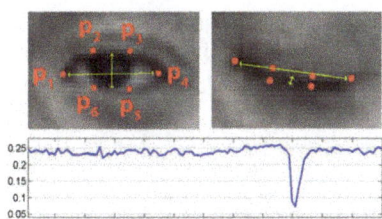

Fig. 4. Eye-frame rate values graph when eyes are closed and open

Looking at the figure above, this is the timed eye-frame rate for a video clip. The eye-frame rate here fluctuated around 0.25 then quickly dropped to 0 then increased again. That showed a blink of an eye had just taken place.

The mouth will be denoted by 12 distinct points (Fig. 5).

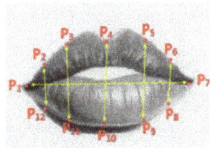

Fig. 5. The coordinates of the points on the mouth

Formula for calculating mouth frame rate

$$MAR = \frac{\|p_2 - p_{12}\| + \|p_3 - p_{11}\| + \|p_4 - p_{10}\| + \|p_5 - p_9\| + \|p_6 - p_8\|}{5\|p_1 - p_7\|}$$

Look at the picture below, this is the oral frame rate over time for a video clip. The oral frame rate here hovers around a 0.3 value when in the normal state then quickly rises to more than 1 and then decreases. This shows that the state of the mouth has changed from normal to open. So we can determine the driver's yawning state (Fig. 6).

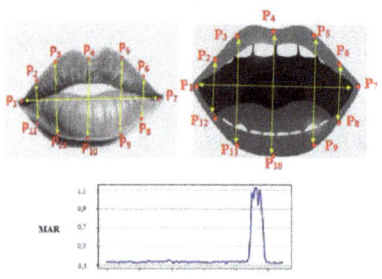

Fig. 6. Mouth frame rate variation chart when close and open

Set threshold standards, to check driver drowsiness: EAR allowed value: 0.25. MAR allowed values: 0.7.

– Warning: During driving, if the ratio of the eyes and mouth do not meet the above two rates. The system will set a timer to provide safety warnings if the ratio of the eyes and mouth.

4 Experiment and Evaluate the Results

a) Experiment
In order to evaluate our proposed method, we used image datasets that are incrementally difficult. We outlined the accuracy of face detection, eye detection, mouth detection and then classification of the eyes as open or closed, yawn mouth or not. The detection results of both the open/closed eyes and yawn mouth are shown in Figs. 7, 8, 9, 10, 11 and 12.

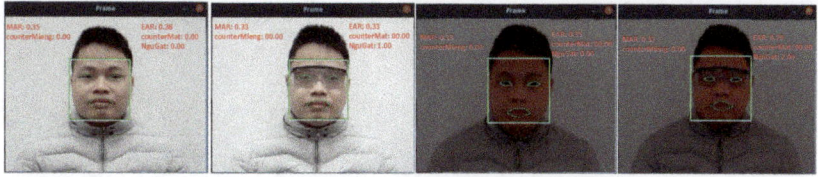

Fig. 7. Front face state, not wearing and wearing glasses, sufficient and underexposed

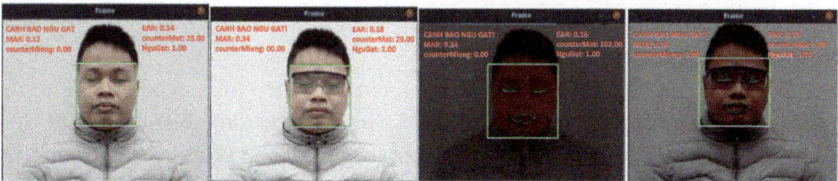

Fig. 8. Eyes closed for more than 1.5 s, not wearing and wearing glasses, sufficient and underexposed

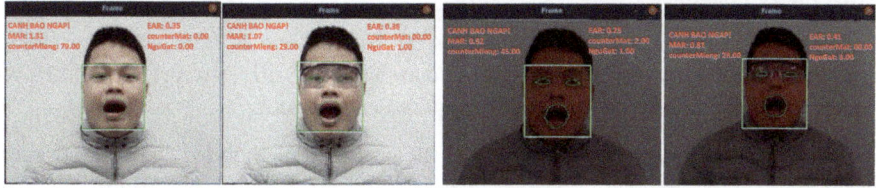

Fig. 9. The state of yawn mouth more than 1.5 s, not wearing and wearing glasses, sufficient and underexposed

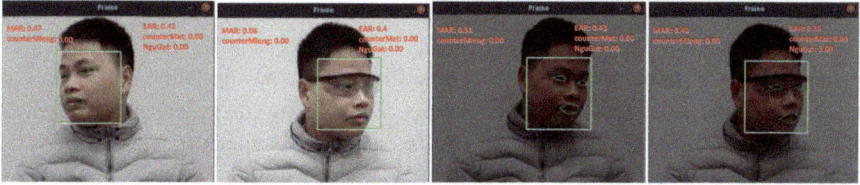

Fig. 10. The face is tilted at an angle of 30°, not wearing and wearing glasses, sufficient and underexposed

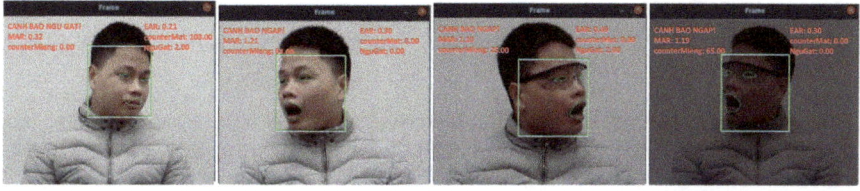

Fig. 11. Eyes closed or yawning, the face is tilted at an angle of 30°, not wearing and wearing glasses, sufficient and underexposed

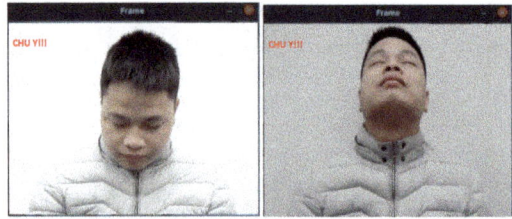

Fig. 12. State head down overslept or head up overslept make the system not recognize

b) Evaluate the Results

The proposed method was evaluated with a real-time video of someone driving a car. The video was recorded in the day time with variations in illumination due to differences in the direction of sunlight. The main purpose of the proposed algorithm is to classify eyes in the detected face and eye images.

- When the driver sitting right posture, the system's exact identification rate is as follows: Face is 97.5%; the eye is 88%; Mouth is 91%. So the system meets the requirements set out.
- When the driver sitting the face is tilted at an angle of 300, the system's exact identification rate is as follows: Face is 97.5%; the eye is 65%; Mouth is 88%. Results are affected by ghosting that occurs when the driver wears glasses. However, the system basically meets the requirements set out.

The above results show the feasibility of the proposed method for driver drowsiness detection. It should be noted, however, that these results were achieved during the day time. For night time, our proposed method can be used on top of face and eye detection algorithms that work during night time.

5   Conclusion

By combining the hardware is the Raspberry Pi 4 processor, the Dlib algorithms with the display function of OpenCV, we conduct design drowsiness warning system for drivers. Experimental results above the image of the driver under light conditions, tilt angle, wear different glasses. Synthesize and compare with the initial requirements of the system, shows: The system works in many different environmental conditions, meet the requirements set out such as: Flexible, convenient, high precision, processing time and alarms are fast enough for a real time request application.

References

1. Dasgupta, A., George, A., Happy, S.L., Routray, A.: A vision based system for monitoring the loss of attention in automotive drivers. IEEE Trans. Intell. Transp. Syst. **14**(4), 1825–1838 (2013)

2. Alioua, N., Amine, A., Rziza, M., Aboutajdine, D.: Eye state analysis using iris detection based on Circular Hough Transform. In: Proceedings of the 2011 International Conference on Multimedia Computing and Systems, ICMCS 2011, pp. 1–5. IEEE (2011)
3. Liu, X., Tan, X., Chen, S.: Eyes closeness detection using appearance based methods. In: Proceedings of the International Conference on Intelligent Information Processing, vol. 385, pp. 398–408. Springer, Berlin (2012)
4. Fa-deng, G., Min-xian, H.: Study on the detection of locomotive driver fatigue based on image. In: Proceedings of the 2010 2nd International Conference on Computer Engineering and Technology, pp. V7-612–V7-615, Chengdu, China (2010)
5. Zhou, L., Wang, H.: Open/closed eye recognition by local binary increasing intensity patterns. In: Proceedings of the 2011 IEEE 5th International Conference on Robotics, Automation and Mechatronics, RAM 2011, pp. 7–11. IEEE, China (2011)
6. Tafreshi, M., Fotouhi, A.M.: A fast and accurate algorithm for eye opening or closing detection based on local maximum vertical derivative pattern. Turkish J. Electr. Eng. Comput. Sci. 24(6), 5124–5134 (2016)
7. Qin, H., Liu, J., Hong, T.: An eye state identification method based on the embedded hidden Markov model. In: Proceedings of the 2012 IEEE International Conference on Vehicular Electronics and Safety, ICVES 2012, pp. 255–260. IEEE, Turkey (2012)
8. Teyeb, I., Jemai, O., Zaied, M., Amar, C.B.: A drowsy driver detection system based on a new method of head posture estimation. In: Proceedings of the International Conference on Intelligent Data Engineering and Automated Learning, pp. 362–369. Springer, Berlin (2014)

An Application of a Method of Weighting Assigning for Attributes Based on the Level of Different Body Structures to Improve the Artificial Neural Network for Diagnosing Hepatitis

Huynh Luong Nghia[1](✉), Cong-Doan Truong[2], Dinh Van Quang[1], Xuan Thu Do[3], and Anh Ngoc Le[1]

[1] Electric Power University, Hanoi, Vietnam
anhngoc@epu.edu.vn
[2] International School, Vietnam National University, Hanoi, Vietnam
doantc@isvnu.vn
[3] University of Transport Technology, Hanoi, Vietnam
thudx@utt.edu.vn

Abstract. Recently, a number of studies on hepatitis using artificial neural networks have attracted a lot of attention. Many methods have been applied, but the results are still limited in terms of the performance of converged network systems and accuracy. In this study, we applied an application of a method of weighting assigning for attributes based on the level of different body structures. Specifically, multi-layer perceptron (MLP) neural networks have been trained for two cases. The first case is that the input of the original features will not be changed. The second case is that the input of attributes corresponding to the levels of body structure was assigned weights. The results show that the quality parameters of the obtained network model in the second case have been significantly improved. To summarize, this approach based on a degree body structures is likely to be able to diagnose hepatitis and other diseases.

Keywords: Artificial neuron networks · Body structure level · Weighting assigning · Hepatitis

1 Introduction

Many studies using artificial neural networks in the classification problem have shown significantly successful results [1–3]. Multi-layer perceptron networks (MLP) have been used effectively to classify the pathology based on symptom information, medical scan images and test results [4] (Fig. 1).

Usually the inputs to this network are symptoms - pathological attributes recorded by different modes and the output as diagnostic results confirmed by the reality used to train the network [5–7]. In general, these networks gave acceptable results, but when the input was too large it could lead to reduced accuracy and increased processing times (slower convergence). To solve this problem, many solutions have been proposed

D.-T. Tran et al. (Eds.): ICISN 2021, LNNS 243, pp. 394–401, 2021.
https://doi.org/10.1007/978-981-16-2094-2_48

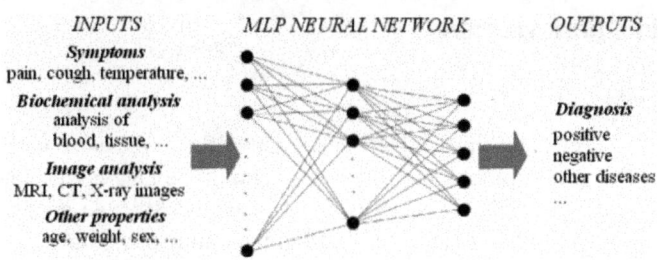

Fig. 1. Medical diagnosis using MLP neural network.

such as selecting the optimal structure of the artificial neural network, reducing the number of input property sizes, and choosing the appropriate lattice algorithm.

Also for this purpose, in [8–10] the paper has initially used a weighting method for the input attributes appropriate to the structural levels of the body for the artificial neural network to support breast cancer diagnosis. The results obtained are significant, especially for diseases that involve changes in the body's microstructure.

The idea of a weighting method corresponding to the body structure level comes from considering the composition of the network input attributes. Specifically, these input parameters/attributes - which are acquired by different biomedical measurement methods - will correspond to the 5 structural levels of the body such as: body $\rightarrow$ functional organs $\rightarrow$ tissue $\rightarrow$ cell $\rightarrow$ biological molecules, as classified in the following Table 1:

Table 1. Structural levels of the body and diagnostic methods.

Level	Structural levels	Diagnostic methods/devices
1	Body level	Pulse, temperature, blood pressure, ...
2	Level of functional organs	Functional diagnostic (Electrocardiogram, Electroencephalogram, Electromyography,...)
3	Tissue Level	CT, MRI, PET SPECT, Nano microscope, analytical test methods
4	Cellular Level: cell types	
5	Bio-Molecular Level	

Intuitively, we understand that the meaning of the symptoms (pathological signs) is different, namely the lower the level of structure, the higher the diagnostic value of the corresponding symptoms (signs), or according to experts, the greater their weight.

Thus, both to check the validity and effectiveness of the above methods and to improve the quality of the network diagnosing some hepatitis diseases, in the following section, we will build and examine the effectiveness of the neural network in turn artificially supported the diagnosis of hepatitis in two unassigned cases and weighted input attributes corresponding to the structural levels of the body.

2 Materials and Methods

2.1 Materials

To build Multi-Layer Perceptron (MLP) neural networks to diagnose pathologies and then improve the quality of the network by assigning weights corresponding to the structural levels of the body, it is necessary to establish suitable databases. In this paper, we select two databases for hepatitis from the online pathology database library of the University of Wisconsin [11].

The Hepatitis dataset was donor by G. Gong from the Carnegie-Mellon University. The data warehouse contained 19 characteristic parameters obtained by calculating and measuring from the characteristics of breast tissue collected from 155 patients. In the database, there are 155 samples, of which 123 are benign and 32 are malignant. The database used to train the network is denoted by Hepatitis Dataset including 2 matrices hepatitisInputs and hepatitisTargets. HepatitisInputs is a 19×155 matrix, which identifies 19 properties of 155 biopsies, hepatitisTargets - a 2×155 matrix, which identifies 2 properties of 155 biopsies.

The matrices of databases will be preliminarily processed to suit the operational requirements of the artificial neural network. Specifically, they will be separated, transposed, and normalized before being used to train the network.

2.2 Methods

- Creating an artificial neural network MLP with NN Toolbox in MatLab.

To ensure the scientific and simplified nature of the research model, we use the Neural Network Pattern Recognition Tool in the MATLAB software for creating diagnostic neural networks [12].

- Artificial Neural Network Design for Hepatitis.

Using two matrices hepatitisInputs and hepatitisTargets, which have been processed as mentioned above and selecting the 3-layer (including input layer) network structure with the corresponding number of neurons 19, 10, 2, it is possible to propose the following artificial neural network solving given task (Fig. 2).

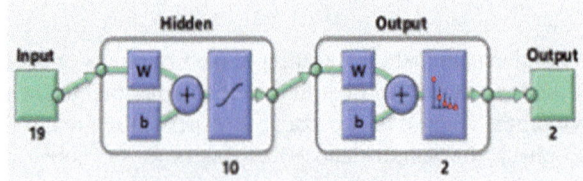

Fig. 2. Artificial neural network for the diagnosis of hepatitis.

For applying the method of assigning weights to body structure levels, it is necessary to classify the input attributes according to the body structural levels for the cases under consideration.

Hepatitis, based on the component size of the 19 input attributes they can be classified into 4 groups corresponding to the 4 structural levels: Body, Level of functional organs, specifically:

- Body Level: attributes 1, 2, 5, 6, 7
- Level of functional organs: attributes 8, 9, 10, 11, 12,13
- Tissue Level: attributes 19
- Bio-Molecular Level: attributes 3, 4, 14, 15, 16, 17, 18

The attributes of these groups will be assigned corresponding W weights that are in the range $0 < W < 1$.

3   Results, Discussion and Conclusion

3.1   Results

After training, validating, and testing networks with a split ratio of 70/15/15, we get the following results.

For the case of hepatitis
When attributes are not assigned by weights.

Fig. 3. The training, validation, testing and all confusion matrix for the hepatitis when attributes are not assigned by weights.

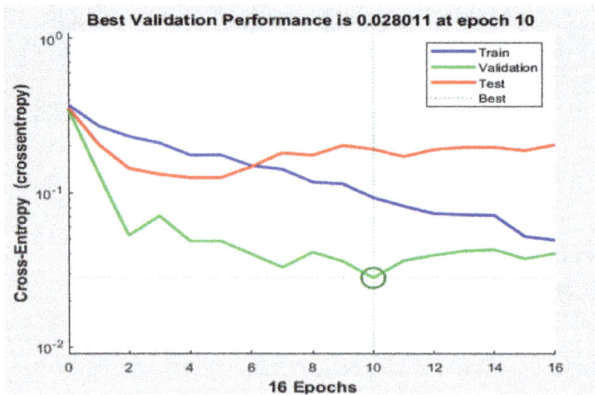

Fig. 4. The best validation performance for hepatitis when attributes are not assigned by weights.

When attributes are assigned by weights.

The results obtained from network training when the attributes are assigned weights corresponding to the 4 groups classified above are shown in Table 2.

Table 2. Table of the best results of neural hepatitis diagnostic neuron network

Case number	Corresponding weight coefficients	Network training results			
		Success rate [%]	Failure rate [%]	Best validation performance	Convergence cycle
0	No weight assigned	92.9	7.1	0.028	10
1	The attributes of the Body Level are weighted with zero	87.7	12.3	0.1434	7
2	The attributes of the Bio-Molecular Level are weighted with zero	87.1	12.9	0.103	7
3	The attributes of the Level of functional organs are weighted with zero	85.2	14.8	0.217	8
4	The attributes of the Tissue Level are weighted with zero	85.8	14.2	0.1293	7
5	The weights 0.2;0.4;0.6;0.8 are assigned correspondingly to attributes of the following groups: Body Level, Bio-Molecular Level, Functional Organ Level, Tissue Level	92.9	7.1	0.028	10

(continued)

Table 2. (*continued*)

Case number	Corresponding weight coefficients	Network training results			
		Success rate [%]	Failure rate [%]	Best validation performance	Convergence cycle
6	The weights 0.8;0.6;0.4;0.2 are assigned correspondingly to attributes of the following groups: Body Level, Bio-Molecular Level, Functional Organ Level, Tissue Level	92.9	7.1	0.028	10
7	The weights 0.6;0.2;0/8;0.4 are assigned correspondingly to attributes of the following groups: Body Level, Bio-Molecular Level, Functional Organ Level, Tissue Level	92.9	7.1	0.028	10
8	The weights 0.4;0.8;0.2;0.6 are assigned correspondingly to attributes of the following groups: Body Level, Bio-Molecular Level, Functional Organ Level, Tissue Level	92.9	7.1	0.028	10
9	The weights 0.1;0.3;0.7;0.9 are assigned correspondingly to attributes of the following groups: Body Level, Bio-Molecular Level, Functional Organ Level, Tissue Level	92.9	7.1	0.028	10
10	The weights 0.9;0.7;0.3;0.1 are assigned correspondingly to attributes of the following groups: Body Level, Bio-Molecular Level, Functional Organ Level, Tissue Level	92.9	7.1	0.028	10
11	The weights 0.7;0.1;0.9;0.3 are assigned correspondingly to attributes of the following groups: Body Level, Bio-Molecular Level, Functional Organ Level, Tissue Level	92.9	7.1	0.028	10
12	The weights 0.3;0.9;0.1;0.7 are assigned correspondingly to attributes of the following groups: Body Level, Bio-Molecular Level, Functional Organ Level, Tissue Level	92.9	7.1	0.028	10

3.2   Discussion

Neural network training results in diagnosing hepatitis in two zero cases and assigned in the number corresponding to the structural levels of the body show:

- The MLP Network Model for hepatitis diagnostics has the same accuracy and convergence as that obtained in [1], so it is suitable for use in medical diagnosis.
- For weighting model:
  - From Table 2 we see that the results obtained in case 2 are better than case 1, 3, 4, it deduces that the group of biomolecules is redundant, so to save computation time, we can ignore them.
  - Case 3 shows a very high error rate, low success rate, so it is unreasonable to remove the attribute at the authority level. This allows to confirm that the group of attributes at the functional organ level is the most important attribute in the process of training the hepatitis network to diagnose hepatitis.
  - It can be seen that case 5 to case 12 are in a special case that the network converges on the overall minimum value by the result of case 0, this implicitly shows that the importance level according to the structure level. In particular, the functional organ-level structure of the lead attribute groups is the same and they differ only in the degree of association between the properties. There is also a supplement to us a method to determine the overall minimum.

3.3   Conclusion

It is significant to assign weight ratios to the input attributes corresponding to the level of body structure. If the weights are specified in the order of the structure levels, efficiency is improved and is better if the reference is not weighted or with a weight of 1 for all input properties. Research results can be applied to the diagnosis of pathology with disease-specific symptoms/attributes obtained at different levels of body structure./
.

References

1. Kajan, S., Pernecký, D., Goga, J.: Application of neural network in medical diagnostics (2015)
2. Rao, D., Zhao, S.: Prediction of Breast cancer (2012)
3. https://gsm672.wikispaces.com/Prediction+of+Breast+cancer, last visit: 5/14/18
4. Al-Shayea, Q.K.: Artificial neural networks in medical diagnosis. Int. J. Comput. Sci. Issues 8(2), 150–154 (2011)
5. Kajan, S.: GUI for classification using multilayer perceptron network, Technical Computing Prague (2009)
6. Amato, F., et al.: Artificial neural networks in medical diagnosis. J. Appl. Biomed. **11.2**, 47–58 (2013). ISSN 1214-0287
7. Papik, Kornel, et al.: Application of neural networks in medicine — a review. Med. Sci. Monit. **4**(3), 538–546 (1998)

8. Nghia H.L., Van Quang, D., Thuy, N.T.: Pathological diagnosis neuron network with inputs corresponding with structure levels of the body. J. Mil. Sci. Technol. **11.17**, 72–78 (2018). ISSN 1859 – 1043

9. Nghĩa, H.L., Văn Quang, D., Ngọc, D.T.B., Thủy, N.T.: Performance improvement of diagnostic neuron network for arrhythmia by method of assigning weights to body structure levels. J. Energy Sci. Technol. – Univ. Electr. (2019) (ISSN: 1859 - 4557)

10. Nghĩa, H.L., Văn Quang, D., Ngọc, D.T.B., Thủy, N.T.: Performance improvement of diagnostic neuron network for dermatology by method of change the number of neuron in a hidden layer. In: Proceedings of National Conference on practical application of high technology - 60 years of development of Military Institute of Science and Technology, vol. 10.2 (2020)

11. UCI: Machine Learning Repository http://archive.ics.uci.edu/ml/

12. https://www.mathworks.com/products/matlab-online.html. Last visit 5/14/18

Development of a Multiple-Sensor Navigation System for Autonomous Guided Vehicle Localization

Anh-Tu Nguyen[(⊠)], Van-Truong Nguyen, Xuan-Thuan Nguyen, and Cong-Thanh Vu

Faculty of Mechanical Engineering, Hanoi University of Industry,
Hanoi 159999, Vietnam
{tuna,nguyenvantruong}@haui.edu.vn

Abstract. In recent years the autonomous guided vehicle (AGV) has been widely applied in the transportation of materials and products in industrial areas where navigation can be done in a constructed environment. The localization of the AGV plays an important role in the control of the AGV for tracking a trajectory. The positioning method based on the inertial navigation system and odometry have acceptable stability and accuracy over the short term. To reduce the errors resulting from the navigation process based on odometry, an additional measurement method is applied using laser measuring technology. This paper presents the development of a multiple-sensor system for localization, in which the data from the encoder, digital compass, and laser scanner are fused using the extended Kalman filter (EKF). The experimental results show remarkable accuracy and stability of the proposed methods concerning positions and orientation.

Keywords: Localization · Triangulation method · Reflector · Extended Kalman Filter (EKF)

1 Introduction

The AGV has been applied in the transportation of materials and products in industrial manufacturing areas as well as in the store. The navigation method of AGV plays an important role in the control of the AGV for following the trajectory [1, 2]. The localization of the AGV is defined by a fixed coordinate system which usually originates from a hall ceiling or one of the corners of the operating area. The position of the AGV is described in terms of two translation coordinate and a radial coordinate. To do this, the navigation system normally works in two basic principles: death reckoning and bearing taking. The dead reckoning is also known as odometry with the fundamental idea is the integration of incremental motion information over time. This method provides good accuracy in the short term, low cost, and high sampling rates. However, the position errors will increase proportionally with the distance traveled by the AGV due to the accumulation of errors [3–5]. The bearing methods use either fixed passive points or active technologies, the AGV might measure the reflection from fixed points to recognize the AGV location. Kim et al. use an RFID system to recognize the position of a system of multiple AGVs, the localization system is composed of RFID readers, RFID tags, and a server, each RFID tag has a unique ID and is fixed at a position in a predefined map. Therefore, this allows the server to determines the AGV position based on the information from the tags [6]. The most previous AGV has guided using magnetic and wire guidance. However, these guidance systems can drive only a pre-determined path and are difficult to change the operating environment or have high costs. Therefore a laser guidance system has been developed for addressing these issues, this method allows obtaining high accuracy in determining vehicle position in indoor environments, such as factories or storages [7]. Another localization method based on laser technology that must be cited here is the work of Śmieszek et al., in the first case, a laser scanner is used to measure the angles and distances to reflectors mounted on the working areas, these data then is used to calculate the AGV position using triangulation method [8]. Many variants of triangulation have been proposed to improve the computational cost, solve the convergence issue, overcome blind spots [9, 10] or event when the AGV receives less than three reflectors [11]. Recently, image processing techniques have been developed for detecting and recognizing the path during the AGV motion. The navigating vision-based system has cameras and computers for image processing. Therefore, reduces the position errors and allows the AGV to change the driving algorithm smoothly while following the desired path [12]. Navigation of the AGV in industrial environments is an important and difficult task with the requirements on the positioning accuracy for docking applications. This paper presents the development of a multiple-sensor system for localization, in which the data from the encoder, digital compass, and laser scanner was fused using the extended Kalman filter (EKF). The experimental results show remarkable accuracy and stability of the position and orientation.

2   Localization Method

2.1   Kinematic Model

In this study, a differential AGV model is considered with two independent driving wheels and two castor freewheels (Fig. 1). The origin of the robot base frame (x_R, y_R) is assigned to the middle of two driving wheels, and θ_R is the mobile robot yaw angle and is the orientation angle from the reference frame (x_{ref}, y_{ref}) relative to the global frame (x_G, y_G). The reflector L_1, L_2, and L_3 located at (x_i, y_i) are added to the working area for the localization using a laser scanner sensor; φ_i is the angle between the reflector number i and the AGV.

The vector state of the AGV is defined $X = [x_R, y_R, \theta_R, v, \omega]^T$, therefore the kinematical model will be as follows:

$$\begin{bmatrix} \dot{x}_R & \dot{y}_R & \dot{\theta}_R \end{bmatrix}^T = R(\theta_R)^{-1} \begin{bmatrix} \dot{x}_{ref} & \dot{y}_{ref} & \dot{\theta}_{ref} \end{bmatrix}^T = R(\theta_R)^{-1} \begin{bmatrix} v & 0 & \omega \end{bmatrix}^T \quad (1)$$

$$\text{In which}: R(\theta_R) = \begin{bmatrix} \cos\theta_R & \sin\theta_R & 0 \\ -\sin\theta_R & \cos\theta_R & 0 \\ 0 & 0 & 1 \end{bmatrix} \quad (2)$$

Therefore : $\dot{X} = \begin{bmatrix} \dot{x}_R & \dot{y}_R & \dot{\theta}_R & \dot{v} & \dot{\omega} \end{bmatrix}^T = \begin{bmatrix} v\cos\theta & v\sin\theta & \omega & 0 & 0 \end{bmatrix}^T$ (3)

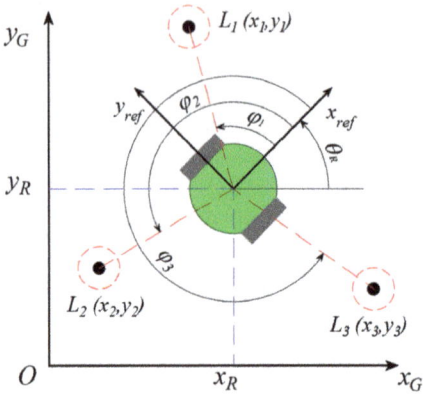

Fig. 1. The description in the robot 2D plane with the respect to a global coordinate system.

By writing in discrete form for the Kalman filter with sampling time Δt, the system model will be:

$$X_k = \begin{bmatrix} (x_R)_k \\ (y_R)_k \\ (\theta_R)_k \\ v_k \\ \omega_k \end{bmatrix} = \begin{bmatrix} (x_R)_{k-1} \\ (y_R)_{k-1} \\ (\theta_R)_{k-1} \\ v_{k-1} \\ \omega_{k-1} \end{bmatrix} + \begin{bmatrix} v_{k-1}\Delta t \cos(\theta_R)_{k-1} \\ v_{k-1}\Delta t \sin(\theta_R)_{k-1} \\ \omega_{k-1}\Delta t \\ 0 \\ 0 \end{bmatrix}$$ (4)

2.2 Extended Kalman Filter

Due to the nonlinearity of an AGV kinematic model and the specificity of each sensor, the EKF is used to fuse multiple types of sensors. EKF requests state system and measurement models, are shown:

$$X_k = f(X_{k-1}) + W_{k-1} \text{ and } z_k = h(X_k) + V_k$$ (5)

In which, f and h are the nonlinear system function and measurement models, respectively. The dynamic system noise W_{k-1} and measurement noise V_k are zero-mean Gaussian noise with associated covariance matrices: $W_k \sim N(0, Q_k)$ and $V_k \sim N(0, R_k)$.

The linearization of functions f and h can be obtained by transforming the Jacobian matrix of f and h at the operating point X_k as follow:

$$F_{k-1} = \frac{\partial f(X)}{\partial X}\bigg|\hat{X}_{k-1}^- \tag{6}$$

$$F_k = \begin{bmatrix} 1 & 0 & -v_k \Delta t \sin \theta_k & \Delta t \cos \theta_k & 0 \\ 0 & 1 & v_k \Delta t \cos \theta_k & \Delta t \sin \theta_k & 0 \\ 0 & 0 & 1 & 0 & \Delta t \\ 0 & 0 & 0 & 1 & 0 \\ 0 & 0 & 0 & 0 & 1 \end{bmatrix} \tag{7}$$

$$H_k = \frac{\partial h(X)}{\partial X}\bigg|\hat{X}_k^- \tag{8}$$

The measurement matrices are established for each sensor as follows:

Using triangulation algorithm [13] for laser to measures the position of the AGV, then the measurement matrix of the laser sensor can be written in the form of:

$$h_{laser} = \begin{bmatrix} x_{laser} \\ y_{laser} \\ \theta_{laser} \end{bmatrix} = \begin{bmatrix} (x_R)_k \\ (y_R)_k \\ (\theta_R)_k \end{bmatrix} \text{ and } H_{laser} = \begin{bmatrix} 1 & 0 & 0 & 0 & 0 \\ 0 & 1 & 0 & 0 & 0 \\ 0 & 0 & 1 & 0 & 0 \end{bmatrix} \tag{9}$$

$$z_k = H_{laser} X_k + V_k \tag{10}$$

Since two encoders are attached to active wheels, applying the kinematic equation of the AGV to calculate v and ω, the measurement matrix of the encoder can be obtained:

$$h_{encoder} = \begin{bmatrix} v_{encoder} \\ \omega_{encoder} \end{bmatrix} = \begin{bmatrix} v_k \\ \omega_k \end{bmatrix} \text{ and } H_{encoder} = \begin{bmatrix} 0 & 0 & 0 & 1 & 0 \\ 0 & 0 & 0 & 0 & 1 \end{bmatrix} \tag{11}$$

$$z_k = H_{encoder} X_k + V_k \tag{12}$$

The digital compass can be used to measures angle, θ, and angular velocity, ω, then the measurement matrix is as follows:

$$h_{compass} = \begin{bmatrix} \theta_{compass} \\ \omega_{compass} \end{bmatrix} = \begin{bmatrix} (\theta_R)_k \\ \omega_k \end{bmatrix} \text{ and } H_{compass} = \begin{bmatrix} 0 & 0 & 1 & 0 & 0 \\ 0 & 0 & 0 & 0 & 1 \end{bmatrix} \tag{13}$$

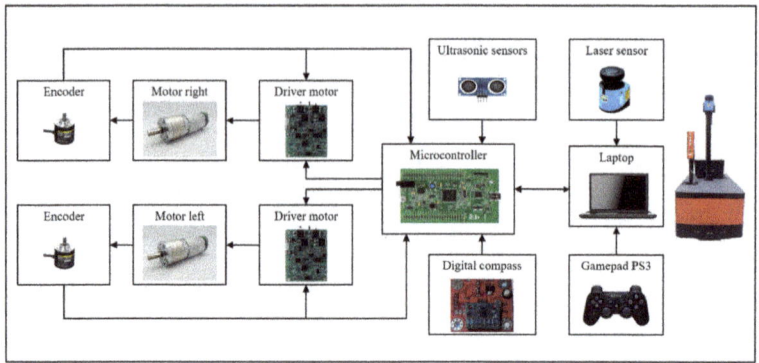

Fig. 2. Schematic of the control system for the experimental AGV.

$$z_k = H_{compass}X_k + V_k \tag{14}$$

3   Experimental Model Description

3.1   Hardware Model Description

To evaluate the accuracy and stability of the proposed approach, an experimental AGV has been designed (Fig. 2). The AGV dimension is $840 \times 540 \times 280$, including four wheels, with two driving wheels being driven by DC motors and two castor freewheels. The control system is developed based on a laptop setting on the AGV. A microcontroller, STM32F407VET6, receives control signals from the laptop and transfers to drive motors. The major sensor of the system is NAV245 installed at the highest position of the AGV. This sensor measures the distance from the AGV to reflectors and the relative angle between the AGV and reflector. A digital compass is added to measure the angle of the AGV. Two encoder sensors are attached to two castor free-wheels to determine the long velocity and angular velocity of the AGV. The speed of two driver motors is controlled using a PID controller. The AGV can operate on the

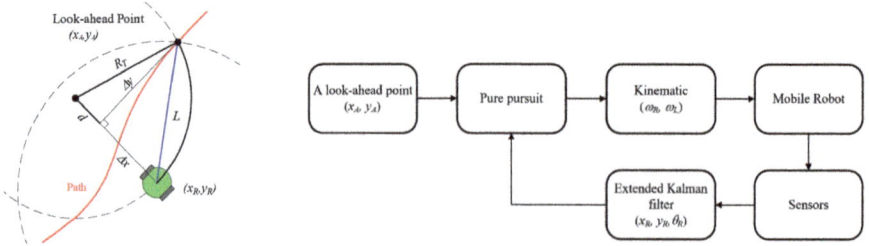

Fig. 3. Pure pursuit path tracking algorithm.

Fig. 4. The control system.

manual mode by using a remote control, which is designed for configuring initial status and establishing positions of reflectors in the environment. Besides, SRF04 sensors are integrated for safety purposes.

3.2 Control Method

In this paper, the pure pursuit path tracking algorithm [14] is used to follow predefined paths. Based on the geometry (Fig. 3), the algorithm calculates a radius (R_T) of the arc connecting between AGV's current position and a look-ahead point (x_A, y_A) with a predefined distance L to path. After that, substituting an angular velocity $\omega = v/R_T$ on an AGV kinematic model [15] to find each wheel velocity, ω_R, ω_L, as in following questions:

$$R_T = \frac{L^2}{2\Delta x} \qquad (15)$$

$$\begin{cases} v = r\frac{\omega_R + \omega_L}{2} \\ \omega = r\frac{\omega_R - \omega_L}{d} \end{cases} \qquad (16)$$

Where r is the radius of a driving wheel, d is the distance between two wheels, v is a long velocity. After every cycle, the algorithm is repeated when the AGV position changes. The whole control system is described in Fig. 4.

4 Result and Discussion

The experimental trajectory has been designed based on a real manufacturing process, the AGV starts from the gathering area and moves in a trajectory of straight lines and arcs. The AGV will move throughout different stations to perform the various tasks and stop at storage. The reflectors were arranged along the two sides of the trajectory to ensure that at least three mirrors can be scanned at any time. The AGV was examined at an average speed of 0.8 m/s and the sampling time to position the AGV for motion control was 40 ms. Figure 5 and 6 show the position and orientation of the AGV in both the designed trajectory and the actual trajectory. Figure 7 and 8 show the position deviation of the AGV. The maximum position deviation in the X direction and Y direction are 17 mm and 16 mm, respectively. It can be seen that when the AGV moves on the straight lines, the deviation is mainly in the direction perpendicular to the direction of translation. The angle deviation of the AGV is described in Fig. 9, the maximum deviation of angle is $0.9°$. Using Eq. 17, the standard deviation of Root Mean Square Error (RMSE) in position and angle are 9.01 mm and $0.34°$, respectively.

$$Disp_{error} = \sqrt{\frac{\sum_{i=1}^{N}(x_{ti} - x_{di})^2 + (y_{ti} - y_{di})^2}{N}} \quad \text{and} \quad \theta_{error} = \sqrt{\frac{\sum_{i=1}^{N}(\theta_{ti} - \theta_{di})^2}{N}} \qquad (17)$$

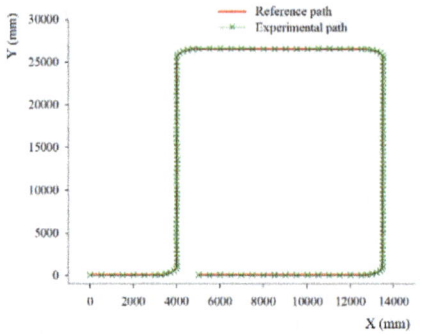

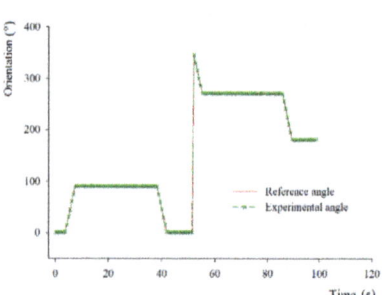

Fig. 5. The position of the AGV.   **Fig. 6.** The orientation of the AGV.

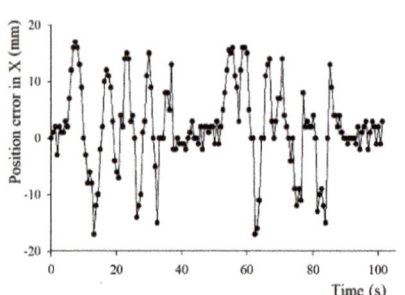

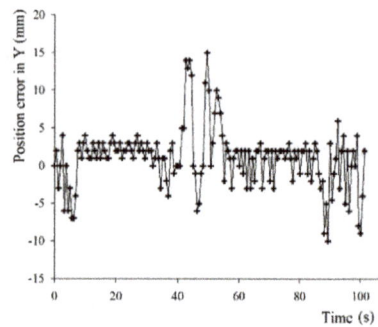

Fig. 7. The deviation of the AGV in X-axis.   **Fig. 8.** The deviation of the AGV in Y-axis.

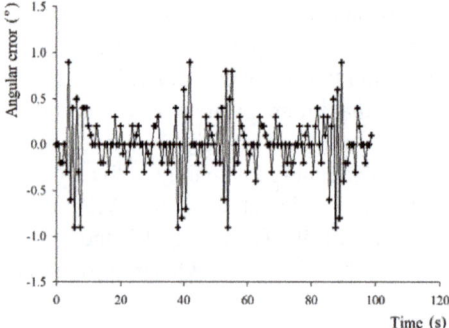

Fig. 9. The deviation of the AGV orientation.

5 Conclusion

This paper addresses the localization of the AGV working in the indoor environment. A multiple-sensor navigation system was proposed to locate the AGV and update reflectors. To obtain the localization accuracy for the AGV in continuous operation, the EKF has been applied for the data fusion. The experiments have been conducted to estimate the deviation of the AGV either in position and orientation which show remarkable accuracy and stability of the method concerning positions and orientation. The proposed navigation system has potential application for AGVs working in indoor industrial environments.

Acknowledgment. The authors would like to acknowledge the financial support provided by the Hanoi University of Industry under Grant No. 04-2020-RD/HĐ-ĐHCN.

References

1. Ferreira, T.P., Gorlach, I.A.: Development of an affordable Automated Guided Cart for material handling tasks. Int. J. Circ. Electr. **1**, 145–150 (2016)
2. Suman, K.D., Pasan, M.K.: Design and methodology of automated guided vehicle-a review. IOSR J. Mech. Civil Eng. 29–36 (2016)
3. Lindsay, K.: Optimal estimation of position and heading for mobile robots using ultrasonic beacons and dead-reckoning. In: IEEE International Conference on Robotics and Automation, pp. 2582–2587 (1992)
4. Cho, B.S., Moon, W.S., Seo, W.J., Baek, K.R.: A death reckoning localization system for mobile robot using inertial sensors and wheel revolution encoding. J. Mech. Sci. Technol. **25**(11), 2907–2917 (2017)
5. Quan, S., Chen, J.: AGV localization based on odometry and LiDAR. In: World Conference on Mechanical Engineering and Intelligent Manufacturing, pp. 483–486 (2019)
6. Kim J., Lee, Y., Moon, K.: Development of RFID-based Localization System for AGV Control and Navigation. In: International Conference on Indoor Positioning and Indoor Navigation (IPIN), pp. 18–21 (2017)
7. Kim, J., Cho, H., Kim, S.: Positioning and driving control of fork-type AGV With laser navigation. Int. J. Fuzzy Logic Intell. Syst. **13**(4), 307–314 (2013)
8. Śmieszek, M., Dobrzańska, M., Dobrzański, P.: Laser navigation application for automated guided vehicles. Meas. Autom. Monit. **61**(11), 503–506 (2015)
9. Vincent, P., Marc, V.D.: A new three object triangulation algorithm based on the power center of three circles. In: Research and Education in Robotics (EUROBOT), Communications in Computer and Information Science vol. 161, pp. 248–262 (2011)
10. Vincent, P., Marc, V.D.: A beacon based angle measurement sensor for mobile robot positioning. IEEE Trans. Rob. **30**(3), 533–549 (2014)
11. Xu, Z., Huang, S., Ding, J.: A new positioning method for indoor laser navigation on underdetermined condition. In: International Conference on Instrumentation & Measurement, Computer, Communication and Control, pp. 703–706 (2016)
12. Nguyen, V.Q., Eum, H.M., Lee, J., Hyun, C.H.: Vision sensor-based driving algorithm for indoor automatic guided vehicles. Int. J. Fuzzy Logic Intell. Syst. **13**(2), 140–146 (2013)
13. Vincent, P., Marc, V.D.: A new three object triangulation algorithm for mobile robot positioning. IEEE Trans. Rob. **30**(3), 566–577 (2014)

14. Jan, B., Frantisek, D., Milan, B., Karol, K., Daniela, P., Sanjeevikuma, P.: Pathfinder—development of automated guided vehicle for hospital logistics. Digital Object Ident. **5**, 26892–26900 (2017)
15. Sandeep, K.M., Jharna, M.: Kinematics, localization and control of differential drive mobile robot. Global J. Res. Eng.: H Robot. Nano-Tech **14**(1), 2–7 (2014)

A Real-Time Human Tracking System Using Convolutional Neural Network and Particle Filter

Van-Truong Nguyen$^{(\boxtimes)}$, Anh-Tu Nguyen, Viet-Thang Nguyen, and Huy-Anh Bui

Department of Mechanical Engineering, Hanoi University of Industry, Hanoi, Vietnam
{nguyenvantruong, tuna}@haui.edu.vn

Abstract. In our paper, a novel approach is proposed for real time tracking human targets in cases of full occlusion and complex environments with a normal webcam. In the proposed methodology, a deep learning model is combined with a particle filter to detect and track targets over frames. In addition, a convolutional neural network is used to have high speed and easy to implement on mobile devices such as automatic guided vehicles or quadcopters. The proposed approach also tracks the human in case of full occlusion and mismatch. The experiment results show that our proposed method achieved quite good FPS (22 FPS) to implement the real-time application or mobile devices. The proposed approach also tracks the human in case of full occlusions.

Keywords: Human tracking · Particle filter · Convolutional neural network

1 Introduction

Human tracking is one of the key problems in computer vision, which has drawn rapid attention from both science and industry. The human tracking motion is widely used in various fields such as military, robotic, or video games, ect. However, tracking human motion is a difficult task when it is conducted with uncontrolled surroundings. The major problem that needs to be resolved in human tracking is the efficient and accurate detection and recognition of targets. In human tracking researches, the main problem that needs to be considered carefully is the accuracy and efficiency of target detection. To be more detailed, the tracking efficiency relies on the speed of finding and identifying the target, whereas the tracking accuracy depends on the robustness of the human target model. Hence, the research and analysis of higher accuracy and better computer vision detection algorithm appears [1–3].

Over the last decades, a large number of methods have been developed to deal with those above problems of human tracking. For instance, in [4], the Camshift algorithm is proposed based on finding the distribution of color histogram. However, the template update ability of Camshift is not appreciated. Kernel-Correlated Filtering (KCF) [5] tracking technique based on correlation filter technique is also a good human tracking algorithm. However, this algorithm could not tackle the full occlusion. Kalmal filter algorithm [6] estimates the coordinate of the target when full occlusion occurs but it could only be used in a linear system when noise is distributed normally. Human tracking using Particle filter [7] is also known as a good method to deal with full occlusion. On the other hand, various approaches try to use Deep Convolutional Neural Network (CNN) to track humans. The problem here is which deep CNN model is suitable to implement on a mobile device. Some architects such as YOLO [8], SSD [9], Faster-RCNN [12] … are famous object detection algorithms applying deep convolution neuron networks. Those algorithms work well on the computer supporting GPU, but it is hard to process on the computer without GPU.

In this article, the Particle Filter is the main method combining with Deep CNN to track along the position of the target. The role of deep CNN in the proposed system is detecting the target automatically when the target comes into the first frame. To address these challenges of real-time, the deep CNN model [10] is adopted in the proposed approach. This model does not take up too many memories and can reach very well speed without the support of a GPU. The experiment shows that our system can track the human target with real-time speed (22 fps). Furthermore, our system is able to solve full occlusion in complex environments.

2   System Architect

2.1   MobileNet-v2 Architect

MobileNet-v2 [10] was created by a team of Google experts. This model is continuing developed based on the previous version Mobile Net [11] with some main modifies the architect such as Linear bottleneck, inverted residual, and Relu6 activation… In Fig. 2, the bottleneck block is a typical struct of MobileNet-v2 built based on Inverted residual block, Linear Bottleneck Block, Batch Normalization, and Relu6 (Fig. 1).

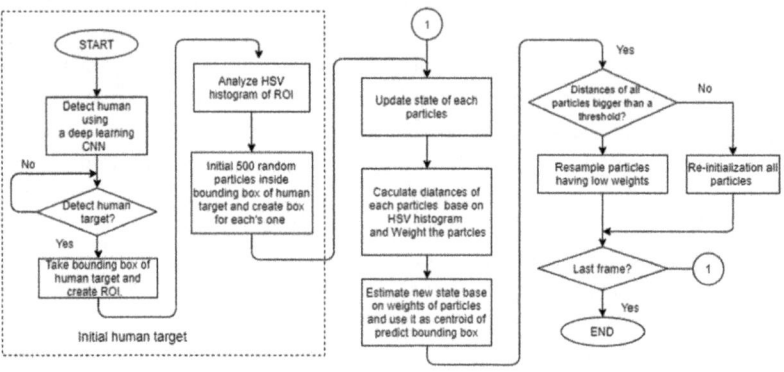

Fig. 1. Flowchart of the proposed system.

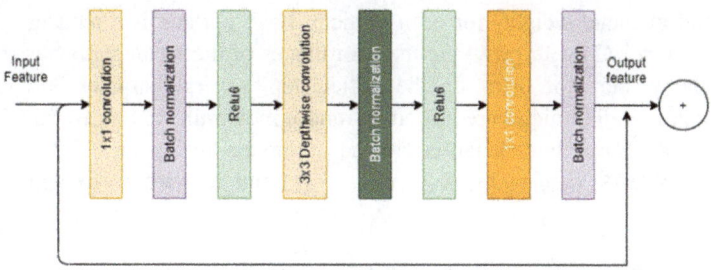

Fig. 2. Bottleneck block.

2.2 One-Stage Detector Deep Learning CNN

For object detection, a CNN model with 6 layers is utilized along with MobileNet-v2. They were combined to get a high effect model. Hence, the model enables to operate on the mobile device precisely. To be more specific, the RGB images with a size of 300×300 are the input of the CNN (6 layers). The expansion of layer 15 (with output stride of 16) of MobileNetv2 is attached to the initialization layer of the model. The rest of the CNN layer is attached to the last feature map of the base model. By using some support layers, the CNN model can detect an object at multiple scales. Through each additional layer, the size of the feature map is reduced four times by using the side equal two.

Like SSD, the input image of the CNN (six layers) model is divided into 8×8 or 4×4 cells. Each cell uses some set of default bounding boxes of different dimensions and aspect ratios. Each bounding box is evaluated by loss value. There are two types of loss: confidence loss and location loss. The total loss is calculated by combining them as:

$$L(x, c, l, g) = \frac{1}{N} \left(L_{conf}(x, c) + \alpha L_{loc}(x, l, g) \right) \tag{1}$$

Where:

- N is the number of default boxes matching with ground truth boxes.
- $L_{conf}(x, c)$ is confidence loss.
- $L_{loc}(x, l, g)$ is location loss.

After predicting, the number of predictions in a frame is not an equal number of actual objects. To tackle this problem, the CNN algorithm applies the non-max suppression. The algorithm sorts the predictions by confidence scores. All boxes which confidence loss less than ct (e.g. 0.012), IoU value less than lt (e.g. 0.47) are eliminated.

2.3 Particle Filter Tracking Method

To track the target, the particle filter is used [7]. This method estimates the position of the human target based on the distribution of a large number of system states called particles and evaluate weights for each particle. Each particle is a rectangle box pre-sented by a vector (2) with (x, y) are the coordinates of the rectangle boxes center and $\frac{dx}{dt}, \frac{dy}{dt}$ are the velocity of them. In the initial step, several particles are defined as N. Coordinates of all particles are picked up randomly within the range of the bounding box (x_0, y_0, W, H) of human target created automatically by the CNN model with $i = 0, 1, 2 \dots N$, $x \in (x_0, x_0, + W)$, $y \in (y_0, y_0, + H)$ and $\frac{dx}{dt}, \frac{dy}{dt}$ are set to zero.

$$s = \left[x, y, \frac{dx}{dt}, \frac{dy}{dt} \right]^T \tag{2}$$

In each frame, all particles are updated by a linear difference Eq. (3). In Eq. (3), s_{t-1}^i is the previous state of each particle, w_{t-1}^i is a multivariate Gaussian random variable as follows:

$$s_t^i = As_{t-1}^i + w_{t-1}^i, \tag{3}$$

To evaluate whether the particles are good or not, each particle has its weight. This weight is considered base on the similarity between the HSV histogram of the human target's template bounding box and the HSV histogram of the particle's rectangle box. This similarity can be computed by using Hellinger distance (4) with H_1, H_2 are HSV histograms that need to be compared, M is the total number of histogram bins. In this paper, the HSV histogram is presented by $8 \times 8 \times 4$ bins to get the best result.

$$d_t^{(i)} = \sqrt{1 - \frac{1}{M\sqrt{\bar{H}_1.\bar{H}_2}} \sum_{u=1}^{M} \sqrt{H_1(u).H_2(u)}} \tag{4}$$

$$\bar{H}_k = \frac{1}{M} \sum_{u=1}^{M} H_k(u) \tag{5}$$

Let σ be a standard deviation of d_t, π_{t-1}^i, be a prior weight of the particle. These weights need normalizing to evaluate which one has a higher change appearance target. The distance values are used to compute the weight for each particle as follow:

$$\pi_t^i = \pi_{t-1}^i \frac{1}{\sigma\sqrt{2\pi}} \exp\left(-\frac{\left(d_t^{(i)}\right)^2}{2\sigma^2} \right) \quad \text{and} \quad \pi_t^i = \frac{\pi_t^i}{\sum_{i=1}^{N} \pi_t^i} \tag{6}$$

With system states and weights of particles, the coordinate of the human target's center can be estimated by Eq. (7).

$$\bar{s} = \sum_{i=1}^{N} s_t^{(i)} \pi_t^{(i)} \tag{7}$$

In each new frame, the particles are resampled based on a distance threshold. As Eq. (4), the rate of similarity between the rectangle box of particle and template is the highest when the distance comes to nearly zero. In particular, the threshold value is chosen sufficiently small to keep the best particles. However, if the distance of threshold is too small, there are none of the satisfying particles. Based on the trial and error method, the most suitable threshold is set as 0.3. After that, the new particles are set around some of the highest weights particles. Nevertheless, in case of all particles have distances bigger than the threshold or the human target is full occlusion, all particles are re-initialization randomly.

3   Experiments and Results

3.1   Training Deep Learning CNN Model

To train robust human detection, the 6000 images of humans are collected from different sources including: 2400 images on Google and 3600 images captured by our camera. For the purpose of reducing the training model time, the transfer learning method is integrated with the pre-trained model on both the COCO dataset and our dataset.

3.2   Running System and Evaluating Results

The developed algorithm is implemented on a computer with 4 GB RAM, INTEL Core i5 8500U CPU 3.0 GHz (6 CPUs), and a 2 Mpx camera. To speed up the process, all images are resized into 270×324 pixels. The algorithm is programmed in Python 3.7 with the support of OpenCV 4.4 library and Numpy Library.

The efficiency of the system is tested with different numbers of particles. To evaluate the accuracy of the number of the particle, the intersection over union (IoU) equation is designed as follows: $\frac{area(R \cap Gr)}{area(R \cup Gr)}$ with R is the rectangular region of the tracking result and Gr denotes the ground truth. If the IoU value is bigger than 0.25, the human target is tracked correctly. In Table 1, it is noticeable that the accuracy rate of the system could reach higher and the number of missing frames is reduced when using more particles. However, when the particles peaks at a certain threshold, the result was almost unchanged. Specifically, in our experiment, when the number of particles reaches 500, the accuracy is still 96.89%.

Table 1. Test algorithm with the different number of particles.

Number of particles	Example	Number of the missing frames	Accuracy (%)	Time per frame
10	677	60	91.1	0.031
50	677	51	92.46	0.033
200	677	41	94.1	0.039
500	677	21	96.89	0.045
1000	677	21	96.89	0.082

In addition, the experiment with significant changes in the color of shirts is implemented to evaluate the precision rate of the proposed system in case of tracking a moving object. Figure 3 illustrates the test experiments with the black and yellow color shirt. The yellow points are the particles used to find the target. The yellow bounding boxes are created for all humans in a frame. The blue bounding box shows the human target. The results indicate that the proposed method still works effectively.

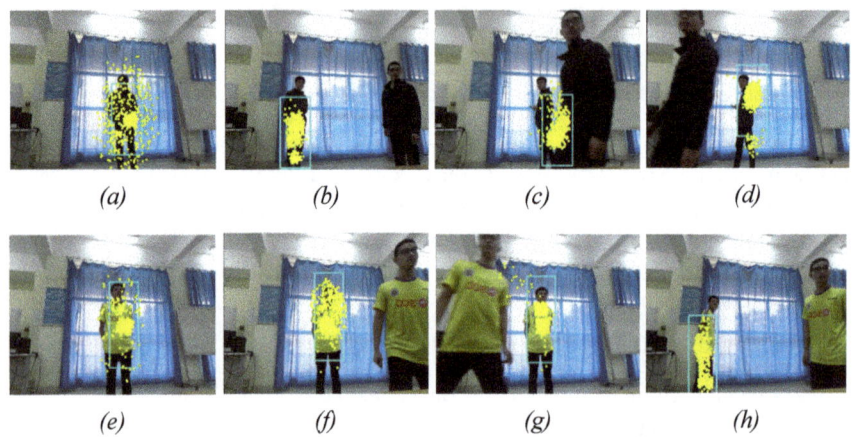

(a)	*(b)*	*(c)*	*(d)*
(e)	*(f)*	*(g)*	*(h)*

Fig. 3. Test algorithm with different color shirts.

4   Conclusion

In this paper, a tracking method that integrates the Particle filter and the convolution neuron network is presented to robustly track human targets. The experimental results under multiple challenge scenes demonstrate that our approach achieves state-of-the-art performance. The detection rate is maintained well even in case of a full occlusion problem. In the future, the authors will keep on developing an algorithm that can track many targets precisely at the same time.

Acknowledgment. This research was funded by Vietnam National Foundation for Science and Technology Development (NAFOSTED) under Grant No. "107.01-2019.311".

References

1. Tran, Q.V., Su, S.F., Nguyen, V.T.: Pyramidal lucas-kanade-based noncontact breath motion detection. IEEE Trans. Syst. Man Cybern.: Syst. **50**(7), 2659–2670 (2020)
2. Nguyen, N.Q., Su, S.F., Tran, Q.V., Nguyen, V.T., Jeng, J.T.: Real time human tracking using improved CAM-shift. In: 2017 Joint 17th World Congress of International Fuzzy Systems Association and 9th International Conference on Soft Computing and Intelligent Systems (IFSA-SCIS), pp. 1–5, IEEE (2017)

3. Tran, Q.V., Su, S.F., Chuang, C.C., Nguyen, V.T., Nguyen, N.Q.: Real-time non-contact breath detection from video using adaboost and Lucas-Kanade algorithm. In: 2017 Joint 17th World Congress of International Fuzzy Systems Association and 9th International Conference on Soft Computing and Intelligent Systems (IFSA-SCIS), pp. 1–4, IEEE (2017)
4. Zhang, Y.: Detection and tracking of human motion targets in video images based on camshift algorithms. IEEE Sens. J. **20**(20), 11887–11893 (2020)
5. Yan, H., Xie, M., Wang, P., Zhang, Y., Luo, C.: Kernel-correlated filtering target tracking algorithm based on multi-features fusion. IEEE Access **7**, 96079–96084 (2019)
6. Deori, B., Thounaojam, D.M.: A survey on moving object tracking in video. Int. J. Inform. Theory **3**(3), 31–46 (2014)
7. Zhou, R., Tang, M., Gong, Z., Hao, M.: FreeTrack: device-free human tracking with deep neural networks and particle filtering. IEEE Syst. J. **14**(2), 2990–3000 (2020)
8. Fang, W., Wang, L., Ren, P.: Tinier-YOLO: a real-time object detection method for constrained environments. IEEE Access **8**, 1935–1944 (2020)
9. Chen, Z., Wu, K., Li, Y., Wang, M., Li, W.: An improved multi-scale object detection network based on SSD. IEEE Access **7**, 80622–80632 (2019)
10. Sandler, M., Howard, A.G., Zhu, M., Zhmoginov, A., Chen, L.: Inverted residuals and linear bottlenecks: mobile networks for classification, detection and segmentation. CoRR, vol. abs/1801.04381 (2018)
11. Howard, A.G., Zhu, M., Chen, B., Kalenichenko, D., Wang, W., Weyand, T., Andreeto, M., Adam, H.: MobileNets: Efficient convolutional neural networks for mobile vision applications. CoRR, vol. abs/1704.04861 (2017)
12. Xu, Q., Zhang, X., Cheng, R., Song, Y., Wang, N.: Occlusion problem-oriented adversarial faster-RCNN Scheme. IEEE Access **7**, 170362–170373 (2019)

A Global Solution for Diversity Data Mining Problem

Nguyen Canh Nam[(✉)]

School of Applied Mathematics and Informatics (SAMI),
Hanoi University of Science and Technology (HUST), Hanoi, Vietnam
`nam.nguyencanh@hust.edu.vn`

Abstract. A new deterministic method based on DC (Difference of Convex function) programming and DC Algorithm (DCA) is presented to solving the maximum diversity problem, which has important applications in Data mining. Most approaches are based on heuristic method while exact solution methods can be applied to only small dimension instances. The globality of solutions given by DCA can be checked by combining it and Semi-Definite Programming (SDP) relaxation techniques in a Branch and Bound (BB) scheme. Numerical experiments on many test problems are presented which show DCA is more efficient than classic algorithms, especially in the large scale setting.

Keywords: DC programming · DCA · SDP relaxation · Data mining · Diversity

1 Introduction

Data mining is one of the most popular technology areas in use today. With the ability to collect, arrange large amounts of data from which to give the most accurate analysis. Data mining process of Data mining is not only limited to data extraction but also used for converting, cleaning, integrating data, and analyzing samples. The info and knowledge built up can be used for applications ranging from finance, healthcare, and telecommunication, to marketing and sale and education. This work investigates a very important problem in Data mining, called the maximum diversity problem. This problem try to recognize, in a community, a part of members, described by some key properties, that show the strongest diversity.

The maximum diversity problem is proved to belong to NP-hard class. In the papers written by Glover et al. [5] and Kuo et al. [9], exact solution algorithms are presented. But the proposed algorithms can only work with instances in small dimensions. So these approaches are not applicable when we work with businesslike tasks in which the size of the community are significant.

Various works have been done for last twenty years [1,3,4,7,8]. All of them are heuristic, authors tried to get the approximated solutions in real life problems

D.-T. Tran et al. (Eds.): ICISN 2021, LNNS 243, pp. 418–428, 2021.
https://doi.org/10.1007/978-981-16-2094-2_51

setting. Therefore, it is always a big objection to propose deterministic methods for this problem with large scale. The algorithm proposed in this work is exactly following this movement.

In a DC optimal problem, the objective functions is a DC one and the set of constraints is convex. The considering problem can be rewritten as an equivalent DC optimization problem by applying the reputable results called exact penalty [11,14]. That is the reason we introduce a method stayed on DC programming called DCA for the resolution of the diversity problem.

DC programming and DCA were first proposed, in the fundamental form, by Tao P.D. in 1985. They then have been broadly grown by his joint works with Hoai An L.T. for the last twenty five years. DCA now is one of the most typical and favored algorithm [10,13,15,16]. Many large scale real life problems were solved by applying DCA. The applicability of DCA in various classes of non-convex and non-smooth optimization problem, even in large-scale setting, as well as the globality of the solutions computed by DCA, even it is a local method, make DCA becomes a scarce method.

In practice, the solutions given by DCA are often global ones. Hence if we can find a good estimation (very near or equal to the value given by DCA), we can then confirm that DCA gives global solutions. Therefore we will integrate DCA into a global approach like BB. In this work, we derive a BB scheme combining DCA and relaxation techniques for general Diversity Maximization Problem. The relaxation techniques used here are named SDP relaxation techniques.

The paper is constructed as follows. Next section is used for the problem statement and applications. Section 3 deals with DC reformulation techniques and SDP relaxation techniques which are all combined in global branch and bound algorithm. Numerical simulations are presented in the Sect. 4 to show the quality of the proposed algorithm. Some conclusions to conclude the paper are presented in the last section.

2   Diversity Maximization Problem and Applications in Data Mining

2.1   The General Diversity Maximization Problem

Let $N = \{1, 2, \ldots, n\}$ and denote $E = \{e_i : i \in N\}$ is a community of n members. Each member e_i in this community has M characteristics. The difference between two members e_i and e_j, denoted by d_{ij} is calculated by $d_{ij} = \sqrt{\sum_{k=1}^{M}(e_{ik} - e_{jk})^2}$, this measure is called *diversity* . The mission is to find a part of size $m, m < n$, from the community in order to maximize the total diversity of this part.

For a subset T of N, we denote the total diversity of T by $\alpha(T) = \sum_{i<j:i,j\in T} d_{ij}$. The Maximum Diversity problem tries to maximize $\alpha(T)$ subject to $|T| = m$.

By introducing a variable $x_j \in \{0, 1\}$ with the convention that $x_j = 1$ if the element e_j is selected otherwise $x_j = 0$, the overall diversity of the subset can be written as $\alpha(T) = \sum_{i=1}^{n} \sum_{j=1}^{i-1} d_{ij} x_i x_j$, or equivalently $\frac{1}{2} x^T D x$ with $d_{jj} = 0$ for all $j = \overline{1, n}$. Finally, we obtain a mathematical model of the Diversity Maximization problem as follows

$$\alpha(T) = \max \quad f(x) = \frac{1}{2} x^T D x \qquad \text{(DDM)}$$

$$\text{subject to} \qquad \sum_{j=1}^{n} x_j = m$$

$$x \in \{0, 1\}^n$$

Remark 1. Although the model (DDM) can represent a wide range of problems. However in certain applications it requires additional constraints. In that case, the heuristic methods normally need to be changed or can not arrive at a solution. Nevertheless our approach presented in this work still works well. In fact, this approach is applicable to the more general model called the Quadratic Binary Constrained Optimization:

$$\alpha = \min \quad f(x) = \frac{1}{2} x^T Q x + q^T x \qquad \text{(QBCP)}$$

$$\text{subject to} \qquad x \in S,$$

$$x_j \in \{0, 1\} \quad \forall j = \overline{1, n}$$

where S is a polyhedral in $\mathbb{R}^n$.

Remark 2. We can suppose that the function f in (QBCP) is convex. Indeed, from the binary property of the variable $x_j, j = \overline{1, n}$, we can smoothly see that the function

$$f_\rho(x) = \frac{1}{2} x^T (Q + \rho I_n) x + q^T x + \frac{1}{2} \rho x^T e_n,$$

where ρ is a number, $I_n \in \mathbb{R}^{n \times n}$ is the identity matrix and $e_n = (1, 1, \ldots, 1)^T \in \mathbb{R}^n$, coincides with f in the feasible set of (QBCP).

As a consequence, by choosing $\rho \leq \lambda_1(Q)$, the function f_ρ becomes convex where $\lambda_1(.)$ denotes the smallest eigenvalue of matrix.

2.2   Data Mining Applications of DDM

Kochenberger et al. in their work [8] presented many important applications of data mining diversity. The applications spread in many areas of business as well as of government such as: Right Sizing the Firm, Agricultural Breeding Stocks, Environmental Balance, Medical Treatment, Genetic Engineering, Molecular Structure Design or Composing Jury Panels...

A solution obtained will give much information about the data and could lead to a very important, maybe death or alive, decision. Let see deeply an example in Composing Jury Panels. In order to make a fair judgment, judge need diverse point of view from the jury panel. So it is perfect if we can maximize the diversity of the selected panel from the qualified citizens. We can go in detail the concrete applications in the context of this paper. The readers can referred to [8] and references therein for more detail.

3   A Global Algorithm

We now take this section to present a method to solve globally Problem (QBCP). The binary property of Problem (QBCP) makes it suitable for a BB scheme. However an efficient algorithm is obtained when we integrate DCA and SDP relaxations.

3.1   DC Program Formulation

A standard DC optimization problem is as follows

$$\alpha = \inf\{f(x) := g(x) - h(x)\}$$

where g and h are lower semi-continuous proper convex functions on $\mathbb{R}^n$.

The vector space of DC functions is quite large to enclose almost real life objective functions and is closed under all the operations usually considered in optimization. Thanks to a well-known result called exact penalty in [11] and its extension in [14] Problem (QBCP) is equivalent to the following one

$$\min \quad \varphi(x) = \frac{1}{2}x^T Q x + q^T x + \tau \sum_{j=1}^{n} \min\{x_j, 1 - x_j\} \tag{1}$$

$$\text{subject to} \qquad x \in S \cap [0,1]^n,$$

where τ is a real number and greater than a $\tau_0 \geq 0$.

The objective function $\varphi(x)$ in Problem (1) is a DC function. Indeed, it comes from the convex property of $f(x)$ and the concave property of $p(x) = \sum_{j=1}^{n} \min\{x_j, 1 - x_j\}$. Precisely

$$\varphi(x) = f(x) + \tau p(x) = f(x) - [-\tau p(x)].$$

3.2   DCA for Solving Problem (1)

The principle of DCA stands on the condition of local optimal solution and dual theory in DC program. DCA generates two sequences $\{x^k\}$ and $\{y^k\}$. They are corresponding solutions to the convex problems (P_k) and (D_k)) given by

$$\inf\{g(x) - h(x^k) - \langle x - x^k, y^k \rangle : x \in \mathbb{R}^n\}, \quad (P_k)$$

$$\inf\{h^*(y) - g^*(y^{k-1}) - \langle y - y^{k-1}, x^k\rangle : y \in \mathbb{R}^n\} \quad (D_k).$$

By twice linearization by means of the subgradients of h and g; we can shorten the scheme of the DCA as following:

$$y^k \in \partial h(x^k); \quad x^{k+1} \in \partial g^*(y^k).$$

Applying DCA to Problem (1) lead to calculate ∂h and $\partial^* g$ where $h(x) = -\tau p(x)$ and $g(x) = f(x)$.

We have $\quad -p(x) = \sum_{j=1}^{n} - \min\{x_j, 1 - x_j\} = \sum_{j=1}^{n} \max\{-x_j, x_j - 1\}.$

Hence

$$\partial h(x) = u : u_j = \begin{cases} \tau & \text{if } x_j < 0.5, \\ -\tau & \text{if } x_j < 0.5 \ , j = 1, 2, \ldots, n. \\ [-\tau, \tau] & \text{otherwise} \end{cases} \quad (2)$$

In the calculation of $\partial^* g(y)$, precisely a subgradient of g^* at y, we need to solve a quadratic convex problem

$$\min\{f(x) - \langle y, x\rangle : x \in S \cap [0, 1]^n\}. \quad (3)$$

We describe below DCA applying to Problem (1)

Algorithm 1. DCA for solving Problem (1)

Step 1 Let $k = 1$ and $\epsilon > 0$. Take x^k as the starting point (may not feasible).
Step 2 Calculate $y^k \in \partial h(x^k)$ as (2).
Step 3 Calculate $x^{k+1} \in \partial g^*(y^k)$ as a global solution of the quadratic convex problem (3).
Step 4 If

$$\|x^{k+1} - x^k\| \leq \epsilon \text{ or } \|\varphi(x^{k+1}) - \varphi(x^k)\| \leq \epsilon$$

then STOP and take x^{k+1} as solution. Otherwise, let $k \leftarrow k + 1$ and return to Step 2.

For a more detail research of DCA the readers are recommended to see [10, 13, 15, 16] and references therein. In order to computed a solution for a non-convex problem, a DCA need to important factor: a convenient DC expression and that for a good starting point.

3.3   SemiDefinite Relaxations

A SemiDefinite relaxation is obtained when we move to the $n \times n$ matrix space. One more time the binary property of $x_j, j = 1, 2, \ldots, n$ gives us a 1-1 corresponding between vector x and matrix $\mathcal{X} = xx^T$ with the constraint $\text{diag}(\mathcal{X}) = x$, where $\text{diag}(\mathcal{X})$ denotes the vector in $\mathbb{R}^n$ whose ith component is x_{ii}.

As a consequence, the matrix $\mathcal{X}$ satisfies the equality $\mathcal{X} = \text{diag}(\mathcal{X})\text{diag}(\mathcal{X})^T$. It is quite strict, we then lessen $\mathcal{X}$ by the condition $0 \preceq \mathcal{X} - \text{diag}(\mathcal{X})\text{diag}(\mathcal{X})^T$ or equivalently, by Schur Complement [6,17],

$$\bar{\mathcal{X}} := \begin{pmatrix} 1 & \text{diag}(\mathcal{X})^T \\ \text{diag}(\mathcal{X}) & \mathcal{X} \end{pmatrix} \succeq 0$$

The representation of the objective function, in term of $\mathcal{X}$, can be done simply by the standard dot product in matrix space. Precisely, the function $\phi(\mathcal{X}) := \frac{1}{2}Q \bullet \mathcal{X} + Diag(q) \bullet \mathcal{X}$, where $Y \bullet \mathcal{X} := \sum_{i,j=1}^{n} y_{ij}x_{ij}$ and $Diag(q)$ stands for the diagonal matrix of order n whose ith component is q_i, is equal to $f(x)$.

The final phase to get a complete a semidefinite relation is to rewrite the constraints in term of X. This phase is called "lifting the constraints". There are four standard lifting [6,17] from simple to complicated but time consuming is also increased along with the complication of obtaining constraints.

Let's describe the lifting with the constraint, $a^T x \leq b$. One again, the natural property of x_i gives us immediately a simple constraint in term of $\mathcal{X}$:

$$\langle Diag(a), \mathcal{X} \rangle \leq b \tag{R1}$$

By squaring both sides of the constraint, provided that $|a^T x| \leq b$, we obtain a constraint in term of $\mathcal{X}$ is as follows:

$$\langle aa^T, \mathcal{X} \rangle \leq b^2 \tag{R2}$$

If $0 \leq a^T x$ we then have $0 \leq a^T x(b - a^T x)$. Or in term of $\mathcal{X}$ we have

$$\left\langle \begin{pmatrix} b \\ -a \end{pmatrix} \begin{pmatrix} 0 \\ a \end{pmatrix}^T, \bar{\mathcal{X}} \right\rangle \geq 0 \tag{R3}$$

The most complicated lifting invented by Lovasz and Schrijver. They did the multiplication of $x_j \geq 0$ and $1 - x_j \geq 0$ for all $j = \overline{1,n}$ to $a^T x \leq b$. As a consequence, in the matrix space we have $2n$ constraints in term of $\mathcal{X}$

$$\begin{cases} \sum_{j=1}^{n} a_j \mathcal{X}_{ij} \leq b\mathcal{X}_{ii} \\ \sum_{j=1}^{n} a_j(\mathcal{X}_{jj} - \mathcal{X}_{ij}) \leq b(1 - \mathcal{X}_{ii}) \end{cases} \quad j = 1, 2, \ldots, n \tag{R4}$$

The following lemma provides the connections between (R1), (R2), (R3) and (R4).

Lemma 1. *(Lemma 3.3.3 [6])   Denoted by S_1, S_2, S_3 and S_4 the feasible sets created by (R1), (R2), (R3), (R4) along with $\mathcal{X} - \text{diag}(\mathcal{X})\text{diag}(\mathcal{X})^T \succeq 0$, respectively. We then have $S_4 \subseteq S_3 \subseteq S_2 \subseteq S_1$* ∎

Table 1. DCA - SDP relaxations

Dim		DCA-R1				DCA-R2				DCA-R3			
n	m	Iter	Value	Time	UB	Iter	Value	Time	UB	Iter	Value	Time	UB
100	40	32	4093	3.92	8868	19	4113	4.14	4416	25	4114	4.78	4225
200	80	52	16066	32.03	35997	43	16086	42.02	16879	39	16103	39.41	16502
300	120	75	35648	124.64	81036	47	35684	151.27	36865	46	35784	146.95	36351
400	160	83	61962	313.67	143751	77	62615	460.73	64134	53	62151	401.73	63400
500	200	88	96724	669.17	224523	69	97056	941.81	99374	54	97042	898.81	98453
120	42	44	4562	7.66	11213	13	4583	6.11	4948	2	4561	4.94	4755
150	55	34	7926	11.44	18733	16	7962	12.70	8408	11	7985	11.91	8152
180	65	33	10768	17.86	26140	32	10793	26.86	11637	31	10797	26.75	11062
240	90	54	20132	53.88	48386	52	20229	77.86	21173	43	20261	72.11	20756
320	115	65	32788	135.39	82662	68	32881	207.33	34185	42	32922	176.95	33681
360	140	62	47845	188.97	113270	93	48037	347.92	49625	87	48109	333.49	49014
450	172	67	71789	400.08	173804	81	71931	686.30	74220	75	71987	662.22	73466
480	180	86	78780	558.63	193892	94	79029	874.84	81203	53	79030	741.36	80391
500	180	85	79090	904.50	202347	64	79436	904.50	81658	70	79546	914.39	80858
560	205	84	101675	922.44	257623	66	102068	1343.33	104795	70	102050	1414.63	103857
600	220	124	116834	1486.00	296572	76	117134	1839.89	120318	89	117166	1992.58	119324
630	220	84	117791	1390.20	311865	64	118201	2057.11	121263	83	118155	2279.00	120266
700	240	109	139031	2371.42	377328	78	139538	3242.80	143130	95	139553	3456.19	142068
720	235	99	133842	2455.67	380209	98	134340	3821.25	138145	109	134297	4073.59	137053
800	275	151	182032	4536.44	494675	112	182542	5846.14	187128	98	182527	5597.38	185859
850	275	141	182275	5355.92	524710	132	183031	7788.16	187686	95	183054	6958.16	186439
900	325	138	252556	6363.98	656986	144	253268	9800.59	258418	106	253436	8955.88	256969
910	310	131	230892	6113.78	635517	124	231525	9452.27	237160	95	231502	9099.63	235753
970	340	172	276590	9353.14	741833	112	277321	11613.00	283348	114	277383	12232.58	281790
1000	370	168	326174	10191.78	832652	77	327115	11595.20	333204	141	327021	14298.13	331521

Remark 3. The resolution of an SDP relaxations gives us a matrix, denoted by X^*, whose diagonal is a vector in $[0,1]^n$. We then can take $\mathrm{diag}(X^*)$ as the starting point for DCA.

Remark 4. At the first launch of DCA, we would get an appropriated starting point for DCA, as techniques proposed in [12], the DCA first used to solve the following concave optimization problem:

$$0 = \min \left\{ \sum_{j=1}^n \min \{x_j, 1 - x_j\} : x \in S \cap [0,1]^n \right\}. \tag{R5}$$

4   Numerical Simulations

We now report the experiment of the proposed algorithm on two sets of data. The first one was used in [3], in which the dimension of problems is limited to 500. The second set of data were randomly generated, as in [3]: the diversities for each instance were taken from the set of values distributed uniformly in $[0..9]$ and their dimensions up to 1000.

Table 2. DCA with initial point given by exact penalty strategy in Remark 4

Dim		SDP relaxation 1		SDP relaxation 2		SDP relaxation 3	
n	m	Value	Time	Value	Time	Value	Time
100	40	3522	1.45	4074	2.72	4074	6.04
200	80	14462	12.24	15994	25.39	15769	40.09
300	120	32264	46.67	35314	100.63	35369	114.51
400	160	57792	128.27	61342	290.91	61284	276.09
500	200	89179	307.49	96091	652.88	95941	601.65
120	42	4012	2.55	4442	4.67	4561	8.02
150	55	6888	4.92	7905	9.89	7974	12.19
180	65	9185	8.73	10652	17.88	10631	30.88
240	90	18369	21.72	20044	46.30	19944	45.80
320	115	29486	57.77	32166	125.28	32198	124.19
360	140	43697	90.14	47165	198.81	47150	188.50
450	172	66243	201.30	70758	441.19	70634	424.55
480	180	71786	254.03	78036	540.41	78336	544.25
500	180	71837	295.09	78282	647.72	78407	626.42
560	205	93990	448.27	100282	962.49	100329	1002.69
600	220	108616	606.81	114958	1279.75	119324	1308.14
630	220	108348	704.77	116191	1511.67	116091	1534.86
700	240	129231	1042.81	136507	2271.72	136632	2247.59
720	235	127173	1159.91	131352	2510.09	131222	2488.36
800	275	170094	1713.77	178620	3725.94	178269	3681.81
850	275	169570	2114.75	178548	4673.52	178361	4657.91
900	325	236372	2628.69	250325	5781.88	250012	5787.52
910	310	215804	2746.38	226285	5983.44	226504	6129.28
970	340	259911	3899.91	273124	7696.08	272687	7916.39
1000	370	307052	3899.91	323015	8674.41	322727	8665.41

The algorithm was written in C and running on Core i7-5200U 2.1GHz, 8G RAM computer. For solving Quadratic convex problems we use CPLEX 11 [18] and all the SDP relaxations are solved by SDPA [2], a callable library in C.

We report the values given by DCA in Table 1 corresponding to the initial point given by the SDP relaxation (R1), (R2), (R3) as described above. We did not carry out the relaxation proposed by Lovasz and Schrijver because it is very time consuming for large scale problems.

The choice of the penalty parameter τ depends on test problems, in our numerical simulations it takes the three value $10^3, 10^4$ and 10^5.

Table 3. Computational results (recapitulative)

Dim		Heuristic - Best		DCA - Best		$UB - Best$	$10^2\epsilon$
n	m	Value	Time	Value	Time		
100	40	4142	234.00	4128	8.38	4225	2.35
200	80	16225	4487.60	16181	123.09	16502	1.98
300	120	35881	17562.01	35738	308.69	36351	1.72
400	160	62454	52056.20	62615	276.09	63400	1.25
500	200	97320	138031.30	97132	2097.19	98453	1.36
120	42			4594	10.95	4755	3.51
150	55			7985	9.78	8152	2.09
180	65			10812	61.53	11062	2.31
240	90			20299	84.27	20756	2.25
320	115			32966	350.11	33681	2.17
360	140			48164	391.34	49014	1.77
450	172			72057	433.33	73466	1.20
480	180			79088	535.72	80391	1.65
500	180			79561	629.45	80858	1.63
560	205			102093	1009.91	103857	1.73
600	220			117847	1328.63	119324	1.25
630	220			118329	1516.80	120266	1.64
700	240			139639	2271.78	142068	1.74
720	235			134494	2541.50	137053	1.90
800	275			182594	3739.67	185859	1.79
850	275			183136	4733.39	186439	1.80
900	325			253544	5964.00	256969	1.35
910	310			231636	5912.70	235753	1.78
970	340			277504	8086.91	281790	1.55
1000	370			327211	8866.89	331521	1.32

In Table 1, the abbreviations are as following:

– Dim: Dimension of testing problems
– Value: Values given by DCA (lower bound for maximum problem)
– UB: Bounds given by SPD relaxations (upper bound for maximum problem)
– time: Computational time in second

We first note that initial point has a strong influence on the qualities of DCAs. All three DCAs (DCA with three initial points given by SDP relaxations) always converge to feasible solutions of Problem (DDM). On the other hand the bounds (upper bound in maximum problem) are improved from (R1) to (R3) as mentioned in Lemma 1.

The most important thing that one should remark is that the bounds provided by SDP relaxations (R2) and (R3) are very close to the values given by the DCA. If we measure the quality of the solution given by DCA by $\frac{UB - Val}{Val}$ we then can see that with (R2) and (R3) we obtained the $\epsilon-$optimal solution with $\epsilon < 2\%$ without doing any iteration in a Brand and Bound scheme.

Table 2 reports the results provided by DCA with initial points given be the exact penalty strategy. In all tests, the four DCA converge to feasible solutions with the same values after 2 iterations.

To assess the efficiency of DCA, we summary the computational experiments in Table 3 where the best values given by DCA and the best upper bounds (UB) obtained are displayed. We also make a direct comparison in values and computation times by quoting the best computational results from the work of Geiza C.S. et al. [3] where they proposed nine strategies for GRASP applying to the Maximum Diversity Problem.

We observe that, DCA give the ε-optimal solution in almost instances with $\varepsilon < 2\%$. Although the values given by GRAPS are a little better than those computed by DCAs the computation times of DCA (+SDP) are much smaller, even for large scale problems.

5   Conclusion

We have proposed a new deterministic algorithm based on DC programming and SDP relaxation technique for the maximum diversity problem (DDM) which has many important applications in Data Mining. Computational experiments show the efficiency of our algorithm. DCA always give feasible solutions of (DDM) which are $\epsilon-$ optimal solutions because their objective values are very close to lower bounds provided by SDP relaxation technique. Evidently this technique is very costly and can not be applied to large scale (DDM) where DCA still works well thanks to its inexpensiveness.

References

1. Andrade, P.M.F., Plastino, A., Ochi, L.S., Martins, S.L.: GRASP for the maximum diversity problem. In: Proceedings of MIC 2003 (2003)
2. Fujisawa, K., Kojima, M., Nakata, K., Yamashita, M.: SDPA - user's manual version 6.00. Research reports on Mathematical and Computing Sciences, Tokyo Institute of Technology (2002)
3. Silva, G.C., Ochi, L.S., Martins, S.L.: Experimental comparison of greedy randomized adaptive search procedures for the maximum diversity problem. J. Heuristic (2006, to appear)
4. Ghosh, J.B.: Computational aspects of the maximum diversity problem. Oper. Res. Lett. **19**, 175–181 (1996)
5. Glover, F., Hersh, G., McMillan, C.: Selecting subsets of maximum diversity, MSIS Report No. 77-9, University of Colorado at Boulder (1977)
6. Helmberg, C.: SemiDefinite Programming for Combinatorial Optimization, ZIB-Report 00-34, October 2000

7. Katayama, K., Naribisa, H.: An evolutionary approach for the maximum diversity problem. Working paper, Department of Information and Computer Engineering, Okayama University of Science (2003)
8. Kochenberger, G., Glover, F.: Diversity datamining. Working paper, The University of Mississippi (1999)
9. Kuo, C.C., Glover, F., Dhir, K.S.: Analyzing and modeling the maximum diversity problem by zero-one programming. Decis. Sci. **24**, 1171–1185 (1993)
10. An, L.T.H., Tao, P.D.: Solving a class of linearly constrained indefinite quadratic problems by DC algorithms. J. Global Optim. **11**(3), 253–285 (1997)
11. An, L.T.H., Tao, P.D., Muu, L.D.: Exact penalty in DC programming. Vietnam J. Math. **27**(2), 169–178 (1999)
12. An, L.T.H., Tao, P.D.: A continuous approach for globally solving linearly constrained quadratic zero - one programming problems. Optimization **50**, 93–120 (2001)
13. An, L.T.H., Tao, P.D.: The DC (difference of convex functions) Programming and DCA revisited with DC models of real world nonconvex optimization problems. Ann. Oper. Res. **133**, 23–46 (2005)
14. An, L.T.H., Tao, P.D., Van Ngai, H.: Exact penalty and error bounds in DC programming. J. Global Optim. **52**(3), 509–535 (2012)
15. Tao, P.D., An, L.T.H.: Convex analysis approach to DC programming: theory, algorithms and applications. Acta Mathematica Vietnamica **22**(1), 289–355 (1997). Dedicated to Professor Hoang Tuy on the occasion of his 70th birthday
16. Tao, P.D., An, L.T.H.: DC optimization algorithms for solving the trust region subproblem. SIAM J. Optim. **8**, 476–505 (1998)
17. Wolkowicz, H., Saigal, R., Vandenberghe, L.: Handbook of SemiDefinite Programming - Theory, Algorithms, and Application. Kluwer Academic Publisher, Dordrecht (2000)
18. ILOG. CPLEX 11.0. Ilog cplex 11.0 reference manual. ILOG CPLEX Division, Gentile, France. http://www.ilog.com/products/cplex

A Study on Anomaly Data Traffic Detection Method for Wireless Sensor Networks

Trong-Minh Hoang[(✉)]

Posts and Telecoms Institute of Technology, Hanoi, Vietnam
hoangtrongminh@ptit.edu.vn

Abstract. The massive growth of the Internet of Things applications recently has brought a lot of benefits to human society. Besides, the increase in the number of sensing devices in wireless sensor network infrastructure has created new challenges in ensuring wireless network security. Specifically, solutions to detect attacks from the edge of the network must be considered to reduce the pressure on the computing elements in core networks. Therefore, approximate approaches to low computational complexity in an Intrusion Detection System (IDS) need to be studied to favor limited-resource devices. In this paper, an intrusion detection model based on hedge algebra is proposed to detect anomaly sensory traffic. The numerical simulation results indicate that our proposed model achieves a higher rate of abnormal patterns than previous studies.

Keywords: Wireless sensor networks · Security · Intrusion Detection System · Anomaly detection · Hedge algebras

1 Introduction

Today, the rapid and robust development of IoT systems has been creating extremely compelling applications for improving the quality of life. By gathering physical environmental phenomena, IoT systems can use them to solve complex intelligence-intensive tasks through effective decision support solutions. In particular, the Wireless Sensor Network (WSN) infrastructure plays a key role in the task of collecting, transmitting, and responding to sensory data. Due to working in a vulnerable open environment like a free space, wireless sensor networks have faced a lot of security issues. Especially, the natural constraints of the IoT working environment make limited resource devices create even more challenges [1, 2].

As the core components of an IoT system, WSN is responsible for sensing environmental physical phenomena and transmitting information to the sink node through various communication links. Hence, unusual sensory traffic is caused by a lot of active or passive attacks such as tampering, Sybil attack, jamming, interception, DDOS, etc. [3, 4]. To determine abnormal traffic behaviors, varied intrusion detection methods have been proposed by exploiting specific sensory data features. Based on statistical rules, anomalous behavior or unusual data is determined through the difference with normal conditions. This solution is highly customizable in IDS satisfying many practical cases. However, the high computation complexity and false alarm are issues that need to be further improved in strongly limited resource environments.

In recent years, the intelligent computing approach for decision support has gained a lot of benefits in varied application areas. This heuristic approach can reduce the calculation complexity while keeping reasonable accuracy by utilizing historical data characteristics. To detect anomaly traffic in wireless sensor network, an IDS based on an artificial intelligence approach or fuzzy logic has become a popular field which attracts more researchers [5–7]. This approach exploits surprising data features to enable accurate identity methods then support a good forecast. Besides their advances, most solutions may require both a large training data set and processing time to reach a demanded accuracy. Otherwise, IDS based fuzzy logic is appropriate in the operation of finite resource environments but needs to continue to improve the accuracy of false alarms.

Originated from the evaluated characteristics of the Shannon entropy principle [8], this study proposes a novel identity system to detect anomaly sensory traffic data. It is the first time, the application of hedge algebra is applied in this security field to support abnormal detection decisions. The proposed model can help reduce false alarm rates while keeping low complexity in operations. The effectiveness of the proposed method is validated by numerical results in the same scenario as other previous proposals. The organization of this paper is structured as follows: The next section presents related work; Sect. 3 briefs our assumptions and base principles; The detailed proposed anomaly detection is illustrated in session 4; in the last section, our conclusions and future works are presented.

2   Related Work

In general, anomaly detection in traffic estimators is proposed by varied methodologies. IDSs are generally categorized into signature-based detection and anomaly-based detection. In which, anomaly-based detection works by estimating a baseline of the normal traffic that compares to present traffic to detect abnormal patterns [9]. To construct IDSs for wireless sensor networks, entropy-based anomaly detection is a traditional approach that has great contributions with varying solutions. This method measures uncertainty in a compact form which summarizing feature distribution. By monitoring sensory traffic in varying time or changing feature aspects, instances created unknown pattern are detected by Entropy thresholds approaches. Several entropy measures were utilized to find out the irregularities that existed in the information content of the normal dataset [10, 11]. To analyze the misbehavior node in wireless sensor networks, the authors in [12] proposed a trust evaluation model with entropy-based weight assignment to mitigate packet dropping or packet modifications. To recognize the DOS attack, several detectors based on the entropy principle have been tested in a real database that provides a variety of performance [13]. With numerous application scenarios in uncertain IoT network conditions, a security scheme of wireless sensor networks is needed to enhance for tailoring with update current requirements such as low and intelligent computations.

IDS based fuzzy logic provides a reasonable computing complexity to estimate data flow characteristics. Since this method has adjusted soft thresholds for outlier events, false alarm issues of the random sensory data can be eliminated. This method has been applied to the Fuzzy Inference System (FIS) model to detect abnormal traffic behaviors in [14]. Focussing on clustering problems, the authors in [8] proposed a fuzzy logic-based jamming detection algorithm to detect the presence of jamming in sensory data. The combined fuzzy-based approaches in [15] were proposed to expose DoS/DDoS or probing attacks over anomalous network traffic in a reasonable manner. These previous works are adapted to limited resource characteristics but the outcome of a fuzzy inference system needs to be enhanced to exploit good entropy features.

3   Premier

3.1   Assumptions

Considering a wireless sensor network utilized to exampled an agriculture IoT system as in Fig. 1, sensor nodes have to collect multiple phenomena environment parameters such as humidity, temperature, or light [16]. These sensory data have created consequent packets that have the inter-arrival time (IAT) followed as a normally distributed random variable [17]. An expected IDS is located at the cluster head location to process abnormal events coming from sensor nodes.

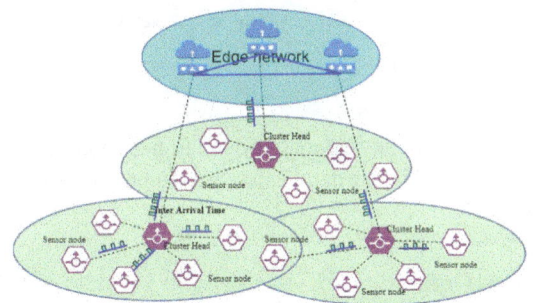

Fig. 1. A typical sensor network

Theoretically, the probability density function of this IAT is an equation below.

$$f(x, \mu, \delta) = \frac{1}{\sqrt{2\pi}} \exp\left[-\frac{1}{2} \left(\frac{x - \mu}{\delta} \right)^2 \right],$$ (1)

where δ is the standard deviation, and $\infty < x < \infty$, $\infty < \mu < \infty$, $\delta > 0$.

In the field of information theory, the entropy Shannon of a message represents the expected information gain when measuring the message. It is a measure of uncertainty associated with a random variable [9]. The Shannon entropy is formulated below.

$$H_s(X) = \sum_{i=1}^{n} p(x_i) \log \frac{1}{p(x_i)}.$$ (2)

Where, $p(x_i)$ is the Probability Mass Function (PMF) of the outcome x_i. An interesting property of Shannon entropy is the assurance of additive features of the main system including subsystems known each other interaction. Hence, it can favor a practical sensor network applicated to IoT application.

Assume that, we already have historical IAT data in the normal mode. Following the Shannon entropy principle, the estimated thresholds of a normal distribution are described below.

$$th = \left[\mu - z_{1-\beta/2}\delta/\sqrt{2}, \mu + z_{1-\beta/2}\delta/\sqrt{2}\right]. \tag{3}$$

Where, (μ, δ) is mean and deviation of the distribution function; n is the size of the sample chain; $z_{1-\beta/2}$ function is the t-student distribution function, $(1 - \beta)100\%$ is a confidence interval which presents the accuracy of the forecast system. With the $\beta = 5\%$, we have z = 1.96. As our previous work, the mean and deviation standard of a sample distribution can be demonstrated by a fuzzy estimator of R.

$$\bar{z}(u) = 2z\left(-\left|\frac{u - \bar{x}}{\delta/\sqrt{n}}\right|\right); u \in R. \tag{4}$$

$$\bar{\delta}^2(u) = \begin{bmatrix} 2F_c\left(\frac{(n-1)\delta^2}{u}\right); 0 < \frac{(n-1)\delta^2}{u} \leq M \\ 2\left(1 - F_c\frac{(n-1)\delta^2}{u}\right); M \leq \frac{(n-1)\delta^2}{u} \end{bmatrix}; u \in R. \tag{5}$$

M is the median value of the estimated deviation standard, u presents a sample value chain. In this case, u is history data which uses to predict abnormal patterns in the current time.

3.2   Hedge Algebras Outline

A hedge algebra is denoted as a set $AX=(X; G; C; H; \leq)$. In which, X and C is a set of values of a linguistic domain (partially ordered set) and a set of constants, $C = \{0, W, 1\}$, (zero, neutral, and unit elements, respectively), G is a set of generators denoted by c^- and c^+, $G = \{c^-, c^+\}$, H is a set of unary operations, $H = H^- \cup H^+$. The operator $\leq$ is a partial ordering relation on set X. Several main core properties are listed below.

To normalize input variables, the linear transformation is used to transform real linguistic variables in scale [a b] into the semantic domain in the normalized interval [0 1].

$$Norm(x) = x_s = \frac{x - a}{b - a}. \tag{6}$$

Two linguistic variables (x_i, x_k) are ordered quantitate values, $A(x_i) < A(x_k)$. A closed metric (η) is defined as the distance correlation for a variable x_* with two previous variables, $A(x_i) \leq A(x_*) \leq A(x_k)$.

$$\eta_k = \frac{A(x_*) - A(x_i)}{A(x_k) - A(x_i)}; \eta_i = \frac{A(x_i) - A(x_*)}{A(x_k) - A(x_i)}. \tag{7}$$

Where, $\eta_i + n_k = 1$, $0 \le \eta_i, \eta_k \le 1$.

The fuzziness measure in the Hegde algebras $(\mu(h))$ domain is presented over the fuzziness measure (fm). We have,

$$fm : X \rightarrow [0, 1]; \ \forall h \in H, \tag{8}$$

$$\sum_{i=-1}^{q} fm(h_i) = \alpha, \ \sum_{i=-1}^{p} fm(h_i) = \beta; \alpha, \beta > 0, \alpha + \beta = 1. \tag{9}$$

The outcome values of this system are presented by real values. Denote $v : X \rightarrow [0, 1]$ is mapped (fm) by the semantically quantifying mapping (SQM). We have the scrip values are presented by the following equation.

$$y_j(HA) = \frac{\sum_{j-1}^{L} \prod_{i=1}^{n} \eta_j(k_i)_j \times \phi_j(x_1...x_n)(p_{1j}...p_{1n})}{\sum_{j-1}^{L} \prod_{i=1}^{n} \eta_j(k_i)_j}. \tag{10}$$

Where, Input set $X = [x_1, x_2, ...x_n]$; a rule set $R = [1, 2, ...L]$; the parameters of rule j^{th} is $(p_{0j}, p_{1j}...p_{nj})$; $A_{ij}(k_i)$ are the linguistic values in the rule j^{th} with its membership $\mu_{ki}(x_i)$; the fuzzy set $K_i = [k_1, k_2, ...k_K]$; $\phi_j(.)$ is a linear function.

4   The Proposed Model

Based on the principles of the Hedge algebras, a proposed anomaly detection model is illustrated in Fig. 2.

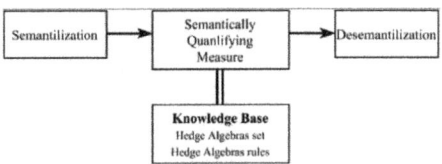

Fig. 2. The proposed HA model

The input data is estimated by the historical normal data from a practical example dataset [12]. In which, the mean and deviation (μ, δ) of statical samples are extracted to examine the limits of two input ranges.

To obey the HA rules from data characteristics, the input real values are transformed into linguistic semantic values. Consequently, these linguistic variables are calculated by closed degree metrics in Eq. (6) (7).

Uncertain conditions are introduced by hedge over the fuzziness measure that is determined in Eq. (8), *fm* is calculated through $\alpha - cut$ parameter as the confidence interval (95%). To set the rules of the HA system, the linguistic variables of inputs are constructed as $\mu = [low, normal, high]$ and $\delta = [small, medium, large]$.

The hedge algebras rules are set in the Semantically Qualifying Measure (SQM) presented in Table 1. Using Eq. (10), the linguistic variables are converted to real outcome values. In our case, these values help to decide in 5 categories such as [anomaly−, suspect−, normal, suspect+, anomaly+]. In more detail, the outcome values are determined to the normal case, anomaly cases, and suspect cases.

Table 1. The system rules

$\mu\backslash\delta$	Small	Medium	Large
Low	Anomaly−	Suspect−	Anomaly+
Normal	Anomaly+	Normal	Suspect+
High	Anomaly+	Suspect+	Anomaly+

To estimate the performance of the proposed model, the IAT data simulated random variables are generated by a random number generator in the Matlab tool. In our previous work, the fuzzy estimator brings a better accuracy than the Shanon estimator. With the same input, the historical data estimation is illustrated in Fig. 3.

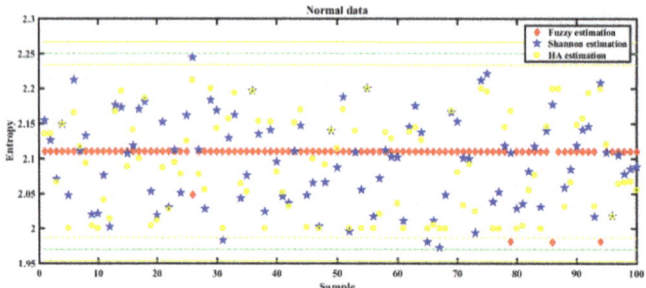

Fig. 3. The IAT data simulation in normal condition.

From input data, the estimated interval of the mean of the IAT data chain is [8 12] and that of the standard deviation is [2 7]. Figure 3 shows that the center point of expected outcome values is 2.1 in normal cases. These parameters are to formulate the closeness degree and linguistic semantic values. As the rules defined, the expected outcome values are a set [1.82 1.98 2.1 2.24 2.5].

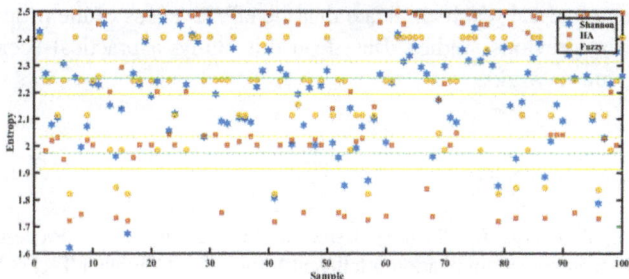

Fig. 4. The IAT data simulation in an abnormal condition.

In Fig. 4, the abnormal patterns are generated by random data numerical generator. The rectangle dot, the circle dot, and the star dot are presented to the detection rate of the proposed HA-based model and other proposals. In this experimental scenario, we enforce randomly 84 real abnormal cases in a whole range of event positions.

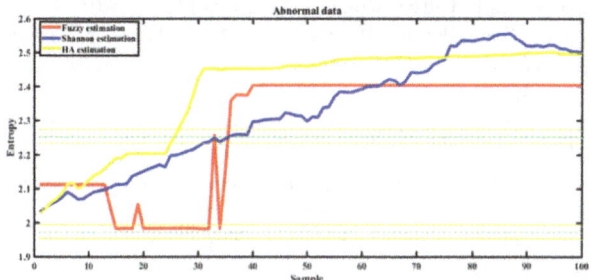

Fig. 5. Comparison of anomaly detection methods.

The outcome detection curves of three methods composed Shanon based detection, fuzzy-based detection, and HA-based detection are shown in Fig. 5. We can see that the proposed model yields results that surpass previous proposals. In detail, the detection rate of the proposed method is 65 anomaly cases and 16 suspect cases. The entropy-based method detects 42 anomaly cases. The fuzzy-based method detects 42 anomaly cases and 41 suspect cases.

5   Conclusion

In this study, the precious properties of the Hedge Algebra have been applied to an anomalous detection pattern in the flow. Uncertainty conditions were formulated by the closeness degree to limit false positive alarms in the proposed model. Hence, the performance of the proposed HA model to detect anomaly IAT patterns has been improved. The numerical results indicate that the effectiveness of the proposed method is better than the previous studies. Based on this study, a practical scenario will be applied in our future works.

References

1. Hameed, A., Alomary, A.: Security issues in IoT: a survey. In: Proceeding of 2019 International Conference Innovation Intelligent Informatics Computer Technology, Bahrain, pp. 1–5. IEEE (2019)
2. Butun, I., Osterberg, P., Song, H.: Security of the Internet of Things: vulnerabilities, attacks, and countermeasures. IEEE Commun. Surv. Tutor. **22**(1), 616–644 (2020)
3. Yang, Y., Wu, L., Yin, G., Li, L., Zhao, H.: A survey on security and privacy issues in Internet-of-Things. IEEE Internet Things J. **4**(5), 1250–1258 (2017)
4. Kaur, P., Kumar, M., Bhandari, A.: A review of detection approaches for distributed denial of service attacks. Syst. Sci. Control Eng. **5**(1), 301–320 (2017)
5. Hamada, Y., Inoue, M., Adachi, N., Ueda, H., Miyashita, Y., Hata, Y.: Intrusion detection system for in-vehicle networks. SEI Tech. Rev. (88), 76–81 (2019)
6. Alnafessah, A., Casale, G.: Artificial neural networks based techniques for anomaly detection in apache spark. Cluster Comput. **23**, 1345–1360 (2020)
7. Vijayakumar, K.P., Ganeshkumar, P., Anandaraj, M., Selvaraj, K., Sivakumar, P.: Fuzzy logic–based jamming detection algorithm for cluster-based wireless sensor network. Int. J. Commun. Syst. **31**(10), 1–21 (2018)
8. Shannon, C.E.: A mathematical theory of communication. Mob. Comput. Commun. Rev. **5** (I), 3–55 (2001)
9. Timothy, J.S., Jonathan, M.S.: Recognition strategies: intrusion detection and prevention. In: Introduction to Information Security, Syngress, pp. 253–274 (2014)
10. Bereziński, P., Jasiul, B., Szpyrka, M.: An entropy-based network anomaly detection method. Entropy **17**(4), 2367–2408 (2015)
11. Callegari, C., Giordano, S., Pagano, M.: Entropy-based network anomaly detection. In: Proceeding of the 2017 International Conference on Computing, Networking and Communication ICNC 2017, pp. 334–340 (2017)
12. Yin, X. Li, S.: Trust evaluation model with entropy-based weight assignment for malicious node's detection in wireless sensor networks. EURASIP J. Wirel. Commun. Netw. **2019**, 198 (2019)
13. Basicevic, I., Blazic, N., Ocovaj, S.: On the use of generalized entropy formulas in the detection of denial-of-service attacks. Secur. Priv. e134 (2020)
14. Nguyen, V.T., Nguyen, T.X., Hoang, T.M., Vu, N.L.: A new anomaly traffic detection based on the fuzzy logic approach in wireless sensor networks. In: Proceedings of ACM International Conference Proceedings Series, pp. 205–209 (2019)
15. Hamamoto, A.H., Carvalho, L.F., Sampaio, L.D.H., Abrão, T., Proença, M.L.: Network anomaly detection system using genetic algorithm and fuzzy logic. Expert Syst. Appl. **92**, 390–402 (2018)
16. Thakur, D., Kumar, Y., Vijendra, S.: Smart irrigation and intrusions detection in agricultural fields using I.o.T. Procedia Comput. Sci. **167**, 154–162 (2020)
17. Doddapaneni, K., et al.: Packet arrival analysis in wireless sensor networks. In: Proceedings of 2015 IEEE 29th International Conference on Advanced Information Networking and Applications Workshops, Gwangju, pp. 164–169 (2015)

Research of Large-Scaled Signalling Networks in Terms of Their Network Features and Sturdiness in Case of Mutations

Cong-Doan Truong[1], Duy-Hieu Vu[2], Duc-Hau Le[3],
and Hung-Cuong Trinh[4(✉)]

[1] International School, Vietnam National University, Hanoi, Vietnam
doantc@isvnu.vn
[2] Department of Information and Communication Technology,
University of Science and Technology of Hanoi, Hanoi, Vietnam
bill-092@st.usth.edu.vn
[3] School of Computer Science and Engineering, Thuyloi University,
175 Tay Son, Dong Da, Hanoi, Vietnam
duchaule@tlu.edu.vn
[4] Faculty of Information Technology, Ton Duc Thang University,
Ho Chi Minh, Vietnam
trinhhungcuong@tdtu.edu.vn

Abstract. The investigations on the connection between dynamics and structural features of networks in biology have been thoroughly explored. Within this research, we suggested how one can use node-based knockout mutation efficiently for researching the robustness of the dynamics of a network. This study was conducted with a randomly generated Boolean network model and a real network of human signalling. The result was that knockout-based robustness had a better correlation with features of the structures in comparison to perturbations in the beginning state. More specifically, it was found that networks with more percentage of coupled feedback loops or positive feedback loops have significantly higher chances of opposing knockout mutations. Furthermore, the higher connectivity or more feedback loops (NuFBL) a node has, the more significant it is. Overall, thanks to measuring knockout-based sturdiness of a network, we found out the connection between network robustness and biological network composition features could be determined.

Keywords: Boolean network models · Network dynamics · Knock-out mutation · Feedback and feedforward loops · Interactions · Human signaling network

1 Introduction

There have been studies about the significant relationship between network composition features and the dynamicity of networks in biology [1, 2]. The sturdiness of networks has been found impacted by feed-forward loops (FFLs) and feedback loops (FBLs), many research shows [3–5]. For instance, a robustly converged network to a set condition has a higher probability of having less negative FBLs and more positive

FBLs, a study shows [3]. Another research illustrated that coherent coupling of FBLs plays a vital role in opposing initial-state mutations [6], while a coherent FFLs is regarded as a fundamental for networks of human signalling, improving networks sturdiness opposing update-rule mutations [5]. In a node, there is a positive correlation in its functional significance and its amount of connections, and the number of FBLs involving it [7]. While the previous researches studied the relationship between dynamics robustness against both initial-state and update-rule mutations, this paper focuses on proposing a new application of using knockout mutation in analyzing the sturdiness of dynamics. This is achieved by using random Boolean network models and network of human signalling, resulting in a correlation in the network structural properties and its knockout-based mutation sturdiness.

The outline of the paper consists of three main parts. First, Sect. 2 talks about the Boolean network model and initial-state, update-rule and knockout mutation definitions. It also introduces the FBLs and FFLs. Next, Sect. 3 illustrates the simulation outcomes from using random Boolean networks and networks of human signalling. Finally, Sect. 4 concludes the study.

2   Preparations and Methodology

2.1   Datasets

Using the Kyoto Encyclopedia of Genes and Genomes (KEGG) [9] - a human signalling network large dataset, we retrieved the large-scale network data [8] of: 7,964 connections and 1,659 genes. These cell-based signalling networks are vital for most procedures in cellular activities.

2.2   Definitions of Boolean Network Models and Sturdiness of Network Dynamics

The sturdiness of any network can be analyzed using a Boolean network model [7, 10, 11]. The network can be written as a $G(V,A)$ directed graph. In which, $V = \{v_0, v_1, \ldots v_{N-1}\}$ with v belonging to V holds value of 0 ("off") or 1 ("on"), correlating to the probable patterns of corresponding genes/proteins. Meanwhile, A called directed links, is a set of 0s and 1s in ordered pairs. This connection (v_i, v_j) can hold the value of one of the following: a positive ("activating") or negative ("inhibiting"), depending on the behavior from v_i to v_j. The Boolean function $f_i : \{0,1\}^{k_i} \rightarrow \{0,1\}$, calculates every v_i at moment $t+1$, using $v_{i_1}, v_{i_2}, \ldots, v_{i_{k_i}}$ with a connection to vi at moment t, and these variables are updated synchronously. Therefore, the update rule can be written as follows: $v_i(t+1) = f_i\left(v_{i_1}(t), v_{i_2}(t), \ldots, v_{i_{k_i}}(t)\right)$.

An array of vectors from v_0 to v_{N-1} can be considered as a state of G. A state trajectory always begins at a state defined as the initial. From this point, it will converge into a limit-cycle attractor or a fixed-point. Using converging dynamics, we can state the sturdiness of the network from the interpretation of the attractor. In this research, we looked at knockout mutations for use in calculating robustness-related dynamics.

Let $s = [v_0(0), v_1(0), \ldots, v_{N-1}(0)]$ be an initial-state array and $F = [f_0, f_1, \ldots, f_{N-1}]$ be an update-rule array; in case a knockout mutation occurs at node v_i, F is then altered into $F^k = \{f_0, \ldots, f_k^i, \ldots, f_{N-1}\}$ with $f_i^k = 0$, meaning the state of v_i is rested to off value, and will remain as such for the remainder of the transition period. Therefore, the for node vi opposing the knock-out mutation, can be written as:

$$\gamma_k(v_i) = \frac{\sum\limits_{s \in S} I(\alpha(s, G, F) = \alpha(s, G, F^k))}{|S|} \tag{1}$$

Where: $\alpha(s, G, F)$ will return the attractor prior to mutation, and $\alpha(s, G, F^k)$ will return the attractor after mutation. The function I compares the two attractors, and will return 1 if they are the same or 0 if not. From this, we can define network G's robustness against knockout mutation for all nodes V of this network, written as $\gamma_k(G)$, with the following formula:

$$\gamma_k(G) = \frac{\sum\limits_{v \in V} \gamma_k(v)}{|V|} \tag{2}$$

We also consider two other mutations that have been studied in other research: initial-state and update-rule. From the array of initial state s and update-rule f, an initial-state mutation will change s into s' where $s' = [v_0(0), \ldots, 1 - v_i(0), \ldots, v_{N-1}(0)]$ meanwhile, a update-rule mutation will alter f into $f' = \{f_0, \ldots, f_i', \ldots, f_{N-1}\}$. Similarly to the knockout based mutation, we define a node's robustness against update-rule as $\gamma_r(v_i)$ and initial-state mutation $\gamma_s(v_i)$, with the following formula:

$$\gamma_s(v_i) = \frac{\sum\limits_{s \in S} I(\alpha(s, G, F) = \alpha(s', G, F))}{|S|} \tag{3}$$

$$\gamma_r(v_i) = \frac{\sum\limits_{s \in S} I(\alpha(s, G, F) = \alpha(s, G, F'))}{|S|} \tag{4}$$

Therefore, the sturdiness of network G opposing the perturbations can be calculated by:

$$\gamma_s(G) = \frac{\sum\limits_{v \in V} \gamma_s(v)}{|V|} \tag{5}$$

$$\gamma_r(G) = \frac{\sum\limits_{v \in V} \gamma_r(v)}{|V|} \tag{6}$$

2.3   Feedback and Feed-Forward Loops Concepts

2.3.1   Feedback Loop and Coupling of Feedback Loops

In a directed graph, a feedback loop is defined as a closed simple cycle $v_0 \rightarrow v_1 \rightarrow v_2 \rightarrow \ldots \rightarrow v_{L-1} \rightarrow v_L$ with $v_L = v_0$, and no nodes are repeated. A FBL has two signs, positive if there are no negative connections or an even amount; otherwise, it's treated as negative.

Given 2 feedback loops $P = u_0 \rightarrow u_1 \rightarrow \ldots u_K$ and $Q = v_0 \rightarrow v_1 \rightarrow \ldots v_L$. If there exists a path of $M = u_{i_1} \rightarrow u_{j_1}, with (i_1, j_1) \in \{0, K\}$, and $N = v_{i_2} \rightarrow v_{j_2}$, $with(i_2, j_2) \in \{0, L\}$, where $M = N$, then we can call M is the subsequence of P and Q. And if P and Q have opposite signs, then M is an incoherent coupling; else, it is coherent.

2.3.2   Feedforward Loop

Consider directed graph $G(V, A)$, if there exists two nodes: one source node v_i and one sink node v_j, which has multiple traversable paths going from the source to the sink, then that array of traversable paths is called a feed-forward loop with v_i start and v_j end. The sign of those paths is found by considering the number of negative connections in said path: positive if zero or even amount of negative connections, and negative if there are odd amounts of negative connections. The feedforward loop is coherent if every path in that set has the same sign, and incoherent if they have different signs.

3   Results

By applying Barabasi-Albert (BA) model [12], an unsystematic Boolean network array of size 1000 was generated. These has $|V| = 30$ and $|A| = 84$. We used these 1000 networks, and the KEGG signalling networks, to study the link between the knockout-based sturdiness with some composition features on the network as a whole and on the node in particular.

3.1   Correlation Between Knockout-Based Mutation Sturdiness Against the Ratio of Positive Feedback Loops

We evaluated the amount of positive FBLs and the sturdiness of the network against knockout-based and initial state mutations. As illustrated in Fig. 1, the higher percentage the positive FBL, the sturdier network is against knockout mutations. On the other hand, against initial state mutations, there was not clear evidence to suggest that the amount of positive FBLs influences the sturdiness of networks (correlation coefficient = −0.00316; p = 0.956 using one sample t test). Therefore, we could determine the robustness of a network against knockout-based mutations using its ratio of positive feedback loops, this result is also verified by previous studies [3].

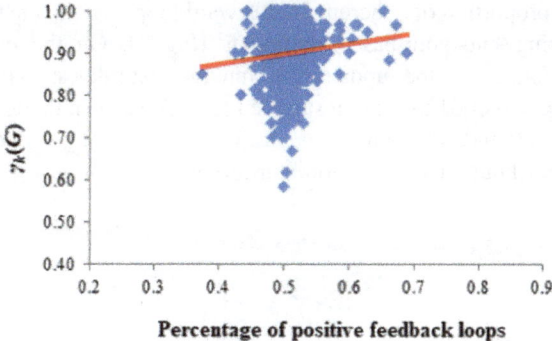

Fig. 1. Graph for correlation between knock-out mutation sturdiness and ratio of positive feedback loops (correlation coefficient = 0.23423, p < 0.01 using one sample t test)

3.2 Correlation Between Knockout-Based Mutation Sturdiness Against the Ratio of Coherent Feedback Loops

We studied the number of coherent FBLs in a network and the sturdiness of it against knockout-based and initial state mutations. Figure 2 shows that the percentage of coherent FBLs has a strong positive correlation with the sturdiness against knockout mutations, but it is not significant against initial-state mutations (correlation coefficient = 0.01302, p = 0.855 using one sample t test for initial-state mutations). From this, we could deduce that the robustness of a network is also dependent on the proportion of coherent feedback loops.

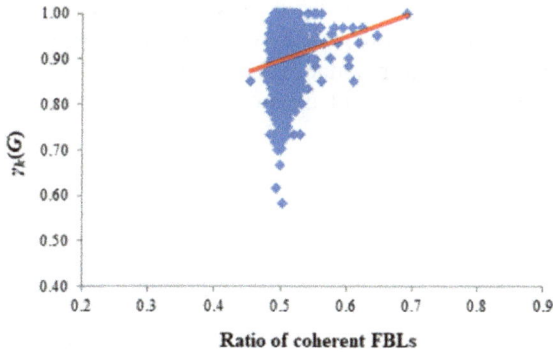

Fig. 2. Graph of correlation between knockout-based mutation sturdiness and ratio of coherent feedback loops (correlation coefficient = 0.52060, p < 0.0001 using one sample t test)

3.3 Correlation Between Knockout-Based Mutation Sturdiness Against the Ratio of Coherent Feedforward Loops

We also considered the amount of coherent FFLs in a network and its relation to the robustness of knockout-based and update-rule mutations. We did not find an important correlation of the proportion of coherent feedforward loops within a network against its robustness opposing knockout-based mutations (Fig. 3). On the contrary, a good correlation was found on the update-rule mutation sturdiness: correlation coefficient = 0.22840, p < 0.0001 using one sample t test. Thus, even though coherent FFLs are seen as a design foundation of networks for human signalling, it does not contribute significantly to knockout-based mutation sturdiness.

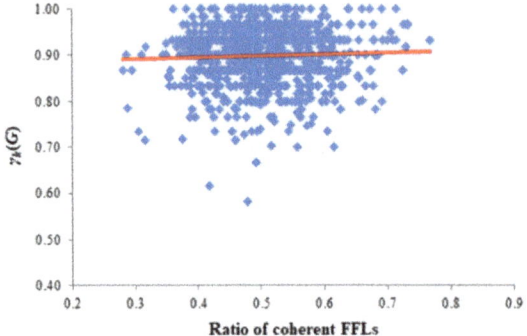

Fig. 3. Graph of correlation between knockout-based mutation sturdiness and ratio of coherent feedforward loops (correlation coefficient = 0.03588, p = 0.169 using one sample t test)

3.4 Correlation Between Knockout-Based Mutation Sturdiness Against Structural Features in Node-Based Level, on Random Generated Networks and KEGG Human Signalling Networks

The importance of the functions of a node in a network is based on the likeliness of the network converging into another attractor if the node alters in a knockout perturbation, in other words, a node has high importance if it has high sturdiness against mutations. Firstly, we studied the relationship of the number of connections to a node and its

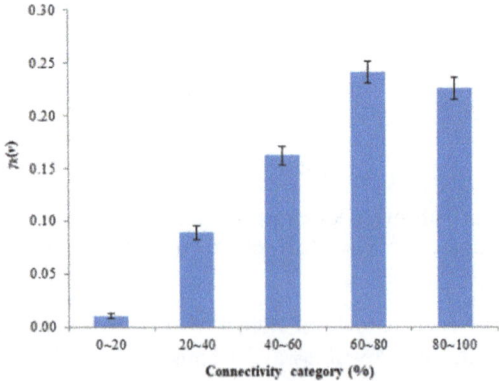

Fig. 4. Relationship between functional importance of a node and its number of connectivities in random generated networks (1000 networks, |V| = 30, |A| = 84)

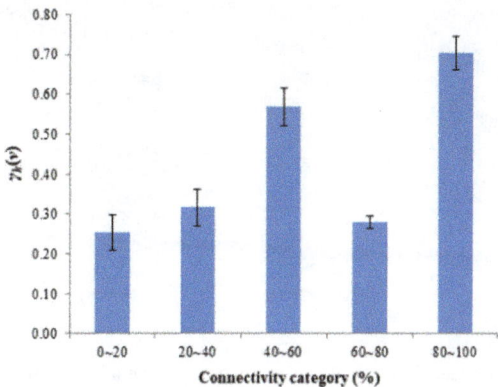

Fig. 5. Relationship between functional importance of a node and its number of connectivities in KEGG human networks

functional importance, by dividing the nodes into five groups depending on their connectivity. We found significant positive correlations in Fig. 4 and Fig. 5 in the random generated networks and KEGG human signalling networks.

Secondly, we also looked at the number of FBLs going through a node and the significance to the functional value of said node. In the random generated network, the nodes were divided into 5 categories depending on their NuFBLs categories; however, KEGG human signalling network, only a few nodes (152 nodes) had feedback loops passing it, so it is divided into 2 categories: "feedback loop" and "no feedback loop". The result is as follow (Fig. 6 and 7):

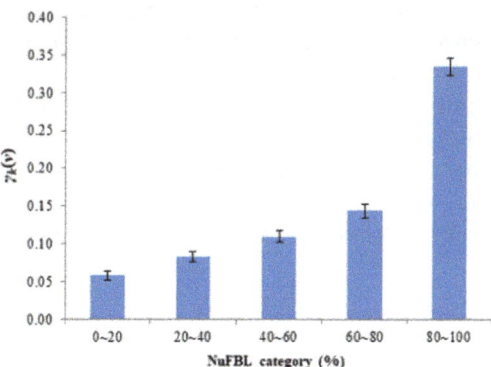

Fig. 6. Relationship between functional importance of a node and its NuFBL category in random generated networks (1000 networks, $|V| = 30$, $|A| = 84$)

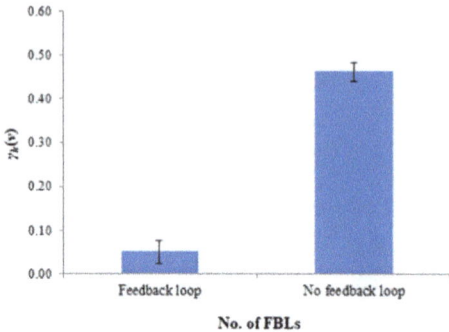

Fig. 7. Relationship between functional importance of a node and its NuFBL category in KEGG human signalling networks

For all four of the graphs above, the confidence level of the functional significance of the nodes are very high (95%). In light of this, we could claim that within a network, nodes with more connections and involved in more FBLs are more significant, because when these nodes are altered by mutation, the network is more likely to converge to a different attractor.

4   Conclusion

Within the scope of this project, we had presented a new functionality of perturbations by knockout: inspecting the sturdiness of the dynamics of a network. By analyzing human signalling networks and random Boolean network models, some interesting correlations between compositional features of a network and its knock-out mutations robustness could be found. Specifically, some significant observations on the effect of positive FBL and coherent couplings of FBLs on the knock-out based mutation robustness were made. Not only that, in the scope of nodes, we found out that there are evidences suggesting that, a node with more connections and amount of feedback loops (NuFBL) passing through, would be more important to the network, since the knockout perturbation of that node makes the network more likely to converge into another attractor. To summarize, the link between signaling network dynamics and structural features could be studied by measuring the sturdiness of that network against knockout mutations.

References

1. Klein, C., Marino, A., Sagot, M.-F., Milreu, P.V., Brilli, M.: Structural and dynamical analysis of biological networks. Briefings Funct. Genomics **11**, 420–433 (2012)
2. Prill, R.J., Iglesias, P.A., Levchenko, A.: Dynamic properties of network motifs contribute to biological network organization. PLoS Biol. **3**, e343 (2005)
3. Kwon, Y.-K., Cho, K.-H.: Quantitative analysis of robustness and fragility in biological networks based on feedback dynamics. Bioinformatics **24**, 987–994 (2008)

4. Kremling, A., Bettenbrock, K., Gilles, E.D.: A feed-forward loop guarantees robust behavior in Escherichia coli carbohydrate uptake. Bioinformatics **24**, 704–710 (2008)
5. Le, D.-H., Kwon, Y.-K.: A coherent feedforward loop design principle to sustain robustness of biological networks. Bioinformatics **29**, 630–637 (2013)
6. Kwon, Y.-K., Cho, K.-H.: Coherent coupling of feedback loops: a design principle of cell signaling networks. Bioinformatics **24**, 1926–1932 (2008)
7. Kwon, Y.-K., Choi, S., Cho, K.-H.: Investigations into the relationship between feedback loops and functional importance of a signal transduction network based on Boolean network modeling. BMC Bioinformatics **8**, 384 (2007)
8. Kim, J.-R., Kim, J., Kwon, Y.-K., Lee, H.-Y., Heslop-Harrison, P., Cho, K.-H.: Reduction of complex signaling networks to a representative kernel. Sci. Sig. **4**, ra35- (2011)
9. Kanehisa, M., Goto, S.: KEGG: kyoto encyclopedia of genes and genomes. Nucleic Acids Res. **28**, 27–30 (2000)
10. Kauffman, S., Peterson, C., Samuelsson, B., Troein, C.: Random Boolean network models and the yeast transcriptional network. Proc. Natl. Acad. Sci. **100**, 14796–14799 (2003)
11. Shmulevich, I., Lähdesmäki, H., Dougherty, E.R., Astola, J., Zhang, W.: The role of certain post classes in Boolean network models of genetic networks. Proc. Natl. Acad. Sci. **100**, 10734–10739 (2003)
12. Barabási, A.-L., Albert, R.: Emergence of scaling in random networks. Science **286**, 509–512 (1999)

Designing System of Receiving and Processing Urine Flow Rate for the Diagnosis of Benign Prostate Enlargement

Vu Tran[1], Huong Pham[2(✉)], Lam Ninh[1], Khanh Le[1], Huy Hoang[1], and Hao Nguyen[3]

[1] Hanoi University of Science and Technology, Hanoi, Vietnam
[2] International School, Vietnam National University, Hanoi, Vietnam
huongptv@isvnu.vn
[3] VietDuc Hospital, Hanoi, Vietnam

Abstract. Uroflowmetry is one of the most useful tools for patients with lower urinary tract symptoms (LUTS). Several tests with uroflowmetry and data analysis have been used. However, these approaches are costly, time consuming and the test cannot be kept confidential. Hence, there is an urgent need for an automatic and confidential test in Vietnam. This paper introduces a prototype system for uroflow-measurements using wireless implementation, which is simple and guarantees patients' privacy. It helps doctors track the symptoms of patients and choose the suitable treatment. The doctor will log into the device and continue to process and display the urinary flow measurement results. This research also includes preliminary test system to evaluate the accuracy of the system.

Keywords: Uroflowmetry · Lower urinary tract symptoms · Preliminary system

1 Introduction

Benign prostate hyperplasia (BPH) is a common condition when men get older, especially after 50 years old. An enlarged prostate gland can cause unpleasant urinary symptoms, such as blocking the flow of urine from the bladder. It can also cause other problems to the bladder, urinary tract or kidney. People are diagnosed with BPH when the volume of the gland exceeds 30 cm^3 and the maximum flow of urine is less than 15 ml/s. It is estimated that BPH occurs in about 50% of men aged 50 years old, and nearly 90% when they are 80–90 years old [1].

Normally, in hospitals or clinics, patients were asked to answer the IPSS (International Prostate Symptom Score) questionnaire about their symptoms. Then, a variety of specialized clinical and laboratory tests are performed (for example, prostate-specific antigen level - PSA). Ultrasound test of the urinary tract is performed to verify the amount of the remaining urine in the bladder. However, with the increasing number of patients in recent years and a wide variety of symptoms, this method has become complicated and time consuming for both doctors and patients. In general, ultrasound diagnosis is an indispensable phase in diagnosis of related diseases, but due to this fact, the risk of cross-infection between patients will be extremely high while measurements are taken.

Uroflowmetry is a first line screening for patient with lower urinary tract symptoms (LUTS) recommended by Good Urodynamic Practice 2002(ICS-GUP2002) [2]. It is a noninvasive urodynamic test which provides many useful information such as maximum flow rate, average flow rate, urination volume, urination time, time to maximum flow, time delay(hesitation) for diagnosis and follow-up of urological disease progression, especially in patients with lower urinary tract symptoms (LUTS) [3]. IPSS is a vital tool for the assessment and ranking of (LUTS). Correlation between IPSS and uroflowmetry data will continue to enhance the symptom identification and treatment of patients. It can also be inferred that IPSS and uroflowmetry may be used to assess and track patients after prostate surgery.

In reality, some devices have been manufactured and marketed with ±5% accuracy with each value over time and ±2% with the whole processes, storing data via SD card; using RS-232; USB or Bluetooth connection to connect to PC through desktop software [2, 4]. However, these devices are accompanied by 80–230 VAC power supply, wire connection, trolleys, and some devices also add an EMG channel that makes the cost of the device very high. Due to the high cost and complexity of the device, not many devices are used, especially in Vietnam. The accessibility of over-numbered patients per device is very limited.

With all the above reasons, a modern solution is required, inheriting the advantages of the classical method and overcoming its drawbacks, which is non-invasive, fast, precise, and minimize unnecessary contact between patients, ensuring privacy with wireless connections between computers and devices. Our paper proposed the urinary flow measurement model with the advantages of collecting data from the patient, and then presenting the calibration methods, including weight calibration, and urine flow simulation. The paper includes four parts. Section 1 is the introduction. Section 2 presents the methodology. Section 3 provides the experiment setting and results. Section 4 concludes the paper.

2 Methods

2.1 Uroflowmetry

Urine flow rate has 3 types based on measurement method: weight transducer, rotary disc and test strip [5, 6]. The weight transducer method uses a weight sensor (Load cell) that measures urine flow over time to calculate the flow rate of urine. The rotary disc method calculates the flow by measuring the difference between the used power when rotating the disc. On the other hand, the test strip method calculates by measuring the capacitance that changes when activating the sensor by the urinary volume [3]. After comparing these methods, we chose to develop our prototype based on the principle of weight transducer, which has high precision of the direction and position of urine when compared to the rotary disc method.

Normal urine flow rate plots are presented as a bell curve (Fig. 1) with a flow rate of about 20–25 ml/s in men [7]. Whereas a maximum flow <10 ml/s would suggest pathology [8]. The parameters used to display details of the urinary flow measurement test are listed below [4, 9].

Maximum flow rate (Q $_{max}$) - the maximum flow rate during urination.
Average Flow (Q $_{ave}$) - total volume of urine divided by the total time to urinate.
Urination time - the total time to urinate including any interruptions.
Time to maximum flow - the time from the start to reach maximum flow
Time delay (hesitation) - the time between the feeling of wanting to urinate and the
time when the urination is started (it usually less than 10 s).

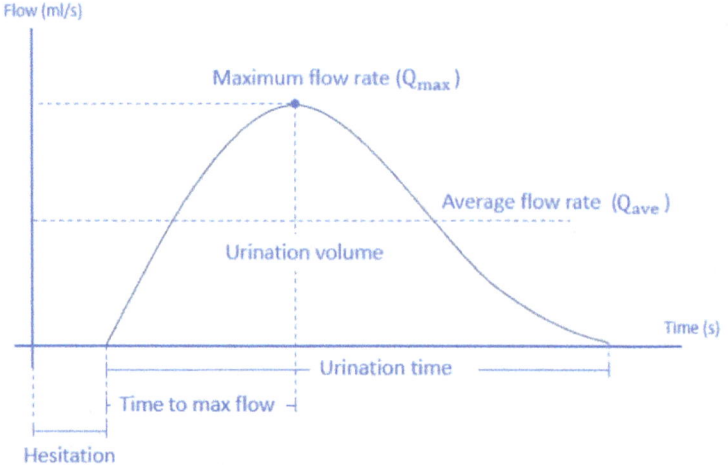

Fig. 1. Typical flow rate graph with parameters

2.2 Block Diagram

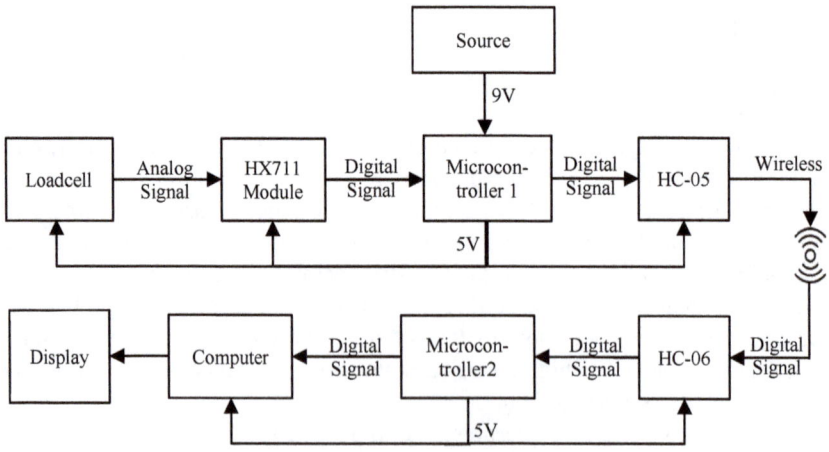

Fig. 2. Block diagram system

Figure 2 shows the block diagram of the system. The analog signal obtained from the sensor is converted to digital signal via the HX711 module. HC-05 is the transferring module while HC-06 is the receiving module. HX711 module sends data to micro-controller, then microcontroller 1 sends data to microcontroller 2 via Bluetooth connection, then microcontroller sends signal to computer and displayed on the application interface. The microcontroller 1 is supplied with the source power of 9 V which then converts to 5 V to provide for each unit left of the device, microcontroller 2 and the Bluetooth HC-06 supplied with 5 V output source from computer.

The Bluetooth coverage area is less than 15 m from the hotspot. The downsides of Bluetooth are the weak connection in an environment with lots of obstacles, long set-up time (about 5–10 s), low speed (max about 720 Kps). In our projects the data is small, so Bluetooth is considered suitable and does not affect the data in the project scale. In the designed prototype unit, we implemented a data transmission model using two HC-05 and HC-06 modules, in which HC-05 is used to send signals from the device (called master); and HC-06 is used to receive signals transmitted from the device and send it to the computer through the connection port (called slave), with a data transmission rate of 9600 bps, 8 data bits, 1 stop bit, there is no parity.

2.3   Prototype Device

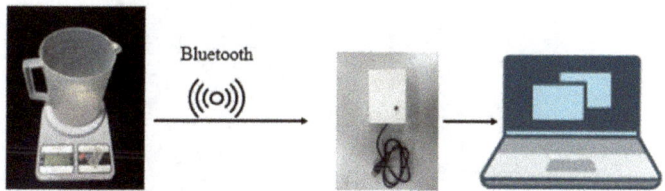

Fig. 3. The hardware components of proposed device

Figure 3 presents the proposed uroflowmeter system. The measurement part includes a big cup (2000 ml) placed on a scale to hold the urine. The scale is designed with components as showed in Fig. 4, including a 5 kg load cell, modul HX711, module Arduino and a Bluetooth sending module Hc-05.

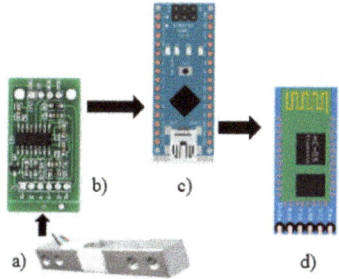

Fig. 4. The hardware components of uroflowmetry equipment. a) Loadcells 5 kg, b) Module HX711, c) Module Arduino, d) Bluetooth sending module HC-05

The load cell is used to measure the weight of urine in the cup. Its weight measurement capacity is 5000 g with 0.3 mg resolution. The sampling rate of the data taken from the scale is 10 Hz. The signal obtained from the loadcell is analog, so we use HX711 to convert into digital signal. The signal is collected from the beginning to the end of the urinate process. The ADC converter has a resolution of 24 bits, it ensures that the data is captured as accurately as possible. Module Arduino receives data from the HX711 and sends to Bluetooth sending module HC-05. All components are placed inside a waterproof case of the electronic scale to ensure safety.

The HC-06 Bluetooth module receives data from the module HC-05 via Bluetooth connection. Figure 5 is the receiving part, including module HC-06 and a microcontroller 2 that communicates with the computer via a USB connection for processing and storing data. When linked to the device, the LED light of the HC-06 unit (the circular component below) will light up 1 s slower than the LED of the microcontroller, while the link between HC-06 and HC-05 is disconnected, the HC-06 LED will flash constantly every 0.1 s. When linked, the LED will slow down approximately 2 cycles every 1 s, and when connected successfully, the LED remains on.

Fig. 5. Receiving module

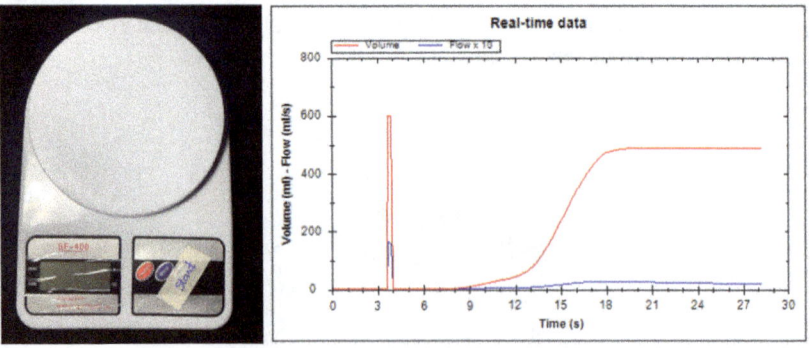

Fig. 6. The complete device is photographed from the top down and graphs tested when a force is applied to the device's face after 6 s

Time from the beginning of the measurement to the beginning of urine is not identical, we know it as hesitation time. The precision in determining exactly that time period is crucial. Therefore, we have designed a "Start" button on the scale to determine when to start the measurement as shown in Fig. 6 (left). When the patient begins to urinate, press this button to start the measurement, then on the application display will appear a pulse with an amplitude of 600 at the 3rd second as shown in Fig. 6 (right). In this experiment we have simulated the normal human urination, the start time is at 3 s and the release of the valve is at 7 s. There is also a button located on the receiver, as shown in Fig. 5 on the left, it's used to re-establish the connection.

3 Calibration

3.1 Calibration Loadcell

The signal from the loadcell after passing the HX711 will show a number in the range 2^{24}, so we have to find the formula for the calibration coefficient with real weight.

Calibration weight includes 1–50 g set weights and 50–200 g set weights. The scale is set to measure up to 5 kg with 0.05 g error. HX711 considered the correlation between displayed results and weights is a linear function.

The calibration weights are placed on the surface of the device, for each weight level of the weights, we record the displayed results in the Excel table. From the data sheet recorded as in Table 1, we calculate the corresponding correction coefficient for loadcell.

Table 1. Recorded data after weight placement

Number of trials	1	2	3	4	5	SUM (ave)
Weight	Ave	Ave	Ave	Ave	Ave	Average
5	1516.5	1516	1539	1529.5	1517	1523.6
15	4264.5	4259.5	4283	4278.5	4271.5	4271.4
20	5850.5	5537	5853	5850.5	5857	5789.6
26	7521.5	7527.5	7539	7568.5	7535	7538.3
30	8604	8603	8580	8747	8561.5	8619.1
41	11770	11770.5	11776.5	11763	11780	11772
48	13825	13802	13823	13800.5	13818	13813.7
50	14451.5	14460.5	14442.5	14453	14460.5	14453.6
62	18119	18130.5	18110	18147	18163	18133.9
69	19926	19919	19948.5	19946.5	19909.5	19929.9
86	24967	24970.5	24992.5	24973	24995	24979.6
97	28254.5	28240	28267	28283.5	28244	28257.8
100	29017	29037.5	29019	29017	29009	29019.9
131	37868.5	37878	37874.5	37870	37882	37874.6
147	42780	42789	42789.5	42795	42749	42780.5
207	59548.5	59560.5	59570	59559.5	59567.5	59561.2

Using the tool "Insert Scatter (X, Y)" we find the linearity of the data in Table 1. The correction factor is calculated through the tool "Display Equation on chart" in the toolkit of Microsoft Excel is in the formula in Fig. 7.

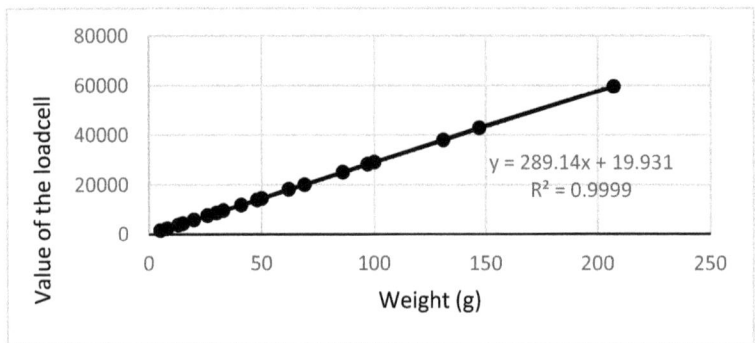

Fig. 7. Calibration formula of loadcell; y: display value of the loadcell; x: actual value of the weight

The chart shows all the data are projected into a linear function. Hence, the calibration factor of the loadcell that we are testing is 289.14. We added the correction factor to the old load code, the calibration experiments were rerun after loading the new code, the results showed that it works well.

3.2   Evaluate the Accuracy of the System

After the calibration of the system is completed, we evaluate the accuracy of the system. We have conducted the same experimental steps as the load cell calibration, the results in Table 2 show error from 0.05%–1.2%. This result is completely suitable for items weighing less than 5 kg.

Table 2. Results obtained when evaluate by the heavy object

Reality (g)	First time		Second time		Third Time		Fourth Time	
	Weight obtained (g)	Error (%)	Weight obtained (g)	Error (%)	Weight obtained (g)	Error (%)	Weight obtained (g)	Error (%)
20	20.22	1.09	19.76	1.2	20.17	0.85	20.19	0.951
50	50.2	0.4	49.89	0.22	49.76	0.48	50.15	0.3
100	100.06	0.06	99.98	0.02	99.95	0.05	100.1	0.1
200	200.1	0.05	200.14	0.07	199.9	0.05	199.94	0.03
250	250.04	0.016	250.09	0.036	249.9	0.04	248.98	0.408
300	300.3	0.1	300.2	0.067	300.25	0.083	300.18	0.06
350	350.15	0.043	350.4	0.114	350.5	0.143	349.8	0.057
400	399.2	0.2	400.2	0.05	400.6	0.151	400.4	0.1
450	450.6	0.134	449.2	0.178	449.4	0.134	450.8	0.178
500	498.9	0.22	500.8	0.16	500.51	0.102	500.9	0.18

3.3   Urine Flow Simulation Measurement

To simulate urine flow, we set up the experiment as shown in Fig. 8. The valve is used to control the flow rate of water from a water tap.

Fig. 8. Urine flow simulation experiment

The urinary flow measurement system was tested in preliminary research that simulated urine flow with tested flow rates of 15 ml/s, 20 ml/s, and 30 ml/s. Table 3 below shows the results of the above test. The collection stream rate is 180 s long, recording the speed over time in the minimum of 0.1 s. Based on the obtained results, there is an error in the measurement process, the speed error is from 0.06%–1.3%. These errors are partly due to the fault of the setting up flow rate and the system itself. In this case, the error is small and acceptable.

Table 3. Results obtained when testing the flow rate by simulation

Flow rate (ml/s)	Velocity obtained (ml/s)	Error (%)
15	15.2	**1.3**
20	20.15	**0.75**
30	30.02	**0.06**

3.4   Preliminary Results

The urinary flow measurement system was tested in the preliminary study between volunteers and team members (n = 2). The figure below (Fig. 9) shows the data after the test run.

The desktop software used in the prototype trials displayed patient identification information and displayed the data as two graphs; volume-time graph (red line) and flow-time graph (flow x 10, blue line). Flow metering test also shows calculated parameters; flow max, average flow, voiding time, time to max flow and total volume. Furthermore, our product is also capable of calculating average velocity, maximum acceleration, number of time interval.

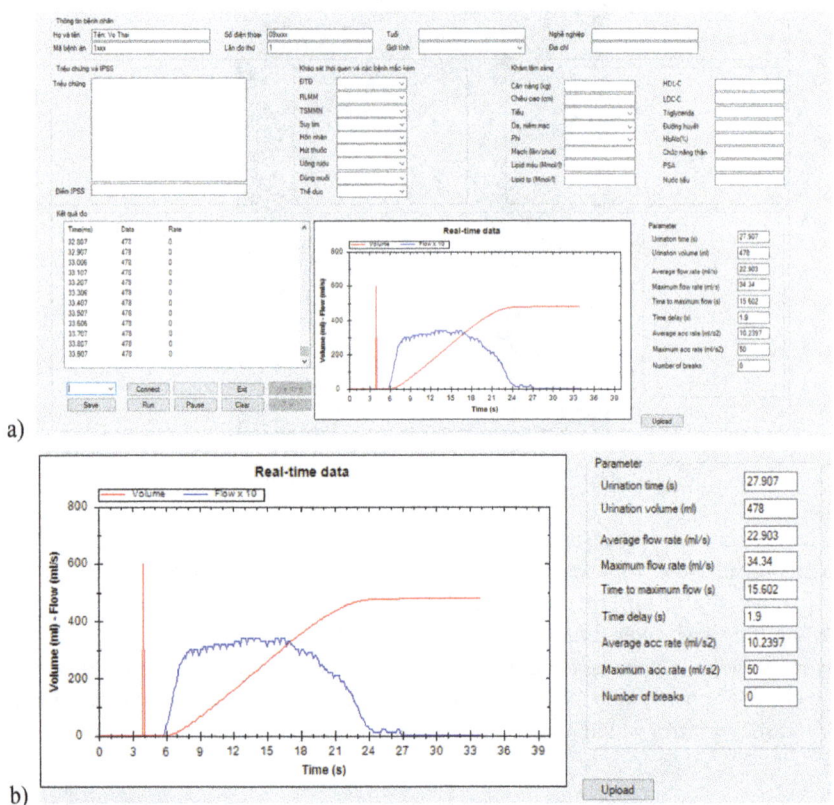

a)

b)

Fig. 9. Data were recorded from preliminary testing. a) Desktop software and data visualization; b) data measured

The parameters obtained from the device as shown in Fig. 9 show that the condition of the volunteer urinary tract is normal, without any abnormalities. The above parameters are very useful for the initial check of the lower urinary tract and related diseases to help doctors classify and detect early to avoid a bad risk.

Prototype testing volunteers noted a few problems with the product as follows: inconvenience in cleaning containers after testing is over, system waiting time starting too long (about 1 min) and the distance of the wireless connection under obstructive environments is less than 10 m, unlike the ideal theoretical conditions of 15 m. Data and signal transmission method with wireless connection tested at 2 points without any communication problems; Building C9, Hanoi University of Technology and Building D8, Hanoi University of Technology.

4 Conclusion

Preliminary systems for urinary flow measurement at Clinics or Hospitals with wireless deployment using Bluetooth technology show promise potential to create low-cost and easy-to-use urinary flow methods.

Our project also showed accurate result of maximum flow rate, average flow rate, urination volume, urination time, time to maximum flow, time delay (hesitation) and with what mentioned it will certainly capable of contributing to the future of health system in Vietnam.

Our coming targets focus on redesigning electronic components and usability of the product, further precise testing, bigger community scale diagnosing and research the correlation of time-based age parameters with IPSS ratings.

References

1. Thomas, A.P., Abrams, P.: Lower urinary tract symptoms, benign prostatic obstruction and the overactive bladder. BJU Int. **85**(3), 57–68 (2000). Discussion 70-1
2. Schafer, W., et al.: Good urodynamic practices: uroflowmetry, filling cystometry, and pressure-flow studies. Neurourol. Urodyn. **21**(3), 261–74 (2002)
3. Gabuev, A., Oelke, M.: Latest trends and recommendations on epidemiology, diagnosis, and treatment of benign prostatic hyperplasia (BPH). Aktuelle Urol. **42**(3), 167–178 (2011)
4. Rattanasomrerk, S., et al.: Development of tele-home uroflowmetry system with NB-IoT implementation. In: 12th Biomedical Engineering International Conference (BMEiCON), Ubon Ratchathani, Thailand, pp. 1–4 (2019)
5. McAninch, J., Lue, T.: Smith and Tanagho's General Urology, 18th edn. McGraw Hill Companies Inc., New York (2013)
6. Wein, A.J., et al.: Campbell-Walsh urology, vol. 3 (2016)
7. Lee, S.W.H., Chan, E.M.C., Lai, Y.K.: The global burden of lower urinary tract symptoms suggestive of benign prostatic hyperplasia: a systematic review and meta-analysis. Sci. Rep. **7**(1), 1–10 (2017)
8. Hann-Chorng Kuo, M.: Interpretation of uroflowmetry. Incont. Pelvic Floor Dysfunct. **2**, 51–55 (2007)
9. Aganovic, D.: The role of uroflowmetry in diagnosis of infravesical obstruction in the patients with benign prostatic enlargement. Med Arh. **58**(1 Suppl 2), 109–11 (2004)

An Improved Relay Selection of Cooperative MIMO Based on Channel Priority Matching Algorithm

Trung Tan Nguyen[✉]

Faculty of Radio-Electronic Engineering, Le Quy Don Technical University, Hanoi, Viet Nam

Abstract. In recent years, cooperative communication as well as relay selection problem have been known as a new research area that may provide many interesting applications in practical networking system. In this paper we consider a cooperative multiple input multiple output (Co-MIMO) system in which multiple antenna of mobile terminals are used. We examine the content of centralized cooperative relay selection and propose a new relay selection method based on channel priority matching algorithm. In the specific algorithm, we presented thresholds for Worst-Link-First Matching (WLF). Theoretical and simulated results indicated that our proposed algorithm can achieve higher performance and lower computational complexity.

Keywords: Cooperative MIMO · Worst-link-first matching · Relay selection

1 Introduction

Cooperative communication is a new research area that may have many interests due to spatial diversity gain, system throughput increase and multi-antenna terminals are not a requirement [1–3]. In cooperative communications network, transmission between source and destination was helped by a relay. There are more than one candidate relays, which makes the question how to select relay to maximize the system performance. Beside that, MIMO technique has known as a effective solution in performance improvement and reducing fading influence on radio communications systems. The combining of cooperative communication and MIMO is a trend.

There are many previous works focused on cooperative relay selection. In reference [4], Zinan Lin considered the concept of the user cooperative area, only when users in this area were chosen to be relay, this method required distance information between different nodes.

This paper examined Decode-and-Forward (DF) scheme and centralized relay selection problem, proposed a new cooperative relay selection method based on WLF algorithm extended to MIMO systems. In the specific algorithm, we presented thresholds for WLF. Threshold 1 was established to prevent channel gain very low or no gain after cooperation. Threshold 2 guaranteed maximum system performance. Relay selection based on channel matrix used in this paper. Thus, it is not important for BS to know location information of users. Analysis and simulation results show that the

D.-T. Tran et al. (Eds.): ICISN 2021, LNNS 243, pp. 456–464, 2021.
https://doi.org/10.1007/978-981-16-2094-2_55

proposed algorithm can yield higher performance and lower computational complexity compared with the traditional WLF algorithm in [5].

The rest of paper is organized as follows. System model is described in Sect. 2. Section 3 presented channel priority matching algorithm between users. Section 4 shows simulation analysis in algorithm performance of the priority matching between users. Finally, a conclusion is provided in Sect. 5.

2   System Model

This paper uses system model shown in Fig. 1 in which the destination (BS of radio cell or an access point in a WLAN) can supports N mobile users.

Any user can cooperate with another user. In our model, each user is equipped with multiple antenna. The numbers of antenna and relay node are optional. For simply purpose, assume all equipment uses 2 antennas, 2-hop transmission via a selected relay. s is the transmitted signal from the source nodes, $s = [s_1, s_2]^T$ in which s_n is the symbol transmitted from the n antenna of the source node and $n = 1, 2, \ldots.$ is the antenna index. The communication between the source and destination is done in 2 phases. Phase 1 is the direct link between two nodes and phase 2 is the relaying link, a selected intermediate node decoded received signal and forwarded it to destination. In this model, we only take interested in link between source and intermediate relay node. The channel matrix between the source and any intermediate node j, $j = 1, 2, \ldots, K$ is defined as

$$H^{sr_j} = \begin{bmatrix} h_{11}^{sr_j} & h_{12}^{sr_j} \\ h_{21}^{sr_j} & h_{22}^{sr_j} \end{bmatrix} \tag{1}$$

y_j are the received signal at the relay node j.

$$y_j = H^{sr_j}s + z_{r_j} \tag{2}$$

where z_{r_j} is the noise vector affecting the receiver of the relay j.

Because cooperation is not always useful, a pair of users can choose not to cooperate if it does not provide energy gain. In that case, they communicate with BS via a conventional non-cooperative scheme.

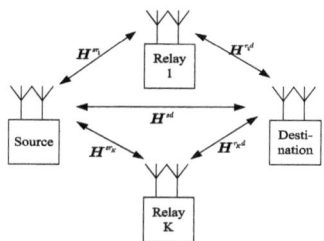

Fig. 1. System model

3   Channel Priority Matching Algorithm Between Users

Traditional WLF algorithm is shown in Fig. 2 [5]. The user with the worst channel quality (furthest from BS) consumes more energy for transmission than the near one. Therefore, these users need a higher priority level because of poor channel quality and energy-consuming quantity. This requires location information for all users. The algorithm selects the user with the best channel quality matches the user with the worst channel quality matches, and then removes those two users, selects the best one and the worst user in the remaining to match each other until the number of users have not matched is less than 2 as shown in Fig. 2.

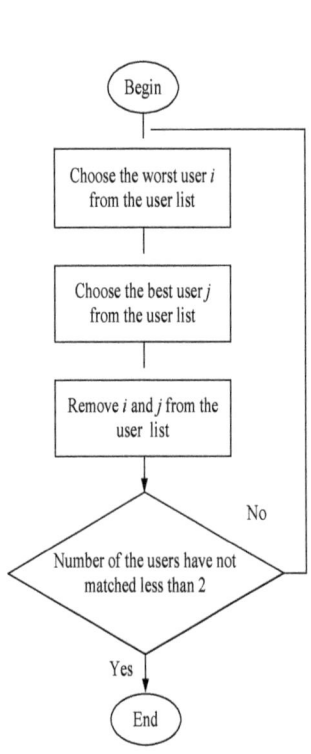

Fig. 2. Classic WLF algorithm process

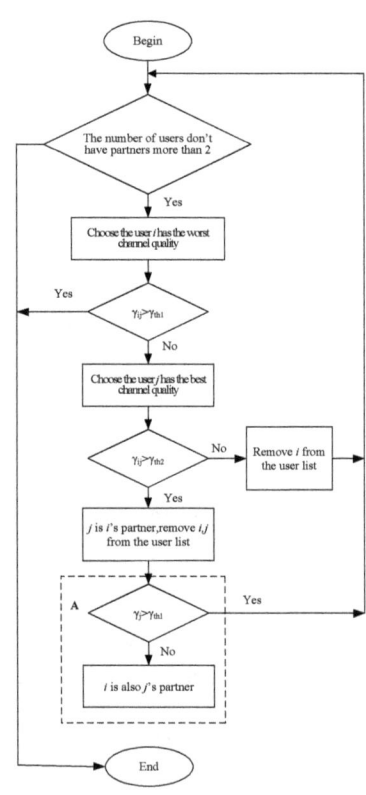

Fig. 3. WLF algorithm with threshold

A new method in Fig. 3 uses channel matrix H to appreciate the channel condition and relay selection instead of the distance energy loss. So it is not important for the base station to control location of all users. Matrix H is included in CSI as basic information provided in MIMO systems use space-time block code. In this algorithm, relay j self-calculated attitude square of channel between relay j, source and destination, respectively. The minimum value is selected. The intermediate relay nodes compare all of these minimum values and choose the maximum one. Selection algorithm can be described by the following equation

$$h_j = \min\left\{ |h_{sj}|^2, |h_{jd}|^2 \right\} \tag{3}$$

The relay j that has h_j maximum is the best selected relay node.

The delay time at j th relay node is $T_j = \frac{\lambda}{h_j}$ with λ is constant.

In the specific algorithm, we presented thresholds for WLF. Threshold 1 is to avoid the situation as channel conditions between the users and the base station are so rich that no necessary to cooperation or cooperation provides immeasurable gain. This threshold implies that relay and relay selection just are applied when extremely necessary to minimize complexity. Threshold 2 was set up to guarantee maximum average energy gain. Threshold 1 and 2 can be expressed in Eq. 4. In Fig. 3, A procedure is to avoid wasted resource when channel station is good so no essential for user j does not need to select a partner.

Assume that space-time coding used in transmitter and maximum likelihood decoding for receiver. Signal-to-noise ratio in the process of selecting should be

$$\gamma = \frac{E\|H\|_F^2}{2N_0} \tag{4}$$

is required energy to send 1bit information. N_0 is one-sided power spectral density of Gaussian white noise. H is channel matrix for the corresponding link, $\|\bullet\|_F^2$ is Frobenius square norm, that is $\|H\|_F^2 = \sum_{i=1}^{M_t} \sum_{j=1}^{M_r} |H_{j,i}|^2$, $H_{j,i}$ is the channel coefficient from the $j - th$ transmitting antenna to the $i - th$ receiving antenna.

Using BPSK modulation, the instantaneous bit error rate in the non-cooperative scheme can be expressed as

$$P_1 = Q(\sqrt{2\gamma_{SR}}); \qquad Q(x) = \frac{1}{\sqrt{2\pi}} \int_x^\infty e^{-\frac{t^2}{2}} dt \tag{5}$$

Similar in the case of cooperation

$$P_{CO} = P_1 \times \alpha + (1 - P_1) \times P_2 \tag{6}$$

P_1 for phase 1, error probability at relay can be $P_1 = Q(\sqrt{2\gamma_{SR}})$.

P_2 for phase 2, when relay decoded received signal from source, error probability at destination is expressed as $P_2 = Q(\sqrt{2(\gamma_{SR} + \gamma_{RD})})$.

α for the error probability at destination when relay decoded incorrectly. When relay decodes incorrectly, which means the source sends BPSK s, while relay forwards $-s$ so when $\gamma_{SD} \geq \gamma_{RD}$, it is equivalent to send s, so

$$\alpha = Q(\sqrt{\frac{(\gamma_{SD} - \gamma_{RD})^2}{\gamma_{SD} + \gamma_{RD}}}) \tag{7}$$

When $\gamma_{SD} < \gamma_{RD}$, it is equivalent to send $-s$, so

$$\alpha = 1 - Q(\sqrt{\frac{(\gamma_{SD} - \gamma_{RD})^2}{\gamma_{SD} + \gamma_{RD}}}) \tag{8}$$

System performance evaluation based on the average BER usually used for the distributed algorithm. In the centralized algorithm, user matching is performed by the base station, so applying it is no longer suitable (because BER of individual users may be high however the average BER of the system is low). Thus this paper appreciated system performance based on the average energy gain. The definition can be written as [5].

$$G = 10\lg\left(\frac{\sum_{i=1}^{N} E_{bi}^{no}}{\sum_{i=1}^{K} (E_{bi}^{S} + E_{bi}^{R}) + \sum_{i=K+1}^{N} E_{bi}^{no}}\right) \tag{9}$$

In the cellular uplink channel, energy loss is a significant problem. Thus, energy gain can be a measurement for the relay selection algorithm. Generally, there is a requirement for minimum bit error rate in each system, under a given acceptable BER, the average energy gain is used in terms of energy saving in cooperation compared to non-cooperation. Assume that a system can support N users, K users is matched to cooperate, the remaining users do not have a partner. E_{bi}^{m}, (m = R, S, no) represents the required energy for source node (user i) to transmit 1 bit in non-cooperative scheme. E_{bi}^{S}, E_{bj}^{S} represents the required energy for source node (user i) and relay node (user j) to transmit 1 bit in cooperative scheme. $\sum_{i=1}^{N} E_{bi}^{no}$ is independent on matching algorithm, from Eq. 9 we can see that when $\sum_{i=1}^{K} (E_{bi}^{S} + E_{bi}^{R})$ is minimum, the system can achieve maximum average energy gain.

4   Simulation Analysis Performance of the Priority Matching Between Users

4.1   Simulation Parameters

The simulation conditions are summarized in Table 1. We assume that there are N mobile users, Rayleigh fading channel, the base station can know CSI between the terminals and the base station.

Setting of the Threshold 1: Figure 4 shows how the average power gain of multi-antenna system changes with threshold 1 (γ_{th1}) when the number of users are 60, 80, 100. Threshold 1 was established to prevent channel gain from being very low or no gain after cooperation. This reduces the complexity of the system and avoids unnecessary cooperation. At this time, threshold 2 (γ_{th2}) is replaced by two theoretical thresholds to guarantee the transmission performance of the cooperation. Two theoretical thresholds are $\gamma_{SR} > \gamma_{SD} \times 2$ and $\gamma_{RD} > \gamma_{SD}$. First assumes that $R \rightarrow D$ link state is ideal, and three-point single-hop model, when $\gamma_{SR} > \gamma_{SD} \times 2$ cooperative gain is occured. Further assumes that the link $S \rightarrow R$ is ideal, the source and relay transmits with unit power. Then the equivalent SNR is $\gamma_{RD} + \gamma_{SD}$, in cooperative scheme total energy is 2. In the non- cooperative scheme, the equivalent SNR is γ_{SD} and the total energy is 1. If the transmit energy increases to 2, the corresponding equivalent SNR increases to $2\gamma_{SD}$. To make sure that the cooperative gain can be achieved, $\gamma_{RD} + \gamma_{SD} > 2\gamma_{SD}$ is satisfied, that is the threshold $\gamma_{RD} > \gamma_{SD}$. From the above analysis, we can see that both of these thresholds are minimum in the theoretical analysis to start achieving energy gain.

Table 1. Simulation parameters

Cooperative model	2 hops, DF
Cooperative scenarios	Time division duplex, transmitting and receiving in the same frequency
Transmit diversity scheme	Alamouti Space-time block code
Modulation	BPSK
Number of antennas in each node	2
System error rate	10^{-3}
Channel	Rayleigh Flat fading channel
Power spectral density for Gaussian white noise	1

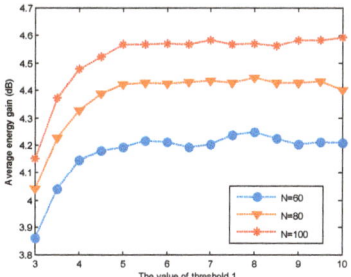

Fig. 4. The determination of threshold 1

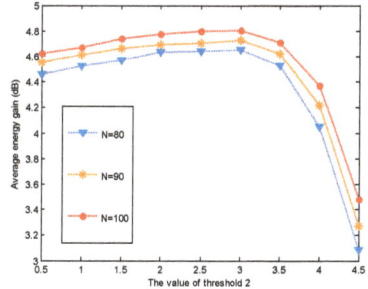

Fig. 5. The determination of threshold 2

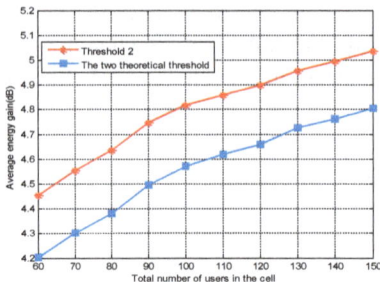

Fig. 6. Performance comparison of channel priority matching algorithm between users using the threshold 2 and using 2 theoretical thresholds.

Figure 4 evidently shows that value of threshold 1 rises, the system performance improves quickly. This is because the threshold 1 was set to decline the requirement of cooperation. But if threshold 1 is greater than 5, the system gain is not obvious. This is due to the excessively high value of threshold 1, users with better self-condition channel still cooperate even when they can achieve significant performance without cooperation. However, the smaller γ_{th1} value can reduce the computational complexity of the system. So the appropriated value of γ_{th1} can be 5.

Setting of the Threshold 2: Figure 5 shows how the average power gain of the MIMO system and threshold 2 (γ_{th2}), when the numbers of users are 80, 90, 100. Obvious from the diagram, in the case of threshold 1 is fixed, average energy gain varies quickly with the threshold 2. As the value of threshold 2 increased, the system performance improves. This is because the adding of threshold 2 increases the quality of cooperation. However when this threshold continues to increase, the average energy gain of the system began to decline sharply. From Fig. 5, maximum average energy gain is obtained when threshold 2 is approximate 3, so select 3 as the value of threshold 2.

Figure 6 is a performance comparison between 2 situations. First, the value of threshold 1 is 5, and threshold 2 is 3. Second, using 2 theoretical thresholds $\gamma_{SR} > \gamma_{SD} \times 2$ and $\gamma_{RD} > \gamma_{SD}$. As can be seen from the figure, more than 0,25 dB in average energy compared with the case using the two theoretical thresholds. So we choose the threshold 2 instead of two theoretical thresholds.

4.2   Analysis of Algorithm Performance

Table 2 indicates that in the channel priority matching algorithm between users using multiple antenna, the number of users has not found a partner is independent on the total number of users. The data are average of 1000 simulation calculation. When the total number of users increases, the number of users has not found partner unchanged. So probability that users are not able to find partners is low when distributed density of users is high (The number of users is large). We vary the number of users for both high density and low density case. Figure 3, 4, 5, 6 all show that the larger number of users, the higher average energy gain of the system can be achieved. It implies the that channel priority matching algorithm can get better performance in a high density system such as in the shopping malls, schools, apartments, stadiums.

Table 2. Number of users can not find partners when total number of users is changed

Total number of users in the cell	60	70	80	90	100	110	120	130	140	150
Users not find partner	5	5	5	5	5	5	6	6	6	6

Figure 7 is a performance comparison between the priority matching algorithms and random selection algorithms applied to the MIMO system in terms of energy consumption. Random selection algorithm is simple but low energy gain. The energy gain is nearly unchanged when the number of users changes. With the channel priority matching algorithm, the larger number of users, the higher distributed density, the more considerable energy gain. This also describes how much important of the selection algorithms is in the cooperative MIMO system.

Figure 8 shows how the average energy gain of the WLF algorithm for the multi-antenna system is influenced by the number of users in case using threshold and without threshold. And it is also the relation between the average energy gain and number of users in MIMO system compared with the single antenna system.

We can obviously see from Fig. 8 that the selection algorithm with threshold provides higher average energy gain. Average power gain of the MIMO system has been lower than the single antenna system. Channel priority matching algorithm for MIMO system has brought average energy gain, but it is less than the single-antenna systems. Because each MIMO terminal is equipped multiple antennas (it is assumed that 2 antennas in this paper), so even if without cooperation, it still yields the diversity gain, under a same certain BER, required energy for MIMO terminal users to transmit 1 bit is less than that for single-antenna system users without cooperation, while the gain from the cooperation is limited, so the average energy gain for the MIMO system with relay selection is not as much as the single antenna system. In simulation process, under the same number of simulation, the performance of the MIMO system is more stable, while performance of the single antenna system is volatile. Only when the number of simulation increases to a magnitude, the performance of the single-antenna system become stable.

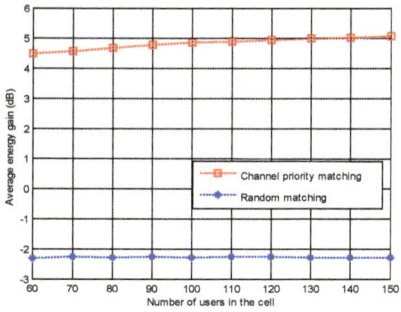

Fig. 7. Performance comparison of channel priority matching and random selection algorithms

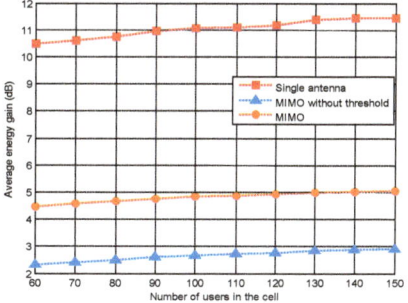

Fig. 8. Performance comparison of performance comparison at different situations

5   Conclusion

In this paper, a channel priority selection algorithm between users for multiple antenna system was proposed. Analysis and theoretical results for average energy gain were put into simulation. The simulation results match the theoretical ones and proves that the channel priority algorithm between users provided a better gain of average energy and better performance in a high distributed density network, and it is more suitable for shopping malls, schools, apartments, and this kind of high-density user system. Threshold setting and selecting a channel priority matching algorithms between users have great impact on the performance of multi-antenna system. Reasonable setting of the threshold gains 2 dB more in average energy.

References

1. Sadek, A.K., Han, Z., Liu, K.J.R.: A distributed relay-assignment algorithm for cooperative communications in wireless network. In: IEEE International Conference on Communications, pp. 1592–1597 (2006)
2. Wang, J., Sha, X., Lu, W., et al: Partner selection strategy for users with high speed in cooperative diversity systems. In: 22nd Canadian Conference on Electrical and Computer Engineering, St. John's, Canada, pp. 852–855 (2009)
3. Torabi, M., Ajib, W., Haccoun, D.: Performance analysis of amplify-and-forward cooperative networks with relay selection over Rayleigh fading channels. In: IEEE 69th Vehicular Technology Conference, Barcelona, Spain, pp. 1–5 (2009)
4. Lin, Z., Erkip, E., Stefanov, A.: Regions and partner choice in coded cooperative systems. IEEE Trans. Commun. **54**(7), 1323–1334 (2006)
5. Mahinthan, V., Cai, L., Mark, J.W., et al.: Maximizing cooperative diversity energy gain for wireless networks. IEEE Trans. Wirel. Commun. **6**(7), 2530–2539 (2007)

Keras Convolutional Neural Network Model in Sequential Build of 4 and 6 Layer Architectures

Praneet Amul Akash Cherukuri[1(✉)], Nguyen Thi Dieu Linh[2],
Sai Kiran Vududala[1], and Rahul Cherukuri[3]

[1] CMR Institute of Technology, Hyderabad, TS, India
[2] University of Industry, Ho Chi Minh, Vietnam
[3] Gokaraju Rangaraju Institute of Engineering and Technology,
Hyderabad, TS, India

Abstract. Machines see pictures utilizing pixels. Pixels in pictures are generally related. A convolution increases a grid of pixels with a channel framework or 'part' and summarizes the duplication esteems. Convolutional Neural Network is one of the most sought over concepts of technology for Image classification. Implementing them through Keras took a step forward and initiated fast processing. In this research paper, the authors intend to explore the different architectures of the convolutional neural networks and the inline layers of the network and understand the influence of training batches and epochs on it. To make things significantly simpler to decipher, we will utilize the 'accuracy' metric to see the ac- curacy score to comprehend the likelihood of cross-entropy.

Keywords: Image classification · Neural network · ReLU · Keras · CNN · Epochs

1 Introduction

1.1 A Subsection Sample

Image classification being the most intuitive domain in the industry depends mostly on the neural network classification to make the process much faster and simpler. In the wide range of neural networks, the convolutional neural network stands out to be the highest performing neural network of all other neural networks. CNN's are a somewhat exceptional neural network for handling information that is known to have a matrix-like topology. This could be a one-dimensional time arrangement information which is a lattice of tests after some time or something like even two-dimensional picture information at the lattice of pixels in space [1]. Three crucial features that decrease the number of boundaries in a neural network are first we have meager cooperation's between layers in a run of the mill feed-forward neural nets is each neuron in one layer is associated with each other in the following. This prompts an enormous number of boundaries that the network needs to realize which thus can cause different issues well for one such a large number of boundary gauges implies we'll require a ton of preparing information intermingling time likewise increments and the authors may wind up with an overfitted model. The authors can lessen the number of boundaries through the round- about, in which neuron and layer 5 is associated with three neurons in the past layer rather than all the neurons. These three neurons in layer four are immediate collaborations for the neuron and layer five since they influence the neuron being referred to legitimately such neurons comprise the open field of the layer five neurons [2]. Being referred to moreover, these layer four neurons have responsive fields of their own in the past layer three. These straightforwardly influence layer four neurons and in a roundabout way influence a couple of the layer five neurons. If the layer is run profound the authors needn't bother with each neuron to be associated with each other to convey data all through the network. At the end of the day, inadequate connections between layers should get the job done. The second advantage of leeway of CNN's is boundary sharing, this further addresses the decreased boundaries as did sparse interactions previously.

CNN makes facial highlights, for instance, consider having a picture which in the wake of going through the convolution layer offers ascend to a volume then a segment of this volume taken through the profundity will speak to highlights of a similar piece of the picture moreover each component in a similar profundity layer is created by a similar channel that convolves x' the picture [3]. The individual focuses on a similar profundity of the component map that is the yield 3d volume is made from a similar part, as such, they are made utilizing a similar arrangement of shared boundaries. This radically lessens the number of boundaries to gain from a common Neural Network. Another element of convolutional neural networks is identical portrayal now a capacity f is set to be equal to another capacity G if F of G of X is equivalent to G of F of X for an information X.

2   Architectures

A basic Convolutional neural network consists of mainly four layers that can be constructed as Convolutional layers, Rectified Linear Unit (ReLU) Layer, Pooling Layer, and a Fully connected layer. Usually, any CNN is structured in the below format such as starting with the input, then onto Convolution, then onto ReLU, then onto the Pooling layer, and finally to the Fully connected layers. In a fundamental sense, Convolutional layers apply a convolution activity to the information. This gives the data to the following layer. Pooling consolidates the yields of groups of neurons into a solitary neuron in the following layer. To build the architecture it could be outlined as follows, the functionality of each layer will attain the workflow as follows. It Starts with an input picture at that point applies various channels to it to make a component map. The following stage incorporates a ReLU capacity to increment non-linearity, at that point on moving to a pooling layer to each element map [4].

It swells the pooled pictures into one long vector. Next, the authors input the vector into a completely connected counterfeit neural network. The last fully-connected layer gives the voting of the classes that the authors are after. Then it successively trains through sending engendering and backpropagation for some, numerous times. This rehashes until we have an all-around characterized neural network with prepared loads and high- light identifiers.

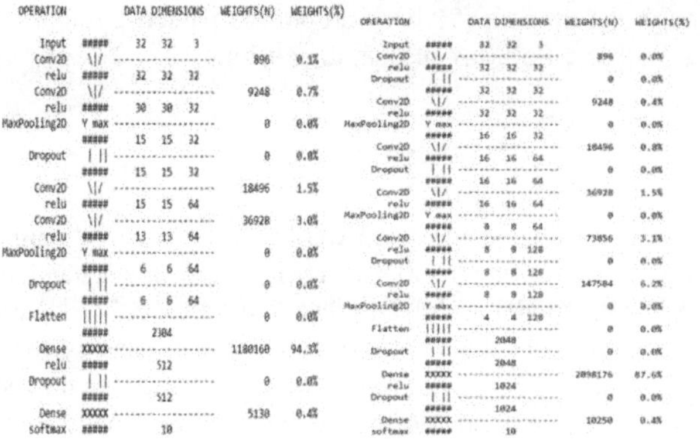

Fig. 1. Visual representation of the 4 & 6 layer architecture

2.1   4 – Layer Architecture

As understood from the architectures, we virtualize the given architectures into 4 and 6 layers. The 4 layer architecture as shown in Fig. 1 is unilaterally divided into 8 parts. The first part being the input layer, and then there is a sort of instruction that includes two Conv2d layers, 2 ReLU layers, and max-pooling layer. After the first sort of instruction, a dropout is involved and the same second sort is processed. The Data dimensions for the first ReLU is $32 \times 32 \times 32$ and then reduced to $32 \times 32 \times 30$. Then in the second instruction data dimensions for the ReLU is $15 \times 15 \times 64$ and which is reduced to $13 \times 13 \times 64$. After dropout, the model is flattened with the dimensions of 2304 after which the dense layer having weights in 1180160 with a percentage of 94.3%. Then the ReLU layer is applied again with a dimension of 512 and passed through the dropout. This dropout is passed through a dense layer of weight 5130 to output a SoftMax of data dimensions 10. In total, the 4 Layer architecture has four Convolutional Layers, five ReLU layers, two max-pooling layers, three dropouts, and two dense layers to give out a SoftMax prediction. In the next phase, the authors will explore the 6 Layer architecture [5].

2.2   6 – Layer Architecture

Following the four-layer architecture, for the six-layer architecture, We visualize the given 6 layer architecture as shown in Fig. 2. In total, the 6 Layer architecture has six Convolutional Layers, seven ReLU layers, three max-pooling layers, five dropouts, and two dense layers to give out a SoftMax prediction. The architecture is unilaterally divided into 11 parts. The first part being the input layer, and there we have pairs of Conv2d and ReLU layers differed by Dropout and max-pooling layers for three times in accordance. In the three pairs, the first pair starts with the dimension depth of $32 \times 32 \times 32$, transforming to $16 \times 16 \times 64$ then on converting to $8 \times 8 \times 128$ finally flattening at $4 \times 4 \times 128$. After dropout, the model is flattened with the dimensions of 2048 after which the dense layer having weights in 2098176 with a percentage of 87.6%. Then the ReLU layer is applied again with a dimension of 1024 and passed through the dropout [5]. This dropout is passed through a dense layer of weight 10250 to output a SoftMax of data dimensions 10.

3 Dataset

After analyzing the given architectures, to explore further a dataset named CIFAR 10 is experimented with the two frameworks to retrieve and understand the trends concerning accuracy for the model at different levels of training. The dataset is contained with 60,000 32 × 32 pixel shading photos of items from 10 classes, for example, frogs, winged creatures, felines, ships, and so forth. The class marks and their standard related number qualities are recorded underneath. These are exceptionally little pictures, a lot littler than a run of the mill photo, and the dataset was planned for PC vision research. CIFAR-10 is a surely known dataset and broadly utilized for benchmarking PC vision calculations in the field of AI [6]. The dataset is divided into 10 classes with the following names airplane, automobile, bird, cat, deer, dog, frog, horse, ship, truck. The structure of the test outfit is measured, and we can build up a different capacity for each piece. This permits a given part of the test saddle to be altered or traded, on the off chance that we want, independently from the rest. We can build up this test saddle with five key components. They are the stacking of the dataset, the arrangement of the dataset, the meaning of the model, the assessment of the model, and the introduction of results.

4 Experimental Framework

After analyzing the given dataset for its complexity. Both the architectures are trained for different frameworks of 50, 75, and 100 epochs, and the training and validation losses and accuracies are plotted onto the line chart to understand the frequencies. A confusion matrix is also plotted for every training set for both the architectures [7].

4.1 4 Layer – 50 Epochs

In this Training Accuracy Vs Validation Accuracy illustration, we will plot Num of Epochs on X-axis and Accuracy on Y-axis, hence from the given graph the training accuracy for CNN 4-layer(50 Epochs) model is 79.87 concerning the validation accuracy of 75.58. Similarly for training loss Vs Validation loss is 59.73 regarding the Validation loss of 74.44.

4.2 4 Layer – 75 Epochs

In this Training Accuracy Vs Validation Accuracy illustration, we will plot Num of Epochs on the X-axis and Accuracy on Y-axis, hence from the given graph the training accuracy for CNN 4-layer(75 Epochs) model is 80.87 for the validation accuracy of 75.87.

4.3   4 Layer – 100 Epochs

In this Training Accuracy Vs Validation Accuracy illustration, we will plot Num of Epochs on X-axis and Accuracy on Y-axis, hence from the given graph the training accuracy for the CNN 4-layer(100 Epochs) model is 81.75 for the validation accuracy of 75.96. Similarly for training loss Vs Validation loss is 56.43 concerning the Validation loss of 79.49.

4.4   6 Layer – 50 Epochs

In this Training Accuracy Vs Validation Accuracy illustration, we will plot Num of Epochs on the X-axis and Accuracy on Y-axis, hence from the given graph the training accuracy for CNN 6-layer(50 Epochs) model is 73.46 concerning the validation accuracy of 66.74. Similarly for training loss Vs Validation loss is 81.41 concerning the Validation loss of 102.53.

4.5   6 Layer – 75 Epochs

In this Training Accuracy Vs Validation Accuracy illustration, we will plot Num of Epochs on the X-axis and Accuracy on the Y-axis, hence from the given graph the training accuracy for CNN 6-layer(75 Epochs) model is 73.29 concerning the validation accuracy of 66.99. Similarly for training loss Vs Validation loss is 82.06 concerning the Validation loss of 102.46.

4.6   6 Layer – 100 Epochs

In this Training Accuracy Vs Validation Accuracy illustration, we will plot Num of Epochs on the X-axis and Accuracy on the Y-axis, hence from the given graph the training accuracy for CNN 6-layer(100 Epochs) model is 72.16 for the validation accuracy of 64.03. Similarly for training loss Vs Validation loss is 86.32 concerning the Validation loss of 110.77.

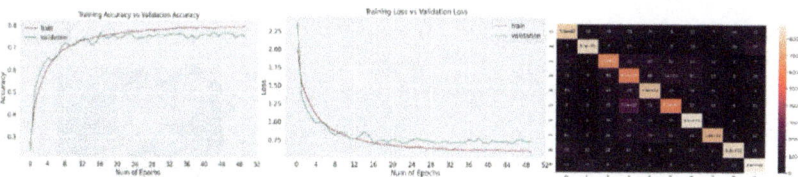

Fig. 2. Visualization of training model for 4 layers 50 epochs

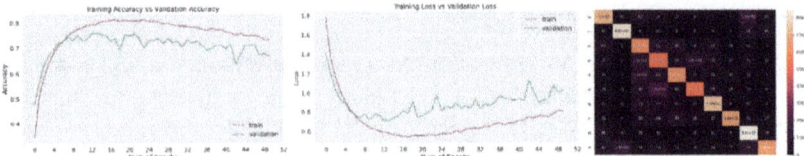

Fig. 3. Visualization of training model for 6 layers 50 epochs

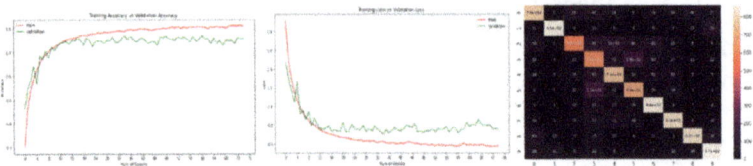

Fig. 4. Visualization of training model for 4 layers 75 epochs

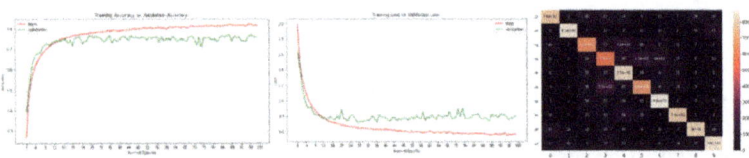

Fig. 5. Visualization of training model for 4 layers 100 epochs

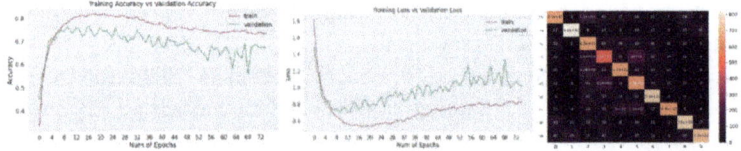

Fig. 6. Visualization of training model for 6 layers 75 epochs

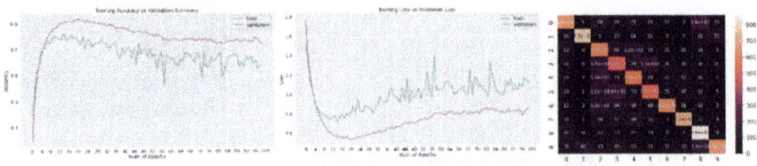

Fig. 7. Visualization of training model for 6 layers 100 epochs

5   Conclusion

After analyzing the 4 and 6 architectures with different epoch values of 50, 75 and 100, the final part of the exploration is performed where the validation accuracies of the model are visualized to understand the linear or exponential nuances of the accuracies for various training batches. After visualizing the accuracies we are provided with a double bar graph model that visualizes the layers, epochs, and their accuracies. It is understood that the values for the 4 layers are represented in blue color and the values for the 6 layers are represented with orange color. The graph is divided into three sets each set representing a different set of epoch values, on the x-axis we have the epoch sets and on the y-axis, we have the accuracy listed.

Analyzing the graph we understand that for the 4 layer architecture there wasn't much of a change in the accuracy changes in the ranges of 50, 75 and 100. It is predicted that for higher epoch values it could be stated to increase more. When taken into consideration the SoftMax predictions, the predictions are much repetitive reducing the entropy.

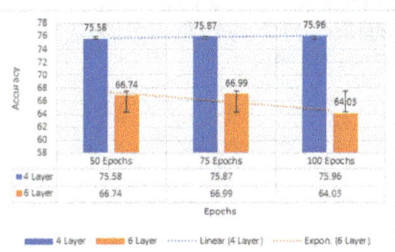

Fig. 8. Bar chart representation for comparison of accuracies for both 4 and 6 layers on 50, 75 and 100 epoch values

For the four-layer architecture, the accuracy for the three sets remained in the 75 range area differing the decimal point areas with an average ranging at 75.804. Transiting into the six-layer architecture the accuracy showed a downfall in the trend for a different set of epoch ranges. Hitting off with 66.7 for 50 epoch range then managing the same decimal change for 75, it is observed that there was an exponential decrease in the accuracy. This exponential decrease could be identified as an anomaly, but when taken the SoftMax predictions they remained constant giving the 6 layer architecture a much understanding of the extra convolutional layer. Since we apply $3 \times 3 \times 3$ filters note that the depth of the filters is the same as that of the input which is 1. In this case for every scaled image, the output of convolution concerning each filter is the width of the image, width of the filter plus 1/4 a stride of 1. This enhances the depth finding and a note of reduction in the validation accuracy.

References

1. Lawrence, S., Giles, C.L., Tsoi, A.C., Back, A.D.: Face recognition: a convolutional neural-network approach. IEEE Trans. Neural Netw. **8**(1), 98–113 (1997). https://doi.org/10.1109/72.554195
2. Li, H., Lin, Z., Shen, X., Brandt, J., Hua, G.: A convolutional neural network cascade for face detection. In: Proceedings of the IEEE Conference on Computer Vision and Pattern Recognition, pp. 5325–5334 (2015)
3. Simard, P.Y., Steinkraus, D., Platt, J.C.: Best practices for convolutional neural networks applied to visual document analysis. In: ICDAR, vol. 3, no. 2003, August 2003
4. Kirana, K.C., Wibawanto, S., Hidayah, N., Cahyono, G.P., Asfani, K.: Improved neural network using integral-RELU based prevention activation for face detection. In: 2019 International Conference on Electrical, Electronics and Information Engineering (ICEEIE), Denpasar, Bali, Indonesia, pp. 260–263 (2019). https://doi.org/10.1109/ICEEIE47180.2019.8981443
5. Dileep, P., Das, D., Bora, P.K.: Dense layer dropout based CNN architecture for automatic modulation classification. In: 2020 National Conference on Communications (NCC), Kharagpur, India, pp. 1–5 (2020). https://doi.org/10.1109/NCC48643.2020.9055989
6. Krizhevsky, A.: The CIFAR-10 Dataset. CIFAR-10 and CIFAR-100 Datasets. https://www.cs.toronto.edu/~kriz/cifar.html. Accessed 16 July 2020
7. Visa, S., Ramsay, B., Ralescu, A.L., Van Der Knaap, E.: Confusion matrix-based feature selection. MAICS **710**, 120–127 (2011)

On the Worst Case User Performance in Milliliter Cellular Networks

Sinh Cong Lam[1,4], Duc-Tan Tran[2,4(✉)], Viet-Huong Pham[3,4],
and Xuan Thu Do[4]

[1] Faculty of Electronics and Telecommunications,
VNU University of Engineering and Technology, Hanoi, Vietnam
congls@vnu.edu.vn
[2] Faculty of Electrical and Electronic Engineering, Phenikaa University,
Hanoi 12116, Vietnam
tan.tranduc@phenikaa-uni.edu.vn
[3] VNU International School, Hanoi, Vietnam
thudx@utt.edu.vn
[4] University of Transport Technology, Hanoi, Vietnam

Abstract. In the cellular networks, the user at the corner of a cell is usually called worst-case user mobile (WMU) experience the lowest performance. This paper studies WMU performance in millimeter wave cellular systems in three aspects such as throughput, Average Packet Delay (APD) and Packet Loss Ratio (PLR). The multi-slope model in which is recommended by 3GPP to simulate wireless communication is utilized to estimate the received signal power at the WMU in downlink. In order to improve the WMU performance, the Carrier Aggregation technique which has been introduced for Long-Term Evolution (LTE) is studied. By utilizing CA scheme, the WMU is allowed to utilize more than one sub-carrier to perform communications. The simulation results in Network Simulation NS3 indicates that the CA technique can significantly improve the WMU performance. For example, maximum achievable throughput of the CA system is 8.25 Mb/s which is nearly double than that of the non-CA system.

Keywords: Carrier Aggregation · 5 generation · Millimeter wave · Throughput · Average packet delay · Packet loss ratio

1 Introduction

The 5G (5G generation) cellular system is being commercially deployed in some countries to replace the current 4G LTE to provide the ultra high Quality of Services, particularly Latency at ultra low values and Reliable at ultra high levels. Theoretically, the 5G network can provide ultra high data speed at 20 Gb/s [2]. In order to obtain these requirements, the Base Stations (BSs) need to be distributed in a service area with ultra high density [8]. Furthermore, the International Telecommunication Union (ITU) recommends that the 5G operational frequency band should have millimeter wave such as the frequency range form 5 GHz–60 GHz.

Research on the feasibility of 5G millimeter wave network has become popular [8]. Most of research works utilized mathematical tools such as stochastic geometry model [14] and simulation tools [9] to do performance analysis and optimization. While the mathematical approach usually analyzed the performance at the physical layer, the simulation tools can examine user performance at a higher layer such as Radio Link Control (RLC) layer [9]. The main performance metrics in RLC layer are user throughput, Average end-to-end Packet Delay (APD) and end-to-end Packet Loss Ratio (PLR). Although these references discussed above provided important approaches to evaluate the 5G millimeter wave system, the performance metrics were derived for the typical user. In other words, each metric was computed by taking the expected value of the corresponding metric value of all users in a given cell. Although this approach can evaluate the overall cell performance, it is not suitable to examine a specific user, especially the user experiences the lowest received signal strength, i.e. Worst-case Mobile User (WMU)[12]. Thus, this paper focuses on the performance analysis of the WMU.

The WMU has studied in some research work in the literature such as [6,7,10]. In these works, the Coordinated Multi-point (CoMP) [4] schemes such as Joint Transmission (JT) and Joint Scheduling (JS) were used to improve the WMU performance. However, the feasibility of CoMP in a practical network is still a question because it requires additional bandwidth and hardware components.

This paper studies Carrier Aggregation [5] as an additional technique to enhance the WMU performance. CA scheme was initially introduced in 4G LTE cellular network systems to increase user data rate. By deploying this scheme, each mobile user can utilize more than one Component Carriers (CC) during its transmission in which one sub-carrier is Primary CC and others are call Secondary CC. The feasibility of CA technique in 5G cellular network systems was discussed in [3,13] in which it was expected that this technique can explore the available spectrum of other wireless systems such as wifi. However, the deployment of CA technique can be challenged by some critical issues such as InterCell Interference and synchronization. The selection of component carriers depends on various network conditions such as devices properties, user buffer and available interface [13]. This paper studies effects of CA on WMU performance. Furthermore, while References [3,13] dealt with the conventional power loss model in which the power loss exponent keeps constant over distances, this paper follows the 3GPP specification by utilizing the multi-slop path loss model [1]. In this model, the path loss exponent is considered as a function of the distance between the transmitter and receiver.

2   System Model

We assume that the wireless link experiences the same path loss model and stochastic fading channel. Thus, WMU has a fixed position at the intersection of three cells We suppose that there is no handover in this system model, so that the user stick with only one BS during its transmission Since the WMU has the same distances to other BSs, the WMU only connects with one BS during its transmission. Hence, the system model can be degraded into a single cell network model and the position of the WMU is at the cell corner.

Path Loss Model. In Document [15], 3GPP highlights that the power loss of a signal over distance in a cellular network system should follow a multi-slop model. Mathematically, the multi-slop model can represented by the variance of the path loss exponent according to the wireless link distance. For simplicity, the path loss exponent α is determined as follows

$$\alpha = \alpha_k \qquad \text{when} \qquad l_{k-1} < l < l_k \tag{1}$$

in which d is the distance between the user and the transmitting BS.

Thus, the power loss over distance d is computed as the following equation

$$PL(dB) = PL_0 + 10 \sum_{m=1}^{M} \alpha_{m-1} \log_{10}\left(\frac{l_m}{l_{m-1}}\right) + 10\alpha_M \log_{10}\left(\frac{l}{l_M}\right) \tag{2}$$

where PL_0 (dB) is power loss at reference point; M is number of slopes.

Carrier Aggregation. CA technique was firstly introduced for LTE-A cellular system to increase the network performance and spectrum efficiency [15]. This technique is standardized for both downlink and uplink channels, and also feasible for both time domain and frequency domain. As stated in 3GPP document, CA technique can explore additional component carriers contiguously or non-contiguously intra-band, or inter-band.

The BS can select the component carrier in different manners such as: contiguously intra-band; non-contiguously intra-band; and non-contiguously inter-band.

Generally, selection of component carriers depends on various conditions such as the available bandwidth, the maximum signaling traffic load in uplink. Besides it also depends on particular purposes such as to increase the network data rate, interference avoidance. In a LTE-A system, the number of component carriers is determined and fixed during the user transmission process. In a later system such as 5G system, the dynamic CA technique was introduced [1] to adjust the number of component carries according to the user demands and network conditions (available bandwidth, interference).

3   Simulation

3.1   System Model

In this work, the WMU performance metrics are evaluated by using Network Simulator 3 (NS3) tool [11]. There is one BS in the service area which has a

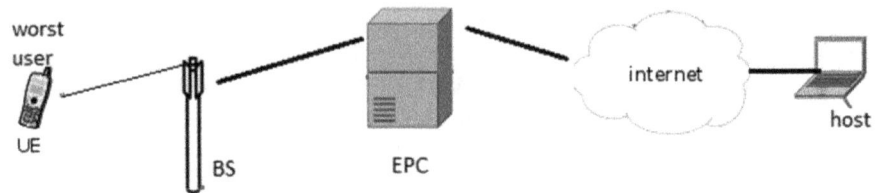

Fig. 1. Simulation model

hexagonal shape. The WMU is located at the cell corner and performance data exchange with the remote host in the Internet. The system model is figured out as in Fig. 1.   The main characteristics of the model are described as follows

- The wireless link between the WMU and its associated BS has a three-slop power loss model. The path loss exponent α at a distance of d is

$$\alpha = \begin{cases} 2 & \text{if } 50 \leq l(m) \leq 100 \\ 2.5 & \text{if } 100 \leq l(m) < 150(m) \\ 3 & \text{if } l(m) \geq 150 \end{cases} \tag{3}$$

- Other links such as the link from BS to EPC, and from EPC to the Internet and the remote host are supposed to be perfect links which have unlimited bandwidth, unlimited queue buffer and negligible delay. Thus, the performance of the data transmission between the WMU and the remote host only depends on the wireless link between WMU and the BS.

The parameters utilized in the simulation program summarized in Table 1

Table 1. Simulation parameters

Simulation parameters	Values
Radius of cell	0.250 (kmm)
Carrier Frequency	5.15 GHz
Number of component carriers	1
Size of Packet	576 byte
Simulation duration	60 (s)

It is assumed that the WMU continuously transmits its data without considering the network conditions. The time duration between two consecutive packets is defined as the packet interval. The packet interval can be used to represent the data rate transmitted by the WMU. A smaller packet interval, a higher number of packets transmitted by WMU and consequently a higher transmitted data rate. For example, when the packet interval is 0.5 ms. Then, 2000 *packets* are generated by the WMU every second. If every packet has a size of 576 byte. Thus, the transmit data rate of WMU at the physical layer is upto 9,216 Mb/s.

For every value of packet interval, the simulation examines WMU performance metrics in terms of downlink throughput, APD and PLR.

3.2   Simulation Results

Figures 2, 3, 4 illustrate the downlink throughput, PLR and ADP of WMU in a millimeter wave cellular system with and without CA technique. It is noted that the cellular system with CA technique can utilize more a wider band than other without this technique. In other words, the user in the former system can be allocated a wider band than other in the later one. Thus, it is possible that the CA system can achieve better performance than the non-CA system. As seen from Figs. 2, 3, 4, the CA technique can significantly increase the performance of WMU in terms of downlink throughput, PLR and APD.

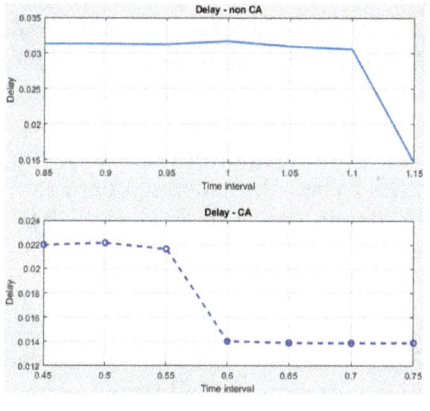

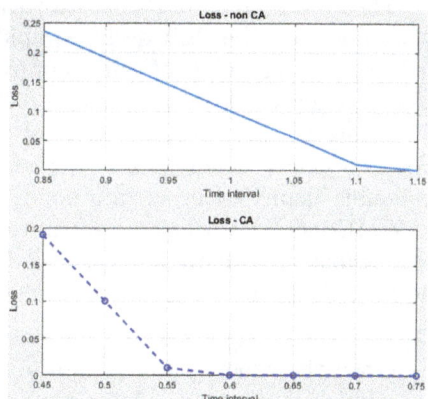

Fig. 2. APD   Fig. 3. PLR

As shown in Fig. 2, the APD in both CA and non-CA system keeps steady at maximum values when the packet interval values are small enough. Take the non-CA system for example, the APD remains at 0.318 s when the interval between two consecutive packets at the WMU is smaller than 0.1100 ms which corresponds to the transmitted data rate of WMU is greater than 4.19 Mb/s. In contrast, in the CA system, the APD remains at 0.022 s when the WMU transmitted data rate is higher than 9.21 Mb/s.

When the packet interval increases, which corresponds to a decline in transmitted data rate of the WMU, the APD falls before keeping constant at the minimum values of 0.014 for both systems. This phenomenon is explained as:

- When the interval between packets is small enough, packets are generated every second may be higher than the transmission capability. Thus, some packets will be dropped while some packets are transferred to the buffer before being transmitted to the remote host. It is also noted the APD only depends on the successfully transmitted packet. When the packet interval slightly reduces, the number of loss packets reduces but the number of packets in buffer does not change. Thus, APD does not change in this case.

- When the packet interval significantly reduces, the number of packets in the buffer also reduces. This reduces the waiting time in the buffer of each packet. Consequently, APD reduces.
- When the packet interval dramatically reduces, all the generated packets of the WMU will be instantly transmitted. Thus, the APD keeps constant at the minimum value.

The figure illustrates that for a specific value of APD, the CA system allows the WMU transmit at a significantly faster data rate than the non-CA system. For an instant, when a APD of 0.014 s is required, the WMU in the CA system is able to transmit at 7.68 Mb/s which is nearly 92% faster than the WMU in the system without CA technique.

The decline trend of PLR with packet interval is presented Fig. 3. The figure illustrates that the PLR declines rapidly and linearly when the packet interval falls in a specific range of value. For example, the range from 0.85 to 0.11 in the case of a non-CA system and from 0.45 to 0.5 in the case of a non-CA system.

Figure 4 represents the effects of packet interval on the WMU throughput. For a specific value of packet interval, the WMU throughput in the CA system is double than that other in a non-CA system. When the generated data rate of WMU increases which corresponds to a decline in packet interval, the WMU throughput in both cases remains at the maximum values before passing a rapid decline. This can be explained as follows:

- That each system has its own maximum network throughput
- When the generated data rate of the WMU is smaller than the maximum network throughput, i.e. for great values of packet interval, all the generated packets will be conveyed successfully.

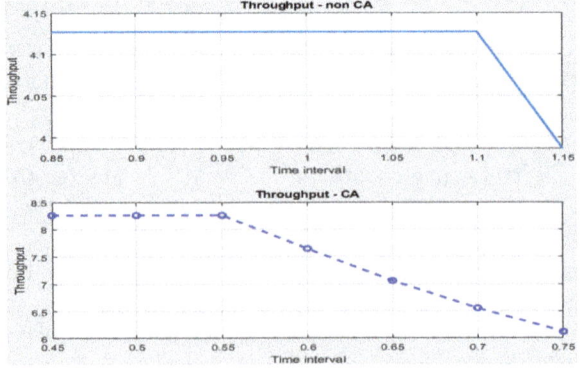

Fig. 4. WMU throughput

– When the generated data rate of the WMU is great than the maximum network throughput, .i.e. for small values of packet interval, apart of packets will be transmitted while others will be dropped as in Fig. 3. Thus, the WMU throughout remains at the maximum network throughput.

4   Conclusion

This work studied WMU which is allocated at the cell corner was analyzed for a 5.15 GHz- milliliter wave system with and without CA technique. The three-slope power loss model was utilized in the NS3 simulation program to represent the wireless link transmission characteristics. The WMU performance metrics in terms of throughput, PRL and APD were analyzed by using NS3 simulation tool The achieved results states that the AC technique provides huge improvement in WMU performance. For example, the system with CA can achieve a double maximum throughput than other without CA.

References

1. 9, GPP TR: Evolved universal terrestrial radio access (E-UTRA); further advancements for E-UTRA physical layer aspects (2010)
2. Agiwal, M., Roy, A., Saxena, N.: Next generation 5G wireless networks: a comprehensive survey. IEEE Commun. Surv. Tutor. **18**(3), 1617–1655 (2016)
3. Alkhansa, R., Artail, H., Gutierrez-Estevez, D.M.: LTE-WiFi carrier aggregation for future 5G systems: a feasibility study and research challenges. Procedia Comput. Sci. **34**, 133–140 (2014). The 9th International Conference on Future Networks and Communications (FNC'14)/The 11th International Conference on Mobile Systems and Pervasive Computing (MobiSPC'14)/Affiliated Workshops
4. Daewon, L., Hanbyul, S., Clerckx, B., Hardouin, E., Mazzarese, D., Nagata, S., Sayana, K.: Coordinated multipoint transmission and reception in LTE-advanced: deployment scenarios and operational challenges. IEEE Commun. Mag. **50**(2), 148–155 (2012). https://doi.org/10.1109/MCOM.2012.6146494
5. ETSI, TS, 136, 104, V13.3.0: LTE; evolved universal terrestrial radio access (E-UTRA); base station (BS) radio transmission and reception (2016)
6. Hashima, S., Elnoubi, S., Alghoniemy, M., Shalaby, H., Muta, O., Mahmoud, I.: Analysis of frequency reuse cellular systems using worst case signal to interference ratio. In: Proceedings of the 12th International Conference on Telecommunications, pp. 185–190 (2013)
7. Jung, S.Y., Lee, H., Kim, S.: Worst-case user analysis in poisson voronoi cells. IEEE Commun. Lett. **17**(8), 1580–1583 (2013)
8. Kamel, M., Hamouda, W., Youssef, A.: Ultra-dense networks: a survey. IEEE Commun. Surv. Tutor. **18**(4), 2522–2545 (2016). https://doi.org/10.1109/COMST.2016.2571730
9. Mezzavilla, M., Zhang, M., Polese, M., Ford, R., Dutta, S., Rangan, S., Zorzi, M.: End-to-end simulation of 5G mmWave networks. IEEE Commun. Surv. Tutor. **20**(3), 2237–2263 (2018)
10. Nigam, G., Minero, P.: Spatiotemporal base station cooperation in a cellular network: the worst-case user. In: 2015 IEEE Global Communications Conference (GLOBECOM), pp. 1–6 (2015)

11. NS3: https://www.nsnam.org
12. Tse, D., Viswanath, P. (eds.): Fundamentals of Wireless Communication. Cambridge (2005)
13. Vidhya, R., Karthik, P.: Dynamic carrier aggregation in 5g network scenario. In: 2015 International Conference on Computing and Network Communications (CoCoNet), pp. 936–940 (2015)
14. Zhang, X., Andrews, J.G.: Downlink cellular network analysis with multi-slope path loss models. IEEE Trans. Commun. **63**(5), 1881–1894 (2015). https://doi.org/10.1109/TCOMM.2015.2413412
15. 3GPP TR 36.814 Release 9: Evolved universal terrestrial radio access (e-utra); further advancements for e-utra physical layer aspects (2010)

Open-Loop Short-Circuit Branch Line Dual-Mode Bandpass Filter for Low-Bands 5G Applications

Trong-Hieu Le[✉] and Manh-Cuong Ho

Faculty of Electronics and Telecommunications, Electric Power University, Hanoi, Vietnam
hieult@epu.edu.vn

Abstract. This paper introduces a new design of microwave bandpass filter based on microstrip open-loop short-circuit dual-mode resonator. The principle of the short circuit branch line of the filter is to split the mode of the fundamental wave or harmonic wave of the quarter wavelength. Two transmission zeros (TZs) poles are formed near the center frequency, which improves the out of band rejection of microstrip filter. The actual test results show that when the frequency range of the filter is 694MHz–806MHz, and the relative bandwidth is 29.4%. This class of proposed filter exhibits the compact size and the good quality factor.

Keywords: Microstrip dual-mode filter · Open-loop resonator · Transmission zeros

1 Introduction

Rapid development of wireless communication present extraordinary demand for RF microwave filters with high selectivity and small size. The band below 1000 MHz in general and the 700 MHz band in particular have a high value of use for mobile communication due to its good transmission ability, bringing about high investment efficiency. These bands can be used to provide broad coverage in rural and suburban areas and have deep indoor coverage in large, high-density, residential and structural areas. Therefore, the 700 MHz band plays an important role in deploying and expanding 4G networks and serving as a foundation for 5G networks in the near future.

Wolff and Knoppik (1971) first analyzed the dual-mode characteristics of ring resonators, dual-mode resonators have been a hot topic research and attracted much attention [2, 3]. According them, the number of resonators when designing the filter with given specifications due to each dual-mode resonator can be realized as a dual tuned resonator, which makes the filter structure more compact. The researches [4–9] show that the microstrip filter based on the microstrip open-loop resonator as well as hybrid resonators can provide multiple transmission paths for signals between the input and output ports, and its stopband has two transmission zeros, which makes the frequency response of the edge of the passband steeper and the out of band rejection better.

In this paper, a new type of RF microstrip bandpass filter using open-loop short-circuit branch line (SCBL) dual-mode resonators is proposed, which works at 694 MHz–806 MHz. The proposed open-loop short-circuit dual-mode BPF exhibits a pair of attenuation poles with reduced insertion loss and miniaturization. The test results are in good agreement with the frequency response simulation results. The central frequency of the passband is 736 MHz, of which the 3 dB bandwidth frequency bandwidth is 217 MHz, and the 3 dB bandwidth insertion loss of the proposed filter is 0.37 dB when the relative bandwidth is 29.4%.

2   Design Theory

2.1   Open-Loop Short-Circuit Branch Line Dual-Mode Resonator

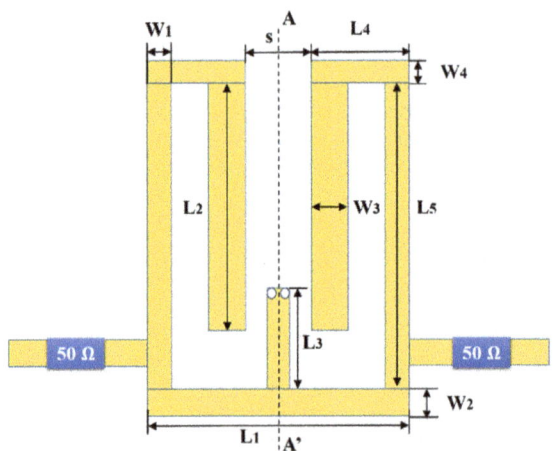

Fig. 1. Open-loop short-circuit branch line dual-mode resonator configuration.

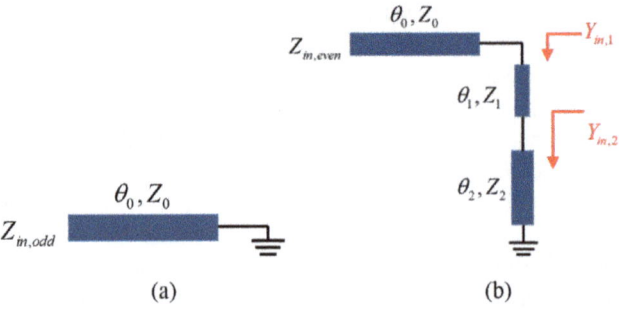

Fig. 2. The proposed resonator equivalent circuit: (a) Odd-mode. (b) Even-mode.

The introduced dual-mode open-loop SCBL resonator configuration is displayed in Fig. 1. A short-circuit branch line stub is setup at the center of the short-circuit branch line resonator. A−A' is a reference plane and two perturbations formed by hole-via to ground, which are set at the center of upper section of the loop. Figure 2 shows the equivalent circuit of short-circuit branch line resonator, which is able to be calculated the odd mode excitation. The value input impedance of short-circuit branch line resonator is calculated as:

$$Z_{in,odd} = jZ_0 \tan \theta_0 \tag{1}$$

According to $Y_{in,odd} = 0$ is the resonant condition, then the first f_{odd} resonant can be derived from:

$$f_{odd} = \frac{c}{4\sqrt{\varepsilon_e}(2L_1 - s/2)} \tag{2}$$

The proposed equivalent circuit in Fig. 3 (b) can be applied for the excitation of even-mode resonant frequency. The value input impedance one-port of proposed short-circuit branch line resonator is calculated as:

$$Y_{in,1} = -j\frac{1}{Z_2 \tan \theta_2} \tag{3}$$

$$Z_{in,even} = Z_0 \frac{Y_{in,2} + jZ_0 \tan \theta_0}{Z_0 + jY_{in,2} \tan \theta_0} \tag{4}$$

According to $Y_{in,even} = 0$ is the resonant condition, the f_{even} resonant is calculated as:

$$f_{even} = \frac{c}{2\sqrt{\varepsilon_e}(2L_1 - s/2 + L_2 + L_3)} \tag{5}$$

2.2   Enhance Out-Band Rejection

Figures 3 (a) and (b) illustrate the mainly affection of the coupling level between the resonator and feed-lines on the attenuation of the out-band bandpass filter. This affection can be observed through the adjustable location of the transmission zeros in the stopband of the proposed filter. It is worth mentioning that the larger L_3, the shorter coupling length between the horizontal feedlines and the short-circuit branch line resonator. In Fig. 3(a), L_3 is tuned from 2.8 to 5.8 mm of length, the magnitude S_{11} is obtained below −20 dB, and the appeared two transmission zeros in the upper stopband realize a higher spurious suppression for proposed filter. Moreover, for a shorter coupling length L_2 between and the short-circuit branch line resonator and the vertical feed-lines, shown in Fig. 3(b), the attenuation of the stopband is enhanced and the transmission zeros are shifted to the high frequencies. Therefore, the high rejection level of the proposed filter stopbands can be achieved by optimally design the coupling level to approach a pair of distributed transmission zeros. Otherwise, by changing the coupling length L_1 as well as the gap s, the larger values L_1, the better out of band rejection from -40 dB to -50 dB is observed as displayed in Fig. 3(c).

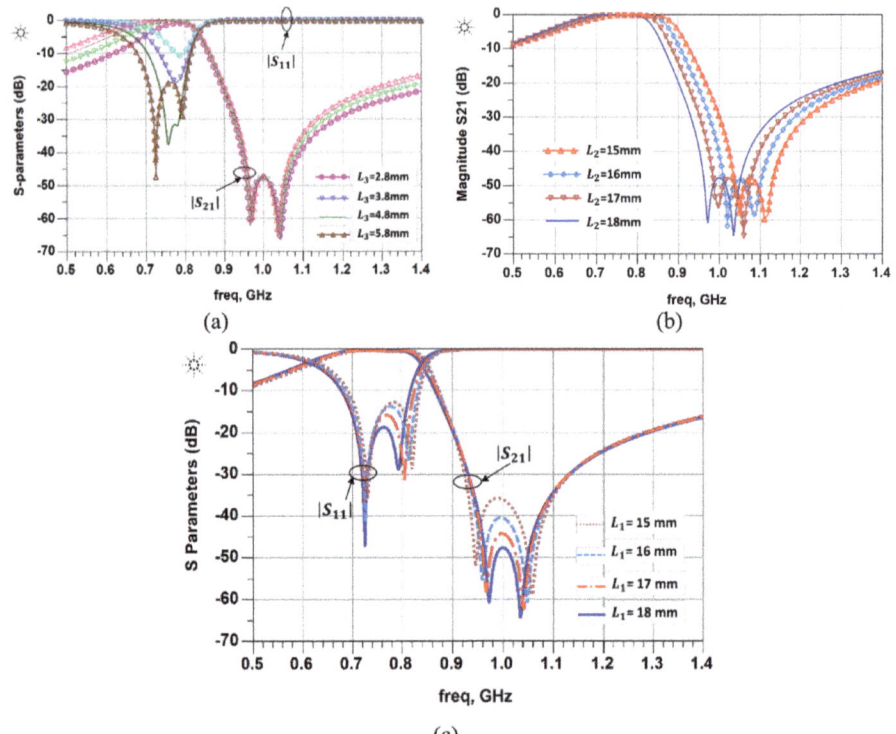

Fig. 3. Calculated frequency response vs. different length L_1, L_2, L_3, respectively.

3 Filter Performance

The structure of the open-loop short-circuit dual-mode band-pass filter proposed in this paper is designed on the Taconic TSM-30 substrate, which has a relative permittivity of 3.0 and a thickness of 0.76 mm. The principle of short-circuit branch line is to split the fundamental wave or harmonic wave of quarter wavelength into two poles near the center frequency. In the design, Sonnet simulation software [10] is used to simulate the filter design. The optimal dimensions (all in mm) of the proposed BPF are as follows: $W_1 = 1.6$, $W_2 = 2$, $W_3 = 3$, $W_4 = 0.9$, $W_5 = 1.4$, $L_1 = 18.4$, $L_2 = 18.7$, $L_3 = 5.8$, $L_4 = 6.6$, $L_5 = 20.6$, $s = 5.2$. The input/output microstrip lines feeders of 50 Ω characteristic impedance are set on both sides of the proposed filter. The fabricated BPF layout with the whole size of 30 $\times$ 30 (mm), is shown in Fig. 4.

Fig. 4. Photograph of open-loop SCBL BPF manufactured.

In the experimental process, the measurement is completed by the Network Analyzer Agilent N5230A. Figure 5 shows the S parameters test and simulation results of open-loop SCBL dual-mode filter. According to the test, the center frequency of passband is 736 MHz, and the 3 dB bandwidth is 217 MHz. When the relative bandwidth is 29.4%, the insertion loss of 3 dB bandwidth is 0.37 dB. It should be noted that, two TZs are realized at 0.98 GHz and 1.07 GHz, respectively, of high-band bandpass response. It is obvious that the closer the transmission zeros to the passband, the sharper fall-off at upper pass-band. Consequently, the high frequency selectivity of the proposed bandpass filter can be achieved. Through the comparison of test results and simulation results, the seen slight frequency discrepancies between the simulation and test results is attributed to the unexpected fabrication tolerances. It can be seen that the design simulation and measurement results are in good agreement.

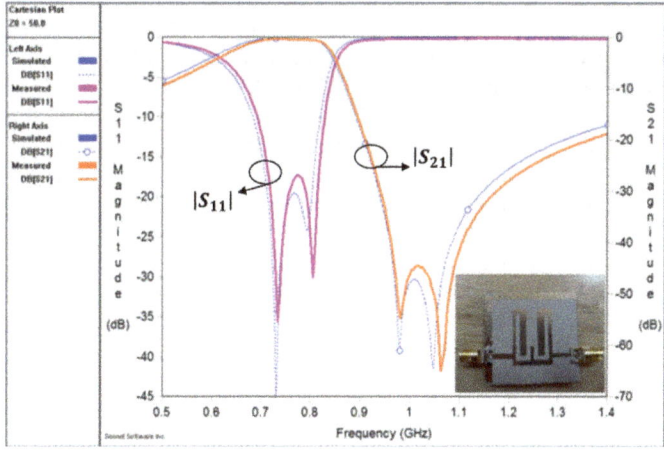

Fig. 5. Comparison between test results and simulation results (inset the photograph of the fabricated bandpass filter).

4 Conclusions

In this paper, a novel bandpass filter with open-loop short-circuit dual-mode resonator structure is designed. It has the characteristics of strengthening out of band suppression, reducing insertion loss and miniaturization, and the structure designed is flexible, which has certain practical value. In wireless communication system, RF receiver can be used to suppress the interference of image frequency. The proposed filter exhibits advantages such as highly-selectivity, compact size, and easy integrated with other planar microwave circuit. This filter is promising as important roles in 5G low-bands transceiver wireless communications systems which helpfully support widespread coverage across urban, suburban, rural areas and as well as Internet of Things (IoT) services.

References

1. Koul, S.: Broadside, edge-coupled, symmetric strip transmission lines. IEEE Trans. Microw. Theory Tech. **30**(11), 1874–1880 (1982)
2. Fiedziuszko, S.J.: Dual-mode dielectric resonator loaded cavity filters. IEEE Trans. Microw. Theory Tech. **30**(9), 1311–1316 (1982)
3. Lin, T.-W., Lok, U.-H., Kuo, J.-T.: New dual-mode dual-band bandpass filter with quasielliptic function passbands and controllable bandwidths. In: Microwave Symposium Digest (MTT), 2010 IEEE MTT-S International, pp. 576–579, May 2010
4. Hong, J.-S., Shaman, H., Chun, Y.-H.: Dual-mode microstrip open-loop resonators and filters. IEEE Trans. Microw. Theory Tech. **55**(8), 1764–1770 (2007)
5. Hong, J.-S., Lancaster, M.J.: Theory and experiment of novel microstrip slow-wave open-loop resonator filters. IEEE Trans. Microw. Theory Tech. **45**(12), 2358–2365 (1997)
6. Le, T.H., Nguyen, L.C., Tran, B.D., Zhu, X.W.: Highly-selective and compact bandpass filters using microstrip - coaxial resonator. In: Proceedings of 2019 26th International Conference on Telecommunications (ICT), Hanoi, April 2019
7. Le, T.-H., Ge, C., Zhu, X.-W.: Two-pole bandpass filter using new hybrid resonator. In: Proceedings of 2015 Asia-Pacific Microwave Conference (APMC), December 2015
8. Le, T.-H., Zhu, X.-W., Ge, C., Duong, T.-V.: A novel diplexer integrated with a shielding case using high Q-factor hybrid resonator bandpass filters. IEEE Microw. Wirel. Compon. Lett. **28**(3), 215–217 (2018)
9. Duong, T.-V., Le, T.-H., Tran, B.-D.: Design of a high-selectivity quad-band bandpass filter using open-ended stub-loaded resonator. Electromagnetics **40**(7), 515–525 (2020)
10. EM User's Manual, ver. 12.56, Sonnet Softw. Inc., Syracuse, NY, 1986–2009

FPGA Hardware Image Encryption Co-simulation Utilizing Hybrid LFSR and Chaos Maps

Yasmine Mustafa Khazaal[1(✉)], Khamis A. Zidan[2],
and Fadhil Sahib Hasan[3]

[1] Computer Engineering Department, Al-Iraqi University College,
Baghdad, Iraq
yasmine_altimmimi@aliraqia.edu.iq
[2] Al-Iraqia University, Baghdad, Iraq
khamis_zidan@aliraqia.edu.iq
[3] Electrical Engineering Department, Mustansiriyah University,
Baghdad, Iraq
fadel_sahib@uomustansiriyah.edu.iq

Abstract. This paper introduces a new version of the chaotic map-based image encryption system that uses Xilinx System Generator (XilinxSG) systems. Two (PRBGs) are proposed and named XOR LFSR Fixed Point Chaotic Maps-PRBG (XORLFPCM-PRBG) plus Combined XOR and XNOR for LFPCM-PRBG (XXLFPCM-PRBG) in which LFSR is combined with FPCM to encrypt an image. The Randomized Test Benchmarks at the (NIST) are applied for texting the randomization suggested PRBGs. Also, the security analysis includes the correlation-coefficient, the information-entropy, and the attacks differential (NPCR and UACI) are being used to test the intelligibility of the suggested encryption system. FPGA co-simulation hardware was also provided via FPGA SP605 XC6SLX45T to check the image encryption method of reality. The outcome showed that both XORLFPCM-PRBG and XXLFPCM-PRBG proposed systems are suitable for ciphering images founded on stream cipher an besides surpass about traditional algorithms.

Keywords: Xilinx system generator (XilinxSG) · Image encryption · Chaotic maps · FPGA hardware co-simulation · LFSR

1 Introduction

The world of digital, the internet is a widely spread channel for transmitting or transmitting information: photos, messages, videos, and audios. This precious channel is weak unlicensed entry during data and network transmission. Cryptography combines knowledge with cryptographic algorithms into unreadable cipher texts. It has been used to pass and secure classified information from unauthorized access. Image cipher processing converts an image to another that can not be seen, while image decryption is a process that recovers an original image from a password or encrypted key image. Encryption of the image is, therefore, one of the ultimate solutions to protect image privacy [1]. In order to secure transmission over the internet, image encryption must be effectively designed with high safety and low complexity [2].

The usage of chaos for encrypting images has come as a potential alternative to several safety issues, since chaotic systems have many benefits for volatile elements, such as vulnerable initial conditions and values for parameters, clear design and aperiodic signal, rendering it suitable for cryptographic systems. [3]. Researchers have to update numerous image encryption algorithms over the last years.

Such algorithms were brought out utilizing picture encryption strategies of the chaotic method. But since chaotic systems have many important characteristics and do more than standard encryption algorithms (DES [4] and AES [5]). Numerous scientists have used the chaotic map to cipher pictures [6, 7]. Amin et al. [8] proposed a chaotic block cipher to cipher an image, using a chaotic system to capture 256 plain image. By mixing linear feedback shift register (LFSR) with chaotic maps, Abd El-Latif et al. [9] increase keyspace and the security of conventional systems. Zhang and Liu [10] used the disorderly tenting map and swapping algorithms to optimize the system 's protection and keyspace. Zhu et al. [11] optimized encrypted compression in which the original images are first mixed with a distinct 2D hyperchaos process, then the remaining Chinese theorem dispersed and compressed the mixed image. Zhang et al. [12], also used a circular replacement box besides key-stream buffer to cipher an image. Tang et al. [13] The primary frame was split into superposed blocks, random-block shuffles were carried out, and Arnold was transformed, and a disordered map was used to produce a safe matrix for block-specific encryption. Via similar investigators but by additional study, Tang et al. [14], Proposed Henon Map, Lozi Map, Tent Map and Duffing Map, together with an LFSR image encryption system.

2 Hybrid LFSR and Fixed Point PRBG-Based Chaotic Maps

The block diagram of the hybrid LFSR and fixed point chaotic maps based P-RBG (LFPCM-P-RBG) is shown in Fig. 1. Firstly, the initial values X_o and Y_0 with N bits are initialized into the discrete chaotic functions $X_{n+1} = f(X_n, Y_n)$ and $Y_{n+1} = g(X_n, Y_n)$, where X_{n+1} and Y_{n+1}, is the next interest of the state of X_n and Y_n, respectively. The sequence Xn is fixed point format (FPF) with N bits word-length (WL), Xn $= (b_1, b_2, .., b_N)_2$, $b_j \in \{0, 1\}$. The FPF has a WL = 32-bit architecture that is selected to be the optimal choice to get the chaotic signal. The signal is then passed through parallel to serial converted and scrambled with suitable pseudo-noise sequence generated by LFSR and feedback after transformed to parallel bits applying sequential to parallel converter. Slicing function is applied to the signal Xn to select the least significant eight bits from it and then Convert LFPCM-PRBG stream bits to sequential bits utilizing parallel to sequential converter [15]. The value of M is 8 bits which are selected to get the synchronization between the PRBG and the stream plain image.

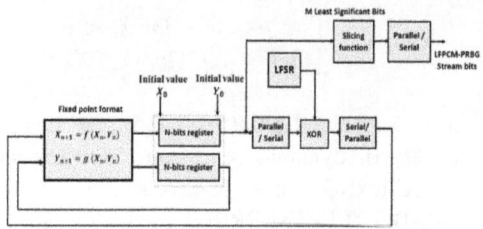

Fig. 1. The overall structure of LFPCM-PRBG.

2.1 Chaotic Logistic Map

The logistic map is given as a simple mathematical equation as [19]:

$$X_n = rX_n(1 - X_n) \tag{1}$$

where X_n is the state variable which is among [0, 1]. The interest values of the parameter r are these in the interval [0,4], [27, 28]. XilinxSG is used for the application of Eq. (1) Added to the LFSR fixed point Logistic Map-PRBG (LFPLoM-PRBG) method in Fig. 1. The defined point presentation parameter for this chart is the integer length (IL) = 2 bits long and the fraction length (FL) = 30 bits. This parameter is a fixed point display. The value of r = 4.

2.2 Chaotic Lozi Map

The muddled map offers an easy two dimensional map [19] within Eq. (2):

$$X_{n+1} = 1 - \alpha|X_n| + Y_n \tag{2}$$

$$Y_{n+1} = \beta X_n$$

Where α and β are the control parameters of the Lozi map, and X_n is the state variable. XSG is used to implement Eq. (2) and apply to the system in Fig. 1 to produce LFSR fixed point Lozi map-PRBG (LFPLM-PRBG). The parameter of the fixed point presentation for this map is the integer length (IL) = 4 bit and the fraction length (FL) = 28 bits. The value of $\alpha = 1.4$ and $\beta = 0.3$.

2.3 Chaotic Tent Map

This map is repeated cycle by mathematics in the shape of a shelter that depicts a concealed complex system over time. It acquires a point Xn on the actual line and draws it to alternative level, shown in Eq. (3), [20].

$$Xn+1 = \begin{cases} \mu X_n & for\ X_n < \frac{1}{2} \\ \mu(1-X_n) & for\ \frac{1}{2} \le X_n \end{cases} \tag{3}$$

Where μ is a positive exact fixed point (this 0.5), the state variable is Xn. The shelter map shows a variety of dynamic behaviour, from expectable to disordered, reliant on the value of μ. XilinxSG is used to execute Eq. (3) and refer to the LFSR Fixed Point Tent Map-PRBG (LFPTM-PRBG) method in Fig. 1. The fixed point display parameter, to this map the integer length (IL) = 4 bit and the fraction length (FL) = 28 bit.

2.4   Chaotic Henon Map

To investigate the dynamics of the Lorenz system, Henon proposed the famous two-dimensional Henon map. The following Eq. (4) gives the Henon map [21].

$$X_{n+1} = 1 - \alpha X_n^2 + \beta Y_n \\ Y_{n+1} = X_n \tag{4}$$

It is a two-dimensional nonlinear diagram that often says that $(\alpha, \beta) = (1.4, 0.3)$ is a unique appeal to Henon's map. XilinxSG is required for Eq. (4) deployment. To use it in the system in Fig. 1 to result in the Henon map-PRBG (LFPHM-PRBG) fixed point LFSR. The parameter for this map is the integer length (IL) = 4 bit plus the fraction length (FL) = 28 bit.

2.5   Chaotic Duffing Map

The Duffing map [22] is a two-dimensional discrete dynamical system. The equation for this type of system is given by (5):

$$X_{n+1} = Y_n \\ Y_{n+1} = \beta X_n + \alpha Y_n - Y_n^3 \tag{5}$$

where $\alpha = 2.75$, $\beta = 0.2$. This algorithm is used to generate pseudo-random bit-streams. For Eq. (5) Implementation XilinxSG is used, and apply the LFSR fixed point DM PRBG (LFPDM-PRBG) for the system under Fig. 1. The parameter in this chart is the integer length (IL) = 4 bit also the FL = 28 bit.

2.6   XOR LFSRFixed Point Chaotic Maps Based PRBG

Within this subset, a proposed XOR LFSR Fixed Point Chaotic Maps-PRBG (XORLFPCM-PRBG), the five PRBGs: LFPLoM-PRBG, LFPLM-PRBG, LFPTM-PRBG, LFPHM-PRBG and LFPDM-PRBG are worked with each other applying XOR, as presented in Fig. 3. The j-th XORLFPCM-PRBG key $K_{FPXORCM,j} = K_{LFPLoM,j} \oplus K_{LFPLM,j} \oplus K_{LFPTM,j} \oplus K_{LFPHM,j} \oplus K_{LFPDM,j}$ where $\oplus$ is XOR operation or calculation with mod-2 and $K_{LFPLoM,j}, K_{LFPLM,j}, K_{LFPTM,j}, K_{LFPHM,j},$ and $K_{LFPDM,j}$ are

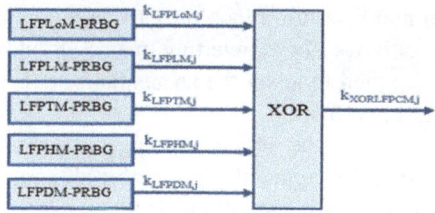

Fig. 2. Block diagram of XORLFPCM-PRBG.

PRBG generated by combined LFSR and Fixed point Logistic, Lozi, Tent, Henon and Duffung map, respectively (Fig. 2).

2.7   Combined XOR and XNOR for LFPCM-PRBG

The novel form of PRBG named XXLFPCM-PRBG may be produced via applying combined four PRBG (LFPHM-PRBG, LFPDM-PRBG, LFPTM-PRBG, and LFPLM-PRBG) using XNOR function then combined the result of this system with LFPLoM-PRBG using XOR function, as shown in Fig. 4. The j-th XNORLFPCM-PRBG key $K_{XNORLFPCM,j} = K_{LFPHM,j} \odot K_{LFPDM,j} \odot K_{LFPTM,j} \odot K_{LFPLM,j}$ where $\odot$ is XNOR operation and $K_{LFPHM,j}, K_{LFPDM,j}, K_{LFPTM,j}$, and $K_{LFPLM,j}$, are PRBG generated by LFSR and fixed point Henon, Duffung, Tent, and Lozi maps, respectively. Then combined $K_{XNORLFPCM,j}$ with $K_{LFPLoM,j}$ are XORed to generate the j-th XXLFPCM-PRBG $K_{XXLFPCM,j} = K_{XNORLFPCM,j} \oplus K_{LFPLoM,j}$ where $\oplus$ is XOR operation and $K_{LFPLoM,j}$ is the stream bits of LFPLoCM-PRBG.

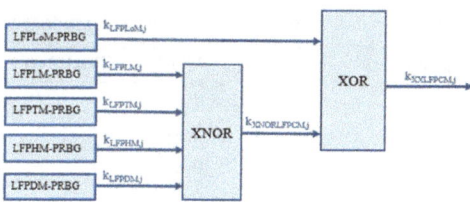

Fig. 3. XXLFPCM-PRBG system.

3   Image Encryption FPGA Framework

The FPGA board SP605 is used with 27 MHz clock frequency. Matlab/Simulink functions are each component before gateway-in and after gateway-out, and each component after gateway-in and before gateway-out are device generator functions that enable the use of fixed-point arithmetic blocks libraries. The system is comprised of stages of encryption and decryption. The representation of the source color is divided into pictures of red, green and blue. This image's data type is Unint8 format. The original image in the first stage of encryption, $I \in L_r \times L_c$ dimension, (where L_r row numbers and L_c column numbers) use block pre-processing transformed to sequential examples. Pre-processing block for converting matrix I into 8-bit serial samples (Unit8). The gateway is applied to adapt the sequential sample shape to a fixed point format with WL = 8 and FL = 0.

Sequential bits after that can be collected using the serial converter XORed on disorderly maps to construct the ciphered message for all sorts of PRBGs. Using the gateway-out, the same PRBG key is used in the decryption stage to retrieve the unique stream bits that pass over the sequential to parallel converter and back into a Matlab/Simulink setting. The postprocessing block is applied for converting the sequential image size to the unique sample (Lr × Lc) to recuperate the unique image with the identical scope.

4   Randomness Test Results

In order to examine the casualness of PRBGs generated by the proposed methods, The (NIST) statistical tests [23] are applied and applied for all proposed PRBGs. In all of the most popular tests, NIST tests contain 15 tests, including frequency, frequency of blocks, runs, cumulative-sums, and discrete Fourier transform. The stream PRBG is considered as random bits if the P-value of each test is equal or greater than 0.01, P-value represent the distribution value that is extracted from each test, and 0.01 threshold value that utilized to decide the stream bits are random or not. If the value of the P-value approach to one, the better randomness occurs, while zero P-value refers to the stream bits are determinant sequence. The NIST suite tests for XORLFPCM-PRBG plus XXLFPCM-PRBG proposed PRBG are presented in Table 1.

Table 1. NIST suite testing for PRBGs proposed.

Number	P-Value of randomness tests/test name	XORLFPCM-PRBG	XXLFPCM-PRBG
1	Frequency	0.1256	0.4074
2	Block-Frequency	0.5151	0.2328
3	Runs Test	0.8173	0.7501
4	Cumulative-Sums	0.4610	0.8148
5	DFT	0.7176	0.7004

5   Performance and Security Analysis

To exam the safety plus the performance of the proposed image encryption system, different tests are applied, containing histogram, correlation coefficients, information entropy, keyspace, plus difference attack analysis. In this section, the analysis is used for various sizes of color images. Experimentations are achieved using Simulink/Matlab and XSG tools.

5.1   Analysis of the Histogram

The histogram analyzed signifies the delivery of every pixel values. The distribution of the plain pixels is usually uneven, plus several pixels have a extraordinary incidence in the plain image. And also worth manipulating. To resist attack, the proposed system must distribute the cipher image to avoid an assault on the cipher image. The original colored image is shown in Fig. 4(1), Lena.jpg, the size 256 to 256, and Fig. 4(2), (3) and (4) respectively with the R, G, and B channels histograms of the real image, and Fig. 5(1) displays a ciphered image of Lena by using XXLFPCM-PRBG system.

Figures 5(2), (3) and (4) display histogram of the RGB chains of cipher picture that provide a stable and outstanding histogram distribution.

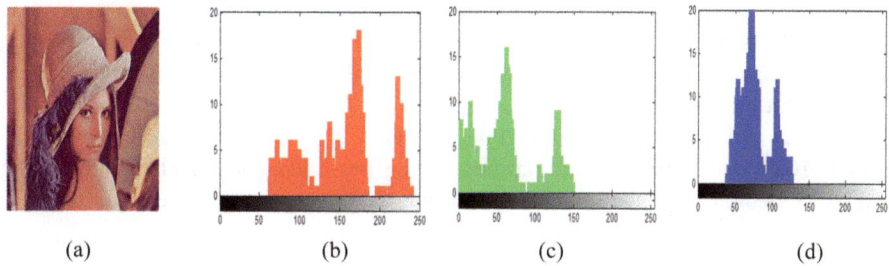

(a)   (b)   (c)   (d)

Fig. 4. Original image. (a) Lena width 256 × 256 original image, (b) Lena Red Channel Histogram, (c) Lena Green Channel Histogram, (d) Lena's Blue Channel Histogram.

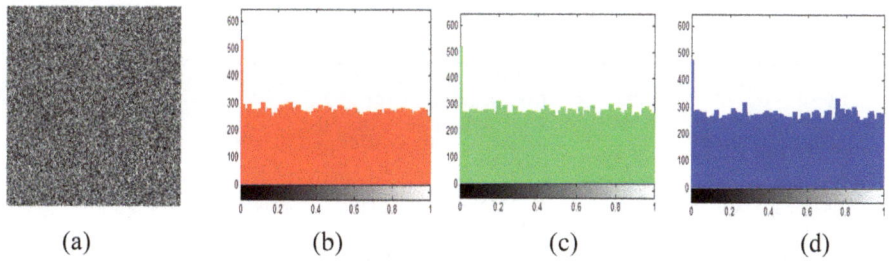

(a)   (b)   (c)   (d)

Fig. 5. XXLFPCM Image Encryption. (1) Ciphered image, (2) Cipher R channel histo, (3) Cipher G channel histo, (4) Cipher B channel histo.

5.2   Correlation Coefficient Analysis

On the encrypted image, the most statistical analysis is the analysis of the correlation coefficient. In the plain image, it tests the correlation between two pixels and encrypted images. The algorithm of image encryption is assumed to be successful if it covers entirely the p image attributes, and encrypted image is completely randomized plus strongly uncorrelated. Unless the coefficient of similarity = 1 otherwise, the two images are the same. And the encryption fails in this situation. The encrypted image is absolutely contrary to the transparent image while it's meaning $= -1$ is. The pixel correlation in horizontal, vertical, Eqs. (6–8) can be measured diagonal directions, in order to compute the correlation coefficient of some color values of two pixels in similar location in original plus cipher images [24].

$$CorCoef = \frac{Covar(x,y)}{\sqrt{Vari(x)} \times \sqrt{Vari(y)}} \tag{6}$$

$$Vari(x) = \frac{1}{N}\sum_{i=1}^{N}\left[(x_i - E(x))^2\right] \tag{7}$$

$$Covar(x,y) = \frac{1}{N}\sum_{i=1}^{N}[(x_i - E(x)) \times (y_i - E(y))] \tag{8}$$

Where CorCef coefficient of correlation in x and y, Vari(x) represents the variance of the real image x, Covar(x, y), x and y, E(x) is predicted, plus N is the entire digit of pixels within the image matrix.

Table 2 and Table 3 for Lena image with a scale of 256×256 as well as 1024×1024 to analyze the similarities of the proposed chaotic charts. Tables 4 besides 5 demonstrate the comparative analysis of correlation coefficients for a

Table 2. Comparative analysis of the Lena pic coefficient 256×256 with a number of chaotic maps and channels.

Channels	LFPLo M-PRBG	LFPL M-PRB G	LF PT M-PR BG	LFP HM - PR BG	LFP DM- PRB G	XORLF PCM-PRBG	XXLFPC M-PRBG	FPXOR CM-PRBG	FPCCM-PRBG
red Horizontal	0.03445	9.23016e-005 5.75745e-014		0.0076 0.0029		0.0096	-0.0029		
Vertical	0.02989		0.00172 0.00149	0.0084 0.0016		0.0078	-0.0016		
Diagonal	0.0336	8.30834e-005 0.00264		0.0074 0.0012		0.0113	-0.0012		
Green Horizontal	0.03372	-0.00268- 0.00508		0.0012 0.0050		0.0066	-0.0050		
Vertical	0.03074	-0.00251- 0.00424		0.0042 0.0050		0.0059	-0.0050		
Diagonal	0.03349	-0.00141 -0.00387		0.0020 0.0056		0.0074	-0.0056		
Blue Horizontal	0.02994	-0.00162- 0.00323		7.2204e-004	-0.0030	-8.8249e-004	0.0030		
Vertical	0.02674	4.30088e-005- 0.00211		0.0042 3.6872e-005		-0.0012		- 3.6872e-005	
Diagonal	0.02812	3.91790e-004 - 0.00274		0.0021 3.9550e-004		2.3581e-004 3.9550e-004		- 3.9550e-004	

Table 3. Lena 512 × 512 ciphered image compare correlation coefficient analysis for various directions between our program and our established structures.

Channels		XORLFPCM-PRBG (our proposed)	XXLFPCM-PRBG (our proposed)	Ref. [26]	Ref. [25]	Ref. [46]
Red	Horizontal	0.0019	−0.0019	0.0033	3.82×10^{-4}	−0.0022
	Vertical	0.0016	−0.0016	0.0155	1.9×10^{-3}	−0.0015
	Diagonal	0.0017	−0.0017	0.0158	4.06×10^{-5}	−0.0017
Green	Horizontal	7.2555e−004	−7.2555e−004	0.0294	-4.97×10^{-4}	0.0017
	Vertical	9.9625e−004	−9.9625e−004	0.0146	2.3×10^{-3}	−8.398e−004
	Diagonal	0.0015	−0.0015	0.0102	-3.4×10^{-3}	−0.0011
Blue	Horizontal	0.0035	−0.0035	0.0086	-1.1×10^{-3}	8.445e−004
	Vertical	0.0030	−0.0030	−0.0229	-1.1×10^{-3}	0.0013
	Diagonal	0.0037	−0.0037	−0.0366	-1.2×10^{-3}	0.0013

Table 4. Lena 512 × 512 ciphered image of our system and existing structures is a comparative analysis of the correlation coefficient.

Channels	XORLFPCM-PRBG (our proposed)	XXLFPCM-PRBG (The proposed)	Ref. [25]	Ref. [17]	Ref. [15]
R	0.00016	−0.0016	0.000626	0.0027	−0.0031
G	5.3665e−014	−5.4626e−004	0.0000219	−0.0019	0.0160
B	0.0027	−0.0027	−0.000475	0.0003	−0.0190

512 × 512 ciphered Lena image between model XORLFPCM-PRBG, XXLFPCM - PRBG and further traditional techniques are [15, 17, 25, 25, 26].

5.3   Analysis of Information Entropy

The entropy review is another security factor of knowledge sometimes used to characterize an image's ambiguity or randomness. The signal s entropy for information is expressed as [24].

$$Entrp(s) = \sum_{n=0}^{2^N - 1} P(s_i) \times \log_2\left(\frac{1}{P(s_i)}\right) bits \qquad (9)$$

where $P(s_i)$ is the probability of the symbol s_i, N is the total of bits on behalf of the basic element of the source s besides d 2^N is entirely the basic unit groupings.

Tables 6 and 7 demonstrate a comparative analysis of the Lena Image Information Entropy for various channels and unstable graphs, sizes 256 × 256 and 1024 × 1024 respectively. The Comparative Knowledge Entropy Knowledge Table 8 provides the

Table 5. Comparative analysis of Lena image size 256 × 256 information entropy with various channels and system of chaotic maps.

Channels		LFPLoM-PRBG	LFPLM-PRBG	LFPTM-PRBG	LFPHM-PRBG	LIPRBG	XORLFPCM-PRBG	XXLFPCM-PRBG	FPXORCM-PRBG	FPCCM-PRBG
R e d	Horizontal	7.3056	7.3056 / 7.3056		7.3056 / 7.3056		7.3056	7.3056		
	Vertical	7.3055	7.3055	7.3055	7.3055 / 7.3055		7.3055	7.3055		
	Diagonal	7.3054	7.3054 / 7.3054		7.3054 / 7.3054		7.3054	7.3054		
G r e e n	Horizontal	7.6043	7.6043 / 7.6043		7.6043 / 7.6043		7.6043	7.6043		
	Vertical	7.6041	7.6041 / 7.6041		7.6041 / 7.6041		7.6041	7.6041		
	Diagonal	7.6043	7.6043 / 7.6043		7.6043 / 7.6043		7.6043	7.6043		
B l u e	Horizontal	7.0654	7.0654 / 7.0654		7.0654 / 7.0654		7.0654	7.0654		
	Vertical	7.0662	7.0662 / 7.0662		7.0662 / 7.0662		7.0662	7.0662		
	Diagonal	7.0659	7.0659 / 7.0659		7.0659 / 7.0659		7.0659	7.0659		

Table 6. Comparison of information entropies of the 512 × 512 Lena image between proposed system and established systems.

Channels	XORLFPCM-PRBG (The proposed)	XXLFPCM-PRBG (The proposed)	Ref. [31]	Ref. [25]	Ref. [17]	Ref. [15]
Red	7.2797	7.2797	7.999328	7.999301	7.9974	7.9993
Green	7.5993	7.5993	7.999322	7.999214	7.9969	7.9993
Blue	7.0041	7.0041	7.999277	7.999336	7.9884	7.9993

information entropy of the 512 = to 512 Lena representation of our XORLFPCM-PRBG, XXLFPCM-PRBG, and the structures identified to us such as: [17, 25, 30] and [15].

6   Conclusion

This paper uses an XilinxSG technique and uses hybrid LFSR and fixed point chaotic map-based stream cipher to implement a color image encryption algorithm. Chaotic maps are transformed into a fixed-point format and these maps are applied to generate new PRBG versions either alone or combined in the XOR process. In the Lena picture protection analysis is performed with different sizes, including histogram, correlation coefficient, entropy and key space analysis, and the effects are compared to previously recorded systems. The systems proposed often have a strong central area that is susceptible to small changes. The results of the FPXXCM-PRBG synthesis shows that the maximum frequency is around 29,179 MHz with an output of approximately 233,432 Mb/s. Eventually, the real period of the planned systems is checked utilizing the hardware co-simulation system of FPGA Xilinx SP605 XC6SLX45 T.

References

1. Leong, M.P., Naziri, S.Z.M., Perng, S.Y.: Image encryption design using FPGA. In: 2013 International Conference on Electrical, Electronics and System Engineering, ICEESE 2013, 27–32 April 2013 (2013). https://doi.org/10.1109/ICEESE.2013.6895037
2. Dang Philip, P., Chau Paul, M.: Image encryption for secure internet multimedia applications. IEEE Trans. Consum. Electron. **46**(3), 395–403 (2000)
3. Khan, K.: Chaotic cryptography and its applications in telecommunication systems. Telecommun. Syst. **52**(2), 513–514 (2013)
4. Banerjee, S.K.: High speed implementation of DES. Comput. Secur. **1**(3), 261–267 (1982)
5. US National Institute of Standards and Technology (NIST): Announcing the advanced encryption standard (AES). Federal Information Processing StandardsPublication **29**(8), 2200–2203 (2001)
6. Fridrich, J.: Symmetric ciphers based on two-dimensional chaotic maps. Int. J. Bifurcat. Chaos **8**(6), 1259–1284 (1998)
7. Chen, G., Mao, Y., Chui, C.K.: A symmetric image encryption scheme based on 3D chaotic cat maps. Chaos Solitons Fractals **21**(3), 749–761 (2004)
8. Amin, M., Faragallah, O., Abd El-Latif, A.A.: A chaotic block cipher algorithm for image cryptosystems. Commun. Nonlinear Sci. Numer. Simul. **15**(11), 3484–3497 (2010)
9. Abd El-Latif, A.A., Niu, X., Amin, M.: A new image cipher in time and frequency domains. Opt. Commun. **285**(21–22), 4241–4251 (2012)
10. Zhang, G., Liu, Q.: A novel image encryption method based on total shuffling scheme. Opt. Commun. **284**(12), 2775–2780 (2011)
11. Zhu, H.G., Zhao, C., Zhang, X.D.: A novel image encryption compression scheme using hyper-chaos and Chinese remainder theorem. Sig. Process.: Image Commun. **28**(2013), 670–680 (2013)
12. Zhang, X., Zhao, Z., Wang, J.: Chaotic image encryption based on circular substitution box and key stream buffer. Sig. Process.: Image Commun. **29**(8), 902–913 (2014)
13. Tang, Z., Zhang, X., Lan, W.: Efficient image encryption with block shuffling and chaotic map. Multimed. Tools Appl. **74**(15), 5429–5448 (2015)
14. Tang, Z., Song, J., Zhang, X., Sun, R.: Multiple-image encryption with bit-plane decomposition and chaotic maps. Opt. Lasers Eng. **80**, 1–11 (2016)
15. Chai, X.L., Gan, Z., Lu, Y., Zhang, M., Chen, Y.: A novel color image encryption algorithm based on genetic recombination and the four-dimensional memristive hyperchaotic system. Chin. Phys. B **25**(10), 100503–100515 (2016)
16. Li, L., Abd-El-Atty, B., El-Latif, A.A., Ghoneim, A.: Quantum color image encryption based on multiple discrete chaotic systems. In: Proceedings of the 2017 Federated Conference on Computer Science and Information Systems, FedCSIS 2017, pp. 555–559 (2017)

17. Wang, W., Si, M., Pang, Y., Ran, P., Wang, H., Jiang, X., Liu, Y., Wu, J., Wu, W., Chilamkurti, N., et al.: An encryption algorithm based on combined chaos in body area networks. Comput. Electr. Eng. **65**, 282–291 (2018)
18. Parvaz, R., Zarebnia, M.: A combination chaotic system and application in color image encryption. Optics Laser Technol. **101**, 30–41 (2018)
19. Abdullah, H.A., Abdullah, H.N.: FPGA implementation of color image encryption using a new chaotic map. Indones. J. Electr. Eng. Comput. Sci. **13**, 129–137 (2019)
20. Rahimov, H., Babaei, M., Farhadi, M.: Cryptographic PRNG based on combination of LFSR and chaotic logistic map. Appl. Math. **2**(12), 1531–1534 (2011)
21. Dridi, M., Hajjaji, M., Mtibaa, A.: Hardware implementation of encryption image using Xilinx system generator. In: 2016 17th International Conference on Sciences and Techniques of Automatic Control and Computer Engineering (STA), pp. 772–775 (2016)
22. Merah, L., Ali-Pacha, A., Hadj-Said, N., Mecheri, B., Dellassi, M.: FPGA hardware co-simulation of new chaos-based stream cipher based on Lozi map. Int. J. Eng. Technol. **9**(5), 420–425 (2017)
23. Sathishkumar, G.A., Bhoopathy Bagan, K., Sriraam, N.: Image encryption based on diffusion and multiple chaotic maps. Int. J. Netw. Secur. Appl. (IJNSA) **3**(2), 181–194 (2011)
24. Wen, H.: A review of the Henon map and its physical interpretations. School of Physics Georgia Institute of Technology, Atlanta, GA, pp. 1–9 (2014). 30332-0430
25. Riaz, M., Ahmed, J., Shah, R.A., Hussain, A.: Novel secure pseudo-random number generator based on duffing map. Wirel. Pers. Commun. **99**(1), 85–93 (2018). https://doi.org/10.1007/s11277-017-5039-9
26. Rukhin, A., et al.: A Statistical Test Suite for Random and Pseudo-random Number Generators for Cryptographic Applications. NIST Special Publication 800-22 Revision 1a, 1–131 (2011)
27. Ahmad, M., Doja, M.N., Beg, M.S.: Security analysis and enhancements of an image cryptosystem based on hyperchaotic system. J. King Saud Univ.-Comput. Inf. Sci. 1–9 (2018)
28. Ahmad, M., Al Solami, E., Wang, X., Doja, M.N., Beg, M.M.S., Alzaidi, A.A.: Cryptanalysis of an image encryption algorithm based on combined chaos for a BAN system, and improved scheme using SHA-512 and hyperchaos. Symmetry **10**(7), 266–283 (2018)
29. Fu, C., Zhang, G., Zhu, M., Chen, Z., Lei, W.: A new chaos-based color image encryption scheme with an efficient substitution keystream generation strategy. Secur. Commun. Netw. **2018**, 1–13 (2018)
30. Rodriguez-Orozco, E., Garcia-Guerrero, E.E., Inzunza-Gonzalez, E., López-Bonilla, O.R., Flores-Vergara, A., Cárdenas-Valdez, J.R., Tlelo-Cuautle, E.: FPGA-based chaotic cryptosystem by using voice recognition as access key. Electronics **7**(12), 414–429 (2018)
31. Zhu, S., Zhu, C., Wang, W.: A new image encryption algorithm based on chaos and secure hash SHA-256. Entropy **20**(9), 716, 22–39 (2018)

Central Multipath Routing to Minimize Congestion in Tcp/Ip Networks Using Neural Networks

Hiba A. Tarish$^{(\boxtimes)}$ and Alaa Q. Raheema

Civil Engineering Department, University of Technology, Baghdad, Iraq
{120046,40345}@uotechnology.edu.iq

Abstract. Neural networks used to resolve the issue of multipath routing in TCP/IP networks. In this paper, the proposed method to design neural networks for solving this problem is centrally described. Through that, this method will be applied to an example of a TCP/IP computer network. Two types of neural networks are used in this work that is Feed Forward Neural Network (FFNN) and Elman Recurrent Neural Network (ERNN). FFNNs are used as recurrent neural networks to make a central multipath routing decision. ERNN is used to solve the same problems that the FFNN solve it for improving the performance of the TCP/IP computer networks. The multipath routing decision strongly depends on the congestion status of nodes or links in TCP/IP computer networks. Then, the treatment of congestion will be included in the proposed method, finally, The outcomes gotten through training plus testing phases of the proposed neural networks, the simulation program display their excellent implementation in making multipath routing decision depending on the pointer of congestion, when they are applied on TCP/IP computer network structures.

Keywords: Multipath routing · TCP/IP computer network · FFNN · ERNN · Central multipath routing decision · Congestion

1 Introduction

People can interconnect effectively when they settle on using a universal language. Computers work in a similar method. Transmission Control Protocol/Internet Protocol (TCP/IP) is a language which makes computers speaks. Nowadays, a network administrator can select one of several protocols. Nevertheless, the TCP/IP protocol is the furthermost extensively operated. One reason is that TCP/IP is the choice protocol used in the Internet, the domain's major network. A computer has to use TCP/IP in order to communicate because it is compatible with nearly all computer available [1]. Present protocols for Internet routing mostly route packets among a specified duos of source plus destination endpoints over the fastest hop-path. Conditional on the traffic load, congestion may ascend on standard links distributed via several quickest path routes, whereas new links can be operated to ease the congestion are underutilized. Within multi-path routing, routers keep a multi-path fixed number of dual or additional paths for every destination subnet [2]. The multipath routing could propose numerous advantages which are load balancing, higher aggregate bandwidth, and lower end-to-end delay; effectively

D.-T. Tran et al. (Eds.): ICISN 2021, LNNS 243, pp. 499–507, 2021.
https://doi.org/10.1007/978-981-16-2094-2_60

alleviate congestion, and bottlenecks and fault-tolerance [3]. Deployment of multipath routing includes important dual tasks that are locating an appropriate group of paths that form a multipath route plus distributing traffic thru a multi-path route. Locating a multipath route may be considered further difficult than Locating the quickest path [4, 5].

2 Related Work

Many researches focusing routing and multipath routing algorithms, some of these researches include:

Dong-Bum Kim and Seung-Ik Lee, 2011, in their paper they said that Ad-hoc network is significant in numerous parts aimed at wireless communications. In network, routing algorithms have a significant effect on communication execution. Towards decreasing network overheads, the paper propositions a mixture routing algorithm, Group Hierarchical Routing (GHR), that is a mixture of proactive plus reactive routing approaches. The evaluation by further mixture routing approaches demonstrations that GHR needs fewer communication difficulties in route detection plus conservation, [6].

S. Bhattacharjee and S. Bandyopadhyay, 2013 suggested interferences plus congestion routing path selection algorithms aimed at the parallel distribution of data packets in wireless networks. The suggested technique discoveries the node disjoint path pair with fewer coupling and congestion level for load balancing scheme. In the path set selection procedure, both the inter path interference and interference of nodes beyond the transmission range taken into consideration. Founded on the congestion level of every path, traffic is spread accordingly on every path pair. Simulation outcomes display that cramming plus interference aware multipath routing advances the network Quality of Services [7].

Zhang Yu-tong, 2017, said that the Computer Network has a significant outcome on progressing the effectiveness of the communication system besides application necessities in a lifetime. In turn, to found a transmission path with movement features plus advance the effectiveness of Computer Network in information transmission, the improved clustering routing protocol founded on node position utilizing the lowest distance routing competition mechanism was suggested within this paper. The clustering routing protocol provides filled thought to the location of computation nodes plus the transmission direction in terms of clustering besides routing path choice. The simulation outcomes display the innovative suggested optimized clustering routing protocol founded on node location has an improved communication presentation in the features of transmission postponement besides network load compared with the traditional routing protocol [8].

Sensor nodes are used in distant wireless sensor networks (WSNs), that are mostly restricted via battery lifetime plus communication variety. Only path routing instruments in WSNs directs to severe power operation besides path malfunction. A multipath routing scheme operating a cooperative neighboring node notion is showed within this study. Grounded on the accessibility of energy at adjacent nodes, multipath routes are known. The simulation outcomes prove the suggested schema improves presentation in terms of multi-path identification delay, end-to-end delay, packet delivery ratio, plus network lifetime [9].

3 Multipath Routing and Congestion Control

Routing in the computer network is critical functionality that affects the network management plus the quality of services in global networks mutually. The aim of the routing algorithm in the usual routine is to discover the minimum-cost path from sender to receiver [8, 8]. A multipath computation method denotes an algorithm that computes multiple paths in the middle of node pairs. The paths computed in the middle of nodes affects the performance gains a node pair can get. Multipath routing can be utilized via end-users to deliver packets to an endpoint over multiple paths. A beneficial extension of the single path-based routing system is a node that contains just one single valuable path to get to the endpoint. Multipath routing is a favorable direction of the present routing system as it can progress the network performance in terms of dependability and throughput.

Edge routers in every network learn numerous paths to grasp a specific endpoint then store them totally in a routing table. From sets of paths, a router later put on a set of policies to choose a particular active route. A router, by choice, promotes the active route to every neighboring network, be subject to the business relationship [5, 12, 12]. A router guides the packets crossways the links with the assistance of bidirectional links. Router chooses on the foundation of evidence gained from the routing table concerning the following continuing endpoint link to where packets travel to [10].

To design a multi-route, the following three components are used:

1. Multi route path construction: The straight path amid the source plus destination is the foundation that the multiple path finding depends on. Afterward, founding the quickest path cost (Reference cost), every alternative path is calculated whose cost is inside a sensible delta of the reference cost. This process confirms that the latency against the throughput tradeoff is valued [11].
2. Network monitoring: The statistics are polled locally via the router later transmitted to adjacent routers; this process is achieved in-band plus not via an exterior monitoring method. A one-bit representation of congestion is used, comparable to, thereby, decreasing to a smallest the scope of the observing information [12].
3. Routing topology representation: every router preserves its routing vector, containing a congestion representation of its paths to changed networks. Routing vectors are later switched with neighboring routers utilizing the In-Network Monitoring. The neighboring router is obligatory in order to clarify this information corresponding to its table; consequently, every router requires to be alert of the forwarding table of every one of its adjacent routers. In order to do this, the perception of the routing mask is introduced, which signifies the forwarding tables structure in multipath permitted routers [13].

4 Proposal System Model

This section shows proposed approaches via neural networks to explain the multipath routing issue locally in the context of TCP/IP networks. Neural networks of these methods might be constructed as a multipath routing system, or as a portion of the integral multipath routing system. It is in charge for constructing the multipath routing choice. Neural networks in these methods determine a complete path to the destination node (so on, all available paths can be found). For that purpose, they are needed general information such as (load, bandwidth) of links, or (throughput, buffer capacity, current buffer size, arrival data rate, congestion status) of nodes. The TCP/IP computer networks which are considered in this work shown in Fig. 1.

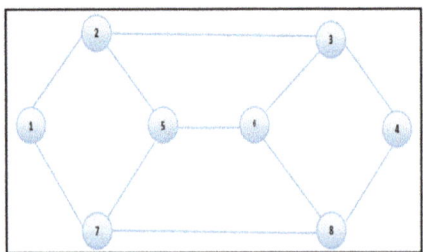

Fig. 1. Computer network CN1

4.1   Central Multipath Routing Decision

Feed Forward Neural Network (FFNN) besides Elman Recurrent Neural Network (ERNN). Are used in this work for making central multipath routing decisions in the proposed method. In this method, a multipath routing decision depends on (load and bandwidth) of links. In this method, neural networks are used for making central multipath routing decision to route a packet starting at the source node and ending at the destination node depending on two metrics which are (load and bandwidth) of links, the imposing values of these metrics for the examples of TCP/IP computer network (CN1) are given in Table 1.

Table 1. Load and bandwidth of links for TCP/IP computer network (CN1)

Link	Load (Kbps)	Bandwidth (Kbps)
(1–2)	40	60
(1–7)	170	235
(2–3)	110	180.3
(2–5)	25	45
(3–4)	205	235
(3–6)	40	50
(5–6)	4	24.6
(5–7)	10	19.6
(6–8)	137	157
(7–8)	100	140
(8–4)	9	29.5

4.2   Using Feed Forward Neural Network

A feed-forward neural network (NN1) is designed for making a central multipath routing decision. This neural network has 2M+2 inputs; which are the source, destination, M of links load, plus M of links bandwidth, where M is amount that represents the number of links of the computer network. The output path has a sequence of nodes that are determined from source to destination in r-steps. For that, top adjacent node for the source node is decided at first step, then this adjacent node converts to the basis node, and the best adjacent node for it is decided and so on till the finest neighbor node to the destination node is determined. This neural network is shown in Fig. 2.

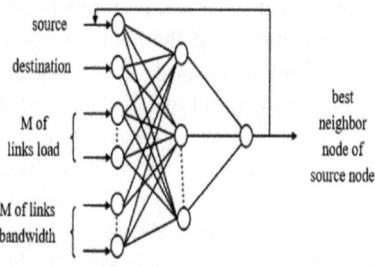

Fig. 2. The feed-forward neural network (NN1)

The TCP/IP computer network has the neural network (NN1) on the central node. This neural network receives information about the source node, the destination node, and (load and bandwidth) of all links in the TCP/IP computer network. It makes central multipath routing decisions depend on this information. First, it determine the finest neighbor node to the source node, which has most considerable bandwidth of link that connecting it with the source node and the load on this link must be lesser than its bandwidth (condition) if two neighbor nodes have same values for bandwidth the best one of them is the node that the load for its link lesser than the load to the other node and the condition is right. Then, this neighbor node becomes the source, and the neural network (NN1) determines the finest neighbor node for this source node in the same way, and so on until it defines the finest neighbor node to the destination node. Router (central-node) performs the following iterative steps for finding multiple paths:

multipath routing algorithm
Begin
1. Index=1;
2. New_Source=Source;
3. Found path by recurrent neural network (NNi);
4. Source= New_Source;
5. Index=Index+1;
6. If (all available paths are found) then jump to (7) else return to (3); // (recurrent steps).
7. Stop;

When the first best path is determined, the router returns to determine the second-best path, and so on until it determines all available paths. In the beginning, the packet forwarded at the first best path, and due to any congestion or fault, the packet will be forwarded on the second-best path to be not lost (where the fault link is the link that has the maximum load and minimum bandwidth, for example, 500 kbps and 0 kbps).

4.3   Using Elman Recurrent Neural Network

An Elman recurrent neural network (NN2) is designed for making the central decisions of the multipath routing similar to NN2.

This neural network has 2+2M+m inputs, and they are the source, destination, M of links load, M of links bandwidth, and m of context units. The output is a path that has a sequence of nodes that is determined from source to destination in r-steps. NN2 works in the same way as NN1. For instance, in the TCP/IP computer network (CN2), for deciding all available paths from node 1 to node 4, the neural network (NN2) at central node accepts source 1, destination 4, (load and bandwidth) of all links in the TCP/IP computer network, this information is given in Table 1, and the content of the context units, while this content is the previous state of hidden units. Then it determines the first best path depending on several iterative steps, the neighbor nodes for the source 1 are (2 and 7) the best one of them is 7. Then, the best neighbor node for this best node is decided, and so on, till the best neighbor node for the destination node (4) is determined; as a result, the best path is (1 7 8 6 3 4). Then, the router returns to determine the second-best path in the same way (by using a multipath routing algorithm that described in the previous subsection, where i = 6) until all available paths are determined, as shown in the Table 2.

5   Simulation Results

The neural networks (NN1 and NN2) have been applied on the example TCP/IP computer network (CN1). The application has been recognized in C++. The applied initiation function is the sigmoid function in Eq. (1). These neural networks using the router to determine all available paths from the source node to the terminus node.

The training set and the test set are arranged and used to train the neural networks and to define the entirely available paths from source to destination. The inputs for the neural network (NN1) are (2M+2), they are source, destination, M of links load and M of links bandwidth, while in the neural network (NN1), there is another input is that m of content of context units, for that the inputs become (2M+m+2). In the neural network (NN1), the number of hidden units is (10) for the TCP/IP computer network (CN1), also in the neural network (NN2), number of hidden units = number of context units = 10 for the TCP/IP computer network (CN1) (these values are selected as the best value by trial and error).

$$f(vi) = \frac{1}{1 + \exp(-\lambda vi(k))} \tag{1}$$

In the training stage, accidental values in the middle of −0.5 and 0.5 use as original weights of connection of the neural networks. Each one of these two neural networks is

Table 2. Sample of the test outcome of the neural networks (NN1 and NN2) for the TCP/IP computer network (CN1)

S	D	No. path	Real paths ordering depend on its precedence	The output of the neural network
1	4	1	(1,7,8,6,3,4)	(7,8,6,3)
		2	(1,7,8,6,5,2,3,4)	(7,8,6,5,2,3)
		3	(1,7,8,4)	(7,8)
		4	(1,7,5,2,3,4)	(7,5,2,3)
		5	(1,7,5,2,3,6,8,4)	(7,5,2,3,6,8)
		6	(1,7,5,6,8,4)	(7,5,6,8)
		7	(1,7,5,6,3,4)	(7,5,6,3)
		8	(1,2,3,4)	(2,3)
		9	(1,2,3,6,8,4)	(2,3,6,8)
		10	(1,2,3,6,5,7,8,4)	(2,3,6,5,7,8)
		11	(1,2,5,6,8,4)	(2,5,6,8)
		12	(1,2,5,6,3,4)	(2,5,6,3)
		13	(1,2,5,7,8,6,3 4)	(2,5,7,8,6,2)
		14	(1,2,5,7,8,4)	(2,5,7,9)
4	5	1	(5,2,3,4)	(2,3)
		2	(5,2,3,6,8,4)	(2,3,6,8)
		3	(5,2,1,7,8,6,3,4)	(2,1,7,8,6,3)
		4	(5,2,1,7,8,4)	(2,1,7,8)
		5	(5,6,8,7,1,2,3,4)	(6,8,7,1,2,3)
		6	(5,6,8,4)	(6,8)
		7	(5,6,3,4)	(6,3)
		8	(5,6,3,2,1,7,8,4)	(6,3,2,1,7,8)
		9	(5,7,1,2,3,4)	(7,1,2,3)
		10	(5. 7, 1, 2, 3, 6, 8,4)	(7,1,2,3,6,8)
		11	(5,7,8,6,3,4)	(7,8, 6,3)
		12	(5,7,8,4)	(7,8)

trained on (10, 15, and 10) training data sets for the TCP/IP computer networks (CN1). The training is nonstop up till the MSE turn into suitable. value of the (MSE) that used for the entire TCP/IP computer networks is (0.5e−7). In the neural network (NN1), the training set consists of inputs of the neural network which includes source, destination, (load and bandwidth) of all links in the TCP/IP computer network, while in the neural network (NN2), in addition to these inputs another input is the content of units of the context layer and desired outputs (for NN1 and NN2), which are set of nodes of path. The values of learning rate (η) are selected by trial and error of the neural network (NN1 and NN2) as (0.9 and 0.7) to the TCP/IP computer network (CN1). The obtained results shown in Figs. 3 and 4 for these two neural networks. Those Figures views mean squared error versus several epochs, the fault is reduced in case the amount of epochs amplified; also the number of epochs of the neural network (NN2) is lesser than the amount of epochs of the neural network (NN1). Results of testing for these two neural networks shown in Table 2 for the TCP/IP computer network (CN1). The success rate of testing these neural networks for the TCP/IP computer networks (CN1) shown in Table 3.

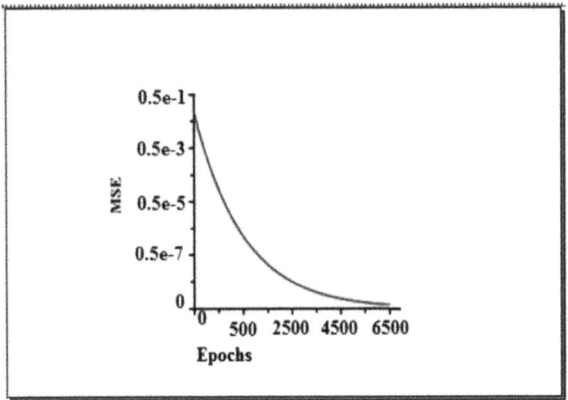

Fig. 3. Error versus number of epochs of the FFNN (NN1) for the TCP/IP computer network (CN1)

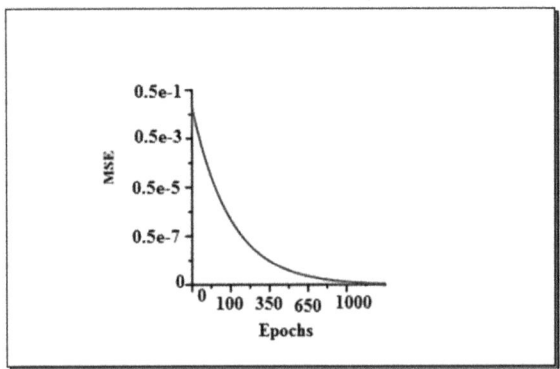

Fig. 4. Error versus number of epochs of the ERNN (NN2) for the TCP/IP computer network (CN1)

5.1   Simulation of Central Methods Using Omnet++

In central methods all available paths from source to destination are determined, these paths ordered depending on its precedence as shown in the results of C++ programming language that used in the previous section, but these results are so significant to simulate it in the OMNet++, so for TCP/IP computer network (CN1) the first six best paths will be simulated as shown in the Table 3.

Table 3. The testing proposed neural networks success rate for the TCP/IP (CN1)

Computer-Network	Neural-Network	The success rate of test on trained sets	The success rate of test on other sets
CN1	NN1	100%	90%
	NN2	100%	93%

6   Conclusions

The central multipath routing decisions are made by two neural network (FFNN and ERNN). In this method, the neural network requires information about all links in the TCP/IP computer network to determine the complete path starting at the basis node and finishing at endpoint node. The metrics used in this method are (load and bandwidth) of links. The neural network defines the first best path, and then the router returns to use the neural network to decide the second-finest path until the router determines all available paths. To decide the first finest path, in the beginning, the source node determines the best neighbor node to it. Then this neighbor node became the source, then this source determines the best adjacent node to it, and so on until it reaches the destination node. 'The finest adjacent node to the source node' is the node that has the largest bandwidth of link that linked it with the source node, and the load on this link must be lesser than its bandwidth. The congestion pointer in this method is a load of link besides, from the experimental at these two types the ERNN is best than the FFNN, because of the number of epochs was lesser and the success rates were more significant than the FFNN. The outcomes gotten through training and testing phases of the proposed neural networks demonstrate the excellent presentation in deciding multipath routing decision depending on the pointer of congestion when applied for example of computer network (CN1) structure.

References

1. Blank, A.: TCP/IP Foundations. Neil Edde, London (2004)
2. Mark, B., Zhang, S.: A Multipath Flow Routing Approach for Increasing Throughput in the Internet. University of George Mason, U.S. National Science Foundation (2008)
3. Yi, J., Cizeron, E., Hamma, S., Parrein, B., Lesage, P.: Implementation of Multipath and Multipath Description Coding in OLSR. University of Nantes (2009)
4. Sohn, S., Mark, B., Brassil, J.: Congestion-Triggered Multipath Routing Based on Shortest Path Information. University of George Mason (2007)
5. Saavedra, J.M., Bustos, B.: Sketch based image retrieval using keyshapes. Multimed. Tools Appl. (2013). https://doi.org/10.1007/s11042-013-1689-0
6. Kim, D.-B., Lee, S.-K.: A new hybrid routing algorithm: GHR (Group Hierarchical Routing). In: International of IEEE
7. Bhattacharjee, S., Bandyopadhyay, S.: Maximally node disjoint congestion aware multipath routing in wireless networks for delay minimization. Int. J. Comput. Appl. (IJCA) (2013)
8. Zhang, Y.: Research of computer network data transmission routing method. In: International Conference on Smart City and Systems Engineering (2017)
9. Chanal, P.M., Kakkasageri, M.S., Shirbur, A.A.: Energy aware multipath routing scheme for wireless sensor networks. In: IEEE 7th International Advance Computing Conference (2017)
10. Ahmad, S., Mustafa, A., Ahmad, B., Bano, A., Hosam, A.S.: Comparative study of congestion control techniques in high speed networks. (IJCSIS). Int. J. Comput. Sci. Inf. Secur. 6(2) (2009)
11. Al-Shabibi, Martin, B.: Multi route a congestion-aware multipath routing protocol. In: International Conference on High Performance Switching and Routing (2010)
12. Alshahen, H.: Neural network based routing in computer networks. University of Basrah, Master thesis (2009)
13. Kurose, J., Ross, K.: Computer Networking: A Top Down Approach Featuring the Internet, 5th edn. United State of America (2010)

Machine Learning System Using Modified Random Forest Algorithm

Dena Kadhim Muhsen[1]([✉]), Teaba Wala Aldeen Khairi[1],
and Noor Imad Abd Alhamza[2]

[1] Computer Science Department, University of Technology, Baghdad, Iraq
{110120, 110053}@uotechnology.edu.iq
[2] Ministry of Education, Baghdad, Iraq

Abstract. Machine learning algorithms are one of the most advanced failed in computer science, so developing this kind of algorithms is very important subject to work on, also many researches are produced papers about developing these algorithms such as decision tree, random forest, this paper produce development on random forest algorithm, the developed algorithm called hybrid feature selection random forest (HFSRF), the algorithm depend on the accuracy percentage that based on weight of two kinds of measures, information gain (IG) and gain index (GI), the algorithm worked on six different datasets such as ionosphere dataset and lung-cancer dataset, finally the experimental results shows that the proposed HFSRF is better than the classical random forest when applied on all the datasets.

Keywords: Machine learning · Random forest · Hybrid feature selection · Decision tree · Information gain · Gain index

1 Introduction

Machine learning is a computer science division that desires to study patterns from information to enhance the execution of different tasks. A used healthcare research comprises typically machine learning to be applied for describing very adaptable, automatized plus computationally strong methods to identifying patterns in multipart data organizations (e.g., interactions, underlying dimensions, nonlinear associations or subgroups). Often this description comes in use opposite to "traditional" parametric approaches that contain a significant number of statistical assumptions, and need the dimensions or subgroups of interest to be pre specified interactions among predictors and the functional form of the association between predictors and the result. Significant range and Multi-modal evaluations of mental disorder phenotypes as well as the related danger and predictive issues (e.g., imaging data, physiological factors and self-report measures) are carried out by researchers in the fields of psychiatry and clinical psychology. Now, it is very common that datasets included a huge number (thousands sometime) of measurements that frequently are evaluated over time.

It is possible to apply machine learning on such complicated data structure to help reaching a better perspective about metal disorders defect. It can also forecast the pattern of symptoms and risks, as well as study action outcomes and difference action reply. The aim of this paper is to present a primer in supervised machine learning (machine learning for prediction) containing usually applied terminology, modeling building, algorithms, validation plus evaluation procedures.

However, before holding a discussion about supervised learning, it is essential initial to recognize its difference from unsupervised learning, and the topics of research tasks for which every can be utilized. In order to determine the method(s) of machine learning to be used, first the research question must be specified. Depending on Hernán and his colleagues' framework, there are three main tasks of data research: prediction, description and causal inference. Each of these three tasks can use machine learning, but depending on the specified question of a research, outdated statistical methods can evidence to be more appropriate besides sufficient [1, 2].

2 Literature Survey

Some related works are addressed by the literature review to present a brief about the pervious valuable works in RF and ML.

Ramón Díaz-Uriarte and Sara Alvarez de Andrés, 2006, [3], the researchers in this paper indicated that the relevant genes selection for sample classification is mutual task in most studies of gene expression; where the attempts of researchers focused on identifying the smallest possible set of genes that is still able to achieve suitable predictive performance (for example, to be used in future in with diagnostic goals in clinical practice). Several gene selection methods utilize univariate (gene-by-gene) rankings of gene relevance plus arbitrary thresholds to choice the amount of genes, can just be used to two-class glitches, and applied gene selection ranking criteria distinct to the classification algorithm. On the other side of this, random forest is a classification algorithm suitable for microarray data: it displays outstanding presentation even when almost all predictive variables are noise, may be utilized when the quantity of variables is superior to the amount of observations and in glitches containing more than two classes, and returns measures of variable significance. Therefore, it is significant to recognize the presentation of random forest with microarray data and its possible usage for gene selection.

Jeffrey S. Evans, Melanie A. Murphy and Zachary A. Holden, 2011, [4], the researchers in this paper indicated that despite the sensitivity of the ecological modeling component, the ultimate goal of ecological studies is balancing between accurate interfaces and predictions (Peters 1991; De'ath 2007). Highly accurate predictions of ecological spatial patterns and process are the core on which the applied applications of ecology in bio-diversity, preservation planning and conservation planning (Millar et al. 2007). Nevertheless, the efforts of obtaining accurate models using traditional frequentist data model are limited by the complicated nature of the ecological systems (Breiman 2001a; Austin 2007). Well-distinct subjects in ecological modeling, such as spatial autocorrelation, non-stationary, complex non-linear interactions, extraordinary-dimensionality, scale, anisotropy and important signal make a contribution in problems that difficultly addressed by the frequentist data model (Olden et al. 2008).

At a critical evaluation of data in ecological models, it is rare that data would fulfill the presumptions of multivariate normality, independence and reevaluation (Breiman 2001a). this resulted in fixed re-evaluation of modeling methods and the special effects of reoccurring issues for example spatial autocorrelation

Seyed Amir Naghibi, Kourosh Ahmadi and Alireza Daneshi, 2017, [5], the researchers in the this paper took in account the constant uprising level of water scarcity problem in various countries, the study planning is applying random forest (RF), optimized random forest (RFGA), genetic algorithm and support vector machine (SVM) methods to evaluate the potentials of groundwater via spring sites. For this result, 14 efficacious variables containing fault-based, river-based, DEM-derived, lithology and land use were provided. The total number of spring locations was (842), 70% (589) of which were applied for model training while the other 30% were used for model evaluation. The presented model were operated and groundwater potential maps (GPMs) were produced.

In order to obtain an evaluation for the methods efficiency, receiver operating features (ROC) curve was plotted. The results of this study indicated that different kernels of SVM model had less effectiveness than RF and RFGA methods. Area under curve (AUC) of ROC value for RFGA and RF was predestined as 85.6% and 84.6% respectively. AUC of ROC was calculated as SVM- linear (78.6%), SVM-polynomial (76.8%), SVM-sigmoid (77.1%), and SVM radial based function (77%). As well as, the outcomes signified a significance slop angel, altitude and TWI in groundwater evaluation. Other places that suffer water scarcity could make use of the methodology proposed in the current study for assessing and managing the groundwater potential.

S. K. Lakshmanaprabu, K. Shankar, M. Ilayaraja, Abdul Wahid Nasir, V. Vijayakumar and Naveen Chilamkurti, 2019, [6], in their paper they said that the internet of things (IoT) is an internet amongst things over progressive communication with no human's process. The medical field can be improved through effectively using data sorting in IoT to discover different hidden truth. The current pater develops big data analytics on IoT founded healthcare system by the use of MapReduce process and Random Forest Classifier (RFC). The E-health data are gathered from the patients with diverse illnesses is considered for analysis. The best attributes are selected through the use of Enhanced Dragonfly Algorithm (IDA) from the database for the improved classification. Lastly, RFC classifier is applied to categorize the E-health data with the aid of ideal features. It is experiential from the application outcomes is that the supreme precision of the proposed technique is 94.2%. In order to confirm the efficiency of the proposed technique, the dissimilar performance measures are analyzed and compared with current approaches.

3   Random Forest Algorithm

It is one of the different types of machine learning algorithms that is utilized for supervised learning. In other words, for learning from labelled data and present forecasts depending on the pattern have been learned. Both regression and classification tasks are able to use RF.

Decision trees are the base on which RF depends. Predictive models in machine learning are created using the technique of decision trees. The name of decision trees came due to the branches of "if …. then…" decision splits followed by prediction like the trees branches. If a class is to be predicted for a sample, for example, the movement will start from the tree bottom up to the body, until reaching the first split-off branch.

This split may be considered as a feature in machine learning, assume it is "age", the decision now is made about what division to be followed: "if a given sample age is more than 30, continue along the left branch, else take the right branch". This process is repeated until reaching the succeeding division and keeps taking similar decision process till there are no more divisions before. The final result a predicted value or class in decision trees is presented by the endpoint which is called a leaf in.

At every division, the feature thresholds that finest split the (residual) samples locally is originate. For tasks or classification and variance reduction for regression, the most widely known metrics to identify the "best split" are information gain and Gini impurity. The similarity between humans' decision making and single decision trees makes the last very simple to competence and visualize: with a chain of easy rules. Though, they are not very strong, that is to say they don't generalize well to unseen sections. Here is where Random Forests come into play [7, 8].

4 Feature Selection

Feature selection is now very vigorous research zone within the arena of data mining. It lets eliminating terminated and unrelated data sets of huge extent. In this segment two equations are utilized within feature selection plus size the relation of the attribute within class.

4.1 Information Gain

Features like Gini index, Chi squared test, Information entropy and Correlation can be scored in many methods. Diversity, for example, can be measured using Entropy among many methods. Also, information entropy can scale the impurity of information to quantify the uncertainty of predicting the value of the aim variable.

The effectiveness of features in classification can be measured by Information Gain (IG) on high dimensional data on a wide range. It is the expected amount of information. The gain of higher information leads to a improved discriminative power for decision-making. IG is considered to be a valuable scale for deciding the feature relevance for classification [9].

Assumed a cosmos of text messages, $t = \{t, 1, 2, \ldots, tn\}$ plus a probability $P(ti)$, in occurrence of every text message, predictable content of information of a message M is specified via:

$$I(t) = \sum_{i=1}^{n} -P(ti)log2\, ni = 1(P(ti)) \tag{1}$$

4.2 Gini Index

In a distribution, the coefficient of Gini is a single number, which is used for scaling the inequality. Commonly, Gini is utilized in economics to measure the deviation degree of the distribution of a country's income or wealth from a total equal distribution. The impurity of D is measured by Gini Index, data partition or else a fixed training tuples, shown in the following Eq. (2) [10].

$$(D) = 1 - \sum_{i-1}^{m} p2i\,mi = 1 \tag{2}$$

5 Proposed Modified Random Forest

In this paper proposed mix features selection Random Forest algorithm are suggested. A selection features in the RF algorithm play a vital part within defining the effectiveness and execution of algorithm. Traditional approaches utilized in features selections offers decent outcomes. Although mix features selections Random Forest algorithm are enhanced and dominance over further traditional algorithms. It is obvious the features are significant within a specific technique and are not significant within other technique. Two approaches (GI and IG). These are decent approaches of range; however there are good features that show in one of them so the proposed algorithm presents Eq. (3) in the middle of the dual approaches.

$$Fx = (w \times info\ gain(Fx)) + ((1 - w) \times (1 - Gini(Fx))) \tag{3}$$

The algorithm of decision tree is shown in Algorithm 1.

Algorithm 1: The DT for training
Input: Sample of Training Dataset
Features Number: Chosen Features Numbers   Depth: Tree Level Depth.
Output: tree

Start
Step.1: build tree (train, n_features, depth)
Step.2: for index=1 to n_features:
          gain Val = gain (dataset, index)
Step.3: For Every Row In Train:
          groups = Split Train dataset on row [index]
          value Gini = Gini Index (groups)
          Value (Fi)= (w * info gain (Fi)) + ((1 - w) *
          (1 – Gini (Fi)))
          if Gini < old Gini then p_index, p_Groups =
          Index, Groups End For
          End For
Step.4: If (depth not equal zero) then
          build tree (left p_groups, n_features)
          build tree (right p_groups, n_features)
     else return {root= (left p_groups, right p_groups)
    end if
End.

Thus, proposed HFSRF is demonstrated in Algorithm 2.

Algorithm 2: HFSRF Algorithm
Input: Train: training dataset, Test: testing dataset, N Trees: Trees Number In Forest, N Features: Selected   Feature Amount
Output: Learned Trees with Accuracy

Start
 	
Step.1: For j = 0 To n_Trees Get Random
 	
       Feature =Random (train, n_Features)
 	
   Train = sub   Train (Train, size sample)
 	
   tree = build tree (train, feature)
 	
   trees. append(tree)
 	
Step.2: For I = 0 to n   Trees
 	
      correct = Acquire vote (test, trees(I)))
 	
 End For
 	
Step.3: Accurse = Correct Divide / Class Number within Test Data
 	
End

5.1   Dataset

In this paper the applied the regular RF algorithm and the modified RF algorithm on two datasets to masseur the accuracy and quality for each algorithm. These datasets are (Ionosphere and Lung Cancer) the descriptions of the datasets are showed in the following Table 1.

Table 1. Explanation of the dataset

Dataset	Instances Num	Attribute Num	Attribute type	The Area of dataset
Ionosphere	400	33	Real, Number	Physical
Lung-cancer	70	55	Number	Natural life

6   Experimental Results

In this paper we applied HFSRF on an actual dataset described in Table 1, we implement mutually the usual Random Forest which utilizes Gini Index also information gain measures also with HFSRF the outcomes are demonstrated in the following Tables (2 and 3). These tables have (0.1....0.9) columns to exemplify the algorithm mix heaviness, likewise there are dual columns that exemplify the GI plus IG. Where rows represent total of trees (5, 10, 20) that utilized in all datasets. In Table 2 the suggested HFSRF reached the uppermost outcomes in circumstance (five trees, weight 0.3), the accuracy = 91.53%, in the next circumstance (ten trees, weight 0.45), achieved 92.46%, and in the next circumstance (twenty trees, weight 0.4) accuracy = 94.22%. In Table 3 the suggested HFSRF reached uppermost outcomes in the instance (five trees, weight 0.4) accuracy = 76.66%, plus in instance (ten trees, weight 0.4, 0.3) where the IG accuracy = 76.66%, instance (fifteen tree, weight 0.8) accuracy = 87.67% (Fig. 1).

Table 2. The accuracy statistical for ionosphere dataset

Tree number	Weights of hybrid									GI	IG
	0.1	0.2	0.3	0.4	0.5	0.6	0.7	0.8	0.9		
5	89.14%	90.22%	**91.53%**	88.68%	90.10%	89.41%	91.00%	88.68%	90.68%	91.14%	87.24%
10	91.71%	90.67%	91.15%	90.68%	**92.46%**	91.41%	92.71%	90.75%	90.29%	90.00%	82.67%
20	92.86%	91.70%	91.34%	**94.22%**	91.41%	93.41%	92.68%	90.06%	91.43%	91.00%	86.67%

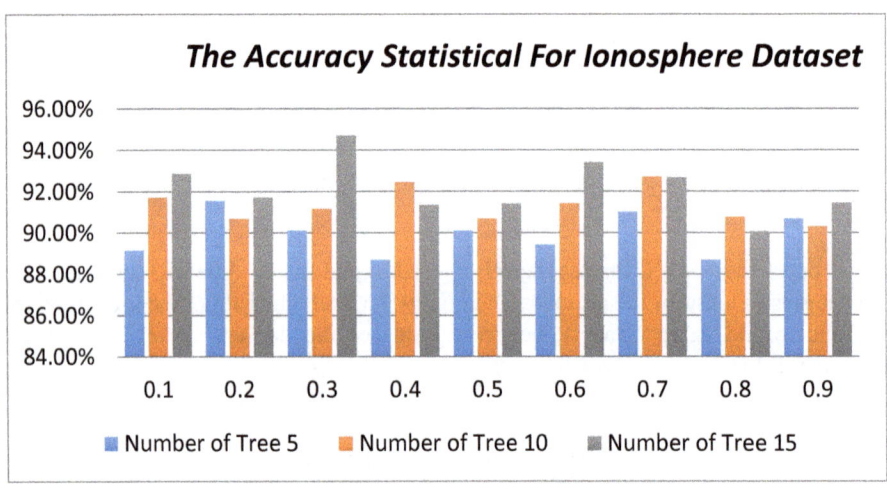

Fig. 1. Accuracy measurement of ionosphere

Table 3. The accuracy statistical for lung-cancer dataset

Tree number	Weights of hybrid									GI	IG
	0.1	0.2	0.3	0.4	0.5	0.6	0.7	0.8	0.9		
5	73.43%	76.57%	73.00%	**76.66%**	**76.66%**	73.04%	74.33%	77.67%	73.03%	73.03%	66.77%
10	66.77%	73.43%	**76.77%**	**76.66%**	74.33%	66.77%	71.00%	**77.67%**	53.03%	70.10%	**76.77%**
20	71.00%	83.23%	71.00%	81.00%	76.66%	66.77%	64.39%	**87.67%**	70.30%	70.10%	73.43%

7 Conclusions

HFS modification is completed in the feature selection stage which is done on RF Algorithm. The mixed measurement of IG and GI in the stage of feature selection in the operation of learning. Enhancements are done on the chief algorithm of feature selection via changing single measurement for feature selection stage with a mixed feature selection measurement that signifies GI measurement and IG measurement. The enhancements are meant to acquire control between the features plus class that in similar circumstance, where the distinct measurement cannot acquire. In the Hybrid proposition dual measurement works simultaneously to select the top feature reliant on the features relation with class that equipped to the succeeding stage in creation evaluation between the standard feature selection and HFS algorithms. The algorithm obtained rules are evaluated. It was concluded that HFS algorithms is more favorable than indusial measurement in utmost of Dataset, since HF has collected the control of dual measurement as it presented within tables from (2) and (3), No each period the growing in numeral of trees in every RF will rise the rate of accuracy approximation, it hinges on the type of dataset, Learning process utilizing HFS algorithms necessitates sum time of dual measurements on contrary one measurement needs a smaller amount. Which incomes Traditional FS Algorithms is enhanced than HFS Algorithms in time difficulty of learning procedure.

References

1. Zhang, C., Ma, Y.: Ensemble Machine Learning, Methods and Applications. Springer, Heidelberg (2012)
2. Rodriguez-Galiano, V., Sanchez-Castillo, M., Chica-Olmo, M., Chica-Rivas, M.: Machine learning predictive models for mineral prospectively: an evaluation of neural networks, random forest, regression trees and support vector machines (2015). www.elsevier.com/locate/oregeorev
3. Díaz-Uriarte, R., de Andrés, S.A.: Gene selection and classification of microarray data using random forest (2006)
4. Evans, J.S., Murphy, M.A., Cushman, S.A., Holden, Z.A.: Modeling species distribution and change using random forest. In: Predictive Species and Habitat Modeling in Landscape Ecology, pp. 139–159 (2011)
5. Naghibi, S.A., Ahmadi, K. , Daneshi, A.: Application of Support Vector Machine, Random Forest, and Genetic Algorithm Optimized Random Forest Models in Groundwater Potential Mapping. Springer, Dordrecht (2017)
6. Lakshmanaprabu, S.K., Shankar, K., Ilayaraja, M., Nasir, A.W., Vijayakumar, V., Chilamkurti, N.: Random forest for big data classification in the internet of things using optimal features. Int. J. Mach. Learn. Cybern. **10**, 2609–2618 (2019)
7. Biau, G., Scornet, E.: A random forest guided tour. Test (2016). https://doi.org/10.1007/s11749-016-0481-7
8. Zhu, Y., Xu, W., Luo, G., Wang, H., Yang, J., Lu, W.: Random forest enhancement using improved artificial fish swarm for the medial knee contact force prediction. Artif. Intell. Med. **103**, 101811 (2020)
9. Jadhav, S., He, H., Jenkins, K.: Information gain directed genetic algorithm wrapper feature selection for credit rating. Appl. Soft Comput. **69**, 541–553 (2018)
10. Clémençon, S., Depecker, M., Vayatis, N.: Ranking forests. Mach. Learn. Res. J. **14**(1), 39–73 (2013)

Robust Speaker Identification System Based on Variational Bayesian Inference Gaussian Mixture Model and Feature Normalization

Aliaa K. Hassan$^{(\boxtimes)}$ and Ahmed M. Ahmed

Department of Computer Science, University of Technology,
10066 Baghdad, Iraq
`110018@uotechnology.edu.iq`

Abstract. Voice is an important human feature for identifying an individual in normal human's communication. Automatic speaker recognition (ASR) mechanisms can, therefore, be considered a client-friendly form of biometric type that is used in applications such as banking, forensics, teleconferencing, and so on. This paper presents a text-independent speaker identification system based on Variational Bayesian Gaussian Mixture Model (VBGMM). Four types of features which are: MFCCs, derivatives of MFCCs, Log Filter-bank Energies and Spectral Sub-band Centroids, in addition to feature normalization have been used in the proposed system. The performance evaluation of proposed system is compared with the traditional Gaussian Mixture Model (GMM). The two modeling techniques with different types of covariance (Diagonal and Tied) are examined using the TIMIT and Arabic corpus. The recognition rates for the two datasets indicate that VBGMM is superior to GMM particularly when using data normalization for the extracted features. The recognition rates that achieved in this experiment were 98.3% and 93.3% for the TIMIT and the Arabic corpus respectively.

Keywords: Speaker identification · Feature normalization · GMM · TIMIT · Variational Bayesian inference

1 Introduction

Speaker Identification is a technique that uses a pre-defined input speech sample compared with voices to pick the one that most closely corresponds to the input sound sample, allowing an unknown speaker to be selected from a collection of different speakers. This is accomplished by the extraction of speaker-dependent characteristics from preset voice recordings. It has two modes of operation: open-set and closed-set. In closed-set, the speaker-test is measured in a closed set to all the speaker models in the dataset and return the speaker ID that has the closed match, there is no rejection. Whereas Open-set can be considered as closed-set with verification task so it considered a complex problem than a closed-set [1, 2].

Speaker recognition can be classified into two main types: identification and verification. Speaker identification is the process of recognizing (identifying) a person from a set of many individuals. Speaker verification (sometimes referred to as

D.-T. Tran et al. (Eds.): ICISN 2021, LNNS 243, pp. 516–525, 2021.
https://doi.org/10.1007/978-981-16-2094-2_62

authentication) finds out whether an individual is who claims to be from speaking samples of that person. Speaker recognition systems can be categorized further as text-dependent or text-independent. In text-dependent speaker recognition, the speaker uses a specific phrase that would be known to the system. On the other hand, in text-independent speaker recognition, the speaker can use any phrase because the system does not have any stored phrase to compare with. Therefore, text-independent speaker recognition is more challenging than Text-dependent speaker recognition, and this what will be focused on in this paper [2].

Usually, a standard speaker recognition program consists of two stages: an enrollment process (or training) and an authentication (or testing) stage. Throughout the training cycle, a person speaks an acceptable sentence in a microphone or related tool connected to the program. The process then derives voice signal basic features that will be used to build a speaker model [3].

The testing phase for the recognition of speakers can be cast as a function of pattern recognition. As such it can be broken down into two parts, a feature extraction stage, and a classification stage. The feature extraction stage is the same as that used in the enrolment process. The classification stage can be divided into two parts; pattern matching and decision making. The pattern matching portion aims to compare the features evaluated with the speaker models. The decision section analyzes the score(s) of correlations that may be either statistical or deterministic [3].

Owing to their vocal tract forms, larynx proportions, and other aspects of their voice-producing organs, speakers have different voices. A speaker's speech has its characteristics which allow us to recognize the speaker. These characteristics are acquired through a stage in the speaker recognition systems called the feature extraction [4].

Speech offers a broader variety of authentication solutions options relative to other biometric technologies like fingerprints, iris, etc. This is because while speech is a non-stationary signal used to relay a message to the listener by words from the speaker, it is often able to provide additional speaker-related information such as nationality, gender, age, emotional status, and so on. Also, though, the speech can be used to gain the speaker's identity. Consequently, the main purpose of a Speaker Recognition Program is to identify and define these sources of knowledge hidden in the speech signal and use it to gain the speaker's identity [5].

This work's main contribution consists of two stages, the first being an effective method of applying the variational inference to speaker models. The experimental results indicate that the proposed VBGMM approach has the potential for a speaker recognition improvement. The second is the benefits to normalize the feature data to have a unit form in an order to decrease the noise and other outlier data. That was achieved by using Normalization. This step improves the recognition rates by for the two used algorithms.

2 Theoretical Background

Figure 1 shows the speaker Identification block diagram. The key elements include feature extraction and modeling of speakers.

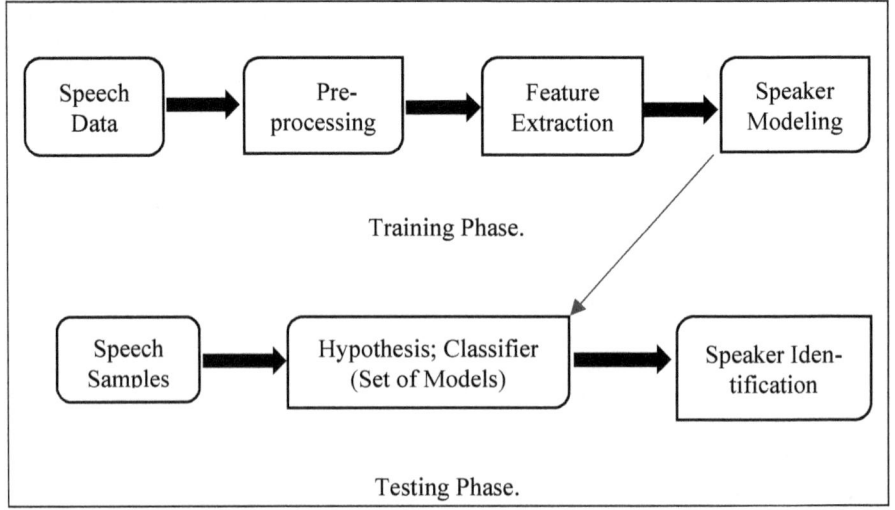

Fig. 1. Speaker identification system block diagram

2.1   Feature Extraction Methods

The purpose of the feature extraction method is to extract a condensed, effective set of parameters that reflect the acoustic impedance observed for subsequent use from the input speech signal. Feature extraction is the tool used to reduce the voice signal data aspect while maintaining the necessary information. Speech signal includes tons of information not all required to identify the speaker. Good features should be resilient against noise and distortion, should appear frequently and of course, in speech, should be easy to determine from voice sound, and should be difficult to mimic. The extracted characteristics can be classified into spectral, Spectro-temporal, speech source, short term, prosodic, etc. Short term spectral features are extracted from speech signals by dividing them into small frames of lengths of 20–30 ms. The speech source utilizes vocal tract features. Here we use the Mel-Frequency Cepstral Coefficients (MFCC) short term feature [6].

MFCC is an audio extraction method that extracts speaker-specific parameters from the speech. (MFCC) is the most common and dominant method of extracting spectral characteristics for the speech by using the Fourier Transformed signal processing of the perceptually dependent Mel spaced filter bank [7].

We apply the following equation to calculate delta functions from MFCCs.

$$d_t = \frac{\sum_{n=1}^{N} n(c_{t+n} - c_{t-n})}{2 \sum_{n=1}^{N} n^2} \tag{1}$$

Where 'N' is summed over by the number of deltas. Commonly taken as 2.

In this work, 92-dimensional features from speech frames are extracted. There are 20 MFCC features, 20 Delta MFCC (derivatives of MFCC) features, 26 Log Filterbank Energies Features and 26 Spectral Sub-band Centroids Features.

MFCC derivatives provide details on the overtime dynamics of MFCCs. It turns out when the delta-MFCC is measured and appended to the original MFCC features, it would have a great impact. They usually substantially increase the accuracy of ASR recognition by incorporating a characterization of temporal dependencies to the models, which are considered to be statistically independent of each other [8].

Filter-bank energies (FBEs) have a real physical meaning that can be beneficially used to integrate the properties of the human speech processing system (such as masking) into the automated speech recognition process. Such energies are computed in a number of methods but it often involves smoothing of standard spectral measurements and compression of non-linear amplitude. The most commonly used non-linear operator would be the logarithm, which has the added benefit of transforming a gain factor conveniently removed in the feature space into an additive part. Though the logarithm may be the most suitable non-linear operator for clean speech recognition [9].

Spectral sub-band centroids (SSC) were used as an additional feature of speech and speaker recognition of cepstral coefficients. SSCs are measured as sub-band centroid frequencies, which capture the dominant short-term spectrum frequencies. SSCs are measured as sub-band spectra centroid frequencies and they give the positions of the power spectrum local maxima. SSCs are used for speaker recognition, audio fingerprinting, and speech recognition. The accuracy of SSC detection in noise-free environments is lower compared with MFCCs. SSCs can, however, outperform MFCCs in noisy environments and can be merged with MFCCs to provide additional information [10].

2.2   Feature Scaling (Normalization)

Is the scaling method for individual samples to have unit norm. This is a tool used for standardizing the set of independent data variables or features. It is also known as data normalization in data processing and is generally done during the preprocessing step of the data. Complicated real-time pattern recognition technologies use features that are created with different sources from many different feature extraction algorithms. There may be various dynamic ranges of these features. For example, common distance measures, the Euclidean distance, implicitly assign features with wide ranges to be weighted more than those with limited ranges. Therefore, Normalization of features is needed to roughly normalize ranges of features and make them have fairly the same effect when measuring similarities [11].

The term "norm" is widely used in Euclidean space for reference to the vector norm. This is known as the "Euclidean norm" formally known as the L2-norm. In Euclidean space, the Euclidean norm maps a vector to its length. Using the L2-norm (also known as spatial sign preprocessing), we applied the normalization on the extracted feature vector. A multivariate subset of the Sign concept is the Spatial Sign. Many studies recently examined multivariate estimators of the spatial sign-dependent covariance structures. These new estimators for peripheral observations are considered robust. From a computational perspective, estimators based on the spatial sign are very easy to accomplish, as they allow the data to be transformed into their spatial signs, from which the standard estimator is then determined [12].

2.3   Speaker Modelling

Classical approaches to speaker modeling and scoring are divided into two main classes: stochastic and template models. In template models, an amount of similarity is specifically computed between the target and the test feature vectors. VQ is a typical example of speaker recognition for speaker models in this context. The desired characteristic vectors are clustered, and then each cluster is described by the cluster centroid. That decreases the target characteristic vectors to a set defined as a codebook. On the other hand, every speaker is modeled in stochastic models by an evaluated stochastic function. In this sense, the GMM was the most common generative model, in which the likelihood of each human acoustic class is. Modeling the speaker is a crucial step in a speaker recognition process. We need even more data on stochastic processes for the model speakers [13].

3   Proposed Methodology

In this paper, we propose an identification scheme based on VBGMM that recognizes speakers from two separate databases (TIMIT and Arabic corpus). The Identification System structure is shown in Fig. 2. It is clear from Fig. 2 That the process of identification consists of three primary phases: extraction of the features, normalization, and Speaker modeling. Two Different models (GMM and VBGMM) will take as input the Combined 4 types of Feature (MFCCs, derivatives of MFCCs, Log Filter-bank Energies and Spectral Sub-band Centroids) Of the speaker's training samples and try to learn how to distribute them, which will be representative of that speaker.

When testing whenever the speaker of a new voice sample is to be identified, it will first extract the sample's 92-dimensional function and then use the trained speaker (GMM or VBGMM) models to measure the scores of features for all models. It is presumed that the speaker model with the highest score is the recognized speaker.

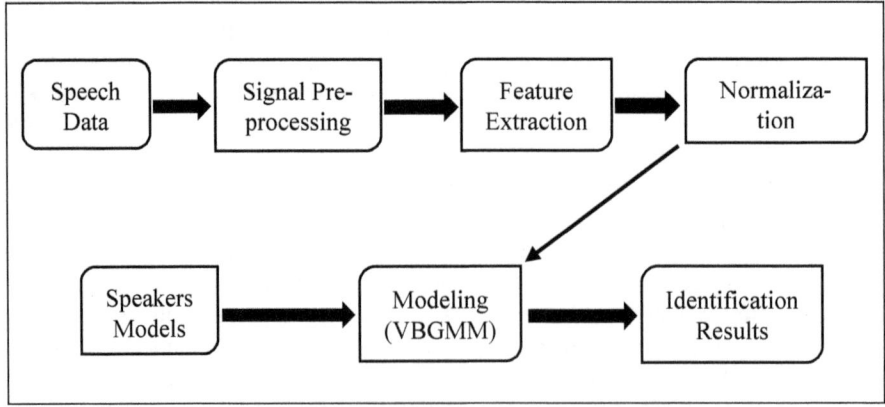

Fig. 2.   Proposed speaker identification system.

4   Experimental Results

4.1   Dataset Discretion

TIMIT

The TIMIT speech corpus was developed to provide voice data for acquiring acoustic-phonetic information and for improving and testing automated systems of speech recognition. TIMIT was the result of collaborative efforts by the Defense Advanced Research Projects Organization-Information Science and Technology Office (DARPA-ISTO) at several facilities under sponsorship. Data corpus architecture was a collaborative project of the Massachusetts Institute of Technology (MIT), Stanford Research Institute (SRI), and Texas Instruments (TI). TIMIT comprises a total of 6300 words, 10 sentences spoken by 630 speakers from 8 major United States dialect regions [14].

It comprises 630 speaker voice messages (438 M/192 F) and reads the speaker's 10 different sentences. Each sentence has about 3 s in length. Due to the 8 kHz bandwidth, acoustic distortion, microphone variability, distortion and lack of intersession differentiation, errors in recognition should result from the overlap of speaker ranges [14].

Arabic Corpus

The Arabic Corpus is a testing dataset was made from the Arabic language. The recording situation is a normal office. 15 speakers participated in the recording sessions which were held in Baghdad, in a quiet room in a governmental institute. The training content comprises distinct 75 voice utterances, 5 for each speaker. The spoken expressions are selected to include most of the Arabic alphabet. The recording is performed with an integrated noise removal microphone with a 16KH sampling rate [15].

4.2   Results

The experiments were applied to different cases, each algorithm (GMM or VBGMM) used with both Diagonal and Tied covariance, and each case utilized with and without the feature scaling (Normalization) technique. The recognition rate (RR) is calculated by:

$$RR = \frac{number\ of\ correctly\ recognized\ speakers}{Total\ Speakers\ Number} * 100 \qquad (2)$$

Case I

In this case, our experiment was implemented for the Arabic dataset. The GMM and VBGMM with both (Diagonal and Tied) covariance were used to model the feature obtained. 3 utterances for each of the 15 speakers were used for training and the remaining 2 utterances for testing. The results for this case are shown in Table 1.

Table 1 RR (in %) for different cases for the Arabic Dataset.

Table 1. RR (in %) for different cases for the Arabic Dataset.

No	Modeling technique	Dataset	Normalization	Recognition rate
1	GMM – Diagonal	Arabic Corpus	Without	80%
2	GMM – Diagonal	Arabic Corpus	With	90%
3	GMM – Tied	Arabic Corpus	Without	53.3%
4	GMM – Tied	Arabic Corpus	With	90%
5	V-GMM – Diagonal	Arabic Corpus	Without	70%
6	V-GMM – Diagonal	Arabic Corpus	With	80%
7	V-GMM – Tied	Arabic Corpus	Without	56.6%
8	V-GMM – Tied	Arabic Corpus	With	93.3%

Case II

Now, in this case, the experiment was implemented for the second dataset, The TIMIT dataset. Again, GMM and VBGMM with both (Diagonal and Tied) covariance were used to model the feature obtained. All speakers (630) were chosen from the TIMIT database. 1 SA, 1 SI, and the first 3 SX phrases are used in a training set and the remaining 5 phrases are used in a test set. The findings for this case are summarized in Table 2.

Table 2. RR (in %) for different cases for the TIMIT Dataset.

No	Modeling technique	Dataset	Normalization	Recognition rate
1	GMM – Diagonal	TIMIT	Without	82,73%
2	GMM – Diagonal	TIMIT	With	92%
3	GMM – Tied	TIMIT	Without	70.5%
4	GMM – Tied	TIMIT	With	80%
5	V-GMM – Diagonal	TIMIT	Without	37.17%
6	V-GMM – Diagonal	TIMIT	With	52.54%
7	V-GMM – Tied	TIMIT	Without	66.9%
8	V-GMM – Tied	TIMIT	With	96.3%

Case III

In the last case, The VBGMM with both (Diagonal and Tied) covariance were used to model the feature obtained from the TIMIT dataset. All speakers (630) were chosen with 7 utterances for training and 3 (1 SA, 1 SI, and 1 SX) for Testing. The results for this case are shown in Table 3.

Table 3. RR (in %) for different cases for the TIMIT Dataset.

No	Modeling technique	Dataset	Normalization	Recognition rate
1	GMM – Diagonal	TIMIT	Without	94.33%
2	GMM – Diagonal	TIMIT	With	98%
3	GMM – Tied	TIMIT	Without	58.94%
4	GMM – Tied	TIMIT	With	97.62%
5	V-GMM – Diagonal	TIMIT	Without	41.7%
6	V-GMM – Diagonal	TIMIT	With	84.34%
7	V-GMM – Tied	TIMIT	Without	66%
8	V-GMM – Tied	TIMIT	With	98.2%

4.3   Discussion

Speaker recognition has significant value in so many applications, particularly in the security area. While several algorithms have been used to identify speakers, it is still not possible to recognize speakers effectively under certain conditions, so new methods of speaker recognition are needed. VBGMM is newly used in this paper for text-independent speech recognition. VBGMM Is a variant of algorithms for variational inference in the Gaussian mixture model. Variational inference is an extension of expectation-maximization, which increases the lower limit on model evidence (including priors) rather than the likelihood of outcomes. Since VBGMM incorporates regularization by integrating information from prior distributions, this avoids the singularities often seen in expectation-maximization results but adds some subtle biases to the model.

From the outcomes in the last section, it is understood that VBGMM performs better than traditional GMM, in particular with the feature normalization. To get superior performance, we combined the four types of features (MFCCs, derivatives of MFCCs, Log Filter-bank Energies and Spectral Sub-band Centroids). By combining, the dimension of the feature vector increased to 92. This also integrates the feature vector with temporal dependencies, understanding of human speech and many other characteristics of talking into the ASR.

The feature scaling (Normalization) was also a key factor in this method, In particular with the tied covariance for VBGMM and the diagonal covariance with GMM. From the results, we have shown that Normalization improved the recognition rates of the proposed method by about 30%, and that is a huge improvement. In this approach, the data normalization applied with the L2-norm (spatial sign preprocessing). Usually, data normalization is a preprocessing technique that performs before the feature selection process. However, here the normalization is applied to the feature vector after the feature extraction process.

For the well-known TIMIT dataset, the VBGMM and feature scaling (Normalization) methods are proposed to improve the efficiency of conventional GMM. We also performed a speaker recognition experiment to test the impact of these techniques using another database, an Arabic database, which was built and documented in a standard environment. GMM with diagonal covariance gives a good recognition rate for both databases when using data normalization based on the results obtained in Case I and III. Cases II and IV indicate that the identification levels usually increase to 3.3 and 4.3 points for Arabic and TIMIT datasets by using the proposed approach for these databases. The difference here is that Tied covariance works better with VBGMM. As expected, although the duration of training for either of the two methods with tied covariance is significantly greater than that of diagonal covariance, the duration of training is not so significant relative to the significance of the improvements in recognition rate. It must be noted that when compared with the diagonal GMM memory requirement, VBGMM with tied covariance requires more memory.

5   Conclusion

This work compares the performance of Variational Bayesian inference GMM and regular GMM with two different datasets (TIMIT and Arabic corpus) for speaker identification system. It showed that the recognition rate can be greatly improved by using the proposed VBGMM with data normalization for four types of features (MFCCs, derivatives of MFCCs, Log Filter-bank Energies and Spectral Sub-band Centroids). It also showed that the Tied covariance works very well with VBGMM, on the contrary for GMM which works better with Diagonal covariance. The system performance comparison for both Arabic and TIMIT datasets has been tested and it shows that the proposed technique, The VBGMM with feature normalization works better than GMM for the two datasets described.

References

1. Sadıç, S., Gülmezoğlu, M.B.: Common vector approach and its combination with GMM for text-independent speaker recognition. Expert Syst. Appl. 38(9), 11394–11400 (2011)
2. Beigi, H.: Fundamentals of Speaker Recognition (2011)
3. Turner, C., Joseph, A.: A wavelet packet and mel-frequency cepstral coefficients-based feature extraction method for speaker identification. Procedia Comput. Sci. 61, 416–421 (2015)
4. Alsulaiman, M., Mahmood, A., Muhammad, G.: Speaker recognition based on Arabic phonemes. Speech Commun. 86, 42–51 (2017)
5. Nayana, P.K., Mathew, D., Thomas, A.: Comparison of text independent speaker identification systems using GMM and i-vector methods. Procedia Comput. Sci. 115, 47–54 (2017)
6. Paulose, S., Mathew, D., Thomas, A.: Performance evaluation of different modeling methods and classifiers with MFCC and IHC features for speaker recognition. Procedia Comput. Sci. 115, 55–62 (2017)
7. Sithara, A., Thomas, A., Mathew, D.: Study of MFCC and IHC feature extraction methods with probabilistic acoustic models for speaker biometric applications. Procedia Comput. Sci. 143, 267–276 (2018)
8. Kim, C., Stern, R.M., Kumar, K.: Delta-spectral cepstral coefficients for robust speech recognition. Science (80), 1–4 (2011)

9. Nadeu, C., Macho, D., Hernando, J.: Time and frequency filtering of filter-bank energies for robust HMM speech recognition. Speech Commun. **34**(1–2), 93–114 (2001)

10. Kinnunen, T., Zhang, B., Zhu, J., Wang, Y.: Speaker verification with adaptive spectral subband centroids. Lecture Notes in Computer Science (including Subseries Lecture Notes in Artificial Intelligence and Lecture Notes in Bioinformatics), vol. 4642, 58–66 LNCS. (2007)

11. KumarSingh, B., Verma, K., Thoke, A.S.: Investigations on impact of feature normalization techniques on classifier's performance in breast tumor classification. Int. J. Comput. Appl. **116**(19), 11–15 (2015)

12. Serneels, S., De Nolf, E., Van Espen, P.J.: Spatial sign preprocessing: a simple way to impart moderate robustness to multivariate estimators. J. Chem. Inf. Model. **46**(3), 1402–1409 (2006)

13. Hourri, S., Kharroubi, J.: A novel scoring method based on distance calculation for similarity measurement in text-independent speaker verification. Procedia Comput. Sci. **148**, 256–265 (2019)

14. TIMIT Acoustic-Phonetic Continuous Speech Corpus - Linguistic Data Consortium. https://catalog.ldc.upenn.edu/LDC93S1

15. Karim Abdul-Hassan, A., Hasson Hadi, I.: Intelligent authentication for identity and access management: a review paper. Iraqi J. Comput. Informatics **45**(1), 6–10 (2019)

Document Retrieval in Text Archives Using Neural Network-Based Embeddings Compared to TFIDF

Sura Khalid Salsal[(✉)] and Wafaa ALhamed

College of Information Technology, University of Babylon, Hillah, Iraq
it.wafaa.mohammed@uobabylon.edu.iq

Abstract. In big text archives, to return the result of a search query to the system, the query needs to be matched with every item in the archive until the best match is found, this can be done faster by using an indexing structure to speed up the process of matching and retrieval, while many solutions exist, most lack either the accuracy or the efficiency, we propose creating such a structure by organizing the data in a text archive into clusters of similar content using an improved Fuzzy-CMeans clustering algorithm, limiting the search only in the most relative clusters, also we use a neural network to encode the words into numerical vectors resulting in words with simulate meaning having similar vector values providing a semantic-based clustering while depends on the context of the document rather than only the frequency of its words, while also comparing this neural network embedding method in terms of speed, memory usage and accuracy to the traditional TFIDF method which is a word frequency based method of creating the document vectors.

Keywords: Data retrieval · Text archives · Text clustering · Word embeddings · Neural networks · Skip-gram · Fuzzy-CMeans · TFIDF · Retrieval evaluation

1 Introduction

In this fast-expanding age of the internet and cloud-based services, data are becoming the modern age gold mine. Companies race to collect user data, organize it, process it to learn the user behavior allowing for better content delivery services; All of it starts with the basic premises of data retrieval. The more data are collected and store, the more difficult it becomes to manage it efficiently and retrieve data quickly. So, it is more critical than ever to find more efficient data delivery methods by lowering the storage cost, providing faster response times while maintaining acceptable accuracy. Feature selection is the process of selecting a subset of the terms. It is a critical step for reducing dimensionality, removing irrelevant data, increasing learning accuracy, and improving result comprehensibility [1]. While some have worked on the feature selection part, like in work done in 2016 by Zhao Li, Wei Lu, Zhanquan Sun, and Weiwei Xing, they propose a feature selection method by combining both TF-IDF and mutual information. First, the TF-IDF values of documents are calculated, and word vectors of the documents were generated. The probability distribution of TF-IDF of each word is estimated

D.-T. Tran et al. (Eds.): ICISN 2021, LNNS 243, pp. 526–537, 2021.
https://doi.org/10.1007/978-981-16-2094-2_63

with the maximum likelihood method. Second, joint mutual information used to measure the relationships between features and categories [1], in the 2019 work of Haojie Huang and Raymond K. Wong [2], they work on relation extraction, which can be expressed as labeling related documents with relevant terms to represent their relation. They describe a framework consisting of an unsupervised training phase and a weakly-supervised extraction phase for relation extraction. It learns relation embeddings through the relation features of the target entity pairs only and in work published in 2020 by S. N. Karpovicha, A. V. Smirnov, and N. N. Teslyab [3], they propose using a probabilistic topic modeling to create a classification model by training the model on only a subset of the data they call the positive examples, that is, documents that belong to the target class. While the proposed method provides high accuracy when it comes to the class definition of documents, it also requires pretraining the models on a humanly labeled dataset. The accuracy of the model is dependent on the accuracy of human labeling.

2   Problem Overview

To explain the solution, we must first explore the problem, retrieving data from a system is usually done by a feature matching methodology; this allows the system to retrieve the most relevant data to the search query; the more semantic the feature extraction is, the more humanly relative is the results.

We can summarize the following:

A good system must emphasize the following characteristics:

1- Fast retrieval times
2- While maintaining low storage cost
3- Semantically accurate retrieval results
4- Allowing the flexibility to use in a wide range of environments
5- Lower the computational complexity for building the indexing structure.

3   Proposed Solution

In text-based retrieval systems, where an item is a document, and a group is a topic:

Data retrieval requires finding a document with minimum effort; using Data segmentation/clustering as a flexible alternative is possible. Instead of searching for the most relevant result matching one document at a time in all the data, we segment the data to related groups and only search for the desired item in the group that better relates to its topic. A One-to-Many relation is created by combining fuzzy logic with clustering systems like fuzzy-CMeans [4]. A computer cannot deal with the text data as we know it, so we have to transform it into a form that it can understand and process. The clustering task's success and accuracy depend highly on the quality of the engineered features.

Feature extraction in text data is known as word-to-vector; such algorithms are TF-IDF [5] and Bag Of words [6]; we propose using a neural network to provide a semantic embedding layer in the system allowing us to device short and accurate relations between documents.

Retrieval systems are used with data storage systems, they are designed in a way to allowing fast data access when data is needed and requested by a query, and a good system is designed to be capable of retrieving not just identical data to the query but also pieces of data that are similar to the query to some predefined degree of similarity,, we propose a method to deal with text data while also comparing the performance, storage cost, and accuracy of when using the traditional way in embedding layer vs. using the neural network-based approach.

3.1   The Preprocessing Layer

Before working on the text data, the text must be preprocessed, removing all the non-Latin characters, ignoring words smalled than three letters, removing stop words, and finally stemming.

3.2   The Embedding Layer

To represent a text document in the system, we must transform it into numerical data then compress the data using some feature engineering methods to produce a feature vector for every document. While documents may have different sized, the doc-vector must have a consistent size across all the documents. Each term is represented by a vector called the word-vector. The doc-vector was created by combining the word-vector for every word in the document using a unique method (Fig. 1).

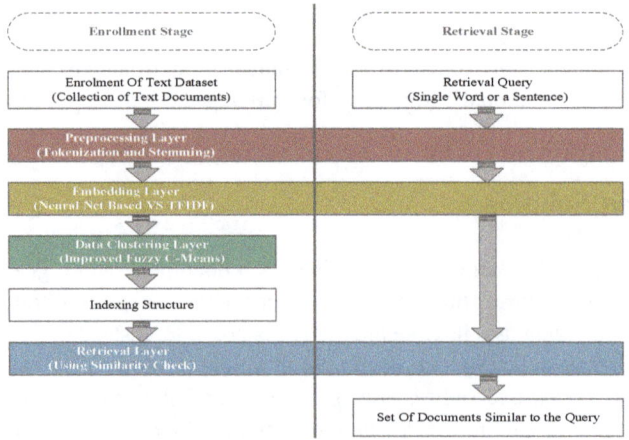

Fig. 1. Proposed system layers and stages

Using Traditional TF-IDF

In traditional TF-IDF [5], the vector must contain a cell to represent each term in the vocabulary, producing vectors with large sizes, sometimes bigger than the document it tries to represent. This happens when the document does not contain all the vocabulary words, which is typical for almost all documents. (no single document contains all the vocabulary). A vector may contain lots of cells, where almost all of them have zero values wasting valuable storage space and slowing down the clustering process. The term value calculated using equations (see Eq. 1 Eq. 2 Eq. 3) [5].

$$TFIDF\ score\ for\ term\ i\ in\ document\ j = TF(i, j) * IDF(i) \qquad (1)$$

$$TF(i, j) = \frac{term\ i\ \ Frequency\ \ in\ \ document}{total\ \ words\ \ in\ \ document} \qquad (2)$$

$$IDF(i) = \log_2 \left(\frac{total\ \ documents}{documents\ with\ term} \right) \qquad (3)$$

Reducing the vector's size can be achieved by a filtering process, where we ignore terms with counts less than the termIgnoreThreshold, while filtering may reduce the size of a vector. Still, it also reduces the accuracy of the system because terms are removed from the vector. this value calculated from the equation:

$$termIgnoreThreshold = total\ tokens * filtering\ threshold \qquad (4)$$

where the total tokens are the count of all the tokens in the whole dataset.

Using Neural Networks and Skip-Grams

Using a neural network as a feature engineering as in the skip-gram model [7] is more suitable because it can find relations between words and represent them in a predefined vector size, which is much smaller TF-IDF. Implementing the skip-gram model as a method to create doc-vectors required the creation of the embedding matrix; The network must train using a fake task [8].

The steps used can be seen in the Fig. 2, and the network architecture is at Fig. 3.

After creating the embedding matrix using the neural network, the matrix is loaded into the system. Each document is preprocessed and tokenized. To calculate the doc-vector for each document, we use the equation:

$$DocVector = \frac{\left(\sum_{i=0}^{n} vec[i] \right)}{n} \qquad (5)$$

Where n is the count of all tokens in the document, vec[i] is the vector representation taken from the skip-gram embedding for the token number i.

3.3 The Clustering Layer

To find the desired object in the system, we need to divide the data into smaller groups, This structure is Achieved using the FCMeans clustering algorithm, allowing items to exist in multiple groups with varying degrees of membership.

One drawback of the traditional c-mean algorithm is the slow convergence time due to the random initialization of cluster centroids. Fuzzy-CMeans Flowchart is in Fig. 4.

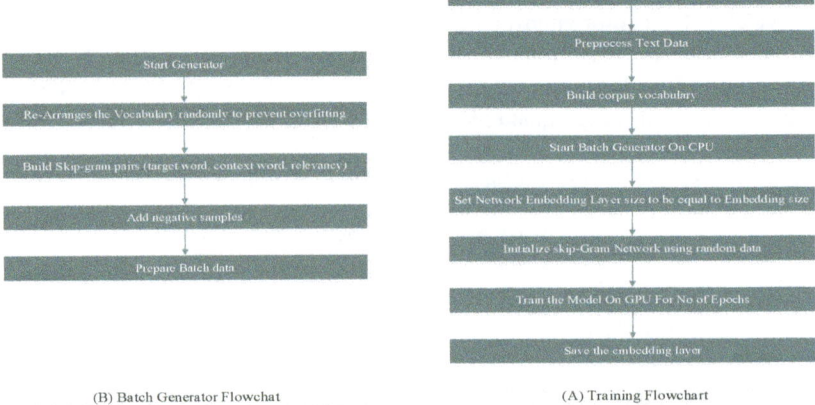

(B) Batch Generator Flowchat   (A) Training Flowchart

Fig. 2. Training flowchart

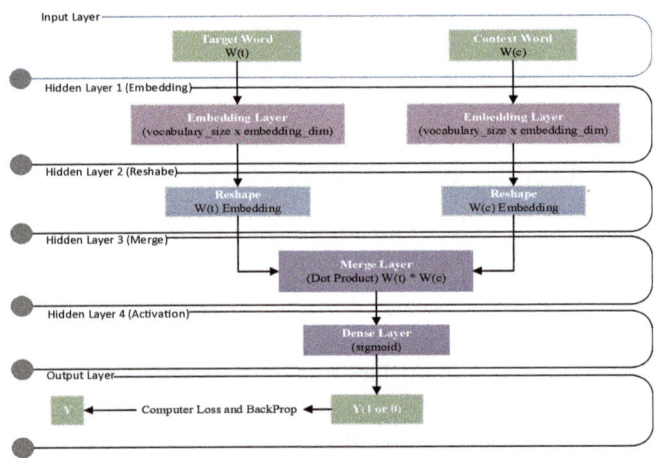

Fig. 3. Deep neural network architecture

Instead, we use Canopy [9] to choose the initial centroids reducing the convergence time, outputting a list of possible clusters. The ones with values above the population threshold calculated in Eq. (6) are chosen.

$$population\ threshold = \frac{NoOfPatarrens}{NoOfClusters/2} \qquad (6)$$

NoOfPatterns are equal to the number of document vectors, and the NoOfClusters are the required cluster numbers.

If the Canopy fails to find adequate clusters, increment the thresholds value by (0.05, 0.005), respectively, and retry.

One of the most critical aspects of the Fuzzy CMeans algorithm is predicting the fuzziness value [10], also known as fuzzifier value, which is a challenge by itself. Predicting the fuzziness value is done with the equation:

$$f(D,N) = 1 + \left(\frac{1418}{N} + 22.05\right)D^{-1} + \left(\frac{12.33}{N} + 0.243\right)D^{-0.0406\ln(N)-0.1134} \quad (7)$$

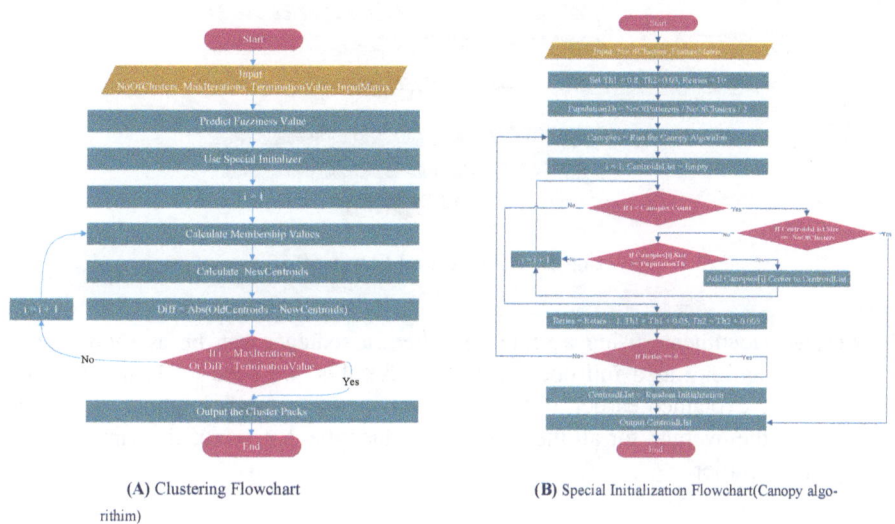

(A) Clustering Flowchart

(B) Special Initialization Flowchart(Canopy algorithim)

Fig. 4. Fuzzy-CMeans flowchart

3.4 The Retrieval Layer

Since a cluster is represented by its center, all we have to do to retrieve an item from the data (as a response to a query) is to calculate the item doc-vector using the embedding matrix created earlier. This process is illustrated in the flowchart in Fig. 5 cluster centroid using cosine similarity, then select the cluster with the most similar centroid as the most similar cluster.

If the cluster contains sub-clusters, we also find the most similar sub-cluster the same way as above.

When there are no more sub-clusters, we compare the item doc-vector with each item inside the cluster using the cosine similarity, then order the results progressively from highest to lowest. The items with the highest values are the most similar to the query item and considered the output result.

The retrieval process can work with two types of input.

1. Retrieve Documents using a single word: in this case, the doc-vector is equal to the word-vector, and it is directly obtained from the embedding matrix.

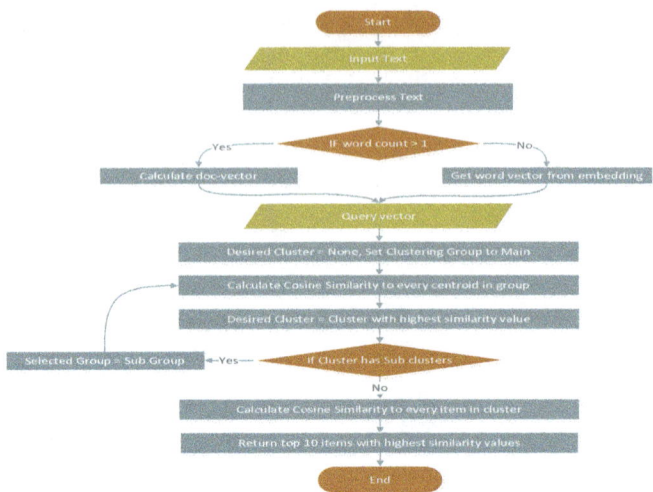

Fig. 5. Data retrieval flowchart

2. Retrieve Documents using a sentence: where a sentence can be as short as two
   words or as long as a full document. To solve this, we need to do the text pre-
   processing explained earlier, obtain the word-vector for each resulting token, and
   calculate the average for all the token to produce the doc-vector the same process
   explained earlier.

4   Testing Method and Environment

To test the system, we used two datasets. Reuters Dataset: it contains 21578 documents
were collected and manually classified by personnel from Reuters Ltd. [11].

This dataset creating a challenge to test the system's capability to classify such hard
data [12].

The Blog Authorship Corpus: This dataset is created in 2004 from the website
(blogger.com); it contains 681228 posts collected from 19320 different bloggers containing over 140 million words [13]. This dataset is more suitable to test the system
because of the wide range of topics it contains representing the true nature of the cloud
data, which is a diverse data content.

We opted to work on each dataset in isolation. Where network training, clustering,
and retrieval are done on one dataset at a time.

The Specifications and parts used in the testing machine and the software versions,
including (OS, Python [14], Keras [15], TensorFlow [16]), are in Table 1.

4.1   Preprocessing

Before using the dataset, preprocessing must be done. The results are in the Table 2.
Since the Blogs dataset is too huge to be processed using the TF-IDF method, we opted

Table 1. Testing system specs

Specs	
CPU	Intel Core i7 6700k
RAM	DDR4 2400MHz 32GB
GPU	Nvidia 980m 8GB GDDR5
OS	Windows 10 Build 17763
Nvidia Driver Version	445.87
Python Version	3.6.10
Keras Version	2.3.1
Tensorflow version	2.1.0

Table 2. Preprocessing result

	Routers	Blog Full	Blog Half
File Size in Byte	27,636,766	806,198,769	---
Avrg Docs Extraction Time in ms	429	386,150	---
No Of Docs	19,043	678,170	380,778
Avrg Tokenig + Steming Time in m	6,180	258,725	120,750
Total Charecters In Dataset	27,636,766	804,721,229	---
Total Charecters Extracted	16,006,391	750,526,207	---
Total Untokenized Words	2,706,337	138,519,314	67,189,680
Total Tockens	1,252,004	49,060,348	23,795,004

to reduce the data size to about half by selecting only the top half of the dataset documents.

4.2   Embedding

The dataset must be loaded and preprocessed then presented in a tokenized format. Creating the document feature matrix is done using both methods:

TF-IDF Embedding

When dealing with a large dataset, the TF-IDF fails because of document size and slow processing time, even when using the filtering threshold because the vector size is enormous. Also TF-IDF suffers from a data size dependency problem, where calculating each doc-vector requires using the count of all terms. By adding a new document, changing dataset term count, requiring the recalculation of all the doc-vectors to compensate for the change. This process of recreating the doc-vector with each change to the dataset reduces the system efficiency.

Solving the dependency and enormous vector size problems is done using the skip-gram model to generate the embeddings; this ensures the fixed short length of all the doc-vectors in the matrix and ensures a static independent word-vector used across the system no matter how many documents to enroll.

Table 3. TF-IDF Steps, time and results

	Routers	Blog	Blog Half
Avrg Frequencies Time in ms	390	13,230	6,124
Filter threshold	0.00005	0.00005	0.00005
Total Terms Count	26,099	456,484	295,988
Avrg Counting Time in ms	296	17,679	10,418
Filtered Terms Count	2,238	2,836	2,856
Avrg Filtering Time in ms	235	2,340	1,850
TFIDF Size (row * col)	19043 x 2238	678170 x 2836	380778 x 2856
TFIDF Size in Ram	170.47 MegaByte	7693.16 MegaByte	4350.01 MegaByte
Avrg TFIDF Time in ms	2,919	1,637,600	72,780

Skip-Gram Embedding

The training parameters used for both datasets are in Table 4, This matrix can be used later to process all the documents enrolled in the system.

Steps to Create the Skip-gram embedding:

1. The dataset must be preprocessed and tokenized.
2. Train the neural network using the preprocessed and tokenized data,
3. Output the embedding layer as a CSV file to be used later
4. Plot sample words as a 2d graph to show the relations between words
5. Use the embedding to create the doc-vector matrix (document feature matrix)

Table 4. Training parameters for both datasets

	Vocabulary Size	Embedding Vector Size	Total Documents	Samples Per epoch	No of Epochs	Avrg Epoch time	Total training time
Reuters Dataset	26100	100	19042	1000	40	15 sec	600 sec
Blog Dataset	456485	100	669932	10000	40	1060 sec	42400 sec

While also worth noting that although the training time for the skip-gram model is longer than running a single TF-IDF calculation, it is only done once, making it more efficient in the long run.

To create the document feature matrix, we apply the steps to both databases:

1. Load embedding matrix for the corresponding dataset.
2. Finding term frequencies for each document
3. Calculate the doc-Vector
4. Create a Matrix by stacking doc-Vectors

The results can be seen in Table 5.

Table 5. Skip-gram results and time.

	Routers	Blog	Blog-Half
Avrg Terms Frequencies Time in ms	390	12,460	12,460
Skipgram Matrix Size (row * col)	19043 x 100	678170 x 100	380778 x 100
Skipgram Matrix Size in Ram	7.62 MegaByte	270.27 MegaByte	152.31 MegaByte
Avrg Creation Time in ms	428	21,920	10,600

4.3   Clustering Layer

We use the doc-vector matrices extracted in previous steps and apply improved Fuzzy CMeans clustering. The results of the clustering can be seen in Table 6. The result is five clusters, each with a list of the documents and a document membership matrix.
   A set of keywords or tags can represent each cluster.

Table 6. Clustering parameters, results, and time

	Routers		Blog		Blog-Half	
	TFIDF	Skip-Gram	TFIDF	Skip-Gram	TFIDF	Skip-Gram
No Of Clusters	5					
Max Allowed Iterations	100					
Termination Threshold	0.00001					
Predicted Fuzziness	1.005840	1.025110	---	1.013915	1.001555	1.015250
Actual Termination Value	0.000010	0.000010	---	0.000010	0.000010	0.000010
Actual Used Iterations	79	28	---	92	53	34
Clustering Time in ms	337,680	7,700	---	1,265,001	15,041,232	258,928
Avrg Building Structure Time	185	140	---	3,450	2,530	3,270

4.4   Evaluation

The system is evaluated using four criteria: Speed, resource usage, precision, and recall.
   The speed the system takes to create the embedding matrices and cluster the data is measured in milliseconds. The memory used to store the embedding matrices is recorded, the results from in the three databases used (Routers, Blogs, and Blog-Half). Are in Table 7. The numbers have been taken from Tables 3, 5 and 6.

Table 7. Comparing the speed and resource usage.

	Ruters		Blog		Blog-Half	
	TFIDF	Skip-Gram	TFIDF	Skip-Gram	TFIDF	Skip-Gram
Embadding Time	3,840	818	1,670,849	34,380	91,172	23,060
Embadding Size	170.47 MB	7.62 MB	7693.16 MB	270.27 MB	4350.01 MB	152.31 MB
Clustering Time	337,680	7,700	---	1,265,001	15,041,232	258,928
Doc-Vector Length	2,238	100	2,836	100	2,856	100

5 Conclusion

1. When using large datasets, the traditional method of building an embedding matrix using the TFIDF fails due to the high computational complexity it requires and, in turn, the long time needed to perform the calculations.
2. Using deep learning to build an encoder to translate a text word to a numeric word-vector is much more efficient and requires fewer resources than the traditional method of TFIDF.
3. The old TFIDF method uses up to 28x times more storage and requires up to 50x times more time to represent the same data in comparison to the new deep learning-based method.
4. The old TFIDF method requires more time to do the calculations due to the long vector used to represent the document, where we need an item to represent each term in the corpus which can be a vast number (the blog's dataset has 456484 different terms), where the Skip-gram requires only a fixed score size (length of 100 in our case) to represent every document regardless of the number of terms the corpus contains.
5. As can be seen in the terms used to represent the clusters and the retrieval examples used, we notice that the result returned using the skip-gram is more semantically correct, due to the nature of the embedding algorithm, it finds the relation between words and giving similar words similar values making them cluster together. Thus the documents that talk about similar content tend to have similar values, where in contrast, the TFIDF method only relies on the frequency of the words to determine the relation between documents, and the frequency of a word does not always represent the topic in discussion.
6. Using the skip-gram method in the retrieval system allows the system to have semantic retrieval results, which are more humanly understandable.

References

1. Li, Z., Lu, W., Sun, Z., Xing, W.: A parallel feature selection method study for text classification. Neural Comput. Appl. **28**, 513–524 (2017). https://doi.org/10.1007/s00521-016-2351-3
2. Huang, H., Wong, R.: Reducing feature embedding data for discovering relations in big text data. In: Proceedings of the 2019 IEEE International Congress on Big Data, BigData Congress 2019 - Part 2019 IEEE World Congress on Services, pp. 179–183 (2019). https://doi.org/10.1109/BigDataCongress.2019.00038
3. Karpovich, S.N., Smirnov, A.V., Teslya, N.N.: Classification of text documents based on a probabilistic topic model. Sci. Tech. Inf. Process. **46**, 314–320 (2019). https://doi.org/10.3103/S0147688219050034
4. Cebeci, Z., Yildiz, F.: Comparison of K-Means and fuzzy CMeans algorithms on different cluster structures. J. Agric. Informatics **6** (2015). https://doi.org/10.17700/jai.2015.6.3.196
5. Beneker, D., Gips, C.: Using clustering for categorization of support tickets. CEUR Workshop Proc. **1917**, 51–62 (2017)

6. Bhoir, S., Ghorpade, T., Mane, V.: Comparative analysis of different word embedding models. In: International Conference on Advances in Computing, Communication and Control 2017, ICAC3 2017, pp. 1–4. Institute of Electrical and Electronics Engineers Inc. (2018). https://doi.org/10.1109/ICAC3.2017.8318770

7. Rong, X.: word2vec Parameter Learning Explained (2014)

8. Mikolov, T., Chen, K., Corrado, G., Dean, J.: Efficient estimation of word representations in vector space. In: 1st International Conference on Learning Representations, ICLR 2013 - Workshop Track Proceedings. International Conference on Learning Representations, ICLR (2013)

9. McCallum, A., Nigam, K., Ungar, L.H.: Efficient clustering of high-dimensional data sets with application to reference matching. In: Proceeding Sixth ACM SIGKDD International Conference on Knowledge Discovery and Data Mining, pp. 169–178 (2000). https://doi.org/10.1145/347090.347123

10. Schwämmle, V., Jensen, O.N.: A simple and fast method to determine the parameters for fuzzy CMeans cluster analysis. Bioinformatics **26**, 2841–2848 (2010). https://doi.org/10.1093/bioinformatics/btq534

11. UCI Machine Learning Repository: Reuters-21578 Text Categorization Collection Data Set. https://archive.ics.uci.edu/ml/datasets/reuters-21578+text+categorization+collection. Accessed 12 Aug 2020

12. Debole, F., Sebastiani, F.: An analysis of the relative hardness of reuters-21578 subsets. J. Am. Soc. Inf. Sci. Technol. **56**, 584–596 (2005). https://doi.org/10.1002/asi.20147

13. Schler, J., Koppel, M., Argamon, S., Pennebaker, J.: Effects of age and gender on blogging. In: AAAI Spring Symposium - Technical Report SS-06-03, pp. 191–197 (2006)

14. Welcome to Python.org. https://www.python.org/. Accessed 12 Aug 2020

15. Keras: the Python deep learning API. https://keras.io/. Accessed 12 Aug 2020

16. TensorFlow. https://www.tensorflow.org/. Accessed 12 Aug 2020

Mobile Robot Path Planning Optimization Based on Integration of Firefly Algorithm and Quadratic Polynomial Equation

Noor Alhuda F. Abbas$^{(\boxtimes)}$

Department of Computer Technologies Engineering,
AL-Esraa University College, Baghdad, Iraq
nooralhuda@esraa.edu.iq

Abstract. The rapid expansion in the industrial revolution has been aroused to utilize of the robot in various high accuracy and productivity applications. Optimal path planning is an important concern in navigation of mobile robot interesting with finding an optimal path. The firefly algorithm is considered as an increasingly promising tool of Swarm Intelligence, which is used in various optimization areas. The objective of the proposed method is to find the free-collision points in mobile robot environment and generate the optimal path based on firefly algorithm and D* algorithm. D* algorithm has been applied to find the shortest path. The essential function of applying the firefly algorithm and Quadratic parametric equations is to iterative globally search and determine the accurate positions of a set of intermediate via points within the free space and generating the corresponding smooth trajectory at a specified time respectively. The collision with obstacles can be efficiently avoided by constructing the free Cartesian space, which guarantee a free collision path planning. The simulation results illustrate the efficiency of proposed methods performance in finding and determining the optimal path and trajectory even in various degrees of environmental complexity.

Keywords: Mobile robot · Optimal path planning · Firefly algorithm

1 Introduction

Much attention has been paid for the utilization of a mobile robot in a variety of automated industrial environments. A mobile robot is increasingly used in a wide range of applications that includes many services like planet exploration, surveillance, landmine detection, etc. In all these applications, the locomotion mechanism is a significant requirement of the mobile robot that makes it navigate unbounded throughout its environment. The most crucial issue to attain autonomous mobile robots is a collision-free path planning. Path planning of the mobile robot concerned with generating a path from a starting to a goal points, with the provision of optimized a performance criterion including avoiding obstacles collision, reducing time interval, decreasing the path traveling cost. The minimum distance is the very commonly adopted criteria has to be attained [1–3]. The essential path-planning problem is to create a path for a moving object that satisfies obstacles-collision avoidance. So, it is

© The Author(s), under exclusive license to Springer Nature Singapore Pte Ltd. 2021
D.-T. Tran et al. (Eds.): ICISN 2021, LNNS 243, pp. 538–547, 2021.
https://doi.org/10.1007/978-981-16-2094-2_64

considered as a nondeterministic polynomial time (NP) complete problem. That means, the required computational time for solving the path planning problem increases dramatically depending on the increase of problem size. So that, it is difficult to apply the traditional optimization methods in finding the shortest robot path in real time [4–6]. Optimization can be defined as the process of searching and determining the best solution or values to given problems according to certain constraints by adjusting the restricted variables to desired characteristics that optimize the objective. The searching process can be achieved using multiple agents that essentially forms an evolving agents system. Moreover, the evolving system can be upgraded by iterations depending on a set of mathematical equations or rules [6–8].

In current work is aimed to generate the optimal trajectory for the mobile robot by providing the optimal method based on integration method of firefly algorithm and Quadratic polynomial equations based on D* algorithm. The objective of the proposed hybrid method is to specify the optimal control effect points for the corresponding shortest trajectory of robot in order to drive it from a stated initial point to a specified desired point while satisfying certain constraints of motion objectives.

2 Point to Point Trajectories Planning Using Quadratic Parametric Equations

Quadratic parametric equations are the type of parametric equations that generate the sets of the points for defining the trajectory at specific interval time u. Moreover, u indicates the independent parametric variable between [0 and 1], where each value of u determines a specific point on the trajectory. Separation of variables, direct computation of points coordinate and easy to express and implementation are the main characteristics of Quadratic parametric equations, which distinguishes them from other equations. The Quadratic parametric equations generate the trajectory by defining the three points representing the initial, singularity intermediate via-point and goal points [9]. However, the points of the trajectory can be generated according to the following equation:

$$r(u) = 2(0.5 - u)(1 - u)a + 4u(1 - u)b + 2u(1 - 0.5)c \tag{1}$$

$$\text{At } u = 0 \Rightarrow r(u) = a \tag{2}$$

$$\text{At } u = 0.5 \Rightarrow r(u) = b \tag{3}$$

$$\text{At } u = 1 \Rightarrow r(u) = c \tag{4}$$

$$\text{By adding time parameter to the equation:} u = \frac{t - t_0}{t_f - t_0} \tag{5}$$

$$\text{If } t_0 = 0 \text{ Then } u = \frac{t}{t_f} \tag{6}$$

The coordinate values of the trajectory can be determined based on time parameter according to the following equations:

$$x(t) = 2 \times \left(0.5 - \frac{t}{t_f}\right)\left(1 - \frac{t}{t_f}\right)x_a + 4 \times \frac{t}{t_f}\left(1 - \frac{t}{t_f}\right)x_b + 2 \times \frac{t}{t_f}(1 - 0.5)x_c \qquad (7)$$

$$y(t) = 2 \times \left(0.5 - \frac{t}{t_f}\right)\left(1 - \frac{t}{t_f}\right)y_a + 4 \times \frac{t}{t_f}\left(1 - \frac{t}{t_f}\right)y_b + 2 \times \frac{t}{t_f}(1 - 0.5)y_c \qquad (8)$$

The result of the applied of quadratic parametric equations with specific time is the smooth trajectory from the initial point to the goal point, passing by a singular intermediate via-point [9].

3 The Proposed Method of Firefly Algorithm for Mobile Robot Path Planning

Finding the shortest path being considered as the optimization problem that was intended to determine a potentially shortest route from start to the goal points. The Integration of firefly algorithm and Quadratic polynomial equations are used to construct the free-collision smooth shortest path for mobile robot. The principle of operation of the proposed method is the planning for the shortest and smooth path within the free Cartesian space according to two iterative processes: construction process of free space followed by generation process of shortest smooth path. Importantly, the proposed method is completely operating within the free space for the provision of the guaranteed collision-free planning, as demonstrated in the block diagram Fig. 1.

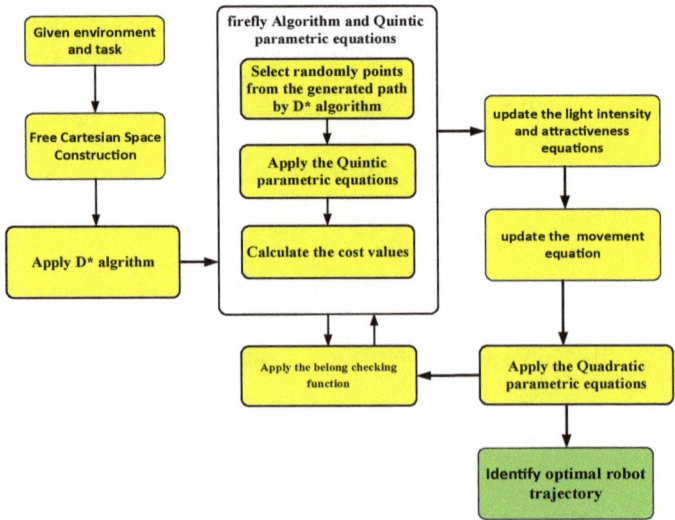

Fig. 1. Block diagram of proposed method.

Initially, the free space of the mobile robot environment is constructed by analyzing the robot environment in order to find collision points and free-collision points. The collision points are found by intersecting the robot and obstacles areas, while the rest points are considered as the free space points. In case of the obstacles are formulated as the circle, the collision checking function is:

$$d_{L1} = \sqrt{(x - x_{c1})^2 + (y - y_{c1})^2} \tag{9}$$

$$d_{L2} = \sqrt{(x - x_{c2})^2 + (y - y_{c2})^2} \tag{10}$$

$$d_{Ln} = \sqrt{(x - x_{cn})^2 + (y - y_{cn})^2} \tag{11}$$

$$v = \max\left(1 - \frac{d_L}{r}, 0\right) \forall d_L \in \{d_{L1}, d_{L2}, \ldots, d_{Ln}\} \tag{12}$$

Where d_{L1}, d_{L2}, d_{Ln} represent the distances between robot and center of the circular obstacle (x_c, y_c). In addition, v represent the verification value of collision checking function. As soon as the free space is constructed, the proposed method of D* and firefly algorithm is the planning for the shortest and smooth path within the free space according to two iterative processes: construction of shortest robot path process followed by smooth trajectory generation process. Importantly, the proposed method is completely operating within the free space for the provision of the guaranteed collision-free planning.

In the process of shortest robot path construction, the D* algorithm is applied to generate the shortest path from start to goal points by avoiding all the obstacles. The principle of operation of proposed D* algorithm is the locally planning for the shortest path within the free space until reach to goal point. Moreover, the D* algorithm is initialized by placing the start node as the "current node" on the open list, which is inserted into the currently planned path. Subsequently, the current node is expanded to eight connected neighborhood nodes for determining the next arm movement "candidate next node". However, only nodes that belong to the free Cartesian are used. Accordingly, the cost value is computed for each neighborhood nodes by implementing the heuristic cost function:

$$F = g(n_f) + h(n_f) \quad \forall n_f \in N^k \tag{13}$$

While in the process of smooth trajectory generation, Quadratic polynomial equations and firefly algorithm are applied for smooth trajectory generating based on generated path of D* algorithm. Initially, the set of free collision paths for the mobile robot are generated by randomly selecting a specific number of intermediate points to formulate the initial firefly population. However, the specific number of intermediate points are selected randomly from the free space. After that, the Quadratic polynomial equations are used to generate the corresponding path that connects each subsequence points together.

$$q_x(t) = 2 \times \left(0.5 - \frac{t}{t_f}\right)\left(1 - \frac{t}{t_f}\right)qx_a + 4 \times \frac{t}{t_f}\left(1 - \frac{t}{t_f}\right)qx_b + 2 \times \frac{t}{t_f}(1 - 0.5)qx_c$$

(14)

$$q_y(t) = 2 \times \left(0.5 - \frac{t}{t_f}\right)\left(1 - \frac{t}{t_f}\right)qy_a + 4 \times \frac{t}{t_f}\left(1 - \frac{t}{t_f}\right)qy_b + 2 \times \frac{t}{t_f}(1 - 0.5)qy_c$$

(15)

Subsequently, the collision checking equation is operated to test the collision of the generated path. In case there is no collision between the path points and obstacles, the generator path is accepted. Otherwise, the generated path is ignored and other free-collision points from the mobile robot environment are randomly selected to generate another path alternatively. Then, the Euclidean distance is applied to calculate the fitness function (distance cost) to each acceptable path according to:

$$\text{Trajectory}_{\text{Cost}} = \sum_{i=1}^{m} \sqrt{(x_i - x_{i+1})^2 + (y_i - y_{i+1})^2}$$

(16)

Subsequently, the specific number of firefly algorithm iterations are iteratively operated to update the light intensity and the attractiveness for each paths. The movement of the *i-th* firefly is attracted to another more attractive firefly *j*, and the movement equation is applied for each x and y coordinate of intermediate via points as following:

$$s_{xi} = s_{xi} + \beta_0 e^{-\gamma r^2} + \alpha \varepsilon_i$$

(17)

$$s_{yi} = s_{yi} + \beta_0 e^{-\gamma r^2} + \alpha \varepsilon_i$$

(18)

Depending on the modified movement of the *i-th* firefly, the Quadratic polynomial equations ate applied to generate the corresponding path that connects each subsequence intermediate via point points together. After that, the collision checking equation is operated to test the collision of the currently generated path. In case there is no collision between the path points and obstacles, the generator path is accepted and added to updated population. Otherwise, the generated path is ignored.

Basing on the updated population of firefly algorithm, firefly algorithm are iteratively update the light intensity, the attractiveness and movement for each paths according to specific number of iterations. Finally, the best path from whole firefly algorithm iterations with minimum cost function is considered free-collision shortest path for mobile robot.

4 Result and Discussion

Simulation and Computer modeling have been done to implement and test the proposed method in finding the optimal trajectory path. In this paper, the proposed method is tested on 2-D difficulties suggested mobile robot environments that have x−y dimensions limited from (−50 to 50) cm and contain various shapes of static obstacles. However, the specification of the suggested environment is that composed of a variety of static obstacles. Initially, the suggested environment was analyzed according to the process of mobile robot free space analysis in order to detect the free space and guarantee a collision-free path planning. After that, the firefly algorithm and quantic polynomial equations operate to find the free collision path for the mobile robot. Moreover, the start and goal points are selected from the free space of mobile robot. However, prior to planning the robot path, the start and goal points must be examined and ensured within the free Cartesian space. Otherwise, there will be no task for a path and trajectory planning.

The Fig. 2 represents the run of suggested mobile robot environment that contain of a variety of static obstacles. In Fig. 2, a, the black circles demonstrate the obstacles space of suggested mobile robot environment. In Fig. 2, b, the green points demonstrate the free space of suggested mobile robot environment. Figure 2, c represent the free-collision shortest path of mobile robot from the start point (2, −1) to the goal point (1, −8) with a fitness function cost equal to 27.313. While the Fig. 2, d represent the optimal smooth path that generated from applied integration method of firefly algorithm and Quadratic polynomial equation with a fitness function cost equal to 25.473.

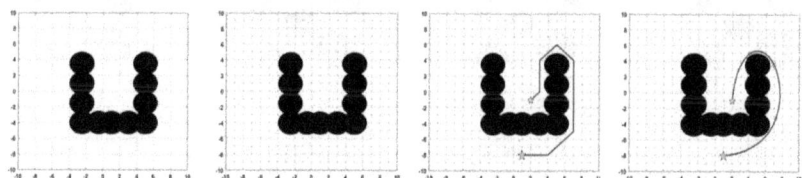

Fig. 2. Run of Fourth Suggested Mobile Robot Environment.

The proposed method also applied to test another suggested mobile robot environment as represented in the Fig. 3. The mobile robot has to move from the start point (3, 7) to the goal point (−3, 7) with a fitness function cost equal to 42.142. While the optimal smooth path that generated from applied integration method of firefly algorithm and Quadratic polynomial equation with a fitness function cost equal to 38.887.

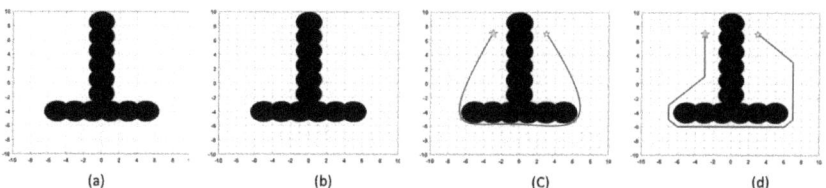

(a)   (b)   (C)   (d)

Fig. 3. Run of five Suggested Mobile Robot Environment.

The Fig. 4 represents the run of suggested mobile robot environment that contain of a variety of static obstacles. The mobile robot has to move from the start point (8, 8) to the goal point (−8, 4) with a fitness function cost equal to 36.485. While the optimal smooth path that generated from applied integration method of firefly algorithm and Quadratic polynomial equation with a fitness function cost equal to 34.05.

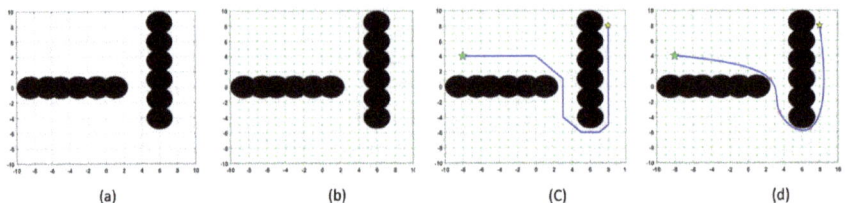

(a)   (b)   (C)   (d)

Fig. 4. Run of Six Suggested Mobile Robot Environment.

The proposed method also applied to test another suggested mobile robot environment as represented in the Fig. 5. The mobile robot has to move from the start point (5, −8) to the goal point (−6, 8) with a fitness function cost equal to 21.142. While optimal smooth path that generated from applied integration method of firefly algorithm and Quadratic polynomial equation with a fitness function cost equal to 20.071.

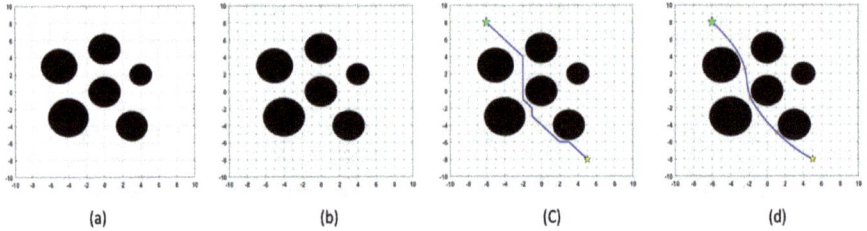

(a)   (b)   (C)   (d)

Fig. 5. Run of Seven Suggested Mobile Robot Environment.

The proposed method also applied to test another suggested mobile robot environment as represented in the Fig. 6. The mobile robot has to move from the start point (−8, −6) to the goal point (8, 2) with a fitness function cost equal to 23.55. While the optimal smooth path that generated from applied integration method of firefly algorithm and Quadratic polynomial equation with a fitness function cost equal to 20.773.

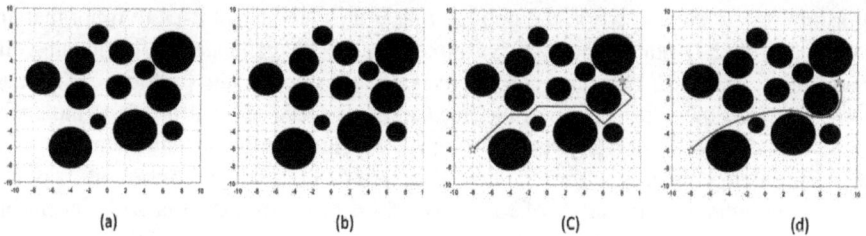

| (a) | (b) | (C) | (d) |

Fig. 6. Run of Eight Suggested Mobile Robot Environment.

Table 1 explain the results of all cases that are tested on the suggested environments.

Table 1. Tested Results of Run of suggested mobile robot environments

No. environment	Test point		D* cost	Firefly cost
	Start	Goal		
1	2, −1	1, −8	27.313	25.473
2	3, 7	−3, 7	42.142	38.887
3	8, 8	−8, 4	36.485	34.05
4	5, −8	−6, 8	21.142	20.071
5	−8, −6	8, 2	23.55	20.773

Modelling of the run of the proposed method with the suggested environment shows that the obstacles are successfully avoided, the free collision optimal path is efficiently planned even in case a variety of mobile robot environment complexity, and mobile robot smoothly follow the optimal path.

5   Conclusion

Optimal path planning is an important concern in navigation of autonomous mobile robots, which is to find an optimal path according to some criteria such as distance, time or energy. However, distance or time being the most commonly adopted criterion. Nature-inspired metaheuristic algorithms are considered powerful in solving modern global optimization problems, especially when dealing with NP-hard optimization such as the travelling salesman problem and path planning. In this paper, a swarm intelligence approach (firefly algorithm), which is inspired by the biological behavior is applied to the robot path optimization problem. The two essential functions of the proposed improvement are used to find the free space of mobile robot environment and to generate the free-collision shortest path from its start to goal points based on firefly algorithm. In the current work, the optimal trajectory planning of 2-DOF robot that are moving in the 2-D static known environment has been generated by proposing the optimization method. This analysis subjects to the three main problems. The first problem aims at the construction of the free Cartesian space of the robot that guarantees a free collision trajectory planning through analyzing the given environments. The second problem concerns within the determination of the optimal control effect points for the corresponding trajectory of robot. The third problem concerns with generating the shortest trajectory that satisfied the motion criteria depending on attained optimal control effect points. Hybridization of the firefly algorithm with Quadratic parametric equations is proposed for solving the problem of offline global trajectory planning of mobile robot in the known static environment. Two essential sequential functions of the hybrid method are the determination of control effect intermediate via points and generation of the corresponding smooth trajectory. However, the firefly algorithm is used to generate and rapidly converge a specific number of intermediate via points. While the Quadratic parametric equations are applied for producing the corresponding smooth trajectory at the specific time. In the proposed method of firefly algorithm and Quintic parametric equations, the hybridization of the fast convergence of the firefly algorithm and the simplicity of implementation of the firefly algorithm with Quintic parametric equations demonstrates the efficiency of the proposed method in finding the shortest smooth trajectory of the mobile robot. The detailed results of implementing the proposed method demonstrate the efficiency of performance in finding the free collision optimal trajectory solution with different complexity of robot environments.

References

1. Giuseppe, C., Fernando, G.: Motion and Operation Planning of Robotic Systems, vol. 29. Springer (2015)
2. Babak, R., Javad, M., Hasan, K., Gholam, D., Ali, O.: Robot manipulator path planning based on intelligent multiresolution potential field. Int. J. U- And E- Serv. Sci. Technol. 8(1), 11–26 (2015)
3. Xu, X., Li, Y., Yang, Y., et al.: A method of trajectory planning for Ground Mobile Robot based on ant colony algorithm. In: IEEE International Conference on Robotics and Biomimetics, pp. 2117–2121. IEEE (2017). https://doi.org/10.1109/ROBIO.2016
4. Raheem, F.A., Sadiq, A.T., Abbas, N.A.F.: Robot arm free Cartesian space analysis for heuristic path planning enhancement. Int. J. Mech. Mechatron. Eng. 19, 29–42 (2019)
5. Sadiq, A.T., Hasan, A.H.: Robot path planning based on PSO and D∗ algorithms in dynamic environment. In: 2017 International Conference on Current Research in Computer Science and Information Technology (ICCIT). IEEE, Slemani, Iraq (2017)
6. Sadiq, A.T., Raheem, F.A., Abbas, N.A.F.: Optimal trajectory planning of 2-DOF robot arm using the integration of PSO based on D∗ algorithm and quadratic polynomial equation. In: 1st Conference of Engineering Researches, Baghdad, Iraq, March 2017

7. Sadiq, A.T., Raheem, F.A., Abbas, N.A.F.: Robot Arm path planning using modified particle swarm optimization based on D* algorithm. Al-Khwarizmi Eng. J. **13**, 27–37 (2017)
8. Sadiq, A.T., Raheem, F.A., Abbas, N.A.F : Robot arm trajectory planning optimization based on integration of particle swarm optimization and A* algorithm. J. Comput. Theoret. Nanosci. **13**(3), 1046–1055 (2019)
9. Raheem, F.A.: Obstacle avoidance and Cartesian trajectory planning of robot manipulator using modified Bezier curve. In: Modeling, Theory, Methods and Facilities X (10th) International Scientific-Practical Conference, Novocherkassk, Russian Federation (2010)

Megakaryocyte Images Segmentation Using Wavelet Transform and Modified Fuzzy C-means Techniques

Shaima Ibraheem Jabbar[1]([⊠]) and Liqaa M. Al Sharifi[2]

[1] Babylon Technical Institute, Al Furat Al Awsat Technical University,
Babylon, Iraq
shaima.jabbar@atu.edu.iq
[2] Collage of Medicine, Pathology Department, Babylon University,
Babylon, Iraq

Abstract. The image segmentation is crucial and effective step in the description of megakaryocyte cell because it allows to identify the main details (cytoplasm and nucleus) and processed it separately from unnecessary details (background). This paper is proposed a state of art technique for the automatic segmentation for megakaryocyte structure image. The proposed method consists of several steps and it combines different tools: wavelet transform technique and modified fuzzy c-means technique. The first step is data collection, data will be collected as photograph of microscope sample which illustrates megakaryocytes of subject's sample. In this work, 25 samples (aspirates stained by gemsa stain) are collected from 10 recruited patients. Wavelet transform technique was used to differentiate between image content and noise as pre-processing step, also to prepare image to the next step. Implementation proposed modified fuzzy c-means technique is the third step. The modification of fuzzy c-means technique deals uncertainty of the image details based on measure of fuzziness and Euclidean distance via fuzzy inference technique. Based on quantitative evaluations tools (sensitivity, specificity and accuracy) the proposed method exhibits remarkably high level of accuracy compared with Otsu's segmentation method in the case of detection nucleus and cytoplasm (8% and 24% respectively).

Keywords: Megakaryocyte images · Wavelet transform · Fuzzy inference technique · Fuzzy c-means technique · Measure of fuzziness

1 Introduction

The disintegration of megakaryocyte cell leads to the production of platelets. Platelets can be defined as natural source for the continuity of life. The accumulation of platelet fibers form blood clots, which are necessary for coagulation therefore stopping the bleeding of an injury [1–4]. Defining the two main structures which make up megakaryocyte cells (the nucleus and cytoplasm) is crucial for when studying the behavior of megakaryocytes which contributes greatly to the processing of platelet production. It is hard to differentiate between the colors of the nucleus and the cytoplasm due to the similarity between them in the colors and textures of megakaryocyte components (the nucleus and the cytoplasm) when observing the megakaryocyte image of bone marrow. After staining and magnifying, it may be possible to label and illustrate the megakaryocyte cell components (the nucleus and the cytoplasm) under a microscope, but it is not possible to delineate or extract these components manually or accurately. Several techniques were presented and applied for detection the morphological properties of the megakaryocyte structure. The first research performed a manual differentiation between different diseases of megakaryocyte of the bone marrow [4]. Recently, researchers have proposed and developed different automated techniques. The objective of this was the delineation of megakaryocyte image structures (nucleus and cytoplasm). Some of these researches involved an unsupervised machine learning tool by combining morphological filters and wavelet techniques [5, 6] to discriminate geometrical features of megakaryocytes diseases. In addition, supervised techniques were performed for the same previous purpose using a region-based active contour method [5, 6]. Moreover, the open source software Cell Profiler was employed to categorize the different morphological features changes of a megakaryocyte cell [7]. In spite of this technique which was an automated approach, it is sometimes difficult to customize the irregular structure of mega images. Therefore, in this work we propose and apply a state of art that is an automated and unbiased nature method to separate megakaryocyte components using amended fuzzy c-means techniques with wavelet transform techniques. Frequently, the objective function of fuzzy c means clustering technique which depends on evaluation of the Euclidean distance between centroid and the points of data and updating the value of the center of a cluster based on this evaluation. However, in this work, we present a novel unbiased nature method for the segmentation of the significant structure of megakaryocyte images. Firstly, the megakaryocyte image is decomposed into a set of components using the wavelet transform technique then we select the most suitable component which shows image content (most low frequency) and rejects high frequency of grey level intensities and noise. This step is carried out as a pre-processing step before isolating the megakaryocyte structure. In the next stage, the megakaryocyte will be fuzzified using the membership function and the measure of fuzziness of the candidate pixel and its neighbourhood will be calculated. The fuzzy structure element is modelled based on the measure of fuzziness used to detect and define megakaryocyte structure. Images are mapped to a fuzzy set using the membership function which depends upon the measure of fuzziness. This approach analyses in depth the ambiguous behavior of the image such as the interaction of grey level intensity which belongs to the same cluster, for example: the presence of cytoplasm pixels in the nucleus region is due to the collaboration of pixels between the two regions.

The rest paper as follows: related techniques are presented in the same section of introduction. Section II illustrates the framework of the proposed method and various techniques which represent fundamental blocks of the proposed method that are presented in the same section. Experiments and results are presented in section III, while section IV and V demonstrate discussions and conclusions respectively. References are in the final section of the paper.

1.1   Related Works

In our work, various techniques are described in this section as follows: technique of wavelet transform, evaluation of fuzziness measure metric and Fuzzy c-means technique (FCT).

1.1.1   Wavelet Transform Technique

The technique of wavelet transform is one of potential technique in many applications of image processing because it is possible to involve it for extraction a unique feature of the image. It is efficient to identify a high level of quality of image content and eradicate unnecessary information [8–10].

1.1.2   Measure of Fuzziness

Measure of fuzziness is a tool which has been involved to evaluate uncertainty degree of fuzzy data [11]. It has many applications in image processing such as image contrast enhancement [12], bioinformatic. [13] and image segmentation [14]. Measure of fuzziness provides a structure about distributing of grey level intensity, low level of measure of fuzziness indicates to the high level of similarity while high level of it illustrate different grey level intensity distribution. Equation (1) illustrates evaluation of measure of fuzziness.

$$MF\left(\mu_B\left(I_{i,j}\right)\right) = \frac{2}{i*j}\sum_i\sum_j min\left(\mu_B\left(I_{ij}\right), 1 - \mu_B\left(I_{ij}\right)\right) \tag{1}$$

where $I_{i,j}$ is input image, i, j are dimensions of this image and $\mu_B\left(I_{ij}\right)$ is image after fuzzification.

1.1.3   Fuzzy C-means Technique (FCT)

Fuzzy c-mean is an efficient tool has been used for labelling data into a set of clusters. The number of clusters depends on the objective of segmentation, for example if the aim is separation image into two objects so we need three cluster two of them depict two object and the third cluster is background. Fuzzy c means technique is firstly proposed by Dunn in 1973 [15] and it was improved by Bezdek 1981 [16]. Then this method become the cornerstone of many applications such as image segmentation [17, 18], bioinformatics [19]. And classification of big data [20]. The basic idea of this method depends on It is based on minimization of the distance between each point of data and centre using the following objective function:

$$H_n = \sum_{i=1}^{k}\sum_{j=1}^{cent} h_{ij}^n\left\|Tx_i - cent_j\right\|^2 \qquad 1 \leq n < \infty \tag{2}$$

In the objective function can measure the similarity between any measured data and the center, where m is a real number > 1, h_{ij} is the membership function of Tx_i in the cluster j, Tx_i is the ith of d-dimensional measured data (image pixels) and $cent_j$ is a set of centre of the cluster. Fuzzy partitioning is performed by an iterative optimization of the objective function with the update of membership h_{ij} and the cluster centres $cent_j$ using Eq. (3) and Eq. (4):

$$h_{ij} = \frac{1}{\sum_{r=1}^{cent} \left(\frac{Tx_i - cent_j}{Tx_i - cent_r}\right)^{\frac{2}{n-1}}} \tag{3}$$

The centre of cluster or centroid indicates to the mean of all points, weighted by their membership of belonging to the cluster; therefore, this method is called fuzzy c-means.

$$cent_j = \frac{\sum_{i=1}^{k} h_{ij}^n . Tx_i}{\sum_{i=1}^{k} h_{ij}^n} \tag{4}$$

A local minimum of H_n indicates to stop the process of clustering This occurs when, where r is iteration number and it is a condition value between 0 and 1.

2 Methodology

The depiction the main structure of megakaryocyte cell (nucleus and cytoplasm) can be obtained using proposed technique. Megakaryocyte images are collected based on particular data collection protocol.

2.1 Data Acquisition

25 photographic image samples were collected from 10 bone marrow samples of patients (5 males and 5 females), age range (6–49) years with normal heterogeneous morphology. Those patients signed from patients to give permission to collect microscope image and finally photographic images were obtained by specific camera for the eye piece of the microscope. The digital images acquired from this configuration are in form of 400 × 430 pixels, and it is color images (R, G, B).

2.2 Planned Technique

The planned technique comprised from several steps; these steps are embodied in the following flowchart which illustrates in the figure below.

2.3 Modified Fuzzy C-Means Technique (MFCT)

Evaluation of Euclidean distance is significant property to perform the usual fuzzy c-means technique. One the other hand, However, due to the complex structure of megakaryocyte images Euclidean distance is not enough to recognize all details which belong to this cluster. So, in this work, we suggest to modify the protocol of traditional fuzzy c-means method to improve its performance. Measure of fuzziness is an efficient

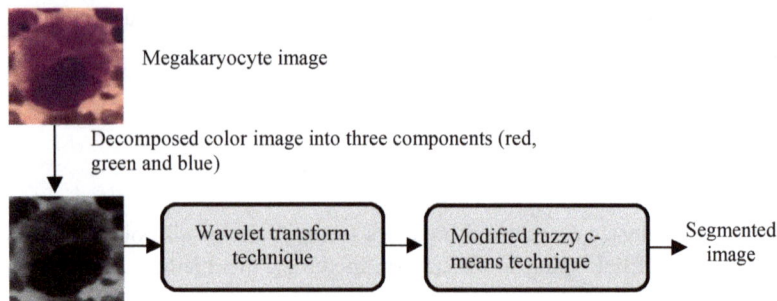

Megakaryocyte image

Decomposed color image into three components (red, green and blue)

Wavelet transform technique → Modified fuzzy c-means technique → Segmented image

Fig. 1. Illustrates the pipeline of proposed technique

metric which illustrates a high level of information from uncertainty data. Figure 2 demonstrates the steps of performing Modified Fuzzy c-means technique (MFCT).

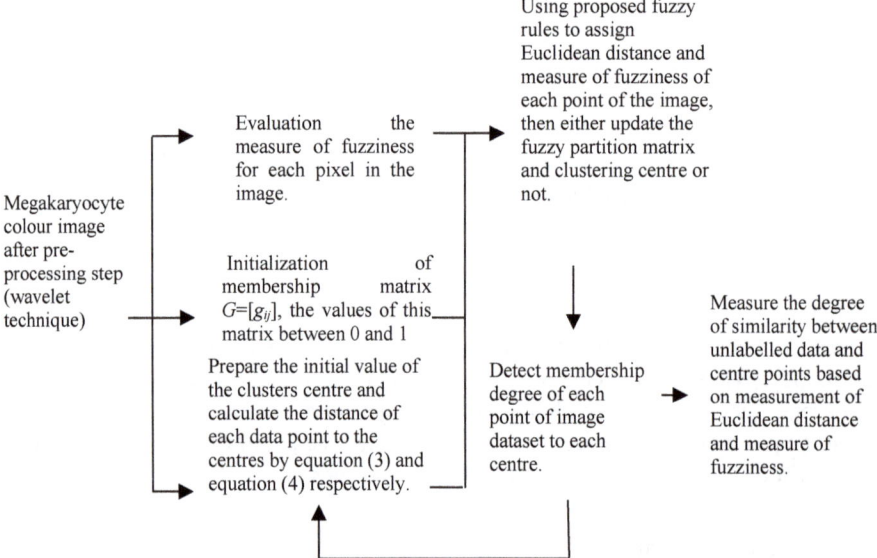

Using proposed fuzzy rules to assign Euclidean distance and measure of fuzziness of each point of the image, then either update the fuzzy partition matrix and clustering centre or not.

Evaluation   the measure of fuzziness for each pixel in the image.

Megakaryocyte colour image after pre-processing step (wavelet technique)

Initialization   of membership   matrix $G=[g_{ij}]$, the values of this matrix between 0 and 1

Prepare the initial value of the clusters centre and calculate the distance of each data point to the centres by equation (3) and equation (4) respectively.

Detect membership degree of each point of image dataset to each centre.

Measure the degree of similarity between unlabelled data and centre points based on measurement of Euclidean distance and measure of fuzziness.

Fig. 2. Flowchart of the steps of performing Modified Fuzzy c-means technique (MFCT).

So, we recruit four fuzzy rules as follows:

Rule 1: If Measure of fuzziness is high & Euclidean distance is low then update of centre of cluster.

Rule 2: If Measure of fuzziness is low & Euclidean distance is high then select of centre of cluster.

Rule 3: If Measure of fuzziness is low & Euclidean distance is low then update of centre of cluster.

Rule 4: If Measure of fuzziness is high & Euclidean distance is high then select of centre of cluster.

3 Results

Based on the following steps in the methodology section, results are illustrated in this section. The first step is decomposing megakaryocyte colour image into three components (red, green and blue). However, the recommended component to select for the rest of processing was red component. The main reason is the clarity of the red image. The second step is applying wavelet transform technique to obtain significant features of the image and prepare it to the clustering step. Separation the most relevant information to the cluster was achieved using proposed technique, it is modified fuzzy c-means clustering method. The results of this method were compared with other traditional method (Otsu's multi-thresholding method) [21–23]. Otsu's segmentation method is a powerful tool which has been used for image segmentation by classifying pixels into groups. Figure 3 present the results after applying two methods.

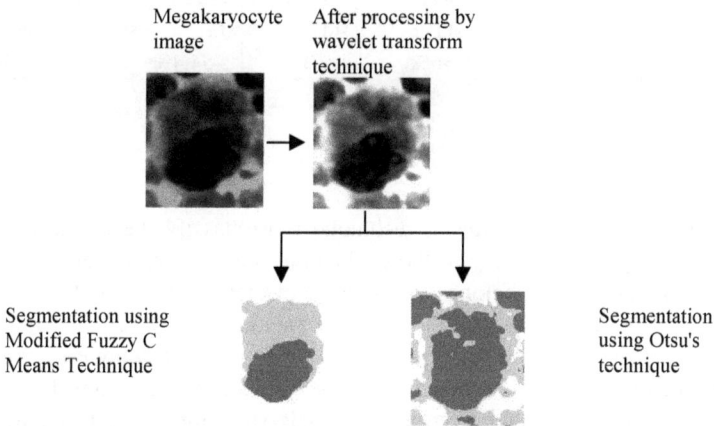

Fig. 3. Represents proposal technique (wavelet technique combined with fuzzy c-means technique)

Figure 3 illustrates input image and approximation component of the image after applying wavelet transform technique.

The last step was quantitative assessment of the output images after performing previous steps which mention above. We have 25 output images, so we need 25 reference images to prove the strength of the performance of proposed method compared with other method. These assessment metrics are sensitivity, specificity, recall, precision and accuracy [24–26]. These metrics.

Table (1) and Table (2) show results of quantitative assessment of segmented images for cytoplasm and nucleus using two methods (proposed method and Otsu) all values represent mean of evaluation of quantitative assessment metrics (sensitivity, specificity, recall, precision and accuracy).

Table 1. Sum up the comparing between Proposed technique and Otsu method of the Cytoplasm region cluster

Assessments	Sensitivity	Specificity	Recall	Precision	Accuracy
Mean & std	mean	mean	mean	mean	mean
Proposed technique against Otsu	>2.1%	>5.2%	>8%	>16%	>24%
	std	std	std	std	std
	>18%	<23%	>3%	>34%	>14%

Table 2. Sum up the comparing between Proposed technique and Otsu method of the Nucleus region cluster

Assessments	Sensitivity	Specificity	Recall	Precision	Accuracy
Mean & std	mean	mean	mean	mean	mean
Proposed technique against Otsu	>2%	>0.8%	>1.88%	>10.2%	>8%
	std	std	std	std	std
	>6%	<3%	>54.4%	>31%	>47%

4 Discussions

The main purpose of fuzzy c means technique is to classify the image into sets of objects or sub regions according to the evaluation of Euclidean distance. Sometimes regions are separated but it belongs to the same cluster nonetheless. For this reason, pixels that might be far away from the centre of the cluster may be labelled to another cluster. However, in this work we modified fuzzy c means technique to explain this limitation. Considering the determination of Euclidean distance alongside the evaluation measures of fuzziness can get more accurate classification of pixels in clusters. The Wavelet Transform technique was utilized as a pre-processing step to remove irrelevant information and keep important data as it is, in order to avoid uncertainty of image information in megakaryocyte cells.

Our results be governed by the determination of quantitative assessment tools (sensitivity, specificity, recall, precision and accuracy). These tools were used on segmented images that were the results of two methods (Modified Fuzzy c-means technique and Otsu's technique). Table 1 and Table 2 reported the percentage of the mean value of assessment metrics evaluation for 25 image samples. The proposed method (the fuzzy c means method) showed the much better accuracy compared to the traditional Otsu's method.

The accuracy of the cytoplasm clustering was around 24% more accurate in comparison to Otsu's method. Using the modified method, we attained approximately 8% more accuracy in comparison to the nucleus cluster in the image using Otsu method. On the other hand, the standard deviation values in most metrics using Otsu's method is greater than the standard deviation values of the modified fuzzy c-means method.

5   Conclusions and Future Work

Applying both the modified fuzzy c-means and wavelet transform techniques for the segmentation of megakaryocyte cells is a relatively new concept. In this paper, we have developed the fuzzy c-means technique and considered the measurement of fuzziness as metric in order to ensure the correct labelling of the image pixels to the right cluster. As a pre-processing step, wavelet transform analysis can present useful recognition and quantification results on the morphological patterns of the megakaryocyte. The proposal technique offers a significant advantage over the commonly used segmentation method (Otsu's method).

The modified fuzzy c-means technique is more advanced than the traditional multi-thresholding method (Otsu's method) is in agreement with the assessment values of statistical analysis. This work could be the starting point for future investigations. In future works, we may be able to examine the morphological features of megakaryocyte cell structures, such as evaluations like area, perimeter a thickness, of the nucleus the cytoplasm. In addition, it will be possible to calculate the number of megakaryocytes.

References

1. Psaila, B.: Single-cell analyses reveal megakaryocyte-biased hematopoiesis in myelofibrosis and identify mutant clone-specific targets. Mol. Cell **78**, 477–492 (2020)
2. Rieger, M.A., Smekal, B.M., Schroeder, T.: Improved prospective identification of megakaryocyte–erythrocyte progenitor cells. Br. J. Haematol. **144**(3), 448–451 (2009)
3. Noetzli, L.J., French, S.L., Machlus, K.R.: New insights into the differentiation of megakaryocytes from hematopoietic progenitors. Arterioscler. Thromb. Vasc. Biol. **39**(7), 1288–1300 (2019)
4. Cramer, E.M.: Megakaryocyte structure and function. Curr. Opin. Hematol. **6**(5), 354–361 (1999)
5. Gorelashvili, M.G., et al.: Megakaryocyte volume modulates bone marrow niche properties and cell migration dynamics. Haematologica **105**, 895–904 (2020)
6. Jung, H., Lodhi, B., Kang, J.: An automatic nuclei segmentation method based on deep convolutional neural networks for histopathology images. BMC Biomed. Eng. **24**(1), 1–12 (2019)
7. Ballaro, B., et al.: An automated image analysis methodology for classifying megakaryocytes in chronic myeloproliferative disorders. Med. Image Anal. **12**, 703–712 (2008)
8. Song, T.H., et al.: A circumscribing active contour model for delineation of nuclei and membranes of megakaryocytes in bone marrow trephine biopsy images. In: SPIE 9420, Medical Imaging 2015: Digital Pathology (2015)
9. Tzu-Hsi, S., Victor, S., Hesham, E., Nasir, M.: Rajpoot, dual-channel active contour model for megakaryocytic cell segmentation in bone marrow trephine histology images. IEEE Trans. Biomed. Eng. **64**(12), 2913–2923 (2017)
10. Salzmann, M., Hoesel, B., Haase, M., Mussbacher, M., Schrottmaier, W.C., Kral-Pointner, J. B., Finsterbusch, M., Mazharian, A., Assinger, A., Schmid, A.J.: A novel method for automated assessment of megakaryocyte differentiation and proplatelet formation. Platelets (2018). https://doi.org/10.1080/09537104.2018.1430359
11. Mallat, S.: A theory for multiresolution signal decomposition: the wavelet representation. IEEE Trans. Pattern Anal. Mach. Intell. **11**, 674–693 (1989)

12. Xizhi, Z.: The application of wavelet transform in digital image processing. In: 2008 International Conference on MultiMedia and Information Technology, pp. 326–329 (2008)
13. Misiti, M., Misiti, Y., Oppenheim, G., Poggi, J.J.: Wavelet toolbox: for use with MATLAB. The Math Works, Inc., Natica (1997)
14. Ebanks, B.R.: On measures of fuzziness and their representations. J. Math. Anal. Appl. **94** (1), 24–37 (1983)
15. Lopes, N.V., Couto, P.M., Pinto, P.M.: Automatic histogram threshold using fuzzy measures. IEEE Trans. Image Process. **19**, 199–204 (2010)
16. Romualdi, C., Campanaro, S., Campagna, D., Celegato, B., Cannata, N., Toppo, S., Valle, G., Lanfranchi, G.: Pattern recognition in gene expression profiling using DNA array: a comparative study of different statistical methods applied to cancer classification. Hum. Mol. Genet. **12**(8), 823–836 (2003)
17. Tobias, O.J., Seara, R.C., Soares, F.A.P.: Automatic image segmentation using fuzzy sets. In: 38th Midwest Symposium on Circuits and Systems Proceedings (1995)
18. Dunn, J.C.: A fuzzy relative of the ISODATA process and its use in detecting compact well-separated clusters. J. Cybern. **3**, 32–57 (1973)
19. Bezdek, J.C.: Pattern Recognition with Fuzzy Objective Function Algorithms. Plenum Press, New York (1981)
20. Banerjee, T., Keller, J.M., Skubic, M., Stone, E.: Day or night activity recognition from video using fuzzy clustering techniques. IEEE Trans. Fuzzy Syst. **22**(3), 483–493 (2014). https://doi.org/10.1109/TFUZZ.2013.2260756
21. Otsu, N.: A threshold selection method from gray-level histograms. IEEE Trans. Syst. Man Cybern. **9**(1), 62–66 (1979). https://doi.org/10.1109/TSMC.1979.4310076
22. Jianzhuang, L., Wenqing, L., Yupeng, T.: Automatic thresholding of gray-level pictures using two-dimension Otsu method. In: 1991 International Conference on Circuits and Systems. Conference Proceedings, China, pp. 325–327 (1991)
23. Dongju, L.: Otsu method and K-means. In: Ninth International Conference on Hybrid Intelligent Systems, vol. 1, pp. 344–349. IEEE (2009)
24. Lopez-Molina, C., De Baets, B., Bustince, H.: Quantitative error measures for edge detection. Pattern Recogn. **46**(4), 1125–1139 (2013)
25. Alani, M.S., Alheeti, K.M.: Precision statistical analysis of images based on brightness distribution. Adv. Sci. Technol. Eng. Syst. **2**(4), 99–104 (2017). https://www.flinders.edu.au/science_engineering/fms/School-CSEM/publications/tech_reps-research_artfcts/TRRA_2007.pdf
26. Gaia, B.F., Pinheiro, L.R., Umetsubo, O.S., Costa, F.M. Cavalcanti, M.: Comparison of precision and accuracy of linear measurements performed by two different imaging software programs and obtained from 3D-CBCT images for Le Fort I osteotomy. Dentmaxillofac. Radiol. **42**, 1–8 (2013)

RESTful API Design for a Real-Time Weather Reporting System

Yasmin Makki Mohialden$^{(\boxtimes)}$, Nadia Mahmood Hussien,
and Hanan Abed AL Wally

Computer Science Department, College of Science,
Mustansiriyah University, Baghdad, Iraq
ymmiraq2009@uomustansiriyah.edu.iq

Abstract. A real-time weather information system is an urgent need in characterizing qualitative data. It can provide accurate weather minute data based on the Internet of Things (IoT). The present weather reporting system is developed using the REST API technique to detect and provide information on climatic changes associated with air quality using different web sources. The used method has the ability to extract the API member set and is supported by an efficient searching interface. The proposed method observes significant searchability of API member sets by using inclusion graphs and is automatically extracted from the source code. The proposed method incorporates four sites to compare and select the proper state of air pollution. Moreover, the system is able to provide an alert message to specify the weather conditions in real time from the Real-Time Air Quality index, IQAir, AirNow, and BreezoMeter. The system has successfully shown real-time information on the weather of a particular place with efficiency and usability responses.

Keywords: RESTful API · Weather Real-Time Reporting System · Internet of Things

1 Introduction

Weather is an air condition that includes conditions of humidity, temperature, pollution content, and barometric pressure [1]. Weather analysis plays an essential role in human activity. Therefore, it needs to involve all information possible to maintain a safe environment. It affects many fields such as the industrial, agriculture, and aviation department [2]. Weather reporting applications generally extract weather information from an accurate weather system. It provides information regarding temperature, humidity, and all time-to-time changes in climate to help in planning daily missions [3]. In many severe weather events, a Weather Real-Time Reporting System was involved. It provides real-time observation and collection of weather information and location [4]. Gathering weather data uses Internet of Things (IoT) devices to measure physical parameters and upload them in real time to cloud storage. The measured data can be observed anywhere using Internet-enabled devices [5]. This facilitates monitoring weather data even in difficult geographical terrains. Moreover, this procedure reduces the manpower needed to collect local data and eliminate risks from visiting sites [6].

Some weather factors can cause results of precipitation to high winds. While some weather conditions can cause "natural disasters". For that, investigation of weather

© The Author(s), under exclusive license to Springer Nature Singapore Pte Ltd. 2021
D.-T. Tran et al. (Eds.): ICISN 2021, LNNS 243, pp. 557–565, 2021.
https://doi.org/10.1007/978-981-16-2094-2_66

conditions influence the air quality. Thus, the air quality considered to be influenced by both emissions and weather conditions [7]. Emissions and air pollution represent a major problem of recent decades, which has a toxicological impact on the environment and human health. There are many factors can cause air pollution such as volcanic activities, motor of engines and industrial activities. It is linked with millions of death and non-healthy effects around the world each year [8, 9]. For that, one of the objective of this work is to provide a Weather Report that involves both of weather conditions and Air Quality based on the specific web pages due to importance of these information.

In order to extract weather information from open source repositories could, Application Programming Interface (API) is used as a search tool. API obtained by mining are mixtures of several irrelevant elements. For that, API needs to have the ability to extract information from source code efficiently and automatically [10, 11]. There are many studies concern with the development of support tools that utilize API applications [12–18]. Some studies help the user of API usages [17] and other studies recommend API methods for specific application or combinations of API methods [13, 17] to develop the software.

The present paper developed a method to let the user efficiently and automatically gathered the weather data with air pollution information collected by IoT sensors and stored in open source repositories. For IoT applications, RESTful Web service architectures developed to create, retrieve, delete, update and query documents and metadata. The process of RESTful API creation considered the following steps are the fundamental steps for creating a schema modeling. The first step is determining business value, next step is choosing metrics, followed by defining use cases finally designing API and creating a schema model. The RESTful API process is presented in [19–22].

NoSQL database is used in this research due to high reliability of storage specific data. NoSQL databases are non-relational, distributed, open source. Also, these databases are horizontally scalable (in linear way) [23–25]. based on the presented knowledge, the authors structured the paper to involve the Related work in Sect. 2, Sect. 3 observe the proposed system, Sect. 4 presents the results and discussion, and Sect. 5 the conclusion.

2   Related Works

RESTful API design (Representational State Transfer) is a process to create a series of small modules. REST usually takes advantage of HTTP in case of using APIs Web. Many researchers discuss and apply many methods for real-time current and forecast weather.

Jones et al., in 2018 proposed a novel approach for real-time forecast weather detection in EPW format. The researchers apply a free online tool chain. The users built a real-time weather files which can directly requested the files for building predictive control system. The developed forecasting observe the next 48-h information. The results observed a high accuracy weather forecast which can be used instead of the data gained from stations. The model shows high real-time performance analysis and benchmarking. The created datasets were rich in information which enable them to study the energy storage optimization. It was feasible due to unique real-time optimization and automatic building simulation based on the machine to machine data communication. Also, it reduces the effects of installing stations for weather processes using large sets of meteorological data. They approved that the error in produced weather data file selection reaches to 30% [26].

Jakaria et al., in 2018 proposed a weather prediction technique that employ historical weather data from multiple stations to train the machine learning models. The proposed method able to provide usable forecasts using weather conditions to predict the near future based on short period of time. Results evaluation observed a high accuracy models which provide a usable state-of-the-art techniques [27].

Ramanathan and Korte in 2014 proposed an architecture of RESTful web services using sensor data. It can be proposed as a service over the web, it takes the location of the user into account which provides nearest weather data from the available sensors. It applies a document oriented NoSQL database [28].

Wang and Haiyan in 2019 developed a system to perform a complex analysis using an original series of sample entropy and data pre-processing. The optimization method is removing noise due to selecting the optimal input structure. Forecasting the hourly AQI series was optimized by using multi-objective multi-verse optimization process and modified using least squares support vector machine and [29].

Based on the presented overview and related works, this paper presents a RESTful API design for finding a specific data based on the correlation of air quality and weather condition involving a signal alert as in the next section.

3   The Proposed System

The proposed system developed to detect and merge the current temperature for the required area or city, in both Arabic and English languages, by using the RESTful API technique. The system uses IOT technology due to the high level of information accuracy in addition to the wide range of stored data in the web repository. The operation of the system starting with sending a shortest message service (SMS) based on the selected weather value for instance 50 °C. also, it provides all the necessary weather and environment descriptions as shows in Fig. 1 and Fig. 2.

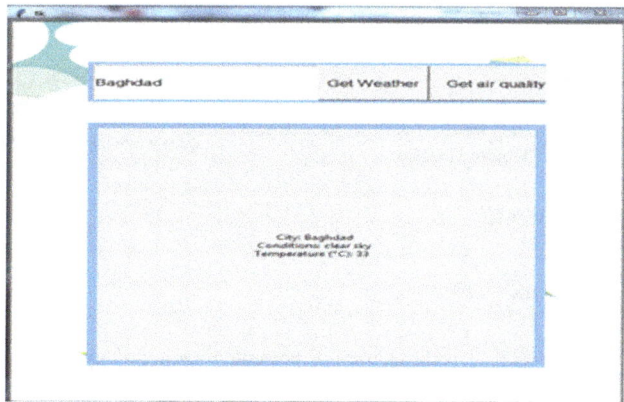

Fig. 1. The English interface of the system

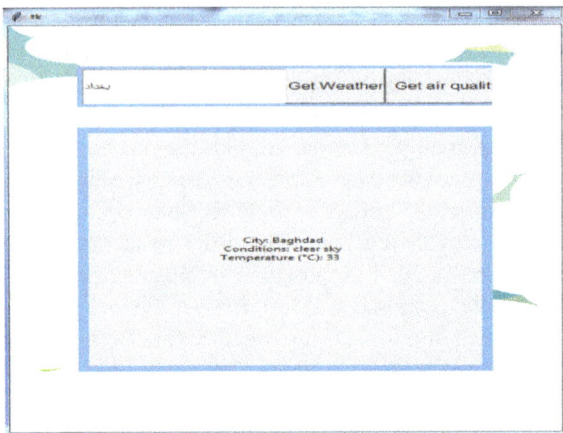

Fig. 2. The Arabic interface of the system

For building the system information, four sites used as resources to merge and compare the useful information based on the web base methods and algorithms. The automatic searching between the different four web repositories and the extracting web information provides a specific air pollution of the costumer location. A real time weather and air quality monitoring method that proposed in the present paper as shown in the block diagram in Fig. 3.

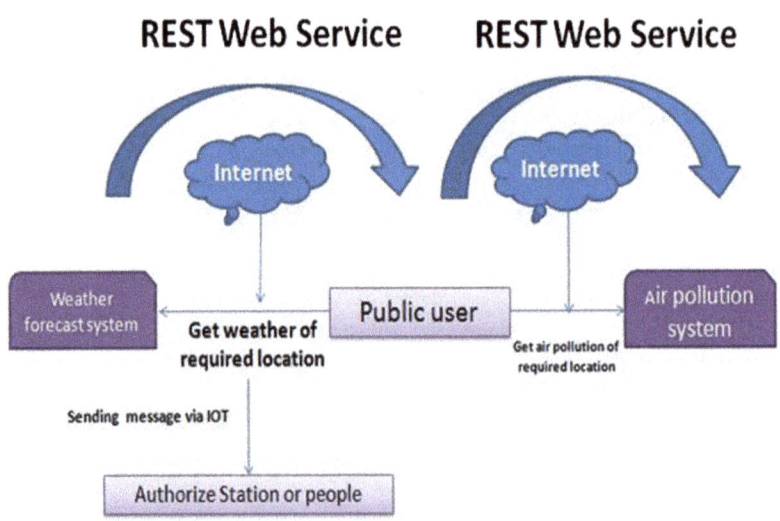

Fig. 3. Block diagram of presented methodology

4   Results and Discussion

The proposed method observe developed inclusive relations in API due to rich obtained information and useful application. It provides the user real-time weather data results correlated with air quality. The system developed a web page application that can get multi sites of air quality sites and can display the air quality or air pollution of any city or country. Figure 4 display the air quality web sites.

Fig. 4. The air quality web sites.

The returned information contains:

A. **air-quality index:**
The first real-time result of proposed system is the air-quality AQI. The repertory data involves more than 12000 stations related to 1000 major cities of one hundred countries. The information available in [30] collected form air quality stations used to monitor the air quality around the world. For high quality level, the published reading is concern with the particular matter (PM2.5/PM10). The AQI standard prepared and developed based on United State Environmental Protection Agency US EPA. One of the contributions in the present work is the estimated data of the non-indexed positions. The propose method able to detect the real-time neighbor stations data to evaluate the missing air quality values in certain position. For that the developed methodology is not only have the ability to merge information automatically, but its also recognize the un-available data.

B. **IQAir**
IQAir Is a Swiss-based air quality company started in 1963 aims to provide clean air for breathing through collaboration, information and technology solution. It is working to change the environment. For that, they operate the world's largest free real-time air quality information platform and engage an ever-growing number of global organizations, citizens, and governments [31].

C. **AirNow**
AirNow is a one-stop source for air quality data. It provides data of air quality at state, national and world views. The collected data is gathered from tribal

monitoring agencies or local measurements based on federal references or methods of equivalent monitoring which approved by EPA. It can provide national reporting consistency, quality control, and the ability to distribute data to the researchers, educators, public and many other data systems. Also, it involves the data of more than 500 cities. Fire conditions and locations, health air quality information. Some locations also provided the air quality of NO2 and CO with the value of PM and zone [31].

D. **BreezoMeter**

It is a source of air quality data used for enhancing human health via making better decision due to air pollution levels at their resident location. It analyzes data based on multiple sources by applying advance algorithms and techniques of machine learning. Continues validation developed the specific BreezoMeter information to provide high level of accuracy. The accuracy of calculation reach to 0.003 miles (5 m/16 ft.) in more than 90 countries. For more details see [32].

Based on the four independent sources above, the present methodology merge and specify the useful data to extract the helpful decision information of air quality cor-related with weather condition. The data sources used IoT government sensors and satellite observation provides all the database needed for costumer life. Based on these data the present work will select, compare and separate the functional numerical response supported with alarm signal.

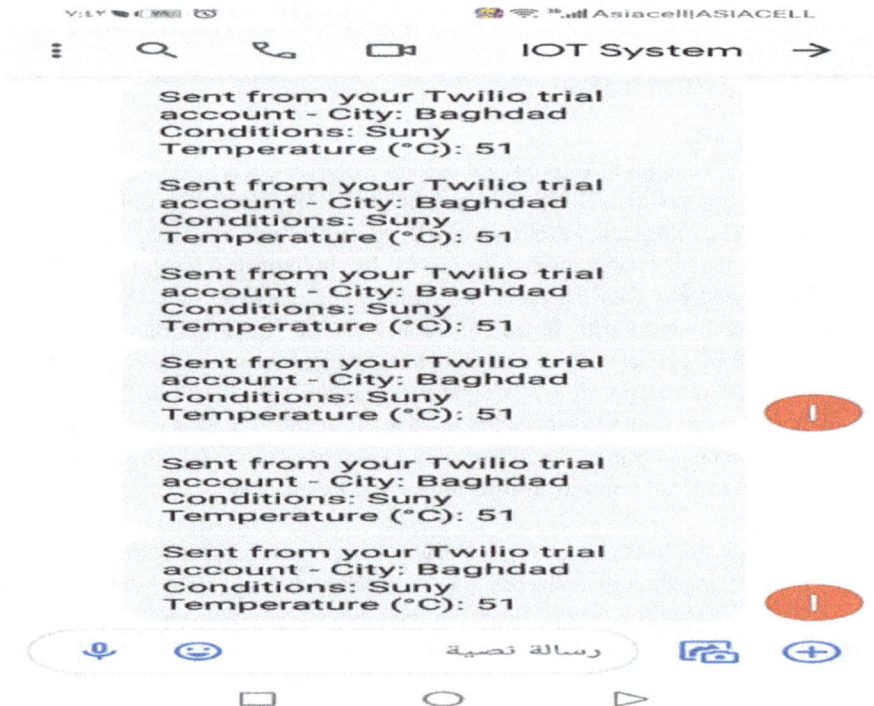

Fig. 5. Alert base SMS message sample

When the temperature greater or equal to 50 °C, the system sends an SMS message to authorize station or monitoring part that shows in Fig. 5.

It can be seen that the system provides an effective integrated information which detect and merge data automatically (superposing the data) for all qualitative feature of selected targets. Also, the data involves air quality sites with air pollutions of each city or specific position.

5   Conclusion

The proposed system presents a useful method to search for API technique that is required to implement a functional base knowledge. The automatic integration of the weather and air quality information based on the costumer keywords is the main contribution in this work. It can provide an accurate specific selected data in real-time as a weather report supported with SMS alert message. The system developed to get the weather description based RESTful API technique, which is used the get operation to fetch the data when the user enter the name of city. The results observe a superior responses by presenting a comparison among four different web sites. The IoT technology provide the present system with all necessary data based on the specific sensors in weather station which helps to meet the working goals. The system observe an easy way to use with fast response to user requires.

Acknowledgment. The Authors would like to thank Department of Computer Science, College of Science, Mustansiriyah University, Baghdad, Iraq, for supporting this study.

References

1. Sipani, J.P., Patel, R.H., Upadhyaya, T., Desai, A.: Wireless sensor network for monitoring & control of environmental factors using arduino. Int. J. Interact. Mob. Technol. **12**(2), 15–26 (2018). https://doi.org/10.3991/ijim.v12i2.7415
2. Sousa, P.J., Tavares, R., Abreu, P., Restivo, M.T.: NSensor–wireless sensor network for environmental monitoring. Int. J. Interact. Mob. Technol. **11**(5), 25–36 (2017). https://doi.org/10.3991/ijim.v11i5.7067
3. Antic, M., Jovanovic, S., Cvetanovic, S.: Development of eStudent iOS mobile application. Int. J. Interact. Mob. Technol. **7**(1), 35–40 (2013). https://doi.org/10.3991/ijim.v7i1.2295
4. Chaganti, S.Y., Nanda, I., Pandi, K.R.: Cloud-based weather monitoring system. Springer (2020). https://doi.org/10.1007/978-981-15-0978-0
5. Bian, Z., Sha, Y., Lu, W., Chenwei: Weather real-time reporting system based on intelligent network (2014)
6. Rao, B.S., Rao, K.S., Ome, N.: Internet of Things (IoT) based weather monitoring system. Int. J. Adv. Res. Comput. Commun. Eng. **5**(9), 312–319 (2016)
7. Nwe, C.M., Htun, Z.M.M.: A smart weather monitoring system using Internet of Things. Int. J. Sci. Eng. Res. (IJSER) (2018). ISSN 2347-3878
8. Jhun, I., Coull, B.A., Schwartz, J., Hubbell, B., Koutrakis, P.: The impact of weather changes on air quality and health in the United States in 1994–2012 (2015). https://doi.org/10.1088/1748-9326/10/8/084009

9. Ghorani-Azam, A., Riahi-Zanjani, B., Balali-Mood, M.: Effects of air pollution on human health and practical measures for prevention in Iran. J. Res. Med. Sci **21**, 65 (2016)

10. Wang, H., Tseng, C., Hsieh, T.: Developing an indoor air quality index system based on the health risk assessment. In: Proceedings of the Indoor Air Conference, Copenhagen, Denmark, 17–22 August, vol. 1 (2008)

11. Kondoh, Y., Nishimoto, M., Nishiyama, K., Kawabata, H., Hironaka, T.: Efficient searching for essential API member sets based on inclusion relation extraction. Int. J. Netw. Distrib. Comput. **7**(4), 149–157 (2019). https://doi.org/10.2991/ijndc.k.190911.002

12. Montandon, J.E., Borges, H., Felix, D., Valente, M.T.: Documenting APIs with examples: lessons learned with the APIMiner platform. In: 2013 20th Working Conference on Reverse Engineering (WCRE), Koblenz, Germany, pp. 401–408. IEEE (2013)

13. Lamba, Y., Khattar, M., Sureka, A.: Pravaaha: mining android applications for discovering API call usage patterns and trends. In: Proceedings of the 8th India Software Engineering Conference (ISEC), Banglore, India, pp. 10–19. ACM (2015). https://doi.org/10.1145/2723742.2723743

14. Zhong, H., Xie, T., Zhang, L., Pei, J., Mei, H.: MAPO: mining and recommending API usage patterns. In: Drossopoulou, S. (ed.) European Conference on Object-Oriented Programming. Lecture Notes in Computer Science, vol. 5653, pp. 318–343. Springer, Heidelberg (2009)

15. Wang, J., Dang, Y., Zhang, H., Chen, K., Xie, T., Zhang, D.: Mining succinct and high-coverage API usage patterns from source code. In: 2013 10th Working Conference on Mining Software Repositories (MSR), pp. 319–328. IEEE, San Francisco (2013). https://doi.org/10.1109/MSR.2013.6624045

16. Nguyen, T.T., Pham, H.V., Vu, P.M., Nguyen, T.T.: Recommending API usages for mobile apps with hidden Markov model. In: 2015 30th IEEE/ACM International Conference on Automated Software Engineering (ASE), pp. 795–800. IEEE, Lincoln (2015). https://doi.org/10.1109/ASE.2015.109

17. Nishimoto, M., Kawabata, H., Hironaka, T.: A system for API set search for supporting application program development. IEICE Trans. Inf. Syst. **J101-D**, 1176–1189 (2018)

18. Hoffmann, R., Fogarty, J., Weld, D.S.: Assieme: finding and leveraging implicit references in a web search interface for programmers. In: Proceedings of the 20th Annual ACM Symposium on User Interface Software and Technology (UIST), Newport, Rhode Island, USA, pp. 13–22. ACM (2007)

19. Barry, D.K.: Web Services, Service-Oriented Architectures, and Cloud Computing: The Savvy Manager's Guide, p. 248. Morgan Kaufmann is an imprint of Elsevier (2013)

20. Pereira, C.R.: Building APIs with Node.js, p. 135. Apress, New York (2016)

21. Doglio, F.: REST API Development with Node.js, p. 323. Apress (2018). https://doi.org/10.1007/978-1-4842-3715-1

22. Malakhov, K., Kurgaev, O., Velychko, V.: Modern RESTful API DLs and frameworks for RESTful web services API schema modeling, documenting, visualizing (2018). arXiv preprint arXiv:1811.04659. https://doi.org/10.15407/pp2018.04.059

23. Sharma, V., Dave, M.: SQL and NoSQL databases. Int. J. Adv. Res. Comput. Sci. Softw. Eng. **2**(8), 20–27 (2012)

24. Moniruzzaman, A.B.M., Hossain, S.A.: Nosql database: new era of databases for big data analytics-classification, characteristics and comparison (2013). arXiv preprint arXiv:1307.0191

25. Chen, J.K., Lee, W.Z.: An Introduction of NoSQL databases based on their categories and application industries. Algorithms **12**(5), 106 (2019). https://doi.org/10.3390/a12050106

26. Du, H., Jones, P., Segarra, E.L., Bandera, C.F.: Development of a rest API for obtaining site-specific historical and near-future weather data in EPW forma. Building Simulation and Optimization, Emmanuel College, University of Cambridge (2018)
27. Jakaria, A.H.M., Hossain, Md.M., Rahman, M.: Smart weather forecasting using machine learning: a case study in Tennessee (2018)
28. Ramanathan, R., Korte, T.: Software service architecture to access weather data using RESTful web services. In: Fifth International Conference on Computing, Communications and Networking Technologies (ICCCNT), pp. 1–8 (2014) https://doi.org/10.1109/ICCCNT.2014.6963122
29. Li, H., Wang, J., Li, R., Haiyan, L.: Novel analysis forecast system based on multi-objective optimization for air quality index. Clean. Prod. **208**, 1365–1383 (2019). https://doi.org/10.1016/j.jclepro.2018.10.129
30. https://waqi.info/
31. https://www.iqair.com/world-air-quality-ranking
32. https://breezometer.com/accurate-realtime-air-quality-data

Filter Based Forwarding Information Base Design for Content Centric Networking

Mohammad Alhisnawi[(✉)]

College of Information Technology, University of Babylon, Hilla, Babylon, Iraq
mohammad.alhisnawi@uobabylon.edu.iq

Abstract. Content Centric Networking (CCN) is a modern suggested networking architecture that aims to utilizing named data rather than named hosts for communication in order to override the limitations of the existing IP architecture. Every content in CCN paradigm is specified by its name and every packet holds a CCN name that defines the required content. Forwarding Information Base (FIB) table is one of the most significant components in any CCN router that have crucial functions in packet forwarding process. Designing a scalable FIB table that have the ability to perform quick look up on changeable-length hierarchical CCN names can be considered as a challenge in CCN paradigm. In this paper, a new design of FIB table for CCN networks has been suggested. Our design depends on modifying a Cuckoo filter (CF), an already proposed query data structure, and utilizing the resulted modified version of CF as a FIB table (we called CF-FIB). Our evaluation show that CF-FIB have the ability to attain perfect search speed and offer very acceptable scalability to large-scale prefix tables.

Keyword: Content centric networking · Forwarding information base · Interest packet · Data packet · Cuckoo filter

1 Introduction

Currently, the traffic magnitude on the Internet is growing noticeably and the target of Internet users is to get their requested contents. Users are not interest about the location of the content that they need to download if they can get it suitably.

Content-Centric Network (CCN) has been proposed [1] to satisfy the purpose of the users be redesigning content distribution mechanism. Instead of the position of the content, CCN is designed to retrieve the content based on its information. Hence, CCN routers have the ability to realize the information of the contents when forwarding them and also they have the ability to cache them by identifying every portion of the content through an identifier (i.e., CCN name) [2]. If the content caching in CCN has been utilized efficiently, then this will minimize noticeably the redundant traffic in the network and also it will enhance the quality of the communication by increasing the sharing of the bandwidth for every user [3]. There exist three fundamental data structures that are utilized in every CCN router [4, 5]:

- Pending Interest Table (PIT): this table is utilized to store unsatisfied Interests. Whenever a new Interest packet reaches the CCN router, a new entry in PIT is generated and this entry will be deleted from PIT when the Data packet arrive.

© The Author(s), under exclusive license to Springer Nature Singapore Pte Ltd. 2021
D.-T. Tran et al. (Eds.): ICISN 2021, LNNS 243, pp. 566–576, 2021.
https://doi.org/10.1007/978-981-16-2094-2_67

- Content Store (CS): this cache memory is utilized to buffer some of the Data packets that already arrive to the CCN router.
- Forwarding Information Base (FIB): this table holds the necessary information that are utilized to forward the arriving Interests.

CCN utilizes variable-length names which are tokenized and have the hierarchical structure rather than utilizing static-length addresses [6]. The forwarding decision in CCN depends on examining the incoming CCN name against the above data structures. The result of this examination will be one of two cases: LPM or exact match [7]. In CCN, FIB table holds CCN name prefixes and their corresponding interface (notice that in CCN the word port and interface are used interchangeably) and forwards the incoming Interest packets by computing the LPM of the CCN name [8].

CCN confronts three fundamental issues for handling the CCN names [9]. First, saving CCN names on CCN routers will need greater storage than saving IP addresses on IP-based routers. This will make FIB table have no place to fit in the current routers and also this will have a great impact on the speed of upgrading these routers online speed. Second, because of the huge size of FIB table, the lookup and update processes will consume considerable amount of time which will result in degrading the performance of packet delivery process [10].

The contribution of this work includes proposing a new design and implementation of forwarding information base table in CCN routers. The basic idea behind this design is to modify a previously suggested data structure named Cuckoo filter (CF) [11] and employing the resulted modified data structure to design FIB table (we called CF-FIB).

The rest of this paper is structured as follows. Section 2 illustrates the former related works. Section 3 presents a general description for Cuckoo filter. Section 4 depicts the suggested FIB design. Section 5 describes our evaluation to the proposed FIB. In Sect. 6, we conclude our work.

2 Related Works

Here, some of the previously suggested works will be presented. In [12], the authors suggested to use Name Component Encoding (NCE) method to encode the CCN name components before looking up through FIB table. NameFilter [13], is another technique that have been proposed later to improve the FIB lookup process by utilizing Bloom filter (BF) with two stages. A data plane with a software-based forwarding mechanism have been suggested in [4]. They utilized hash tables with a good hash computation in order to overcome the collision. A greedy technique [14] have been proposed to enhance the FIB lookup process. In [15] another CCN name lookup technique has been suggested through reorganizing the architecture of FIB table to prop greedy lookup mechanism and enhancing the hash table. A binary Patricia trie structure have been employed in [16] to perform FIB lookup and forwarding. Utilizing Bloom filter to perform a pre-searching lookup process has been proposed in [17]. Another encoding technique (RaCE) has been proposed in [18] that employs radix trie structure to implement FIB table. To implement CCN FIB, [18] suggested memory-effective and scalable radix trie based name component encoding technique, RaCE. An ordered trie structure have been suggested by [10], called Patricia trie, to implement FIB table.

3   Cuckoo Filter Background

Here, a brief description for Cuckoo filter [11], which represents basic for the suggested FIB table, will be presented. The main reason for suggesting Cuckoo filter is to overcome the short comes that are arise in the earlier suggested query data structures Bloom filter (BF) and Quotient filter (QF) [19, 20]. The main advantage of CF is the ability to accomplish the fundamental operations (i.e., insertion, query, and deletion) in $O(1)$ time. Rather than storing the complete string, CF stores solely the element's fingerprint [21]. Many networking applications make use of Cuckoo filter because of its efficiency. Deep packet inspection [22], IP lookup [23], and packet classification [24] are some examples for these applications. The standard cuckoo hashing [18] represents the fundamental concept for the general work of Cuckoo filter. Inserting an element in cuckoo hashing technique may need to apply the relocation on some of the earlier inserted elements to prepare vacant position for this element. In the standard cuckoo hashing, the original elements are inserted to the hash table and, thus, the relocations process is possible but, in the CF, solely the fingerprints of elements are inserted. So, there must be some way in Cuckoo filter to perform the relocation. The latter issue has been addressed by utilizing partial-key cuckoo hashing technique that work in the following way: Suppose that there is an item x, the two candidate buckets are computed as follow:

$$bucket1(x) = calculatehash(x); \tag{1}$$

$$bucket2(x) = bucket1(x) \oplus alculatehash(calculatefingerprint(x)) \tag{2}$$

4   Proposed Forwarding Information Base Design

In our proposed FIB design, we will modify the standard Cuckoo filter in order to make it suitable for our FIB design (we called CF-FIB). So, first, in this section, we will describe our modification for CF in the following way: The standard Cuckoo filter, as mentioned in Sect. 4, consists of a number of buckets with equal number of slots in

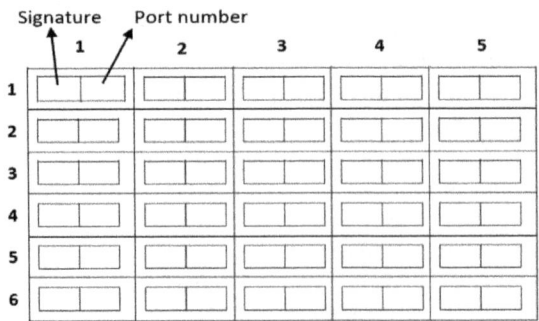

Fig. 1. General structure of CF-FIB.

every bucket. These slots contain the fingerprints of the inserted items. In our modification for CF, we will expand these slots such that every one of them will hold two values: the first one is the item's fingerprint and the second one is the output interface (as will be explained later). Figure 1 depicts the general structure for CF-FIB.

Algorithm 1 depicts the main steps to insert an item into CF-FIB. The algorithm starts by taking the input item (CCN name as will be explained later) and the output face. Then, it calculates the two possible buckets to insert this item. If any of these possible buckets have an empty entry, then the item (with its associated interface) will be inserted into CF-FIB. Otherwise, the relocation operation must take place to prepare an empty slot to insert this item.

Algorithm 1: Item insertion
1 Input: item(x), output face(f);
2 finger = calculatefingerprint(x);
3 bucket1 = calculatehash(x);
4 bucket2 = bucket1 $\oplus$ calculatehash(finger);
5 if (bucket1 or bucket2 holds not occupied entry) then
6   put finger and f to that bucket;
7   //process finished;
8   quit;
9 // Else, perform a relocation;
10 bucket = pick randomly between (bucket1 and bucket2);
11 for i ← 0 to allowed number of attempts do
12   pick an entry (e) randomly from bucket;
13   swap finger and its associated f and the fingerprint and its associated f stored in entry e;
14   bucket = bucket $\oplus$ calculatehash(finger);
15   if bucket holds not occupied entry then
16   put finger and f to bucket;
17   //process finished;
18   quit;
19 endfor

Algorithm 2: Item examination
1 Input : item(x);
2 finger = calculatefingerprint(x);
3 bucket1 = calculatehash(x);
4 bucket2 = bucket1 $\oplus$ calculatehash(finger);
5 if (bucket1 or bucket2 holds finger) then
6   return (output face(f));
7 else
8   return (negative query);

Algorithm 3 depicts how to remove an item from CF-FIB. First, the algorithm calculates the two possible buckets for the item x. Then, these buckets will be checked seeking for x. If x is already mapped into CF-FIB, then it will be removed with its associated output face.

Algorithm 3: Item deletion
```
1 Input : item(x);
2 finger = calculatefingerprint(x);
3 bucket1 = calculatehash(x);
4 bucket2 = bucket1 ⊕ calculatehash(finger);
5 if (bucket1 or bucket2 holds fin) then
6               remove fing and f;
7               return (task accomplished);
8 else
9               return (error task);
```

Now we will explain how the proposed CF-FIB works. CF-FIB works in two main stages: configuration stage and lookup stage. The configuration stage works as a pre-processing operation in which all the required CCN names with their associated output faces are added into the CF-FIB. After this stage, the CCN router will be able to receive the Interest packets and doing packet processing operations.

Algorithm 4 depicts the configuration stage which include inserting all the name parts into CF-FIB. It requires two values as input: the CCN name and its output face. It begins by extracting the number of portions in the incoming CCN name. Then, it will traverse through each part of the CCN name beginning from the shortest part to the longest one (full CCN name).

Algorithm 4: Configuration stage
```
1 Input: CCN name(x), output face(f);
2 //extract number of name's parts;
3 p ← extractnumberofpartitions(x);
4 for i ← 1 to p do
5               //treat name's parts individually from short to long;
6               cp ← getnamepart(x; 1..i);
7               //function checkpart return true if the current name part is not
already inserted;
8 if (checkpart(cp,CF-FIB)) then
9               //reach final part or not;
10     if (i=p) then
11                         insert-FIB(cp,f);
12              else
13                         insert-FIB(cp,null);
```

For every part, the algorithm examines if the current part is already inserted or not, because every part of the CCN name must be inserted once, and if it is not already inserted then it will be inserted into CF-FIB. If the current part is not the final part (i.e., full CCN name) then "null" will be inserted with it into the CF-FIB as its output face. Otherwise, the output face (f) will be inserted. It can be noticed from the way of insertion in this algorithm that the insertion of CCN name parts in the FIB will be done without duplication. As instance, suppose that there exist two various CCN names (com/yahoo/content1,com/yahoo/content2) both of them will be inserted into CF-FIB but their portions (com/yahoo) and (com) will be inserted solely one time. This will be ensured by utilizing (checkpart) procedure. The purpose of last feature is to minimize the size of CF-FIB and giving it the hierarchical style and, also, it will make the lookup operation extra precise.

Algorithm 5 illustrates the lookup stage which is concentrate on finding the LPM for the incoming CCN name in our proposed CF-FIB. First, the algorithm specifies the default port as its final port then it will extract the number of name's portions (p). Then, the CF-FIB is queried by utilizing the concatenated string of i portions (i = 1..p). If the CF-FIB announce a positive query, the CF-FIB is queried continuously using i + 1 parts and so on. So, here will be two scenarios: first the entire CCN name parts will return positive query and the algorithm will reach the condition (i=p). Second, the CF-FIB will respond with a negative query, and, in this situation the lookup is terminated. In both scenarios the algorithm will return the currently obtained port. Note that in our CF-FIB design, if there is no matching between CF-FIB and the current concatenated parts of CCN name then there is no matching (definitely) between CF-FIB and the longer concatenated parts. The latter property has been ensured by configuration stage.

Algorithm 5: Lookup stage
1 Input: CCN name(x);
2 finalport ← default port;
3 //extract number of name's parts;
4 p ← extractnumberofpartitions(x);
5 for i ← 1 to p do
6   cp ← getnamepart(x; 1. .i);
7   //Query CF-FIB;
8   if (checkCF-FIB(cp,f)) then
9   if (f<> null) then
10   finalport ← f
11   else
12   break;

5   Evaluation

Here, we will discuss our evaluation for the proposed CF-FIB. We utilized ndnSIM package [25] which performs CCN protocol stack for NS-3 network simulator.

In our evaluation, a standard dataset has been utilized. It has been gotten from an open source store [26] which holds datasets of CCN names in different size and formats which are gathered from URLs or by tracing web pages. Three previously proposed FIB design techniques: Components-Trie (CT) [12], Name prefix trie structure with BF (NPT-BF) [17], Greedy-PHT [14] have been utilized in our evaluation.

5.1 Memory Usage

The memory usage for all techniques on various prefix table sizes have been depicted in Fig. 2. The memory consumption for all methods is increased proportionally with the increasing number of prefixes. It can be noticed that NPT-BF consumes the greater memory, and this is the effect of utilizing NPT, CBF, and a hash table. Component trie utilizes lower memory than name prefix trie but greater than the others. Additional data need to be saved in FIB in the Greedy-PHT technique and, thus, it expends larger memory than CF-FIB. CF-FIB needs the minimum memory requirements because of its simple structure.

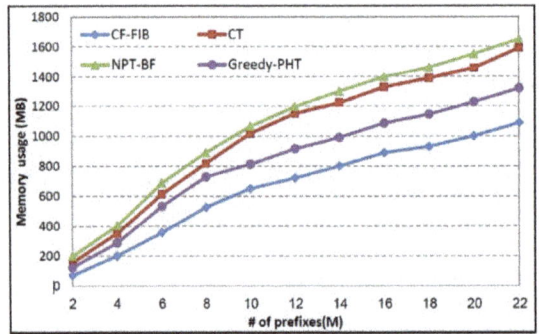

Fig. 2. Memory usage for CT, NPT-BF, greedy-PHT, and CF-FIB.

5.2 Insertion Performance

Figure 3 illustrates the insertion throughput for all methods. Among all of them, NPT-BF has the minimum insertion throughput because of its complicated structures: NPT, CBF, and a hash table. Greedy-PHT has better insertion performance than NPT-BF but lower insertion performance than CT. CF-FIB has the greater insertion performance and this is a result of its simple structure that have the ability to perform the insertion directly and quickly.

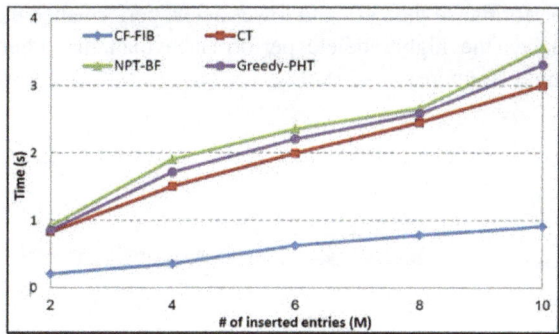

Fig. 3. Insertion performance for CT, NPT-BF, greedy-PHT, and CF-FIB.

5.3 Lookup Performance

The lookup performance for all four methods have been depicted in Fig. 4. The experiment has been accomplished by performing lookup operations on up to 12 million entries and registering the elapsed time to do these operations. CF-FIB has the highest insertion performance, and this is a result of its simple structure that have the ability to perform the lookup operations directly and quickly. Since the lookup process is accomplished sequentially beginning from the root in CT, it will have the lowest achievement. Greedy-PHT has more achievement than CT. NPT-BF showed higher lookup performance than both CT and Greedy-PHT and this is a result of utilizing Bloom filter as a pre-search process which have the ability to eliminate the superfluous accesses to the off-chip memory (i.e., hash table).

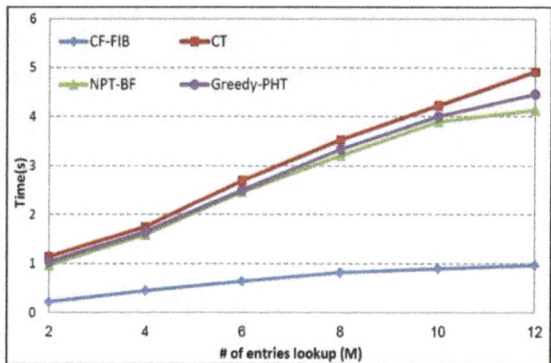

Fig. 4. Lookup performance for CT, NPT-BF, greedy-PHT, and CF-FIB.

5.4 Delete Performance

Figure 5 depicts the delete performance for the four methods. It can be concluded that CT has the worst delete performance because of its complicated structure. Greedy-PHT has better delete performance than CT and lower delete performance than the other two methods. CF-FIB has the higher delete performance than the other three methods because of its simple structure.

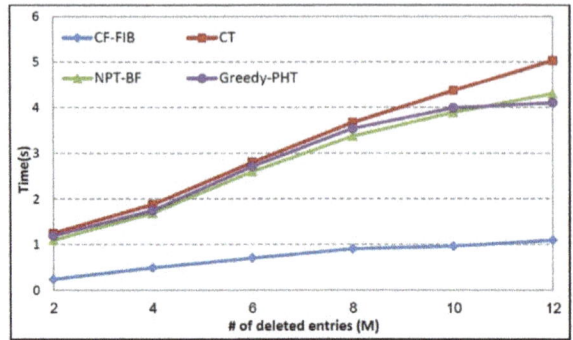

Fig. 5. Delete performance for CT, NPT-BF, greedy-PHT, and CF-FIB.

6   Conclusion

This paper suggested novel design to enhance the lookup achievement of FIB table for packet forwarding in CCN networks. The proposed technique depends on the verity that dealing with an on-chip memory consumes less time than dealing with an off-chip memory. Therefore, we suggested to utilize modified version of Cuckoo filter (CF-FIB) as an on-chip memory to exclude the need of accessing to an off-chip memory. Our choice for Cuckoo filter comes from its ability to perform high speed insertion/query/delete operations with very acceptable memory consumption and false positive rate. Comprehensive experiments on wide range of name prefixes confirm that CF-FIB can considerably minimize the memory consumption, enhance the scalability, and offer gradual updates while accomplishing fast search operation.

References

1. Jacobson, V., Smetters, D.K., Thornton, J.D., Plass, M., Briggs, N., Braynard, R.: Networking named content. In: Proceedings of ACM CoNEXT 2009, pp. 1–12 (2009)
2. Endo, T., Yokotani, A., Ohzahata, S., Yamamoto, R., Kato, T.: An adaptive bandwidth reservation method for a content-centric network. In: 2018 IEEE 42nd Annual Computer Software and Applications Conference (COMPSAC), vol. 2, pp. 763–768. IEEE (2018)
3. Zhang, M., Luo, H., Zhang, H.: A survey of caching mechanisms in information centric networking. IEEE Commun. Surv. Tutor. **17**(3), 1473–1499 (2015)
4. Won, W., Narayanan, A., Oran, D.: Named data networking on a router: fast and dos-resistant forwarding with hash tables. In: Proceedings of the Ninth ACM/IEEE Symposium on Architectures for Networking and Communications Systems, pp. 215–226. IEEE Press (2013)

5. Li, Z., Liu, K., Zhao, Y., Ma, Y.: MaPIT: an enhanced pending interest table for NDN with mapping bloom filter. IEEE Commun. Lett. **18**(11), 1915–1918 (2014)
6. Fan, K., Ren, Y., Wang, Y., Li, H., Yang, Y.: Blockchain-based efficient privacy preserving and data sharing scheme of content-centric network in 5G. IET Commun. **12**(5), 527–532 (2018)
7. Wang, K., Gu, L., Guo, S., Chen, H., Leung, V., Sun, Y.: Crowdsourcing-based content-centric network: a social perspective. IEEE Netw. **31**(5), 28–34 (2017)
8. Xing, L., Zhang, Z., Lin, H., Gao, F.: Content centric network with label aided user modeling and cellular partition. IEEE Access **5**, 12576–12583 (2017)
9. Jmal, R., Fourati, L.: An OpenFlow Architecture for Managing Content-Centric-Network (OFAM-CCN) based on popularity caching strategy. Comput. Stand. Interfaces **51**, 22–29 (2017)
10. Saxena, D., Raychoudhury, V.: N-FIB: scalable, memory efficient name-based forwarding. J. Netw. Comput. Appl. **76**, 101–109 (2016)
11. Fan, B., Andersen, D., Kaminsky, M., Mitzenmacher, M.: Cuckoo filter: practically better than bloom. In: Proceedings of the 10th ACM International on Conference on emerging Networking Experiments and Technologies, pp. 75–88 (2014)
12. Wang, Y., He, K., Dai, H., Meng, W., Jiang, J., Liu, B., Chen, Y.: Scalable name lookup in NDN using effective name component encoding. In: 2012 IEEE 32nd International Conference on Distributed Computing Systems (ICDCS), pp. 688–697. IEEE (2012)
13. Wang, Y., Pan, T., Mi, Z., Dai, H., Guo, X., Zhang, T., Liu, B., Dong, Q.: NameFilter: achieving fast name lookup with low memory cost via applying two-stage Bloom filters. In: 2013 Proceedings IEEE INFOCOM, pp. 95–99. IEEE (2013)
14. Wang, Y., Tai, D., Zhang, T., Lu, J., Xu, B., Dai, H., Liu, B.: Greedy name lookup for named data networking. In: ACM SIGMETRICS Performance Evaluation Review, vol. 41, no. 1, pp. 359–360. ACM (2013)
15. Wang, Y., Xu, B., Tai D., Lu, J., Zhang, T., Dai, H., Zhang, B., Liu, B.: Fast name lookup for named data networking. In: 2014 IEEE 22nd International Symposium of Quality of Service (IWQoS), pp. 198–207. IEEE (2014)
16. Song, T., Yuan, H., Crowley, P., Zhang, B.: Scalable name-based packet forwarding: from millions to billions. In: Proceedings of the 2nd International Conference on Information-Centric Networking, pp. 19–28. ACM (2015)
17. Lee, J., Shim, M., Lim, H.: Name prefix matching using Bloom filter pre-searching for content centric network. J. Netw. Comput. Appl. **65**, 36–47 (2016)
18. Saxena, D., Raychoudhury, V., Becker, C., Suri, N.: Reliable memory efficient name forwarding in named data networking. In: EUC (2016)
19. Bender, M., Farach-Colton, M., Johnson, R., Kuszmaul, B., Medjedovic, D., Montes, P., Shetty, P., Spillane, R., Zadok, E.: Don't thrash: how to cache your hash on flash. Proc. VLDB Endow. **5**(11), 1627–1637 (2012)
20. Alhisnawi, M., Ahmadi, M.: QCF for deep packet inspection. IET Netw. **7**(5), 346–352 (2018)
21. Pagh, R., Rodler, F.: Cuckoo hashing. In: European Symposium on Algorithms, pp. 121–133. Springer, Heidelberg (2001)
22. Alhisnawi, M., Ahmadi, M.: Deep packet inspection using cuckoo filter. In: 2017 Annual Conference on New Trends in Information and Communications Technology Applications (NTICT), pp. 197–202 (2017)
23. Kwon, M., Reviriego, P., Pontarelli, S.: A length-aware cuckoo filter for faster IP lookup. In: 2016 IEEE Conference on Computer Communications Workshops (INFOCOM WKSHPS), pp. 1071–1072, 10 April 2016

24. Abdulhassan, A., Ahmadi, M.: Cuckoo filter-based many-field packet classification using X-tree. J. Supercomput. **75**, 1–21 (2019)
25. Afanasyev, A., Moiseenko, A., Zhang, L.: ndnSIM: NDN simulator for NS-3. University of California, Los Angeles, Technical Report 4, (2012)
26. Schnurrenberger, U.: The Content Name Collection. CNC. https://www.icn-names.net/. Accessed 26 June 2019

Proposed Method to Generated Values of the RADG Design by Lattice and Logistic Map

Sundus Q. Habeeb[(⊠)] and Salah A. K. Albermany

Faculty of Computer Science and Mathematics, University of Kufa, Kufa, Iraq
Salah.albermany@uokufa.edu.iq

Abstract. The typical scheme used to generate values of RADG (Reaction Automata Directed Graph) are lattice based cryptography and chaos logistic map. Lattice based cryptography is efficient method against post quantum computers consists of set of points of n-dimension $\in \mathbb{R}^n$ in lattice space. Logistic map is a mathematical model used to describes the growth of populations by give chaos value between 0 and 1. In this paper, lattice basis $\in \mathbb{R}^2$ is used to generate set of points consider as values of RADG design, based on Gaussian lattice cryptography, Babai's algorithm. Logistic map used to generate and distribution point on RADG states.

Keywords: RADG · Lattice · Gaussian lattice cryptography · Babai's algorithm · Logistic map · Hadamard ratio Cryptography and Network

1 Introduction

Reaction Automata Direct Graph (RADG), which is based on automata directed graph and reaction state, is the main aspect of RADG that gives more than one ciphertext to the same plaintext [1–3]. Lattice cryptography is one of the important and fastest moving area in mathematical cryptography today, lattice supported by strong worst case to average case security. The lattice is new type of hard problem arising in the theory of lattice can be used as the basis for the public key cryptosystem [4, 5]. Lattice based cryptography is set of points in n-dimensional space with periodic structure, consists of set of vectors belong to $\mathbb{R}^n$ that is consider as basis of lattice and must be linear independent vector [6]. Lattice-based cryptographic constructions hold a great promise for post-quantum cryptography, as they have quite strong security proofs based on worst-case hardness, relatively efficient implementations, as well as great simplicity. Lattice-based cryptography is believed to be secure against quantum computers. The concentration here will be mainly on the practical aspects of lattice-based cryptography and less on the methods used to establish their security [7].

Logistic map is mathematical model used to generate chaos value, that is very sensitive for initial condition, simple different in condition will produce high different in value, in the proposed method can be used to generate and distribution points values among RADG states except jump states [8–10]. This paper using a novel keyless security schema RADG which is based on Reaction Automata Direct Graph [1], the

D.-T. Tran et al. (Eds.): ICISN 2021, LNNS 243, pp. 577–588, 2021.
https://doi.org/10.1007/978-981-16-2094-2_68

new trend in security generate set of points belong to lattice space that is considered as a values of RADG states depended on the key of size 448 bit between sender and receiver, the set of points in RADG states useful to increase complexity against post quantum computers since points belong to lattice space.

2 Reaction Automata Directed Graph (RADG)

Reaction Automata Directed Graph (RADG), which is based on automated directed graph and reaction state. The start of the RADG cryptography does not require any key to perform encryption and decryption operation. The first Modeling of RADG is influenced by graph theory, the RADG design can be represented by a set of characters as $\{Q, R, \Psi, \Sigma, J, T, \lambda\}$, such that Q stands for standard states, R acts as reaction states, Σ represents non-empty finite set of input data, Ψ stands for output transitions, J represents non-empty finite set called jump which is subset of Q, T represents transition Function and λ represents the number of value in each state, except jump state does not take any value. RADG cryptography is based on transition between states of the RADG design depended on the value of n, m, k and λ, where $n = |Q|$, $m = |R|$ and $k = |j|$. The encryption starts from standard state if standard state goes to jump state then jump is given random address to any state in reaction to encrypt the input then return back to standard state to continue in encryption process until all value in message is used [1].

3 Gaussian Lattice Reduction in Dimension Two

Gaussian reduction algorithm is used for solving the problem of shortest vector (SVP) and finding optimal basis belong to $\mathbb{R}^2$ lattice which works in polynomial time. SVP has a polynomial solution in the case of two dimensions [11].

Proposition (1): Let $L \in \mathbb{R}^2$ is a 2-dimensional lattice with basis v_1 and v_2, the algorithm terminates and yields a good basis for lattice [5].

Algorithm (1): Gaussian lattice algorithm
Input :two basis for lattice v_1 and $v_2 \in \mathbb{R}^2$
Output: optimal basis for lattice
Loop
If $\| v_2 \| < \| v_1 \|$, swap v_1 and v_2
Compute m=$\left\lfloor \frac{v_1 . v_2}{\| v_1 \|^2} \right\rceil$
If m=0 , return the basis vector v_1 and v_2
Replace v_2 with $v_2 - m\, v_1$
Continue Loop

Assume L $\in \mathbb{R}^n$ be a lattice with basis $v_1, ..., v_n$ and W $\in \mathbb{R}^n$ be a random vector. If the vectors in the basis are sufficiently orthogonal to one another, then the following algorithm solves CVP (closest vector problem) [5]. Algorithm solves some versions of CVP if the vectors in the basis are reasonably orthogonal to one another, but if the basis vectors are highly nonorthogonal, then the vector returned by the algorithm is generally far from the closest lattice vector to **W**.

Algorithm (2): Babai's algorithm
Input: set of lattice basis $v_1, \ldots, v_n \in \mathbb{R}^n$ and arbitrary vector $W \in \mathbb{R}^n$
Output: vector close to W in lattice space
Write w= $t_1 v_1 + t_2 v_2 + ... + t_n v_n$ with $t_1 .... t_n \in \mathbb{R}$
Set $a_i = \lfloor t_i \rfloor$, for i=1 ,2 ,....,n

---

4   Generate Points as a Values for the RADG States in the RADG Design

To generate a set of points as values in states of RADG design, the optimal basis from the lattice based cryptography can be used to generate points for the RADG design, the sender and the receiver choose any two points randomly each point with two components belong to $\mathbb{R}^2$, each component with size 64 bit. The two points are processed by using lattice rules and Gaussian lattice cryptography to get optimal basis of lattice based cryptography, that basis satisfies the two conditions of lattice must be linear independent when Hadamard ratio is greater than or equal 0.5, that makes it difficult to break by post quantum computers [7], this basis of lattice with vector of chaos value from logistic map can generate a set of points, which are tested and processed by Babia's algorithm to ensure that points belong to lattice space. The points in the lattice space can be distributed among RADG states by using chaos values from logistic map. The Algorithm 3 explains generation and distribution of points, Fig. 1 lattice basis generation. Figure 2 generate and distribution point on RADG.

Algorithm (3): Generate Points for The RADG Values in The RADG design
Input: key between sender and receiver of size 448 bit as
n: Size of the set Q, 42 bit and n≥ 8.
m: Size of the set R, 34 bit and m≥ 4.
k: Size of the set J (jump state), 24 bit and k≥ 2.
λ: The number of value in each state, 4 bit and λ ≥ 2.
w: The number of transition to jump state, 24 bit and w≥ 2.
v_1 and v_2: Two points its components ∈ $\mathbb{R}^2$, the size of each component 64 bit, two component in the same point do not has zero values.
r: Values between 3.569 and 4, it sizes 24 bit.
x_0: Initial value between 0 and 1, it sizes 40 bit.
Output: p_{sv} matrix with value of p_{ij} set of point, s_i the address of states and v_i the values of states

A-   Generate a Lattice Basis by Using the Input Points
1- Convert the two point into matrix of size 2×2 call m_v
2- Choose matrix A of size 2×2 randomly with determine ±1 to satisfy the conditions of lattice basis must be linear independent.
3- Testing the lattice basis $W = Am_v$ by using Hadamard ratio to ensure that W contain good basis for lattice when ratio≥0.5.

$$Hd = \sqrt{\frac{det(W)}{\|W_{11}\|\|W_{12}\|\|W_{21}\|\|W_{22}\|}} \ , \ where \ \|W\| = \sqrt{W_1^2 W_2^2}$$

4- convert the matrix in step3 into two vector $b_1 = [W_{11}.W_{12}]$ and $b_2 = [W_{21}.W_{22}]$
5- Process the two point in (4) by Gaussian lattice algorithm to obtain the optimal basis when basis ∈ $\mathbb{R}^2$, see algorithm (1) and save result in $b_1 and\ b_2$.

B- Generate Set of Point Based on Lattice Basis and Logistic Map

1- Compute the value of $\tau = \lambda(n + m - k)$. Where τ mean the number of values in all states.
2- Compute logistic map with value between 0 and 1 by using the following rule $x_{i+1} = rx_i(1 - x_i)$. where $i = 0.1. ... .(2\tau) - 1$.
3- Convert the value of x_i to integer value between 1 and 2τ by using the suggested relationship $g_i = 1 + \lfloor 2\tau x_i \rfloor$.
4- Remove the repetition in g_i and replace that with value between 1 and 2τ, the result is vector with distinct values.
5- Convert the result of g_i into matrix of size 2×τ, denoted by G=[g_{ij}]
6-   Generate set of point based on that relationship

$$p_{ij} = g_{1j}b_1 + g_{2j}b_2 \ .where \ j = 1.2. ... .\tau$$

7- Check the result from (6) by Babia's algorithm to ensure that point within lattice space environment, see algorithm (2).

C- Distribution of the Points on RADG values of states

1. Take vector of size τ from x_i denoted by y_i.
2. Convert the value of y_i to integer value between 0 and $\tau - 1$ by using the suggested relationship $dv_i = \lfloor y_i u \rfloor.where \ u = \lambda(n - k)$
3. Remove the repetition in dv_i and replace that with value between 0 and $\tau - 1$, the result is vector with distinct value
4. Compute $s \ and \ v$ the address and value of state and convert from row to vector , $s = \lfloor \frac{dv_i}{\lambda} \rfloor$ and $v = dv_i \ mod \ \lambda$
5. Create matrix of size τ×4 called psv where the first and second column points values, third Column the addresses of states and last column is values of states.

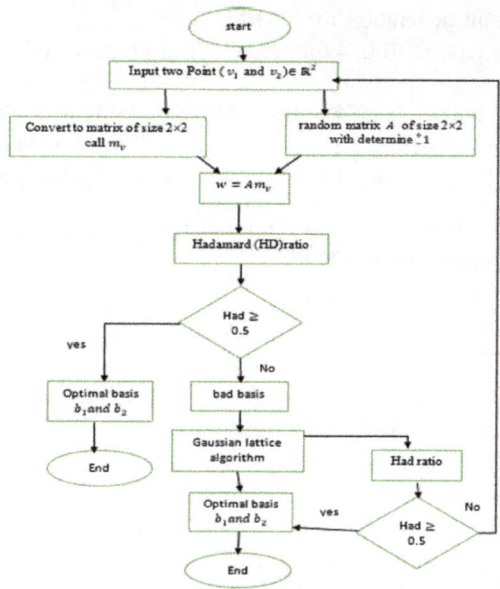

Fig. 1. Lattice basis generation

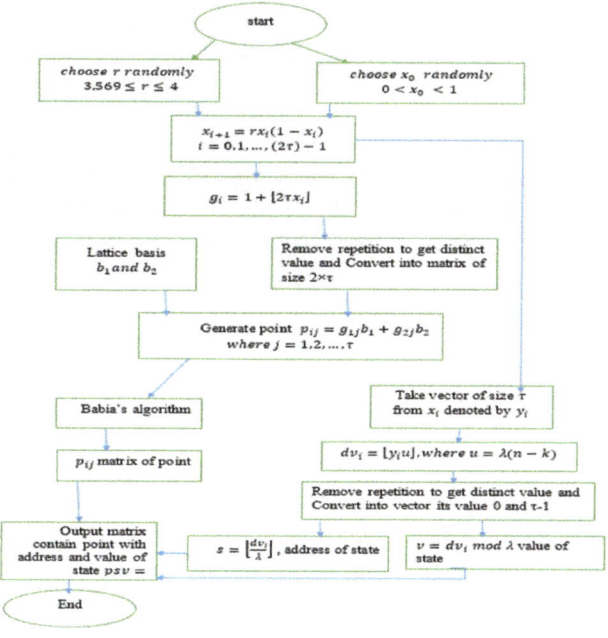

Fig. 2. Generate and distribution point on RADG

Example (1): point generation for RADG.

Input: n = 000……00000100, 42bit, convert from binary to decimal $nd = 4\, then\, n = nd + minimum(n)$,

$n = 4 + 8 = 12.m = 0000\ldots0000.k = 00\ldots0001.\lambda = 0001.w = 00\ldots\ldots001$

From above get $n = 12.m = 4.k = 3.\lambda = 3.w = 3$, v_1 = [66586820,65354729] and v_1 = [6513996,6393464], the Fig. 3 and Fig. 4 show point generation and distribution.

A Generate a Lattice Basis by Using the Input Point

1 $m_v = \begin{bmatrix} 66586820 & 65354729 \\ 6513996 & 6393464 \end{bmatrix}$

2 Assume choose $a = \begin{bmatrix} 0 & -1 \\ -1 & 2 \end{bmatrix}$, the determine of $a = -1$

3 Compute $W = \begin{bmatrix} -6513996 & -6393464 \\ -53558828 & -52567801 \end{bmatrix}$

4 Hd = 9.9215e05 less than 0.5 continuous process

5 b_1 = [−6513996 −6393464], b_2 = [−53558828 − 52567801]

6 Apply Gaussian lattice cryptography b_1= [−2280 1001], b_2= [1324 2376], is the optimal basis for lattice.

7 Hd = 0.9955 ≥ 0.5 that ensure the b_1 and b_2 is optimal basis for lattice

B Generate Set of Point Based on Lattice Basis and Logistic Map

1 Assume choose $r = 3.8\, and\, x_0 = 0.345$

2 Compute $\tau = 39$

3 x_i = {0.3450, 0.8587, 0.4611, 0.9442, 0.2001, 0.6082, 0.9055, 0.3251, 0.8338, 0.5267, 0.9473, 0.1897, 0.5842, 0.9231, 0.2699, 0.7488, 0.7148, 0.7746, 0.6635, 0.8485, 0.4885, 0.9495, 0.1822, 0.5662, 0.9333, 0.2364, 0.6860, 0.8185, 0.5645, 0.9342, 0.2337, 0.6805, 0.8262, 0.5457, 0.9421, 0.2074, 0.6247, 0.8909, 0.3694, 0.8852, 0.3863, 0.9009, 0.3394, 0.8520, 0.4792, 0.9484, 0.1861, 0.5756, 0.9283, 0.2530, 0.7181, 0.7692, 0.6747, 0.8340, 0.5260, 0.9474, 0.1893, 0.5831, 0.9237, 0.2677, 0.7449, 0.7221, 0.7625, 0.6881, 0.8156, 0.5716, 0.9305, 0.2457, 0.7043, 0.7915, 0.6272, 0.8885, 0.3764, 0.8919, 0.3663, 0.8820, 0.3954, 0.9084}.

4 g_i = {27, 67, 36, 74, 16, 48, 71, 26, 66, 42, 74, 15, 46, 72, 22, 59, 56, 61, 52, 67, 39, 75, 15, 45, 73, 19, 54, 64, 45, 73, 19, 54, 65, 43, 74, 17, 49, 70, 29, 70, 31, 71, 27, 67, 38, 74, 15, 45, 73, 20, 57, 60, 53, 66, 42, 74, 15, 46, 73, 21, 59, 57, 60, 54, 64, 45, 73, 20, 55, 62, 49, 70, 30, 70, 29, 69, 31, 71}, with repetition.

5 g_i = {27, 67, 36, 74, 16, 48, 71, 26, 66, 42, 4, 15, 46, 72, 22, 59, 56, 61, 52, 2, 39, 75, 12, 45, 73, 19, 54, 64, 24, 32, 37, 40, 65, 43, 5, 17, 49, 70, 29, 50, 31, 8, 1, 3, 38, 6, 13, 25, 33, 20, 57, 60, 53, 10, 11, 7, 14, 18, 34, 21, 23, 77, 78, 41, 44, 28, 35, 76, 55, 62, 47, 51, 30, 58, 63, 69, 68, 9}, when remove repetition

6 Change size of g_i to $2 \times \tau$

7 p_{ij} =

p1-p13		p14-p26		p27-p39	
27148	186219	-38384	206118	-107600	76813
15896	211860	-12352	100056	-15812	27643
27072	130064	-31400	132077	-8088	56782
-127456	132847	-91268	167233	-49716	83930
-94872	165858	11108	45397	49508	205975
10740	39644	-19040	215369	-123556	175494
-9552	217118	80	147829	-63248	110572
27956	162206	-60088	50039	20824	215611
-46916	200992	1692	8129	-43312	202367
-115912	56804	-78696	52294	-39636	168223
10380	217239	3460	72413	8392	167838
32220	118932	-48760	80553	-52284	227007
-141284	118217	-50520	199617	-143124	89452

8   Check result in step7 by Babia's algorithm see given the same result that ensure the point generated belong to the lattice space environment.

C   Distribution of the Point on RADG

1   y_i = {0.3450, 0.8587, 0.4611, 0.9442, 0.2001, 0.6082, 0.9055, 0.3251, 0.8338, 0.5267, 0.9473, 0.1897, 0.5842, 0.9231, 0.2699, 0.7488, 0.7148, 0.7746, 0.6635, 0.8485, 0.4885, 0.9495, 0.1822, 0.5662, 0.9333, 0.2364, 0.6860, 0.8185, 0.5645, 0.9342, 0.2337, 0.6805, 0.8262, 0.5457, 0.9421, 0.2074, 0.6247, 0.8909, 0.3694}

2   dv_i = {9, 23, 12, 25, 5, 16, 24, 8, 22, 14, 25, 5, 15, 24, 7, 20, 19, 20, 17, 22, 13, 25, 4, 15, 25, 6, 18, 22, 15, 25, 6, 18, 22, 14, 25, 5, 16, 24, 9}, with repetition.

3   dv_i = {9, 23, 12, 25, 5, 16, 24, 8, 22, 14, 1, 21, 15, 28, 7, 20, 19, 36, 17, 30, 13, 2, 4, 34, 3, 6, 18, 31, 35, 10, 37, 38, 32, 33, 11, 26, 27, 29, 0}, when remove repetition

4   s_i = {3, 7, 4, 8, 1, 5, 8, 2, 7, 4, 0, 7, 5, 9, 2, 6, 6, 12, 5, 10, 4, 0, 1, 11, 1, 2, 6, 10, 11, 3, 12, 12, 10, 11, 3, 8, 9, 9, 0}, address of state.

v_i = {0, 2, 0, 1, 2, 1, 0, 2, 1, 2, 1, 0, 0, 1, 1, 2, 1, 0, 2, 0, 1, 2, 1, 1, 0, 0, 0, 1, 2, 1, 1, 2, 2, 0, 2, 2, 0, 2, 0}, value of state.

5   psv is final result of point distribution with address and value of the state.

Point		state	value	Point		state	value
27148	186219	3	0	-60088	50039	4	1
15896	211860	7	2	1692	8129	0	2
27072	130064	4	0	-78696	52294	1	1
-127456	132847	8	1	3460	72413	11	1
-94872	165858	1	2	-48760	80553	1	0
10740	39644	5	1	-50520	199617	2	0
-9552	217118	8	0	-107600	76813	6	0
27956	162206	2	2	-15812	27643	10	1
-46916	200992	7	1	-8088	56782	11	2
-115912	56804	4	2	-49716	83930	3	1
10380	217239	0	1	49508	205975	12	1
32220	118932	7	0	-123556	175494	12	2
-141284	118217	5	0	-63248	110572	10	2
-38384	206118	9	1	20824	215611	11	0
-12352	100056	2	1	-43312	202367	3	2
-31400	132077	6	2	-39636	168223	8	2
-91268	167233	6	1	8392	167838	9	0
11108	45397	12	0	-52284	227007	9	2
-19040	215369	5	2	-143124	89452	0	0
80	147829	10	0				

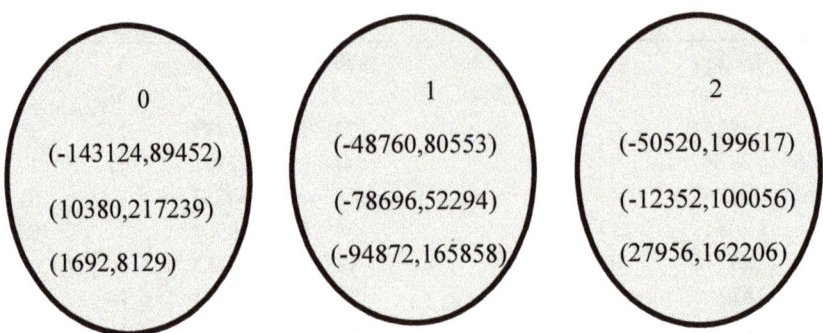

Fig. 3. Some states for point generation and distribution in example (1)

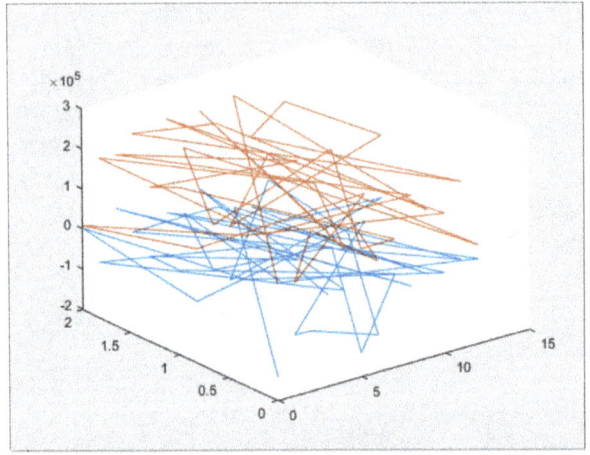

Fig. 4. States and values for point generation and distribution in example (1)

Can take another example and see generation and distribution for points when n = 100, m = 20, k = 15, λ = 3 and w = 15, with the same points in example (1) see Fig. 5, and points when n = 1000, m = 200, k = 100, λ = 3 and w = 100, with same point in example (1) see Fig. 6, when x- axis equal to states, y-axis equal to values and z-axis mean points values with two components.

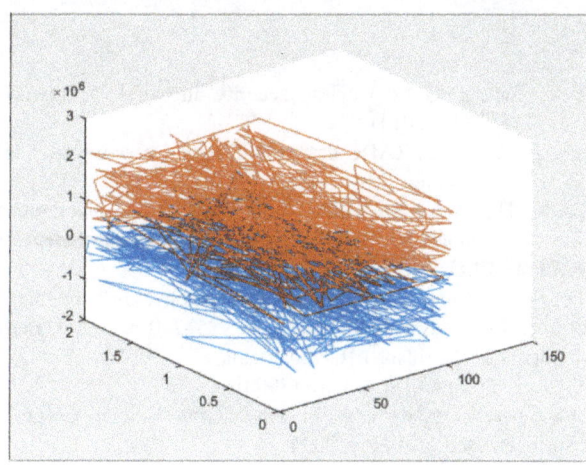

Fig. 5. States and values for point generation and distribution

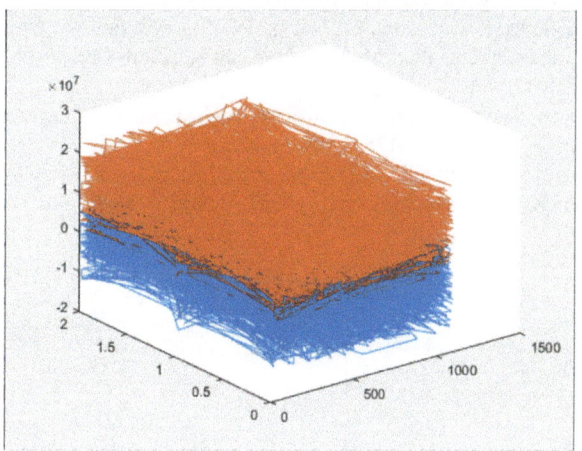

Fig. 6. States and values for point generation and distribution

5   Conclusion

The general system applied on two point $\in \mathbb{R}^n$, with key of the size 448 bit between sender and receiver. The purpose behind this paper is to generate a set of points belong to the lattice space based on lattice basis and chaos from logistic map. These points can be considered as values for the RADG states and can be distributed on the RADG design by using chaos from logistic map. And that achieves the goal of this paper.

References

1. Albermany, S.A., Safdar, G.A.: Keyless security in wireless networks. Wirel. Pers. Commun. **79**(3), 1713–1731 (2014)
2. Albermany, S.A., Alwan, A.H.: RADG design on Elliptic Curve Cryptography. In: ICCIIDT Conference UK, 11–12 October 2016
3. Albermany, S.A.K., Hamade, F.R., Safdar, G.A.: New random block cipher algorithm. In: 2017 International Conference on Current Research in Computer Science and Information Technology (ICCIT). IEEE (2017)
4. Micciancio, D.: The geometry of lattice cryptography. In: International School on Foundations of Security Analysis and Design, pp. 185–210. Springer, Berlin, August 2011
5. Hoffstein, J., Pipher, J., Silverman, J.H., Silverman, J.H.: An Introduction to Mathematical Cryptography, vol. 1. Springer, New York (2008)
6. Regev, O.: Lattice-based cryptography. In: Annual International Cryptology Conference, pp. 131–141. Springer, Berlin, August 2006
7. Micciancio, D., Regev, O.: Lattice-based cryptography. In: Post-quantum Cryptography, pp. 147–191. Springer, Berlin (2009)
8. Patidar, V., Sud, K.K., Pareek, N.K.: A pseudo random bit generator based on chaotic logistic map and its statistical testing. Informatica **33**(4) (2009)
9. Senkerik, R., Zelinka, I., Davendra, D., Oplatkova, Z.: Utilization of SOMA and differential evolution for robust stabilization of chaotic logistic equation. Comput. Math. Appl. **60**(4), 1026–1037 (2010)
10. Levy, D.: Chaos theory and strategy: theory, application, and managerial implications. Strateg. Manag. J. **15**(S2), 167–178 (1994)
11. Mandangan, A., Kamarulhaili, H., Asbullah, M.A.: On the smallest-basis problem underlying the GGH lattice-based cryptosystem. Malays. J. Math. Sci. **13**, 1–11 (2019)

Matrix Application in Engineering Problems

Ahmed Abdulkareem Hadi[(⊠)]

Mathematics Science, Ministry of Education, Directorate of Education, Baghdad, Al-Rusafa, Iraq

Abstract. Matrix is one of the most important pillars of mathematics. Matrix theories are used to solve many engineering problems in different fields such as Steganography, Cryptography, and Wireless Communication. In this paper, the general concept of matrices and their theories that contribute many engineering sciences were presented. The history of matrix is very important where it is as the cornerstone of the work on the matrices in their current form, so review of the most important developments in the science of matrices in this research is our concern. There are various applications of engineering science in many directions in daily life. Therefore, the use of concepts mathematical theories in solving engineering problems of the most important features of applied mathematics. It is necessary to apply the mathematical concepts of integration, differentiation, matrices and determinants, which are useful in helping other sciences to solve linear and simultaneous equations.

Keywords: Cryptography · MIMO · OFDM · Determinants · Matrix · Wireless communication

1 Introduction

Many mathematical expressions and theories are used in many life applications and contribute to a lot of science. Such as partial differential equations which have a vital role in Heat Equation, Wave Equation, Laplace's Equation, a) Helmholtz's equation, b) Schrodinger's equation, and etc. Matrices are also used in many applications, especially engineering applications, since engineering is one of the pillars of life in general. Matrices contribute to, but are not limited to, software engineering where encryption and decryption, mechanical engineering in equations of quantum mechanics, electrical engineering in electrical circuits, voltage laws, wireless communications, graph theory, analysis and geometry etc. Matrix science and its applications are the most important rules that provide science with many practices and principles that help in solving them problems. The concept of the matrix must be clarified, and the definition to represent the meaning of matrices. Matrices with their theories and rules acquire many definitions according to their uses. The word matrix is of Greek origin and its meaning in Greek is very similar to its meaning in English where it means the thing that produced or formed it [1].

Simultaneous equations are the most important ranges in which matrices are used. It made the scientists emphasize that the simultaneous equations are best applied to the science of matrices. Matrices are the solution method for studying simultaneous

D.-T. Tran et al. (Eds.): ICISN 2021, LNNS 243, pp. 589–598, 2021.
https://doi.org/10.1007/978-981-16-2094-2_69

equations and obtaining values for solution. The history of matrices goes back to the Babylonian scientists were able to solve linear equations in two unknowns (system 2X2) and then later the Chinese came in a solution in system 3X3. However, the use of matrices was not applied optimally in that period until the late 19th century [2].

In order for scientists to develop algebra of matrices, it was necessary to rely on the documentation of the algebraic process used in the solution. The notation of the process helps to describe algebraic operations on equations and thus prepares the way to the beginning of the existence of mathematical operations on the matrix such as adding, subtracting or multiplying two or more matrices. Array science came with a view to a broader understanding of algebraic science in order to study the determinants and perform algebraic calculations. In general, algebraic science became more important and influential after the discovery of matrices. In 1848 Sylvester launched the term matrix and its definition [3].

Euler applied conditions to solve equations containing more than one unknown and began with square determinants in which the number of unknowns is equal to the number of equations. With the passage of time came the work on linear equations to address the matrices and their symbols was done through Gauss. Gauss's efforts focused on dealing with variables, numbers, and different equations, which came to us in the form of the Gaussian elimination. The concepts for this method are based on swapping and multiplication and the addition of rows of matrices. All mathematical operations on columns and rows of matrices are aimed at solving equations and finding values for unknowns [8].

After reaching the form of the determinants and making calculations on them Tucker reviewed it and considered that the matrix is the engine and generator of the determinants. Later Arthur Kylie worked on a lot of concepts and principles related to the matrix algebra. One of the most important definitions reached by Kylie and the clearest way to implement them on the matrices was multiplication and how to multiply the matrix in the other and rely on columns and rows to get a true multiplication of the matrices [10].

With the development of technology, thanks to the contributions of the scientific community, the general principles and theories of matrices have become clearer. Mathematical application began to spread in many other sciences related to physics, engineering and chemistry and matrices had an important and effective role in those sciences. Matrices are a system for solving many scientific problems and are prepared for many modern scientific techniques such as coding and linear coordination systems and their role in solving equations of wireless communication.

On the other hand, the matrices also existed in recreational works that rely on games such as the game of Sudoku. The key to all those applications in which the practices and principles of matrices are used was to understand the algebraic history of them. The history of matrices and their applications is the main focus on which many developments in linear algebra depend, the existence of a firm and a clear history of science contributes to the positive contributions that occur in its use in the present era [7].

Encryption is one of the most important achievements in modern technologies where it ensures the passage of important and sensitive information safely. The equations of wireless communication is one of the most important questions posed by the scientific community continuously as it is used in many applications, therefore, the role of matrices in wireless communication is important and effective. The aims of this paper is to demonstrate the role of matrices in wireless communication, matrix application in cryptography and steganography and highlight the role of mathematics in contributing to the solution of engineering problems.

2   Matrix and Uses

Matrices are the order of elements whether numbers or symbols in the form of rows and columns the matrix name (such as 3×2, 3×4 and so on) depends on the number of rows and the number of columns it contain. The matrices have a vital and active role in many applications especially in engineering problems such as electrical circuits, optics and quantum mechanics. The power outputs in the battery are calculated using matrix solution and it has a role in the equations of Kirchhoff's law of current and voltage. A new role for matrices has emerged in dimensional triple projection into two dimensions, which is used to create a realistic decomposition motion. In addition, Google Inc. uses matrices to rank web pages. Matrices have many benefits in areas of geology such as the matrix system is used for seismic surveys and graphs. Furthermore, it is used by many scientists in scientific research in the parts of statistical analysis. The calculations of the matrices have an important role in the processes of automation and robotics where reliance on the matrices components of the columns and rows in the control panel of the movement of robots [4].

3   Matrix Role in Cryptography

Cryptography began by Lester Hill in 1929 where he invented Hill Cipher. Hill Cipher used to deal with more than one code, but by increasing the number of codes, the decryption process becomes more difficult. Hill relied on the configuration of the encryption process on the key matrix. The key matrix was used to encrypt text messages and its inverse to decrypt. It is important to maintain the confidentiality of the key matrix and its inverse between senders and recipients because by knowing the key matrix anyone can decrypt or send incorrect encrypted messages [16].

Not all matrices serve as a key matrix in the Hill encryption process. First, the matrix must have an inverse, in other words, the determinant is not equal to zero. Key matrix operators (factors) must not be shared with the modular base factors for encryption and decryption. The modular rule is determined by the number of symbols in the encryption process. For example, if the encoding process targets English latters in this case, the modular rule is 26, the number of letters from A to Z. Therefore, the key matrix determinant is not divisible by 13 or 2 and must add a number of other symbols to the extent that the modulus becomes prime. It should be noted that it is possible to use any square matrix in encryption, but if the previous conditions are not met in the matrix, it is not suitable to be a key matrix, even if it is square and is not suitable for decryption.

To show how to use matrices to encode and decode messages. The key matrix must be square like 2×2 or 3×3, and meet the above conditions. For example, the key matrix 2×2 is used with message want to encrypt is: SELL NOW. Every letter is equal a number as A = 1, B = 2 and so on as shown in Table 1:

Table 1. Numbers of each Alphabetic letters

Letter	Number	Letter	Number	Letter	Number
A	1	K	11	U	21
B	2	L	12	V	22
C	3	M	13	W	23
D	4	N	14	X	24
E	5	O	15	Y	25
F	6	P	16	Z	26
G	7	Q	17		27
H	8	R	18	-	0
I	9	S	19		
J	10	T	20		

Table 2. The corresponding numerical values for each letter in the message

S	E	L	L	-	N	O	W
19	5	12	12	0	14	15	23

From Table 2 these numerical values are not a fixed form that expresses the letters in the sense that any numerical value can be changed provided the sender and consignee are familiar with it. The message is in the form of a matrix:

$$\begin{bmatrix} 19 & 12 & 0 & 15 \\ 5 & 12 & 14 & 23 \end{bmatrix}$$

Both the sender and the consignee agree on a key matrix that has the aforementioned attributes and contains any numerical values (random) such as:

$$\begin{bmatrix} 3 & 5 \\ 1 & 2 \end{bmatrix}$$

Later the message matrix is multiplying in the key matrix (coding matrix) as:

$$\begin{bmatrix} 3 & 5 \\ 1 & 2 \end{bmatrix} \begin{bmatrix} 19 & 12 & 0 & 15 \\ 5 & 12 & 14 & 23 \end{bmatrix} = \begin{bmatrix} 82 & 96 & 70 & 160 \\ 29 & 36 & 28 & 61 \end{bmatrix}$$

This is sending in list as: 82, 29, 96, 36, 70, 28, 160, and 61 (this converts to encryption matrix). In decrypting a message, the inverse of the key matrix or code matrix are used and presented in Matrix below:

$$\begin{bmatrix} 2 & -5 \\ -1 & 3 \end{bmatrix}$$

Then, the inverse matrix is multiplying in encryption matrix

$$\begin{bmatrix} 2 & -5 \\ -1 & 3 \end{bmatrix} \begin{bmatrix} 82 & 96 & 70 & 160 \\ 29 & 36 & 28 & 61 \end{bmatrix} = \begin{bmatrix} 19 & 12 & 0 & 15 \\ 5 & 12 & 14 & 23 \end{bmatrix}$$

$$\begin{bmatrix} 19 & 12 & 0 & 15 \\ 5 & 12 & 14 & 23 \end{bmatrix} = \begin{bmatrix} S & L & - & O \\ E & L & N & W \end{bmatrix} = \text{SELL NOW}$$

In the previous example, the simple mechanism are used to apply the key matrix and also the size of the message is not large, but as the number of rows and columns increases for both the code matrix and the message matrix, the process becomes more complex. So working on an encrypted message has a key matrix 10×10 become complex. In addition, increasing the rank of the matrix makes the encryption process complicated, making it more secure due to the simple addition to the key matrix offers a large number of options. For example, a matrix 3×3 can provide 5 trillion coded codes, making it difficult to track them.

Despite the complexity associated with the encryption process, it was not difficult to penetrate, especially with the development of technology and the presence of computers that allow the testing of many key arrays until the correct matrix is found. With the recognition of the key matrix, any entity can track and penetrate encrypted messages by saving a lot of time and effort. It is worth mentioning that this nucleus on which cryptography was built helped to preserve the confidentiality of the information and the transparency of sources. Nowadays, Hill's method is no longer the best way to encrypt, but it was the beginning. Recently, modern software and techniques have used the matrix system and calculations and algebraic, which specializes in the expansion and development of cryptography [9].

4 Matrix Role in Wireless Communications

Wireless communication is one of the fastest growing and expanding sectors. Cellular communication is one of the most important forms of wireless communication used by many around the world and is built on a lot of applications on phones and tablets and so on. Mobile phones have a range of sensors and micro devices that process a huge amount of different data with high accuracy and less time. Therefore, the requirements of these wireless systems pose a lot of technical challenges, especially in the signal schemes. So in this section the role of matrices will be discussed in two of the signal schemes they are multiple-input and multiple-output (MIMO) and Orthogonal frequency-division multiplexing (OFDM). Both MIMO and OFDM are the most important signal schemes as they are relied upon in many wireless connections to improve quality and performance [6].

Shannon defined channel capacity as the maximum rate at which data can be transmitted with a small error rate. With the development arrived in the current days that the rate of data transmission increases with acceptable error rates and is disastrous. Those are important points in wireless communication technology [12]. For example, the evolution of Long Term Evolution LTE is two 4G Long Term Evolution LTE – Advanced increase in downlink and uplink data rates. In the downlink, data rates have increased from 150 Mbps to 1 Gbps. For the uplink, data rates have increased from 75 Mbps to 500 Mbps [11].

In theory, many ways to improve channel capacity, for example, reduce noise power, multiplex, increase signal power, exploit frequency diversity, area, etc. For the practice it is not so easy or diverse, due to the limited energy and noise that increase their sources and is expected to change their characteristics in the future. Therefore, the scientific community began to focus on diversity and multiplexing as a means of improving channel capacity. It was found that the most effective way to achieve distinct spatial transmission was to rely on multiple transmit and multiple receivers. The two systems can be discussed as follow (Fig. 1).

4.1   MIMO

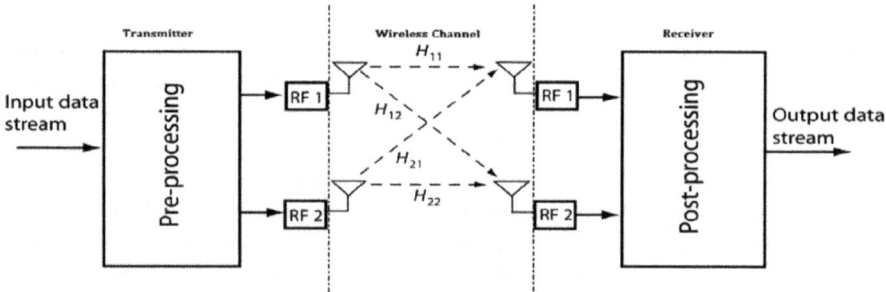

Fig. 1. MIMO system

MIMO can be defined as a system that receives multiple inputs and sends multiple outputs, it is a system based on multiple antennas. In the MIMO model, the transmit system will be expressed in At and the receive system will be expressed in Ar. MIMO model is depicted in an algebraic matrix as follows:

$$
\begin{bmatrix} y_1 \\ \vdots \\ yA_r \end{bmatrix} = \begin{bmatrix} b_{11} & \cdots & b_{1N_t} \\ \vdots & \ddots & \vdots \\ b_{A_r1} & \cdots & b_{A_rA_t} \end{bmatrix} \begin{bmatrix} x_1 \\ \vdots \\ x_{A_t} \end{bmatrix} + \begin{bmatrix} v_1 \\ \vdots \\ v_{A_r} \end{bmatrix}
$$

Where:

$$y = Bx + v \tag{1}$$

B: $[b_{ij}]$ propagation channel, gain from j to i (transmitter to receiver).
x: $A_t * 1$, transmitted data dimensional vector.
y: $A_r * 1$, received data dimensional vector.
V: noise vector.
The following algebraic equation can therefore be inferred:

$$y_i = \sum_{j=1}^{A_t} b_{ij} x_j + v_i, i = 1, 2, \ldots, A_r \tag{2}$$

Hence, it is difficult to retrieve the symbols for the transmitted data where the output symbol (yi) is an input symbol linear combination. It is referred to as interference between symbols (ISI - intersymbol interference), which causes the aliasing of signal. Depending on the singular value decomposition (SVD), the MIMO channel can decompose into independent paths from signals. This feature ensures channel enhancements and successful multicast transmission. Matrix-based software has made solving equations aimed at channel optimization easier [13, 14].

4.2 OFDM

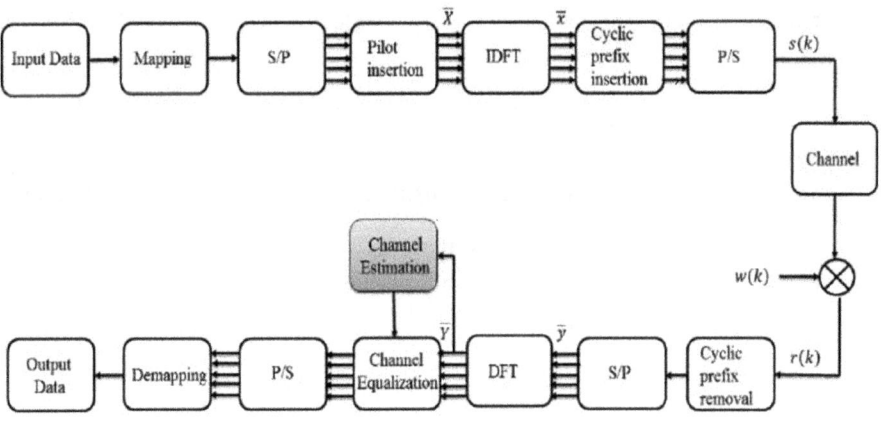

Fig. 2. OFDM system

Orthogonal frequency-division multiplexing is one of the most important technologies on which wireless communication depends. It was the key signaling technology for the recently approved long-term evolution LTE. It appeared to be relied on in many standards for wireless communication systems such as [15] (Fig. 2):

- Digital Video Broadcasting (DVB).
- Wireless Local Area Network (WLAN).
- Wireless Personal Area Network (WPAN)
- Wireless Metropolitan Area Network (WMAN)
- Digital Audio Broadcasting (DAB).
- Wireless Regional Area Network (WRAN) [5].

OFDM is a multi-band system that allows it to divide the spectrum into various narrow frequency bands, through which the data is transmitted by a parallel way. To represent an OFDM system must be based on several concepts, including:

- H: it is transfer matrix.
- x: it is input vector.
- y: it is output vector.
- v: it is noise vector.

$$\text{so,} \qquad y = Hx + v \tag{3}$$

The system is concerned with the implementation of a discrete Fourier transform (DFT) on the receiver aspect and an inverse discrete Fourier transform (IDFT) on the transmitter aspect. In practice, a fast Fourier transform (FFT) algorithm is adopted to implement (DFT) and an inverse fast Fourier transform (IFFT) algorithm is adopted to implement (IDFT). It can be representing in the follow equations:

$$\text{For } x[n] \text{ point in DFT: } X[k] = \frac{1}{\sqrt{N}} \sum_{n=0}^{N-1} x[n] e^{\frac{-j2\pi nk}{N}}, \ k = 0, 1, \ldots, N-1 \tag{4}$$

$$\text{In IDFT : } x[k] = \frac{1}{\sqrt{N}} \sum_{k=0}^{N-1} x[n] e^{\frac{j2\pi nk}{N}}, \ n = 0, 1, \ldots, N-1 \tag{5}$$

So we can describe the DFT process as matrix multiplication in the following form:

$$X = Wx, \qquad \text{where W is an unitary matrix}$$

$$W = \frac{1}{\sqrt{N}} \begin{bmatrix} 1 & 1 & 1 & \cdots & 1 \\ 1 & w & w^2 & \cdots & w^{N-1} \\ 1 & w^2 & w^4 & \cdots & w^{2(N-1)} \\ \vdots & \vdots & \vdots & \ddots & \vdots \\ 1 & N^{N-1} & w^2(N-1) & \vdots & w^{(N-1)^2} \end{bmatrix}$$

So we can describe the IDFT process in the following form:

$$x = W^{-1}X = W^H X, \tag{6}$$

Where H: it is discrete channel.

Referring to the overall form of the inputs and outputs in the form of a matrix as follows:

$$
\begin{bmatrix} y_{N-1} \\ y_{N-2} \\ \vdots \\ y_0 \end{bmatrix} = \begin{bmatrix} h_0 & h_1 & \cdots & h_M & 0 & \cdots & 0 \\ 0 & h_0 & \cdots & h_{M-1} & h_M & \cdots & 0 \\ \vdots & \ddots & \ddots & \ddots & \ddots & \ddots & \vdots \\ 0 & 0 & 0 & h_0 & \cdots & h_{M-1} & h_M \end{bmatrix} \begin{bmatrix} \widetilde{x}_{N-1} \\ \vdots \\ \widetilde{x}_0 \\ \widetilde{x}_{-1} \\ \vdots \\ \widetilde{x}_{-M} \end{bmatrix} + \begin{bmatrix} v_{N-1} \\ v_{N-2} \\ \vdots \\ v_0 \end{bmatrix}
$$

From the above we conduct the following matrix of OFDM system:

$$
\begin{bmatrix} y_{N-1} \\ y_{N-2} \\ \vdots \\ y_0 \end{bmatrix} = \begin{bmatrix} h_0 & h_1 & \cdots & h_M & 0 & \cdots & 0 \\ 0 & h_0 & \cdots & h_{M-1} & h_M & \cdots & 0 \\ \vdots & \ddots & \ddots & \ddots & \ddots & \ddots & \vdots \\ 0 & \cdots & 0 & h_0 & \cdots & h_{M-1} & h_M \\ \vdots & \ddots & \ddots & \ddots & \ddots & \ddots & \vdots \\ h_2 & h_3 & \cdots & h_{M-2} & \cdots & h_0 & h_1 \\ h_1 & h_2 & \cdots & h_{M-1} & \cdots & 0 & h_0 \end{bmatrix} \begin{bmatrix} x_{N-1} \\ x_{N-2} \\ \vdots \\ x_0 \end{bmatrix} + \begin{bmatrix} v_{N-1} \\ v_{N-2} \\ \vdots \\ v_0 \end{bmatrix}
$$

Wireless communication invades the world with modern technologies and its ability to send and receive a huge amount of data and processing accurately and high speed with a small error rate. So this technology (OFDM) creates a wide range of precise performance and enhanced channel capacity that are useful in wireless communications.

It is worth mentioning that there are a wide range of types of matrices that contribute many applications, these types include:

1. unit matrix
2. inverse matrix
3. diagonal matrix
4. complex matrix
5. orthogonal matrix
6. symmetric matrix
7. triangular matrix.

5 Application of Matrices in Engineering Science and Other Science

- Statistics and Probability.
- Vectors of sun which use in solar energy generation.
- Vibrating system.
- Optimal quantum measurements.
- Gaussian Markov property.

6 Conclusion

Mathematical theories of all kinds, whether in integration, differentiation, algebra or any other branch of mathematics, have many applications and diverse in various sciences. Since matrices have been discovered and the evolution of their types and the mastery of algebraic processes that occur on them has become one of the most important pillars of modern science. Matrices are the cornerstones of encryption and decryption, which built on the idea of working a lot of current software that contributed significantly to the security, confidentiality and transparency of information circulated between two parties, institutions or two countries.

Matrices have an important and effective role in wireless communication, which is one of the most important strengths of modern times. The matrix contributes to the most important channel capacity development systems represented by the two systems MIMO and OFDM. These system use matrix to improve performance and reach the best possible power in sending and receiving data. It Infer, Matrices are important algebraic rules and are stand-alone sciences involving many life and practice applications.

References

1. Ando, T.: Concavity of certain maps on positive definite matrices and applications to Hadamard products. Linear Algebra Appl. **26**, 203–241 (1979)
2. Ando, T.: Totally positive matrices. Linear Algebra Appl. **90**, 165–219 (1987)
3. Bhatia, R.: Matrix Analysis, vol. 169. Springer (2013)
4. Bôcher, M., Duval, E.P.R.: Introduction to Higher Algebra. Macmillan (1922)
5. Aarab, M.N., Chakkor, O.: MIMO-OFDM for wireless systems: an overview. In: International Conference on Artificial Intelligence and Symbolic Computation, pp. 185–196 (2019)
6. Goldsmith, A.: Wireless Communications. Cambridge University Press (2005)
7. Hiai, F., Petz, D.: Riemannian geometry on positive definite matrices (2008)
8. Horn, R.A., Johnson, C.R.: Matrix Analysis. Cambridge University Press (2012)
9. Knobloch, E.: From Gauss to Weierstrass: determinant theory and its historical evaluations. In: The Intersection of History and Mathematics, pp. 51–66. Springer (1994)
10. Petz, D., Hasegawa, H.: On the Riemannian metric of α-entropies of density matrices. Lett. Math. Phys. **38**(2), 221–225 (1996)
11. Sesia, S., Toufik, I., Baker, M.: Introduction to LTE-Advanced (2011)
12. Shannon, C.E.: A mathematical theory of communication. Bell Syst. Tech. J. **27**(3), 379–423 (1948)
13. Wang, X., Serpedin, E., Qaraqe, K.: A variational approach for assessing the capacity of a memoryless nonlinear MIMO channel. IEEE Commun. Lett. **18**(8), 1315–1318 (2014)
14. Wang, X., Serpedin, E.: An overview on the applications of matrix theory in wireless communications and signal processing. Algorithms 9(4), 68 (2016)
15. Glover, I.A., Atkinson, R.: Overview of wireless techniques. In: Wireless MEMS Networks and Applications, pp. 1–33 (2017)
16. Zhan, X.: Matrix Inequalities. Springer (2004)

Online Failure Prediction Model for Open-Source Software System Based on CNN

Sundos Abdulameer Hameed. Alazawi[1,2(✉)]
and Mohammed Najm Abdullah[1,2]

[1] Department of Computer Engineering, University of Technology,
Baghdad, Iraq
120002@uotechnology.edu.iq
[2] Department of Computer Sciences, Mustansiriyah University, Baghdad, Iraq
ss.aa.cs@uomustansiriyah.edu.iq

Abstract. Fault injection tools are one of the most important methods used in assessment software dependability, as the fault injection tool speeds up the presence of a fault that causes software and systems failure, which helps in evaluating fault tolerance or the possibility of predicting system failure before the failure occurs. Defects and bugs classification are one of the most significant phases followed in assessing the system and software reliability as the accuracy of the defects classification can affect the evaluation of the reliability of the systems, improve the software performance, and reduce the cost of evaluating the software by regular methods. Based on this, the FIBR model was designed and used in order to evaluate the reliability of open source software, and to build a CNN model for failure software prediction based on bug reports of the open source software referred to in this work. An approach to predict online failure is presented in this work for some open-source software as a target system using the Convolution Neural Network. The steps required to simultaneously predict bug report injections are performed to simulate open source software and then collect the data that is generated by monitoring system behavior states. Data on system behavior during bug report injection operations is added to the original Data Set.

Keywords: Online failure prediction · Open-source software · Convolutional neural networks · Bug classification · Fault injection

1 Introduction

Users are increasingly relying on open source software, which have become one of the most important daily communication software in the world. The serious impact on the basic services of open source software and the great spread of these systems made an urgent need to assess the dependability of these software and their ability to perform tasks and services in an efficient with more reliable manner, and to deal with the fault tolerance principle without affecting the reliability of the performance of the tasks provided by these software and systems [1, 2].

© The Author(s), under exclusive license to Springer Nature Singapore Pte Ltd. 2021
D.-T. Tran et al. (Eds.): ICISN 2021, LNNS 243, pp. 599–608, 2021.
https://doi.org/10.1007/978-981-16-2094-2_70

Errors detection and failure prediction algorithms play important roles through the software dependability assessment because one of the goals of dependability assessment are to find defects or bugs. The detecting and repairing faults costs appears one of the most expensive software development operations [3, 4] Therefore, fault injection techniques are used in order to accelerate the occurrence of failure, thus reducing the time required in testing the system and its behaviour during the development phases. [5, 6].

The most recent research conducted in the field of failure prediction based on the system that has been tested are the methods and machine learning techniques used, and certainly the type of data used by predictors. The studied research and literature based on the type of data used in the prediction, for example, data of a failure tracking type is used. Singh and Shrish (2014) [7] have used classification based on clustering for fault software prediction systems on the dataset collected by failure tracking for NASA MDP, and evaluate this method by performing a general relative analysis in software fault prediction results for the same datasets. They have suggested a model to gain a probability of fault detection equal to 83.3% and balance rates equal to 68.5%.

Ivano (2016) [8], use symptom data for failure prediction goal depend on an existing fault injection tool (GSWFIT) implemented at Coimbra University. The SVM technique is implemented for the failure prediction using a sliding window of 3 s. In his thesis, the model for online failure prediction is presented depending on failure data generation continuously by a virtualized copy of the target system; the data collected is a Symptom data type. In his studies, the target system depends on the Windows XP OS running several workloads for simple file. F-measures of Apache Tomcat webserver workload are 0.549, 0.611, 0.747, and 0.854 when sliding windows are 10, 20, 30, 40, and 50 sequentially.

Detected errors reports data type is used in many research, Davari (2016) [9] builds a prediction framework and he employs some machine learning algorithms such as Naive Bayes, Random Forest, Decision Tree, and Logistic Regression. The dataset of his approach detects error reports type. Evaluation of his approach model is implemented on Mozilla Firefox and its defects file security. Davari takes the Bohrbug and Mandelbug terms and then introduce two terms: BohrVulnerability (BV) and MandelVulnerability (MV), his experiment shows that approach is capable to classify with an accuracy of 67% for 65% of BV files, while accuracy is equal to 75% to predict 84% of the MV files using Random Forest algorithm.

Based on bug report data type, Tóth et al. (2016) select 15 software of java from GitHub to created their dataset by statistical, the authors collected Lines of Code, Commits, and Bug Reports data type. They correspond to bugs that are familiar with classes and files of source code elements and measure a set of product metrics on it. They use several machine learning algorithms and obtained a good result for Random Forest, and Random Tree methods in F-measure among 0.7–0.8. Similarly, Kukkar et al. (2019) [10], present their prediction method using CNN and Random Forest. The method of classification proposed named BCR, F-measures of their method is equal to 96.43%. The results show that the approach increases performance for bug classification with state-of-the-art techniques.

In (2018) Luanzheng et al. [11], Bhardwaj [12] use most of machine learning methods for failure predication. Luanzheng et al. propose PARIS, a machine learning approach based on the rate prediction of manifestations of transient faults. They use pattern-based method to collect reporting of detected error data type and train 100 small computation kernels and test on 25 big benchmarks such as Myocyte, Backprop, Kmeans, etc. They test SV, Gradient Boosting, Random Forest, Kneighbors, and NuSVR regression models and find the Gradient Boosting Regression the best machine learning model for predicting the rate of manifestations of transient faults in terms of prediction accuracy. PARIS provides accuracy among 82% and 77% on prediction average, while the state-of-the-art prediction model cannot predict them (38% vs. -273%). Bhardwaj uses Random forest, linear regression, Neural network, and SVM for number of faults prediction that implemented on PROMISE repository of detected error reports data. RF and SVM predict moderate results. With respect to ARE measure, SVM and Linear regression generates the lowest median values, Random forest, and Neural Network produce moderate minimum values, SVM and Linear regression produces the highest minimum values, Linear regression produces the moderate maximum values (0.84).

Different approaches exist in the literature regarding failure prediction using deep learning algorithms such as Convolution Neural Network and Recurrent Neural Network. Stagge (2018) [13] has discussed CNN and RNN techniques implemented on Hierarchical Attention Network. In Hierarchical Attention Network the data is organized of word by word format, and the result calculates by word and line together. Crash reports dataset are gained for information prediction; this approach can be done using the users learning idea that currently works with software defect classification through crash decision. When the information is not specific, Stagge's system has accuracy about 0.51.

At last based on pattern data of CK features that extracted by statically test method, Fan et al. (2019) [14] suggest an approach based on deep learning technique to the prediction of defects through a recurrent neural network concept. Software and code are tested to predict code defects possibility in software, they use the attention technique for feature extraction to increase the performance of defect prediction. They have experimented on seven software as open-source: Camel, Lucene, Poi, Xerces, Jedit, Xalan, and Synapse. By implementing their method, F-measure for those open-source software sequentially is 0.515, 0.721, 0.764, 0.270, 0.560, 0.644, and 0.477. Similarly, Ke et al. (2016) [15] suggest a new approach based on Recurrent Neural Network. Their system detects early alert signals based on the log driven for information technology system failure prediction. Random Forest, SVM, and RNN are applied on the dataset that are collected from web server cluster system and mailer server cluster system. RNN has high confidence comparing to Random Forest and SVM, where Random Forest has 0.361, SVM has 0.513, while RNN has 0.662.

While an injection of real error data or code is very expensive, we have relied on training CNN model on open-source software bugs reports and their preparations to receive faults and predict them through these bugs reports during bug injected operations Therefore, new model for Fault Injection of Bug Report (FIBR-OSS) is designed for Open-source software depend on the bug reports database as fault library. Our approach of online failure prediction according to the reporting bugs dataset type that injected into four open-source software by our suggested software injector model (FIBR) in [16]. A new dataset that has been updated during monitoring, data generation, and bug reports dataset building, online failure prediction model is done by training of deep neural network represented by convolutional neural networks on failure states being predicted.

2   Failure Prediction

A software bug is what software engineers commonly use to term the release of an error in the software system. A fault or defect is then sorted as a failure which raises the behaviour of software differently from its normalization [17, 18]. Errors detection and faults prediction algorithms play important roles through the software dependability assessment because one of the goals of dependability assessment is to find defects or bugs [19, 20]. Software faults are more costly and time consuming, therefore, fault injection techniques are used [6, 21].

2.1   Online Failure Prediction

Online failure prediction is a mechanism that permits forecasting active failures occurred in a nigh future via system monitoring at runtime (that case is online) such as operating system state. Moreover, it uses prior information around the behaviour of a system (normal, failure, and so on), it is an approach that aims to foresee imminent problems by analysing the monitoring data collected from the system at runtime [8, 22].

Reporting of detected errors, evaluating a big number of faults and error reports in process in real software system or system services is a very interesting field [19]. Error reports are analysed to detect errors that have not still improved to turn out a failure [23, 24].

An Analysis of the error log can be done for:

Past event: Defining an event cause.
Future event: Significant events Prediction.

Several methods can be divided into two sets [19, 25]: Rule-based failure prediction, such as inferring some of the rules, every rule depends on error reports. And pattern-based methods, such as Co-occurrence of errors, Pattern recognition techniques, Statistical test, and classifiers.

2.2   (FIBR-OSS) Model

During the research process for study of injection tools that are published in our survey of related works [26], FIBR-OSS model for bug reports injection as faults, is proposed and implemented in our previous research [16]. Figure 1 shows the system environment FIBR model for bug report injection and new data generation using CNN model proposed.

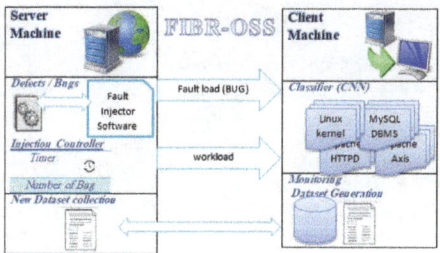

Fig. 1. System Environment for bug report injection and new data generation.

FIBR-OSS model is established using bug reports datasets dependent on a main study of bugs identify that can rationally be predictable to occur the failures states in OSS's as Linux kernel, MySQL DBMS, Apache HTTPD web server, and Apache AXIS WS.

3 Proposed Methodology

In this research, the online failure prediction method has been suggested, the method can be implemented by using the FIBR-OSS prototype as an injection model based on the dataset of bug's reports collected by the Mantis tracking system. There are two basic procedures that work in conjunction with the bug injection times that are system state classification and system response analysis, general structure of generation and building for new dataset during the failure prediction stages.

The failure prediction model collects a new dataset that is generated by the FIBR-OSS model. An online failure prediction process is done by training of deep neural network represented by convolutional neural networks on failure states being predicted. In specific, the monitored data is related to the failures state observed in the monitoring procedures with the failure prediction requirements.

3.1 Building Convolutional Neural Network (CNN)

The parameters used in the current work are Rectified Linear Unit (ReLU) activation function for all convolution layers and Softmax activation function for output prediction [27, 28]. For gradient descent optimization algorithms, Adam [29] is used to enhance the weights in each filter. The overall architecture is shown in Fig. 2. A flattening layer and dense layer with Softmax activation function to predict whether a source file is any of four classes of System State

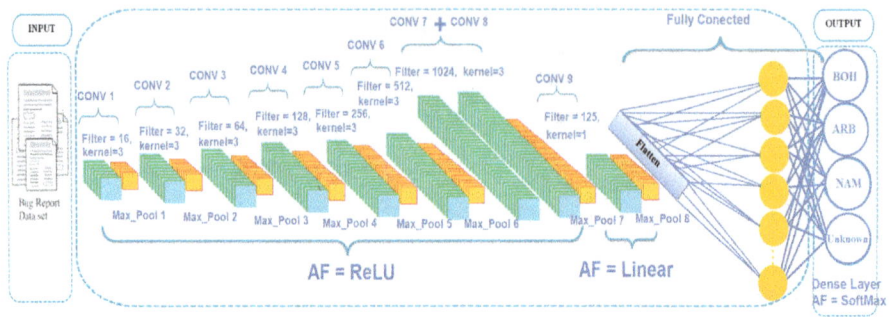

Fig. 2. CNN architecture in current work

3.2   New Dataset Generation

The monitored data are related to the failures state observed in the previous phase with the failure prediction requirements. The first stride at failure prediction is ability to correlate the monitored data with the effect of bugs on system states. Several type of system states may occur through the data generation and collection phase, Failure, Critical, Normal, and Unknown states.

During FIBR running, while CNN model predicted of failure states, the new attributes are fixed:

1. System State,

   - Failure, if fund complex defect makes a system fail
   - Critical, if fund less complex defect make a system in a critical state that must be doing some solution.
   - Normal, if fund a simple defect that not effect system execution,
   - Unknown, fund unknown bug.

2. Response of System based on system state, bug class, and subclass of bug.
3. Expected Solution is determined based on the system state and response of the system.

4   Excremental Results

New data is collected over different time periods, and the monitoring system collects the values of the variables that describe the state of the target system. Here, the prediction process by CNN works simultaneously with the bug report injection operations via the injector FIBR_OSS, and this is a basic stage in this approach, where data is collected during the implementation of the target system and the creation of new dataset and database contains of some reliability metrics that will be used in future. This coincides with the prediction of one of the system cases, extracting the system response to the detected failure with proposed solutions previously collected by monitoring system behavior and finding possible solutions to avoid that behavior so that it does not affect the workflow order as much as possible.

As shown in Fig. 3 bugs setting vary in each dataset for open-source software whose behavior has been analyzed and studied through its bug reports. We find that Axis_soap has the highest number of BOH type followed by MySql and then Appach_httpd_Web Server and Linux Kernel. As for the bug type NAM is more likely than other software, when Linux Kernel is followed by MySql until it is almost non-existent in Axis_soap and Appach_httpd_Web Server. The ARB bug type has the same effect as NAM. The type of bug is not defined in terms of the system response and consequently the lack of logical solutions unless its source is (known) UNK, whose effect is evident in the Linux Kernel and also Appach_httpd_Web Server.

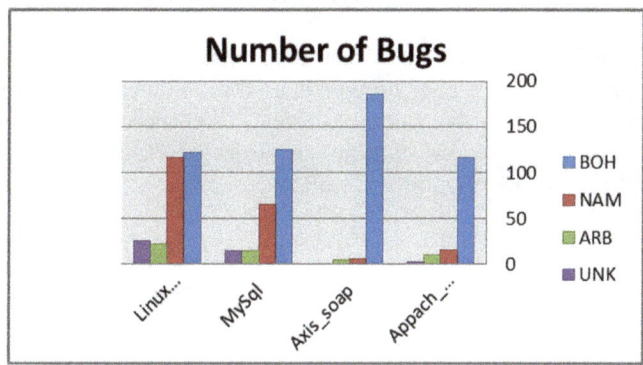

Fig. 3. Bug dataset analysis.

Over 117-min (7020 s) fault-injection campaign carried out within 30-day at random times, more than 1700 variables are collected, representing the system states and new attributes of expected behavior of that system from fault injection, injected a total of 1800 bug report, and collected 217 failures state. Given that the number of data collected and updated during data generation, forecasting of failure, system response, and expected solution are relatively large. We a new dataset results of four open-source software.

When system states are predicated for Linux Kernel, the generation of new dataset for Linux_Ext3-fs, Linux_ Ext3-fs has a little "unknown" state, while the "critical" state of the system is more if compared to cases of a "unknown" state, where "critical" state is causes by NAM and ARB bugs with subclass of TIM, MEM, and STO. Linux_ scsi drivers for example, has a little failures state in the case if the Bug-type is NAM with subclass is SEQ, while the "critical" state is caused by ARB bug with subclass is MEM, and by NAM bug with subclass is NAU.

While in Apache HTTPD_core an example, did not reach a state of "Failure", whereas, the critical situation is repeated when NAM bug with subclass NAU is inject. The rest of the states were of a "Normal" type despite the presence of a defects. In Mysql_replication dataset, failures state is not fund for all Bug-types, but the critical state of the system is a little in both bug type NAM and ARB, while the normal state is more compared to cases of a critical state, which is often caused by loading memory more than the real absorption. It is noted, here, that most states are normal when the Bug-type is BOH and this means exactly the principle of fault tolerance as the system continues to run despite the presence of faults until a fault occurs in which the system should be stopped. Finally, Axis_soap has one "critical" state caused by NAM bug type with subclass is NAU, while the normal state is a most if compared to cases of failure and critical states, which often causes by loading memory more than the real absorption.

In Failure prediction, the set of familiar measures depending on the generated confusion matrixes are used to calculate the values of these measures to find the disparity between actual and predicted value, the three most popular measures evaluating a quality of prediction are Precision, Recall, and F-measure [30]. Accuracy for our model has 0.88 for all bug reports types, and without Unknown bug reports the model have 0.965.

Studied research and literature is based on the type of data used in the prediction, for example, data of a failure tracking type is used in the prediction, failure tracking data, detected error reporting, symptom data. Table 1 appears compares results between current work and previous works depending on errors reports data because it's similar to bug report dataset that is used in this work.

Table 1. Comparing results between current work and previous works depending on errors reports dataset

Related work	Accuracy (%)	Machine learning algorithm
Davari (2016) [9]	0.75	Random Forest
T´oth et al. (2016) [31]	0.82	Random Forest
Kukkar et al.(2019) [10]	0.963	Convolution Neural Network
Luanzheng et al. (2018) [11]	0.85	Support Vector Machine
Bhardwaj, Stagge (2018) [13]	0.85	Linear Regression
Stagge (2018) [13]	0.51	Convolution Neural Network
Fan et al. (2019) [14]	0.764	Recurrent Neural Network
Ke et al. (2016) [15]	0.662	Recurrent Neural Network
Current work	0.965	Convolution Neural Network

5   Conclusion

Online failure prediction models based on software behaviour observations during fault injection operations, which faults can be expressed by bug class and subclass of bugs dataset file.

The choice of data type used in this work depends on the mechanism of our fault injection model for bug reports in open-source software (FIBR-OSS).

To predict failure, a new CNN model were trained. The use of general fault report data to classify defects is a 96.25%. CNN Execute a failover concurrently with bug reports using the FIBR-OSS model on CentOS Linux using a CNN model with a new dataset showing states of system behaviour, it is an acceptable model for building an online failure prediction model.

Acknowledgment. The author's thankful Department of Computer Science/Collage of Science in Mustansiriyah University, and Department of Software Engineering in University of Technology, for supporting this work.

References

1. Ullah, N., Morisio, M., Vetro, A.: A comparative analysis of software reliability growth models using defects data of closed and open source software. In: 2012 35th Annual IEEE Software Engineering Workshop, pp. 187–192 (2012)
2. Natella, R.: Achieving Representative Faultloads in Software Fault Injection. University of Naples Federico II, Italy (2011)
3. Naidu, M.S., Geethanjali, N.: Classification of defects in software using decision tree algorithm. Int. J. Eng. Sci. Technol. **5**(6), 1332 (2013)
4. Kalaivani, N., Beena, R.: Overview of software defect prediction using machine learning algorithms. Int. J. Pure Appl. Math. **118**(20), 3863–3873 (2018)
5. Tamura, Y., Yamada, S.: Reliability and maintainability analysis and its toolbased on deep learning for fault big data. In: 2017 6th International Conference on Reliability, Infocom Technologies and Optimization (Trends and Future Directions) (ICRITO), pp. 106–111. IEEE (2017)
6. Feinbube, L., Pirl, L., Polze, A.: Software Fault Injection: A Practical Perspective, Dependability Engineering, IntechOpen (2017)
7. Singh, P., Verma, S.: An efficient software fault prediction model using cluster based classification. Int. J. Appl. Inf. Syst **7**(3), 35–41 (2014)
8. Irrera, I.: Fault Injection for Online Failure Prediction Assessment and Improvement (2016)
9. Davari, M.: Classifying and Predicting Software Security Vulnerabilities Based on Reproducibility. Queen's University, Canada (2016)
10. Kukkar, A., Mohana, R., Nayyar, A., Kim, J., Kang, B.-G., Chilamkurti, N.: A novel deep-learning-based bug severity classification technique using convolutional neural networks and random forest with boosting. Sensors **19**(13), 2964 (2019)
11. Guo, L., Li, D., Laguna, I.: PARIS: Predicting Application Resilience Using Machine Learning. arXiv preprint arXiv:1812.02944 (2018)
12. Bhardwaj, H.: Software Fault Prediction using Machine Learning Techniques, Computer Science and Engineering Department, Thapar Institute of Engineering and Technology (2018)
13. Stagge, V.: Categorizing Software Defects using Machine Learning. LU-CS-EX 2018-17 (2018)
14. Fan, G., Diao, X., Yu, H., Yang, K., Chen, L.: Software defect prediction via attention-based recurrent neural network. Sci. Program. **2019**, 1–15 (2019)
15. Zhang, K., Xu, J., Min, M.R., Jiang, G., Pelechrinis, K., Zhang, H.: Automated IT system failure prediction: a deep learning approach. In: 2016 IEEE International Conference on Big Data (Big Data), pp. 1291–1300. IEEE (2016)
16. Al-Salam, M.N., Alazawi, S.A.: FIBR-OSS: fault injection model for bug reports in open-source software. Indones. J. Electr. Eng. Comput. Sci. (IAES Publisher) **20**(1), 465–474 (2020)

17. Pushpalatha, M., Mrunalini, M.: Predicting the Severity of Closed Source Bug Reports Using Ensemble Methods, Smart Intelligent Computing and Applications, pp. 589–597. Springer, Heidelberg (2019)
18. Otoom, A.F., Al-Shdaifat, D., Hammad, M., Abdallah, E.E., Aljammal, A.: Automated labelling and severity prediction of software bug reports. Int. J. Comput. Sci. Eng. **19**(3), 334–342 (2019)
19. Tran, H.M., Le, S.T., Van Nguyen, S., Ho, P.T.: An analysis of software bug reports using machine learning techniques. SN Comput. Sci. **1**(1), 4 (2020)
20. Subbiah, U., Ramachandran, M., Mahmood, Z.: Software engineering framework for software defect management using machine learning techniques with Azure. In: Software Engineering in the Era of Cloud Computing, pp. 155–183. Springer (2020)
21. Tamura, Y., Yamada, S.: Reliability and maintainability analysis and its toolbased on deep learning for fault big data. In: 2017 6th International Conference on Reliability, Infocom Technologies and Optimization (Trends and Future Directions) (ICRITO), pp. 106–111 (2017)
22. Pitakrat, T.: Hora: online failure prediction framework for component-based software systems based on Kieker and Palladio. In: KPDAYS, pp. 39–48 (2013)
23. Hernández-González, J., Rodriguez, D., Inza, I., Harrison, R., Lozano, J.A.: Two datasets of defect reports labeled by a crowd of annotators of unknown reliability. Data Brief **18**, 840–845 (2018)
24. Wu, L.L., Xie, B.G., Kaiser, E., Passonneau, R.: BugMiner: software reliability analysis via data mining of bug reports (2011)
25. Svendsen, P.A.: Online failure prediction in UNIX systems. Universitetet i Agder/University of Agder (2011)
26. Alazawi, S.A., Al-Salam, M.N.: Review of dependability assessment of computing system with software fault-injection tools. J. Southwest Jiaotong Univ. **54**(4), 1–15 (2019)
27. Skansi, S.: Introduction to Deep Learning: From Logical Calculus to Artificial Intelligence. Springer (2018)
28. Aggarwal, C.C.: Neural networks and deep learning, vol. 10. Springer (2018). 978-3
29. Jais, I.K.M., Ismail, A.R., Nisa, S.Q.: Adam optimization algorithm for wide and deep neural network. Knowl. Eng. Data Sci. **2**(1), 41–46 (2019)
30. Tanwar, S., Tyagi, S., Kumar, N.: Multimedia Big Data Computing for IoT Applications: Concepts, Paradigms and Solutions. Springer, Heidelberg (2019)
31. Tóth, Z., Gyimesi, P., Ferenc, R.: A public bug database of GitHub projects and its application in bug prediction. In: International Conference on Computational Science and Its Applications, pp. 625–638 (2016)

Mobile Cloud GIS Encrypted System for Emergency Issues

Ameer Yawoz Abd Al-Muna(✉), Abbas Abdulazeez Abdulahmeed, and Karim Q. Hussein

Computer Sciences Department, College of Science, University of Mustansiriyah, Baghdad, Iraq

Abstract. Today, there is significant growth and development in many technologies, especially computing and geographic information systems (GIS). Where these technologies enabled the capabilities to work in the field of maps, smart device applications and positioning systems. These capabilities, including the processing of huge amounts of various types of data through the use of cloud computing and enabling the identification of location information through smartphones, enabled GIS to spread very significantly. In our research, we propose a modern mobile GIS cloud computing system, also this system can shield and protect the data by one of the famous defence algorithms Advanced Encryption Standard (AES) 128 bit by the easiest way and consume little time. Also, we used Google spreadsheet as a cloud, which is easy in terms of use and programming, and it is locked, given its link to a Google account. We also presented some capabilities and limitations regarding the mobile device, as the application that we are working on is compatible with all types of devices and their various capabilities and takes into account the limited memory size, battery usage and many other things.

Keywords: Mobile · GIS · Cloud · Encryption · Management · Google spreadsheet

1 Introduction

Cloud computing for mobile geographic information systems is a kind of promising technologies in our time and it enables the cooperation of both mobile computing and geographic information system and according to the opinion of many researchers and developers that this technology will be used by many people and institutions via smart devices shortly, But such technologies need to be monitored and developed continuously. Cloud computing is a technology that has evolved dramatically and rapidly in our time and has been used by many fields, for example, industrial and commercial, to benefit from its services and the great potential this technology has. We mentioned previously that cloud computing and its services were used by many fields, most notably technology and architecture, especially in the areas of development and research and their integration with the geographic information system. This development caused a great era in the field of developing and updating applications and networks like 3G, which are quicker and secure [1]. AES algorithm is an example of

© The Author(s), under exclusive license to Springer Nature Singapore Pte Ltd. 2021
D.-T. Tran et al. (Eds.): ICISN 2021, LNNS 243, pp. 609–619, 2021.
https://doi.org/10.1007/978-981-16-2094-2_71

the common important and popular algorithms that are most significant used in various areas in the world. This algorithm allows the most active than DES and many other algorithms to protect the data. It is practised in several cryptography rules like Secret Socket Layer (SSL) to give the most powerful messages security among customer, Cryptography and Network Security [2].

The spreadsheet is created through an existing tab in a Google account, which is a similar interface that exists in the Microsoft group, which is an Excel sheet. After completing the settings for programming, we can fill in the fields or enter the data by linking to a specific application by contacting with a link and there is an important point which must be the name The document is identical to the name in the application or the program associated with it and it is important that the same document is not shared with several people to preserve sensitive data, but it can only be shared if the owner of the paper is allowed only to the people who want to send the document to them [3].

2   Related Works

The researchers built and developed applications in which the geographic information system was used in the field of dealing with emergency and critical cases in the health sector, and that was in the city of Mokattam in Cairo, Egypt, where the workers of this project relied on the proximity of the distance between the accident site and the nearest hospital and they used a set of databases and linked them With satellites, all of this is used through a website where all people can use it easily and from anywhere [4]. After lengthy studies, the researchers established systems that deal with different types of accidents based on geographic information systems in New Delhi India. The goal of this project is to be able to deal with the accident as soon as possible and to take an appropriate decision where traffic databases were used at a specific time and the goal is not to choose a long road And there is a lot of movement, which hinders the speed of the paramedics advance and reaching the accident site as soon as possible [5]. The researchers tried to plan a new way to solve sudden accident problems using GIS in Kumasi Metropolis, Ghana so that the rescue team could reach the accident site in the fastest and shortest way [6]. Water is the first source of life in the world, but at the same time, it is considered a major cause of death in the case of polluted water due to factories, waste, waste, etc. Dead pollution causes many dangerous and deadly diseases, including severe diarrhoea, poisoning and skin diseases, such things happen especially in poor regions and countries and because of the lack of necessary services and supplies, there are people who use river water for bathing and drinking and this water may be contaminated and cause them many diseases, so that is why The researchers and those concerned are collecting data and storing them in databases, as this data includes population sites near rivers and monitoring the sources of rivers and the causes of pollution through the use of geographical information systems for the purpose of monitoring and monitoring the sites more accurately Details. The collected data is studied and analyzed for the purpose of finding appropriate solutions and taking the necessary measures [7].

3   Cloud Computing

Cloud computing A modern technology that includes a group of types of computing and has been used in many areas recently, as this technology enables the possibility of storage and use of applications in the Internet and this leads to a reduction in costs and resources as it works regardless of the infrastructure (meaning it is separately). In this case, this technology (cloud computing technology) provides great services for mobile devices in terms of supporting applications such as video games, online commerce, image and video processing, and many other services. The more the number and requirements of devices increase, the greater the need for cloud computing [8]. But, unfortunately, such an evolution of these smart devices lack some of the main sources, for example, the processor speed, size and capacity of the battery and the lack of storage space, which makes such things hamper the handling of cloud computing capabilities that need great capabilities and capabilities [9].

3.1   Google Spreadsheet

Cloud computing is characterized by its ability to deal with various technologies related to the Internet, as there is a set of statistical techniques available at present, including cloud spreadsheets called Google Spreadsheet, Microsoft Excel-Web Application, and Zoho-Sheet. Where some outcome was explained after research and work by researchers and developers not to monitor the performance of these technologies and this leads to showing an error in the results as some techniques are difficult to find information and details that explain the way of working and the cause of this problem are the developers even if there is an update or development in A certain technique but without disclosing it and such things cross a big problem Spreadsheets It is a set of tools used in various types of calculations, statistics and data processing, as this method appears in the form of two-dimensional tables and usually. The results demonstrated that the collaboration between cloud computing and spreadsheet techniques led to great benefits in this area. The emergence of such a modern technology (cloud computing in spreadsheets) has an important purpose in the community and updating of data in the actual moment and the ability of users to conduct various operations on files at the same time. Users, including sharing information with great confidence. When a group of users shares information, especially in the case of sensitive information, this is a great risk if a simple error occurs, so there are such technologies that provide high security and reliable capabilities that are important in making decisions regarding certain questions [10].

3.2 Sheetgo Tool

It is one of the tools associated with the spreadsheet, which in turn is created through a Google Drive linked to a Google account (G-mail) and it is an automated tool that relies on the cloud where it enables users to share and transfer data between many spreadsheets, which is considered one of the basic features of the spreadsheet It has the ability to link more than one table together at the same time.

An important feature of this tool is that it is able to update data automatically and without any interference, which makes it one of the tools for real-time systems. In addition to the ability to send information to many tables associated with different accounts, it is one of the advantages of multiple clouds. This tool first appeared in the year 2010 and it is easy to use and does not need any code (just select condition and simple setting). One of the advantages of this tool is that it is able to filter and classify data in an easy and simple way based on certain conditions that can be controlled from the tool settings and this achieved automatically.

4 Geographic Information System

A GIS It is a system that provides a set of tools used in the field of different applications such as data analysis, dealing with maps and management, etc. It is used in spatial data and other and many applications where it can deal with data of different sources and provides the required treatment in advance and the presence of such an advanced technology has led to facilitating many tasks and shortening with time, reduce costs, etc. Location-based services (LBS) exploit the capabilities of some modern and widespread technologies such as (GPS) technology used in locating, routing and maps, but the existence of specific constraints restricts the progress of such technologies as the limited capabilities of devices and large amounts of data in networks that need to be addressed and security issues [11]. GPS uses spy-in-the-sky that round Earth to transfer information to GPS receivers that are on the earth. The information assists people to determine their place. Geographical Information System (GIS) is a software program that supports people to apply the information that is obtained from the (GPS) satellites. Google Maps has its geo-analytics abilities because Google Maps can deliver network analysis accurately, Google Maps facilitates transactions and driving questions. Nevertheless, network analysis is a popular GIS ability that Google is acting.

5 Mobile

The recent emergence of a large and varied group of portable smart devices and used by huge numbers of people in record time due to the services provided by these devices and they are easy to use and have different platforms such as iPhone and Android[1]. Among the most popular and widespread platforms among the rest of the brands is the Android platform [8]. Industrial analysts show that there were 2.9 billion Android apps in 2016 [9]. The success of Android platform is largely due to the open-source environment for developers, enabling them to sell Android powered devices in low costs

[11]. There are some data endurance paths to collect user and applications data in Android program. One of the advantages is SQLite, which is a light-weight relational [12, 13]. There are almost 3.6 billion smartphones in current use with millions of copies of SQLite in the world. SQLite is regularly used associated with other database generators [14].

6 Encryption Algorithm

The Advanced Encryption Standard (AES) algorithm is one of the block cypher encryption algorithms that was distributed by the National Institute of Standards and Technology (NIST) in 2000. (AES) an algorithm is one on the numerous common and broadly symmetric block cypher algorithm used in the world. The (AES) algorithm has the benefit of staying capable to protect extremely sensitive data located in several kinds of devices and applications. It is very hard to defeat by hackers to get the data when encrypted by AES algorithm. AES can deal with three various kinds of keys with size such as AES 128, 192 and 256 bit and all of this cyphers has 128-bit block size. The symmetric key is enough more powerful and quicker than Asymmetric. Some of the popular symmetric algorithms are the Advanced Encryption Standard (AES) [2]. In our research, we proposed AES 128 bit encryption to encrypt and decrypt data in the simplest way.

6.1 Comparison Between AES and Other Algorithms

There are many proposed algorithms such as Blowfish, AES, DES and 3DES. As one of the important things in testing the performance of the specialized algorithms in data security, encryption, and decryption of different types of files, a collection of methods for that was accepted similarly (P/II-266 MHz) and (P/4–2.4-GHz). According to the outcomes, Blow-fish can present the most reliable production matched to other algorithms and AES has a more dependable performance than 3-DES and DES. It furthermore produces that 3-DES 1/3 throughput of DES [15].

6.2 Algorithm's Architecture

To evaluate the performance of the algorithms, there are criteria set by a group of experts that are very important in measuring mission factors. For example, the algorithm must be uncomplicated and correspond to a large set of programs and is characterized by flexibility and efficient performance [16]. Range of the key, there are three various key sizes are used by AES algorithm to hide the data like (128bit, 192bit, 256bit). The key sizes choose to the number of rounds similarly AES with 10 rounds for 128-bit keys, with 12 rounds for 192-bit keys and with 14 rounds for 256-bit keys [17]. Figure 1 display the AES operations.

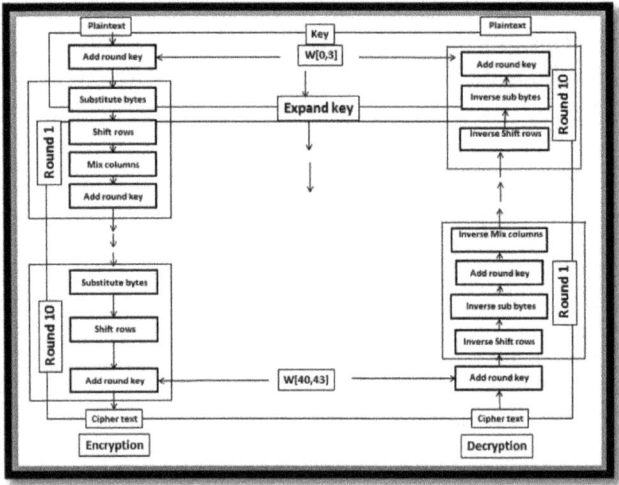

Fig. 1. Encrypt and decrypt process of AES

7   The Mobile Cloud and GIS System

This system depends on the same context as the systems that operate at several levels, they are similar in structure and engineering construction. Figure 2 illustrates the proposed mobile cloud and GIS system [12].

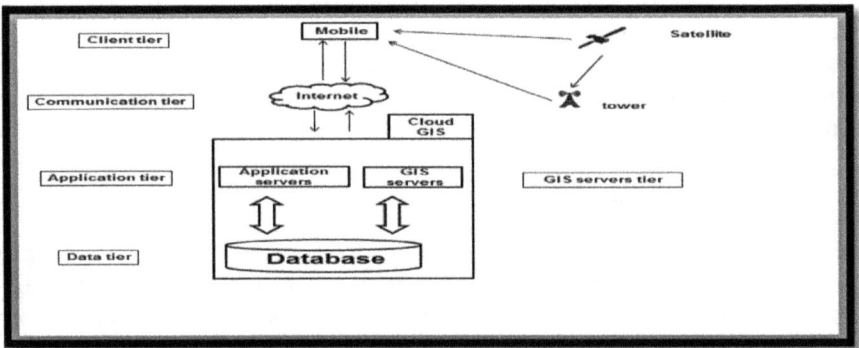

Fig. 2. Mobile Cloud and GIS system

8   Prototype Implementation

We have built an Android application, which is used for emergency, security and health purposes. Anyone can use this application with ease and report the accident status by simply sending a photo, phone number and the status of the situation and the coordinates of the accident to the competent authority in order to be Take appropriate action and solve the problem as soon as possible.

The information is sent to the cloud that is the spreadsheet technology, in this paper we used the spreadsheet technology as a cloud to store and manage the information. This technology is characterized by high security due to its association with a Google account (G-mail) and is known to be unable to break. In addition, we have encrypted the sent information using the advanced encryption standards (AES) algorithm as a precaution in case the information was stolen or it is tampered with or penetrated, or it is viewed by unauthorized persons and this causes major problems and affects the speed of responding to emergency situations, which leads to severe consequences such as death during transmission. It is possible through the management center to control the transmission of information to the categories of the competent authorities, and this means that cases related to injuries or diseases or health matters will be sent to the hospital. As for cases such as theft, kidnapping or security violations, they are sent to police stations etc. This is done through the use of a tool called (Sheetgo), where the received information can be classified, filtered and sent again to the specialized authority, depending on the type of complaint. This application is compatible with various types of Android devices, whether they are modern or old or with high capabilities or low and other things related to smart devices, in addition to that it does not drain a lot of memory due to its small size and does not need large processing operations which cost the CPU time For the phone, such applications are considered important to be established and developed due to the services they provide to the

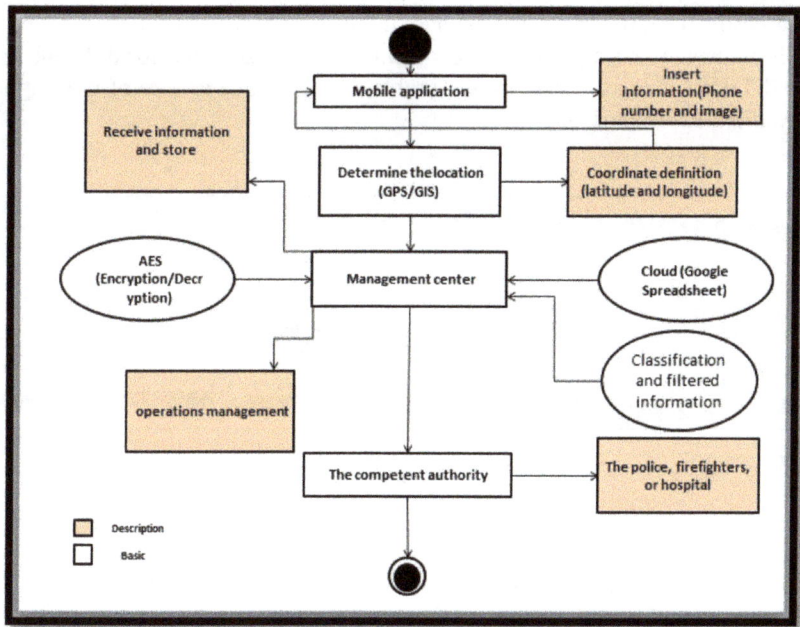

Fig. 3. Methodology diagram

community, especially in emergency situations and health issue. This means providing a humanitarian service as it is used in real life and in saving people's lives and reducing the amount of accidents that happen every day and everywhere. Figure 3 show Methodology Diagram.

Comparison between Sheetgo tool and other algorithms in terms of data classifications show in Table 1.

Table 1. Show comparison between Sheetgo tool and other algorithms in terms of data classifications.

Features	Sheetgo tool	K-means	KNN
Need to select cluster number (k)	No	Yes	Yes
Need to writing code	No	Yes	Yes
Complexity	Low	Medium	Medium
Need to inserted values manually	No	Yes	Yes
Determine the type of classification	Yes	Yes	Yes
Working automatically	Yes	No	No

9   Results and Discussion

The results of this method showed a speed of transmission of information with high accuracy, and all of this is under very powerful protection given the number of security layers and we were able to classify the information automatically and send it to several different destinations (multiple clouds) and at the same time (real time), and we have tried this application on a group of users from very far places and for different issues and we did not notice any delay or any problem at work due to ease of use. Figure 4 show Android application interfaces.

Fig. 4. Android application interfaces

The user uses the application to send information to the competent authorities.

The information is stored in the database in the cloud as shown in Fig. 5. The data are classified into two groups using the Sheetgo tool as shown in Fig. 6 and 7.

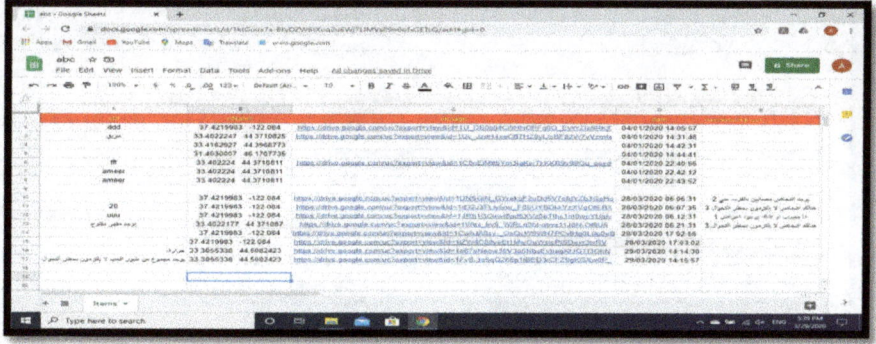

Fig. 5. Cloud database

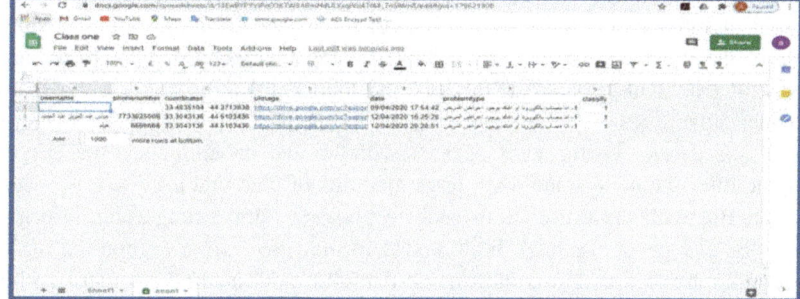

Fig. 6. Class one

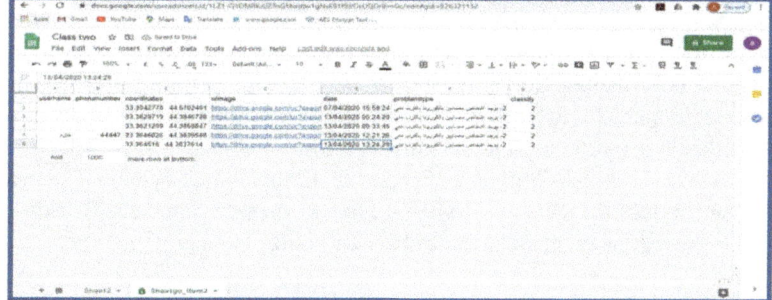

Fig. 7. Class two

The data is encrypted to ensure unauthorized access using the AES algorithm as shown in Fig. 8.

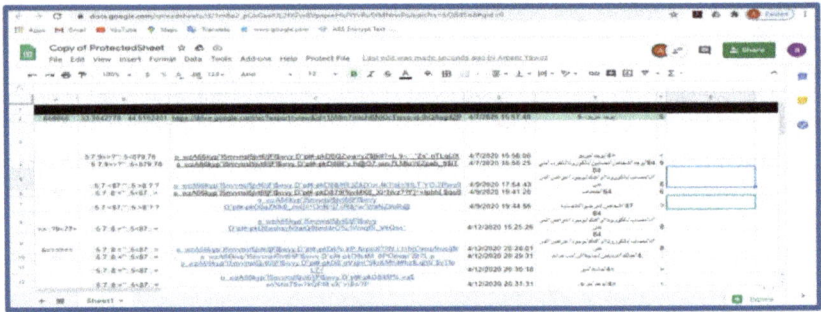

Fig. 8. Encrypt data

10   Conclusion

Geographical Information Systems (GIS) is distinguished by its multiple capabilities and one of the most important goals achieved using this technology is to reduce the large size of data sent from smart devices and their applications to servers. Unfortunately, the capabilities of mobile devices are limited in terms of battery, memory, processor speed, etc. Using such devices with limited capabilities, especially with geographic information systems with large amounts of data that need to be processed, is difficult. But with the presence of such technology (cloud geographic information systems). In this paper we have built a system that uses cloud computing and GIS through an application that was built using Android. And it turns out that despite the limited capabilities of smart phones, but it can deal with such systems as servers of GIS and provide specialized services by locating. As this application facilitates the use of such technologies and systems as GIS. This application can determine the coordinates (latitude and longitude) using the global positioning system (GPS) and map view. And we use spreadsheet as a cloud that is connected with Google drive and we programed it by using Java Script programming language to do the operations of receiving and encryption the information. Certainly, such important data as position-related data are sensitive and vulnerable to theft and penetration. In this paper we use advanced encryption standard (AES) algorithm. This algorithm facilitates to work with various sizes of keys like 128bit, 192bit, and 256bit with 128bits block cypher. The proposed approach can protect the data via (AES) 128bit to preserve the sensitive data from hackers. All operations of this algorithm happen within the cloud.

11   Future Work

In the future work, we suggest to use Dijkstra algorithm and QGIS program to determine the best path and little time to arrive between the respond points and the locations of accident.

Acknowledgment. The Authors would like to express their acknowledgment to the Computer Science Department/College of Science/Mustansiriyha University for the encouragement and support to perform the current research and hoping to carry out more next related researches.

References

1. Ge, X., Wang, H.: Cloud-Based Service for Big Spatial Data Technology in Emergence Management, pp. 126–129 (2007)
2. Singh, G., Supriya, S.: A study of encryption algorithms (RSA, DES, 3DES and AES) for information security. Int. J. Comput. Appl. **67**, 33–38 (2013)
3. Wiemken, T.L., et al.: Googling your hand hygiene data: using google forms, google sheets, and R to collect and automate analysis of hand hygiene compliance monitoring. Am. J. Infect. Control **46**, 617–619 (2018)
4. Gubara, A., Amasha, A., Ahmed, Z., El Ghazali, S.: Decision support system network analysis for emergency applications. In: 2014 9th International Conference Informatics System INFOS 2014 ORDS40–ORDS44 (2015). https://doi.org/10.1109/INFOS.2014.7036694
5. Ahmed, S., Ibrahim, R.F., Hefny, H.A.: GIS-based network analysis for the roads network of the Greater Cairo area. In: CEUR Workshop Proceedings, vol. 2144 (2017)
6. Forkuo, E.K., Quaye-ballard, J.A.: GIS based fire emergency response system. Int. J. Remote Sens. GIS **2**, 32–40 (2013)
7. Hamilton, S.E., Talbot, J., Flint, C.: The use of open source GIS algorithms, big geographic data, and cluster computing techniques to compile a geospatial database that can be used to evaluate upstream bathing and sanitation behaviours on downstream health outcomes in Indonesia, 2000–2008. Int. J. Health Geogr. **17**, 1–11 (2018)
8. Vallina-rodriguez, N., Crowcroft, J.: Modern mobile handsets. Context **15**, 1–20 (2012)
9. Satyanarayanan, M., Bahl, P., Cáceres, R., Davies, N.: The case for VM-based cloudlets in mobile computing. IEEE Pervasive Comput. **8**, 14–23 (2009)
10. McCullough, B.D., Talha Yalta, A.: Spreadsheets in the cloud - Not ready yet. J. Stat. Softw. **52**, 1–14 (2013)
11. Shah, R.M., Bhat, M.A., Ahmad, B.: Cloud computing : a solution to geographical information systems (GIS). Int. J. **3**, 594–600 (2011)
12. Rao, S., Vinay, S.: Choosing the right GIS framework for an informed enterprise web GIS solution. In: 13th Annual International Conference and Exhibition on Geospatial Information Technology (2010)
13. Grundy, J.: Conformance, pp. 102–109 (2017)
14. Flaten, H.K., St. Claire, C., Schlager, E., Dunnick, C.A., Dellavalle, R.P.: Growth of mobile applications in dermatology - 2017 update. Dermatol. Online J. **24** (2018)
15. Khedr, M.M., Mahmoud, W.H., Sallam, F.A. Elmelegy, N.: Comparison of Nd. Ann. Plast. Surg. **1** (2019). https://doi.org/10.1097/sap.0000000000002086
16. Yenuguvanilanka, J., Elkeelany, O.: Performance evaluation of hardware models of advanced encryption standard (AES) algorithm. In: Conference Proceedings - IEEE SOUTHEASTCON, pp. 222–225 (2008). https://doi.org/10.1109/SECON.2008.4494289
17. Pramstaller, N., et al.: Towards an AES crypto-chip resistant to differential power analysis. In: ESSCIRC 2004 – Proceedings of 30th European Solid-State Circuits Conference, pp. 307–310 (2004). https://doi.org/10.1109/esscir.2004.1356679

Developing a Reliable System for Real-Life Emails Classification Using Machine Learning Approach

Sally D. Hamdi[(✉)] and Abdulkareem Merhej Radhi

College of Information Engineering, Al-Nahrain University, Baghdad, Iraq
sally.dakhel@coie-nahrain.edu.iq

Abstract. Cyber World has become accessible, public and commonly used to distribute and exchange messages between malicious actors, terrorists, and illegally motivated persons. Electronic mail is one of the most frequently used transfers of information on internet media. E-mails are the most important digital proof that courts in various countries and communities use to condemn and that enables researchers to work continually to improve e-mail analysis using state-of-the-art technology to find digital evidence from e-mails. This work introduces a distinctive technology to analyze emails. It is based on consecutive phases, starting with data processing, extraction, compilation, then implementing the SWARM algorithm to adjust the output and to transfer these electronic mails for realistic and precise results by adjusting the support algorithm of vector machines. For email forensic analysis this system includes all the sentiment terms plus positives and negatives cases. It can deal with the machine learning algorithm (Sent WordNet 3.0). Enron Data set is used to test the proposed framework. In the best case, a high accuracy rate is 92%.

Keywords: Cyber forensic · Enron · Swarm · Sent WordNet · Email

1 Introduction

The Internet is a huge source of information, connecting millions of computers to share information around the world and produce a large number of electronic messages. As a result, email appears as a very important application on the Internet for data communication, which is utilized not only by computers but also by numerous electronic devices [1]. Email is a common way to communicate between parties. It transfers information between servers on a specified port number. Typically, email composed using client-side applications with identity sender, then store as a file and delivered to the destination user through one or more servers. Some individuals have found ways to exploit Email for malicious purposes although e-mail connections are designed to make things simple, powerful and effective [2]. The popularity and the low cost of e-mail made it the medium of choice by criminals or persons having mischievous intent [3]. The problem of gathering significant evidence against adversaries by examining suspected e-mail accounts to identify the most appropriate author from group of potential suspects [4]. Cyber Penetrator is approving sophistical apparatus and plans cause dangers to procedure of inclusive phenomenon. These attackers use anti-forensic technique to hide data of cybercrime. Cyber forensics apparatus is enhancing its hardiness and offset these inexorable threats [5]. In the most recent many years advanced legal sciences have become a noticeable movement in current examinations. Because of the unpredictability of this inquisitive movement and to the huge measure of the information to be dissected, the decision of proper computerized devices to help the examination speaks to a focal concern [6]. Data Mining is an application of algorithms to extract patterns of information and to make the useful information available in management. Data Mining has a number of applications in Digital cyber forensics [7].

1.1   Aims of the Proposed Method

- Deal with real-life e-mail dataset (Enron corpus).
- Design and Implement a reliable system that able to classify emails into different categories (malicious or normal).
- Optimize of the feature selection method.
- Achieve better performance of the proposed system.

1.2   Related Works

The researchers provide several of literature of similar work to analysis and classify emails or texts, which focused on classifying in to different categories by using different ways. The researchers' work that related to the email analysis such as: Farkhund Iqbal and et al., in 2010, [8], demonstrated a problem of how to extract writing style from set of email messages written by many anonymous authors. Sobiya R. Khan et al., in 2012, [9], discussed apply data mining technique to realize many functions for the implementation of the statistical analysis of e-mail, and clustering. In the field of identifications of an e-mail author and the analysis of the social network of email, Sobiya R. Khan, and et al., in 2013, [10], proposed a method to analyze and classify emails using a decision tree classifier that showed promising results. Emad E. Abdallah, in 2013, [11], presented an analysis and investigation of anonymous mining email content and suspect writing style. In this work used features stylometric and Machine Learning technique. A major contribution of this work is to reduce the training time by extracting some effective features. Harsh Vrajesh Thakkar 2013, [12], combined the English dictionary SentiWordNet3.0 with existing machine learning Naïve Bayes classifier algorithm which that classifies tweets in positive and negative classes respectively. Abbas.2018, [13], applied a method for analyzing official documents on a computer. V. Sreenivasulu et al., 2015, [14], proposed semantic ontologies and the "Gaussian Mixture Model" of data mining models to investigate cybercrime, and to analyze massive email forensic clusters. The purpose of the semantic tool (WORDNET) is to analyze grammatical rules and determine the term index and abbreviations. This methodology can be very important in investigation from emails. Nirkhi, S., et al. 2016, [15] explored the application of unsupervised techniques (multidimensional scaling techniques and cluster analysis) to solve the problem of author verification. Sofea Azizan and et al. 2017, [16], proposed improvise of the sentiment analysis by using machine learning to detect the acts of terrorism more accurately.

2   Proposed System

For E-mail Mining, overall, the learning count of this technique is used. The structure is made out of a couple of novel stages, each stage has a specific limit: Data preprocessing, Feature extraction, Clustering, Feature decision, Optimization, Classification, and a while later Prediction results. Figure 1, Summarize the packaging work times of our proposed research as follows:

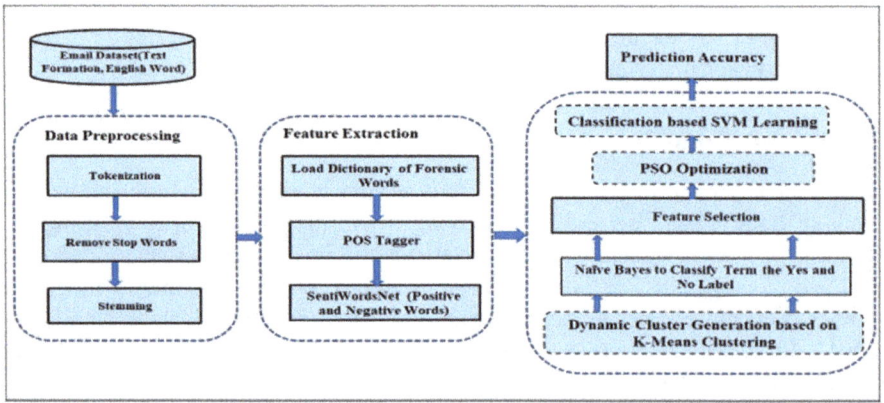

Fig. 1. Architecture of the proposed model

The system is composed of several different phases each phase has a specific function.

A. *E-Mail Dataset*

Enron was distributed during Enron's Corporation's lawful examination and ended up having various honesty issues. The current variant contains 619446 instant messages in their unique structure having a place with 150 workers, generally senior administration of Enron Corporation, composed into organizers. Processed dataset is accessible in the website site (https://www.cs.cmu.edu/~enron/). Figures 2, depicts information choice, client subtleties rundown and butleries content for every client.

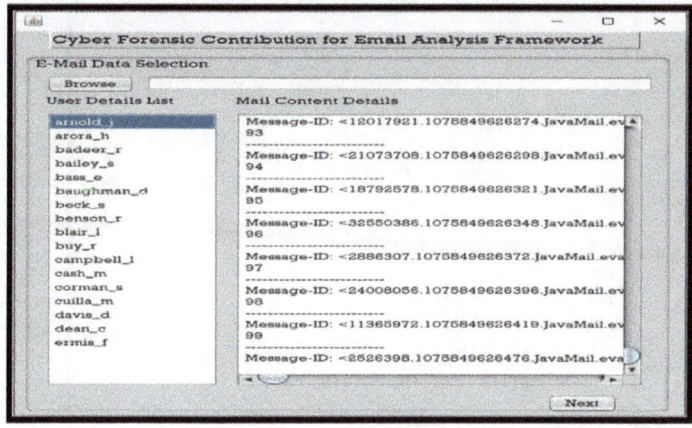

Fig. 2. Data selection

B. *Preprocessing*

Preprocessing is a significant phase in the data mining technique. This is done to prepare data in a form that can be used in the next stages. There are many types of document representation, such as vector-model, graphical mode and so on. Many measurements are also be used for document weighting. Thus, the representation of data is first and foremost before the implementation of the machine learning algorithms. Often, data preprocessing is the main phase in a project using machine learning. If there is much redundant and irrelevant information or data confusion and unreliable data, then discovering knowledge during the training stage will be more difficult. Data preprocessing stages are described in Fig. 3. The following sections present data preprocessing stages:

C. *Tokenization Process*

The fundamental goal of the tokenization stage is separating the text of message into smaller components. The term "tokenization" is referred to the process of phase sentences to be divided into text streams to its constituent meaning, as units called" tokens" or "words". The proposed system is using java tokenizer such that each email message is converted into distinct words or tokens. The list of tokens becomes input for the next steps.

D. *Removal of Stop Words*

The next stage after the tokenization process is the removal of stop words. Stop words are common words that can be found in almost all text scripts. There is a need to remove these words because these words do not hold any useful infor-mation to help determine whether the e-mail message belongs to a specific clas-sification or not. Elimination of stop words from the email message will reduce the feature space dimensions. The list contains about 488 stop words that are used in this work. "Stop words available in this Website (https://github.com/arc12/Text-Mining-Weak-Signals/wiki/Standard-set-of-english-stopwords)" [22]. Data pre-processing process shown in Fig. 3.

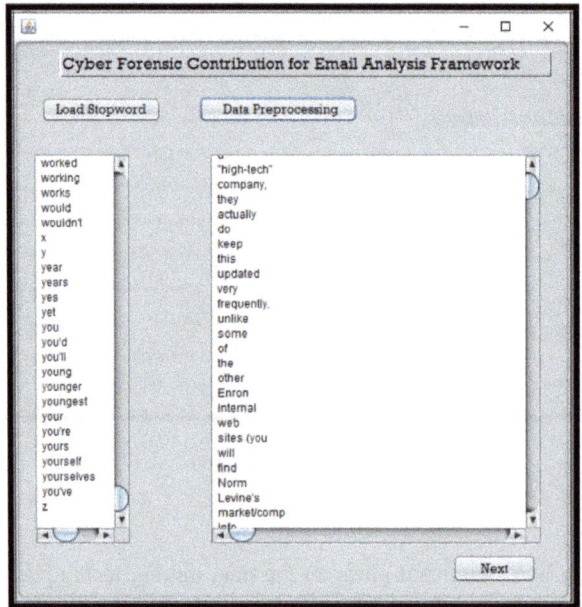

Fig. 3. Data preprocessing phase

E. *Token-Frequency*

Token Frequency can be measured by Eq. (13)

$$TF - idf_i = t_{i,j} \times log\left(\frac{N}{df_i}\right). \tag{13}$$

Thus, TF-idfi is the weight of a Token. N is the number of samples.

F. *Chi-Squared {Selection method}*

For the steps, the deviation from the measured circulation was required to involve an occurrence that by usage condition (14) [23],

$$\aleph^2(f,c) = \frac{N(WZ - YX)^2}{(W+Y)(X+Z)(W+X)(Y+Z)} \tag{14}$$

Where *W, X, Y, and Z* represents occurrences, specifies the attendance or nonappearance of a feature.

G. *Information Gain {Features Reduction}*

As shown in Eq. (15), the entropy drop for a feature equivalent to a grade of the features accountable on their IG score.

$$IG(f,c) = -\sum P(c)logP(c) + \sum P(c|f)logP(c|f) \tag{15}$$

H. *Evaluation*

To assess the degree of subsequent clusters and approve exploratory outcomes, the habitually utilized plan is "F-Measure". It is ensuing from exactness and review. It is the precision methodology utilized in the territory of (IR)" as follows:

$$Recall\,(N_p, C_q) = \frac{O_{pq}}{|N_p|} \tag{16}$$

$$Precision\,(N_p, C_q) = \frac{O_{pq}}{|C_q|} \tag{17}$$

$$F\,(N_p, C_q) = \frac{2 * recall\,(N_p, C_q) * precision\,(N_p, C_q)}{ecall\,(N_p, C_q) + precision\,(N_p, C_q)} \tag{18}$$

2.1   Feature Extraction

Feature extraction is the selection of those data attributes that best characterized a predicted variable. The forensic words are searched in the email dataset and POS Tagging categories then the score is calculated for each term by using SentiWordNet 3.0. The following sections presents features extraction stages:

- **Forensic Words**
  There are a 647 "forensic vocabulary words" has been compiled. "Forensic words available in this Website (https://myvocabulary.com/word-list/crime-vocabulary/)".
- **POS Tagging**
  These tags are used by dividing text scripts into sentences and giving a part-of-speech tag for each token whether it is a "name", "verb", "adverb" or "adjective". SentiWordNet3.0 modeled Part-of-Speech. Example word has POS tagging "(JJ, JJR, JJS, VB, VBD, VBG, VBN, VBP, and VBZ)" represent of "an adjective" score,"verb" score and so as.
- **SentiWordNet3.0**
  Every token is identified with mathematical scores speaking to "positive and negative" assessment data "SentiWordNet3.0". A score determined utilizing "SentiWordNet3.0". "SentiWordNet3.0" word reference is accessible on this "Website (http://sentiwordnet.isti.cnr.it/)" [24, 25]. Figure 4 represents feature extraction process.

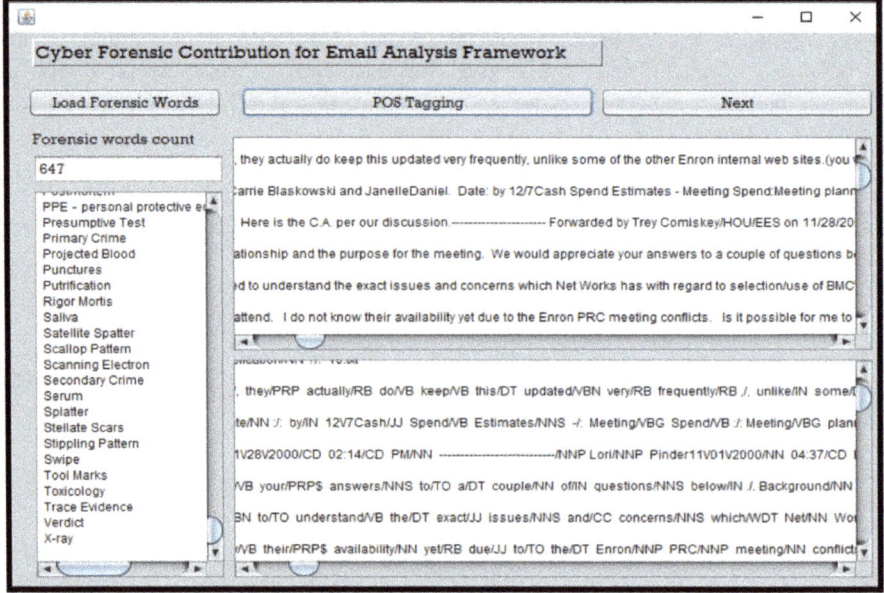

Fig. 4. Feature extraction phase

2.2   Clustering

The scores of each term were accomplished. In light of scores bunching achieved by utilizing the k-implies "clustering algorithm". "In k-means grouping", the middle point is characterized. It isn't progressively produced in the process with the end goal that makes the "middle point node in k-means" powerfully as portrayed in Fig. 5.

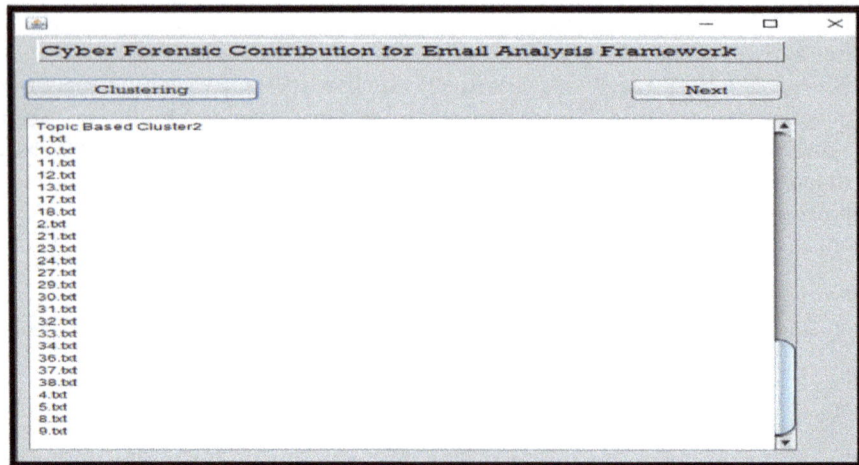

Fig. 5. Clustering phase

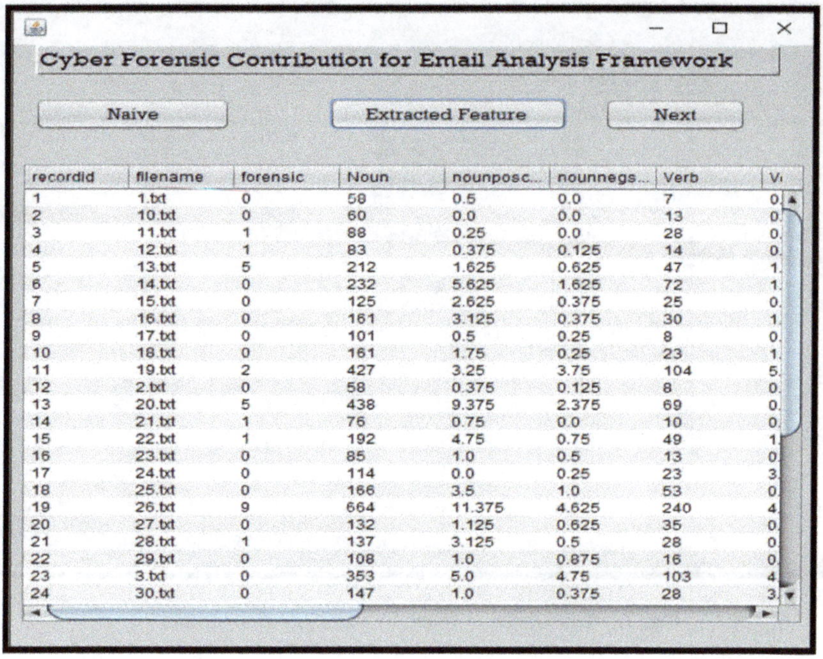

recordid	filename	forensic	Noun	nounposc...	nounnegs...	Verb	V
1	1.txt	0	58	0.5	0.0	7	0
2	10.txt	0	60	0.0	0.0	13	0
3	11.txt	1	88	0.25	0.0	28	0
4	12.txt	1	83	0.375	0.125	14	0
5	13.txt	5	212	1.625	0.625	47	1
6	14.txt	0	232	5.625	1.625	72	1
7	15.txt	0	125	2.625	0.375	25	0
8	16.txt	0	161	3.125	0.375	30	1
9	17.txt	0	101	0.5	0.25	8	0
10	18.txt	0	161	1.75	0.25	23	1
11	19.txt	2	427	3.25	3.75	104	5
12	2.txt	0	58	0.375	0.125	8	0
13	20.txt	5	438	9.375	2.375	69	0
14	21.txt	1	76	0.75	0.0	10	0
15	22.txt	1	192	4.75	0.75	49	1
16	23.txt	0	85	1.0	0.5	13	0
17	24.txt	0	114	0.0	0.25	23	3
18	25.txt	0	166	3.5	1.0	53	0
19	26.txt	9	664	11.375	4.625	240	4
20	27.txt	0	132	1.125	0.625	35	0
21	28.txt	1	137	3.125	0.5	28	0
22	29.txt	0	108	1.0	0.875	16	0
23	3.txt	0	353	5.0	4.75	103	4
24	30.txt	0	147	1.0	0.375	28	3

Fig. 6. Naïve Bayes and extracted features phase

- **"Naïve Bayes Algorithm"**
  Bayes calculation "for any class" likelihood assessment with includes mining. The credulous "Bayes" characterization calculation is utilized for arranging the "yes and no name|. Truly, "it represents to positive scores while No, represent to a negative score [26, 27]". Figure 6 shows Naïve Bayes and Extracted Features.
- **Optimization**
  PSO used to have the best forecast advancement for the best feature [28]. PSO starts by arbitrarily introducing the molecule populace (information ascribes that best portray an anticipated variable). An entire PS transfers "in the pursuit space to locate the best arrangement "(fitness)" by refreshing the position" at that point compute the speed of every molecule. The yield from these stage best highlights (characteristic) are legal, thing, action word, modifier, and descriptor credits as appeared in Fig. 7.

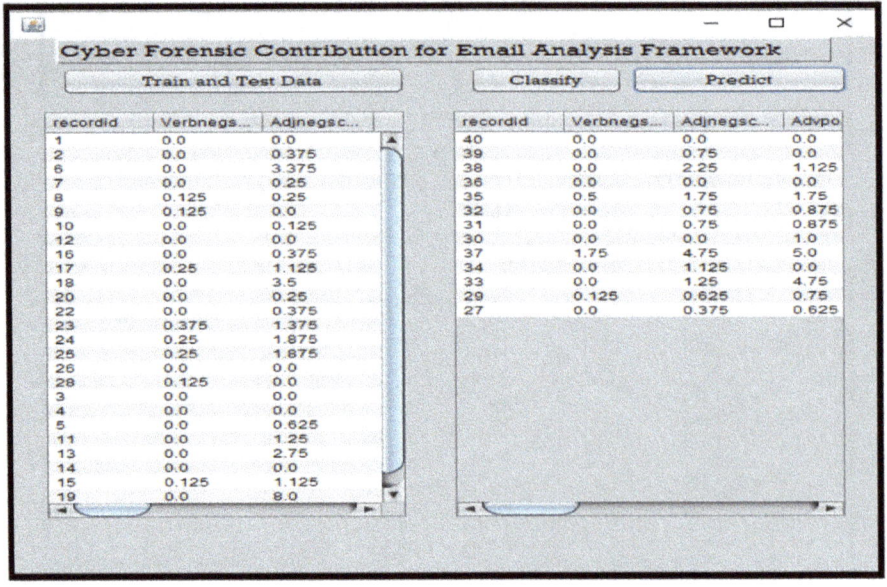

Fig. 7. "Classification frame by using SVM algorithm".

3   Classification

The objective of preparing a classifier is to make divisions between gatherings of various class classifications. "For the email dataset", the arrangement of messages is partitioned by choosing arbitrarily to a preparation 70% of absolute messages and "testing set 30% "of all out messages. In this research, usage of "SVM "by utilizing "LIBSVM" which includes two stages: first, preparing an informational collection and another, utilizing the proposed system to anticipate data of a "testing dataset". The edge for the grouping cycle as appeared in Fig. 7.

The testing stage incorporates every one of these cycles were completed similarly as the preparation period of the model exhibition.

4   Discussion of the Proposed System Results

When using the proposed algorithm for different subsets of databases, the accuracy of the classification, determined by the measurement of acceptably "classified emails" in the test collection, is provided in Table 1. We found that the best precision of classification is 92%.

Table 1. Results accuracy of classification.

Dataset from enron email dataset	No. of samples	Precision %
Subset 1	450	80.7%
Subset 2	950	85.5%
Subset 3	1425	76.9%
Subset 4	1900	87.1%
Subset 5	2345	76.02%
Subset 6	2850	79.02%
Subset 7	3325	92.12%
Subset 8	5000	91.79%
Subset 9	10000	87.7%
Subset 10	20000	85.5%
Subset 11	30000	88.9%
Subset 12	40000	87.1%
Subset 13	50000	89.02%
Subset 14	60000	92.97%
Subset 15	70000	90.12%
Subset 16	80000	92.92%
Subset 17	90000	92.52%
Subset 18	100000	92.72%
Subset 19	200000	90.12%
Subset 20	300000	91.12%

5 Conclusion

E-mails are one of the big data trading mechanisms and are commonly used on the Internet as a protected and powerless means. Email messages are advanced evidence that in many countries a substantial number of courts and social orders have acquired one of the crucial methods to prove them. In order to compel and group such messages, it is necessary to find a useful and efficient method. A study of the suggested method was planned to test the suitability of unknown messages in an official courtroom to obtain evidence to arraign hoodlums. This paper presents an unmistakable procedure for ordering messages dependent on information preparing, managing, refinement, and afterward adjust a few calculations to characterize these messages and afterward utilizing the SWARM calculation to acquire pragmatic and precise outcomes. To test the proposed framework will be select accessible information which is "Enron Data set". A high characterization rate is (92%), which is greater than the order rates referenced in related works introduced in segment II in this research.

References

1. Charalambou, E., Bratskas, R., Karkas, G., Anastasiades, A.: Email forensic tools: a roadmap to email header analysis through a cybercrime use case. J. Pol. Saf. Reliab. Assoc. Summer Saf. Reliab. Semin. 7(1) (2016)
2. Meghanathan, N., Allam, S.R., Moore, L.A.: Tools and techniques for network forensics (2010). arXiv preprint: arXiv:1004.0570
3. Tsochataridou, C., Arampatzis, A., Katos, V.: Improving digital forensics through data mining. In: IMMM 2014, The Fourth International Conference on Advances in Information Mining and Management, September 2016
4. Maria, K.A.: Authorship Attribution Forensics: Feature Selection Methods in Authorship Identification Using A Small E-mail Dataset. M.Sc. thesis, University of Athens (2016)
5. Sridhar, N., Lalitha Bhaskari, D., Avadhani, P.S.: Survey paper on cyber forensics. Int. J. Comput. Sci. Syst. Eng. Inf. Technol. pp. 113–118 (2011)
6. Haggerty, J., Karran, A.J., Lamb, D.J., Taylor, M.: A framework for the forensic investigation of unstructured email relationship data. Int. J. Digital Crime Forensics (IJDCF) 3(3), 1–18 (2011)
7. Bhardwaj, A.K., Singh, M.: Data mining- based integrated network traffic visualization framework for threat detection. Neural Comput. Appl. 26(1), 117–130 (2015)
8. Iqbal, F., Binsalleeh, H., Fung, B.C., Debbabi, M.: Mining write prints from anonymous e-mails for forensic investigation. Digital Investigation 7(1– 2), 56–64 (2010)
9. Khan, S.R., Nirkhi, S.M., Dharaskar, R.V.: Mining e-mail for cyber forensic investigation. Int. Conf. Adv. Comput. Electron. Electr. Eng. ICACEEE, Int. J. Comput. Sci. Appl. (UACEE), 2(2), 112–116 (2012)
10. Khan, S.R., Nirkishi, S.M., Dharaskar, R.V.: E-mail data analysis for application to cyber forensic investigation using data mining. In: Proceedings of the 2nd National Conference on Innovative Paradigms in Engineering and Technology (NCIPET), New York, USA (2013)
11. Abdallah, E.E., et al.: Simplified features for email authorship identification. Int. J. Secur. Netw. 8(2), 72–81 (2013)
12. Thakkar, M.H.V.: Twitter sentiment analysis using hybrid naive Bayes. M.Sc. thesis, Department of Computer Engineering, Sardar Vallabhbhai National Institute of Technology, Surat (2013)
13. Abbas, S.F.: Cybercrime system to identification the author of instance message using chat messages. M.Sc. thesis, The University of Technology, Department of Computer Science, Baghdad, Iraq (2018)
14. Sreenivasulu, V., Prasad, R.S.: A methodology for cybercrime identification using email corpus based on gaussian mixture model. Int. J. Comput. Appl. 117, 13 (2015)
15. Nirkhi, S., Dharaskar, R.V., Thakare, V.M.: Authorship verification of online messages for forensic investigation. Proc. Comput. Sci. 78, 640–645 (2016)
16. Azizan, S.A.: Terrorism detection based on sentiment analysis using machine learning. J. Eng. Appl. Sci. Univ. Technol. Petronas, Bandar Seri Iskandar, Peerak, Malaysia 12(3), 691–698 (2017)
17. Khairkar, P.K., Phalke, D.A.: Enhanced document clustering using K- Means with Support Vector Machine (SVM) approach. Int. J. Recent Innov. Trends Comput. Commun. (IJRITCC), 3(6), ISSN: 2321-8169, pp. 4112–4116, June 2015
18. Sailaja, D., Kishore, M., Jyothi, B., Prasad, N.R.G.K.: An overview of pre-processing text clustering methods. Int. J. Comput. Sci. Inform. Tech. 6, 3119–3124 (2015)
19. Gidofalvi, G., Zadrozny, B.: Feature selection for class probability estimation (2002)

20. Support Vector Machine (2019). http://www.statistics4u.com/fundstat_eng/cc_data_structure.html
21. Sharma, T., Tomar, G.S., Kumar, B., Berry, I.: Particle swarm optimization based cluster head election approach for wireless sensor network. Int. J. Smart Device Appl. **2.2** (2014)
22. Stop Words (2019). https://github.com/arc12/Text-Mining-WeakSignals/wiki/Standard-set-of-english-stopwords
23. Oludare, O., Stephen, O., Ayodele, O., Temitayo, F.: An optimized feature selection technique for email classification. Int. J. Sci. Technol. Res. **3.10**, 286–293 (2014). (48-[24] Firmanto, A., Sarno, R.: Prediction of movie sentiment based on reviews and score on rotten tomatoes using SentiWordnet. In: 2018, International Seminar on Application for Technology of Information and Communication, IEEE (2018))
24. Soni, V., Patel, M.R.: Unsupervised opinion mining from text reviews using SentiWordNet. Int. J. Comput. Trends Technol. (IJCTT) **11** (2014)
25. Hamdi, S.D., Radhi, A.M.: Cyber forensic email analysis and detection based on intelligent techniques. M.Sc. thesis, Al-Nahrain University, College of Information Engineering, Baghdad, Iraq (2019)
26. Howedi, F., Mohd, M.: Text classification for authorship attribution using Naive Bayes classifier with limited training data. Comput. Eng. Intell. Syst. **5**(4), 48-56 (2014). ([27] Amer, A.: Real Time for Arabic Speech Emotion Recognition. M.Sc. thesis, AL-Mustansiriyah University, Iraq)
27. Howedi, F., Mohd, M.: Text classification for authorship attribution using Naive Bayes classifier with limited training data. Comput. Eng. Intell. Syst. **5**(4), 48–56 (2014)
28. Wang, Y., Kim, K., Lee, B., Youn, H.Y.: Word clustering based on POS feature for efficient twitter sentiment analysis. Hum. Comput. Inf. Sci. **8.1,** 17 (2018)

Breast Cancer Classification Based on Unsupervised Linear Transformation Along with Cos Similarity Machine Learning

Maha S. Altememe, Ashwan A. Abdulmunem[(✉)],
and Zinah Abulridha Abutiheen

University of Kerbala, Karbala, Iraq
Ashwan.a@uokerbala.edu.iq

Abstract. In this research, we propose applying machine learning techniques on Breast Cancer classification problem that based on combination of unsupervised linear transformation along with Cos similarity machine learning. In unsupervised transformation we use principal components analysis (PCA). PCA used in analysing Breast cancer dataset to select effective features. In addition, we propose using classification model, Cos similarity to test accuracy of selected features. We prove the performance acceptability of the suggested PCA-Cos technique by conducting extensive practical experiments on binary-class Breast cancer dataset. The experimental results prove that PCA-Cos technique is promising method for manipulating features selection and cancer classification challenges and achieves the significant accuracy. Compared to previously suggested methods.

Keywords: Breast cancer dataset · Classification · Machine learning · Feature selection

1 Introduction

With the accelerated growth of the information technologies, we have witnessed a proliferation tremendous amount of data stored in database, files, and other repositories, it is mounting importance in the development of more effective analysis, processing, accurate, readable, and ease of understanding that could assist in decision-making [1]. Data mining is a process of mechanically looking out warehouses of massive data to detect patterns and objects that exceed minor analysis and assess the possibility of future events. Data mining which is also called as "Knowledge Discovery in Data (KDD)" is a powerful technique that can be used to get patterns relationships inside a store data. Data mining include two kinds: Supervised and Unsupervised. Supervised data mining is a process directed by a formerly known based on feature and explains the behaviour of objective acting as a function of a range of independent features or predictable. This is contrast to unsupervised data mining where the objective is pattern discovery [2, 3].

The real problem in the way data is handled is to choose the right algorithm to find the best solution. There are many ways to analyse and drill data using multiple

D.-T. Tran et al. (Eds.): ICISN 2021, LNNS 243, pp. 632–640, 2021.
https://doi.org/10.1007/978-981-16-2094-2_73

methods. In this study, we used two algorithms to find the best solution using medical data, a Principal component analysis (PCA) algorithm that filters for data then implements the Cos similarity algorithm to find the correct class.

The Scope of this research will be utilizing medical dataset obtained from UCI Machine Learning Repository [4]. Cos similarity algorithm will be applied for medical data sets and select breast cancer data to classify the correct instance by finding the closest sample from training dataset with a predefined target.

The basic objective of this research is to utilize two algorithms for prospecting of medical dataset. Specifically, to design method integrate PCA and Cos algorithms to classify data and evaluate the algorithm by comparing its performance against other optimal techniques. This research will provide researchers with simple methods to analysis and filter data. It presents a novel combination approach based on PCA and Cos similarity algorithm for classification tasks.

This paper organize as follows first section to presents an overview of related work, second section explain the proposed approach while the results and conclusion will be conducted in last two sections respectively.

2 Literature Review

Data mining [5–7] have been applied to several of medical fields to improve decision-making. In [8] a work has been present comparative review of data mining classifiers in cardiovascular illness expectation. Data mining have numerous techniques that can be applied to assurance and prevention of cardiovascular illness among patients. Whereas the study uses four classifiers in data mining to predict cardiovascular illness in patients are compared: rule rely upon schemes, decision tree, artificial neural networks and support vector machine. These techniques are compared on principle of sensitivity, specificity, error rate, accuracy, true positive rate and false positive rate.

Shah, Kusiak, and Dixon [9] have been implemented data mining predicting the survival kidney dialysis patients using fixed object to extract knowledge about the interaction between these variables and patient survival. Tow completely different data mining algorithms are utilized for extracting information within the variety of decision rules. The total classification precision for all data mining algorithms became significantly higher using the independent sample dataset over the variety datasets. The prediction accuracy of individual visit depend rule sets magnified over the combination. These enhancements were despite the exclusion defines important attributes e.g. diagnosis, time for dialysis, target weight, etc.

Principal components analysis (PCA) [9] is a famous technique for unsupervised linear transformation. Given a set of data on n dimensions, PCA aims to find a linear subspace of dimension d lower than n such that the data points lie in the fundamental on this linear subspace. Such a reduced subspace tries to take care of most of the variability of the data.

Many works have been suggested to deal with Breast cancer identification [10–12]. Cabrera [13] introduced a scheme to manipulating the classification based on feature extraction through texture analysis. The classification based on X-ray images. In recent works, machine learning [7, 14] has a vital role in solving the problems associated with breast cancer classification. Several works proposed to use a machine learning algorithm to obtain an acceptable result. One of these works, Chtihrakkannan [15] used Grey-Level Co-occurrence Matrix (GLCM) along with Multi-layer Perceptron (MLP) classifier. This method can classify the breast cancer images with reasonable findings. Moreover, Deep learning methods also taken a clear place in cancer classification. The researchers [16–18], employed this technique to improve the results.

Pirouzbakht [18] used multi-layered Convolutional Neural Network [8, 19–22]. This network consists of several different layered, pooling, convolutional and fully connected layers.

In Shen [16], the authors improved the accuracy rate by using a deep learning machine learning. The deep learning based on "end-to-end" training approach. While [17] suggested method based on three basics steps namely: Region of interest extraction, secondly, features selection and finally classification step uses Convolutional Neural Networks (CNN).

3   Proposed Approach

3.1   Breast Cancer Dataset

The dataset used in this analysis is the breast-cancer-Wisconsin. Data file obtained from the machine learning repository of UCI [4]. The dataset consists of 569 breast instants, 32 attributes for each instant. It is organized in 32 columns each row reflects findings belonging to the breast FNA obtained through medical examination of a patient. The first column is a corresponding identification code for each patient; The second column represent Diagnosis (M = malignant, B = benign). From 3–32 ten unique values are determined for every cell nucleus: radius (distances mean from centre to points on the perimeter), texture (standard deviation of gray-scale values), perimeter, area, smoothness (variety of neighbourhood in radius lengths), compactness (perimeter^2/area - 1.0), concavity (concave parts severity of the contour), concave points (number of contour concave parts), symmetry fractal dimension ("coastline approximation" - 1). In this dataset 357 are diagnose as B (benign) and 212 as M (malignant). In this work, Intelligent architecture has been proposed based on machine learning to classify B and M based on Cos similarity machine learning. To deal with, the dataset divided into training (80% of whole dataset) and testing (20%). As a result, 455 (300 for B, 155 for M) instances for training and 114 (57 for each B and M) instances for test. Table 1 show the training and testing sets.

Table 1. Breast cancer dataset

Class	Total data	Training data	Testing data
B	357	300	57
M	212	155	57

3.2   PCA Algorithm

We propose to use a combination of PCA with Cos similarity algorithms to find best features of Cancer dataset named PCA-Cos algorithm. Principal Component Analysis (PCA), is well known for reduction dimensional and statistical measurements in big data manipulating [9, 23] concerned with elucidating the variance structure of a collection of variables and uses symmetrical adjustment to convert over that arrangement of perceptions of conceivably associated factors into an arrangement of estimations of directly unrelated factors referred to as the most segments [24]. Processing large data matrix is time consuming, to deal with issue PCA can be used to decrease large data matrix and keep the important information in that matrix which is required for next processing steps. As far as math, the fundamental step of the standard principal component analysis accomplished by constructing a covariance matrix and compute its eigenvalues and eigenvectors. This method concerned representing of N × N data matrix as a one-dimensional vector of N * N elements, by replacing the rows of the data matrix one after another. At that point figure the covariance matrix of the whole data set and calculate the eigenvector of this covariance matrix. The eigenvector with the biggest eigenvalue is the direction of greatest variation.

3.3   Cos Similarity

A famous method used to coordinate comparable documents depends on counting the maximum value. It is a measurement used to determine how comparative the data set are regardless of their size. Numerically, the angle cosine is measured between two vectors projected in a multi-dimensional space. Equation below is cosine similarity between vector a and vector b [25].

$$\cos(a, b) = \frac{\bar{a}.\bar{b}}{||\bar{a}||\,||\bar{b}||} = \frac{\sum_1^n a_i b_i}{\sqrt{\sum_1^n a_i^2}\sqrt{\sum_1^n b_i^2}} \tag{1}$$

Where:

$\bar{a}$ and $\bar{b}$: are the vectors.

$\bar{a}.\bar{b} = \sum_1^n a_i b_i = a_1 b_1 + a_2 b_2 + \ldots + a_n b_n$ is the dot product of the two vectors.

$||\bar{a}||\,||\bar{b}||$: vector lengths.

4   Results

In the classification step, the cos similarity has been used on the data set with and without using PCA algorithm (30 status for each person), and after applied PCA (extract the ten best of statuses), the results improved from 78.9% to 99.12%. Where each test sample compared with data training in class B (infected person) and class M (healthy person). Table 2 shows samples of results without using PCA algorithm and using PCA algorithm where: Bmattest i: sample one in testing data and Bmattraining j: each raw in data set where j is 1 to 300 (number data in class B) and Mmattraining k: each raw in data set where k is 1 to 155 (number data in class M).

Table 2. Classification results based on Cos similarity by using PCA and without using PCA algorithm (as example for test instances (B = 10, M = 10 with 20 training instances)

Test_instance	Trainging_instances	Cos similarity results using PCA	Cos Similarity results without using PCA
Bmattest 1	Bmattraining 1	0.998754768	0.999530267
	Bmattraining 2	0.997255312	0.999359561
	Bmattraining 3	0.996654603	0.998275044
	Bmattraining 4	0.999229521	0.999390815
	Bmattraining 5	0.999606428	0.999944628
	Bmattraining 6	0.999312281	0.999181341
	Bmattraining 7	0.999939843	0.999640034
	Bmattraining 8	0.99987498	0.999920666
	Bmattraining 9	0.99962273	0.999970027
	Bmattraining 10	0.998273543	0.99997943
Mmattest 1	Bmattraining 1	0.986961699	0.997848967
	Bmattraining 2	0.996204724	0.997518979
	Bmattraining 3	0.980220255	0.995642654
	Bmattraining 4	0.984848399	0.997566792
	Bmattraining 5	0.988693648	0.998969335
	Bmattraining 6	0.993268248	0.997157773
	Bmattraining 7	0.989574891	0.998099965
	Bmattraining 8	0.987201474	0.999676505
	Bmattraining 9	0.99084738	0.999622772
	Bmattraining 10	0.982269626	0.999410016

Based on these calculations, the largest value is chosen according to cos similarity algorithm as a classified class. By using PCA algorithm, the best 10 features for each data person are selected. Table 3 shows test results by using PCA algorithm and without PCA algorithm using cos similarity for sample of testing set, as follow.

Table 3. Classification results based on Cos similarity by using PCA algorithm and without using PCA (as example for test instances (10 for B and 10 for M with 20 training instances)

Testing instance	Closest sample using PCA	Classification results using PCA	Closest sample without PCA	Classification results without PCA
Bmattest1	Bmattrain68	True	Bmattraining 105	True
Bmattest2	Bmattrain8	True	Mmattraining 10	False
Bmattest3	Bmattrain100	True	Bmattraining 3	True
Bmattest4	Bmattrain221	True	Bmattraining 167	True
Bmattest5	Bmattrain82	True	Bmattraining 110	True

(continued)

Table 3. (*continued*)

Testing instance	Closest sample using PCA	Classification results using PCA	Closest sample without PCA	Classification results without PCA
Bmattest6	Bmattrain9	True	Bmattraining 82	True
Bmattest7	Bmattrain247	True	Bmattraining 72	True
Bmattest8	Bmattrain129	True	Bmattraining 157	True
Bmattest9	Bmattrain220	True	Bmattraining 273	True
Bmattest10	Bmattrain28	True	Mmattraining 89	True
Mmattest1	Mmattrain33	True	Bmattraining 243	False
Mmattest2	Mmattrain107	True	Mmattraining 111	True
Mmattest3	Mmattrain83	True	Mmattraining 101	True
Mmattest4	Mmattrain56	True	Mmattraining 112	True
Mmattest5	Mmattrain7	True	Mmattraining 110	True
Mmattest6	Mmattrain106	True	Bmattraining 82	True
Mmattest7	Mmattrain102	True	Mmattraining 144	False
Mmattest8	Mmattrain13	True	Bmattraining 245	True
Mmattest9	Mmattrain115	True	Bmattraining 124	False
Mmattest10	Mmattrain155	True	Bmattraining 157	False

As a whole of the accuracy of the dataset. Figures 1, 2 illustrate fusion matrices the TP, TN, FP and FN values for each B and M with and without using PCA respectively.

	B	M
B	57	0
M	1	56

Fig. 1. Fusion matrix with PCA

	B	M
B	52	5
M	19	38

Fig. 2. Fusion matrix without PCA

Based on these matrices the Precision and recall metrics will be calculated. The following Table 4 will represent these metrices:

Table 4. Represent metrices

	Recall	Precision	Recall (PCA)	Precision (PCA)
Test set	0.7323	0.91228	0.9827	1.0

From the metrices above, it is clearly to see that the Recall and precision increased by approximately 0.2 and 0.1 receptively. To validate the results, we used 3-fold cross validation training data 300 for class B and 155 for class M and testing data with 57 for each B and M as Fig. 3.

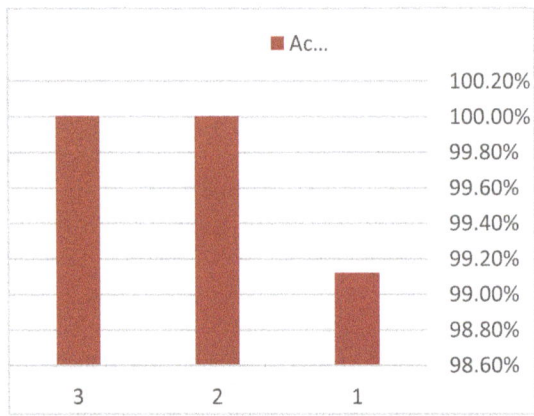

Fig. 3. 3-fold cross validation

It is explained the PCA has improved the results above 20% percentage.

5   Conclusion

Based on the experiments we can conclude that, The Cos similarity learning can work effectively along with PCA algorithm. By using this combination, the results obviously improved. The results without PCA is 78.9% about 24 false negatives values from whole testing instances. While when using PCA the accuracy increased to 99.12% give more acceptable findings to justify this combination. As a result, a machine learning with effective feature selection give a reliable outcome in more vital problem in the health community.

Acknowledgements. We are happy to acknowledge University of Kerbala for supporting our work.

References

1. Zaïane, O.: Chapter I: Introduction to Data Mining. Prince Knowledge Discovery Databases, pp. 1–15 (1999)
2. Azuaje, F.: Review of "data mining: practical machine learning tools and techniques" by Witten and Frank. Biomed. Eng. Online 5(1), 51 (2006)
3. Agarwal, S.: Data mining: data mining concepts and techniques. In: Proceedings - 2013 International Conference on Machine Intelligence Research and Advancement, ICMIRA 2013, pp. 203–207 (2014)
4. UCI Website. https://archive.ics.uci.edu/ml/datasets/Breast+Cancer+Wisconsin+%28Diagnostic%29
5. Yoo, I., Alafaireet, P., Marinov, M., Pena-Hernandez, K., Gopidi, R., Chang, J.F., et al.: Data mining in healthcare and biomedicine: a survey of the literature. J. Med. Syst. 36(4), 2431–48 (2012)
6. Han, J., Kamber, M., Pei, J.: Data Mining: Concepts and Techniques, 3rd edn. Solution Manual.
7. Sudharshan, P.J., Petitjean, C., Spanhol, F., Oliveira, L.E., Heutte, L., Honeine, P.: Multiple instance learning for histopathological breast cancer image classification. Expert Syst Appl. 117, 103–111 (2019). https://doi.org/10.1016/j.eswa.2018.09.049
8. Chaitrali, M., Dangare, S., Sulabha, M., Apte, S.: A data mining approach for prediction of heart disease using neural networks. Int. J. Comput. Eng. Technol. 3(3), 30–40 (2012). www.iaeme.com/ijcet.asp
9. Shlens, J.: A tutorial on principal component analysis derivation, discussion and singular value decomposition (2003)
10. Id, S.: Breast Cancer Classification based on Unsupervised Linear Transformation along with Cos Similarity Machine Learning (2020)
11. Nilashi, M., Ibrahim, O., Ahmadi, H., Shahmoradi, L.: A knowledge-based system for breast cancer classification using fuzzy logic method. Telemat. Inform. 34(4), 133–144 (2017). https://doi.org/10.1016/j.tele.2017.01.007
12. Nasser, A.M., Raheem, H.A.: Image compression based on data folding and principal component 4(2), 67–78 (2016)
13. Guzmán-Cabrera, R., Guzmán-Sepúlveda, J.R., Torres-Cisneros, M., May-Arrioja, D.A., Ruiz-Pinales, J., Ibarra-Manzano, O.G., et al.: Digital image processing technique for breast cancer detection. Int. J. Thermophys. 34(8–9), 1519–1531 (2013)
14. Chtihrakkannan, R., Kavitha, P., Mangayarkarasi, T., Karthikeyan, R.: Breast cancer detection using machine learning. Int. J. Innov. Technol. Explor. Eng. 8(11), 3123–3126 (2019)
15. Chtihrakkannan, R., Kavitha, P., Mangayarkarasi, T., Karthikeyan, R.: Breast cancer detection using machine learning (11), 3123–3126 (2019)
16. Shen, L., Margolies, L.R., Rothstein, J.H., Fluder, E., Mcbride, R.: Deep learning to improve breast cancer detection on screening mammography, pp. 1–12 (2019)
17. Shamy, S., Dheeba, J.: A research on detection and classification of breast cancer using k-means GMM & CNN algorithms (6), 501–515 (2019)
18. Pirouzbakht, N., Mej, J.: Algorithm for the detection of breast cancer in digital mammograms using deep learning, pp. 46–49 (2017)
19. Pirouzbakht, N.: Algorithm for the Detection of Breast Cancer in Digital Mammograms Using Deep Learning

20. Tsuchida, T., Negishi, T., Takahashi, Y., Nishimura, R.: Dense-breast classification using image similarity. Radiol. Phys Technol. **13**(2), 177–186 (2020). https://doi.org/10.1007/s12194-020-00566-3

21. Khan, S.U., Islam, N., Jan, Z., Din, I.U., Rodrigues, J.J.P.C.: A novel deep learning based framework for the detection and classification of breast cancer using transfer learning. Pattern Recogn. Lett. **125**, 1–6 (2019). https://doi.org/10.1016/j.patrec.2019.03.022

22. Shen, L., Margolies, L.R., Rothstein, J.H., Fluder, E., McBride, R., Sieh, W.: Deep learning to improve breast cancer detection on screening mammography. Sci. Rep. **9**(1), 1–12 (2019)

23. Naseriparsa, M.: Combination of PCA with SMOTE resampling to boost the prediction rate in lung cancer dataset **77**(3), 33–38 (2013)

24. Leen, T.K.: Dimension reduction by local principal component analysis, pp. 1493–1516 (1997)

25. Sohangir, S., Wang, D.: Improved sqrt_cosine similarity measurement. J. Big Data **4**, 1–13 (2017)

A Comparison of Handwriting Systems Based on Different Style Characteristics

Huda Abdulaali Abdulbaq$^{(\boxtimes)}$, Yasmin Makki Mohialden,
and Qabas Abdal Zahraa Jabbar

Mustansiriyah University, Baghdad, Iraq
{huda.it,ymmiraq2009,qabas_a}@uomustansiriyah.edu.iq

Abstract. In the current age of digitalization, recognizing the writer of a handwritten text plays an important role in information proceeding. It has been attractive research problem in document analysis. The handwriting recognition aims to enable the machine to recognize the handwritten characteristics. For that it is used in the forensics, document examination and paleography. This paper present an automatic writer recognition method based on digitalized images of unconstrained writing and describes the different techniques used in text-dependent and text-independent methods to recognize the writer.

Keywords: Handwriting identification · Cookbook · Writer recognition · Text-dependent writer

1 Introduction

Writer's automatic identification of handwriting digitalized samples became an interested research subject during the last decade. Many researchers created an effective writer recognition techniques and tested different handwritten secrets (languages). They considered the specific characteristics of each script which enable them to develop specific approach for handwritten recognition. The researchers extracted the characteristics of writer script from individual writing samples. The developed methodologies compared the detected characteristics across different samples in order to specify the writer given samples [1]. Hand writing recognition is related to the human and personalized identification field [2]. The concept of handwriting recognition is to eliminate the writers dependent variations by extracting a set of features depended on different allographs of the same character. This technique facilitate the recognition of writing style. On the other hand, writer identification depends on the differences in handwriting styles and derive the benefit of the variations in characteristics and identify the writer of given sample.

Writing samples in the writer recognition task, as well as other related problems, can be categorized as online or offline. In online writer identification, the researcher uses a digitizer such as a pad, digital pen, or a touch screen to capture handwriting. In addition to the shapes of words and characters, these digitized online samples contain additional information about the writing style, including the writing speed, sequence of strokes, pressure at different points, among other characteristics. The offline writing samples, on the other hand, are available in the form of scanned images making offline writer identification are more challenging task as opposed to the online scenario [3].

One can differentiate two broad classes in automatic offline writer identification methods: text independent and text dependent [4]. The requirements of text-dependent methods compared with the same fixed text based on different writer's samples. These methods are very similar to signature verification, where each writer provides a particular predefined text for his/her authentication. In general, text-dependent methods require the segmentation of text into characters or words, which are then compared with the same characters or words in the training data to recognize its writer.

In recent years, scientists investigated writer identification across multiple languages as well as historical manuscripts. In addition to writing samples in English, several researchers have addressed writer identification in Arabic [5–7] and different ancient languages

In this paper, we will be presenting a broad review of the techniques developed for writer identification on offline English writing samples. The next section presents outlines of the existing methods of text-dependent writer identification methods. another section discusses the significant contributions in the text-independent writer recognition domain. Finally, analyzing the used methods is presented and compared with discussion on their performance.

2   Text-Dependent Writer Recognition

The earlier research on writer identification was mainly a fixed given text, also the present system compares the hand written characteristics to identify the writer. The present methods required a segmentation process for handwritten elements in order to compare these elements and observe the similarity between them. The used methods like signature verification methods able to recognize the characteristics and of handwritten in high rates, even with the small amounts of available handwriting [8]. compared the classical writer identification approaches mainly, including text-dependent methods. These classical approaches were mostly evaluated on under-study number of writers [9, 10]; performed evaluations on 100 writers and each writer contributed a single word 10 times, achieving an overall identification rate of 98%.

A set of macro and micro features were computed for feature extraction from the writing samples. The writing letters macro features were extracted on the word level, paragraph and document while micro features were extracted at character level or allograph. Macro features contained a set of eleven features which were grouped into three categories namely: dark features (a measure of grey values, the threshold value of grey level and the amount of black pixels); contour features (stroke formation and slope); and averaged line-level features (height and slant). The two(aspect ratio and indentation) features were extracted at the paragraph level, while three (word length, upper and lower zone ratios) features were extracted at the word level. Micro features consisted of 512 binary features calculated at the character level, including concavity, structural, gradient and features.

After feature extraction from the samples, their effectiveness to characterize the writer was evaluated on writer verification and identification. For macro features, the specification of two samples distance of documents based on feature vectors A and B as defined by the Euclidean distance.

$$\text{Euclidean distance} = \sum_{i=1}^{d} \sqrt{(A_i + B_i)^2} \tag{1}$$

Where d represents the number of attributes

The distance between two characters (allographs) represented by the micro binary feature vectors A and B was calculated as

$$\Sigma\tau\,(A,\,B) = A\tau\,B + -\_(A-\_B\,)/2 \tag{2}$$

The system was evaluated using the nearest neighbor rule by randomly selecting n writers from a total of 1000, selecting one sample written by one of these writers as a query document, and the rest $(3 \times n - 1)$ sample as reference. For each n, the evaluation was carried out 1000 times. Macro and micro features sets were first tested separately and then combined by using the macro feature as a filter that selected 100 similar samples from 1000 samples and then micro feature identified writer among 100 writers. Writer identification yielded accuracy in the range of 98% for two writers and 87% for 900 writers. The writer verification task reported an accuracy of 96% using an artificial neural network with combined sets of features.

Furthermore, some studies focused on the comparison between individual characters. [11, 12] ranked the discriminatory power of each handwritten individual character based on the area under the receiver operating characteristic (ROC) curve. The analysis showed that the characters 'G', 'b,' 'N,' 'I,' 'K,' 'J,' 'W,' 'D,' 'h,' 'f' were the ten most discriminating characters. In numbers, '4' was found to be the most discriminating numeral while '1' was the least discriminating numeral. [13] extended the research by using the same set of features by comparing the individuality of characters with four words 'been,' 'Cohen,' 'Medical' and 'referred.' Later, the same study was then extended to 25 words by adding WMR (Word Model Recognizer) Characters ranked based on the area under the ROC curve [12, 14] introduced a feature obtained by morphologically transforming the projection functions of a word. After the banalization and thinning process (generating one-pixel full words), the horizontal projection of each word was computed. Spaces between the letters were accounted for by two projection functions, with and without zero bins. The morphological opening was then successively applied (Fig. 1) by increasing the size of the structuring element to compute the feature vector representing the writing.

The performance was evaluated using word databases of (50) writers. Each writer contributes (45) samples of two different words having similar length in English and Greek script. The authors employed two different classification schemes neural networks and the Bayesian classifier. Around 96% identification rate was achieved by using a neural network classifier, and a 92% identification rate was achieved by using the Bayesian network classifier for both databases (English and Greek) separately.

[15] proposed a method based on text-dependent writer identification by Kannada. The Kannada database consisted of collected handwritten words which is gathered from 25 writers. This collection represent the independent dataset which represents a training and testing input representation for writer identification. The specific feature

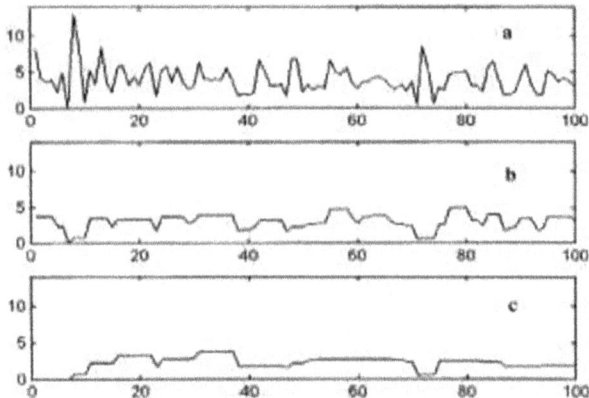

Fig. 1. Projection function for morphological opening: (a) histogram using primary projection function on the word 'characteristic' (b) the line structuring element is opening with length 3 (c) the line structuring element is opening with length 7. [1]

vector involves a spatial direction features based on resolution variety which is used the Discreet Cosine Transform and Random Transform. The word image used to extract some structural features such as on-pixel ration and aspect ratio. The method uses a combination of the features of several words. The method presents performance analysis using the nearest neighbor classifier and 5-fold cross-validation. Using single words only the method has an average identification accuracy of 93.2582%. The accuracy reaches 100% by features a combination of three or more words. The results show a higher impact on writer identification using vectors extracted from longer words or words having more structural variation.[16] studied different feature classification techniques and extraction methods required for writer identification on both Indic and non-Indic scripts. They found out that the writer identification systems have limited accuracy rates in non-Indic scripts compared to Indic scripts, while [17] proposed a new approach for offline and text-independent writer identification. The approach is based on image mining techniques using the writer's training samples. The image mining-specific techniques include genetic algorithms and SVM classifiers.

To evaluate this method performance in different languages, Other databases such as IAM English database, CASIA Chinese database, and two others for Persian and Kannada were examined. The results for these languages demonstrate that the method has over 99% accuracy.

The above paragraphs discussed the well-known writer recognition techniques in the text-dependent mode. Some of the features employed for text-dependent writer identification, for example, slope and slant, have also been successfully applied to text-independent writer recognition. Although these text-dependent methods report high recognition rates, they are not very useful for many practical applications where the text in question is different from the text in the training database. Therefore, most of the research carried out on writer identification.

3   Text-Independent Writer Recognition

This section presents a review of different approaches for text-independent writer recognition. The existing paper has been categorized into three groups; global approaches, local approaches, and approaches based on handwriting recognition.

4   Global Approaches

Global approaches for writer identification characterize the writer of a given document based on the general look and feel of the writing. Two main techniques that have been effectively applied in this category include texture and fractal analysis of handwriting, as presented in the following.

[18] Present a text-independent writer identification using contour-directional feature (CDF) and the SIFT descriptor. The method includes two stages—a SIFT descriptor codebook of local texture patterns that are extracted from images. to verify the similarities between different images an Occurrence histograms are calculated and utilized. A list of reference images is obtained for each image. This rundown is refined, a short time later, utilizing the form directional element and SIFT descriptor. Assessment is finished utilizing the ICFHR2012-Latin dataset and the ICDAR2013 dataset. Results display that the technique beats the modern calculations and files.

[19, 20] Proposed a grapheme-based approach to offline Arabic writer identification. Rather than extricating common graphemes, it creates its graphemes. The technique creates and tests one full and four incomplete codebooks. The technique extricates (60) element vectors utilizing coordinating of a layout with (411) individual scholars from the IFN/ENIT data set. The outcomes indicated the wide representatively and the great speculation capacity of engineered codebooks. We got a top1 rate = 90.02% and a top5 rate = 96.35% for essayist distinguishing proof, and an EER = 2.1% for author check.

[21] proposed a system that addresses the robustness aspect of automatic hand writer identification. The framework utilizes sacked discrete cosine change (BDCT) descriptors. Multiple predictor models to obtain a decision are generated by Universal codebooks. The BDCT approach allows exploited for identification by DCT features. The system has been evaluated on the original form of handwritten documents of various datasets with comparable performance with state-of-the-art systems. Indistinct documents of two dissimilar datasets have been tested with well results. The system is suitable for digital forensics, where the documents may not be in ideal conditions.

This section presents a review of different approaches for text-independent writer recognition. Based on the methodology employed, the existing approaches have been categorized into three groups; global approaches, local approaches, and approaches based on handwriting recognition. Each of these is discussed in the following subsections.

[22] examined writer identification from Arabic handwriting samples. Their procedure rely upon separating pieces of composing utilizing textural descriptors, Gray Level Run Length (GLRL) Matrices and Histogram of Oriented Gradients (HOG).The system is evaluated with the IFN/ENIT Arabic handwriting databases, database (411

writers), the QUWI database (1,017 writers) and the KHATT database (1000 writers). Fusion using the 'sum' rule give the maximum identification rates reading: 96.86, 85.40, and 76.27% using QUWI databases, and the IFN/ENIT, KHATT, respectively. n. Each of these is discussed in the following subsections.

5   Texture Analysis

[23], propose a text-independent approach for writer identification to extract texture features using multichannel spatial filtering techniques. The system first determines and corrects the skew in the documents and normalizes the space between different words and lines to a predefined size. After normalization, random non-overlapping text blocks are extracted, and features are computed from each block using gray-scale co-occurrence matrices (GSCM). [24] For each image, 60 features were extracted from the gray-scale co-occurrence matrices (GSCM), and 32 features were extracted by the multichannel Gabor filtering. Two sets of 20 writers each were grouped, and 25 samples of non- overlapping blocks per writer were extracted for evaluations. For experiments, among each group of 20 writers, 10 samples were used for training and 15 samples for testing followed by 15 for training and 10 for testing. Two classifiers weighted Euclidean distance (WED) and closest neighbor classifier (K-NN) were considered for ID. Various mixes of highlights under the two classifiers were evaluated, and a maximum identification rate of about 96% was achieved by using Gabor filters underweighted Euclidean distance (WED).

Another significant contribution is presented in [22, 26], which characterizes the writing style of an author by using contour-based directional probability distribution functions (PDFs). Contour direction distribution is for each point on the contour fragment. Angles are determined concerning the horizontal axis and are counted in a histogram quantized into 8, 12 or 16 directions. Then the histogram is normalized to convert it into a probability distribution. Two additional features, "bivariate contour-angle probability distribution" and "joint probability distribution," were also introduced, which realized significant improvements in writer identification rates. The system was evaluated on 250 writers of the Firemaker dataset using a leave-one-out strategy.

Multi-stream (CNN) in depth learn to reorganizing writers was represented as a proposed by [27] called Deep Writer that uses a pieces of regional handwritten as input and with SoftMax trained classification loss.

This proposal have a many contribution like:
The multi-stream structure for the essayist distinguishing proof undertaking.
Information expansion figuring out how to improve execution.
Fix checking system to deal with text pictures with various lengths

Trial of significant essayist found that dialects, for example, English and Chinese may share standard highlights for author recognizable proof and that joint preparing yields improved execution. Exploratory outcomes on IAM and HWDB datasets show high acknowledgment exactness:

6 Fractal Dimension

Mandelbrot defines a fractal dimension as "a number which measures the degree of irregularity or of fragmentation of a set" Many problems in different fields have been tackled using fractal dimension. The study of Handwriting on fractal analysis of handwriting has been presented in [28], where the authors aim to identify the writer based on fractal compression. The compression process involves partitioning the original image into sub-images R called ranges, which must be considered as a end result of an affine contractive transformation of domains D also belonging to the same image. The fractal compression process changes the image by the best transformations, and the position of associated R and D. Based on the extraction of fundamental similarities in writing a reference base is generated for each writer [29].

During the decompression, the sub-image is using transformation to regenerate the handwriting. For writer identification, the quality of the questioned document after fractal compression and decompression is compared with the original image. The reference base of each writer is used in the decompression process, and the one reporting the highest correspondence between the actual and decompressed writing is chosen as the author of the query document [27]. Experiments carried out on 20 different writers reported an identification rate of 85%. The approach is interesting but is not likely to perform on databases with a more significant number of writers (Fig. 2).

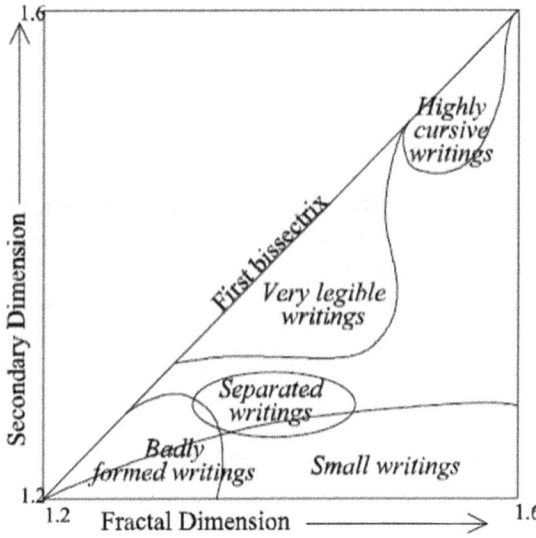

Fig. 2. Legibility graph. [1]

After presenting the global (texture and fractal-based) approaches to writer identification, the next section details the local approaches proposed in the literature.

[39] presented provided a taxonomy of dataset and a review of writer identification methods, feature extraction methods, as well as classification (conventional and deep learning-based).

7   Local Approaches

This section discusses the writer's recognition techniques based on localized features of handwriting. In local approaches, the main focus lies on characters or allographs in writing. The subsequent subsections first present the low-level features followed by a detailed discussion of the codebook-based approaches for writer identification.

Low-Level Features

Based on the visibility characteristics of lines of handwritten text [30], a writer identification system is presented, which is implemented by segmenting the text into lines and extracting features from each line of text. A k-nearest neighbor classifier is employed to compare the features from the input text and with the features coming from writers with a known identity.

The proposed system evaluated on a database of 50 writers realizes an identification rate of 90% when using single lines of text. The identification rate rises to approximately 100% when using the whole page of each writer. The determination process is to detect the edges taken from certain distance and the angles that each fragment makes. The next section presents to most popular and recent methods of writer recognition.

8   Codebook Based Methods

The idea of the codebook (writer invariants) was extended in [1, 31] by generating a universal codebook of all the writers in the database and hence representing a common feature space. The system evaluated on 88 writers and 150 writers of PSI and IAM database respectively. The rate of identification is 93% and 86% respectively. [32] Apply the same idea of writing pattern in order to characterize the writer. The authors are binarize the handwriting and the process starting with dividing the text into small fragments based on small windows as shown in Fig. 3. The same operation will group the codebook [1, 33]. The first step is building the codebook based in the writing samples by extracting the lines features into elements and labeled each element into the codebook. The used features for specifying each word elements are the direction of the line and the angle. The extracting technique is by a specific guide [22].

Fig. 3.   Division of the word 'headlines' in small windows [1, 33]

The complement of the codebook-based attribute, the researchers proposed a set of counter-based features to discover the curvature information of writing [34]. The counter and codebook are combined and evaluated on IAM (650 writers), RIMES (375 writers), and also a combination of both the databases (1,025 writers) achieving identification rates of 89%, 85%, and 86% respectively. An example codebook generated by this method is illustrated in Fig. 4(a), Fig. 4(b).

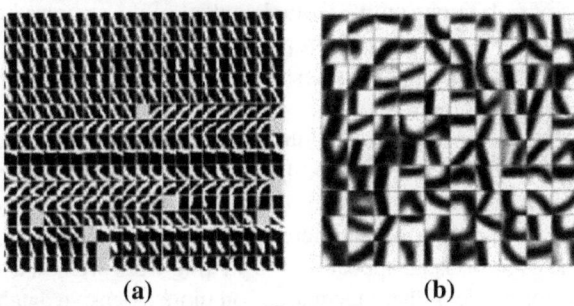

(a)   **(b)**

Fig. 4. Writer specific and universal codebook (a) Writer specific codebook obtained from a sample (Siddiqi and. (b) Universal codebook of size 100 generated from 50 samples [1, 33].

Another unusual codebook-based method was presented in [37], which segment the text based on the K-adjacent segments (KAS) method. KAS feature is an ordered sequence of Kline segments in the writing image, which determines the relationship between sets of neighboring edges where each pair shares an endpoint to extract the object. Once the text is segmented, similar KAS features were group together to generate a universal codebook to characterize, the writers. To generate the universal codebook, authors use Arabic script, and the system is evaluated on a database of 650 writers of IAM database realized identification rates of 92.1%. The study also claims that the proposed segmentation scheme is a language-independent approach that can segment any script.

[38] Presents a strategy to identify authorship of handwritten document based on online applications. They used a codebook-based Vector of Local Aggregate Descriptor (VLAD) which is considered as image object retrieval. The codebook involves a set of code vectors correlated with Voronoi cells detected from clustering algorithms and using online features vector. However, VLAD formulation cannot specifying the writers effectively. To overcome this problem, the VLAD formulation improved by adding a novel descriptor and the normalization process for features vector has been correlated with VLAD. The performance of this descriptor is considered based on two Handwriting Databases-the IBM-UBI and IAM. The presented method observes a noticeable increase over the VLAD.

9 Conclusion

The discussion of text-independent and text-dependent writer recognition methods observed over the years, it can be concluded that in general, text-dependent methods produce better results, but naturally, they require the same text to for comparison, a constraint which may not always be met in most of the practical applications. Text-independent methods for writer identification are much more useful, capable of finding a writer without any restrictions on the textual content of the samples being compared. These methods, therefore, are more appropriate for many practical applications, but naturally, they require a certain minimum amount of text for training and testing to achieve appreciable recognition rates.

The earlier methods of writer recognition were mostly text-dependent and considered only a limited number of writers. Later, more sophisticated text-independent methods were proposed which were able to realize acceptable identification rates on relatively large databases. Among the different types of methods, the global techniques which consider the overall visual appearance of writing work well on a smaller number of writers. The performance of these techniques on more extensive databases, however, is debatable. On databases with a significant number of writers (of the order of hundreds), local techniques have shown better performances. Recently, codebook-based writer identification has emerged as an effective method for characterizing the writer of a document, and the results of these methods are found to be better than other features.

The discussion of stat-of-the-art methods observe a quantitative comparison which represent the main contributions if the offline writer and summarized in Table 1.

Table 1. offline writer contribution to date papers identification.

Authors	Years	Contribution	Result
Jehanzeb, S.	2013	Apply the adaptive windows positioning method based on the writing meaning and individuality	the adaptive windows positioning present an efficient method due to the high accuracy in signature detection and this method enable to detect the emotional effect on the signature
B.V. Dhandra	2015	The method identify handwriting based on Kannada handwriting	The identification accuracy was 93.2583% based on simple words and 100% based on combination features of three or more words
Mohsen Mohammadi	2015	The used techniques are the SVM correlated with genetic algorithm which is mining the image and extracting the features base on training process	The results observe that the accuracy was over than 99% and the methodology shows a good results in different

(*continued*)

Table 1. (*continued*)

Authors	Years	Contribution	Result
Linjie Xing, Yu Qiao	2016	optimization and Design a multi-stream of writing structure and writer identification. Also, based on different image length its introduce the deep writer with a patch scanning strategy	The aftereffects of IAM and HWDB see that the pre-owned model accomplish high precision scopes to 99.01% dependent on 301 scholars. what's more, 97.03% on 657 authors with one English sentence input, 93.85% on 300 journalists with one Chinese character input
Faraz AhmadKhan	2017	in place of conventional DCT Bagged discrete cosine transforms (BDCT) descriptors is used	(two English and two Arabic) datasets System tested on four challenging BDCT has shown robustness to noise and blurring

Acknowledgment. The author thankful Department of Computer Science, College of Science Mustansiriyah University/Baghdad.

References

1. Jehanzeb, S.M.: Off-line Writer Recognition for English Handwritten Text Documents Based on Extended Codebook Approach (Doctoral dissertation, Universiti Teknologi Malaysia), (2013). https://doi.org/10.1016/j.patrec.2015.03.004
2. Bunke, H.: Recognition of cursive Roman handwriting: past, present, and future. In: Proceedings of the Seventh International Conference on 2003 Document Analysis and Recognition 2003, pp. 448–459, 3–6 August 2003. https://doi.org/10.1109/icdar.2003.1227707
3. Leclerc, F., Plamondon, R.: Automatic signature verification: the state of the art 1989–1993. Int. J. Pattern Recogn. Artif. Intell. **8**(2), 643–660 (1994). https://doi.org/10.1142/s0218001494000346
4. Abdi, M.N., Khemakhem, M., Ben-Abdallah, H.: A novel approach for off-line arabic writer identification based on stroke feature combination. In: 24th IEEE International Symposium on Computer and Information Sciences (2009). doi:https://doi.org/10.1109/iscis.2009.5291888
5. Gazzah, S., Amara, N.E.B.: Writer identification using modular MLP classifier and genetic algorithm for optimal features selection. In: Proceedings of Third International Conference Advances Neural Network, vol. 2, pp. 271–276. Springer-Verlag, Chengdu, China (2006). doi:https://doi.org/10.1007/11760023_
6. Br-Yosef, I., Beckman, I., Kedem, K., Dinstein, I.: Binarization, character extraction, and writer identification of historical hebrew 145 calligraphy documents. Int. J. Doc. Anal. Recogn. **9**(2–4), 89–99 (2007). https://doi.org/10.1007/s10032-007-0041-5

7. Schuckers, M.: Receiver operating characteristic curve and equal error rate computational methods in biometric authentication, 202-5_5, pp. 155–204. Springer, London (2010). https://doi.org/10.1007/978-1-84996-202-5_5

8. Plamondon, R., Lorette, G.: Automatic signature verification and writer identification - state of the art. Pattern Recogn. **22**(2), 107–131 (1989). https://doi.org/10.1016/0031-3203(89) 90059-9

9. Kuckuck, W.: Writer recognition by spectrum analysis. In: Proceedings of 1980 International Conference on Security through Scientific Engineering, West Berlin, Germany, pp. 1–3 (R47) (1980)

10. Naske, R.D.: Writer recognition by prototype related deformation of handprinted characters. In: Proceedings of 7th International Conference on Document Analysis and Recognition, vol. 2, pp. 819–822 (1982)

11. Schlapbach, A., Bunke, H.: Off-line writer identification and verification using Gaussian mixture models. In: Marinai, S., Fujisawa, H., (eds.) Machine Learning in Document Analysis and Recognition, vol. 90, Studies in Computational Intelligence, pp. 409–428. Springer (2008). https://doi.org/10.1007/978-3-540-76280-5_16

12. Bin, Z., Srihari, S.N., Sangjik, L.: The individuality of handwritten characters. In: Proceedings of the 2003 Document Analysis and Recognition 2003, pp. 1086–1090 (2003)

13. Laulederkind, S.J., Hayman, G.T., Wang, S.J., Smith, J.R., Lowry, T.F., Nigam, R., Munzenmaier, D.H.: The rat genome database, 2013—data, tools, and users. Brief. Bioinf. **14**(4), 520–526 (2013). https://doi.org/10.1093/bib/bbt007

14. Zois, E.N., Anastassopoulos, V.: Morphological waveform coding for writer identification. Pattern Recogn. **33**, 385–398 (2000). https://doi.org/10.1016/s0031-3203(99)00063-1

15. Dhandra, B.V., Vijayalaxmi, M.B.: A novel approach to text dependent writer identification of Kannada handwriting. Proc. Comput. Sci. **49**, 33–41 (2015). https://doi.org/10.1016/j. procs.2015.04.224

16. Dargan, S., Kumar, M.: Writer identification system for Indic and Non-Indic scripts: state-of-the-art survey. Arch. Comput. Methods Eng. 1–29 (2019). https://doi.org/10.1007/s11831-018-9278-z

17. Mohammadi, M., Moghaddam, M.E., Saadat, S.: A multi-language writer identification method based on image mining and genetic algorithm techniques. Soft. Comput. **23**(17), 7655–7669 (2019). https://doi.org/10.1007/s00500-018-3393-5

18. Xiong, Y.J., Wen, Y., Wang, P.S., Lu, Y.: Text-independent writer identification using the SIFT descriptor and contour-directional feature, In: 2015 13th International Conference on Document Analysis and Recognition (ICDAR), pp. 91–95. IEEE (2015). https://doi.org/10. 1109/icdar.2015.7333732

19. Abdi, M.N., Khemakhem, M.: A model-based approach to offline text-independent Arabic writer identification and verification. Pattern Recogn. **48**(5), 1890–1903 (2015). https://doi. org/10.1016/j.patcog.2014.10.027

20. Khan, F.A., Tahir, M.A., Khelifi, F., Bouridane, A., Almotaeryi, R.: Robust off-line text-independent writer identification using bagged discrete cosine transform features. Expert Syst. Appl. **71**, 404–415 (2017). https://doi.org/10.1016/j.eswa.2016.11.012

21. Hannad, Y., Siddiqi, I., Djeddi, C., El-Kettani, M.E.Y.: Improving arabic writer identification using a score-level fusion of textural descriptors. IET Biometrics **8**(3), 221–229 (2019)

22. Schomaker, L., Franke, K., Bulacu, M.: Using codebooks of fragmented connected-component contours in forensic and historic writer identification. Pattern Recogn. Lett. **28** (6), 719–727 (2007). https://doi.org/10.1016/j.patrec.2006.08.005

23. Bulacu, M., Schomaker, L.: Text-independent writer identification and verification using textural and allographic features. IEEE Trans. Pattern Anal. Mach. Intell. **29**(4), 701–717 (2007). https://doi.org/10.1109/tpami.2007.1009
24. Ma, L., Wang, Y., Tan, T.: Iris recognition based on multichannel Gabor filtering. In: Proceedings of Fifth Asian Conference on Computer Vision, vol. 1, pp. 279–283, January 2002
25. Bulacu, M., Schomaker, L., Vuurpijl, L.: Writer identification using edge-based directional features. In: Proceedings of ICDAR 2003 International Conference on Document Analysis and Recognition, pp. 937–941 (2003). http://citeseerx.ist.psu.edu/viewdoc/download?doi=10.1.1.58.2438&rep=rep1&type=pdf
26. Li, Q., Qiu, Z., Yao, T., Mei, T., Rui, Y., Luo, J.: Action recognition by learning deep multi-granular spatio-temporal video representation. ICMR (2016). https://doi.org/10.1145/2911996.2912001
27. Xing, L., Qiao, Y.: DeepWriter: a multi-stream deep CNN for text-independent writer identification. In: Proceedings of the 15th International Conference on Frontiers in Handwriting Recognition, ICFHR 2016, pp. 584–589, China, October 2016. https://doi.org/10.1109/icfhr.2016.0112
28. Rehman, A., Naz, S., Razzak, M.I.: Writer identification using machine learning approaches: a comprehensive review. Multimedia Tools Applications, pp. 1–43, September 2018. https://doi.org/10.1007/s11042-018-6577-1
29. Fisher, Y., et al.: Fractal Image Compression: Theory and Application. Springer Verlag, New York (1995). https://doi.org/10.1007/978-1-4612-2472-3
30. Hertel, C., Bunke, H.: A set of novel features for writer identification, in audio- and video-based biometric person authentication, Series Lecture Notes in Computer Science, Kittler, J., Nixon, M., (eds.) Springer, Berlin/Heidelberg, vol. 2688, pp. 1058–1058 (2003). https://doi.org/10.1007/3-540-44887-x_79
31. Bensefia, A., Paquet, T., Heutte, L.: A writer identification and verification system. Pattern Recogn. Lett. **26**(13), 2080–2092 (2005). https://doi.org/10.1016/j.patrec.2005.03.024
32. Siddiqi, I., Vincent, N.: Combining global and local features for writer identification. In: Proceedings of 11th International Conference on Frontiers in Handwriting Recognition (ICFHR), Canada (2008). https://www.researchgate.net/profile/Nicole_Vincent/publication/251566062_Combining_Global_and_Local_Features_for_Writer_Identification/links/54059fcc0cf23d9765a7134c/Combining-Global-and-Local-Features-for-Writer-Identification.pdf
33. Siddiqi, I., Vincent, N.: Stroke width independent feature for writer identification and handwriting classification. In: Proceedings of the 11th International Conference on Frontiers in Handwriting Recognition (ICFHR), Montreal, Canada, pp. 1–6 (2008). http://www.iapr-tc11.org/archive/icfhr2008/Proceedings/papers/cr1014.pdf
34. Korolchuk, V.I., Saiki, S., Lichtenberg, M., Siddiqi, F.H., Roberts, E.A., Imarisio, S., O'Kane, C.J.: Lysosomal positioning coordinates cellular nutrient responses. Nat. Cell Biol. **13**(4), 453 (2011). https://doi.org/10.1038/ncb2204

Designing and Implementation of Mobile Application for Healthcare Monitoring

Zahra Qassim Hami[(⊠)] and Boshra F. Zopon Al Bayaty

Computer Science Department, College of Sciences, Mustansiriyah University,
Baghdad, Iraq
boshraalbayatymbu@uomustansiriyah.edu.iq

Abstract. With the Corona virus, mobile applications are becoming more important especially these days because all people stay at home and cannot visit the hospital, and the clinic abroad becomes a danger. Mobile technology has the potential to impact healthcare. In the paper we use the mobile application as a platform Remote monitoring to measure various biological variables such as the ECG signal and heart rate of patients. The main contribution of this paper is to use the fuzzy algorithm to analyze patient body parameters, transmit them to a smart cell phone, and send an alarm in case of danger. This helps save heart patients before the worst cases happen. Finally, the results obtained from this project are displayed on a smartphone or PC as well. The accuracy of the results is 90% was proven by comparing the project readings with the medical devices used by doctors.

Keywords: Health care monitoring · Mobile application · Fuzzy algorithm

1 Introduction

With the development of technology, the evolution of home health care has become possible, while it is possible to care and follow the patient's condition from home. in the past two decades, Mobile phone technologies have greatly advanced. In many mobile applications, these technologies are now widely used, especially for health [1]. Remote health monitoring, including Portable sensor and many health apps, draws interest from the communications and health divisions, decreases patient monitoring costs and increases the quality of treatment in hospitals or at home during the recovery time, thus enhancing patient's quality of life [2].

Especially in the case of the difficulty of patient going to the hospital, as happened with the current situation in the spread of Corona virus and curfews that affected periodic reviews and the doctor's follow-up of the patient's condition, which increased the need to use programs to care for the patient from home. In this paper, a system is designed and implemented to track the patient's condition from home and based on smartphone and web service to patients [3]. To monitor physiological information, the portable terminal integrates vital signal sensors. Smartphones utilize as an intuitive interface between man and machine and a platform for transmitting information so that the patient and doctor can easily follow the patient's health status and give the

D.-T. Tran et al. (Eds.): ICISN 2021, LNNS 243, pp. 654–665, 2021.
https://doi.org/10.1007/978-981-16-2094-2_75

appropriate treatment without the need to leave the home. From that we can conclude the following:

Firstly, "Reducing the time and effort that the patient can spend to reach the hospital or the specialist doctor".

Secondly, "the patient does not make an appointment with the doctor, as he can obtain information about his medical condition at any time. Third, "The condition can be monitored by any other family member who is allowed to enter the application and this is very useful if the patient is in a state of danger or unconscious." Fourth, "an economical idea, as electronic advice is cheaper than consulting a doctor, as subscribing to the application is either free or for a nominal amount per month." Finally, the patient's location can be obtained via an external GPS or an internal Wi-Fi signal [4].

2 E-health

The use of information and communication technology (ICT) is used when delivering health care services. This means that (ICT) can be used for multiple health operations, such as clinical and educational, search and administration. m-Health is a medical procedure and mobile devices such as cell phones, patients with instrument-watching and cellular devices support public health. Approximately 58% of Member States have a World Health Organization (WHO) e-Health program, 55% of the country's e-Health Strategy in order to protect electronic patient records, 87% of countries admit that they have one or more national m-Health programs [88%] [5] .

3 Related Works

In the past 10 years, healthcare surveillance systems have attracted great interest from researchers. The primary aim is to create a reliable patient observation system in order to healthcare professionals could observer their patients, whether they were in the hospital or performing their regular daily activities. In this paper, we provide a wireless healthcare observing system based on mobile devices that can provide real-time online information about a patient's physiological conditions [6]. Among the most important works that have been accomplished and presented as follows:

Amna Abdullah, et al. (2015) proposed a Wireless Healthcare Monitoring System (WHMS) based on a smartphone that offers real-time online data about the medical condition of a patient. In addition, sending alarming messages and reminders about the patient's health status to those responsible for monitoring patients' condition for the needful medical diagnosis and immediate care. The system is made up of sensors, information acquisition unit, smartphone and LabVIEW software. This system is capable of viewing, register and transmit the patient's physiological information. In addition, the proposed WHMS system also supports internet connection so that patient data can be viewed and accessed from anywhere in the world at any time by healthcare professionals. The patient has medical sensors that convert changes in the observed physiological amounts into electronic information that can be measured and recorded.

LabVIEW software helps observer and display information. Can observe patient temperature, heart rate, muscle, blood pressure, blood sugar level, and ECG data [7].

Mohammed Al-khafajiy, et al. (2019) this work allow the elderly, the disabled and people with chronic conditions to live in their living environment safely, Ambient Assistive Living (AAL) programs Cooperatively allow assistance for an environment that is controlled by health care providers (for example, families, friends, and medical staff). These systems consist of several types of wearable medical sensors, smart phones, portable devices, servers, computer networks and software applications, and are connected to each other in an auxiliary environment for the transfer of knowledge and the presentation of services. Sensors and actuators are linked to AAL applications and patient gateways to send medical information to the information center, which then becomes ready for health observing purposes [8].

R Bulent Cobanoglu, et al. (2018) Remote health monitoring is explored alongside conduct change assist features and persuasion strategies using in mobile health applications [9].

Ranjeet Kumar, et al. (2017) proposed to create an IOT based patient observing app. The progress of the body sensor network in healthcare applications has made patient care more meaningful. they suggestion an Android Mobile based healthcare system using Body Sensor Network (BSN). BSN nodes include "temperature, humidity, and pulse rate sensor" [10].

Yunzhou Zhang, et al. (2015) proposed to provides a comprehensive solution; specially, Physiological factor's, inclusive respiratory rate and heart rate, are measured through sensors can worn and registered by a mobile phone that supplies a graphical interface for the user to monitor his health status easier. By providing the doctors and families with required information via a web interface and staff to monitor certified patient's condition and facilitate remote diagnosis; And supports real-time warning and GPS services during urgent situations, such as a fall or heart attack, so that unexpected events can be dealt with in a timely fashion [11].

Soumya S. Kenganal1, et al. (2016) Vital parameters such as ECG, temperature, BP, heart rate and so on are controlled by the proposed device and transmitted wirelessly via ZigBee technology. n a computer-based program called a nurse's central station, the sent data is interpreted where the computer acts as the center with the ZigBee receiver. This information is updated repeatedly in the database. The Android app brings all the modified information from the database and shows it. This assist the doctor in real time to obtain the current condition of the patient. If any given patient parameter exceeds the preset threshold value, an automatic notification will appear in the Doctor android mobile app [12].

Yvette E. Gelogo, et al. (2015) focus on the integration of wearable monitoring device and Android smart phone applications for healthcare monitoring system. This study proposes a healthcare system monitoring system that is convenient and within reach of users. Integration of wearable watch [13].

Mustafa A Al-Sheikh, et al. (2020) proposed an experimental model for achieving wearable, compact, low-control power Consumption, real-time remote monitoring device for bio-signals based on Internet of Things technology. Trusted and checked bio-signs are achieved through this implementation. The evaluation is conducted through the output of Comparisons between the readings of this project and the medical instruments that physicians and health professionals use [14].

4   Experimental Setup

4.1   Requirement Engineering

The Requirements Engineering phase consists of a set of research details that should be completed with each part of the work being specified in details and the purpose for which it is used. These requirements are multifaceted, here will explain them in detail as follow:

4.2   Functional Requirements

There are several points during this part:

- Preparing an environment
- Choosing of algorithm
- Select data set (readings)

4.3   Non-Functional Requirements

In a system requirement, points of interest of software and hardware are specified. This system prerequisite stage I partitioned into two groups.

Hardware Components

- **Heart Pulse Sensor:** Pulse sensors have become popular due to their use in health monitors-. The sensors used are simple, cheap and very reliable in obtaining reasonable evidence of heart rate in daily use. It works by sensing the change in light absorption or reflection through the blood as pulses through the arteries, a technique

Fig. 1. Heart pulse sensor

called gracefully photoplethysmography (PPG). The light signal rising and falling can be used to determine the pulse, and then calculate the heart rate. Figure 1 Displays the pulse sensor in use [15].

- **ECG** (is a complete representation of the heart's electrical activity on the surface of the human body. It is widely applied in the clinical diagnosis of heart disease and can be reliably used as a procedure to monitor cardiovascular function. ECG signals have been widely used to detect heart disease due to its simplicity and non-invasive nature Node [16]. Displays in Fig. 2.

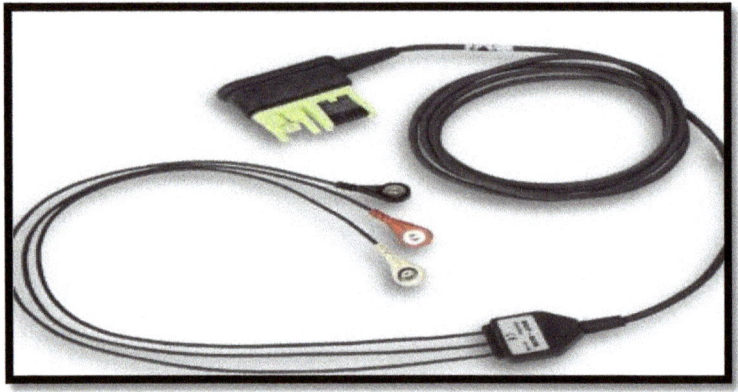

Fig. 2. ECG

- **MCU:-** Node MCU Development board is featured with wi-fi capacity, serial communication protocols, digital pins and analog pins. Node MCU is very important for IoT applications. By attaching it to the pulse heart sensor, will be able to present the result on website or mobile phone because of the property of wi-fi in it. This device will be shown in Fig. 3.

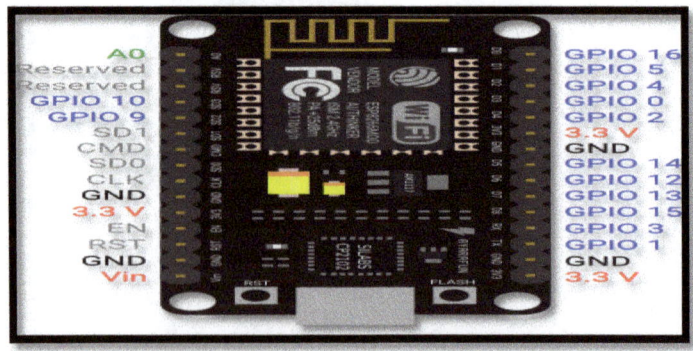

Fig. 3. Node MCU.

- **Computer:-** if the user wants to present the result in the computer, he/she must enter the website link in any web browser then login to this website to be able from seeing the results.
- **Mobile Phone:-** if the user wants to watch the result in his/her own mobile phone he/she must download the application and then also login as a user in a purpose of getting the result.

Software Components

- **Arduino 1.8.12:-** The heartbeat sensor is programmed using Arduino. Arduino is an open-source platform which using to construct electronic projects. Arduino comprises of both of hardware and software pieces, hardware.
- **ASP.NET:-** a developer platform consists of tools, programming languages, and libraries to build many different types of applications.
- **C#:-** Pronounced "C-Sharp" It is Microsoft's "object-oriented programming language" that runs on the.NET Platform. We can build different types of stable and strong applications with the aid of the C# programming language, such as window applications, web applications, distributed applications, web server applications and database applications [17].
- **MySQL:-** A properly designed database is fundamental to achieving the goal with developing a software system. Some principles guide the process of designing the database.

4.4   Design and Implementation

The main objective of the proposed application is monitoring the situation of heart patients by the specialist doctor. This is done by setting a sensor to read the patient's heartbeat. This sensor is programmed using Arduino, and the results are displayed

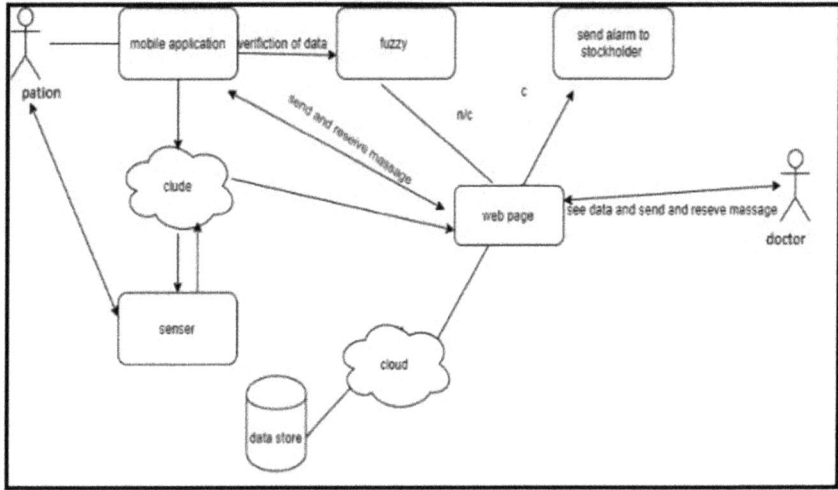

Fig. 4. Proposed system architecture

either on the website or on the mobile phone as an application. Here will be talking in detail about the system architecture in addition to the advantages of using fuzzy algorithm which used in the analysis of the results of readings. Figure 4 Illustrate proposed system architecture.

4.5   The Methodology

Fuzzy algorithm used Determine the critical or normal condition, so that the patient can know his health status by the doctor supervising the case. According to the National Institute of Health and American Heart Association, the average resting heart rate for adults is between 60 and 100 beats per minute. We put the value of threshold and if the readings larger or smaller of normal it can determine whether the patient is in a stable condition or not, the algorithm of fuzzy used shown in Algorithm (1):

<table>
<tr><td colspan="1" align="center">Algorithm (1) Fuzzy</td></tr>
<tr><td>Input: - pulse Readings</td></tr>
<tr><td>Output: - Report either normal or crucial state</td></tr>
<tr><td>Begin
Step1: Put heart pulse sensor on the patient finger
Step2: Readings result from previous step will be sending to cloud as a URL
Step3:   heart   pulse readings will present in a range between 58-125
• Give each reading its equivalent fuzzy value.
• If value between (0.6 -0.9) send report "normal state"
• Else
• Send report "crucial state".
End</td></tr>
</table>

Fuzzy values ranged between 0–1 and the heart pulse must be ranged between 72–82 pulses per minute, Table 1 present each value of heart beat and the corresponding values of supposed fuzzy value.

Table 1. Heart beat value VS fuzzy value

Heart beat value	Fuzzy value
0–10	0.0
11–20	0.1
21–30	0.2
31–40	0.3
41–50	0.4
51–60	0.5
61–70	0.6
71–80	0.7
81–90	0.8
91-100	0.9
101 and above	1

Some of readings will be tacking and then analyze according fuzzy:

Table 2. Sensor readings

Id	Sensor name	Reading	Date time	Patient
1	Heart Beating	98	7/18/2020 3:14:33 AM	1
2	Heart Beating	55	7/18/2020 3:16:55 AM	2
3	Heart Beating	97	7/18/2020 3:19:26 AM	1
4	Heart Beating	62	7/20/2020 11:13:22 AM	2
5	Heart Beating	84	7/20/2020 11:14:33 AM	3
6	Heart Beating	95	7/21/2020 3:00:54 AM	1
7	Heart Beating	88	7/21/2020 4:15:33 AM	3
8	Heart Beating	70	7/26/2020 1:20:6 AM	2

If readings from 1–8 will be tacking and fuzzy values were given to it, all results are recorded in the Table 3:

Table 3. Result analyze

Id	Readings values of sensor	Fuzzy values
1	98	0.9
2	84	0.8
3	97	0.9
4	70	0.7
5	95	0.9
6	62	0.6
7	88	0.8
8	55	0.5

Table 4. Comparison between this paper heart rate sensor and three other methods

Id	This paper (BPM)	Pulse oximeter fingertip (BPM)	Digital blood pressure (BPM)
No.1	88	89	87
No.2	79	78	77
No.3	67	68	68

By tacking the average value of fuzzy values as follow:
Average = Fuzzy value/no. of readings [18]
From Table 3 the summation of fuzzy value equal "0.5–0.9" and the number of readings is 8, So the average will be 0.6. Since the values between (72–82) are the best, the result should be compared to the fuzzy value of this range. Fuzzy value for 72–82 equals to 0.7, this meaning 0.7 is the best value. 0.6 is close to the good, so it is considered to be good state of patient now, the chart of heart pulse will be illustrating for the example in paper:

Chart (1): - Heart Pulse "example 1"

To determine the accuracy of the project three people tested heart rate sensor. The first is a young male, the second is a middle-aged woman, and the third is an elderly woman. The heart rhythm of each of these individuals. It is compared in two ways, as seen in the table. The first method is to use a fingertip pulse oximeter to record your heart rate. The second method is to use a digital heart rate measurement blood pressure monitor as it is illustrated in Table 4 [14].

Table 5. Compared with some previous research

Authors	Methodology	Limitations	Ref.
Amna Abdullah, et al. (2015)	Consists of sensors, data acquisition unit, smartphone and LabVIEW software.	The basic limited that the system able of only real-time observing of the patient's status, unprofessional analysis and directive. test made only when the smartphone charged	7
Ranjeet Kumar, et al. (2017)	Created an IOT based patient monitoring app. Android Mobile based healthcare system using Body Sensor Network (BSN). BSN nodes include "temperature, humidity, and pulse rate sensor	The details "from a single patient, history of patient care, recent improvements and potential recommendations can be made for a short record interval. This project can be extended with more sensors connected to more patients and by providing unique Health mate credential to each patient we can increase functionality. Others extending this project can also use GSM module to get text message to get patients status and GPRS module to check patient's location".	10
Soumya.S. Kenganal, et al. (2016)	Monitors vital parameters such as ECG, temperature, BP, heart rate etc., and transmits wirelessly via ZigBee technology.	Need to measuring some more parameters of the patients and also increasing the number of patients for monitoring simultaneously. Data uploading can be done using WIFI8266 module that may result faster uploading of data into the database and also enhancing the android application functionality	12
Mustafa A Al-Sheikh et al. (2020)	Remote monitoring of the patient's live ECG signal, heart rate, SPO2 and body temperature The signals are monitored and analyzed by means of measurements (Arduino)	There is a need to add a feature to send phone calls and SMS messages in the event of an emergency to physicians, patients, relatives and emergency centers	14
The proposed system	We used ASP.NET with Heart Pulse Sensor, ECG and fuzzy Algorithm Determine the critical or normal condition	Difficulty finding cases in order to test the system on them and find high-precision sensors	

5 Compared with Other Research

When we compare our system with other related system we find that the different is that the system can Ignores the inaccurate results of the sensors, and with the help of fazzy algorithm it can determine if the case is critical, and inform the patient and the treating doctor and that will help save the patient.

The project can be validated by research like Ranjeet Kumar et al., Mohammed Al-khafajiy et al., and Mustafa A Al-Sheikh et al. which is Close to the project we are doing.

6 Conclusions

The health care system is very important, especially for the elderly. In this work, the focus on cardiac patients from the elderly, where technological advantage was gained at the present time and with the help of the Internet network, the patient can communicate with the doctor remotely, when The results are normal within the range (60–100), but when it is within the (40–60 or 100–120) range, the situation is critical and the device alerts the stoke holder to take action, and the system can Ignores the inaccurate results of the sensors. The use of Fuzzy algorithm helps in determining the state of health more accurately as it classifies readings in a way that makes the result clearer.

Acknowledgments. The authors would like to thank University of Mustansiriyah (www.uomustansiriyah.edu.iq) Baghdad-Iraq for its support in present work.

References

1. Conejar, J.R., Kim, H.K.: A study for home and mobile U-healthcare system. Int. J. Softw. Eng. App. **9**, 255–260 (2015)
2. Ajiroghene, O., Obiei-uyoyou, O., Chuks, M., Ogaga, A., et al.: Development of an autonomous mobile health monitoring system for medical workers with high volume cases. Adv. Ro. Au. **S2**, 1 (2015). https://doi.org/10.4172/2168-9695.1000s2-005
3. Sachin, P., Kaustubh, G., Vishal, D., Akash, B., et al.: Heal. Monit. Mobile Appl. Wearable Sens. M. J. Re. Eng. Techn. **4**(3), 1257–1265 (2017)
4. Bulent, C.I.F., Atmaca, O.: Design and implementation of mobile patient monitoring system: G.S.J. **6**, 7 (2018)
5. Sreenivasa, V., Rao, T., Murali, K.: A design of mobile health for android applications. Am. J. Eng. Res. (AJER) **3**, 20–29 (2014)
6. Rameswari, R., Divya, N.: Smart health care monitoring system using android application: a review. Int. J. R. Techn. Eng. (IJRTE) -**7** 4S (2018)
7. Amna, A., Asma, I., Aisha, R., et al.: Real time wireless health monitoring application using mobile devices. I.J.C.N.C. **7**, 3 (2015)
8. Al-khafajiy, M., Baker, T., Chalmer, C.: Remote health monitoring of elderly through wearable sensors. MuL. T. Ap. **78**, 24681–24706 (2019). https://doi.org/10.1007/s11042-018-7134-7

9. Bulent, C.I.F., Oguzhan, A.: Design and implementation of mobile patient monitoring system. GSJ. **6**, 7 (2018)
10. Ranjeet, K., Rajat, M., Amit, A.M. et al.: IOT based health monitoring system using android App. ARPN J. Eng. Appl. Sci. **12**,19 (2017)
11. Zhang, Y., Liu, H., Su, X., Jiang, P., We, D.: Remote mobile health monitoring system based on smart phone and browser/server structure. J. Heal. Eng. **6**,717–7384 (2015)
12. Soumya, S.K., Rengaprabhu, P.: Real time health care monitoring system using android mobile. Int. J. Adv. Res. Electr. Instrum. Eng. **5**, 5 (2016)
13. Yvette, E.G., Haeng-Kon, K., Youn-Ky, C.: A design of CBD intelligent framework for U-healthcare application. Int. J. SW. Eng. App. **9**(6), 247–256 (2015) doi: https://doi.org/10.13140/rg.2.1.1062.0567
14. Mustafa, A., Ibrahim, A.: Design of mobile healthcare monitoring system using IoT technology and cloud computing. In: 3rd International Conference on Sustainable Engineering Techniques, ICSET (2020). https://doi.org/10.1088/1757-899x/881/1/012113
15. Maradugu, A.K., Ravi, Y.S.: Android based health care monitoring system. In: 2nd International Conference on Innovations in Information, Embedded and Communication Systems ICIIECS, vol. 5 (2015)
16. Rameswari, R., Divya, N.: Smart health care monitoring system using android application: a review. Int. J. Re. Tec. Eng. (IJRTE) ISSN.7 4S (2018)
17. Math open reference. https://www.mathopenref.com/calcaveval.html
18. Lakmini, P.M., Naeem, R., Keshav, D.: Remote patient monitoring: a comprehensive study: J. Ambient. Intell. Human. Comput. (2019). https://doi.org/10.1007/s12652-017-0598-x

Enhanced 2-DPR Replacement Algorithm

Rusul Fadhil[1] and Israa Hussain[2(✉)]

[1] Adhamiya, Baghdad, Iraq
[2] Atifiyah, Baghdad, Iraq

Abstract. Caching is a common term used in many systems to maximize the performance of the system such as a web cache. This concept has been utilized in processor's speed to lower the latency of fetching data from the main memory throw system bus long way to the processor. Cache Miss and Cache Hit is main events occurred when any block is being referenced by the processor from cache. Replacement algorithm works to make the decision to evict the block when Cache Miss occurred, as well as, increase the possibility of keeping block when Cache Hit occurred. The objective of replacement algorithm is to utilize the cache memory as efficient as possible with limit size of cache to minify cache miss and maximize the cache hit to get better efficiency for the system. Two-Dimensional Pyramid Replacement algorithm, (2-DPR), is an algorithm that combines the advantages (LRU) Least Recently Used and (LFU) Least Frequently Used (LRU) , and eliminates their disadvantages. The problem of this algorithm is giving high performance, low complexity, implementable in hardware, but not high hit ratio. In this paper, an enhanced method of 2-DPR is used to take advantage of Low Inter-reference Recency Set (LIR) to calculate the distance between the last referenced index and the current reference index in the cache by setting a counter called Inter-Reference Recency (IRR) in each block. By this method, we are attempting to gain high performance and high hit ratio.

Keyword: Cache memory · Replacement Algorithms · Cache miss · Cache Hit · Least Recently Used

1 Introduction

Since decades, processor cache memory was known as a very small memory placed between the processor and the main memory and used to hold data that a processor mostly need it. Cache is used to minify the period of time that takes to move data between the processor and the main memory and it makes the system more efficient [1]. Management of ordinal memory organization. Computer science is constantly getting more efficient and this number of technologies is being deployed for this reason and its importance [2].

Processor's cache is often smaller than the main memory because the more size of cache the more expensive will be. This limit size of cache can increase the efficiency of the system by placing mostly needed data in the cache memory and keep the rarely needed data in the main memory. So, specific techniques must be used to utilize the size of cache to get better results with limited size. The term used to measure the efficiency of cache is "Hit Ratio" or "Hit Rate", the higher Hit Ratio the better system performance [3].

© The Author(s), under exclusive license to Springer Nature Singapore Pte Ltd. 2021
D.-T. Tran et al. (Eds.): ICISN 2021, LNNS 243, pp. 666–673, 2021.
https://doi.org/10.1007/978-981-16-2094-2_76

In lifetime of memory cache manufacturing, researchers created many algorithms to manage cache's buffer, these algorithms known as "Replacement Algorithms", such algorithms are: LRU, LFU, LIRS, and ARC.In all environments, caches are often organized as multi-level hierarchies [4]. One of the issues faced in cache memory is "Cache Miss". Cache miss means that one of the blocks in cache is not found when it's needed by the CPU. Cache Miss is occur if the referenced block is not found in main memory and must to be transferred from secondary memory.

As Fig. 1, a cache memory is a fast and small piece between the main memory and CPU that acts as a buffer for frequently used data.one of the problems in cache memory called "Cache Miss" [5].

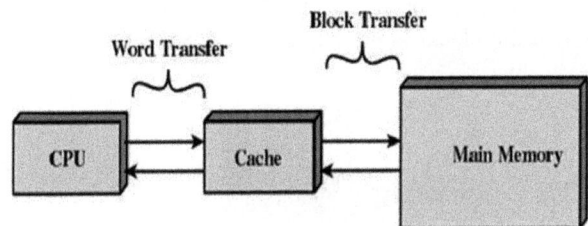

Fig. 1. CPU, Cache and Main Memory.

In the other case, if a block is found in the cache when it's needed then it's called "Cache Hit". The role of replacement algorithms is to lower Cache Miss and maximize Cache Hit as much as possible. If any Cache Miss appeared when the CPU requests block, a replacement strategy is implemented to switch blocks between the main memory and Cache Memory to replace unneeded block with the needed one, this decision is made by replacement algorithms [6].

These algorithms decides which blocks remains and which leaves the cache. So, different the goal of designing Replacement Algorithms is to increase the hit ratio for better performance replacement algorithm designs will cause different effects on the cache performance, and that will lead to the effect of performance of the system [6, 7].

Several replacement algorithms had been developed since the beginning of cache memory creation. Some of these algorithms are:

Optimal: replaces the mostly needed block. It's not recommended to implement in practical, but can be used measure the other replacement techniques [8]. Random: this method places the block within the cache in totally random order [9].

First-In-First-Out (FIFO): this algorithm works just like queue structures; the oldest block is chosen when a new block must be placed [10].

Least Recently Used (LRU): this algorithm works by setting a counter in each block to measure how this block is being referenced again in the future [11].

Least Frequently Used (LFU): this method also works by setting a counter for each block, but this time, this counter is used to measure how this block is rarely used and must be evicted from the cache when a new block is being requested by the CPU [12, 13].

2   Replacement Algorithm

The DPR supports a variety of functions by reconfiguring the dynamic region as needed however it does not use a lot area [14].

Salam A. and Hussein [15, 16] suggested that the algorithm is a hybrid of Two-Dimensional pyramid cache, as seen in Fig. 1. The structure of this pyramid consists of four levels (L1, L2, L3, L4). The insertion starts at the beginning of L1, and the eviction is done through L4.

In this system, each block has two counters:

1- Reference counter (called R): it holds the recency for all blocks and it's been controlled as follows:
   a. R = 0, if the block has been referenced or re-referenced.
   b. R = R + 1, if the blocks is not being referenced.

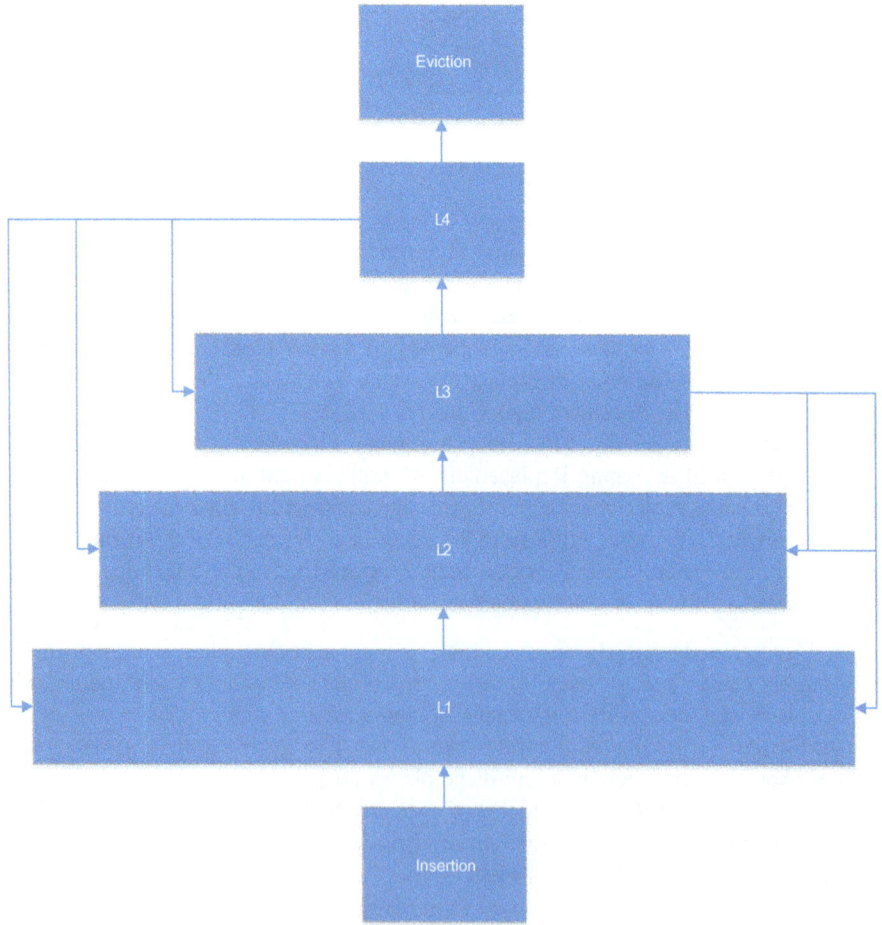

Fig. 2. 2-DPR Algorithm Cache Hierarchy.

2- Frequency counter (called F): it holds the frequency for all blocks within the cache, and it works as follows:

   a. $F = F + 1$, if the block has been referenced or re-referenced.

   b. $F = F$, if the blocks is not being referenced (Fig. 2).

3 The Proposed Method

As mentioned earlier, the original 2-DPR has two counters; the first one is R counter (Reference Counter) which is increment each time the block is not been referenced and set to zero when it's been referenced. The second one is F counter (Frequency Counter) which is increment each time the block is been referenced. This proposed method enhances the hit ratio of the original 2-DPR by adding additional counter to the algorithm. The new counter is called IRR (Inter-Reference Recency) counter inspired by the LIRS (Low Inter-reference Recency Set) caching replacement algorithm. The F counter is used to remain the most used blocks, while R is used to evacuate the least used blocks. The role of IRR counter is to calculate the distance between the last referenced block index and the current reference index, then, we use the IRR counter with R counter for evacuation purposes. With that way, the least used blocks will be discovered more than using R counter only, for this reason, the IRR counter will enhance the algorithm to gain higher hit ratio.

The IRR will be set to infinity if the block is first referenced, then after that, whenever the block is being referenced, the IRR is calculated by subtract the last referenced index of block the current value of IRR within the same level of cache. If block position has been set on higher or lower level within the cache, the IRR will be reset.

The first part algorithm pseudo code is mentioned as following:

```
INPUT: Blocks[x]

PROCEDURE

BEGIN

FOR LOOP Block IN Blocks

IF Block in cache THEN

        SWITH (block.Level)

        CASE 1:

Block,R -0;

Block.F++;

IF Block is First Reference THEN

Block.IRR – Infinity;

ELSE

Block.IRR – Block.LastReferencedIndex – IRR;

ENDIF

FOR LOOP Block2 IN Cache

IF Block2  ISNOT Block THEN

Block2.R++;

Block2.F- F;

ENDIF

NEXT

Level1.Push(Block);

                ENDCASE
```

4   Experimental Results

This work has maintained the features of 2-DPR that increases the performance of the computer system as showed in the following efficiency criteria:

1- The simplicity of the design to be implementable as hardware.
2- Ease of the design work steps.

3- Scan-resistant that makes the algorithm doesn't scan the whole cache blocks in evection.

4- High hit ratio.

In addition to all these features, our work enhanced the algorithm to get much higher hit ratio than the original 2-DPR.

To evaluate our new replacement algorithm, we made a simulation process to test our policy. We obtained the following results:

A. 2-DPR can be implemented as a hardware cache that can be attached to the CPU (on-chip) or detached from it (off-chip). This feature helps in achieving higher processing speed than other modern algorithms such as LRU (which needs an expensive hardware), or ARC, CAR, CART, CLOCK-Pro and Dueling CLOCK (which can be implemented only by the operating system) as a software.

B. Although this technique is highly dynamic, but the 2-DPR working steps are easy to be performed, remembered and programmed [6].

C. When a cache miss occurs, and there is no empty space to insert the new data block, the eviction is done without scanning all cache blocks. The cache block in (L4) will be deleted (because it is the best candidate for eviction at any clock tick), since it will be balanced between the least recently and frequently used. So, this means that 2-DPR is scan-resistant, and can achieve a high processing speed comparing with (LRU) and (LFU).

D. Note that the 2-DPR frequency threshold had removed the (LFU) problem which 'keeps the older blocks in the cache even if they are no longer in use'. Because they capture a higher frequency than other blocks. It is evident in the workings of the algorithm implementer.

E. The two parameters of 2-DPR, (R and F), at every cache referencing, fluctuates over its entire range of the existing cache blocks to a specific position and control their movement by balancing between 'recency' and 'frequency'. So, this proves that 2-DPR is a Self-Tuning and a high dynamic technique.

F. After testing 2-DPR, and the enhanced 2-DPR in cache size of (10) for a 3 datasets on a cache size of 10 blocks (2 datasets picked up from this paper [13] and one dataset randomly generated) (Table 1):

The previous table shows the dataset used in applying the work of the algorithm, and accordingly it was found that this new enhanced 2-DPR has higher ratio than the original one. We used the following equation to calculate the hit ratio for both algorithms:

$$HitRatios\ \% = \frac{Number\ of\ cache\ hits}{Total\ number\ of\ memory\ references} * 100$$

Table 1. The dataset

Dataset1	100,110,120,130,1,2,3,4,1,2,4,140,150,3,2,1,4,6,5,170,180,190,200,5,6,1,7,3,4,2,5,67,4, 210,220,1,3,5,7,2,4,6,1,8,6,7,2,3,4,2,3,6,7,8,230,240,250,260,270,280,290,300,7,8,1,3,5,6, 2,4,1,8,6,310,320,330,340,350,1,8,7,2,4,3,8,6,5,7,1,360,370,380,390,400,410,6,5,4,3,2,1,6
Dataset2	1,15,2,3,15,17,16,4,17,16,7,6,5,15,17,8,5,21,16,15,17,2,3,16,21,17,15,16,4,7,6,7,3,2,17,2 1,23,24,8,15,12,17,19,21,16,4,1,2,3,8,19,8,7,6,1,2,3,4,5,6,7,19,18,17,16,2,13,1,2,3,4,5,21, 22,18,16,14,12,2,4,6,1,2,3,4,5,18,19,20,21,22,23,24,10,11,12,13,14,15,9,6,7,6,5,4,3,2,1
Dataset3	38,38,40,7,11,19,22,2,8,38,46,28,35,49,5,8,49,45,4,6,23,27,17,10,7, 3,9,3,41,22,13,16,16,20,8,7,38,25,39,10,20,28,45,46,6,45,50,47,30, 43,15,16,41,48,21,46,19,19,18,8,8,48,14,46,22,27,12,13,36,21,25, 35,41,49,49,43,2,50,16,24,41,33,27,10,43,10,16,32,11,30,38,39,29, 10,6,31,46,43,13,17,2,46,29,26,13,30,48,19,15,31,22,31,9,7,39,47, 28,39,31,18,17,41,21,36,38,35,11,31,3,9,27,47,44,4,21,27,9,43,16, 39,24,3,9,34,13,28,38,24,46,41,3,47,29,4,50,46,23,49,21,12,13,14, 12,8,20,18,27,1,37,44,40,44,41,47,19,31,5,19,48,27,39,19,27,29,50, 34,23,36,42,35,14,8,14,40,45,29,50,36,12,44 [14]

In Table 2 the result of lustrate higher ratio for the Enhanced 2-DPR Algorithm and the original 2-DPR the difference for the increase is clear in the resurrected ratios (Fig. 3).

Table 2. New enhanced 2-DPR

	2-DPR	Enhanced 2-DPR
Dataset1	43%	55%
Dataset2	34%	42%
Dataset3	19%	20%

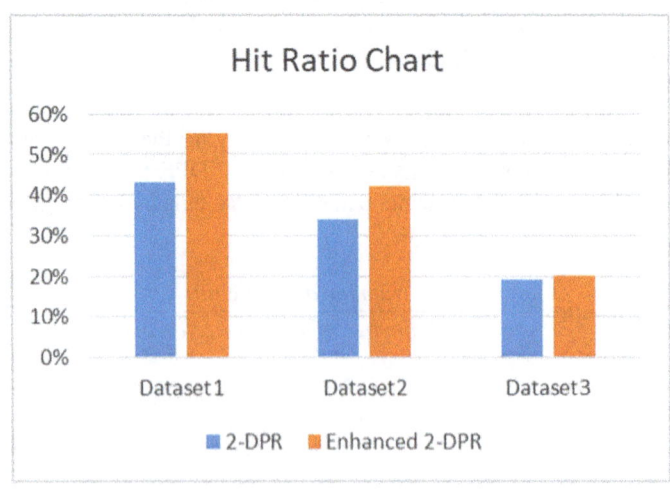

Fig. 3. Hit Ratio Chart

The diagram shows the clear difference to the algorithm optimization for each of the dataset used.

5 Conclusion

In this paper by maintaining the features of 2-DPR, we were able to enhance the algorithm by adding another counter to increase the hit ratio. So, now a new and more powerful algorithm is introduced and it has many advantages:

- High performance (by obtaining a higher hit ratio than the original 2-DPR).
- Remarkably fast speed (because it is scan resistant).
- Easy working steps.
- Implementable as a hardware, which is also High speed
- Self-tuned.

References

1. Bala, I., Raina, V., Sharma, O., Alam, A.: Analytical study of CAR, CLOCK and LRU page replacement algorithms in operating system. Int. J. Adv. Res. Comput. Sci. Softw. Eng. **4**, 747–751 (2014)
2. Shi, X., An, X., Zhao, Q., Li, H., Xia, L., Sun, X., Guo, Y.: State-of-the-art internet of things in protected agriculture. Sensors **19**(8), 1833 (2019)
3. Samiee, K.: A replacement algorithm based on weighting and ranking cache objects. Int. J. Hybrid Inf. Technol. **2**(2), 93–104 (2009)
4. He, X., Kosa, M.J., Scott, S.L., Engelmann, C.: A unified multiple-level cache for high performance storage systems. Int. J. High Perform. Comput. Netw. **5**(1–2), 97–109 (2007)
5. Danappa, R.: DAS page replacement algorithm. Doctoral dissertation (2020)
6. Abbas, S.H., Hussein, S.A.: 2-DPR: a novel, high performance cache replacement algorithm. Diyala J. Pure Sci. **12**(3-part 2), 98–113 (2016)
7. Aho, A.V., Denning, P.J., Ullman, J.D.: Principles of optimal page replacement. J. ACM (JACM) **18**(1), 80–93 (1971)
8. Belady, L.A.: A study of replacement algorithms for a virtual-storage computer. IBM Syst. J. **5**(2), 78–101 (1966)
9. Tanenbaum, A.S.: Modern Operating System, 3rd edn, pp. 199–213. Pearson Prentice Hall, New Jersey (2009)
10. Shah, K., Mitra, A., Matani, D.: An O (1) algorithm for implementing the LFU cache eviction scheme. (1), 1–8 (2010)
11. Alzakari, N., Dris, A.B., Alahmadi, S.: Randomized least frequently used cache replacement strategy for named data networking. In: 2020 3rd International Conference on Computer Applications and Information Security (ICCAIS), pp. 1–6, March 2020
12. Lee, D., Moon, H., Sejong, O., Park, D.: mIoT: metamorphic IoT platform for on-demand hardware replacement in large-scaled IoT applications. Sensors **20**(12), 3337 (2020). https://doi.org/10.3390/s20123337
13. Hussein, S.A.: A Proposed Multilevel Replacement Algorithm for Cache Memory M.Sc. Al-Mustansiriyah University (2016)
14. Salam, A., Mohsin, R.: Compression and analysis between classic and modern cache replacement techniques J. Iraqi (2018)

Short Interval Forecasting Model of COVID-19 in IRAQ Using State Space Modelling

Amina Atiya Dawood$^{(\boxtimes)}$ and Balasem Allawi Al-Isawi

University of Babylon, Hillah, Iraq
{amina, balasem}@uobabylon.edu.iq

Abstract. In December 2019 COVID-19, a viral disease originated in Wuhan (China), spread through China, and break out through the rest of the world, the disease is called coronavirus. The world health organisation (WHO) declared this disease as pandemic in March 2020. The recommended measures in almost every infected country was to enforce social distance. In 17 March 2020 Iraq declared curfew "Social distance" and provinces lockdown for the entire country as a precautious measure to control the spread of the disease. The daily reports data collected from JHU can be represented as time series, thus it can be analysed using time series analysis methods. The challenges in this pandemic include the short in data points for each vital pointer, the current continuous exponential trend in confirmed cases, and the randomness as the dominant phenomenon in observations. Those reasons made it very hard to predict or forecast long periods in the future. This paper suggests short periods for forecasting and an online updated state space model. The model records high accuracy with low error percentages comparing with other statistical models, which indicate that this rank of predictors is more suitable than other statistical models. The proposed model based on state space model, which is a technique to forecast time series with unobserved parameter, seasonal and unseasonal. The exponential behaviour of the time series assists in the good performance of the predictors.

Keywords: State space models · Statistical models · COVID-19 · Time series · Time series analysis

1 Introduction

COVID-19 is a very infectious viral disease originated from Wuhan province in mainland China in December 2019. The fast spread of this disease raises a number of challenges to governments and the whole world population [1]. Till the writing of this paper there was no actual medication for this virus. COVID-19 virus takes about 4.4 to 7.5 for an infected person to infect other people [2], which known as the series interval of the disease.

The learned lessons from the Chinese social distancing indicated the possibility to control the epidemic at many levels; globally, country wise, provinces, and even individual voluntarily isolation. Beside that governments across the world facing series

D.-T. Tran et al. (Eds.): ICISN 2021, LNNS 243, pp. 674–685, 2021.
https://doi.org/10.1007/978-981-16-2094-2_77

challenges in two, hard to balance, responsibilities to minimising fatalities and the catastrophic economic impact [3].

The data gathered by major world body, WHO, and other institutes like the Centre for Systems Science and Engineering at Johns Hopkins University (JHU CSSE), and the medical network DXY.cn. The data types delivered by these communities varies depending on their view and perspective. The world health organisation (WHO) delivers their data as online daily reports in PDF files format, this method can be due to the official nature of the organisation. JHU repository is now hosted as a GIT repository with daily updates and it also implemented in an online graphical platform [4]. The DXY.cn medical network provides a tabular real-time tracker for COVID-19 worldwide and accompanied GIT repository. For data scientists these data provide rich content that can be used to analyse to produce qualitative and quantitative insights.

When dependent data points represented on an independent temporal dimension then they can be visualised as time series. This resolution opens the possibility to use statistical models, among other approaches, to forecast the behaviour of the virus. This paper uses state-space model (SSM) as the intended statistical model to forecast the development of COVID-19 cases in Iraq. The type of state-space model used is the local trend model, which is a rank of state-space models that take into account the trend of the time series to assess in forecasting the effect of the unobserved parameters.

In the next section we will discuss the data sources used in this paper, followed by viewing the current state of COVID-19 in Iraq. A description of the SS model used and the fitting process then results are shown. Finally, a conclusion section for the outcomes of the paper.

2 The JH Dataset

The main data source for this paper is the publicly available repository [5] from the centre for Systems Science and Engineering at Johns Hopkins University (JHU CSSE). A COVID-19 real-time tracker [4] was developed in [6] which shows the accumulative counts for confirmed, deaths and recovered cases among other administrative and geospatial information, a screen shot is illustrated in Fig. 1. The demographic accuracy for the tracker is province level for mainland China; city level for the USA, Australia, and Canada; country level for the rest of the world. Other supplementary data sources include WHO coronavirus reports [7], and DXY dataset [8].

This repository creates a comma separated values data file (.csv) for each day of observation. The rows contain the administrative level, and each row contains several columns, including country name, province (for countries like China, USA, and Australia), geo location, number of confirmed cases, number of deaths, number of recovered case, and number of active cases.

The followed approach to pre-process the collected data is illustrated in the below pseudo-algorithm (Algorithm 1).

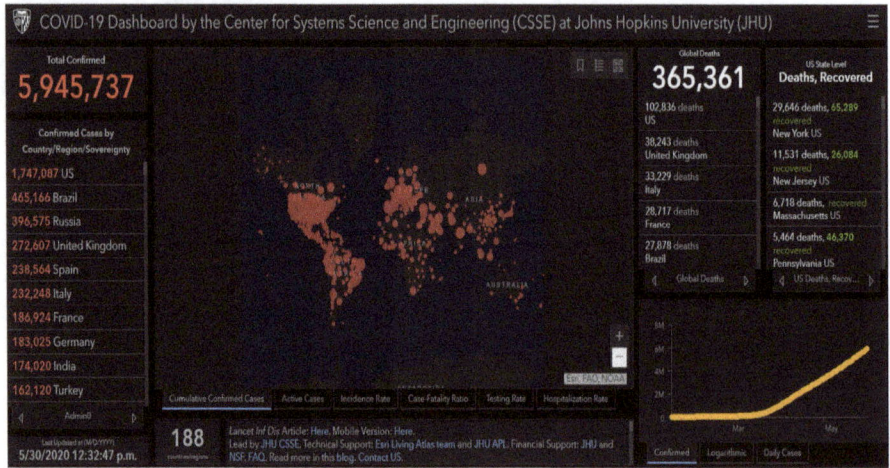

Fig. 1. COVID-19 real-time tracker dashboard [4]

3   COVID-19 IRAQ Data Stats

The first case in Iraq reported by [5] was in 25th of February 2020 after more than a month from the starting of the pandemic. At that date the world total statistics was about 80,000 confirmed (97% in China) and about 2,700 deaths (99% in China) [9]. Figure 2 shows the development of COVID-19 cases in Iraq.

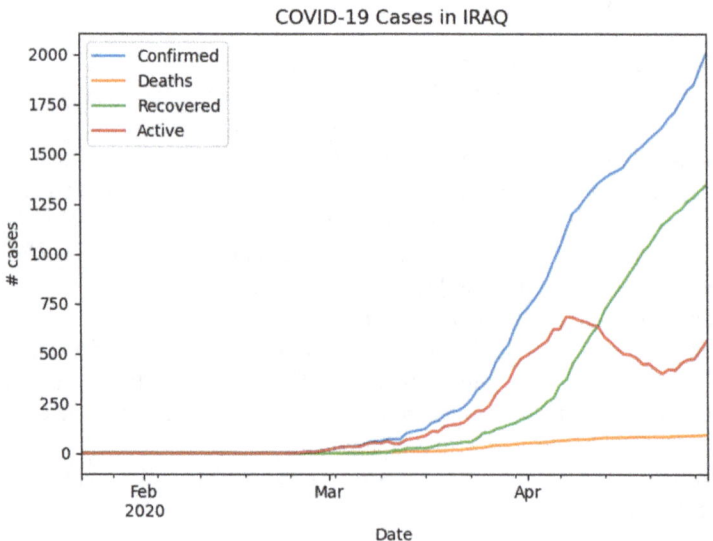

Fig. 2. Daily accumulated cases of COVID-19 in IRAQ

Figure 3 shows the weakly growth in mean confirmed, recovered, and deaths cases in Iraq respectively, and Table 1 shows the main descriptive statistics of cases in Iraq. The overall trend for all series shows an exponential positive trend which indicate a consistent incremental behaviour. The ideal expectation for the future data-points should show the following behaviour to indicate drawback of the COVID-19 cases in general. Confirmed cases would reach a climax and plot a zero-slop horizontal line indicating (no new cases). Recovered cases will continuously shows exponential growth (number of recovered cases overpower both confirmed and deaths cases). Death will drop as the virous lifecycle ended or vaccines will be developed (death from the virous drop to zero as no new cases and recovered cases will rise). The active cases eventually would be zero, active cases are calculated by the formula

$$Active\ Cases = Confirmed - Deaths - Recovered \tag{1}$$

Optimally Active cases will be zero when Confirmed = (Deaths + Recovered).

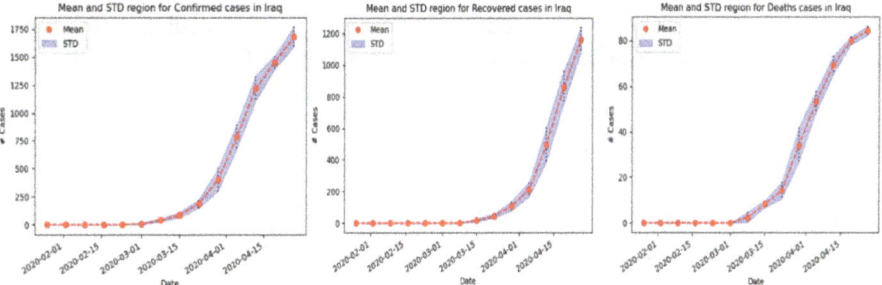

Fig. 3. Weakly mean values (with STD) for Confirmed, Recovered, and Deaths cases in IRAQ, respectively.

4   Model Design

Our SSM has the following hyperparameters, $\theta\left(\sigma_\varepsilon^2, \sigma_\xi^2, \sigma_\zeta^2\right)$, taking the initial values of the time series standard deviation. The optimizer of this model uses maximum likelihood estimator with Kalman filter. When parameters are selected for optimization, the optimizer unbound the values of the parameters, that mean they can take any value. Some statistics cannot accept certain values, like negative values for variances, therefore a transformation function is used to overcome this issue. The transformation function used is the square value of the unbound parameter, with square root as its inverse to retrieve the bounded parameter value.

Kalman filter is used to calculate the mean and variance parameters for the unobserved states of the given observations. The whole states distribution is Gaussian and parameterised by the calculated mean and variance. Kalman filter is a recursive process with updates given from newly inserted observations.

A Kalman filtering method can be derived from the model in (4), starting with some assumptions. First, Y_{t-1} is a vector of observations $(y_1, \ldots, y_{t-1})'$ where t = 2,3,..., the conditional distribution of the random walk $\alpha_t|Y_{t-1}$ is $N(a_t, P_t)$, $\alpha_t|Y_t$ is $N(a_{t|t}, P_{t|t})$ and α_t $_{+1}|Y_t$ is $N(a_{t+1}, P_{t+1})$. The object of the filter is to calculate the estimator $a_{t|t}$ of state α_t and the one-step ahead predictor a_{t+1} of state α_{t+1} and their correspondence variances $P_{t|t}$ and P_{t+1}. For optimization purposes it is important to calculate the one-step ahead error v_t of y_t.

$$v_t = y_t - a_t \qquad \text{for } t = 1, \ldots, n \qquad (2)$$

And to calculate the expected value and variance of vt

$$
\begin{aligned}
E(v_t|Y_{t-1}) &= E(\alpha_t + \varepsilon_t - a_t|Y_{t-1}) = a_t - a_t = 0, \\
\text{Var}(v_t|Y_{t-1}) &= \text{Var}(\alpha_t + \varepsilon_t - a_t|Y_{t-1}) = P_t + \sigma_\varepsilon^2, \\
E(v_t|\alpha_t, Y_{t-1}) &= E(\alpha_t + \varepsilon_t - a_t|\alpha_t, Y_{t-1}) = \alpha_t - a_t, \\
\text{Var}(v_t|\alpha_t, Y_{t-1}) &= \text{Var}(\alpha_t + \varepsilon_t - a_t|\alpha_t, Y_{t-1}) = \sigma_\varepsilon^2,
\end{aligned}
\qquad (3)
$$

For t = 2, ..., n. The above filtering constrains in equations set (3) can help to handle the general linear state space model and produce the following notations

$$
\begin{aligned}
F_t &= \text{Var}(v_t|Y_{t-1}) = P_t + \sigma_\varepsilon^2, \\
K_t &= \frac{P_t}{F_t},
\end{aligned}
\qquad (4)
$$

Where F_t is the variance of the prediction error v_t and K_t is Kalman gain. Now we can define the optimization updating parameters from time t to time $t + 1$

$$
\begin{aligned}
v_t &= y_t - a_t, & F_t &= P_t + \sigma_\varepsilon^2, \\
a_{t|t} &= a_t + K_t v_t, & P_{t|t} &= P_t(1 - K_t), \\
a_{t+1} &= a_t + K_t v_t, & P_{t+1} &= P_t(1 - K_t) + \sigma_\varepsilon^2,
\end{aligned}
\qquad (5)
$$

For t = 1, ..., n. Equations set (5) represent Kalman filter for the local level model. Note that the terms a_{t+1} and P_{t+1} are only for forecasting purposes to forecast $t + 1$ given t. The forecast errors $\{v_t\}$ are called innovations because here they summaries the unpredictable section of y_t that cannot be predicted from the past at $t = 1, \ldots, n$. The filter calculates the conditional mean and variance for Y_t.

The other part of Kalman filter is the smoother, which calculates the global conditional mean and variance on Y_n. The smoother takes the estimated outcome from the filter and recursively run through the observations from the last to the first. The parameters accounted for the smoother are

$$
\begin{aligned}
\hat{\mu}_t &= E(\mu_t|Y_n), \\
V_t &= \text{var}(\mu_t|Y_t), \\
r_t &= \text{weighted sum of future innovations}, \\
N_t &= \text{var}(r_t), \\
L_t &= 1 - K_t.
\end{aligned}
\qquad (6)
$$

The smoothing model in (6) runs recursively by the following rules:

1- Initial values: $r_n = 0$, $N_n = 0$.
2- Recursion update rules:

$$r_{t-1} = F_t^{-1} v_t + L_t r_t, \quad N_{t-1} = F_t^{-1} + L_t^2 N_t,$$
$$\hat{\mu}_t = a_t + P_t r_{t-1}, \quad\quad V_t = P_t - P_t^2 N_{t-1}, \tag{7}$$

The next step is to fit the model around the given time series. The fitting process aim to optimize the model hyperparameters to minimize the estimator error, maximize their likelihood to simulate the behaviour of the original observations and to give better forecasts.

The model selection will be based on the performance when forecasting one week in the future with minimum error rates.

5   Results

The total data points for each time series is 80. The paper considers only the non-zero cases which started from 24th of February 2020.

The model will fit data by choosing a weekly increment in the time series length starting from March 2020, the dates taken along with their descriptive statistics are shown in Tables 1, 2 and 3 and illustrated in Fig. 4.

Table. 1. Descriptive statistic for COVID-19 in Iraq aggregated weekly for confirmed case states.

Week commencing	Weekly statistics		Statistics from 24th of February 2020	
	Mean	STD	Mean	STD
COVID-19 case state	Confirmed			
2020-03-01	7.571	6.503	7.571	6.503
2020-03-08	40.286	12.284	23.929	19.424
2020-03-15	85.714	22.552	44.524	35.897
2020-03-22	184.143	38.386	79.429	71.220
2020-03-29	403.000	103.600	144.143	152.206
2020-04-05	783.286	112.955	250.667	281.423
2020-04-12	1219.429	112.768	389.061	431.916
2020-04-19	1451.571	60.638	521.875	537.523
2020-04-26	1682.143	88.479	650.794	626.236
2020-05-03	2075.857	160.485	793.300	734.879
2020-05-10	2549.857	145.639	952.987	866.219

Table. 2. Descriptive statistic for COVID-19 in Iraq aggregated weekly for deaths case states.

Week commencing	Weekly statistics		Statistics from 24th of February 2020	
	Mean	STD	Mean	STD
COVID-19 case state	Deaths			
2020-03-08	3.400	1.673	3.400	1.673
2020-03-15	8.143	1.574	6.167	2.887
2020-03-22	14.286	3.729	9.158	5.091
2020-03-29	34.143	7.777	15.885	12.685
2020-04-05	53.286	4.716	23.818	19.260
2020-04-12	69.286	4.071	31.775	24.760
2020-04-19	80.000	1.732	38.957	28.659
2020-04-26	84.286	1.976	44.833	30.814
2020-05-03	92.714	3.039	50.328	32.809
2020-05-10	103.429	3.645	55.794	35.063

Table 3. Descriptive statistic for COVID-19 in Iraq aggregated weekly for recovered case states.

Week commencing	Weekly Statistics		Statistics from 24th of February 2020	
	Mean	STD	Mean	STD
COVID-19 case state	Recovered			
2020-03-15	16.000	10.033	16.000	10.033
2020-03-22	43.000	10.847	29.500	17.235
2020-03-29	105.143	30.596	54.714	42.533
2020-04-05	210.000	46.957	93.536	80.739
2020-04-12	493.714	111.661	173.571	183.721
2020-04-19	859.857	103.739	287.952	310.767
2020-04-26	1163.857	75.981	413.082	423.220
2020-05-03	1386.143	76.789	534.714	512.256
2020-05-10	1634.286	68.932	656.889	595.465

Figure 4 shows that the weekly mean statistic can represent the behaviour of the time series, on the other hand, global mean shows great divergent from the actual observations.

The main experiments were committed on the SS model with the original observations and the logged observations. For each main set, starting from the first week of March 2020, we systematically capture the time series from the first data point ending with the end of the week ending on 16th May 2020. The captured data are used to fit the model using maximum likelihood optimiser, then the model will be used to forecast the last week which is a hold-out subset to compute the forecasting goodness, from 17 May to 23 May 2020.

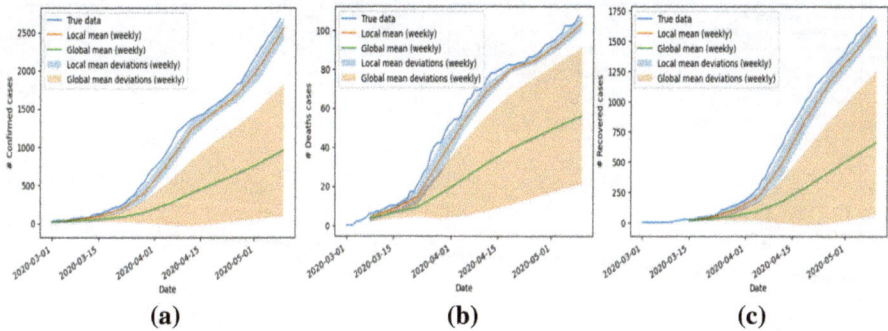

Fig. 4. Local and global descriptive statistics illustration for COVID-19 Iraq in all cases states; (a-Confirmed cases, b-Death cases, c-Recovered cases)

The number of data points varies depending on the cases label. Confirmed cases have 83 readings (11 weeks), death cases 74 readings (10 weeks), and for recovered cases we have 69 (9 weeks). That can be seen in the data illustrated in Table 3.

The metrics used in this paper are mean absolute error (MAE), mean squared error (MSE), root mean squared error (RMSE), mean absolute percentage error (*MAPE*), median absolute percentage error (*MdAPE*), root means squared percentage error (*RMSPA*), root median squared percentage error (*RMdSPE*). These metrics are scale independent as they are percentages, hence they will not be biased by the number of observations. The equations for those metrics are summarised in the following set of functions.

$$
\begin{aligned}
MAE &= mean(|Y_t - F_t|) \\
MSE &= means\left((Y_t - F_t)^2\right), \\
RMSE &= \sqrt{means\left((Y_t - F_t)^2\right)}, \\
MAPE &= mean\left(\left|\frac{Y_t - F_t}{Y} \times 100\right|\right), \\
MdAPE &= medean\left(\left|\frac{Y_t - F_t}{Y_t} \times 100\right|\right), \\
RMSPA &= \sqrt{mean\left(\left(\frac{Y_t - F_t}{Y_t} \times 100\right)^2\right)}, \\
RMdSPA &= \sqrt{medean\left(\left(\frac{Y_t - F_t}{Y_t} \times 100\right)^2\right)},
\end{aligned}
\tag{8}
$$

6   SS Model Performance

The resulted models from original and logged observations shows better fitting on the logged observations. The logged observations obtained from obtaining log function on the original observations. Figure 5 shows the general shapes for the original and logged observations.

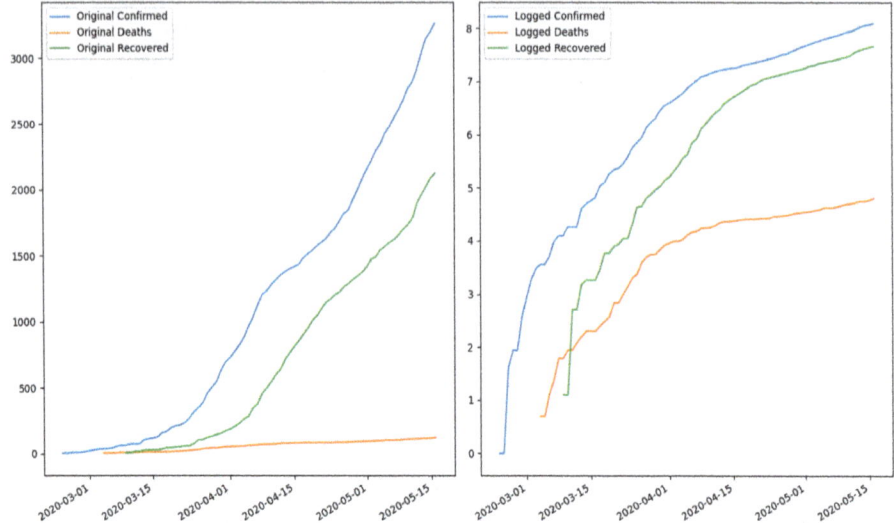

Fig. 5. COVID-19 Iraq observations shapes; original (left), logged (right).

Table 4 shows the comparison in fitting performance between original and logged observations. In addition, Figs. 6 and 7 illustrate the fitting performance for original and logged modelled data, respectively.

Table 4. Statistics of the fitted models for Confirmed cases.

Observations	#	AIC	BIC	MAE	MSE	RMSE
Confirmed						
Original	83	702	709	12.736	306.691	17.513
Logged	83	−153	−146	1.077	1.043	1.227
Deaths						
Original	74	265	272	1.132	2.137	1.462
Logged	74	−163	−157	1.053	1.013	1.120
Recovered						
Original	69	564	571	10.409	237.784	15.420
Logged	69	−23	−16	1.112	1.079	1.316

From Table 3, we can conclude that transforming the observations using the log function makes the time series more predictable. The observations are represented as an accumulated values curve which is easily fitted as log curve. The low coefficients of errors indicate the small distance between the fitting residuals and their corresponding true observations. The previous assumption is also supported in Figs. 6 and 7.

From the above discussion we conclude that the model obtained from logged observation produce a model has stronger ability to forecast future observations. The next set of experiments will be a comparison between our SS model and other commonly used statistical models.

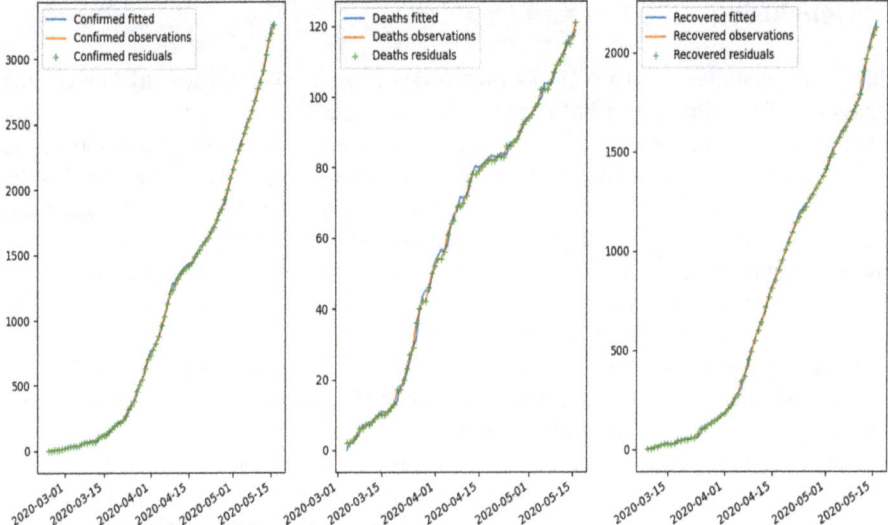

Fig. 6. Fitting performance for SS model on original observations.

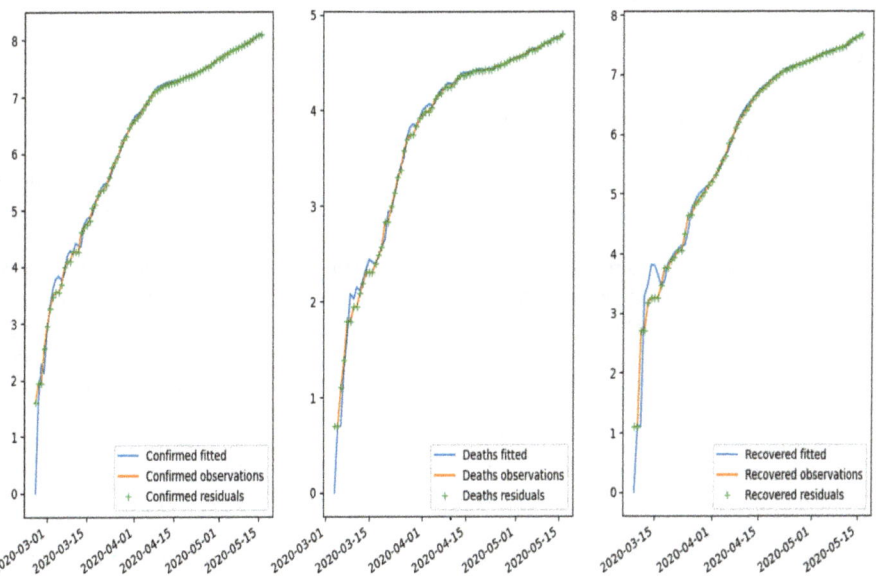

Fig. 7. Fitting performance for SS model on logged observations.

7 Conclusion

This paper examines the COVID-19 time series in Iraq. The aim was to forecast the behaviour of the time series in future to adapt suitable situation and services by Iraqi government. The paper used a state-space model method to fit the observations, the method used was linear trend model. The reason for choosing a state space model is the ability of such models to fit time series with random walk of an unobserved parameter. The forecasting error percentages recorded the lowest percentage for SSM when comparing with other models (Table 4), but the univariant nature of the considered observations time series omitted other unobserved factors and combine them in the noise term inside the adopted statistical model.

Results of the model can be considered as guidelines for the near future status of the virus spread rates. Another advantage is that the model can be retrained regularly to get new insights from the updated observations.

The current exponential nature of the recorded cases does not indicate any considerable constant decline in short interval forecasts; hence this type of models cannot predict the end of the epidemic to any intense. This paper shows the feasibility of using state-space models to forecast time series with epidemic data and get very good results to forecast short intervals in the future.

References

1. Buheji, M., Buheji, A.R.: Designing intelligent system for stratification of COVID-19 asymptomatic patients. Am. J. Med. Med. Sci. **10**(4), 246–257 (2020)
2. WHO: Coronavirus disease (COVID-2019) situation report—30, 19 February 2020. https://www.who.int/docs/default-source/coronaviruse/situation-reports/20200219-sitrep-30-covid-19.pdf?sfvrsn=3346b04f_2. Accessed 2 May 2020
3. Anderson, R.M., Heesterbeek, H., Klinkenberg, D., Hollingsworth, T.D.: How will country-based mitigation measures influence the course of the COVID-19 epidemic. Lancet **395** (10228), 931–934 (2020)
4. C. JHU: COVID-19 Dashboard by the Center for Systems Science and Engineering (CSSE) at Johns Hopkins University (JHU) (2020). https://www.arcgis.com/apps/opsdashboard/index.html#/bda7594740fd40299423467b48e9ecf6. Accessed 23 Apr 2020
5. J. CSSE: Novel Coronavirus (COVID-19) Cases, provided by JHU CSSE (2020). https://github.com/CSSEGISandData/COVID-19. Accessed 23 Apr 2020
6. Dong, E., Du, H., Gardner, L.: An interactive web-based dashboard to track COVID-19 in real time. Lancet Infectious Diseases (2020)
7. WHO: Coronavirus disease (COVID-2019) situation reports (2020). https://www.who.int/emergencies/diseases/novel-coronavirus-2019/situation-reports. Accessed 23 Apr 2020
8. DXY: COVID-19 Global Pandemic Real-Time Report (2020). https://ncov.dxy.cn/ncovh5/view/en_pneumonia. Accessed 23 Apr 2020
9. WHO: Coronavirus disease (COVID-2019) situation reports, 24 February 2020. https://www.who.int/docs/default-source/coronaviruse/situation-reports/20200224-sitrep-35-covid-19.pdf?sfvrsn=1ac4218d_2. Accessed 27 Apr 2020
10. Codling, E.A., Plank, M.J., Benhamou, S.: Random walk models in biology. J. R. Soc. Interface **5**(25), 813–834 (2008)

11. Kalman, R.E.: A new approach to linear filtering and prediction problems. Trans. ASME–J. Basic Eng. (82), 35–45 (1960)
12. Auger-Methe, M., Newman, K., Cole, D., Empacher, F., Gryba, R., King, A., Thomas, L.: An introduction to state-space modeling of ecological time series. arXiv:2002.02001 (2020)
13. Adejumo, A.O., Daniel, J.: Time series anaalysis of brent crude oil prices per barrel: a Box-Jenkins approach. Ann. Comput. Sci. Ser. **17**(1), 87–101 (2019)
14. Shapiro, S.S., Wilk, M.B.: An analysis of variance test for normality (complete samples). Biometrika **52**(3/4), 591–611 (1965)
15. d'Agostino, R.B.: An omnibus test of normality for moderate and large sample size. Biometrika **58**(2), 341–348 (1971)
16. Stephens, M.A.: EDF statistics for goodness of fit and some comparisons. J. Am. Stat. Assoc. **69**(347), 730–737 (1974)
17. Ivanovski, Z., Milenkovski, A., Narasanov, Z.: Time series forecasting using a moving average model for extrapolation of number of tourist. UTMS J. Econ. **9**(2), 121–132 (2018)
18. Sumitra, I.D.: Comparison of forecasting the number of outpatients visitors based on naïve method and exponential smoothing. In: Materials Science and Engineering (2019)

Enhance a Cloud-Based Distance Learning Computing Management System (LCMS)

Muqdad Abdulraheem Hayder[1(✉)] and Ekhlas Ghaleb Abdulkadhim[2]

[1] College of Education for Human Sciences, University of Kerbala,
Kerbala, Iraq
moqdad.a@uokerbala.edu.iq
[2] Collage of Tourism Sciences, University of Kerbala, Kerbala, Iraq
ekhlas.g@uokerbala.edu.iq

Abstract. In the age of information and communication, older methods of teaching and learning are less effective. In order to keep pace with the changing environment, humans need to look for new ways and means of transferring knowledge and raising awareness. A new approach to education that emphasizes inclusive education, as well as one of the quickest and shortest possible ways to achieve this, is E-learning. E-learning has been growing exponentially with the increased number of students, volume of teaching materials, and provision of the available educational resources and services. The main challenge in this regard lies in the optimization of computing resources, data storage, communications demands, and dynamic concurrency through cost-efficient platforms with proper scalability. However, numerous educational institutions struggle to constantly sustain their resources and infrastructures with the rapid growth of e-learning. Cloud computing is considered to be an effectual strategy for the mentioned issues as it is a computation paradigm, by which IT resources could be presented as services and made available to the users via the Internet. Cloud computing provides great opportunities for higher education institutions to use e-learning solutions without investing in hardware, software, or infrastructure. In this research, we are trying to expanded a cloud based model called Moodle. Moodle has the features that allow it to be scaled for very large extensions. Many institutions use Moodle as their platform for online courses. The present study aimed to propose a model for the provision of IaaS, PaaS, and SaaS services for e-learning purposes and educational.

Keywords: E-learning · Distance learning · Cloud computing · Moodle · OpenStack

1 Introduction

E-learning is implemented using different software on every education level, including universities, middle schools, high schools, and other educational institutions. Compared to conventional schooling, e-learning has prominent advantages, the most significant of which is cost-efficiency owing to the use of physical environments based on information technology (IT). Such education is accessible everywhere at any given time, while it also suits learners it terms of convenience. As for educators, e-learning

D.-T. Tran et al. (Eds.): ICISN 2021, LNNS 243, pp. 686–693, 2021.
https://doi.org/10.1007/978-981-16-2094-2_78

enables them to update the educational materials at hand and benefit from integrated multimedia teaching techniques; such example is the use of user-friendly educational models, which facilitates the learning process. Another advantage of e-learning is that educators will be able to exchange educational materials for contribution and the improvement of their own tools [1]. On the other hand, the limitations of e-learning should also be assessed before the integration of these models into the academic framework. The main limitation of e-learning is its scalability in primary and intermediate levels as some of the tools used in e-learning are only applicable for certain tasks. Consequently, the installed educational tools should be constantly re-configured and updated in the case of extended workload, which in turn increases the costs of education. Therefore, the provided resources for e-learning must be exploited properly. For instance, learners tend to use laboratory computers and servers less frequently in the evening or during holidays, while during the semester, there is increased demand for educational tools and resources. Furthermore, the use of educational software imposes significant costs in terms of the license purchase and maintenance.

Currently, e-learning systems have been used per individual in various educational organizations as this method has lower efficacy in educational resources and is more cost-efficient, while there is the possibility of higher investment in computing infrastructures, scalability improvement, and system maintenance. As such, the public applications of e-learning should be further assessed based on the cloud computing technology. Cloud computing is a paradigm that provides IT resources through various services and is available to the users via the Internet. This novel technology provides proper platforms, infrastructures, and software through services such as IaaS, PaaS, and SaaS, and the main basis is payment per use [1–3]. Cloud computing is an IT service model, which encompasses catalogs regarding user demands, addressing their needs flexibly and adaptively; in this method, users are only charged for the actual usage of the system. The two main features of this paradigm are the transparent scalability and on-demand resource use, which result in the accurate determination of resource needs by requiring no details. Cloud computing is developed to target staff, students, faculty members, and external parties for collaboration.

2 Problem Description Learning Computing Management System (LCMS)

E-learning refers to the learning processes that are enabled via the Internet. The main components of e-learning are multiple-format contents, management of learning experience, and the online community of learners, content developers, and experts. In the present study, the main features of e-learning were classified as flexibility, convenience, accessibility, consistency, and repeatability. Cloud computing is considered to be an effectual strategy for the mentioned issues [5, 6] as it is a computation paradigm, by which IT resources could be presented as services and made available to the users via the Internet. This paradigm is offered as a model of IT service provision with a catalog that responds to the needs and demands of the users flexibly and adaptively, only billing for the actual usage. Therefore, the distinctive features of cloud computing could be referred to as the resource use based on demand and transparent scalability, which help users to exploit the computational resources dynamically and accurately only for necessities without the need to thoroughly recognize the infrastructure. Figure 1 shows the Architecture for E-Learning Cloud computing. Cloud-based e-learning is also used as education service software, which is installed quickly owing to the low hardware requirements. Furthermore, the vendor supports the maintenance and provides regular system updates, thereby allowing the users to merely concentrate on core learning.

According to Masood-Huang [7], the development of e-learning in the cloud environment has been associated with some implications and outcomes. First, it is accessible via the Internet worldwide by any user. Second, there is no software requirement for the user, which minimizes the costs of applications as there is no need for installation, software maintenance, deployment, and server administration costs. As a result, the total ownership costs are comparatively lower, and the system is implemented within a shorter time with fewer staff required.

The other key features of this paradigm include the pay-per-use function and the ability of the SaaS Server to support several educational settings. In addition, the scalability could be maintained as the related applications are run in different server groups; for instance, the perforce is not degraded with the increased number of learners. In the context of the advantages, the paradigm provides the client data in the SaaS Server, which in turn increases the security level, and the users could easily trust the erudite software [8].

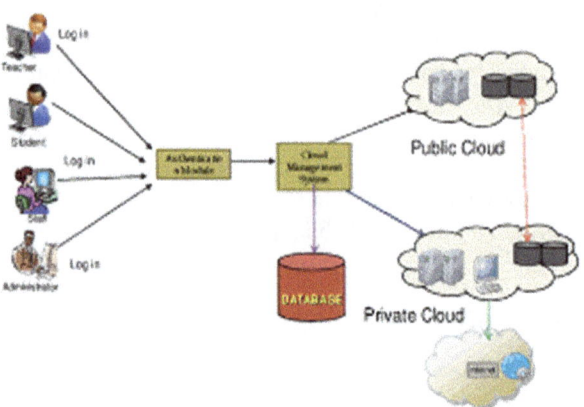

Fig. 1. Architecture for E-Learning Cloud computing

Figure 2 depicts the infrastructures of the IaaS services and the software as SaaS services. The IaaS services enable the storage management and load balance of all the educational systems, scaling management of virtual machines, and back-up and restoring of educational applications. On the other hand, the SaaS services enable the management of the application registry and server application for management and use of the learning content by the clients, while an account management system is also provided to authorized users. Furthermore, these services have virtual desktop deployment on the personal desktop for the learning content, while session management is available to ensure the use of the virtual desktop by the authorized users. Finally, the personalized management of popular learning contents is available for registration management.

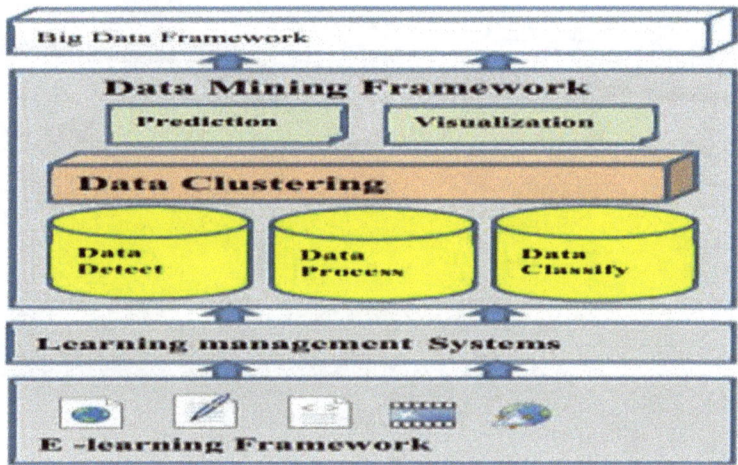

Fig. 2. E-learning environment

3   Research Methodology

We extended on the Moodle that add two fields. Firstly, class online that students and teachers can taking together video and voice by protocol RTSP (real time streaming protocol). Also, the teacher can give lecturers for a group of students and can make experiments inside the lab such as chemistry, biology and physical experiments. In addition, students without teachers can make a group study that can discuss anything that is related to the class that has been taking. Secondly, linked two cables and more in buildings or schools by Openstack. That makes all the data can be saved in the server Openstack. In this way, we are secure our data and in the same time it can be very fast arriving to the users.

3.1   OpenStack

OpenStack is the open-source cloud computing software, which is used for the development of a reliable cloud infrastructure. The main goal of the software is to create the opportunity for an organization to employ cloud computing services based on open-source software using standard hardware. OpenStack is classified as OpenStack compute and OpenStack storage. The former is used for automation in the development and management of large-scale virtual private servers. The latter is used for the propagation of object storage with adequate scalability and redundancy via clusters for data storage in terabytes or petabytes, Figure 3 illustrates the OpenStack. The OpenStack code is licensed by the Apache version 2.0, which enables the users to run and develop other software on OpenStack or transfer modified codes in the form of a patch or new features. OpenStack has recently attracted the attention of large-scale hosting companies, such as Rackspace Hosting and NASA, which benefit from this tool for the management of numerous compute instances and data storage in petabytes [9].

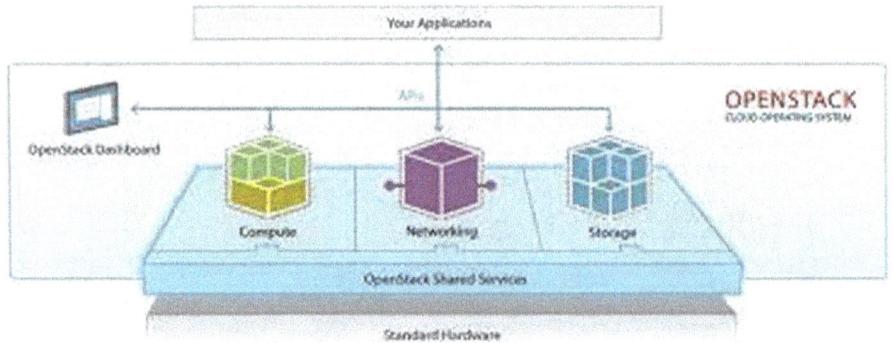

Fig. 3. OpenStack

3.2   Data Base

Database management is a complex process, which has been significantly addressed with programming and programming languages. The computer program used to manage, ask, and answer questions between databases is called the database system administrator, or DBMS. Typically in a database there is a structured description for the entities stored in the database. This description is known by a pattern. The descriptive model shows the objects of the databases and the relationship between them. There are different ways to organize these models, which we call database models [9, 10]. The most widely used model today is the relationship model. Each database in SQL is made up of different parts, these include tables, views, store procedures, functions, and so on. The database used in this project is SQL Server 2012. The types of data used in the project database are:

- Bit, Define Boolean value
- VARCHAR(n),, the maximum variable length string n.
- INT, definition of integer.
- Date, indicates the date with the order of day, month and year.
- NText, save text.

3.3   Installed Moodle on OpenStack

Today, cloud computing and e-learning present a new arena of learning and education. Cloud hosting and the virtualization hardware features could bring about the enhancement and maintenance of learning resources in a cost-efficient manner. IaaS cloud technology has been frequently incorporated into various learning processes, with an emphasis on VM reservation for learners. In the current research, we applied the Moodle technology for e-learning applications through cloud installation (Fig. 4). This software package could be applied for web-based training, which is available in the form of learning management systems, course management systems, and virtual learning environment. Moodle is free of charge as it is open- source and licensed by the GNU Public.

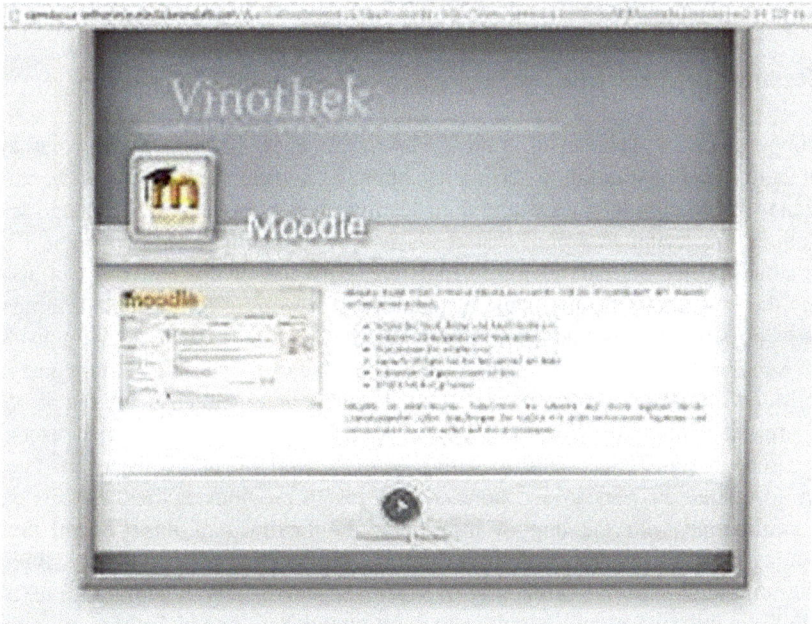

Fig. 4. Moodle Installed On Openstack

Moodle has seven layers of users, including the administrator, who is responsible for the site management (e.g., menus, website display, user privileges and roles), and course creator, who is responsible for the development and teaching of the courses, assignment of others teachers, and monitoring of the unpublished educational contents, and these responsibilities are often covered by the head of the educational program. Another major user is the teacher, who is responsible for the management of the subjects that have been taught, such as effecting changes in the course activity, giving the grades, student dropouts, and appointing non-editing teachers for the subjects [6, 9]. In addition, the software supports a non-editing teacher, who is responsible for teaching the subject and giving grades, while they are not assigned with changing the activities that have been set by the teacher. In other words, a non-editing teacher is appointed as an assistant teacher/lecturer. The fifth layer of the Moodle users is students, who seek the subjects of their need. Student must register in advance for any subjects they are seeking before its presentation. At the end of the course, the teacher assigns a grade to the students based on their accomplishment of the subject. Guests are another layer of the users of the software, who are given the read-only access. The users that have not been listed on the Moodle are considered guests, who are free to join any course that they are allowed to select, while they cannot follow course activities. Finally, there are the authenticated users, which apply to the users who log in. While users could also be the teachers of educational subjects, they are only considered to be authenticated users when it comes to other subjects, in which they could also have the guest status. It is notable that guests and authenticated users are distinctly different as the latter is able to select any subjects, while the former cannot [10].

4   Results

The importance of websites is more visible than ever. In such a way that every organization and institution is trying to build a website. On the other hand, the importance of foreign languages and their learning has increased, and language school institutions are trying to increase students by providing more services. The use of professional professors and advanced tools are examples of this activity. But another feature that can attract people is the availability of a website for students to access as easily as possible. In this study, we tried to add the key features that are not considered in the websites by reviewing the previous works. One of the most important past tasks is the Moodle system. This system is a virtual system that includes many features. The most important of these features are the ability to chat between different users, the forum, send assignments, send questions with the aim of taking the exam and the ability to upload answers to questions. According to previous studies, various works have been done with the aim of improving the facilities of this system, and by examining these works and carefully examining the Moodle system, the possibility of holding online classes, the possibility of voice and video calls in addition to text chat can improve this system and People who can't attend class can be helpful. As a result, in addition to the facilities available in the Moodle system, this research has tried to include two possibilities of holding online classes and the possibility of voice and video calls between different users. The system designed to allow voice and video calls between a group of users, students and professors.

5   Conclusion

According to the results of the study, the design of an e-learning system based on the OpenStack cloud computing has several key components, including easy access, flexibility, ease of use, repeatability, and uniformity. Therefore, the e-learning system could be readily delivered based on the features of cloud computing platforms using the OpenStack software. This method could be used in proper proportion to efficient learning environments through the personalization of educational contents and facilitate adaptation to the educational model that is being presented. Furthermore, the incorporation of cloud computing into e-learning systems could enhance the scalability and flexibility of learning resources, such as computing, network access, and data storage in a cost-efficient manner through the pay-per-use function, thereby limiting the use of new software and hardware that may not be allowed for educational programs. Also, by extended the Moodle we will be able to have the users connected online together. Moreover, will be able to connect buildings together by using OpenStack.

References

1. Kumar, T.P.: A private cloud-based smart learning environment using moodle for universities. In: Cases on Smart Learning Environments, pp. 188–202. IGI Global (2019)
2. Popel, M.V., Shyshkina, M.P.: The areas of educational studies of the cloud-based learning systems. In: Proceedings of the 6th Workshop on Cloud Technologies in Education (CTE 2018), Kryvyi Rih, Ukraine, 21 December 2018. CEUR Workshop Proceedings, no. 2433, pp. 159–172 (2019)
3. Arora, S., Maini, J., Mallick, P.: Efficient e-learning management system through web socket (2017)
4. Chang, W.Y., Abu-Amara, H., Sanford, J.F.: Transforming Enterprise Cloud Services. Springer, Heidelberg (2010)
5. Zhou, R., Song, Y., He, T., Fahad, A.: Applying augmented reality technology to e-learning (2016)
6. Mayer, R., Clark, R.: E-Learning and the Science of Instruction: Proven Guidelines for Consumers and Designers of Multimedia Learning, 3rd edn. Pfeiffer, Hoboken (2011)
7. Masud, A.H., Huang, X.: ESaaS: a new education software model in e-learning systems. In: Zhu, M. (ed.) ICCIC 2011, Part V. CCIS, vol. 235, pp. 468–475. Springer, Heidelberg (2011)
8. Buyya, R., Broberg, J., Goscinsky, A.: Cloud Computing: Principles and Paradigms. Wiley, Hoboken (2011)
9. Sosinksy, B.: Cloud Computing Bible. Wiley, Hoboken (2011)
10. Pecchia, A., Russo, S., Sarkar, S.: Assessing invariant mining techniques for cloud-based utility computing systems. IEEE Trans. Serv. Comput. **13**, 44–58 (2017)

Author Index

Lightning Source UK Ltd.
Milton Keynes UK
UKHW020736170522
403097UK00002B/25